T0390531

Nanotechnology for Hematology, Blood Transfusion, and Artificial Blood

Nanotechnology for Hematology, Blood Transfusion, and Artificial Blood

Edited by

Adil Denizli
Department of Chemistry, Hacettepe University, Ankara, Turkey

Tuan Anh Nguyen
Institute for Tropical Technology, Vietnam Academy of Science and Technology, Hanoi, Vietnam

Mariappan Rajan
Department of Natural Products Chemistry, School of Chemistry, Madurai Kamaraj University, Madurai, Tamil Nadu, India

Mohammad Feroz Alam
Department of Pathology, JNMC, AMU, Aligarh, India

Khaliqur Rahman
Department of Hematology, Sanjay Gandhi Post Graduate Institute of Medical Sciences, Lucknow, India

ELSEVIER

Elsevier
Radarweg 29, PO Box 211, 1000 AE Amsterdam, Netherlands
The Boulevard, Langford Lane, Kidlington, Oxford OX5 1GB, United Kingdom
50 Hampshire Street, 5th Floor, Cambridge, MA 02139, United States

MATLAB® is a trademark of The MathWorks, Inc. and is used with permission. The MathWorks does not warrant the accuracy of the text or exercises in this book. This book's use or discussion of MATLAB® software or related products does not constitute endorsement or sponsorship by The MathWorks of a particular pedagogical approach or particular use of the MATLAB® software.

Notices

Knowledge and best practice in this field are constantly changing. As new research and experience broaden our understanding, changes in research methods, professional practices, or medical treatment may become necessary.

Practitioners and researchers must always rely on their own experience and knowledge in evaluating and using any information, methods, compounds, or experiments described herein. In using such information or methods they should be mindful of their own safety and the safety of others, including parties for whom they have a professional responsibility.

To the fullest extent of the law, neither the Publisher nor the authors, contributors, or editors, assume any liability for any injury and/or damage to persons or property as a matter of products liability, negligence or otherwise, or from any use or operation of any methods, products, instructions, or ideas contained in the material herein.

British Library Cataloguing-in-Publication Data
A catalogue record for this book is available from the British Library

Library of Congress Cataloging-in-Publication Data
A catalog record for this book is available from the Library of Congress

ISBN: 978-0-12-823971-1

For Information on all Elsevier publications
visit our website at https://www.elsevier.com/books-and-journals

Publisher: Matthew Deans
Acquisitions Editor: Simon Holt
Editorial Project Manager: Gabriela D. Capille
Production Project Manager: Debasish Ghosh
Cover Designer: Greg Harris

Typeset by MPS Limited, Chennai, India

Working together
to grow libraries in
developing countries

www.elsevier.com • www.bookaid.org

Contents

List of contributors

Hayder A. Abdulbari
Center of Excellence for Advanced Research in Fluid Flow, Faculty of Chemical and Natural Resources Engineering, University Malaysia Pahang, Pahang, Malaysia

Dalia Samir Ahmed Mahdy
Fraunhofer Institute for Silicate Research ISC, Würzburg, Germany

Semra Akgönüllü
Department of Chemistry, Hacettepe University, Ankara, Turkey

Fahima Akther
Queensland Micro- and Nanotechnology, Griffith University, Nathan, QLD, Australia; Australian Institute for Bioengineering and Nanotechnology, University of Queensland, St Lucia, QLD, Australia

Mohammad Feroz Alam
Department of Pathology, JN Medical College, Aligarh, India

Ibrahim M. Alarifi
Department of Mechanical and Industrial Engineering, College of Engineering, Majmaah University, Al-Majmaah, Saudi Arabia; Engineering and Applied Science Research Center, Majmaah University, Al-Majmaah, Saudi Arabia

Sidra Amir
Department of Biological Sciences, ILM Group of Colleges, Sargodha, Pakistan

Tatiana Avsievich
Optoelectronics and Measurement Techniques Laboratory, University of Oulu, Oulu, Finland

Nilay Bereli
Biochemistry Division, Department of Chemistry, Hacettepe University, Ankara, Turkey

Muhammad Bilal
School of Life Science and Food Engineering, Huaiyin Institute of Technology, Huaian, P.R. China

Alexander Bykov
Optoelectronics and Measurement Techniques Laboratory, University of Oulu, Oulu, Finland

Santhosh Chidangil
Centre of Excellence for Biophotonics, Department of Atomic and Molecular Physics, Manipal Academy of Higher Education, Manipal, India

Adil Denizli
Biochemistry Division, Department of Chemistry, Hacettepe University, Ankara, Turkey

Katja Ferenz
Institut für Physiologie, Universität Duisburg-Essen Universitätsklinikum Essen, Essen, Deutschland

Sowmya Hari
Department of Bio-Engineering, School of Engineering, Vels Institute of Science, Technology and Advanced Studies, Chennai, India

Sumreen Hayat
Department of Microbiology, Government College University Faisalabad, Faisalabad, Pakistan

Dzuliana Fatin Jamil
Department of Mathematics and Statistics, Faculty of Applied Sciences and Technology, Universiti Tun Hussein Onn Malaysia, Pagoh, Malaysia

Ozan Karaman
Institut für Physiologie, Universität Duisburg-Essen Universitätsklinikum Essen, Essen, Deutschland

Mohsin Khurshid
Department of Microbiology, Government College University Faisalabad, Faisalabad, Pakistan

Jijo Lukose
Centre of Excellence for Biophotonics, Department of Atomic and Molecular Physics, Manipal Academy of Higher Education, Manipal, India

Christian Mayer
Department of Chemistry, Physical Chemistry, University of Duisburg-Essen, Essen, Germany

Setti Sudharsan Meenambiga
Department of Bio-Engineering, School of Engineering, Vels Institute of Science, Technology and Advanced Studies, Chennai, India

Igor Meglinski
Optoelectronics and Measurement Techniques Laboratory, University of Oulu, Oulu, Finland; Interdisciplinary Laboratory of Biophotonics, National Research Tomsk State University, 634050, Tomsk, Russia; Institute of Engineering Physics for Biomedicine (PhysBio), National Research Nuclear University (MEPhI), 115409, Moscow, Russia; Department of Histology, Cytology and Embryology, Institute of Clinical Medicine N.V. Sklifosovsky, I.M. Sechenov First Moscow State Medical University, Moscow, Russia; College of Engineering and Physical Sciences, Aston University, Birmingham, B4 7ET, UK

Ganesh Mohan
Department of Immunohematology and Blood Transfusion, Kasturba Medical College, Manipal Academy of Higher Education, Manipal, India

Saima Muzammil
Department of Microbiology, Government College University Faisalabad, Faisalabad, Pakistan

Nam-Trung Nguyen
Queensland Micro- and Nanotechnology, Griffith University, Nathan, QLD, Australia

Yoong Sheng Phang
Department of Physics and Astronomy, The University of Georgia, Athens, GA, United States

Alexey Popov
VTT Technical Research Centre of Finland, Oulu, Finland

Khaliqur Rahman
Department of Hematology, Sanjay Gandhi Post Graduate Institute of Medical Sciences, Lucknow, India

Rozaini Roslan
Department of Mathematics and Statistics, Faculty of Applied Sciences and Technology, Universiti Tun Hussein Onn Malaysia, Pagoh, Malaysia

Punniavan Sakthiselvan
Department of Bio-Engineering, School of Engineering, Vels Institute of Science, Technology and Advanced Studies, Chennai, India

Yeşeren Saylan
Department of Chemistry, Hacettepe University, Ankara, Turkey

Shah Bahrullah Shah
Institut für Physiologie, Universität Duisburg-Essen Universitätsklinikum Essen, Essen, Deutschland

Aqsa Shahid
Department of Microbiology, Government College University Faisalabad, Faisalabad, Pakistan

Shamee Shastry
Department of Immunohematology and Blood Transfusion, Kasturba Medical College, Manipal Academy of Higher Education, Manipal, India

Hang T. Ta
Queensland Micro- and Nanotechnology, Griffith University, Nathan, QLD, Australia; Australian Institute for Bioengineering and Nanotechnology, University of Queensland, St Lucia, QLD, Australia; School of Environment and Science, Griffith University, Nathan, QLD, Australia

Ahmet Fatih Tabak
Mechatronics Engineering Department, Kadir Has University, Istanbul, Turkey

Huong D.N. Tran
Queensland Micro- and Nanotechnology, Griffith University, Nathan, QLD, Australia; Australian Institute for Bioengineering and Nanotechnology, University of Queensland, St Lucia, QLD, Australia

Salah Uddin
Department of Physical and Numerical Sciences, Qurtuba University of Science and Information Technology, Peshawar, Pakistan

Devasena Umai
Department of Bio-Engineering, School of Engineering, Vels Institute of Science, Technology and Advanced Studies, Chennai, India

Zhi Ping Xu
Australian Institute for Bioengineering and Nanotechnology, University of Queensland, St Lucia, QLD, Australia

Sunishtha S. Yadav
Center for Medical Biotechnology, Amity Institute of Biotechnology, Amity University, Noida, India

Yanjun Yang
School of Electrical and Computer Engineering, College of Engineering, The University of Georgia, Athens, GA, United States

Humaira Yasmeen
Department of Microbiology and Molecular Genetics, The Women University Multan, Multan, Pakistan

Handan Yavuz
Biochemistry Division, Department of Chemistry, Hacettepe University, Ankara, Turkey

Maryam Zain
Department of Biochemistry and Biotechnology, The Women University Multan, Multan, Pakistan

Jun Zhang
Queensland Micro- and Nanotechnology, Griffith University, Nathan, QLD, Australia

Yiping Zhao
Department of Physics and Astronomy, The University of Georgia, Athens, GA, United States

Ruixue Zhu
Optoelectronics and Measurement Techniques Laboratory, University of Oulu, Oulu, Finland

Aimen Zulfiqar
Department of Chemistry, Government College University, Faisalabad, Pakistan

Foreword

The discovery of blood circulation in the human body by William Harvey sparked a new trend of investigation in the scientific community into blood transfusion that could save precious human lives. Early on, human-to-human blood transfusion was substantially practiced, but the outbreak of HIV, hepatitis, and other transmutable diseases elevated the blood prices due to the increased detection tests required before transfusion. Also, the low shelf life of blood adds more complications to the process. Due to these reasons, a worldwide investigation has been triggered for substituting blood with synthetic or semisynthetic materials that can work as efficiently as blood. Recently, the development in medical research allowed the application of nanotechnology for the production of nanomaterials. The main hurdle in producing an alternative material is the toxicity due to the breaking of hemoglobin (a tetramer) into dimers, which are filtered out from the kidney and concentrated in renal tubules, causing toxicity. Until now, various semisynthetic and new techniques have been developed, which involve the alteration of hemoglobin and encapsulation of hemoglobin. Furthermore, perfluorochemicals have also been investigated to synthesize alternatives for blood. Despite the intense ongoing research, the Food and Drug Administration has not authorized a single product for human applications in the United States.

Recently, Wei Zhu and colleagues reported the successful production of duplicate red blood cells (RBCs) that have all the characteristics of natural RBCs and also offers some other benefits. The produced replicas are flexible enough to squeeze through the narrow capillaries and have high circulation time (Guo et al., 2020). The team packed the synthesized cells with hemoglobin, toxin sensors, anticancer drugs, or magnetic nanoparticles (NPs) to reveal the cargo carrier characteristic of the cell. Further investigation will examine the medical application of the artificial cell for toxin biosensing and cancer treatment.

The contribution of nanotechnology in the medical field is unprecedented. It has been considered as a potential technique for drug delivery, tumor imaging, and cancer therapy. However, numerous researches are underway to prepare NPs-based devices for more advanced applications without any toxicity. In view thereof, the present book is very welcome.

This book covers the fundamental concepts of NPs' interfaces, interactions, immunity, transportation, toxicity, and interactions with coagulated blood in the bloodstream. Much attention is given to the techniques that involve the application of NPs for predicting blood diseases and disorders, blood purification, bioreactors for blood component replacement, and drug delivery systems. Furthermore, an extended section addressing the current development in nanotechnology for blood transfusion and artificial blood, along with future scopes, has been presented. Overall, this book provides a detailed insight into the significant advancements in nanotechnology for hematology.

Thanks.

Ram Gupta
Department of Chemistry, Pittsburg State University,
Pittsburg, KA, United States

Reference

Guo, J., Agola, J. O., Serda, R., Franco, S., Lei, Q., Wang, L., Minster, J., Croissant, J. G., Butler, K. S., Zhu, W., & Brinker, C. J. (2020). Biomimetic rebuilding of multifunctional red blood cells: Modular design using functional components. *ACS Nano*, *14*, 7847−7859. Available from https://doi.org/10.1021/acsnano.9b08714.

Blood–nanomaterials interactions

Tatiana Avsievich[1], Ruixue Zhu[1], Alexey Popov[2], Alexander Bykov[1] and Igor Meglinski[1,3,4,5,6]

[1]Optoelectronics and Measurement Techniques Laboratory, University of Oulu, Oulu, Finland [2]VTT Technical Research Centre of Finland, Oulu, Finland [3]Interdisciplinary Laboratory of Biophotonics, National Research Tomsk State University, 634050, Tomsk, Russia [4]Institute of Engineering Physics for Biomedicine (PhysBio), National Research Nuclear University (MEPhI), 115409, Moscow, Russia [5]Department of Histology, Cytology and Embryology, Institute of Clinical Medicine N.V. Sklifosovsky, I.M. Sechenov First Moscow State Medical University, Moscow, Russia [6]College of Engineering and Physical Sciences, Aston University, Birmingham, B4 7ET, UK

1.1 Introduction

Fast and widespread introduction of nanoparticles (NPs) into various areas of industry and medicine has led to the enormous amount of NP-containing products in our everyday life. Modern humanity is surrounded by NPs emerging from food products (Carrillo-Inungaray et al., 2018), construction materials (Mohajerani et al., 2019), electronics, medical drugs (Rudramurthy & Swamy, 2018), and appliances etc., making NPs' interactions with the human body inevitable. The current state of NP development and application raises concerns on how to make NP applications beneficial, while avoiding the possible health-related risks.

According to the National Nanotechnology Initiative (NNI) annual report (The National Nanotechnology Initiative Supplement to the President's, 2020), the national nanotechnology initiative supplement to the president's 2020 budget requests over $1.4 billion support. NNI investments are categorized by Program Component Area, though each of the components indirectly involves biomedical aspects of NPs, they are mainly related to the "Foundational research" (39%) and "Applications, Devices, Systems" (28%) components, while only 4% is assigned for the "Environment, Health, and Safety" component.

The number of products registered in the Nanodatabase (Consumer Products—The Nanodatabase, 2020) (Fig. 1.1) demonstrates the steady growth within the past decade. Most of these products are from the "Health and fitness" category. Surprisingly, the majority of nanomaterials used in the products are categorized as "unknown" (77%), meaning that their composition was not declared, and consequently these products have an "unknown hazardous" effect on humans.

The unique physicochemical properties of NPs (optical, electronic, magnetic, mechanical, thermal, etc.) originate from their parameters, such as nanoscale size, shape, geometrical structure, chemical composition, surface chemistry, and charge (Khan et al., 2019). These properties make NPs a powerful tool for medical therapy, imaging, or drug delivery. New engineered NPs offer

Nanotechnology for Hematology, Blood Transfusion, and Artificial Blood. DOI: https://doi.org/10.1016/B978-0-12-823971-1.00002-7

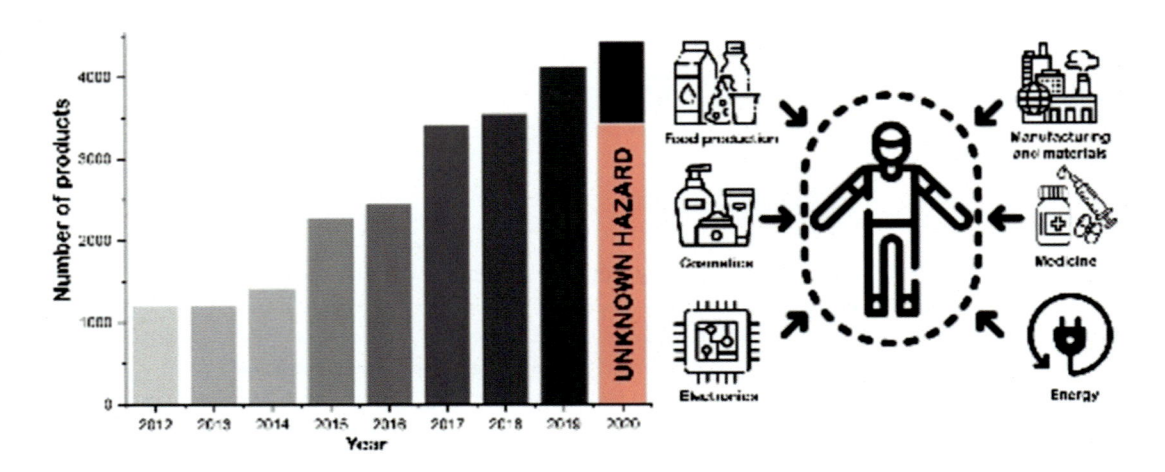

FIGURE 1.1

Registered growth of the nanoparticle (NP)-containing products by Nanodatabase. Main applications of NPs: food, cosmetics, and electronics products, manufacturing and materials, drugs and medications, and energy industries.

novel approaches to prevent and treat difficult cardiovascular conditions (Sarmah et al., 2017), infections, cancer (Revia & Zhang, 2016), and Alzheimer's disease (Leszek et al., 2017), allow imaging and diagnostic modalities (Man et al., 2018), and advance personalized medicine with control drug delivery (Singh & Lillard, 2009; De Jong & Borm, 2008) without off-target toxicities. On the other hand, despite all the trending advantages of nanomaterials, there is a high risk emerging from the shallow understanding of NP-related adverse effects on humans. For instance, some observations demonstrated harmful effects, such as developing neurodegenerative disorders (Lucchini et al., 2012), stroke, myocardial infarction, and inflammation of the upper and lower respiratory tracts (Madl et al., 2014).

The adverse effects arising from NP exposure on living system are called nanotoxicity; on the contrary, NPs are biocompatible if they do not produce any adverse effects, such as immune response or inflammation. The mode of NPs interactions with living systems is defined by the NPs' intrinsic properties: a single change in NP parameters can impact its properties and hence the interaction mechanism. High surface/volume aspect ratio makes NPs highly reactive regarding biological systems, forcing the undesirable interactions with physiological environment.

NP-related risk assessment is complicated because it involves a large number of factors that influence the interaction mode between the NPs and living systems. The complexity of NP behavior in vivo has led to many contradictory observations. At the moment, the generalized scientific understanding of NP impact on living systems is lacking. The prediction of NP behavior upon the interaction with living systems starting from the initial uptake to physiological endpoints requires time and a huge effort from the scientific community. Therefore, the understanding of NPs' influence on the living system is a multistage and time-consuming process to attempt to associate NP intrinsic properties and their impact on biological objects. Only a systematic approach to versatile

NP properties study can help researchers to identify environment-friendly nanomaterials and reduce the risk for human health.

Compatibility with blood or "hemocompatibility" is one of the primary factors to be understood for the smart design and development of nanosystems as therapeutics (Brash, 2018). Since the blood system often serves as an administration and distribution route for medical NPs, for example, in tumor targeting (Lazarovits et al., 2015), blood—NP interactions is the first determinant of the further NP efficiency.

The study of blood—NP interactions at the moment comprises specific cases of individual blood components interfacing NPs. The main challenge is in the significant implications of NP behavior occurring in the multicomponent complex system, which cannot be ignored, but yet are too hard to be understood in in vivo conditions, when the system is working as a whole in a synergetic way. Blood—nanomaterials interactions have been extensively reviewed previously (Brash, 2018; Cruz La et al., 2018; Harpe et al., 2019). However, the lack of understanding of NP—blood interactions and the poor prediction models set the limits on the development of the effective nanodrugs.

This chapter discusses the known mechanisms of blood—NP interactions, the reported risks related to the interaction of blood with NPs, and the future beneficial prospects of nanomedicine.

1.2 **Nanoparticle uptake pathways**

Nanomaterials are easily internalized by the human body on the organ, cell, and molecular levels due to their extremely small size. The overwhelming number of NP sources increase risks of NP uptake after exposure. NPs may enter the human body via a few potential exposure pathways, each of them is eventually escorting NPs to the blood circulatory system

1. Inhalation. NPs inhaled from the nose or mouth are spread in the respiratory system to the lungs. Due to the connections of the system, through the inhalation, NPs can also enter the gastrointestinal tract (Buzea et al., 2007). At the endpoint of the respiratory tract, NPs diffuse fast in the capillary tubes of alveoli, and NPs start entering the lymphatic and the blood circulatory system (Miller et al., 2017). Nemmar, Hoet, et al. (2002) detected the inhaled carbon NPs in blood of the volunteers 1 minutes after inhalation, reaching the maximum within 10—20 minutes.
2. Ingestion. Ingestion of NPs is leading them through the gastrointestinal tract. Epithelial cells in the stomach perform their regular function effectively to absorb NPs, and translocate them into the other organs and the circulatory system (Bergin & Witzmann, 2013).
3. Transdermal exposure. Being in contact with skin, NPs from the cosmetic and pharmaceutical products can be absorbed through the different layers of the skin (Larese Filon et al., 2016). Multilayered skin composition is presented in particular by the blood and lymphatic microvessels (Buzea et al., 2007), dendritic cells, and nerve endings, which are the further routes for NPs.
4. Intravenous administration. Biomedical applications, such as diagnostics, imaging, and therapy, often require an intravenous injection of NP-based pharmaceutics (Man et al., 2018). The blood circulatory system has access to the majority of the organs and can spread the NPs throughout

the body to vital organs such as heart, brain, liver, kidneys, spleen, bone marrow, and nervous system (Dukhin & Labib, 2013).

1.2.1 Nanoparticles and blood

The evolution of NPs is based on the understanding of the correlation between their properties and the corresponding efficacy in vivo (Albanese et al., 2012; Li et al., 2016). NPs of the first generation were functionalized with basic chemistries to estimate biocompatibility and toxicity, however, they were mostly unstable, subject to fast clearance, and caused high toxicity. In the second generation the better solubility, stability, targeting, and decreased adsorption of biomolecules were achieved by polymer grafting. The third generation are environment-responsive NPs for improvement of specific targeting. NPs from all the generations are still actively investigated, and some of them have been already approved by Food and Drug Administration (FDA).

NPs developed for medical applications can be classified by the chemical composition into three main groups (Limongi et al., 2019):

1. Carbon-based NPs: fullerenes, carbon nanotubes (CNTs), and graphene.
2. Organic or lipid-based NPs: polymeric NPs (PNPs), liposomes, dendrimers, extracellular vesicles etc.
3. Inorganic NPs: metal-based (silver (Ag NPs), gold (Au NPs), cadmium (Cd NPs) etc.), metal oxide [zinc oxide (ZnO), titanium dioxide (TiO_2), iron oxide or magnetite (Fe_3O_4), silicon dioxide (SiO_2)], quantum dots (QDs), upconversion NPs.

Known as the "4S parameters" of NPs, the size, shape, surface functionality, and stiffness were found to define their interaction with living systems (Li et al., 2016). Besides, NP fate also depends on the experimental design including the exposure time and the route, the dose, and the personal health state of the patient. Finally, the physiological environment influences the NPs, defining the further possible routes, such as toxicity, clearance, or the achievement of the intended therapeutic effect (Fig. 1.2).

There are various methods of NP production, which depend on the NP type and application. NP synthesis for biomedical applications requires control over NP properties, which define their synthetic identity, that is, size, shape, composition, and surface functionalization (Fig. 1.2), and strongly affect NP initial uptake, distribution, accumulation, and associated toxic effects.

Though the physicochemical properties of engineered NPs can be controlled in the production phase due to known environment conditions, they will be highly influenced by the complex in vivo environment, including biomolecules, cells, and organs, affecting NP properties in the uncontrolled and unpredictable way. Therefore to minimize the impact of the physiological environment of NPs' synthetic identity they go through a prior characterization and testing to ensure proper functioning when applied in vivo. The most common techniques used in NP characterization are (Titus et al., 2019):

1. Atomic force microscopy, scanning (SEM) and transmission (TEM) electron microscopy: size, shape, and size distribution.
2. Dynamic light scattering: size in colloids, hydrodynamic diameter, stability (agglomeration).

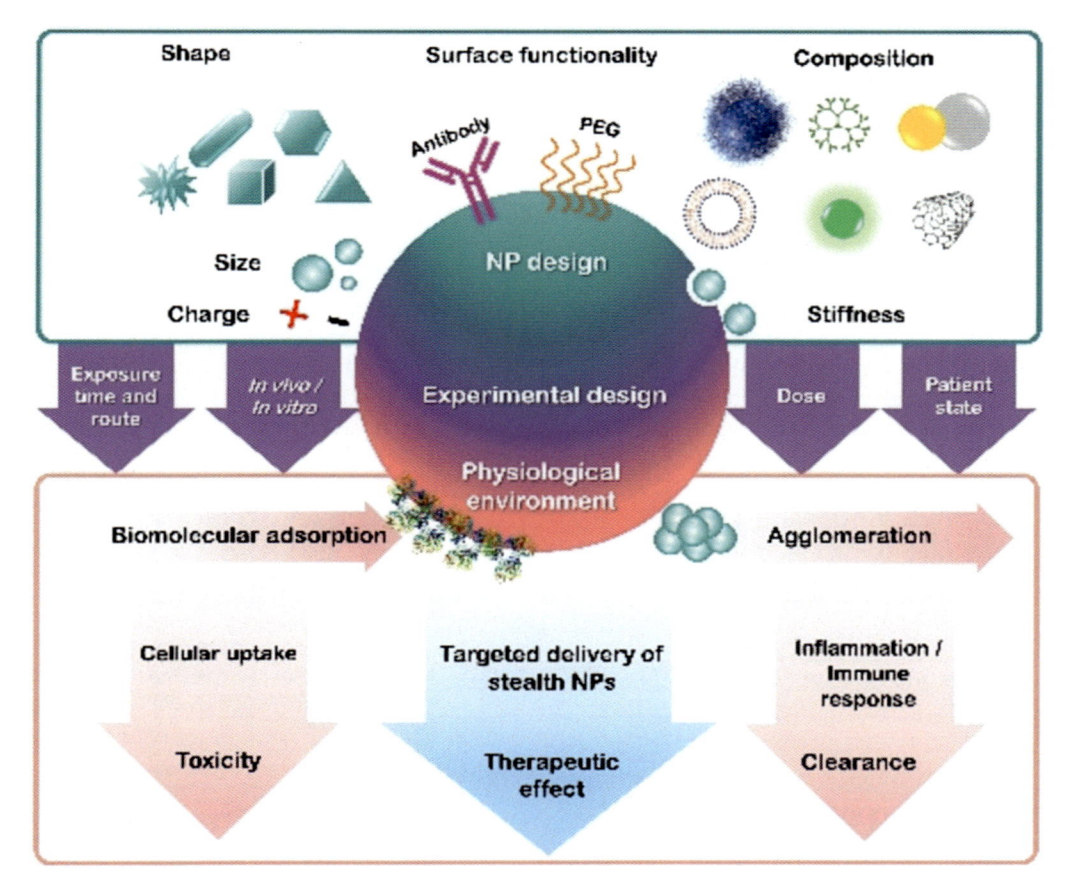

FIGURE 1.2

Nanoparticle (NP) design features and experimental design influence the NP fate in vivo. Variety of factors on the way to the desired therapeutic effect: the design of NP properties, the experimental conditions, and the influence of the physiological environment.

3. X-ray diffraction (XRD), X-ray photoelectron spectroscopy: chemical composition, crystal structure, surface chemistry.
4. Energy dispersive X-ray spectroscopy: elemental analysis of composition.
5. (UV-VIS) absorbance spectroscopy: monodispersity, absorption spectrum.
6. Zeta potential: charge, stability.

NPs in physiological environment undergo fast biomolecules adsorption and obtain so-called biological identity. The latter from now on determines the future fate of NPs, however, its identity is hard to predict because it depends on both the NP properties and the properties of the environment. The complexity of nanomedicine lies in the difficulty to draw generalized conclusions, since the NPs behavior in physiological conditions is always a function of too many factors, such as

NPs' intrinsic properties, concentration, local environment, individual state of the patient, etc. The effective nanodrug should perform its function with minimal disturbance of the physiological parameters.

For example, coating the surface with polyethylene glycol [PEGylated (PEG) NPs], or carbohydrates (Pelaz et al., 2013) allows them to maintain their monodispersity, prevent agglomeration, and ensure biocompatibility. PEG can drastically increase the circulation time of NPs by preventing the biomolecules adsorption. The coating of NPs with organic ligands (antibodies, peptides, DNA, etc.) facilitates the active targeting applications and provides the invisibility of NPs for reticuloendothelial system (RES). Therefore the evaluation of NP physiochemical properties is necessary for deriving the correlation of these properties with physiological response.

Usually, regardless of the exposure route into the human body, NPs enter the bloodstream and encounter blood components. Blood is a complex fluid composed of cellular and acellular elements. Most of the blood volume $\approx 55\%$ is occupied by liquid plasma consisting of 91% water with 9% dissolved proteins and other biomolecules. The remaining 45% is cell components, 99.1% of which are red blood cells (RBCs) and only 0.9% are white blood cells (WBCs) or leukocytes and platelets (Patton & Thibodeau, 2018) (Fig. 1.3). The ratio of the cellular blood constituents is one leukocyte

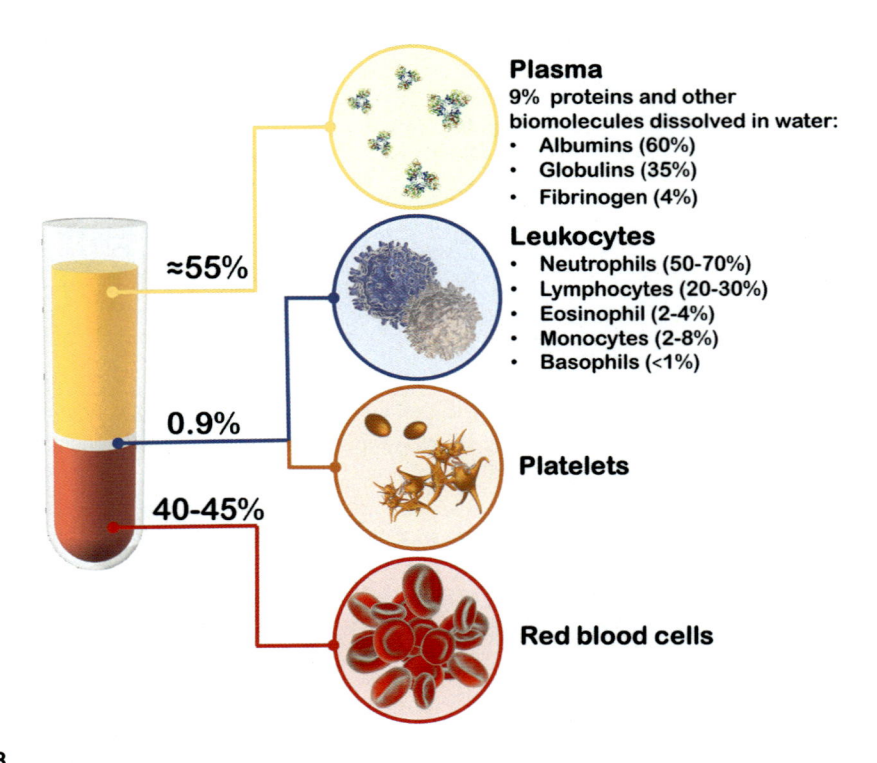

FIGURE 1.3

Blood composition: $\approx 55\%$ of blood plasma, consisting of 9% proteins dissolved in water, and cellular components: leukocytes and platelets (0.9%), and red blood cells (40%—45%).

per 25 thrombocytes per 500 RBCs. Each of the cell types has a specific function: RBCs transport the gases (O_2 and CO_2 exchange), WBCs fight against infection in the immune response, platelets close wounds, and endothelial cells participate in homeostasis and translocate biomolecules from blood to other cells.

Once NPs enter the bloodstream special protective biological barriers are trying to eliminate xenobiotic material. NPs' properties are strongly affected by the blood components leading to NP clearance by RES organs (Jeevanandam et al., 2018) and immune system cells. Blood—NPs interactions define their future fate, and pose a threat of undesirable effects such as inappropriate biodistribution and clearance, toxicity (Cruz La et al., 2018), and low efficacy of nanomedicine.

There are two interchangeable terms describing the possible adverse effects caused by NPs' interaction with blood constituents: hemocompatibility and hemotoxicity. To evaluate the influence of NPs on blood, different hemocompatibility tests are used (Szebeni & Haima, 2013), which estimate the critical interactions with adverse effects between xenobiotic materials and whole blood or its components. The mechanisms leading to the hemotoxicity include the immune response, the activation of the complement system and coagulation factors, and the cell components damage, such as an increased production of the reactive oxygen species (ROS) (Yu et al., 2020).

At first, plasma proteins absorbed on the NP surface form a complex named protein corona (PC) (Pederzoli et al., 2017), and then these "physiologically modified" NPs interact with blood cells. The next processes commonly occurring in blood are complement activation, platelet activation and aggregation, immune cells chemotaxis and activation, and cell damage (Chen et al., 2015), which eventually result in inflammation, thrombosis, or hemolysis (RBC lysis). NPs can be internalized inside the blood cells: via the passive transport in RBCs, due to the lack of the uptake machinery, while WBCs, specially designed to protect organism from the xenobiotic materials, can effectively engulf NPs by the active receptor-mediated phagocytosis (Shang et al., 2014). The most probable outcomes that NPs can cause in blood are presented on Fig. 1.4.

The physicochemical characteristics of the nanomaterial [size (Kwon et al., 2012), shape, composition, surface functional groups, and surface charges (Zhao et al., 2011)], the features of the physiological surroundings (blood, interstitial fluid, cell cytoplasm, etc.), the route of administration, the duration of exposure (Qualhato et al., 2017), and the dose (Fonseca et al., 2018) are the most significant factors influencing the toxicity of NPs.

1. Size. NP size is one of the prime physicochemical properties affecting the toxicity of materials (Kwon et al., 2012; Sukhanova et al., 2018). NPs' large surface area to volume ratio facilitates their high reactivity regarding the environment, from the biomolecule adsorption to the following cell uptake. Many studies have verified that the decreased size is the main cause of toxicity induction (Barshtein et al., 2016), for example, ROS increase in macrophages treated by Ag NPs (Park et al., 2011), or some materials that are nontoxic in bulk exhibit intense toxicity as their size reduces (Liao et al., 2011). The size was found to be a key parameter in NP internalization inside RBCs, compared to NP charge or material (Rothen-Rutishauser et al., 2006). Cytotoxic effects caused by the NPs were reviewed by Shang et al. (2014), underlying the importance of the size in the interaction mode of NP. Theoretical and experimental study of cellular uptake by Zhang et al. (2015) revealed that NPs smaller 5 nm can nonspecifically translocate through the membrane, while bigger NPs internalized via phagocytosis, macropinocytosis, and specific and nonspecific transport mechanisms.

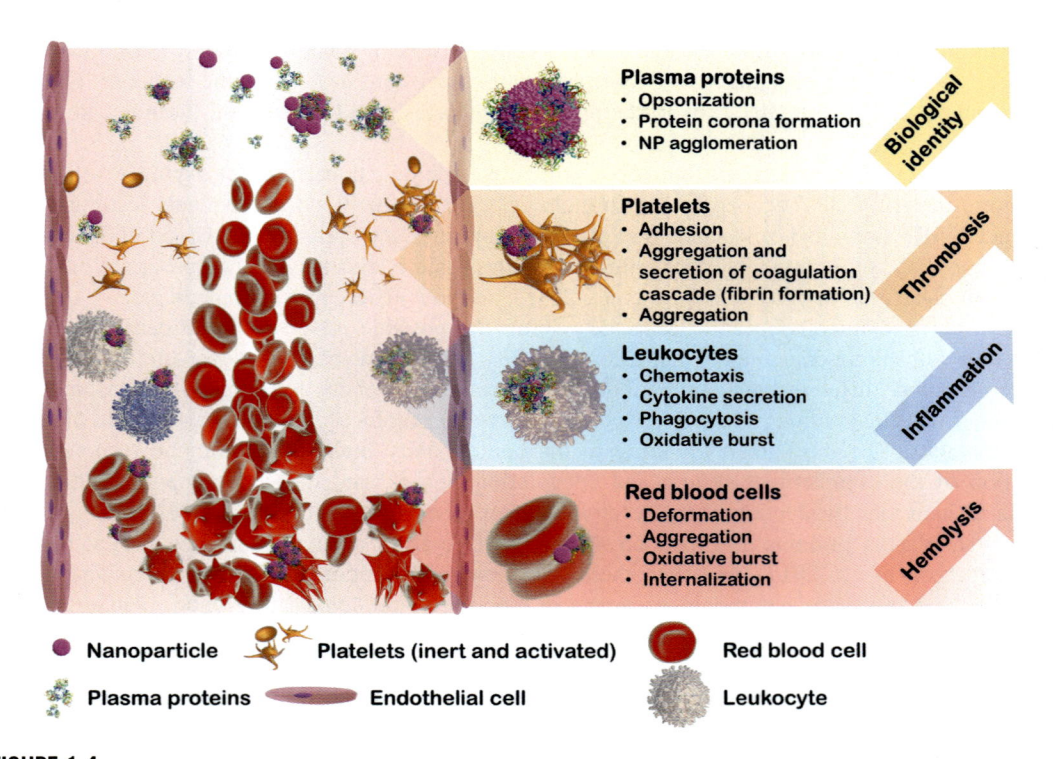

FIGURE 1.4

The most probable outcomes of nanoparticles' interaction with the main blood components inside a blood vessel.

2. Shape. NP shape is an important factor in the interaction with a cell membrane and internalization, and margination process in blood vessels (Cooley et al., 2018). Having the contradictory findings for the role of shape in the toxicity of NPs, Wang et al. (2019) by analyzing available studies, made the conclusion, that rod-shaped NPs in general showed a higher level of cell internalization, and higher drug delivery efficiencies than spherical NPs.
3. Surface charge. The main consideration of the NP surface charge in the interactions with cells is a negative membrane charge, favoring the electrostatic interaction with positively charged NPs, or repulsion from the negative NPs (Lin et al., 2010). Cationic NPs in the blood system are prone to strong interactions with RBCs disrupting cell membranes and causing hemolysis (Aillon et al., 2009). In work by Das et al. (2017) surface functionalization of Ag NPs significantly decreased the inflammation and hemolytic levels.
4. Composition and surface functionality. NP surface engineering is a way to dictate the NP interactions with a living system. Nanomedicine is pursuing NPs' biocompatibility and targeted interaction with specific sites of the body, which can be improved by using coatings to decrease the toxicity levels, stabilize the particles, prevent them from agglomeration, and increase the targeted cellular uptake (Mout et al., 2012).

1.2.2 Evaluation of nanomaterials hemocompatibility

Hemocompatibility is the primary criteria of NP applicability for in vivo and clinical applications. Adverse NP interactions with blood components result in inflammation activation of coagulation, and/or fibrinolysis leading to thrombosis and hemolysis, which are considered interdependent. The guidance of the International Organization for Standardization (ISO 10993-4) highlights five main categories of hemocompatibility evaluation: thrombosis, coagulation, platelets, hematology, and immunology (complement system and leukocytes) for hemocompatibility evaluation (ISO 10993-4, 2017; Weber et al., 2018).

Hemocompatibility tests in vitro rely on whole blood or blood components incubation with NPs with an estimation of the cell count, hemolysis, activation of the coagulation and complement, activation of platelets and leukocytes, and comparison with unmodified controls. Incubation can be done in static, agitated, or shear flow conditions. In vivo animal studies are required for preclinical evaluation of NPs.

Various methods are applied for hemocompatibility estimation. Hematology analyzers perform the blood cell count based on the electrical impedance. Microscopy methods, for example, hemocytometer, are also applied for testing. The decreased number of RBCs due to the membrane rupture (hemolysis) results in hemoglobin and macrovesicles release, which can induce toxicity or thrombus activation. The decrease in platelets number is associated with the thrombogenic effects of NPs. Coagulation (factor XII, factor XI), plasma prekallikrein (PK), and the nonenzymatic cofactor high-molecular-weight kininogen, complement (over 30 proteins), and platelet activation (α-granule release, integrin $\alpha_{IIb}\beta_{III}$ conformational change) are mediated by the complex cascade of the proteins, which are usually quantified using enzyme linked immunosorbent assay (ELISA). ELISA is also efficient for the quantification of leukocyte activation mainly mediated by ROS generation and polymorphonuclear (PMN) elastase. Platelet shape deformation can be assessed with SEM imaging. For more details on hemocompatibility assessment please refer to the review by Weber et al. (2018) and Evani and Ramasubramanian (2016).

NP hemocompatibilty testing should include various aspects. For example, Huang, Lai, et al. (2016) performed diverse investigation of Ag NPs—the most common NP type in medicine. According to the observations, hemolysis and lymphocyte proliferation occurred even at the lowest Ag NPs concentrations (10 mg/mL), but the platelet aggregation, coagulation process, or complement were not activated until the concentration reached $\sim 40\ \mu g/mL$.

1.3 Nanoparticles in blood plasma: biological identity

The biggest barrier NPs experience is at the very beginning after administration into the blood. On being exposed to blood, NPs immediately undergo the spontaneous formation of the plasma biomolecule layer—PC—consisting of proteins, sugars, lipids, and nucleic acids (Rampado et al., 2020; Cedervall et al., 2007). The NP—protein attachment involves different kinds of intermolecular interactions: electrostatic, hydrogen bonding (hydrophobic and hydrophilic), and Van der Waals (Pareek et al., 2018). Hydrophobic or charged NPs are more likely to interact with proteins, than hydrophobic and neutral ones.

The attachment of proteins follows the Vroman effect (Vroman et al., 1980). Biomolecules in plasma are competing to get adsorbed on the surface of NPs, the smallest and most abundant proteins get adsorbed first, but then the affinity becomes the driven force in PC formation, and the high-affinity proteins gradually substitute the low-affinity ones. Among the almost 3700 plasma proteins (Sun et al., 2018), only 100—300 were found to form the PC within less than 5 minutes (Tenzer et al., 2013). On the contrary, PC formation in the study by Allémann et al. (1997) took 1 hour. Interestingly, the abundance of plasma proteins does not correlate with their abundance in PC composition or their affinity (Zhang et al., 2011; Dobrovolskaia et al., 2009; Dufort et al., 2012).

The observations of PC formation at the present time are ambiguous. For example, PC was found to be formed as one monolayer (Röcker et al., 2009), while another study suggested the multilayered structure of PC (Goy-López et al., 2008). Most of the researchers have accepted the model, where the "hard corona," is surrounded by the "soft corona"—the relatively unstable complex of fast exchanging and low-affinity proteins (Hadjidemetriou & Kostarelos, 2017). Docter et al. (2015), to avoid the confusion, suggested the generalized PC term referring to the analytically accessible NP—protein complex. It was shown, that even after forming the "hard corona," the NP—protein complex translocated to another environment, for example, cell compartment, can be subjected to the alterations in protein composition, and on reaching the final site, PC can have the traces of the previously visited environments (Monopoli et al., 2012; Bonvin et al., 2017). Therefore the route of NP administration should be carefully considered in targeted drug delivery to minimize the unpredictable biomolecule adsorption on NP.

Protein—NP interfacing is a mutual process with both sides influencing each other's properties. Protein conformational changes or complete denaturation are able to make the proteins unrecognizable, which will trigger the immune response. The binding of proteins facilitates the dynamic change of NP properties, mainly size and charge, which subsequently define their further routes: the uptake by the cells, the biodistribution, the clearance by the immune cells, and the accumulation sites in the body. Proteins related to physiological response on the toxicological processes were identified in PC of various nanomaterials. Albumin, fibrinogen, apolipoproteins, immunoglobulins, and complement proteins (e.g., C1q, C3b proteins), are the most common proteins found in the PC of various NPs (metal oxide, carbon, polymeric, and liposomes) (Vu et al., 2019). Albumin, being the most abundant plasma protein, due to the highest concentration rapidly absorbs on NPs and reduces their cellular uptake by masking them from the mononuclear phagocyte system (MPS). At the same time, the apolipoprotein, complement, and immunoglobulins "mark" NPs to be recognized and removed by the phagocytic cells (Lee et al., 2014). Complement and coagulation factors have also been identified in forming the PC, which is reducing the target cellular interactions, by making the surface ligands inaccessible for cells (Pearson et al., 2014).

PC composition depends on the inherent NP properties, size, surface properties (Lundqvist et al., 2008), shape (Madathiparambil Visalakshan et al., 2020), charge, and on the extrinsic properties of biological environment, its physiological condition, temperature, localization, etc. The time of exposure (incubation time) has a significant influence on PC, affecting the composition both qualitatively and quantitatively (Palchetti et al., 2017). In general, the correlation between in vivo and in vitro experiments is very weak (García-Álvarez et al., 2018; Amici et al., 2017), underlying the importance of factors influencing PC formation.

Specifically, in the bloodstream in vivo the blood composition, the flow rate, and the presence of pathophysiological processes are crucial for PC formation. For instance, the number of proteins of

NP corona in the blood flow mimicking conditions was shown to change upon the increasing rate of the flow (Bonvin et al., 2017); the microfluidic in vitro methods were proposed by Mahmoudi (2018). Braun et al. (2016) hypothesized, that an increased number of encounters between NPs and proteins in blood flows of a higher rate is the reason for higher protein adsorption on NPs. In the experiment by Jayaram et al. the loosely attached proteins can be removed from NPs by the shear stress at high flow rates (Jayaram et al., 2018). Moreover, the structural changes of the protein were affected by the flow, resulting in different protein−NP interactions and PC composition.

The responses of blood components to NPs are very closely interrelated and can activate and enhance each other. The initial protein layer adsorbed on NP may trigger activation of a coagulation cascade and activation of leukocytes and platelets, which result in inflammation and thrombosis. Even being in the blood system for a short time, pristine NPs were able to induce moderate toxicity by triggering the platelet activation, and to affect vitality of the endothelial cells and RBCs (Tenzer et al., 2013).

PC usually prevents the cellular uptake of NPs (Smith et al., 2012; Lesniak et al., 2012) by decreasing the adhesion to the cell membrane. However, there are also observations reported otherwise (Tenzer et al., 2013; Chithrani et al., 2006). It is expected that corona affects the hemocompatibility by coating the NPs to prevent the direct contacts between NPs and blood cells. Just 3%−4% of human serum albumin applied to cover the NPs allows a reduction in the side effects and enhances the therapeutic effects (Cucinotto et al., 2013) in the first NP-based drug for cancer treatment a NP albumin-bound paclitaxel. Besides, the surface of implantable devices with increased albumin adsorption led to a better hemocompatibility (Keogh et al., 1992). However, the composition of corona on the surface of PEGylated (PEG) gold NPs did not correlate with hemocompatibility (Dobrovolskaia et al., 2009).

Synthetic coating is an effective strategy to create the stealth (passive) NPs. PEGylation is a widely used method to reduce the PC formation around the NPs (Suk et al., 2016). However, in the case of high grafting, it can decrease the uptake efficiency by the target cells and fail to hinder phagocytosis, therefore some of the aspects of the procedure should be carefully considered. Zwitterionic coating not only decreased the adsorption of bovine serum albumin (BSA) and stabilized NPs in the physiological environment, but prevented hemolysis (Loiola et al., 2019).

The first human in vivo PC characterization of PEGylated doxorubicin−encapsulated liposomes (Caelyx) performed by Hadjidemetriou et al. (2019) revealed the unexpected result: the most abundant component was the cDNA clone CS0DD006YL02 protein, which was not previously reported in the PC composition of any NPs.

Contradictory observations are not rare at the current stage of PC studies. The described phenomena are rather highly specific cases of intrinsically complicated systems interaction. An improved proteomic analysis workflow offered recently by Blume et al. (2020) provided the efficient proteomic profiling. The large volume of data is hard to analyze, therefore the progress is expected with the help of computational (quantitative structure-activity relationship) models (Berrecoso et al., 2020).

1.3.1 Opsonization

An important part of PC formation taking place within seconds after NP exposure to blood is the opsonization process (Owens & Peppas, 2006). Special opsonin proteins attach to NPs, in this way labeling them to be recognized by phagocytic cells of the MPS followed by destruction and elimination. This protective function of the RES system promotes fast NP elimination and accumulation preferably in the liver and spleen (Aillon et al., 2009), such an outcome, however, is undesirable if the prolonged circulation time of NPs in the bloodstream is required. Opsonization, adhesion, and internalization of NPs by the MPS are still poorly understood.

Opsonin proteins are mostly presented by the serum proteins (laminin, fibronectin, type-I collagen, albumin, etc.), immunoglobulins (i.e., antibodies) (IgG and IgM), and complement system components (C3, C4, and C5). Opsonized NPs, "marked" for specific ligand—receptor interactions, are identified and engulfed by phagocytic cells. While the opsonins facilitate the fast NP clearance from the circulation system, therefore decreasing the NP circulation time, dysopsonins (apolipoprotiens) exhibit the opposite effect by decreasing the NP elimination.

Complement proteins aim to destroy NPs. Certain PC composition can facilitate the complement activation and cause inflammation (Sim & Tsiftsoglou, 2004). Therefore the NP surface should be designed to minimize the immune response and to improve the NP performance.

Hulander et al. (2011) demonstrated differential complement activation of Au NPs of different surface hydrophobicity by evaluating the complement protein (C1) bound to IgG adsorbed on the nanosurface. Inclusion of the hydrophobic groups (NH_2, $-OH$, or $-COOH$) usually enhances the complement activation (Rybak-Smith et al., 2011). Observed complement activation was significantly lower for hydrophilic than for hydrophobic ones. Yu et al. discovered that conformational states of glycopolymer chains covering NP can act as a "switch" for activation and amplification of the complement system.

1.4 Platelets and coagulation system

Platelets are the main contributors to homeostasis along with coagulation factors and vascular endothelium. In the case of injury platelets accumulate at the site of injury and create the fibrin links, forming a clot to cease the bleeding. Moreover, platelets also actively manifest themselves in immune response, vascular inflammation and repair, angiogenesis, and diseases (Mancuso & Santagostino, 2017). Platelets are highly reactive anucleate cells of $1.5-3\,\mu m$, which can perform versatile functions via controlled release of storage granules and procoagulant membrane extracellular vesicles, and undergo structural transformations, possessing a very dynamic cytoskeleton.

Any xenobiotic has at least a minimum ability to induce thrombogenicity (Evani & Ramasubramanian, 2016). NPs can encounter platelets in the vicinity of the vascular endothelium with the highest concentration of platelets or start a cascade of reactions by adhering to the vascular wall. Responding to the interruption of hemostasis, platelets undergo a serious structural transformation mediated by the complex network of signaling pathways: activation, adhesion, secretion, and aggregation.

According to the available data, most NPs possess procoagulant activity. Since NPs can affect the function of platelets by activating or inhibiting them, the thrombogenic activity of NPs is one of the important factors to be estimated, in order to prevent the side effects in a form of thrombosis or bleeding. CdTe QDs (Samuel et al., 2015) promoted the activation and CdSe QDs (Dunpall et al., 2012) the aggregation of platelets. Rutile TiO_2 nanorods caused the acute inflammation-induced thrombosis in rats in vivo and in vitro in a dose-dependent manner (Nemmar et al., 2008).

An unambiguous correlation between the platelet activation and NP size was reported by many studies, but generally suggested that toxicity is not defined solely by the NP size. Platelet activation was not affected by the gold nanospheres up to 60 nm, and was inhibited by bigger size NPs (Aseychev et al., 2013), confirming their biological safety. However, in the study by Deb et al. Sanfins et al. (2014), in contrast to 68 nm NPs, 18 nm Au NPs activated the platelets. Carboxyl-modified polystyrene NPs were reported to have a size-dependent effect on the platelets activation: large NPs (60−220 nm) triggered the activation of the intrinsic pathway, while smaller NPs (26 and 24 nm) inhibited the activation (Deb et al., 2011). Fifty nanometer SiO_2 NPs caused more pronounced platelet aggregation in mouse blood, than 500 nm NPs (Nemmar et al., 2014).

NP surface charge can mediate platelet−NP interactions, as the negatively charged platelets (sialic acid) could interact with positively charged NPs. Platelet activation by the positively charged amine−polystyrene particles was demonstrated in vivo (Nemmar, Hoylaerts, et al., 2002). However, negatively charged surfaces can more easily initiate thrombotic events (Ilinskaya & Dobrovolskaia, 2013). Platelet aggregation of human platelets in vitro was also induced by other cationic nanomaterials, such as dendrimers (Dobrovolskaia et al., 2012) and carboxyl- and amine-modified polystyrene NPs (McGuinnes et al., 2011). However, anionic, neutral, or small cationic dendrimers had no effect on platelet aggregation.

In general, regarding surface properties amino, phosphate, hydroxyls, and carboxyls on the NP surface result in platelet activation (Fröhlich, 2016). NP shape properties also influenced platelets in the study by Ferraz et al. (2010). Alumina membranes with 200 nm pores promoted the generation of platelet microparticles, while 20 nm porous membrane did not have any affect.

Studying the effect of silica NPs on platelet adhesion in flow conditions, Saikia et al. (2018) demonstrated the overexpression of platelet−endothelial cell adhesion molecules (PECAM) under the influence of NPs, which led to enhanced platelet−endothelial cell interactions. PEGylated Au NPs were found to attach to the platelet membrane, but also localized in small grooves, probably opening the membrane canalicular system (Shah et al., 2012). Interestingly, negatively charged Au NPs did not interact with platelets in vitro, but showed great uptake by circulating cells in vivo (Fig. 1.5A).

Still, many NPs have a promising potential for thrombosis treatment: liposomes, PNPs, dendrimers, PEGylated NPs, magnetic and iron oxide NPs (Karagkiozaki et al., 2016). In the study by Shabanova et al. (2018) human thrombin was entrapped in a matrix of Fe_3O_4 NPs, which after injection into a model vessel were guided to the site of bleeding through a magnetic field, where they caused coagulation and accelerated local hemostasis. Au NPs coated with chitosan demonstrated anticoagulant properties (Ehmann et al., 2015), thrombin-binding aptamer-conjugated Au NPs controlled by irradiation with green laser inhibited thrombin activity (Huang, Wei, et al., 2016). Ag NPs (Lateef et al., 2017) showed thrombolytic effect, and ZnO NPs reduced the thrombin generation potential regardless of the NP size (Yang et al., 2017).

Aiming to reduce NP-induced toxicity, Hu et al. (2015) developed an elegant solution—the platelet membrane-cloaked NPs (Fig. 1.5B). Biodegradable polymeric nanoparticle coated with the plasma membrane of human platelets and functionalized with immunomodulatory and adhesion antigens allowed not only to decrease the cellular uptake and avoid the complement activation, but also could mimic some platelet functions. Preliminary incubation of silica NPs in human plasma in

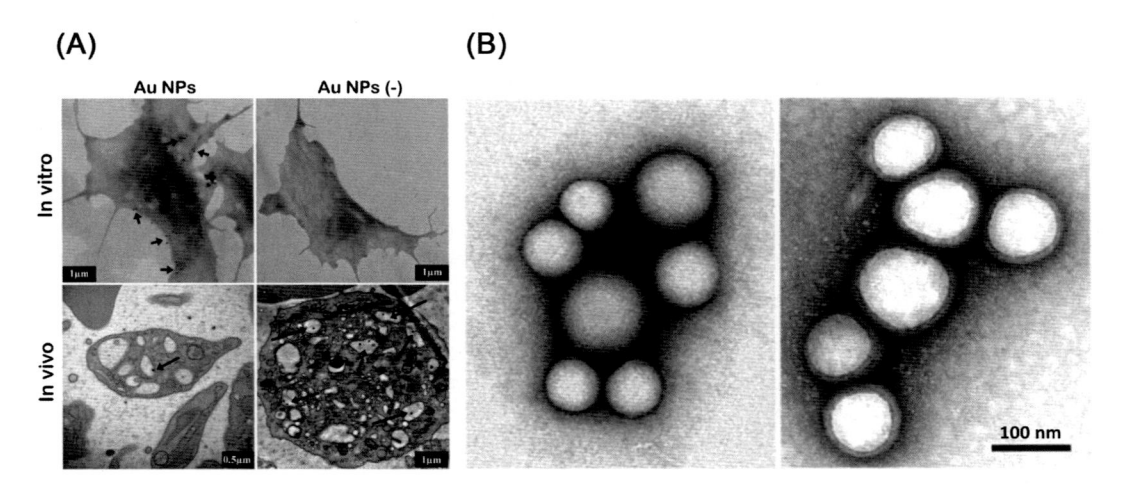

FIGURE 1.5

(A) Au nanoparticles (NPs) bind to the surface of washed platelets after 15 min of incubation. Circulating platelets show Au NPs uptake in 1 h postinjection. Anionic Au NPs (-) do not interact with platelet in vitro, but show the uptake in circulating blood monocytes in vivo.

Adapted with permission from Neha B. S., Vercellotti, G. M., White, J. G., Fegan, A., Wagner, C. R., & Bischof, J. C. (2012). Blood-nanoparticle interactions and in vivo biodistribution: Impact of surface peg and ligand properties. Molecular Pharmaceutics, 9(8), 2146–2155. https://doi.org/10.1021/mp200626j. Copyright 2012 American Chemical Society. (B) TEM images of bare NPs (left) and platelet membrane-cloaked NPs (right) negatively stained with uranyl acetate. *Adapted with permission from Hu, C. M. J., Fang, R. H., Wang, K. C., Luk, B. T., Thamphiwatana, S., Dehaini, D., et al. (2015). Nanoparticle biointerfacing by platelet membrane cloaking. Nature, 526(7571), 118–121. https://doi.org/10.1038/nature15373. Copyright © 2015, Springer Nature.*

the study by Tenzer et al. (2013) helped to avoid platelet activation, which is easily triggered by bare NPs. PC composition of CNTs (De Paoli et al., 2014) had an influence on the platelet activation: albumin PC could decrease the aggregation, while histone H1 and gamma globulins PCs induced platelet aggregation and fragmentation.

The mechanisms behind NP—platelet interactions are orchestrated by numerous factors complicating the interpretation of the observed effects. Further study of the inflammation and immune responses to NPs is required.

1.5 Immune system response: leukocytes

Leukocytes or white blood cells (WBCs) are cell protectors in blood against the foreign substances. WBCs are more complex than RBCs, and are represented by five cell types: neutrophils (12−14 μm), eosinophils (12−17 μm), basophils (12 μm), monocytes (20 μm), and lymphocytes (T cells and B cells) (7 μm). Neutrophils are the most abundant WBC type, constituting over 50% of the circulating leukocytes; thus the majority of functions are provided by neutrophils.

Leukocyte activation can be related to the NP inflammatory profiling as a part of the hemocompatibility testing. Due to the same activating mediators, the important role of leukocytes in coagulation was confirmed along with platelets and RBCs (Swystun & Liaw, 2016). Adhered platelets recruit leukocytes and encourage their migration and activation (Ghasemzadeh et al., 2013).

Reacting to the inflammation, leukocytes functions include chemotaxis, cytokine secretion, neutrophil extracellular traps (NET) formation, oxidative burst, and phagocytosis. Monocytes (biggest leukocytes) mediate inflammatory response by releasing cytokines (IL-6, IL-8) and small chemokines signaling for activation of other cells. Finally, this cascade can lead to an acute inflammation.

NETs formation, which is the host neutrophil defense, was observed as the immune response to the treatment by the cationic solid lipid NPs (cSLNs) (Hwang et al., 2015) (Fig. 1.6). By capturing the platelets, NETs can then stimulate thrombus formation.

After NP opsonization leukocytes can effectively accumulate NPs. Different effects caused by NPs on neutrophils are described in the review by Lin et al. (2018). Ex vivo human neutrophils perform NPs uptake within 15 minutes, preferentially of larger NPs (up to 200 nm in size). Ingested particles reside in intracellular compartments that are retained during activation and degranulation (Bisso et al., 2018). Fasbender et al. (2017) demonstrated concentration-dependent uptake of 1−3 nm graphene QDs, which was faster and hence greater for granulocytes and monocytes, in

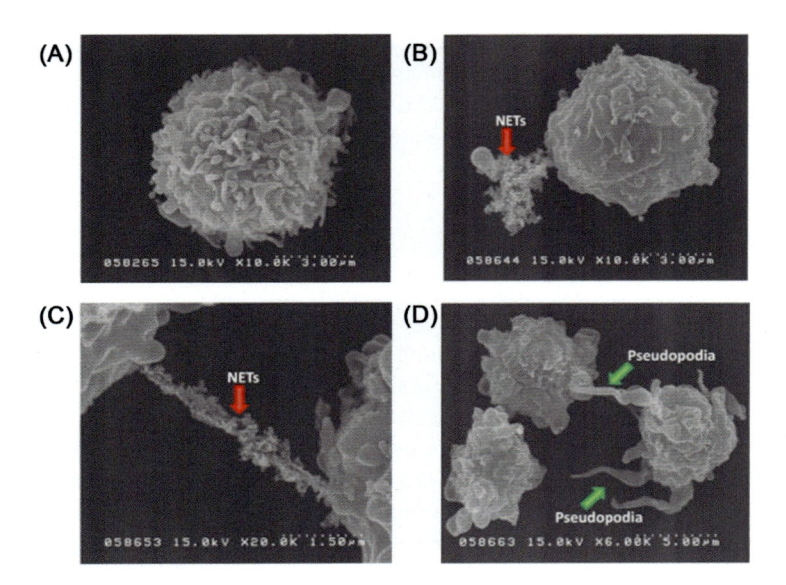

FIGURE 1.6

SEM imaging of human neutrophils treated with phosphate buffered saline (PBS) at a magnification of 10,000 (A), cationic SLNs (55 µg/mL) at a magnification of 10,000 (B), cationic SLNs (55 µg/mL) at a magnification of 20,000 (C), and cationic SLNs (55 µg/mL) at a magnification of 6000 (D).

Adapted with permission from Hwang, T. L., Aljuffali, I. A., Hung, C. F., Chen, C. H., Fang, J. Y. (2015). The impact of cationic solid lipid nanoparticles on human neutrophil activation and formation of neutrophil extracellular traps (NETs). Chemico-Biological Interactions, 235, *106−114. https://doi.org/10.1016/j.cbi.2015.04.011. Copyright © 2015 Elsevier Ireland Ltd.*

comparison to lymphocytes. Ninety percent of cells stayed viable even when concentration reached 500 μg/mL.

The uptake of 30 nm PEGylated Au NPs by leukocytes was observed in vitro, but not in vivo (Shah et al., 2012). The 50- and 100 nm cNDs injected into an animal model did not change the level of proinflammatory cytokine TNF-α production, meaning there was no immune response (Tsai et al., 2016).

According to the review on immunotoxicity of Si NPs by Chen et al. (2018), Si NPs may result in proinflammatory responses, ROS, and autophagy, which are mitigated if NPs are functionalized with amino and phosphate groups. Surface modification of silica NPs with vinyl and aminopropyl diminishes the cytotoxic and genotoxic properties, which are actively manifested for human lymphocytes treated with unmodified NPs (Lankoff et al., 2013).

Chemotactic activity of leukocytes was increased in response to negatively and positively charged Au NPs in work by Durocher et al. (2017), however cationic Au NPs-induced activation was more pronounced. At the same time, negatively charged Au NPs led to a considerable leukocyte apoptosis in contrast to the positive NPs. This finding is in agreement with another study that found cationic Au NPs of 20 and 70 nm to be proapoptotic toward leukocytes (Noël et al., 2016). TiO_2 NPs were shown to cause genotoxicity (DNA damage) in human lymphocytes, interestingly it was more severe at low concentration (0.25 mM) (Ghosh et al., 2010).

There are many advantages of using the leukocytes for NP delivery. They not only inherently circulate in the system, but are able to interact with xenobiotic materials, adhere to blood vessels, and extravasate to surrounding tissue. Macrophage-derived biomimetic NPs (leukosomes) prolong survival in the sepsis model (Molinaro et al., 2019). The great increase in accumulation of NPs in tumors and vascular lesions was achieved by using the leukosomes loaded with rhodamine and gadolinium (Abello et al., 2019).

1.6 Red blood cell—nanoparticle interaction

RBCs (erythrocytes) are the most abundant blood cells, occupying 40%—45% of whole blood, and 99% of formed blood components. The uniqueness of RBCs is the lack of a nucleus and other intracellular organelles, which makes the cell extra deformable to efficiently travel through the narrowest capillaries. The inner cell composition is essentially a viscous fluid—cytoplasm filled with hemoglobin (about 33% of the cell contents; Steck, 1974), which reversibly binds and transports oxygen and carbon dioxide. Human RBCs are biconcave discotic cells of 6—8 μm in diameter and 0.8—1 μm thickness in the center and 2—2.5 μm at the edges. Such a shape favors effective gas diffusion into the cell and facilitates its transport and exchange. The RBC phospholipid membrane integrity is supported by the cytoskeleton of spectrin and actin (Mohandas & Gallagher, 2008). Among the bilayer of lipids, phospholipids, and cholesterol, there are almost 50% of integral membrane proteins incorporated into the membrane, which play roles in transport, cell adhesion, and structure of the RBC.

The relative structural simplicity and availability make RBCs a great cellular model in studies of cell—NPs interactions (Wadhwa et al., 2019). For the normal functioning RBCs should preserve the shape, deformability, and integrity, which can be easily affected by NPs. These changes directly affect the main factors of blood viscosity: RBC aggregation, RBC deformation, and hematocrit [the percentage of RBCs in a volume of whole blood (vol%)] (Harpe et al., 2019; Wu et al., 2018).

NPs-related toxic effects include RBC morphological changes and functionality impairment caused by the oxidative stress or inflammatory response increase (Baranwal et al., 2018). NP toxicity largely depends on the time of exposure, dose (Fonseca et al., 2018), size (Shang et al., 2014; Kwon et al., 2012), surface (Zhao et al., 2011), composition, and the route of administration.

One should bear in mind, that toxic processes induced by NP-RBC interactions are closely interrelated. Change of RBC deformability will affect aggregation, which, except for the NP attachment, does not exclude internalization, or ROS activation. The strong manifestation of these effects, separately or in synergy, usually leads to hemolysis. The most common outcomes of RBC—NP interactions (Harpe et al., 2019) are described below.

1.6.1 RBC aggregation, adhesion, and agglutination induced by nanoparticles

NPs acting as bridges can induce RBC aggregation and clumping. Negatively charged sialylated glycolipids on the RBC membrane result in electrostatic repulsion between RBCs, creating an electric zeta potential (ζ) (Eylar et al., 1962). In other words, zeta potential represents RBC surface charge, which prevents RBC aggregation (Fernandes et al., 2011).

RBCs under the low-shear stress have a tendency to form linear aggregates (rouleaux), which easily disintegrate upon shear rate increase (Baskurt et al., 2011). RBC aggregation is dependent on plasma proteins, mostly fibrinogen (Neu & Meiselman, 2002), as RBCs do not aggregate in the phosphate buffer solution, which does not contain any natural or synthetic polymers (Meiselman, 1993; Avsievich et al., 2018). At the moment, there are two mechanisms proposed to explain RBC aggregation (Avsievich, Zhu, et al., 2020; Zhu et al., 2020). One of them (Asakura & Oosawa, 1954) suggests that the depletion layer between adjacent RBCs forces them to interact via osmotic pressure, while the other theory relies on the protein cross-bridges formation between RBC membranes (Brooks, 1988). While the experimental confirmation was found in favor of both hypotheses for specific cases, it is possible that both of them can be merged in to a hybrid model, such as described by Avsievich et al. (2018). Interaction energy dependence between individual RBCs in dextran solutions differed from the one observed in plasma, however, in the mixture of dextrans, where the molecular weights and fractions matched those of proteins in plasma, the interaction mode was almost identical (Fig. 1.7A). SEM imaging demonstrated the difference between RBC binding in plasma and dextran solutions (Fig. 1.7B, C).

RBC aggregation can indicate pathophysiological processes affecting the blood microcirculation, usually manifested in hematological disorders or metabolic disorders (Baskurt & Meiselman, 2008; Simmonds et al., 2013). Fig. 1.8 demonstrates RBC aggregates formed 30 minutes after incubation in control sample in blood plasma, and a sample treated with 0.01 mg/mL of carboxylated NDs.

Han et al. (2012) observed that positively charged hydroxyapatite NPs induced aggregation in RBC suspensions, forming the "caves" by attaching to RBC membrane. Formation of electrostatic bridges during NPs attachment to RBC membrane was suggested to be a driven force of aggregation. Kim and Shin (2014), observing the higher RBC aggregation after treatment with smaller (~ 30 nm) Ag NPs comparing to the large ones (~ 100 nm), suggested that RBC aggregation should be considered along with changes in RBC shape and deformability.

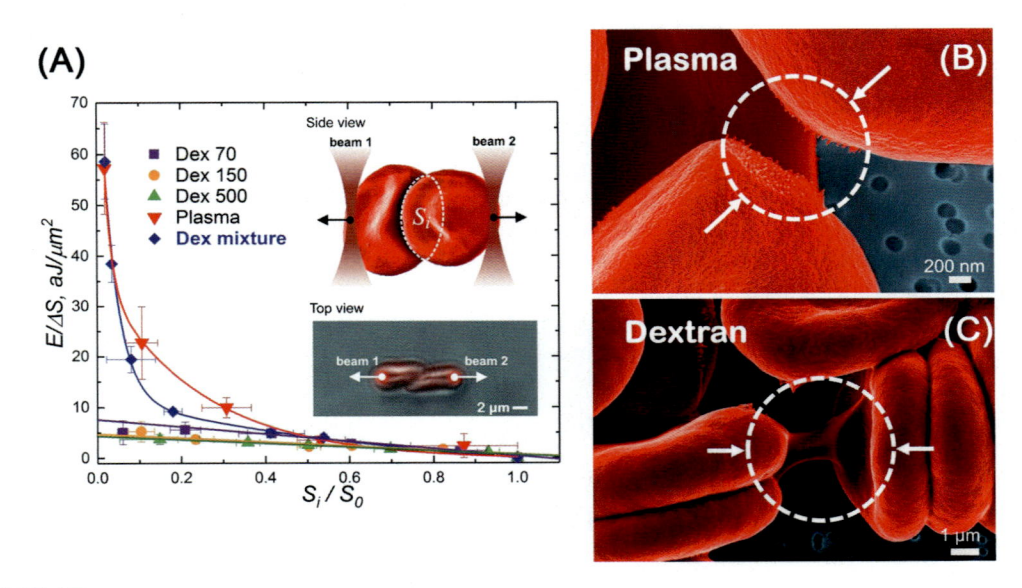

FIGURE 1.7

(A) Relative interaction energy of the adhering RBC as a function of relative conjugated surface area S_i to the total initially overlapping area S_0. False-colored scanning electron microscope images of the bridges between RBC formed in (B) blood plasma, and (C) dextran 500 kDa.

Adapted with permission from Avsievich, T., Popov, A., Bykov, A., & Meglinski, I. (2018). Mutual interaction of red blood cells assessed by optical tweezers and scanning electron microscopy imaging. Optics Letters, 43, 3921. *https://doi.org/10.1364/ ol.43.003921. © The Optical Society.*

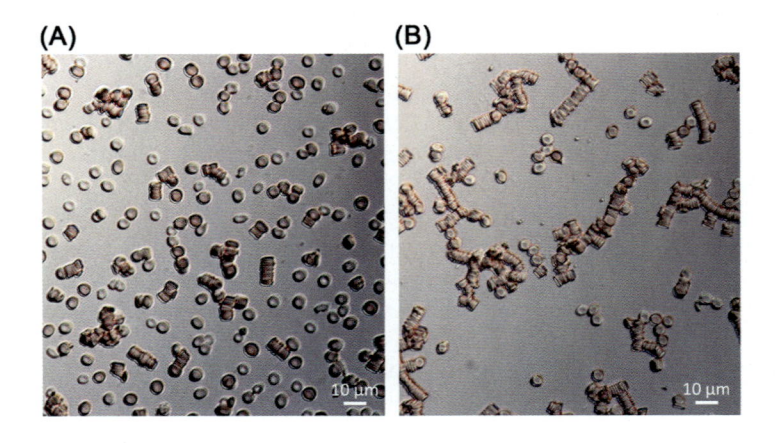

FIGURE 1.8

(A) RBC aggregates in a blood plasma sample after 30 min of incubation and (B) sample treated with carboxylated NDs.

Since plasma proteins significantly affect NPs, the influence of different types of NPs on single-cell RBC aggregation was estimated in protein-rich plasma with optical tweezers (Avsievich et al., 2019). Mutual RBC interactions between individual cells were affected by 100 nm carboxylated NDs, leading to the double increase of the adhesion force between the RBCs, compared to other NP types (Fig. 1.9). The analog study confirmed the hemocompatibility of 600 nm polymeric nanocapsules (Avsievich, Tarakanchikova, et al., 2020). Carboxylated graphene QDs, in work by Kim et al. (2016), turned out to be more toxic at high concentrations (750 and 1000 µg/mL), compared to nonfunctionalized, hydroxylated-caused QDs, as they had a more pronounced effect on the activation of aggregation and hemolysis. Carboxylated NP toxic effects can be a result of NP synthesis, they often cause the generation of radical oxidative stress.

RBC adhesion usually describes RBC interactions with endothelial cells. Barshtein et al. noted the enhancement of RBC aggregation and RBC adhesion to endothelial cells in a concentration-dependent manner for polystyrene NPs (Barshtein et al., 2016).

RBC agglutination, compared to RBC aggregation, is an irreversible process of severe irregular RBC clumping, often promoted by antibodies on the RBC surface. The stronger promotion of RBC agglutination and adhesion by smaller NPs, rather than larger ones, was repeatedly demonstrated in different studies (Barshtein et al., 2016; Han et al., 2012; Li et al., 2008).

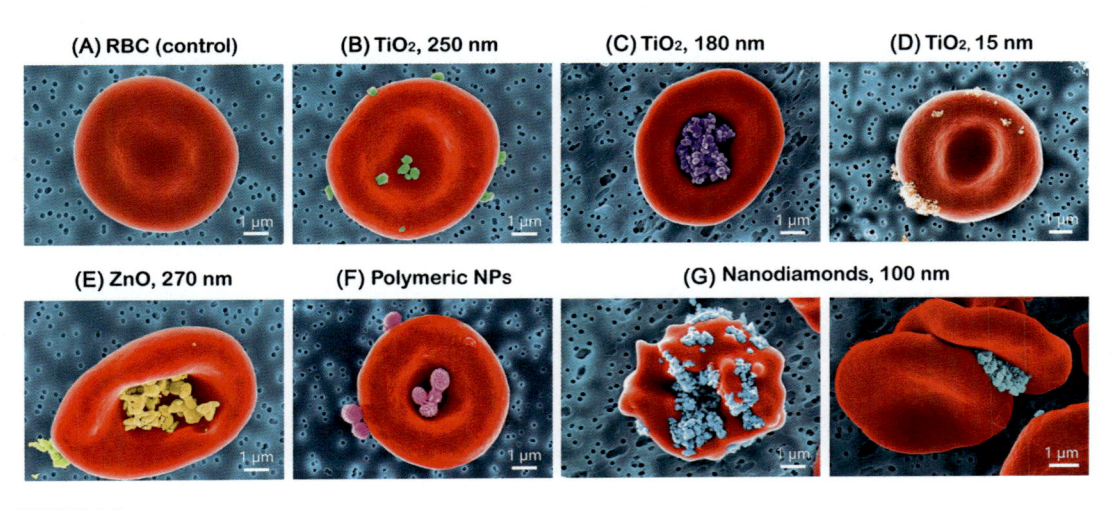

FIGURE 1.9

Colored scanning electron microscope images representing a diversity of observed nanoparticle (NP) localizations on the RBC surface: (A) normal conditions (control); RBC incubated with (B) TiO$_2$ 250 nm, (C) TiO$_2$ 180 nm, (D) TiO$_2$ 15 nm, (E) ZnO NPs 270 nm, (F) polymeric NPs 600 nm, and NDs 100 nm (G) echinocyte formed due to adhesion of NDs and two RBCs adhered together.

Adapted with permission from Avsievich, T., Popov, A., Bykov, A., & Meglinski, I. (2019). Mutual interaction of red blood cells influenced by nanoparticles. Science Reports, *9(1), 1–6. https://doi.org/10.1038/s41598-019-41643-x. Copyright © 2019, Springer Nature.*

1.6.2 RBC deformation and membrane deformability change induced by nanoparticles

RBC deformability is a vital property for effective oxygen delivery. Changes in RBC rigidity can impair RBC oxygenation/deoxygenation function and rheology of microcirculation.

Zhao et al. (2011) studied the interaction of mesoporous Si NPs (MSNs) with RBCs, and concluded that 100 nm MSNs attachment to membrane does not influence the membrane properties, in contrast to 600 nm MSNs causing the RBC shape transformation to echinocyte (Fig. 1.10). The effect was associated with a decreased deformability, due to the reduction in the ratio of surface area to volume. In the study by Kim et al. Kim and Shin (2014), however, the size of Ag NPs had the opposite relation to the deformability as the smaller Ag NPs induced the greater reduction of deformability. The decrease of the membrane fluidity was also initiated by 100 nm cNDs in the concentration-dependent manner in the study by Lin (2012). Graphene QDs at high concentration (750 µg/mL) (Kim et al., 2016), and PEGylated Au NPs (He, Liu, et al., 2014) reduced deformability of RBCs.

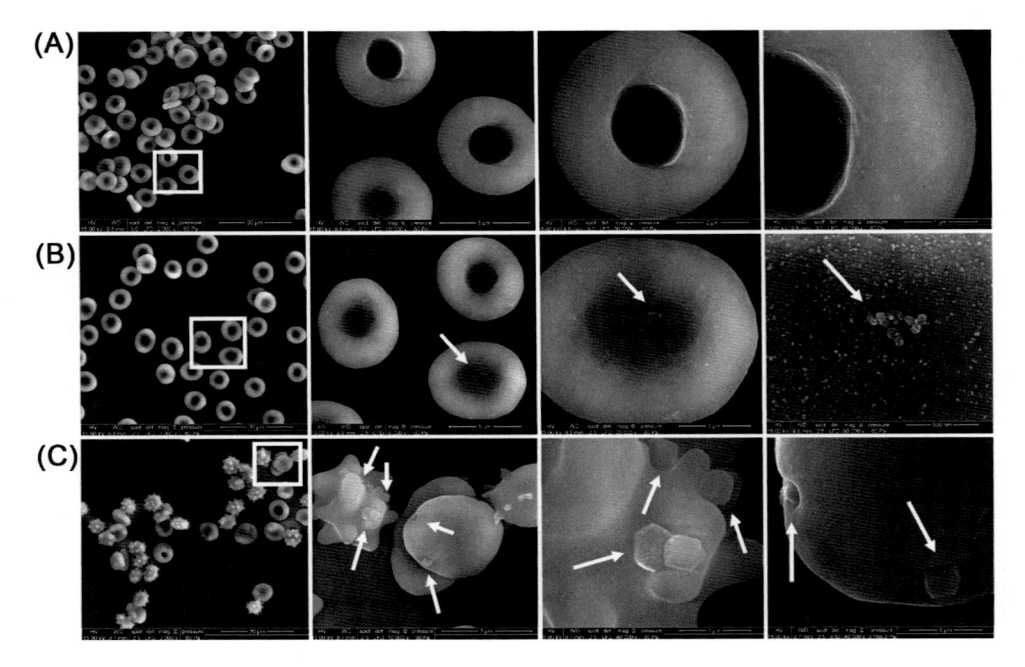

FIGURE 1.10

Scanning electron microscope images of RBCs (5% hematocrit) incubated for 2 h at room temperature with (A) Phosphate buffered saline (PBS) as control, (B) 100 µg/L of 100 nm mesoporous Si NP (MSN), and (C) 100 µg/mL of 600 nm MSN. Images increase in magnification from left to right with features highlighted with white squares or arrows. The nanoparticles attached on the cell surface are distinguished by the particle shape and surface textural difference between the particles and RBCs.

Reprinted with permission from Zhao, Y., Sun, X., Zhang, G., Trewyn, B. G., Slowing, I. I., & Lin, V. S. Y. (2011). Interaction of mesoporous silica nanoparticles with human red blood cell membranes: Size and surface effects. ACS Nano, 5(2), 1366–1375. https://doi.org/10.1021/nn103077k. Copyright 2011 American Chemical Society.

The membrane fluidity increase was also confirmed in some studies. Wang et al. studied passive transport of 4 nm zwitterionic QDs through the RBC membrane and found significant membrane softening, when QDs were incorporated into RBCs (Wang et al., 2012), while the overall membrane structure remained completely intact.

The concentration-dependent abnormal morphology (spherocytosis, echinocytosis), agglutination, and hemolysis were observed for TiO_2 (Li et al., 2008), Ag NPs (Osborne et al., 2015), Au NPs (Shrivastava et al., 2016), platinum NPs (Asharani et al., 2010). Ghosh et al. revealed the TiO_2 NPs interaction with Hb by the change in Hb spectra (Ghosh et al., 2013). High Hb concentration occurred in rats treated with graphene QDs (5 mg/kg) (Wang et al., 2013), but RBC antioxidant defense (glutathione, catalase, and peroxiredoxin-2) is able to slow down the damage.

1.6.3 Nanoparticle induced hemolysis

The lysis of RBCs is a highly undesirable effect causing cell death and affecting the hematocrit by decreasing the RBCs count. Hemolysis assay is a standard procedure in the NP testing to set safety limits for NP concentration. RBC destruction results in hemoglobin (Hb) release, and the absorption spectroscopy is used to reveal the conformation of Hb. Hemolysis is the most obvious and the important parameter in the hemocompatibility estimation, however, the exact mechanism of the process is not completely understood yet. In general, hemolysis may result from the critical cases of NP-RBC interaction.

RBC negative charge was associated with hemolytic activity of Ag NPs arising from strong interactions with RBCs (Das et al., 2017). NP size decrease is associated with increased hemolysis (Chen et al., 2015; Shang et al., 2014). Graphene NPs demonstrated the greatest hemolytic activity at the smallest size, compared to much lower hemolysis caused by graphene sheets (Liao et al., 2011). Coating graphene oxide with chitosan nearly eliminated hemolytic activity.

The first report on carboxylated carbon nanotubes (CNTs) interaction with RBCs showed no hemolysis or internalization of NPs (Donkor et al., 2009), and the same was later observed for the pristine CNTs. Acid functionalization of CNTs, however, caused the dose-dependent hemolysis (Sachar & Saxena, 2011).

Hemolysis assays are common in clinical practice, and often require only visual observation and the count of intact cells. As an example, visual inspection by a trained clinician confirmed the hemocompatibility of three NDs types: unmodified, oxygenated, and hydrogenated (Wasowicz et al., 2017). In the case of mesoporous Si NPs, hemolytic activity is activated by 600 nm Si NPs resulting in fatal RBC deformations, while 100 nm Si NPs absorbed on the membrane did not cause adverse influence (Zhao et al., 2011).

1.6.4 Nanoparticle uptake by RBCs

Due to relative functional and structural simplicity of RBCs, they perform the nonphagocytic passive uptake of NPs. Rothen-Rutishauser et al. (2006) used different microscopy methods and quantitative analysis to reveal the role of NP size, charge, and material in the uptake by RBCs. The maximum size of internalized NP aggregates was 200 nm, and internalization only depended on the NP size.

However, there are some controversial observations, for instance, 100 nm cNDs were not able to penetrate the RBCs (Lin, 2012). Herewith, internalized 5 nm cNDs did not promote Hb conformational changes or deoxygenation dynamics at high enough concentrations (up to 50 µg/mL). At

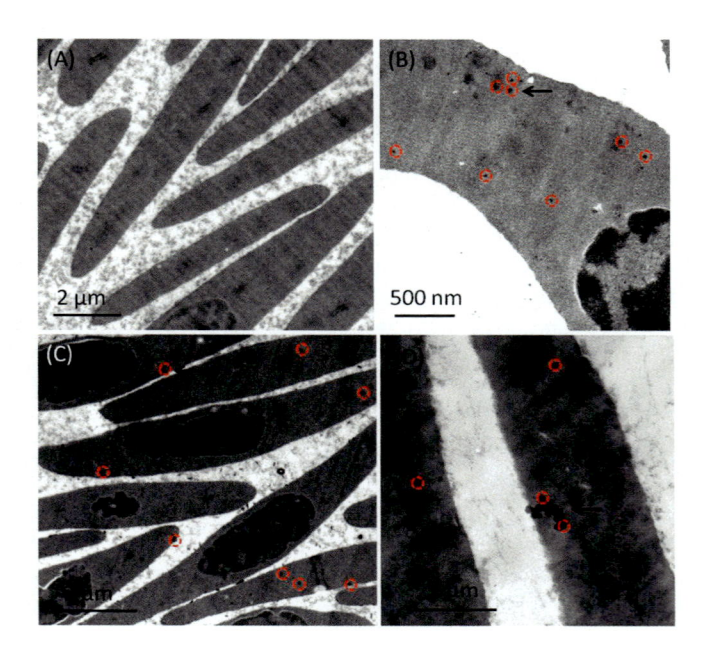

FIGURE 1.11

Transmission electron microscopy images of the Ag nanoparticles (NPs) adsorption and uptake by RBCs. (A) RBCs control; (B) Ag NPs 15 nm treated RBCs; (C) Ag NPs 50 nm treated RBCs; (D) Ag NPs 100 nm treated RBCs. All three sizes of Ag NPs were incubated with RBCs for 2 h. Individual Ag NPs are indicated by red circles, and the aggregate of Ag NPs are indicated by black arrows.

Reprinted with permission from Chen, L. Q., Fang, L., Ling, J., Ding, C. Z., Kang, B., & Huang, C. Z. (2015). Nanotoxicity of silver nanoparticles to red blood cells: Size dependent adsorption, uptake, and hemolytic activity. Chemical Research in Toxicology, 28(3), 501–509. https://doi.org/10.1021/tx500479m. Copyright 2015 American Chemical Society.

the same time, 4 nm QDs owing to zwitterionic nature of the ligands on the surface were internalized without pore formation (Wang et al., 2012), demonstrating the functionalization capacity in reducing toxicity.

Chen et al. found that among Ag NPs of different sizes the most effective adsorption and RBC uptake was observed for 50 nm Ag NPs, compared to 15- and 100 nm, suggesting that 50 nm NP size is the optimal for the passive uptake (Chen et al., 2015) (Fig. 1.11).

RBC uptake of NPs without disturbance of cell's functionality became a powerful method for drug delivery. The next section describes methods of NP encapsulation for RBCs use as carriers of NPs.

1.6.5 RBC-based nanoparticle carriers

The motivation for RBC use is their abundance in blood, ease of obtaining and purification, structural simplicity, and the presence of special proteins, like CD47, acting to prevent RBC engulfment by macrophages. Prone to prolong circulation, RBCs can increase the circulation time of NPs (over twofold increase compared to PEGylated NPs; Hu et al., 2011), however, the additional active targeting is required to deliver the drug to a specific site.

The RBC membrane incorporates special ligands facilitating the adsorption of NPs: antigen−antibody, RBC-specific surface ligands, avidin−biotin bridge, carbohydrate moiety oxidized by periodate, and other chemical linkers (Sun et al., 2017). Methods of RBC employment as a nanocarrier were described in the extensive review by Villa et al. (2016), and the recent review by Wadhwa et al. (2019). There are two approaches to prepare the RBC-based NP carriers—surface and internal loading (Villa et al., 2016).

1. Surface loading is a NP binding to RBC via affinity ligands by specific or nonspecific binding.
   - Nonspecific NP binding implies the NP adsorption to RBC membrane domains, for instance, hydrophobic PNPs independent of charge were attracted to hydrophobic areas on RBCs (Chambers & Mitragotri, 2007). Electrostatic attraction facilitated the binding of positively charged amino groups on magnetic NPs to negatively charged sialic acid groups of RBC (Mai et al., 2013).
   - Specific NP binding is achieved by the ligand−receptor or chemical conjugation on RBC. The effective binding was demonstrated by coating NPs with antibodies of proteins (Villa et al., 2015), and NP detachment and vascular endothelium (Villa et al., 2015), which is critical for the drug delivery. Noncovalent adsorption of NPs to RBCs by Anselmo et al. (2013) allowed to reduce their accumulation in liver and spleen. NPs detached from RBCs left indents fixed with glutaraldehyde after shear stress exposure, but no indents were observed if fixation was done after shear stress (Fig. 1.12A−D). Avidin-functionalized NPs specifically bind to biotin on RBCs (Wang et al., 2014; Muzykantov & Taylor, 1994).
   Brenner et al. (2018) demonstrated that intravascularly administrated RBC-hitchhiked NPs absorbed to the first organ downstream without any toxic effects. The stream flow upon adsorption of NPs to RBC surface was shown to be improved (Pan et al., 2018). NPs attached to the glycocalyx of RBC membrane masked attached NPs and demonstrated the increased lifetime of NPs in the blood circulatory system (Zaitsev et al., 2012). Conjugation of NPs with RBCs was shown to be an effective way to bypass the immune system (Villa et al., 2016; Zhang, 2016). Self-recognizing markers such as CD47 present on the surface of the RBC membrane prevent its disturbance and show a signal for macrophages (Tsai et al., 2010). Fifty- and 100 nm cNDs injected in the animal model at the concentration below 100 μg/mL attached to RBCs and circulated in vivo for at least 30 minutes without any effect on RBC oxygenation state (Tsai et al., 2016).
2. Internal loading of NP (encapsulation) through the RBC membrane. Encapsulation employs different methods to load NPs inside the RBC. Fabrication methods were comparatively described in the review by Xia et al. (2019), and Villa et al. (2016). Hypotonic (osmotic) treatment is based on temporal pore opening, allowing NPs to translocate within the cells (Rossi et al., 2016; Wu et al., 2014). Wu et al. (2014) loaded Fe_3O_4 NPs into RBC to turn them into "functional micromotors," which can be guided under acoustic propulsion (Fig. 1.12E−G). Method requires <100 nm NPs. Fe_3O_4 (Antonelli et al., 2011) and Au NPs (Ahn et al., 2011) were encapsulated in RBCs for magnetic resonance imaging (MRI). Hypotonic treatment, however, can irreversibly damage the membrane, disrupting its functionality. The alternative

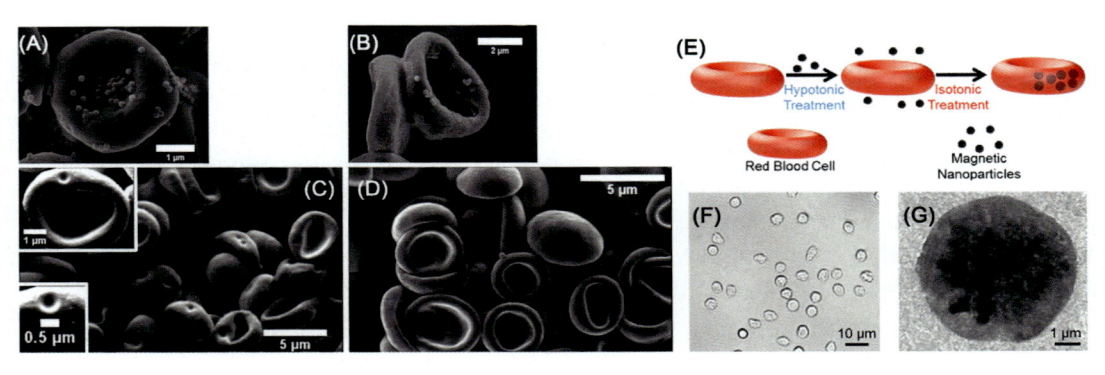

FIGURE 1.12

Controlled detachment of nanoparticles (NPs) from the RBC surface via applied shear stress using a rheometer: (A) no induced shear, and (B) 5 Pa shear stress for 15 min at 37°C. (C) scanning electron microscope (SEM) images of 500 nm polystyrene spheres detached from RBCs after fixation with glutaraldehyde showing indents caused by particles attachment and (D) SEM images of 500 nm polystyrene spheres detached from RBCs prior to fixation with glutaraldehyde showing that indents are reversible and do not permanently deform RBCs.

Reprinted with permission from Anselmo, A. C., Gupta, V., Zern, B. J., Pan, D., Zakrewsky, M., Muzykantov, V. et al. (2013). Delivering nanoparticles to lungs while avoiding liver and spleen through adsorption on red blood cells. ACS Nano, 7(12), 11129–11137. https://doi.org/10.1021/nn404853z. Copyright © 2013, American Chemical Society. Preparation of the RBC motors: (E) magnetic NPs are loaded into regular RBCs by using a hypotonic dilution encapsulation method. Optical (F) and (G) TEM images of the RBC motors. *Reprinted with permission from Wu, Z., Li, T., Li, J., Gao, W., Xu, T., Christianson, C. et al. (2014). Turning erythrocytes into functional micromotors. ACS Nano, 8(12), 12041–12048. https://doi.org/10.1021/nn506200x. Copyright © 2014 American Chemical Society.*

method of RBC encapsulation by He, Ye, et al. (2014) is based on cell translocating peptide—low molecular weight protamine—where RBCs remain intact. Fe_3O_4 NPs encapsulation by electroporation method was described in work by Rao et al. (2017). Electrical pulses create transient pores in the cell membrane, providing an entrance route for NPs. Hydrophobic QDs infused into the RBC membrane preserved optical properties in the experiment by Guo et al. (2016), as a proof-of-concept of an effective membrane nanoprobe. One more effective approach is lipid fusion, where RBC-derived vesicles (Ren et al., 2016) were used to camouflage NPs. Ren et al. designed magnetic NPs coated with RBC membranes and observed the enhanced therapeutic efficacy.

1.7 Factors influencing nanoparticles biodistribution

1.7.1 Nanoparticles margination in the bloodstream

Blood flow itself affects NP circulation times, pharmacokinetics, and biodistribution. NPs in the bloodstream are subject to a margination, which forces them to interact with vascular walls. When the adhesive force exceeds hydrodynamic flow, NPs firmly attach to endothelial cells and can be translocated to target tissues. Understanding of the main principle of the margination process will allow control over NP circulation time and targeted delivery by the NP adherence and translocation through a vessel wall. NP size, shape, surface, and stiffness were found to be the main determinants of margination.

Small NPs were found to flow among the RBCs, which facilitate their longer circulation, while larger NPs tend to marginate more easily (Müller et al., 2014; Lee et al., 2013). Lee et al. (2013) used intravital microscopy and computer simulation methods to reveal that NPs <100 nm are radially distributed in the capillary, but margination increases with increase of NP size.

Nonspherical NPs were found to marginate faster, and adhere stronger to the vessel wall, compared to spherical particles (Cooley et al., 2018). 3D computational modeling by Vahidkhah and Bagchi (2015) for spherical, oblate, and prolate NPs in the suspension of deformable RBCs demonstrated the highest near-wall accumulation for oblate NPs, then spherical, and then prolate NPs. The adherence of oblate ellipsoid NPs was higher than for prolate and spherical NPs.

Stiff NPs also tend to higher margination propensity, comparing to soft elastic capsules (Kumar & Graham, 2011). Müller et al. (2016) demonstrated in the numerical experiment, that margination of the rigid carriers is better than that for deformable ones. But the margination of deformable carriers improves at a low shear rate in the narrow capillaries, so they can adhere better due to the natural stretching providing the larger area for interactions.

1.7.2 Vascular endothelium

Vascular endothelium serves as a pathway for NPs translocation to other parts of the body. The circulation time of NPs and targeted delivery effectiveness depends on the NP−endothelium interaction. Endothelial cells are flat-shaped cells, about $1-2\,\mu m$ thick and $10-20\,\mu m$ in diameter. Vascular endothelium plays a pivotal role in homeostasis by controlling the passage of the molecules going through in and out of the bloodstream, participating in platelet and leukocyte adhesion, thrombosis, and inflammation.

The widespread of Ag NPs can produce severe toxicity during interaction with endothelial cells and result in thromboembolic problems (Sun et al., 2016). Among observed reactions are decreased cell viability, ROS production, and increased production of interleukins and adhesion proteins, which can promote inflammation (Shi et al., 2014; Grosse et al., 2013).

Damage of endothelial cells induced by Si NPs followed by increased expression of the adhesion factor caused hypercoagulation (Feng et al., 2019). Even a very low concentration of Fe_3O_4 NPs (0.1 mM) caused cytoskeleton disruptions and endocytosis with following apoptosis (Wu et al., 2010). Fe_2O_3 and ZnO NPs elicit a pronounced inflammatory response above a threshold concentration of $10\,\mu g/mL$, while Fe_2O_3 NPs did not cause an inflammatory response even at the highest concentration (Gojova et al., 2007).

1.7.3 Biodistribution

Toxic effects and biodistribution in vivo are usually studied in experiments on mice and rats (Ghasemzadeh et al., 2013). Size-dependent biodistribution was found in the study (De Jong et al., 2008). NPs of 10 nm in size were rapidly distributed to organs of the in vivo model, whilst the large NPs (50—250 nm) were accumulated in liver, spleen, and blood (De Jong et al., 2008), suggesting that smaller NPs can pass through the defensive system of the organism.

Biodistribution of Ag NPs with different sizes (10, 40, and 100 nm) and coatings was studied by intravenous injection at the concentration 10 mg/kg. Although each type of NP was found to cause toxic damage of tissues, larger particles were less toxic, probably due to their lower penetration capacity (Recordati et al., 2016). Gold NPs were shown to be toxic for mice, causing weight loss, decrease in the hematocrit, and reduction of the red blood cell count (Zhang et al., 2010). The study by Fabian et al. (2008) suggested hemocompatibility of TiO_2 NPs intravenously injected in rats (5 mg/kg). Within 28 days of experiments no signs of inflammation were registered.

Cd-based QDs were distributed throughout a mice body within 15 minutes after injection to the caudal vein, and then accumulated in liver, spleen, kidneys, red bone marrow, and lymph nodes. Two years later none of the QDs were detected by any means, except for the fluorescence in lymph nodes (Hwang et al., 2015). The change of the fluorescence spectrum demonstrated that degradation of QDs went slowly, therefore no toxic effects occurred. Similar results were obtained by Lin et al. (2018). Sim and Tsiftsoglou (2004) showed that CdTe QDs predominantly accumulated in the liver, decreasing the amount of antioxidants in it and inducing oxidative stress in liver cells.

1.8 Summary

Blood is the primary system of the body in terms of NP interactions. Any exposure pathway inevitably leads to blood—NP interactions.

The interaction of NPs with blood is highly context-dependent, including NP properties, experiment design, and physiological environment properties, implicating the interpretation of the occurred effects and drawing of general conclusions. According to the numerous studies, the biological identity of NPs given by the physiological environment determines their future fate such as biodistribution and biokinetics rather than synthetic identity. Blood, being the multicomponential system, can provide a variety of responses to NP introduction, and each of them is governed by hemocompatibility. NP interactions are currently studied with separate blood components, as well as with whole blood in vitro and in vivo. Different experimental conditions starting from the in vivo and in vitro approaches, in vivo models, cell types, dose, and incubation times often lead to inconsistency of the obtained results, underlying the need for standardized NP testing.

With a growing number of findings, which often appear ambivalent, the database resource summarizing the current knowledge will help to classify the NP behavior in physiological environment and to generate a correlation between NPs and biological response. Further computer simulation approaches will guide a subsequent development of nanomedicine, enabling the prediction, control, and optimization of their properties.

Understanding blood—NPs interactions is essential not only for the effective translation of new NPs into clinical application, but also for the development of new approaches on the way to the synthesis of artificial blood. Synthetic and biomimetic approaches, such as blood cells-cloaked NPs hold great potential to improve the hemocompatibilty and pharmacokinetics of the nanodrugs.

References

Abello, J., Nguyen, T. D. T., Marasini, R., Aryal, S., & Weiss, M. L. (2019). Biodistribution of gadolinium- and near infrared-labeled human umbilical cord mesenchymal stromal cell-derived exosomes in tumor bearing mice. *Theranostics*, *9*, 2325−2345. Available from https://doi.org/10.7150/thno.30030.

Ahn, S., Jung, S. Y., Seo, E., & Lee, S. J. (2011). Gold nanoparticle-incorporated human red blood cells (RBCs) for X-ray dynamic imaging. *Biomaterials*, *32*(29), 7191−7199. Available from https://doi.org/10.1016/j.biomaterials.2011.05.023.

Aillon, K. L., Xie, Y., El-Gendy, N., Berkland, C. J., & Forrest, M. L. (2009). Effects of nanomaterial physico-chemical properties on in vivo toxicity. *Advanced Drug Delivery Reviews*, *61*(6), 457−466. Available from https://doi.org/10.1016/j.addr.2009.03.010.

Albanese, A., Tang, P. S., & Chan, W. C. W. (2012). The effect of nanoparticle size, shape, and surface chemistry on biological systems. *Annual Review of Biomedical Engineering*, *14*, 1−16. Available from https://doi.org/10.1146/annurev-bioeng-071811-150124.

Allémann, E., Gravel, P., Leroux, J. C., Balant, L., & Gurny, R. (1997). Kinetics of blood component adsorption on poly(D,L-lactic acid) nanoparticles: Evidence of complement C3 component involvement. *Journal of Biomedical Materials Research*, *37*(2), 229−234, https://doi.org/10.1002/(SICI)1097-4636(199711)37:2 < 229::AID-JBM12 > 3.0.CO;2-9.

Amici, A., Caracciolo, G., Digiacomo, L., Gambini, V., Marchini, C., Tilio, M., et al. (2017). In vivo protein corona patterns of lipid nanoparticles. *RSC Advances*, *7*(2), 1137−1145. Available from https://doi.org/10.1039/c6ra25493d.

Anselmo, A. C., Gupta, V., Zern, B. J., Pan, D., Zakrewsky, M., Muzykantov, V., et al. (2013). Delivering nanoparticles to lungs while avoiding liver and spleen through adsorption on red blood cells. *ACS Nano*, *7*(12), 11129−11137. Available from https://doi.org/10.1021/nn404853z.

Antonelli, A., Sfara, C., Manuali, E., Bruce, I. J., & Magnani, M. (2011). Encapsulation of superparamagnetic nanoparticles into red blood cells as new carriers of MRI contrast agents. *Nanomedicine: Nanotechnology, Biology, and Medicine*, *6*(2), 211−223. Available from https://doi.org/10.2217/nnm.10.163.

Asakura, S., & Oosawa, F. (1954). On interaction between two bodies immersed in a solution of macromolecules. *The Journal of Chemical Physics*, *22*(7), 1255−1256. Available from https://doi.org/10.1063/1.1740347.

Aseychev, A. V., Azizova, O. A., Beckman, E. M., Dudnik, L. B., & Sergienko, V. I. (2013). Effect of gold nanoparticles coated with plasma components on ADP-induced platelet aggregation. *Bulletin of Experimental Biology and Medicine*, *155*(5), 685−688. Available from https://doi.org/10.1007/s10517-013-2226-x.

Asharani, P. V., Sethu, S., Vadukumpully, S., Zhong, S., Lim, C. T., Hande, M. P., et al. (2010). Investigations on the structural damage in human erythrocytes exposed to silver, gold, and platinum nanoparticles. *Advanced Functional Materials*, *20*(8), 1233−1242. Available from https://doi.org/10.1002/adfm.200901846.

Avsievich, T., Popov, A., Bykov, A., & Meglinski, I. (2018). Mutual interaction of red blood cells assessed by optical tweezers and scanning electron microscopy imaging. *Optics Letters*, *43*, 3921. Available from https://doi.org/10.1364/ol.43.003921.

Avsievich, T., Popov, A., Bykov, A., & Meglinski, I. (2019). Mutual interaction of red blood cells influenced by nanoparticles. *Science Reports*, *9*(1), 1−6. Available from https://doi.org/10.1038/s41598-019-41643-x.

Avsievich, T., Tarakanchikova, Y., Zhu, R., Popov, A., Bykov, A., Skovorodkin, I., et al. (2020). Impact of nanocapsules on red blood cells interplay jointly assessed by optical tweezers and microscopy. *Micromachines*, *11*(1), 19. Available from https://doi.org/10.3390/mi11010019.

Avsievich, T., Zhu, R., Popov, A., Bykov, A., & Meglinski, I. (2020). The advancement of blood cell research by optical tweezers. *Review Physics*, 100043. Available from https://doi.org/10.1016/j.revip.2020.100043.

Baranwal, A., Srivastava, A., Kumar, P., Bajpai, V. K., Maurya, P. K., & Chandra, P. (2018). Prospects of nanostructure materials and their composites as antimicrobial agents. *Frontiers in Microbiology*, 9, 422. Available from https://doi.org/10.3389/fmicb.2018.00422.

Barshtein, G., Livshits, L., Shvartsman, L. D., Shlomai, N. O., Yedgar, S., & Arbell, D. (2016). Polystyrene nanoparticles activate erythrocyte aggregation and adhesion to endothelial cells. *Cell Biochemistry and Biophysics*, 74(1), 19—27. Available from https://doi.org/10.1007/s12013-015-0705-6.

Baskurt, O., Neu, B., & Meiselman, H. J. (2011). *Red blood cell aggregation*. CRC Press.

Baskurt, O. K., & Meiselman, H. J. (2008). RBC aggregation: More important than RBC adhesion to endothelial cells as a determinant of in vivo blood flow in health and disease. *Microcirculation*, 15(7), 585—590. Available from https://doi.org/10.1080/10739680802107447, New York, N.Y. 1994.

Bergin, I. L., & Witzmann, F. A. (2013). Nanoparticle toxicity by the gastrointestinal route: Evidence and knowledge gaps. *International Journal of Biomedical Nanoscience and Nanotechnology*, 3(1—2), 163—210. Available from https://doi.org/10.1504/ijbnn.2013.054515.

Berrecoso, G., Crecente-Campo, J., & Alonso, M. J. (2020). Unveiling the pitfalls of the protein corona of polymeric drug nanocarriers. *Drug Delivery and Translational Research*, 1—21. Available from https://doi.org/10.1007/s13346-020-00745-0.

Bisso, P. W., Gaglione, S., Guimarães, P. P. G., Mitchell, M. J., & Langer, R. (2018). Nanomaterial interactions with human neutrophils. *ACS Biomaterials Science and Engineering*, 4(12), 4255—4265. Available from https://doi.org/10.1021/acsbiomaterials.8b01062.

Blume, J. E., Manning, W. C., Troiano, G., Hornburg, D., Figa, M., Hesterberg, L., et al. (2020). Rapid, deep and precise profiling of the plasma proteome with multi-nanoparticle protein corona. *Nature Communications*, 11(1), 1—14. Available from https://doi.org/10.1038/s41467-020-17033-7.

Bonvin, D., Aschauer, U., Alexander, D. T. L., Chiappe, D., Moniatte, M., Hofmann, H., et al. (2017). Protein corona: Impact of lymph versus blood in a complex in vitro environment. *Small (Weinheim an der Bergstrasse, Germany)*, 13(29), 1700409. Available from https://doi.org/10.1002/smll.201700409.

Brash, J. L. (2018). Blood compatibility of nanomaterials. *Drug delivery nanosystems for biomedical applications*, 13—31. Available from https://doi.org/10.1016/B978-0-323-50922-0.00002-X.

Braun, N. J., Debrosse, M. C., Hussain, S. M., & Comfort, K. K. (2016). Modification of the protein corona-nanoparticle complex by physiological factors. *Materials Science and Engineering C*, 64, 34—42. Available from https://doi.org/10.1016/j.msec.2016.03.059.

Brenner, J. S., Pan, D. C., Myerson, J. W., Marcos-Contreras, O. A., Villa, C. H., Patel, P., et al. (2018). Red blood cell-hitchhiking boosts delivery of nanocarriers to chosen organs by orders of magnitude. *Nature Communications*, 9(1), 1—14. Available from https://doi.org/10.1038/s41467-018-05079-7.

Brooks, D. E. (1988). *Mechanism of red cell aggregation. Blood cells, rheology, and aging* (pp. 158—162). Berlin/Heidelberg: Springer. Available from https://doi.org/10.1007/978-3-642-71790-1_16.

Buzea, C., Pacheco, I. I., & Robbie, K. (2007). Nanomaterials and nanoparticles: Sources and toxicity. *Biointerphases*, 2(4), MR17—MR71. Available from https://doi.org/10.1116/1.2815690.

Carrillo-Inungaray, M. L., Trejo-Ramirez, J. A., Reyes-Munguia, A., & Carranza-Alvarez, C. (2018). Use of nanoparticles in the food industry: Advances and perspectives. *Impact Nanoscience and Food Industry*, 419—444. Available from https://doi.org/10.1016/B978-0-12-811441-4.00015-7.

Cedervall, T., Lynch, I., Lindman, S., Berggård, T., Thulin, E., Nilsson, H., et al. (2007). Understanding the nanoparticle-protein corona using methods to quntify exchange rates and affinities of proteins for nanoparticles. *Proceedings of the National Academy Science USA*, 104(7), 2050—2055. Available from https://doi.org/10.1073/pnas.0608582104.

Chambers, E., & Mitragotri, S. (2007). Long circulating nanoparticles via adhesion on red blood cells: Mechanism and extended circulation. *Experimental Biology and Medicine*, 232(7), 958—966. Available from https://doi.org/10.3181/00379727-232-2320958.

Chen, L., Liu, J., Zhang, Y., Zhang, G., Kang, Y., Chen, A., et al. (2018). The toxicity of silica nanoparticles to the immune system. *Nanomedicine: Nanotechnology, Biology, and Medicine*, *13*(15), 1939−1962. Available from https://doi.org/10.2217/nnm-2018-0076.

Chen, L. Q., Fang, L., Ling, J., Ding, C. Z., Kang, B., & Huang, C. Z. (2015). Nanotoxicity of silver nanoparticles to red blood cells: Size dependent adsorption, uptake, and hemolytic activity. *Chemical Research in Toxicology*, *28*(3)), 501−509. Available from https://doi.org/10.1021/tx500479m.

Chithrani, B. D., Ghazani, A. A., & Chan, W. C. (2006). Determining the size and shape dependence of gold nanoparticle uptake into mammalian cells. *Nano Letters*, *6*(4), 662−668. Available from https://doi.org/10.1021/nl052396o.

Consumer Products—The Nanodatabase. (2020). Available from: https://nanodb.dk/en/analysis/consumer-products/#chartHashsection (accessed October 11, 2020).

Cooley, M., Sarode, A., Hoore, M., Fedosov, D. A., Mitragotri, S., & Sen Gupta, A. (2018). Influence of particle size and shape on their margination and wall-adhesion: Implications in drug delivery vehicle design across nano-to-micro scale. *Nanoscale*, *10*(32), 15350−15364. Available from https://doi.org/10.1039/c8nr04042g.

Cruz La, G. G. D., Rodríguez-Fragoso, P., Reyes-Esparza, J., Rodríguez-López, A., Gómez-Cansino, R., & Rodriguez-Fragoso, L. (2018). Interaction of nanoparticles with blood components and associated pathophysiological effects. Unraveling the safety profile of nanoscale particles and materials-from biomedical to environmental applications. https://doi.org/10.5772/intechopen.69386.

Cucinotto, I., Fiorillo, L., Gualtieri, S., Arbitrio, M., Ciliberto, D., Staropoli, N., et al. (2013). Nanoparticle albumin bound paclitaxel in the treatment of human cancer: Nanodelivery reaches prime-time? *Journal of Drug Delivery*, 905091. Available from https://doi.org/10.1155/2013/905091.

Das, B., Tripathy, S., Adhikary, J., Chattopadhyay, S., Mandal, D., Dash, S. K., et al. (2017). Surface modification minimizes the toxicity of silver nanoparticles: An in vitro and in vivo study. *Journal of Biological Inorganic Chemistry: JBIC: A Publication of the Society of Biological Inorganic Chemistry*, *22*(6)), 893−918. Available from https://doi.org/10.1007/s00775-017-1468-x.

De Jong, W. H., & Borm, P. J. A. (2008). Drug delivery and nanoparticles: Applications and hazards. *International Journal of Nanomedicine*, *3*(2), 133. Available from https://doi.org/10.2147/ijn.s596.

De Jong, W. H., Hagens, W. I., Krystek, P., Burger, M. C., Sips, A. J. A. M., & Geertsma, R. E. (2008). Particle size-dependent organ distribution of gold nanoparticles after intravenous administration. *Biomaterials*, *29*(12), 1912−1919. Available from https://doi.org/10.1016/j.biomaterials.2007.12.037.

De Paoli, S. H., Diduch, L. L., Tegegn, T. Z., Orecna, M., Strader, M. B., Karnaukhova, E., et al. (2014). The effect of protein corona composition on the interaction of carbon nanotubes with human blood platelets. *Biomaterials*, *35*(24), 6182−6194. Available from https://doi.org/10.1016/j.biomaterials.2014.04.067.

Deb, S., Patra, H. K., Lahiri, P., Dasgupta, A. K., Chakrabarti, K., & Chaudhuri, U. (2011). Multistability in platelets and their response to gold nanoparticles. *Nanomedicine Nanotechnology, Biology and Medicine*, *7*(4), 376−384. Available from https://doi.org/10.1016/j.nano.2011.01.007.

Dobrovolskaia, M. A., Patri, A. K., Simak, J., Hall, J. B., Semberova, J., De Paoli Lacerda, S. H., et al. (2012). Nanoparticle size and surface charge determine effects of PAMAM dendrimers on human platelets in vitro. *Molecular Pharmaceutics*, *9*(3), 382−393. Available from https://doi.org/10.1021/mp200463e.

Dobrovolskaia, M. A., Patri, A. K., Zheng, J., Clogston, J. D., Ayub, N., Aggarwal, P., et al. (2009). Interaction of colloidal gold nanoparticles with human blood: Effects on particle size and analysis of plasma protein binding profiles. *Nanomedicine Nanotechnology, Biology and Medicine*, *5*(2), 106−117. Available from https://doi.org/10.1016/j.nano.2008.08.001.

Docter, D., Westmeier, D., Markiewicz, M., Stolte, S., Knauer, S. K., & Stauber, R. H. (2015). The nanoparticle biomolecule corona: Lessons learned - challenge accepted? *Chemical Society Reviews*, *44*(17), 6094−6121. Available from https://doi.org/10.1039/c5cs00217f.

Donkor, A. D., Su, Z., Mandal, H. S., Jin, X., & Tang, X. S. (2009). Carbon nanotubes inhibit the hemolytic activity of the pore-forming toxin pyolysin. *Nano Research*, *2*(7), 517—525. Available from https://doi.org/10.1007/s12274-009-9049-0.

Dufort, S., Sancey, L., & Coll, J. L. (2012). Physico-chemical parameters that govern nanoparticles fate also dictate rules for their molecular evolution. *Advanced Drug Delivery Reviews*, *64*(2), 179—189. Available from https://doi.org/10.1016/j.addr.2011.09.009.

Dukhin, S. S., & Labib, M. E. (2013). Convective diffusion of nanoparticles from the epithelial barrier toward regional lymph nodes. *Advances in Colloid and Interface Science*, *199*, 23—43. Available from https://doi.org/10.1016/j.cis.2013.06.002.

Dunpall, R., Nejo, A. A., Pullabhotla, V. S. R., Opoku, A. R., Revaprasadu, N., & Shonhai, A. (2012). An in vitro assessment of the interaction of cadmium selenide quantum dots with DNA, iron, and blood platelets. *IUBMB Life*, *64*(12), 995—1002. Available from https://doi.org/10.1002/iub.1100.

Durocher, I., Noël, C., Lavastre, V., & Girard, D. (2017). Evaluation of the in vitro and in vivo proinflammatory activities of gold (+) and gold (−) nanoparticles. *Inflammation Research: Official Journal of the European Histamine Research Society*, *66*(11), 981—992. Available from https://doi.org/10.1007/s00011-017-1078-7.

Ehmann, H. M. A., Breitwieser, D., Winter, S., Gspan, C., Koraimann, G., Maver, U., et al. (2015). Gold nanoparticles in the engineering of antibacterial and anticoagulant surfaces. *Carbohydrate Polymers*, *117*, 34—42. Available from https://doi.org/10.1016/j.carbpol.2014.08.116.

Evani, S. J., & Ramasubramanian, A. K. (2016). *Hemocompatibility of nanoparticles, Nanobiomaterials handbooks* (31). London: Taylor & Francis Group. Available from https://doi.org/10.1201/b10970-32.

Eylar, E. H., Madoff, M. A., Brody, O. V., & Oncley, J. L. (1962). The contribution of sialic acid to the surface charge of the erythrocyte. *The Journal of Biological Chemistry*, *237*(6), 1992—2000.

Fabian, E., Landsiedel, R., Ma-Hock, L., Wiench, K., Wohlleben, W., & Van Ravenzwaay, B. (2008). Tissue distribution and toxicity of intravenously administered titanium dioxide nanoparticles in rats. *Archives of Toxicology*, *82*(3), 151—157. Available from https://doi.org/10.1007/s00204-007-0253-y.

Fasbender, S., Allani, S., Wimmenauer, C., Cadeddu, R. P., Raba, K., Fischer, J. C., et al. (2017). Uptake dynamics of graphene quantum dots into primary human blood cells following in vitro exposure. *RSC Advances*, *7*(20), 12208—12216. Available from https://doi.org/10.1039/c6ra27829a.

Feng, L., Yang, X., Liang, S., Xu, Q., Miller, M. R., Duan, J., et al. (2019). Silica nanoparticles trigger the vascular endothelial dysfunction and prethrombotic state via miR-451 directly regulating the IL6R signaling pathway. *Particle and Fibre Toxicology*, *16*(1), 1—13. Available from https://doi.org/10.1186/s12989-019-0300-x.

Fernandes, H. P., Cesar, C. L., & de Barjas-Castro, M. L. (2011). Electrical properties of the red blood cell membrane and immunohematological investigation. *Revista Brasileria de Hematologia Hemoterapia*, *33*(4), 297—301. Available from https://doi.org/10.5581/1516-8484.20110080.

Ferraz, N., Hong, J., & Karlsson Ott, M. (2010). Procoagulant behavior and platelet microparticle generation on nanoporous alumina. *Journal of Biomaterials Applications*, *24*(8), 675—692. Available from https://doi.org/10.1177/0885328209338639.

Fonseca, L. C., de Araújo, M. M., de Moraes, A. C. M., da Silva, D. S., Ferreira, A. G., Franqui, L. S., et al. (2018). Nanocomposites based on graphene oxide and mesoporous silica nanoparticles: Preparation, characterization and nanobiointeractions with red blood cells and human plasma proteins. *Applied Surface Science*, *437*, 110—121. Available from https://doi.org/10.1016/j.apsusc.2017.12.082.

Fröhlich, E. (2016). Action of nanoparticles on platelet activation and plasmatic coagulation. *Current Medicinal Chemistry*, *23*(5), 408—430. Available from https://doi.org/10.2174/0929867323666160106151428.

García-Álvarez, R., Hadjidemetriou, M., Sánchez-Iglesias, A., Liz-Marzán, L. M., & Kostarelos, K. (2018). In vivo formation of protein corona on gold nanoparticles. The effect of their size and shape. *Nanoscale, 10* (3), 1256−1264. Available from https://doi.org/10.1039/c7nr08322j.

Ghasemzadeh, M., Kaplan, Z. S., Alwis, I., Schoenwaelder, S. M., Ashworth, K. J., Westein, E., et al. (2013). The CXCR1/2 ligand NAP-2 promotes directed intravascular leukocyte migration through platelet thrombi. *Blood, 121*(22), 4555−4566. Available from https://doi.org/10.1182/blood-2012-09-459636.

Ghosh, M., Bandyopadhyay, M., & Mukherjee, A. (2010). Genotoxicity of titanium dioxide (TiO$_2$) nanoparticles at two trophic levels: Plant and human lymphocytes. *Chemosphere, 81*(10), 1253−1262. Available from https://doi.org/10.1016/j.chemosphere.2010.09.022.

Ghosh, M., Chakraborty, A., & Mukherjee, A. (2013). Cytotoxic, genotoxic and the hemolytic effect of titanium dioxide (TiO$_2$) nanoparticles on human erythrocyte and lymphocyte cells in vitro. *Journal of Applied Toxicology: JAT, 33*(10), 1097−1110. Available from https://doi.org/10.1002/jat.2863.

Gojova, A., Guo, B., Kota, R. S., Rutledge, J. C., Kennedy, I. M., & Barakat, A. I. (2007). Induction of inflammation in vascular endothelial cells by metal oxide nanoparticles: Effect of particle composition. *Environmental Health Perspectives, 115*(3), 403−409. Available from https://doi.org/10.1289/ehp.8497.

Goy-López, S., Castro, E., Taboada, P., & Mosquera, V. (2008). Block copolymer-mediated synthesis of size-tunable gold nanospheres and nanoplates. *Langmuir: The ACS Journal of Surfaces and Colloids, 24*(22), 13186−13196. Available from https://doi.org/10.1021/la802279j.

Grosse, S., Evje, L., & Syversen, T. (2013). Silver nanoparticle-induced cytotoxicity in rat brain endothelial cell culture. *Toxicology In Vitro: An International Journal Published in Association with BIBRA, 27*(1), 305−313. Available from https://doi.org/10.1016/j.tiv.2012.08.024.

Guo, X., Zhang, Y., Liu, J., Yang, X., Huang, J., Li, L., et al. (2016). Red blood cell membrane-mediated fusion of hydrophobic quantum dots with living cell membranes for cell imaging. *Journal of Materials Chemistry B, 4*(23), 4191−4197. Available from https://doi.org/10.1039/c6tb01067a.

Hadjidemetriou, M., & Kostarelos, K. (2017). Nanomedicine: Evolution of the nanoparticle corona. *Nature Nanotechnology, 12*(4), 288−290. Available from https://doi.org/10.1038/nnano.2017.61.

Hadjidemetriou, M., McAdam, S., Garner, G., Thackeray, C., Knight, D., Smith, D., et al. (2019). The human in vivo biomolecule corona onto PEGylated liposomes: A proof-of-concept clinical study. *Advanced Materials, 31*(4), 1803335. Available from https://doi.org/10.1002/adma.201803335.

Han, Y., Wang, X., Dai, H., & Li, S. (2012). Nanosize and surface charge effects of hydroxyapatite nanoparticles on red blood cell suspensions. *ACS Applied Materials Interfaces, 4*(9), 4616−4622. Available from https://doi.org/10.1021/am300992x.

Harpe, K., Kondiah, P., Choonara, Y., Marimuthu, T., du Toit, L., & Pillay, V. (2019). The hemocompatibility of nanoparticles: A review of cell−nanoparticle interactions and hemostasis. *Cells, 8*, 1209. Available from https://doi.org/10.3390/cells8101209.

He, H., Ye, J., Wang, Y., Liu, Q., Chung, H. S., Kwon, Y. M., et al. (2014). Cell-penetrating peptides mediated encapsulation of protein therapeutics into intact red blood cells and its application. *Journal of Controlled Release: Official Journal of the Controlled Release Society, 176*, 123−132. Available from https://doi.org/10.1016/j.jconrel.2013.12.019.

He, Z., Liu, J., & Du, L. (2014). The unexpected effect of PEGylated gold nanoparticles on the primary function of erythrocytes. *Nanoscale, 6*(15), 9017−9024. Available from https://doi.org/10.1039/c4nr01857e.

Hu, C. M. J., Fang, R. H., Wang, K. C., Luk, B. T., Thamphiwatana, S., Dehaini, D., et al. (2015). Nanoparticle biointerfacing by platelet membrane cloaking. *Nature, 526*(7571), 118−121. Available from https://doi.org/10.1038/nature15373.

Hu, C. M. J., Zhang, L., Aryal, S., Cheung, C., Fang, R. H., & Zhang, L. (2011). Erythrocyte membrane-camouflaged polymeric nanoparticles as a biomimetic delivery platform. *Proceedings of theNational Academic Science USA, 108*(27), 10980−10985. Available from https://doi.org/10.1073/pnas.1106634108.

Huang, H., Lai, W., Cui, M., Liang, L., Lin, Y., Fang, Q., et al. (2016). An evaluation of blood compatibility of silver nanoparticles. *Scietific Reports*, *6*(1), 25518. Available from https://doi.org/10.1038/srep25518.

Huang, S. S., Wei, S. C., Chang, H. T., Lin, H. J., & Huang, C. C. (2016). Gold nanoparticles modified with self-assembled hybrid monolayer of triblock aptamers as a photoreversible anticoagulant. *Journal of Controlled Release: Official Journal of the Controlled Release Society*, *221*, 9—17. Available from https://doi.org/10.1016/j.jconrel.2015.11.028.

Hulander, M., Lundgren, A., Berglin, M., Ohrlander, M., Lausmaa, J., & Elwing, H. (2011). Immune complement activation is attenuated by surface nanotopography. *International Journal of Nanomedicine*, *6*, 2653. Available from https://doi.org/10.2147/ijn.s24578.

Hwang, T. L., Aljuffali, I. A., Hung, C. F., Chen, C. H., & Fang, J. Y. (2015). The impact of cationic solid lipid nanoparticles on human neutrophil activation and formation of neutrophil extracellular traps (NETs). *Chemico-Biological Interactions*, *235*, 106—114. Available from https://doi.org/10.1016/j.cbi.2015.04.011.

Ilinskaya, A. N., & Dobrovolskaia, M. A. (2013). Nanoparticles and the blood coagulation system. Part I: Benefits of nanotechnology. *Nanomedicine: Nanotechnology, Biology, and Medicine*, *8*(5), 773—784. Available from https://doi.org/10.2217/nnm.13.48.

ISO 10993-4. (2017). Biological evaluation of medical devices—Part 4: Selection of tests for interactions with blood. Int Stand 10993-4.

Jayaram, D. T., Pustulka, S. M., Mannino, R. G., Lam, W. A., & Payne, C. K. (2018). Protein corona in response to flow: Effect on protein concentration and structure. *Biophysical Journal*, *115*(2), 209—216. Available from https://doi.org/10.1016/j.bpj.2018.02.036.

Jeevanandam, J., Barhoum, A., Chan, Y. S., Dufresne, A., & Danquah, M. K. (2018). Review on nanoparticles and nanostructured materials: History, sources, toxicity and regulations. *Beilstein Journal of Nanotechnology*, *9*(1), 1050—1074. Available from https://doi.org/10.3762/bjnano.9.98.

Karagkiozaki, V., Pappa, F., Arvaniti, D., Moumkas, A., Konstantinou, D., & Logothetidis, S. (2016). The melding of nanomedicine in thrombosis imaging and treatment: A review. *Future Science*, *2*(2). Available from https://doi.org/10.4155/fso.16.3.

Keogh, J. R., Velander, F. F., & Eaton, J. W. (1992). Albumin-binding surfaces for implantable devices. *Journal of Biomedical Materials Research*, *26*(4), 441—456. Available from https://doi.org/10.1002/jbm.820260403.

Khan, I., Saeed, K., & Khan, I. (2019). Nanoparticles: Properties, applications and toxicities. *Arabian Journal of Chemistry*, *12*, 908—931. Available from https://doi.org/10.1016/j.arabjc.2017.05.011.

Kim, J., Nafiujjaman, M., Nurunnabi, M., Lee, Y. K., & Park, H. K. (2016). Hemorheological characteristics of red blood cells exposed to surface functionalized graphene quantum dots. *Food and Chemical Toxicology: An International Journal Published for the British Industrial Biological Research Association*, *97*, 346—353. Available from https://doi.org/10.1016/j.fct.2016.09.034.

Kim, M. J., & Shin, S. (2014). Toxic effects of silver nanoparticles and nanowires on erythrocyte rheology. *Food and Chemical Toxicology: An International Journal Published for the British Industrial Biological Research Association*, *67*, 80—86. Available from https://doi.org/10.1016/j.fct.2014.02.006.

Kumar, A., & Graham, M. D. (2011). Segregation by membrane rigidity in flowing binary suspensions of elastic capsules. *Physical Review E: Statistical Nonlinear, Soft Matter Physics*, *84*(6), 066316. Available from https://doi.org/10.1103/PhysRevE.84.066316.

Kwon, T. W., Woo, H. J., Kim, Y. H., Lee, H. J., Park, K. H., Park, S. K., et al. (2012). Optimizing hemocompatibility of surfactant-coated silver nanoparticles in human erythrocytes. *JNN*, *12*(8), 6168—6175. Available from https://doi.org/10.1166/jnn.2012.6433.

Lankoff, A., Arabski, M., Wegierek-Ciuk, A., Kruszewski, M., Lisowska, H., Banasik-Nowak, A., et al. (2013). Effect of surface modification of silica nanoparticles on toxicity and cellular uptake by human peripheral blood lymphocytes in vitro. *Nanotoxicology*, *7*(3), 235−250. Available from https://doi.org/10.3109/17435390.2011.649796.

Larese Filon, F., Bello, D., Cherrie, J. W., Sleeuwenhoek, A., Spaan, S., & Brouwer, D. H. (2016). Occupational dermal exposure to nanoparticles and nano-enabled products: Part I—Factors affecting skin absorption. *International Journal of Hygiene and Environmental Health*, *219*(6), 536−544. Available from https://doi.org/10.1016/j.ijheh.2016.05.009.

Lateef, A., Ojo, S. A., Elegbede, J. A., Azeez, M. A., Yekeen, T. A., & Akinboro, A. (2017). Evaluation of some biosynthesized silver nanoparticles for biomedical applications: Hydrogen peroxide scavenging, anticoagulant and thrombolytic activities. *Journal of Cluster Science*, *28*(3), 1379−1392. Available from https://doi.org/10.1007/s10876-016-1146-0.

Lazarovits, J., Chen, Y. Y., Sykes, E. A., & Chan, W. C. W. (2015). Nanoparticle-blood interactions: The implications on solid tumour targeting. *Chemical Communication*, *51*(14), 2756−2767. Available from https://doi.org/10.1039/c4cc07644c.

Lee, T. R., Choi, M., Kopacz, A. M., Yun, S. H., Liu, W. K., & Decuzzi, P. (2013). On the near-wall accumulation of injectable particles in the microcirculation: Smaller is not better. *Scientific Reports*, *3*, 2079. Available from https://doi.org/10.1038/srep02079.

Lee, Y. K., Choi, E. J., Webster, T. J., Kim, S. H., & Khang, D. (2014). Effect of the protein corona on nanoparticles for modulating cytotoxicity and immunotoxicity. *International Journal of Nanomedicine*, *10*, 97−113. Available from https://doi.org/10.2147/IJN.S72998.

Lesniak, A., Fenaroli, F., Monopoli, M. P., Åberg, C., Dawson, K. A., & Salvati, A. (2012). Effects of the presence or absence of a protein corona on silica nanoparticle uptake and impact on cells. *ACS Nano*, *6*(7), 5845−5857. Available from https://doi.org/10.1021/nn300223w.

Leszek, J., Md Ashraf, G., Tse, W. H., Zhang, J., Gasiorowski, K., Avila-Rodriguez, M. F., et al. (2017). Nanotechnology for Alzheimer disease. *Current Alzheimer Research*, *14*(11), 1182−1189. Available from https://doi.org/10.2174/1567205014666170203125008.

Li, S. Q., Zhu, R. R., Zhu, H., Xue, M., Sun, X. Y., Yao, S. D., et al. (2008). Nanotoxicity of TiO_2 nanoparticles to erythrocyte in vitro. *Food and Chemical Toxicology: An International Journal Published for the British Industrial Biological Research Association*, *46*(12), 3626−3631. Available from https://doi.org/10.1016/j.fct.2008.09.012.

Li, Y., Lian, Y., Zhang, L. T., Aldousari, S. M., Hedia, H. S., Asiri, S. A., et al. (2016). Cell and nanoparticle transport in tumour microvasculature: The role of size, shape and surface functionality of nanoparticles. *Interface Focus*, *6*(1), 20150086. Available from https://doi.org/10.1098/rsfs.2015.0086.

Liao, K. H., Lin, Y. S., MacOsko, C. W., & Haynes, C. L. (2011). Cytotoxicity of graphene oxide and graphene in human erythrocytes and skin fibroblasts. *ACS Applied Materials Interfaces*, *3*(7), 2607−2615. Available from https://doi.org/10.1021/am200428v.

Limongi, T., Susa, F., & Cauda, V. (2019). Nanoparticles for hematologic diseases detection and treatment. *Hematology and Medical Oncology*, *4*(3), 1015761. Available from https://doi.org/10.15761/hmo.1000183.

Lin, J., Zhang, H., Chen, Z., & Zheng, Y. (2010). Penetration of lipid membranes by gold nanoparticles: Insights into cellular uptake, cytotoxicity, and their relationship. *ACS Nano*, *4*(9), 5421−5429. Available from https://doi.org/10.1021/nn1010792.

Lin, M. H., Lin, C. F., Yang, S. C., Hung, C. F., & Fang, J. Y. (2018). The interplay between nanoparticles and neutrophils. *Journal of Biomedical Nanotechnology*, *14*(1), 66−85. Available from https://doi.org/10.1166/jbn.2018.2459.

Lin, Y.-C. (2012). The influence of nanodiamond on the oxygenation states and micro rheological properties of human red blood cells in vitro. *Journal of Biomedical Optics*, *17*(10), 101512. Available from https://doi.org/10.1117/1.jbo.17.10.101512.

Loiola, L. M., Batista, M., Capeletti, L. B., Mondo, G. B., Rosa, R. S. M., Marques, R. E., et al. (2019). Shielding and stealth effects of zwitterion moieties in double-functionalized silica nanoparticles. *Journal of Colloid and Interface Science*, *553*, 540–548. Available from https://doi.org/10.1016/j.jcis.2019.06.044.

Lucchini, R. G., Dorman, D. C., Elder, A., & Veronesi, B. (2012). Neurological impacts from inhalation of pollutants and the nose-brain connection. *Neurotoxicology*, *33*(4), 838–841. Available from https://doi.org/10.1016/j.neuro.2011.12.001.

Lundqvist, M., Stigler, J., Elia, G., Lynch, I., Cedervall, T., & Dawson, K. A. (2008). Nanoparticle size and surface properties determine the protein corona with possible implications for biological impacts. *Proceedings of the National Academy Science USA*, *105*(38), 14265–14270. Available from https://doi.org/10.1073/pnas.0805135105.

Madathiparambil Visalakshan, R., González García, L. E., Benzigar, M. R., Ghazaryan, A., Simon, J., Mierczynska-Vasilev, A., et al. (2020). The influence of nanoparticle shape on protein corona formation. *Small (Weinheim an der Bergstrasse, Germany)*, 2000285. Available from https://doi.org/10.1002/smll.202000285.

Madl, A. K., Plummer, L. E., Carosino, C., & Pinkerton, K. E. (2014). Nanoparticles, lung injury, and the role of oxidant stress. *Annual Review of Physiology*, *76*, 447–465. Available from https://doi.org/10.1146/annurev-physiol-030212-183735.

Mahmoudi, M. (2018). Debugging nano–bio interfaces: Systematic strategies to accelerate clinical translation of nanotechnologies. *Trends in Biotechnology*, *36*(8), 755–769. Available from https://doi.org/10.1016/j.tibtech.2018.02.014.

Mai, T. D., d'orly, F., M'nager, C., Varenne, A., & Siaugue, J. M. (2013). Red blood cells decorated with functionalized core–shell magnetic nanoparticles: Elucidation of the adsorption mechanism. *Chemical Communication*, *49*(47), 5393–5395. Available from https://doi.org/10.1039/c3cc41513a.

Man, F., Lammers, T., & de Rosales, T. M. (2018). Imaging nanomedicine-based drug delivery: A review of clinical studies. *Molecular Imaging and Biology: MIB: The Official Publication of the Academy of Molecular Imaging*, *20*(5), 683–695. Available from https://doi.org/10.1007/s11307-018-1255-2.

Mancuso, M. E., & Santagostino, E. (2017). Platelets: Much more than bricks in a breached wall. *British Journal of Haematology*, *178*(2), 209–219. Available from https://doi.org/10.1111/bjh.14653.

McGuinnes, C., Duffin, R., Brown, S., Mills, N. L., Megson, I. L., MacNee, W., et al. (2011). Surface derivatization state of polystyrene latex nanoparticles determines both their potency and their mechanism of causing human platelet aggregation in vitro. *Toxicological Sciences: An Official Journal of the Society of Toxicology*, *119*(2), 359–368. Available from https://doi.org/10.1093/toxsci/kfq349.

Meiselman, H. J. (1993). Red blood cell role in Rbc aggregation: 1963-1993 and beyond. *Clinical Hemorheology and Microcirculation*, *13*(5), 575–592. Available from https://doi.org/10.3233/CH-1993-13504.

Miller, M. R., Raftis, J. B., Langrish, J. P., McLean, S. G., Samutrtai, P., Connell, S. P., et al. (2017). Inhaled nanoparticles accumulate at sites of vascular disease. *ACS Nano*, *11*(5), 4542–4552. Available from https://doi.org/10.1021/acsnano.6b08551.

Mohajerani, A., Burnett, L., Smith, J. V., Kurmus, H., Milas, J., Arulrajah, A., et al. (2019). Nanoparticles in construction materials and other applications, and implications of nanoparticle use. *Materials*, *12*(19), 3052. Available from https://doi.org/10.3390/ma12193052, Basel.

Mohandas, N., & Gallagher, P. G. (2008). Red cell membrane: Past, present, and future. *Blood*, *112*(10), 3939–3948. Available from https://doi.org/10.1182/blood-2008-07-161166.

Molinaro, R., Pastò, A., Corbo, C., Taraballi, F., Giordano, F., Martinez, J. O., et al. (2019). Macrophage-derived nanovesicles exert intrinsic anti-inflammatory properties and prolong survival in sepsis through a direct interaction with macrophages. *Nanoscale*, *11*(28), 13576–13586. Available from https://doi.org/10.1039/c9nr04253a.

Monopoli, M. P., Åberg, C., Salvati, A., & Dawson, K. A. (2012). Biomolecular coronas provide the biological identity of nanosized materials. *Nature Nanotechnology*, *7*(12), 779−786. Available from https://doi.org/10.1038/nnano.2012.207.

Mout, R., Moyano, D. F., Rana, S., & Rotello, V. M. (2012). Surface functionalization of nanoparticles for nanomedicine. *Chemical Society Reviews*, *41*(7), 2539−2544. Available from https://doi.org/10.1039/c2cs15294k.

Müller, K., Fedosov, D. A., & Gompper, G. (2014). Margination of micro- and nano-particles in blood flow and its effect on drug delivery. *Scientific Reports*, *4*, 4871. Available from https://doi.org/10.1038/srep04871.

Müller, K., Fedosov, D. A., & Gompper, G. (2016). Understanding particle margination in blood flow—A step toward optimized drug delivery systems. *Medical Engineering & Physics*, *38*(1), 2−10. Available from https://doi.org/10.1016/j.medengphy.2015.08.009.

Muzykantov, V. R., & Taylor, R. P. (1994). Attachment of biotinylated antibody to red blood cells: Antigen-binding capacity of immunoerythrocytes and their susceptibility to lysis by complement. *Analytical Biochemistry*, *223*(1), 142−148. Available from https://doi.org/10.1006/abio.1994.1559.

Nemmar, A., Albarwani, S., Beegam, S., Yuvaraju, P., Yasin, J., Attoub, S., et al. (2014). Amorphous silica nanoparticles impair vascular homeostasis and induce systemic inflammation. *International Journal of Nanomedicine*, *9*, 2779. Available from https://doi.org/10.2147/IJN.S52818.

Nemmar, A., Hoet, P. H. M., Vanquickenborne, B., Dinsdale, D., Thomeer, M., Hoylaerts, M. F., et al. (2002). Passage of inhaled particles into the blood circulation in humans. *Circulation*, *105*(4), 411−414. Available from https://doi.org/10.1161/hc0402.104118.

Nemmar, A., Hoylaerts, M. F., Hoet, P. H. M., Dinsdale, D., Smith, T., Xu, H., et al. (2002). Ultrafine particles affect experimental thrombosis in an in vivo hamster model. *American Journal of Respiratory and Critical Care Medicine*, *166*(7), 998−1004. Available from https://doi.org/10.1164/rccm.200110-026OC.

Nemmar, A., Melghit, K., & Ali, B. H. (2008). The acute proinflammatory and prothrombotic effects of pulmonary exposure to rutile TiO_2 nanorods in rats. *Experimental Biology and Medicine*, *233*(5), 610−619. Available from https://doi.org/10.3181/0706-RM-165.

Neu, B., & Meiselman, H. J. (2002). Depletion-mediated red blood cell aggregation in polymer solutions. *Biophysical Journal*, *83*(5), 2482−2490. Available from https://doi.org/10.1016/S0006-3495(02)75259-4.

Noël, C., Simard, J. C., & Girard, D. (2016). Gold nanoparticles induce apoptosis, endoplasmic reticulum stress events and cleavage of cytoskeletal proteins in human neutrophils. *Toxicology in Vitro*, *31*, 12−22. Available from https://doi.org/10.1016/j.tiv.2015.11.003.

Osborne, O. J., Lin, S., Chang, C. H., Ji, Z., Yu, X., Wang, X., et al. (2015). Organ-specific and size-dependent Ag nanoparticle toxicity in gills and intestines of adult zebrafish. *ACS Nano*, *9*(10), 9573−9584. Available from https://doi.org/10.1021/acsnano.5b04583.

Owens, D. E., & Peppas, N. A. (2006). Opsonization, biodistribution, and pharmacokinetics of polymeric nanoparticles. *International Journal of Pharmaceutics*, *307*(1), 93−102. Available from https://doi.org/10.1016/j.ijpharm.2005.10.010.

Palchetti, S., Pozzi, D., Capriotti, A. L., Barbera, G. L., Chiozzi, R. Z., Digiacomo, L., et al. (2017). Influence of dynamic flow environment on nanoparticle-protein corona: From protein patterns to uptake in cancer cells. *Colloids Surfaces B Biointerfaces*, *153*, 263−271. Available from https://doi.org/10.1016/j.colsurfb.2017.02.037.

Pan, D. C., Myerson, J. W., Brenner, J. S., Patel, P. N., Anselmo, A. C., Mitragotri, S., et al. (2018). Nanoparticle properties modulate their attachment and effect on carrier red blood cells. *Scientific Reports*, *8*(1), 1−12. Available from https://doi.org/10.1038/s41598-018-19897-8.

Pareek, V., Bhargava, A., Bhanot, V., Gupta, R., Jain, N., & Panwar, J. (2018). Formation and characterization of protein corona around nanoparticles: A review. *Journal of Nanoscience and Nanotechnology*, *18*(10), 6653−6670. Available from https://doi.org/10.1166/jnn.2018.15766.

Park, J., Lim, D. H., Lim, H. J., Kwon, T., Choi, J. S., Jeong, S., et al. (2011). Size dependent macrophage responses and toxicological effects of Ag nanoparticles. *Chemical Communication*, *47*(15), 4382−4384. Available from https://doi.org/10.1039/c1cc10357a.

Patton, K., & Thibodeau, G. A. (2018). *Anthony's textbook of anatomy and physiology* (21st ed.). Elsevier Health Sciences.

Pearson, R. M., Juettner, V. V., & Hong, S. (2014). Biomolecular corona on nanoparticles: A survey of recent literature and its implications in targeted drug delivery. *Frontiers in Chemistry*, *2*, 108. Available from https://doi.org/10.3389/fchem.2014.00108.

Pederzoli, F., Tosi, G., Vandelli, M. A., Belletti, D., Forni, F., & Ruozi, B. (2017). Protein corona and nanoparticles: How can we investigate on? *Wiley Interdisciplinary Reviews: Nanomedicine and Nanobiotechnology*, *9*(6), e1467. Available from https://doi.org/10.1002/wnan.1467.

Pelaz, B., Charron, G., Pfeiffer, C., Zhao, Y., De La Fuente, J. M., Liang, X. J., et al. (2013). Interfacing engineered nanoparticles with biological systems: Anticipating adverse nano-bio interactions. *Small (Weinheim an der Bergstrasse, Germany)*, *9*(9−10), 1573−1584. Available from https://doi.org/10.1002/smll.201201229.

Qualhato, G., Rocha, T. L., de Oliveira Lima, E. C., e Silva, D. M., Cardoso, J. R., Koppe Grisolia, C., et al. (2017). Genotoxic and mutagenic assessment of iron oxide (maghemite-Γ-Fe_2O_3) nanoparticle in the guppy Poecilia reticulata. *Chemosphere*, *183*, 305−314. Available from https://doi.org/10.1016/j.chemosphere.2017.05.061.

Rampado, R., Crotti, S., Caliceti, P., Pucciarelli, S., & Agostini, M. (2020). Recent advances in understanding the protein corona of nanoparticles and in the formulation of "Stealthy" nanomaterials. *Frontiers in Bioengineering and Biotechnology*, 8. Available from https://doi.org/10.3389/fbioe.2020.00166.

Rao, L., Cai, B., Bu, L. L., Liao, Q. Q., Guo, S. S., Zhao, X. Z., et al. (2017). Microfluidic electroporation-facilitated synthesis of erythrocyte membrane-coated magnetic nanoparticles for enhanced imaging-guided cancer therapy. *ACS Nano*, *1*(4), 3496−3505. Available from https://doi.org/10.1021/acsnano.7b00133.

Recordati, C., De Maglie, M., Bianchessi, S., Argentiere, S., Cella, C., Mattiello, S., et al. (2016). Tissue distribution and acute toxicity of silver after single intravenous administration in mice: Nano-specific and size-dependent effects. *Particle and Fibre Toxicology*, *13*(1), 12. Available from https://doi.org/10.1186/s12989-016-0124-x.

Ren, X., Zheng, R., Fang, X., Wang, X., Zhang, X., Yang, W., et al. (2016). Red blood cell membrane camouflaged magnetic nanoclusters for imaging-guided photothermal therapy. *Biomaterials*, *92*, 13−24. Available from https://doi.org/10.1016/j.biomaterials.2016.03.026.

Revia, R. A., & Zhang, M. (2016). Magnetite nanoparticles for cancer diagnosis, treatment, and treatment monitoring: Recent advances. *Materials Today*, *19*(3)), 157−168. Available from https://doi.org/10.1016/j.mattod.2015.08.022.

Röcker, C., Pötzl, M., Zhang, F., Parak, W. J., & Nienhaus, G. U. (2009). A quantitative fluorescence study of protein monolayer formation on colloidal nanoparticles. *Nature Nanotechnology*, *4*(9), 577−580. Available from https://doi.org/10.1038/nnano.2009.195.

Rossi, L., Pierigè, F., Antonelli, A., Bigini, N., Gabucci, C., Peiretti, E., et al. (2016). Engineering erythrocytes for the modulation of drugs' and contrasting agents' pharmacokinetics and biodistribution. *Advanced Drug Delivery Reviews*, *106*, 73−87. Available from https://doi.org/10.1016/j.addr.2016.05.008.

Rothen-Rutishauser, B. M., Schürch, S., Haenni, B., Kapp, N., & Gehr, P. (2006). Interaction of fine particles and nanoparticles with red blood cells visualized with advanced microscopic techniques. *Environmental Science & Technology*, *40*(14), 4353−4359. Available from https://doi.org/10.1021/es0522635.

Rudramurthy, G. R., & Swamy, M. K. (2018). Potential applications of engineered nanoparticles in medicine and biology: An update. *Journal of Biological Inorganic Chemistry: JBIC: A Publication of the Society of Biological Inorganic Chemistry*, 23(8), 1185−1204. Available from https://doi.org/10.1007/s00775-018-1600-6.

Rybak-Smith, M. J., Tripisciano, C., Borowiak-Palen, E., Lamprecht, C., & Sim, R. B. (2011). Effect of functionalization of carbon nanotubes with psychosine on complement activation and protein adsorption. *Journal of Biomedical Nanotechnology*, 7(6), 830−839. Available from https://doi.org/10.1166/jbn.2011.1347.

Sachar, S., & Saxena, R. K. (2011). Cytotoxic effect of poly-dispersed single walled carbon nanotubes on erythrocytes in vitro and in vivo. *PLoS One*, 6(7), e22032. Available from https://doi.org/10.1371/journal.pone.0022032.

Saikia, J., Mohammadpour, R., Yazdimamaghani, M., Northrup, H., Hlady, V., & Ghandehari, H. (2018). Silica nanoparticle-endothelial interaction: Uptake and effect on platelet adhesion under flow conditions. *ACS Applied Bio Materials*, 1, 1620−1627. Available from https://doi.org/10.1021/acsabm.8b00466.

Samuel, S. P., Santos-Martinez, M. J., Medina, C., Jain, N., Radomski, M. W., Prina-Mello, A., et al. (2015). CdTe quantum dots induce activation of human platelets: Implications for nanoparticle hemocompatibility. *International Journal of Nanomedicine*, 10, 2723. Available from https://doi.org/10.2147/IJN.S78281.

Sanfins, E., Augustsson, C., Dahlbäck, B., Linse, S., & Cedervall, T. (2014). Size-dependent effects of nanoparticles on enzymes in the blood coagulation cascade. *Nano Letters*, 14(8), 4736−4744. Available from https://doi.org/10.1021/nl501863u.

Sarmah, D., Saraf, J., Kaur, H., Pravalika, K., Tekade, R. K., Borah, A., et al. (2017). Stroke management: An emerging role of nanotechnology. *Micromachines*, 8(9), 262. Available from https://doi.org/10.3390/mi8090262.

Shabanova, E. M., Drozdov, A. S., Fakhardo, A. F., Dudanov, I. P., Kovalchuk, M. S., & Vinogradov, V. V. (2018). Thrombin@Fe_3O_4 nanoparticles for use as a hemostatic agent in internal bleeding. *Scientific Reports*, 8(1), 1−10. Available from https://doi.org/10.1038/s41598-017-18665-4.

Shah, N. B., Vercellotti, G. M., White, J. G., Fegan, A., Wagner, C. R., & Bischof, J. C. (2012). Blood-nanoparticle interactions and in vivo biodistribution: Impact of surface peg and ligand properties. *Molecular Pharmaceutics*, 9(8), 2146−2155. Available from https://doi.org/10.1021/mp200626j.

Shang, L., Nienhaus, K., & Nienhaus, G. U. (2014). Engineered nanoparticles interacting with cells: Size matters. *Journal of Nanobiotechnology*, 12(1), 5. Available from https://doi.org/10.1186/1477-3155-12-5.

Shi, J., Sun, X., Lin, Y., Zou, X., Li, Z., Liao, Y., et al. (2014). Endothelial cell injury and dysfunction induced by silver nanoparticles through oxidative stress via IKK/NF-κB pathways. *Biomaterials*, 35(24), 6657−6666. Available from https://doi.org/10.1016/j.biomaterials.2014.04.093.

Shrivastava, R., Kushwaha, P., Bhutia, Y. C., & Flora, S. J. S. (2016). Oxidative stress following exposure to silver and gold nanoparticles in mice. *Toxicology and Industrial Health*, 32(8), 1391−1404. Available from https://doi.org/10.1177/0748233714562623.

Sim, R. B., & Tsiftsoglou, S. A. (2004). Proteases of the complement system. *Biochemical Society Transactions*, 32(Pt 1), 21−27. Available from https://doi.org/10.1042/BST0320021.

Simmonds, M. J., Meiselman, H. J., & Baskurt, O. K. (2013). Blood rheology and aging. *Journal of Geriatric Cardiology*, 10(3), 291. Available from https://doi.org/10.3969/j.issn.1671-5411.2013.03.010.

Singh, R., & Lillard, J. W. (2009). Nanoparticle-based targeted drug delivery. *Experimental and Molecular Pathology*, 86(3), 215−223. Available from https://doi.org/10.1016/j.yexmp.2008.12.004.

Smith, P. J., Giroud, M., Wiggins, H. L., Gower, F., Thorley, J. A., Stolpe, B., et al. (2012). Cellular entry of nanoparticles via serum sensitive clathrin-mediated endocytosis, and plasma membrane permeabilization. *International Journal of Nanomedicine*, 7, 2045. Available from https://doi.org/10.2147/IJN.S29334.

Steck, T. L. (1974). The organization of proteins in the human red blood cell membrane: A review. *The Journal of Cell Biology*, 62(1), 1−19. Available from https://doi.org/10.1083/jcb.62.1.1.

Suk, J. S., Xu, Q., Kim, N., Hanes, J., & Ensign, L. M. (2016). PEGylation as a strategy for improving nanoparticle-based drug and gene delivery. *Advanced Drug Delivery Reviews*, *99*, 28–51. Available from https://doi.org/10.1016/j.addr.2015.09.012.

Sukhanova, A., Bozrova, S., Sokolov, P., Berestovoy, M., Karaulov, A., & Nabiev, I. (2018). Dependence of nanoparticle toxicity on their physical and chemical properties. *Nanoscale Research Letters*, *13*(1), 44. Available from https://doi.org/10.1186/s11671-018-2457-x.

Sun, B. B., Maranville, J. C., Peters, J. E., Stacey, D., Staley, J. R., Blackshaw, J., et al. (2018). Genomic atlas of the human plasma proteome. *Nature*, *558*(7708), 73–79. Available from https://doi.org/10.1038/s41586-018-0175-2.

Sun, X., Shi, J., Zou, X., Wang, C., Yang, Y., & Zhang, H. (2016). Silver nanoparticles interact with the cell membrane and increase endothelial permeability by promoting VE-cadherin internalization. *Journal of Hazardous Materials*, *317*, 570–578. Available from https://doi.org/10.1016/j.jhazmat.2016.06.023.

Sun, Y., Su, J., Liu, G., Chen, J., Zhang, X., Zhang, R., et al. (2017). Advances of blood cell-based drug delivery systems. *European Journal of Pharmaceutical Sciences: Official Journal of the European Federation for Pharmaceutical Sciences*, *96*, 115–128. Available from https://doi.org/10.1016/j.ejps.2016.07.021.

Swystun, L. L., & Liaw, P. C. (2016). The role of leukocytes in thrombosis. *Blood*, *128*(6), 753–762. Available from https://doi.org/10.1182/blood-2016-05-718114.

Szebeni, J., & Haima, P. (2013). Hemocompatibility of medical devices, blood products, nanomedicines and biologicals. TECOmedical Clinical & Technical Review.

Tenzer, S., Docter, D., Kuharev, J., Musyanovych, A., Fetz, V., Hecht, R., et al. (2013). Rapid formation of plasma protein corona critically affects nanoparticle pathophysiology. *Nature Nanotechnology*, *8*(10), 772–781. Available from https://doi.org/10.1038/nnano.2013.181.

The National Nanotechnology Initiative Supplement to the President's. (2020). Budget Product of the National Science and Technology Council 2019, pp. 1–56.

Titus, D., James Jebaseelan Samuel, E., & Roopan, S. M. (2019). *Nanoparticle characterization techniques. Green synthesis, characterization and applications of nanoparticles*. Elsevier. [chapter 12]. https://doi.org/10.1016/b978-0-08-102579-6.00012-5.

Tsai, L. W., Lin, Y. C., Perevedentseva, E., Lugovtsov, A., Priezzhev, A., & Cheng, C. L. (2016). Nanodiamonds for medical applications: Interaction with blood in vitro and in vivo. *International Journal of Molecular Science*, *17*(7)), 1111. Available from https://doi.org/10.3390/ijms17071111.

Tsai, R. K., Rodriguez, P. L., & Discher, D. E. (2010). Self inhibition of phagocytosis: The affinity of "marker of self" CD47 for SIRPα dictates potency of inhibition but only at low expression levels. *Blood Cells, Molecules and Diseases*, *45*(1), 67–74. Available from https://doi.org/10.1016/j.bcmd.2010.02.016.

Vahidkhah, K., & Bagchi, P. (2015). Microparticle shape effects on margination, near-wall dynamics and adhesion in a three-dimensional simulation of red blood cell suspension. *Soft Matter*, *11*(11), 2097–2109. Available from https://doi.org/10.1039/c4sm02686a.

Villa, C. H., Anselmo, A. C., Mitragotri, S., & Muzykantov, V. (2016). Red blood cells: Supercarriers for drugs, biologicals, and nanoparticles and inspiration for advanced delivery systems. *Advanced Drug Delivery Reviews*, *106*, 88–103. Available from https://doi.org/10.1016/j.addr.2016.02.007.

Villa, C. H., Pan, D. C., Zaitsev, S., Cines, D. B., Siegel, D. L., & Muzykantov, V. R. (2015). Delivery of drugs bound to erythrocytes: New avenues for an old intravascular carrier. *Therapeutic Delivery*, *6*(7)), 795–826. Available from https://doi.org/10.4155/tde.15.34.

Vroman, L., Adams, A. L., Fischer, G. C., & Munoz, P. C. (1980). Interaction of high molecular weight kininogen, factor XII, and fibrinogen in plasma at interfaces. *Blood*. Available from https://doi.org/10.1182/blood.v55.1.156.bloodjournal551156.

Vu, V. P., Gifford, G. B., Chen, F., Benasutti, H., Wang, G., Groman, E. V., et al. (2019). Immunoglobulin deposition on biomolecule corona determines complement opsonization efficiency of preclinical and clinical nanoparticles. *Nature Nanotechnology*, *14*(3), 260−268. Available from https://doi.org/10.1038/s41565-018-0344-3.

Wadhwa, R., Aggarwal, T., Thapliyal, N., Kumar, A., Priya., Yadav, P., et al. (2019). Red blood cells as an efficient in vitro model for evaluating the efficacy of metallic nanoparticles. *3 Biotech*, *9*(7), 279. Available from https://doi.org/10.1007/s13205-019-1807-4.

Wang, C., Sun, X., Cheng, L., Yin, S., Yang, G., Li, Y., et al. (2014). Multifunctional theranostic red blood cells for magnetic-field-enhanced in vivo combination therapy of cancer. *Advanced Materials*, *26*(28), 4794−4802. Available from https://doi.org/10.1002/adma.201400158.

Wang, K., Gao, Z., Gao, G., Wo, Y., Wang, Y., Shen, G., et al. (2013). Systematic safety evaluation on photoluminescent carbon dots. *Nanoscale Research Letters*, *8*(1), 122. Available from https://doi.org/10.1186/1556-276X-8-122.

Wang, T., Bai, J., Jiang, X., & Nienhaus, G. U. (2012). Cellular uptake of nanoparticles by membrane penetration: A study combining confocal microscopy with FTIR spectroelectrochemistry. *ACS Nano*, *6*(2), 1251−1259.

Wang, W., Gaus, K., Tilley, R. D., & Gooding, J. J. (2019). The impact of nanoparticle shape on cellular internalisation and transport: What do the different analysis methods tell us? *Materials Horizons*, *6*(8), 1538−1547. Available from https://doi.org/10.1039/c9mh00664h.

Wasowicz, M., Ficek, M., Wróbel, M. S., Chakraborty, R., Fixler, D., Wierzba, P., et al. (2017). Haemocompatibility of modified nanodiamonds. *Materials*, *10*(4), 352. Available from https://doi.org/10.3390/ma10040352, Basel.

Weber, M., Steinle, H., Golombek, S., Hann, L., Schlensak, C., Wendel, H. P., et al. (2018). Blood-contacting biomaterials: In vitro evaluation of the hemocompatibility. *Frontiers in Bioengineering and Biotechnology*, *6*, 99. Available from https://doi.org/10.3389/fbioe.2018.00099.

Wu, X., Tan, Y., Mao, H., & Zhang, M. (2010). Toxic effects of iron oxide nanoparticles on human umbilical vein endothelial cells. *International Journal of Nanomedicine*, *5*, 385. Available from https://doi.org/10.2147/ijn.s10458.

Wu, Y. F., Hsu, P. S., Tsai, C. S., Pan, P. C., & Chen, Y. L. (2018). Significantly increased low shear rate viscosity, blood elastic modulus, and RBC aggregation in adults following cardiac surgery. *Scientific Reports*, *8*(1), 1−10. Available from https://doi.org/10.1038/s41598-018-25317-8.

Wu, Z., Li, T., Li, J., Gao, W., Xu, T., Christianson, C., et al. (2014). Turning erythrocytes into functional micromotors. *ACS Nano*, *8*(12), 12041−12048. Available from https://doi.org/10.1021/nn506200x.

Xia, Q., Zhang, Y., Li, Z., Hou, X., & Feng, N. (2019). Red blood cell membrane-camouflaged nanoparticles: A novel drug delivery system for antitumor application. *Acta Pharmaceutica Sinica B*, *9*(4), 675−689. Available from https://doi.org/10.1016/j.apsb.2019.01.011.

Yang, J. Y., Bae, J., Jung, A., Park, S., Chung, S., Seok, J., et al. (2017). Surface functionalization-specific binding of coagulation factors by zinc oxide nanoparticles delays coagulation time and reduces thrombin generation potential in vitro. *PLoS One*, *12*(7), e0181634. Available from https://doi.org/10.1371/journal.pone.0181634.

Yu, Z., Li, Q., Wang, J., Yu, Y., Wang, Y., Zhou, Q., et al. (2020). Reactive oxygen species-related nanoparticle toxicity in the biomedical field. *Nanoscale Research Letters*, *15*(1), 1−14. Available from https://doi.org/10.1186/s11671-020-03344-7.

Zaitsev, S., Kowalska, M. A., Neyman, M., Carnemolla, R., Tliba, S., Ding, B. S., et al. (2012). Targeting recombinant thrombomodulin fusion protein to red blood cells provides multifaceted thromboprophylaxis. *Blood*, *119*(20), 4779−4785. Available from https://doi.org/10.1182/blood-2011-12-398149.

Zhang, H. (2016). Erythrocytes in nanomedicine: An optimal blend of natural and synthetic materials. *Biomaterials Science*, *4*(7), 1024−1031. Available from https://doi.org/10.1039/c6bm00072j.

Zhang, H., Burnum, K. E., Luna, M. L., Petritis, B. O., Kim, J. S., Qian, W. J., et al. (2011). Quantitative proteomics analysis of adsorbed plasma proteins classifies nanoparticles with different surface properties and size. *Proteomics*, *11*(23), 4569—4577. Available from https://doi.org/10.1002/pmic.201100037.

Zhang, S., Gao, H., & Bao, G. (2015). Physical principles of nanoparticle cellular endocytosis. *ACS Nano*, *9*(9), 8655—8671. Available from https://doi.org/10.1021/acsnano.5b03184.

Zhang, X. D., Wu, H. Y., Wu, D., Wang, Y. Y., Chang, J. H., Zhai, Z. B., et al. (2010). Toxicologic effects of gold nanoparticles in vivo by different administration routes. *International Journal of Nanomedicine*, *5*, 771. Available from https://doi.org/10.2147/IJN.S8428.

Zhao, Y., Sun, X., Zhang, G., Trewyn, B. G., Slowing, I. I., & Lin, V. S. Y. (2011). Interaction of mesoporous silica nanoparticles with human red blood cell membranes: Size and surface effects. *ACS Nano*, *5*(2), 1366—1375. Available from https://doi.org/10.1021/nn103077k.

Zhu, R., Avsievich, T., Popov, A., & Meglinski, I. (2020). Optical tweezers in studies of red blood cells. *Cells*, *9*(3), 545. Available from https://doi.org/10.3390/cells9030545.

Extracorporeal affinity systems and immunoadsorption therapies

Handan Yavuz, Nilay Bereli and Adil Denizli

Biochemistry Division, Department of Chemistry, Hacettepe University, Ankara, Turkey

2.1 Introduction

Extracorporeal therapies are directed toward the removal of potential toxic substances from plasma. Being one of the conventional extracorporeal therapies, hemoperfusion is a method in which the blood from the patient is circulated through an adsorption column in order to remove exogenous and endogenous toxin substances from the blood. Extracorporeal methods are particularly suitable for the treatment of atherosclerosis, cancer, acute poisonings with metals or other substances, and degenerative and autoimmune diseases (Table 2.1) (Abrams et al., 2016; Blaha et al., 2011; Denizli, 2011; Dhokia et al., 2019; Gemelli et al., 2019; Jung et al., 2009; King et al., 2019; Klingele et al., 2020; Lee et al., 2002; Nakaji, 2001; Olbricht, 1993; Schettler et al., 2012; Zeitler et al., 2012).

Conventional techniques like plasma exchange, hemodialysis, hemofiltration, and hemoperfusion are nonselective and require plasma substitutes such as albumin. This may also increases the risk of plasma products related immune reactions. The use of nanoparticle, microsphere, membrane, monolith, hollow fiber, and cryogel-based affinity adsorbents in hemoperfusion systems allows specific adsorption and removal of pathogens (Fig. 2.1).

The patient's own plasma is returned with no need for substitution materials and minimized risk for infectious diseases are some of the benefits of the extracorporeal affinity therapy.

Table 2.1 Diseases potentially treated by extracorporeal affinity therapy.

Disease	References
Hyperbilirubinemia	Dhokia et al. (2019), Gemelli et al. (2019), Jung et al. (2009), Lee et al. (2002)
Familial hypercholesterolemia	Blaha et al. (2011), Olbricht (1993), Schettler et al. (2012)
Autoimmune diseases	Denizli (2011), Nakaji (2001)
Hemophilias (complicated by antibodies to clotting factors VIII or IX)	Zeitler et al. (2012)
Thrombocytopenia	Abrams et al. (2016)
Neurological diseases	Klingele et al. (2020)
Acute poisoning	King et al. (2019)

Nanotechnology for Hematology, Blood Transfusion, and Artificial Blood. DOI: https://doi.org/10.1016/B978-0-12-823971-1.00001-5

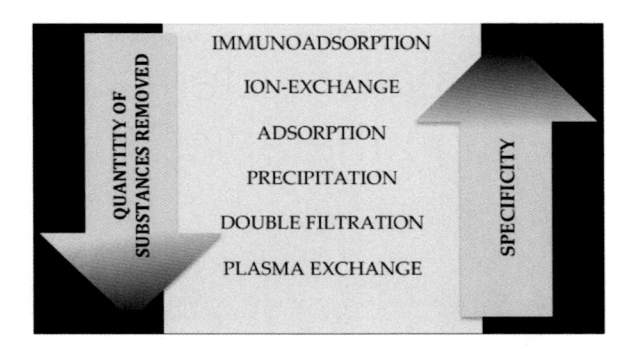

FIGURE 2.1

Brief comparison of therapeutic extracorporeal techniques.

Affinity carriers may be used in a hemoperfusion system. This type of application makes the system more effective, simple, specific, and inexpensive. Many adsorbents such as nanoparticles, microbeads, membranes, fibers, hollow fibers, monoliths, and cryogels have been derivatized with specific ligands and investigated for potential applications to remove unwanted substances directly from plasma (Altintaş et al., 2011; Aşir et al., 2005; Demirçelik et al., 2009; Han & Zhang, 2009; Ma et al., 2005; Rad et al., 2003; Şenel et al., 2002; Shi et al., 2005; Wang et al., 1997; Yavuz et al., 2001, 2016).

2.2 Extracorporeal therapy for autoimmune diseases

The question of "how do we discriminate ourselves from other people?" seems a nonsense question for a person with healthy brain activity, since our brain recognizes and discriminates information belonging to others and ourselves. Like the brain system, the immune system recognizes and distinguishes different structures such as viruses, bacteria, chemicals, toxins, transplanted organs, and tumor cells. The immune system sorts out these functions with a network of cells, tissues, and organs working together. It is amazingly complex and can recognize and remember millions of different structures and produce specific molecules and cells to match and wipe them out. The primary organs of the immune system are the thymus and bone marrow where immune-active cells are made. The secondary organs are tonsils, adenoids, spleen, and lymph nodes where immune-active cells attack and destroy the enemy. White blood cells, or leukocytes, are the most important individual cells of the immune system, which have different appearance and function. These cells are made from the stem cells found in the marrow of long bones, like red blood cells (Sompayrac, 2012).

At the heart of the powerful defense against foreign materials the humoral immune system exists and the basic functional units of the humoral immune system are the special and very specific proteins called antibodies or immunoglobulins, which are made by B-cells. There are five classes of antibodies, as shown in the Fig. 2.2.

The most common type of antibody in the blood is IgG. It is a Y-shaped molecule made up of two identical large proteins and two identical smaller proteins joined together. The ends of large

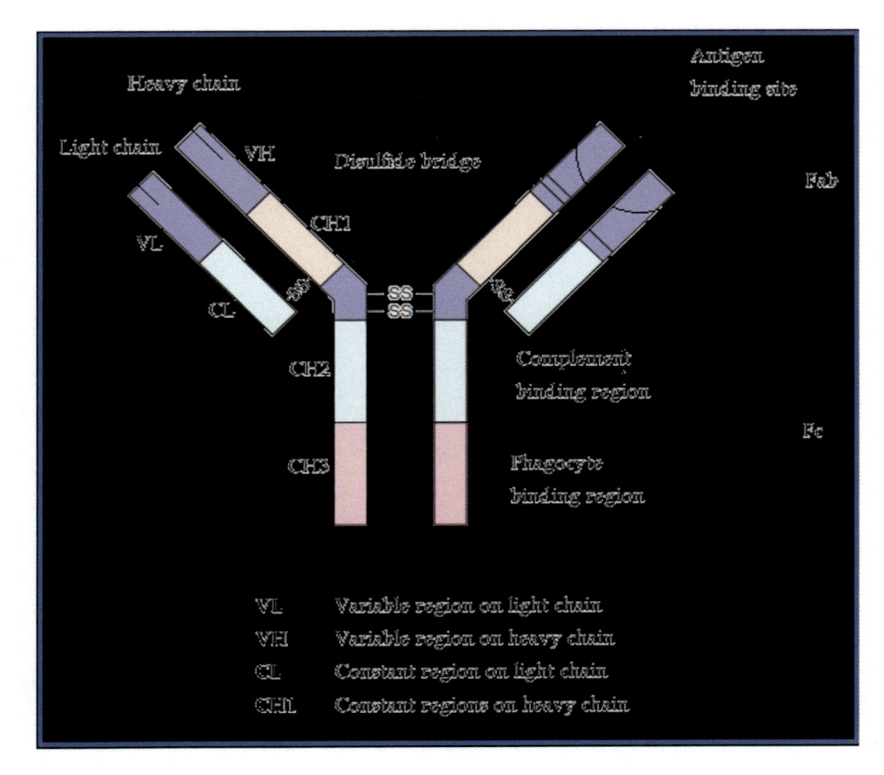

FIGURE 2.2

Classification and properties of antibodies.

molecules have exactly the same amino acids and are called constant regions. The small proteins line around the arms of the Y shape and they are all bonded to each other. The half of the small proteins toward the base is constant for all IgG molecules. The upper ends of the large and small proteins on the tips are called variable regions that make antibodies specific for a certain antigen (Sompayrac, 2012) (Fig. 2.3).

When the underlying defects of the immune system lead to attack on its own organs, tissues, and cells the case of autoimmune diseases arises. There are more than 100 illnesses characterized and according to the Autoimmune Diseases Coordinating Committee (ADCC) of National Institute of Health, USA, their prevalence is rising in the population. Autoimmune diseases are a heterogeneous group of disorders, which can be classified according to how the damage is induced, that is, humoral-mediated autoimmune diseases (myasthenia gravis, autoimmune hemolytic anemia, immune thrombocytopenic purpura, etc.) and cell-mediated autoimmune diseases (rheumatoid arthritis, systemic lupus erythematosus (SLE), thyrodinitis, insulin-dependent diabetes mellitus, etc.) or based on the number of organs effected, such as organ-specific disorders and systemic disorders (Rose & Mackay, 2014; Shoenfeld et al., 2008).

Antibody		Heavy Chain	Molecular weight (Da)	% of total antibody in serum	Function
	Name				
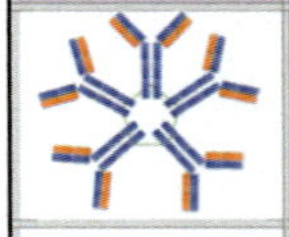	IgM	μ	900 k	6 %	. Primary response . Fixes complement . Monomer serves as B-cell receptor
	IgG	γ	150 k	80 %	. Main blood antibody . Neutralizes toxin . Opsonization
	IgA	α	385 k	13 %	Secreted into mucus, tears, saliva
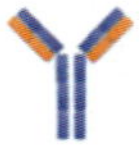	IgE	ε	200 k	0.002 %	. Allergic response anti-parasitic activity
	IgD	δ	180 k	1 %	B-cell receptor

FIGURE 2.3

Structure of an IgG molecule.

There is no complete cure available for these diseases due to the lack of greater understanding of the principles and progression of the diseases. Conversely, despite the diversity of the diseases, autoimmune diseases share common underlying mechanisms and properties, for example, autoantibodies are at the center of nearly all systemic autoimmune diseases. This allows the use of related therapies such as immunosuppressive and immunomodulatory agents, B-cells targeting therapies, and direct removal of circulating autoantibodies (Nakaji, 2001).

Immunoadsorption for the treatment of SLE with the DNA attached charcoal to remove anti-DNA antibodies from rabbit plasma. The first clinical attempt of immunoadsorption was made in 1979 by Terman et al. to treat a female patient suffering from SLE. Then successful treatments were reported by several authors of various diseases including cancer, chronic lymphocytic leukemia, and autoimmune hemolytic anemia (Nilsson et al., 1981; Ray, 1984; Terman et al., 1976, 1979). Ray et al. used protein A-attached adsorbents to remove pathogenic antibodies and immune complexes in various diseases and cancers. Protein A-attached agarose-based adsorbents were also used to treat hemophilia patients (Nilsson et al., 1981; Ray, 1984; Terman et al., 1976, 1979). A dextran-sulfate-based immunoadsorption column was used by Hoshimoto et al. for SLE patients with nephritis together with drug therapy (Hoshimoto et al., 1990). Palmer et al. reported efficiency of protein A immunosorption combined with corticosteroid and cyclophosphamide for the treatment of glomerulonephritis (Palmer et al., 1991).

While protein A has high affinity for the IgG, antigen-bound IgG and IgM complexes, it has reported that the efficiency of protein A immunoadsorption column was affected negatively by antirheumatic drugs. In addition protein A immunoadsorption has common adverse side effects such as joint pain and swelling, abdominal pain, headache, nausea, and dizziness. Several groups reported the performance of different adsorbents including dextran sulfate, phenylalanine, tryptophan, and antibody as a ligand. The mode of interaction of these ligands with antibodies is summarized in Table 2.2 (Denizli, 2011).

Commercially available immunoadsorption columns to treat immune diseases are summarized in Table 2.3.

The adsorbents used in these columns are hydrophobic amino acids, tryptophan, and phenylalanine, which interact through hydrophobic and electrostatic interactions with their targets; dextran sulfate, which binds and removes targets with electrostatic interactions owing to its negative charges; Protein A, which is a bacterial membrane protein that binds Fc region of antibodies specifically; synthetic peptides and poly- or monoclonal antibodies against pathogenic antibodies, etc. Some of the clinical applications of immunoadsorption columns, commercially available or prepared in laboratory, are summarized in the following table (Table 2.4) (Aotsuka et al., 1990; Bhardwaj et al., 2017; Biesenbach et al., 2009; Blaha et al., 2011; Braun et al., 2000; Darnige et al., 1999; Hershko & Naparstek, 2005; Kato & Ikada, 1996; Kong et al., 2002; Müller et al., 2000; Özgür et al., 2011; Pitiot et al., 2000; Rech et al., 2006; Schneider et al., 1990; Süfke et al., 2017; Sugimoto et al., 2006; Suzuki, 2002; Türkmen et al., 2006; Uzun et al., 2010; Ventura et al., 2001; Yan et al., 1988; Yılmaz et al., 2008; Yu & He, 1999; Zhu et al., 1999)

Table 2.2 Immunoadsorbents: mode of action (Denizli, 2011).

Ligand	Binding force	Principle
Anti-IgG	Antigen–antibody	Biological interaction
Protein A	Fc	Biological interaction
Phenylalanine, tryptophan	Hydrophobic	Physicochemical interaction
Dextran sulfate	Ionic	Physicochemical interaction

Table 2.3 Commercial immunoadsorption columns for clinical treatment (Denizli, 2011).

Adsorbent	Commercial name	Manufacturer
Phenylalanine-PVA gel	Immusorba PH	Ashai Medical (Japan)
Tryptophan-PVA gel	Immusorba TR	Ashai Medical (Japan)
Dextran sulfate-cellulose beads	Selesorb	Kaneka (Japan)
Protein A-silica beads	Prosorba	Cypress Bioscience (USA)
Protein A- agarose beads	Immunosorba	Excorim (Sweden)
Aceltylcholine receptor- cellulose beads	Medisorba MG	Kuraray (Japan)
Polyclonal antibodies-cellulose beads	Ig-Therasorb	Therasorb Medical Systems (Germany)

Table 2.4 Clinical applications of some immunoadsorption columns.

Adsorbent	Ligand	Disease	References
Poly(ethylene vinyl alcohol)	L-histidine	Systemic lupus erythematosus (SLE)	Ventura et al. (2001)
Nonwoven Poly(ethylene terephthalate)	DNA	SLE	Zhu et al. (1999)
Chitosan particles	DNA	SLE	Zhu et al. (1999)
Poly(ethylene terephthalate) microfibers	DNA	SLE	Kato and Ikada (1996)
Silica	Protein A	Rheumatoid arthritis	Hershko and Naparstek (2005)
Cellulose gels	Polyanionic ligands	SLE	Aotsuka et al. (1990)
	Anti-IgG	SLE	Biesenbach et al. (2009)
Sepharose	Synthetic peptide (GAM146)	SLE	Rech et al. (2006)
Spherical carbon	DNA	SLE	Yan et al. (1988)
Poly(ethylene vinyl alcohol) hollow fiber	L-histidine	SLE and primary antiphospholipid syndrome	Darnige et al. (1999), Pitiot et al. (2000)
PHEMA-MAAL membrane	DNA	SLE	Uzun et al. (2010)
Cellulose	Dextran sulfate	SLE	Suzuki (2002)
Commercial M-P column	Phenylalanine Tryptophan	SLE	Schneider et al. (1990)
Immusorba PH-35	Phenylalanine	Lupus nephritis	Sugimoto et al. (2006)
	Anti-IgG	Idiopathic dilated cardiomyopathy	Müller et al. (2000)
Sepharose (Immonosorba)	Protein A	SLE	Braun et al. (2000)
Cellulose	DNA	SLE	Kong et al. (2002)
PHEMA cryogel	Polyethylene imine	SLE	Türkmen et al. (2006)
PHEMA cryogel	DNA	SLE	Özgür et al. (2011)
PHEMAH particles	Histidine	Rheumatoid arthritis	Yılmaz et al. (2008)
Polymethacrylate	Protein A-LIGASORB	Multiple sclerosis	Süfke et al. (2017)
Immunosorba ®	Proetin A	Dilated cardiomyophaty	Bhardwaj et al. (2017)
Sepharose 4FF Adsopack 200	Anti human IgG	Myasthenia gravis	Blaha et al. (2011)

2.3 Extracorporeal affinity adsorbents for bilirubin removal

A negatively charged pigment formed in the normal metabolism of heme proteins, bilirubin, is non-soluble in water and transported predominantly by albumin in circulation. When the serum concentration of bilirubin exceeds 1 mg/dL hyperbilirubinemia condition occurs. Hyperbilirubinemia can be classified as conjugated or unconjugated depending on the type of bilirubin present in plasma.

Free bilirubin is highly toxic for organisms. Because of its hydrophobicity it aggregates on phospholipid membranes and damages their integrity. Unconjugated bilirubin can cross the blood−brain barrier and cause encephalopathy and cannot be excreted into urine. Several conditions can result in unconjugated hyperbilirubinemia, these are neonatal or physiologic jaundice, hemolytic anemia, hereditary bilirubin conjugation defects, and acute liver dysfunction due to ingestion of toxins (Francini et al., 2010).

Many bilirubin removal methods have been reported in the literature. Phototherapy is the most common one, however, it is limited because the light can only penetrate a few millimeters of skin and also may induce DNA damage (Tiribelli & Ostrow, 2005).

Plasma exchange, on the other hand, requires large volumes of plasma, which are associated with the risks of hypoglycemia, hypocalcemia, acidosis, and infectious diseases (Aotsuka et al., 1990; Bhardwaj et al., 2017; Biesenbach et al., 2009; Blaha et al., 2011; Braun et al., 2000; Darnige et al., 1999; Hershko & Naparstek, 2005; Kato & Ikada, 1996; Kong et al., 2002; Müller et al., 2000; Özgür et al., 2011; Pitiot et al., 2000; Rech et al., 2006; Schneider et al., 1990; Süfke et al., 2017; Sugimoto et al., 2006; Suzuki, 2002; Türkmen et al., 2006; Uzun et al., 2010; Ventura et al., 2001; Yan et al., 1988; Yılmaz et al., 2008; Yu & He, 1999; Zhu et al., 1999). Hemoperfusion with affinity carriers has become the most promising technique. The first clinical trial for bilirubin removal was reported by Idezuki et al. in 1981, in which anion exchange synthetic fibers were used (Idezuki et al., 1981). Since then a lot of studies have been published on a variety of ligands for bilirubin adsorption (Table 2.5) (Altıntaş et al., 2011; Aşir et al., 2005; Demirçelik et al., 2009; Han & Zhang, 2009; Ma et al., 2005; Rad et al., 2003; Şenel et al., 2002; Shi et al., 2005; Wang et al., 1997; Yavuz et al., 2001, 2016).

As seen from the table, bilirubin adsorption capacity of the adsorbents is quite variable depending on the type of sorbent and the ligand including different amino acids, reactive dyes especially Cibacron Blue F3GA and albumin. Recently molecular imprinting technology also has been used for this purpose and present promising results.

Table 2.5 Some affinity adsorbents for the bilirubin removal from plasma.

Support	Ligand	Bilirubin adsorption capacity	References
Polyacrylamide particles	L-lysine, L-ornithine	0.2−75	Zhu et al. (1990)
Macroreticular resin	Albumin	1−24	Sideman et al. (1981)
Chitosan particles	Poly-L-lysine	1.5	Chandy and Sharma (1992)
Poly(ethylene vinyl alcohol)	Bovine serum albumin	25.0	Avramescu et al. (2004)
Poly(GMA-DVB)	Albumin	30	Kuroda et al. (1996)
PHEMA particles	Dye molecules	6.8−32.5	Denizli et al. (1998), Denizli and Pişkin (2001), Kocakulak et al. (1997)

(Continued)

Table 2.5 Some affinity adsorbents for the bilirubin removal from plasma. *Continued*

Support	Ligand	Bilirubin adsorption capacity	References
Polyamide hollow fiber	Cibacron Blue F3GA	48.9	Kassab, et al. (2000)
mPHEMA particles (MSFB)	Human serum albumin	88.3	Uzun and Denizli (2006)
mPHEMA particles (batch)	Human serum albumin	64.7	Rad et al. (2003)
Chitosan coupled nylon membrane	Cibacron Blue F3GA	64.7	Xia et al. (2003)
Poly(styrene-divinyl benzene)	Quaternary ammonium salt	4.0−8.0	Davies et al. (1990)
Poly(tetrafluoro ethylene) membrane	Cibacron Blue F3GA	76.2	Zhang and Jin (2005a, 2005b)
Poly(butadiene-HEMA) gels	Bovine serum albumin	3.1	Alvarez et al. (2001)
Cellulose acetate fiber	Cibacron Blue F3GA	4.0	Ma et al. (2005)
Polyamide/chitosan membrane	Polylysine	28.6	Shi et al. (2005)
Poly(GMA-AAm-MBA)	Polyethylene imine	16.6	Gao et al. (2007)
Poly(tetrafluoroethylene) membrane	Human serum albumin	71.2	Zhang and Jin (2005a)
Poly(tetrafluoroethylene) fiber	Bovine serum albumin	9.6	Han and Zhang (2009)
Aluminum oxide-silica-membrane	Lysine	17.6	Shi et al. (2010)
Poly(pyrrole)-alumina membrane	Lysine	32.4	Shi et al. (2010)
Poly(glycidyl methacrylate)	Cibacron Blue F3GA	241.5	Altintaş et al. (2011)
Poly(hydroxyethyl methacrylate-N-methacryloyl-(L)-tyrosine methyl ester) [poly(HEMA-MATyr) cryogel]	Molecular imprinting	3.6	Baydemir et al. (2009a, 2009b)
PHEMA/Poly(HEMA-MATyr) bead embedded composite cryogel	Molecular imprinting	10.3	Baydemir et al. (2009a, 2009b)
P(HEMA-MATrp) cryogel disk	L-tryptophan	22.2	Perçin et al. (2013)

2.4 Affinity adsorbents for metal ion removal

Metals have unique biological roles for human metabolism in catalysis, hormone actions, regulation of genes and other regulators, contraction of muscles, nerve function, muscle contraction, enzyme function, and macromolecule stabilization. Even some metals such as cobalt, copper, iron, manganese, molybdenum, nickel, zinc, etc. are essential for the above functions, the adverse effects occur above certain threshold levels for all metal ions. Their toxicity is due to the blocking of an essential functional group of biomolecules, changing the functional conformation and disrupting the integrity of membranes and so on (Clayton & Clayton, 1981).

In several diseases, such as thalassemia and aplastic anemia in which parental iron administration is necessary, chronic iron overload may occur. In addition, acute iron poisoning is frequent among young children. Excess iron has been shown to be related with an increased risk of cancer, oxidative stress in cells, etc. (Mahoney et al., 1989). Cadmium is another toxic metal and its compounds result in kidney damage, impaired regulation of calcium and phosphorus, bone defects, and more (Himeno & Aoshima, 2019). Aluminum, the most widely distributed metal in the environment, has long been considered unsafe for humans at increased levels. It results in oxidative stress, lipid peroxidation, protein denaturation and transformation, neurotoxicity, bone defects, and more (Onyebuchi et al. 2019). Chelating agents are often used to treat metal poisoning nonspecifically and there are few reports on the extracorporeal affinity adsorbents for the removal of various metal ions (Table 2.6).

Table 2.6 Some affinity adsorbents for metal removal from plasma.

Support	Ligand	Capacity	References
P(HEMA-EGDMA) microsphere	Cibacron Blue F3GA Cibacron Blue F3GA/Thionein	17.5 mg/g 38.0 mg/g	98
PHEMA membrane	Thionein	1.72 umol/g	99
Magnetic poly(HEMA-MAC) beads	Cd(II) imprinting	48.8 umol/g	100
Poly(HEMA-MAC) monolith	Cd(II) imprinting	26.6 umol/g	101
PHEMA cryogel	Cibacron Blue F3GA Reactive Green 19 Congo Red	25.5 mg/g 48.0 mg/g 28.5 mg/g	102
PHEMA beads	Cysteinylhexapeptide	11.8 mg/g	103
Polysulfome fibers	Desferrioxamine	-	104
PHEMA film	Cibacron Blue F3GACongo Red Ferritin	3.80 ug/cm^2 4.41 ug/cm^2 8.1 ug/cm^2	105
Poly(HEMA-MAGA) beads	Fe(III) imprinting	92.6 umol/g	106
Poly(HEMA-MAGA) membrane	Fe(III) imprinting	164.2 umol/g	107
Poly(HEMA-MAC)	Fe(III) imprinting	150 ug/g	108
Poly(HEMA-MAGA) cryogel	Fe(III) imprinting	19.8 umol/g	109
Poly(HEMA-MAGA) beads	Al(III) imprinting	0.76 mg/g	110
Polysulfone fiber	Desferrioxamine	94−628 ug	111
Poly(HEMA-MAC)	Hg(II) imprinting	0.45 mg/g	112

A number of polymeric materials with various bulk structures and surface chemistries have been tested in vitro for heavy metal removal from biological fluids. The results suggest that extracorporeal metal removal therapies may be achieved through high efficacy adsorbents.

2.5 Conclusion

Hemoperfusion is an accepted process for the removal of exogeneous and endogeneous substances from plasma for a variety of diseases and for a variety of substances ranging from small metal ions to high-molecular-weight biological molecules. Scientists are still examining more specific and high-performance affinity adsorbents for extracorporeal affinity therapies. Besides specificity and performance, personalization and miniaturization of the extracorporeal setup can be expected in the near future (Perçin et al., 2013).

References

Abrams, D., Baldwin, M. R., Champion, M., Agerstrand, C., Eisenberger, A., Bacchetta, M., & Brodie, D. (2016). Thrombocytopenia and extracorporeal membrane oxygenation in adults with acute respiratory failure: A cohort study. *Intensive Care Medicine*, *42*(5), 844−852. Available from https://doi.org/10.1007/s00134-016-4312-9.

Altintaş, E. B., Türkmen, D., Karakoç, V., & Denizli, A. (2011). Efficient removal of bilirubin from human serum by monosize dye affinity beads. *Journal of Biomaterials Science, Polymer Edition*, *22*(7), 957−971. Available from https://doi.org/10.1163/092050610X496594.

Alvarez, C. I., Gomez, C. G., & Strumia, M. C. (2001). Influence of the polymeric morphology of sorbents on their properties in affinity chromatography. *Journal of Biochemical and Biophysical Methods*, *49*(1−3), 141−151. Available from https://doi.org/10.1016/S0165-022X(01)00194-4.

Aotsuka, S., Funahashi, T., Tani, N., Okawa-Takatsuji, M., Kinoshita, M., & Yokohari, R. (1990). Adsorption of anti-dsDNA antibodies by immobilized polyanionic compounds. *Clinical & Experimental Immunology*, 215−220. Available from https://doi.org/10.1111/j.1365-2249.1990.tb05181.x.

Aşir, S., Uzun, L., Türkmen, D., Say, R., & Denizli, A. (2005). Ion-selective imprinted superporous monolith for cadmium removal from human plasma. *Separation Science and Technology*, *40*(15), 3167−3185. Available from https://doi.org/10.1080/01496390500385376.

Avramescu, M. E., Sager, W. F. C., Borneman, Z., & Wessling, M. (2004). Adsorptive membranes for bilirubin removal. *Journal of Chromatography B: Analytical Technologies in the Biomedical and Life Sciences*, *803*(2), 215−223. Available from https://doi.org/10.1016/j.jchromb.2003.12.020.

Baydemir, G., Bereli, N., Andaç, M., Say, R., Galaev, I. Y., & Denizli, A. (2009a). Bilirubin recognition via molecularly imprinted supermacroporous cryogels. *Colloids and Surfaces B: Biointerfaces*, *68*(1), 33−38. Available from https://doi.org/10.1016/j.colsurfb.2008.09.008.

Baydemir, G., Bereli, N., Andaç, M., Say, R., Galaev, I. Y., & Denizli, A. (2009b). Supermacroporous poly (hydroxyethyl methacrylate) based cryogel with embedded bilirubin imprinted particles. *Reactive and Functional Polymers*, *69*(1), 36−42. Available from https://doi.org/10.1016/j.reactfunctpolym.2008.10.007.

Bhardwaj, G., Dörr, M., Sappa, P. K., Ameling, S., Dhople, V., Steil, L., Klingel, K., Empen, K., Beug, D., Völker, U., Felix, S. B., & Hammer, E. (2017). Endomyocardial proteomic signature corresponding to the response of patients with dilated cardiomyopathy to immunoadsorption therapy. *Journal of Proteomics*, *150*, 121−129. Available from https://doi.org/10.1016/j.jprot.2016.09.001.

Biesenbach, P., Schmaldienst, S., Smolen, J. S., Horl, W. H., Derfler, K., & Stummvoll, G. H. (2009). Immuno-adsorption in SLE: Three different high affinity columns are adequately effective in removing autoantibodies and controlling disease activity. *Atherosclerosis.*

Blaha, M., Pitha, J., Blaha, V., Lanska, M., Maly, J., Filip, S., Brndiar, M., & Langrova, H. (2011). Experience with extracorporeal elimination therapy in myasthenia gravis. *Transfusion and Apheresis Science*, *45*(3), 251−256. Available from https://doi.org/10.1016/j.transci.2011.10.003.

Braun, N., Erley, C., Klein, R., Kötter, I., Saal, J., & Risler, T. (2000). Immunoadsorption onto protein A induces remission in severe systemic lupus erythematosus. *Nephrology Dialysis Transplantation*, *15*(9), 1367−1372. Available from https://doi.org/10.1093/ndt/15.9.1367.

Chandy, T., & Sharma, C.P. (1992). Polylysine immobilized chitosan beads as adsorbents for bilirubin (Vol. 16).

Clayton, G. D., & Clayton, F. E. (Eds.), (1981). *Patty's Industrial Hygiene and Toxicology.* Wiley.

Darnige, L., Legallais, C., Arvieux, J., Pitiot, O., & Vijayalakshmi, M. A. (1999). Functionalized hollow fiber membrane cartridge for adsorption of anticofactor/antiphospholipid antibodies: A potential tool for treatment. *Artificial Organs*, *23*(9), 834−839. Available from https://doi.org/10.1046/j.1525-1594.1999.06250.x.

Davies, C. R., Malchesky, P. S., & Saidel, G. M. (1990). Temperature and albumin effects on adsorption of bilirubin from standard solution using anion-exchange resin. *Artificial Organs*, *14*(1), 14−19. Available from https://doi.org/10.1111/j.1525-1594.1990.tb01587.x.

Demirçelik, A. H., Andaç, M., Andaç, A. C., Say, R., & Denizli, A. (2009). Molecular recognition based detoxification of aluminum in human plasma. *Journal of Biomaterials Science. Polymer Edition*, *20*, 1235−1258.

Denizli, A. (2011). Autoimmune diseases and immunoadsorption therapy. *Hacettepe Journal of Biology and Chemistry*, *39*(3), 213−229.

Denizli, A., Kocakulak, M., & Pişkin, E. (1998). Bilirubin removal from human plasma in a packed-bed column system with dye-affinity microbeads. *Journal of Chromatography B: Biomedical Applications*, *707*(1−2), 25−31. Available from https://doi.org/10.1016/S0378-4347(97)00612-9.

Denizli, A., & Pişkin, E. (2001). Dye-ligand affinity systems. *Journal of Biochemical and Biophysical Methods*, *49*(1−3), 391−416. Available from https://doi.org/10.1016/S0165-022X(01)00209-3.

Dhokia, V. D., Madhavan, D., Austin, A., & Morris, C. G. (2019). Novel use of Cytosorb™ haemadsorption to provide biochemical control in liver impairment. *Journal of the Intensive Care Society*, *20*(2), 174−181. Available from https://doi.org/10.1177/1751143718772789.

Francini, M., Targher, G., & Lippi, G. (2010). Serum bilirubin leves and cardiovascular disease risk: a Janus Bifrons? *Advances in Clinical Chemistry.*

Gao, B., Lei, H., Jiang, L., & Zhu, Y. (2007). Studies on preparing and adsorption property of grafting terpolymer microbeads of PEI-GMA/AM/MBA for bilirubin. *Journal of Chromatography B: Analytical Technologies in the Biomedical and Life Sciences*, *853*(1−2), 62−69. Available from https://doi.org/10.1016/j.jchromb.2007.02.055.

Gemelli, C., Cuoghi, A., Magnani, S., Atti, M., Ricci, D., Siniscalchi, A., Mancini, E., & Faenza, S. (2019). Removal of bilirubin with a new adsorbent system: in vitro kinetics. *Blood Purification*, *47*(1−3), 10−15. Available from https://doi.org/10.1159/000492378.

Han, Xy, & Zhang, Zp (2009). Preparation of grafted polytetrafluoroethylene fibers and adsorption of bilirubin. *Polymer International*, *58*(10), 1126−1133. Available from https://doi.org/10.1002/pi.2640.

Hershko, A. Y., & Naparstek, Y. (2005). *Removal of pathogenic autoantibodies by immunoadsorption, . Annals of the New York Academy of Sciences* (Vol. 1051, pp. 635−646). New York Academy of Sciences. Available from https://doi.org/10.1196/annals.1361.108.

Himeno, S., & Aoshima, K. (2019). Cadmium toxicity: New aspects in human disease.

Hoshimoto, H., Tsuda, H., Kanai, Y., Kobayashi, S., Hirose, S., Shinoura, H., Yokahari, R., & Kinoshita, M. (1990). Selective removal of anti-DNA and anticardiolipin antibodies by adsorbent plasmapheresis using dextran sulfate columns in patients with SLE. *The Journal of Rheumatology, 18.*

Idezuki, Y., Hamaguchi, M., Hamabe, S., Moriya, H., Nagashima, T., Watanabe, H., Sonoda, T., Teramoto, K., Kikuchi, T., & Tanzawa, H. (1981). Removal of bilirubin and bile acid with a new anion exchange resin: Experimental background and clinical experiences. *Transactions - American Society for Artificial Internal Organs*, *27*(1), 428−433.

Jung, A., Krisper, P., Zadora, M., Haditsch, B., Stauber, R., Holzer, H., & Schneditz, D. (2009). Clearance and rate of conjugated bilirubin removal during extracorporeal liver support therapy with FPSA Prometheus™. In *IFMBE proceedings* (Vol. 25, Issue 7, pp. 620−623). Springer Verlag. <https://doi.org/10.1007/978−3−642−03885-3_172>

Kassab, A., Yavuz, H., Odabaşi, M., & Denizli, A. (2000). Human serum albumin chromatography by Cibacron Blue F3GA-derived microporous polyamide hollow-fiber affinity membranes. *Journal of Chromatography B: Biomedical Sciences and Applications*, *746*(2), 123−132. Available from https://doi.org/10.1016/S0378-4347(00)00311-X.

Kato, K., & Ikada, Y. (1996). Immobilization of DNA onto a polymer support and its potentiality as immunoadsorbent. *Biotechnology and Bioengineering*, *51*(5), 581−590, https://doi.org/10.1002/(SICI)1097−0290(19960905)51:5 < 581::AID-BIT10 > 3.0.CO;2-L.

King, J. D., Kern, M. H., & Jaar, B. G. (2019). Extracorporeal removal of poisons and toxins. *Clinical Journal of the American Society of Nephrology*, *14*(9), 1408−1415. Available from https://doi.org/10.2215/CJN.02560319.

Klingele, M., Allmendinger, C., Thieme, S., Baerens, L., Fliser, D., & Jan, B. (2020). Therapeutic apheresis within immune-mediated neurological disorders: Dosing and its effectiveness. *Scientific Reports*, *10*(1). Available from https://doi.org/10.1038/s41598-020-64744-4.

Kocakulak, M., Denizli, A., Rad, A. Y., & Pişkin, E. (1997). New sorbent for bilirubin removal from human plasma: cibacron blue-immobilized poly(EGDMA-HEMA) beads. *Journal of Chromatography B*, *693*.

Kong, D. L., Schuett, W., Dai, J., Kunkel, M., Holtz, M., Yamada, R., Yu, H., & Klinkmann, H. (2002). Development of cellulose-DNA immunosorbent. *Artificial Organs*, *26*.

Kuroda, H., Ranaka, T., & Osawa, Z. (1996). Selective adsorption of bilirubin by macroporous poly(GMA-DVB) beads. *Die Angewandte Makromolekulare Chemie*, *237*.

Lee, K. H., Wendon, J., Lee, M., Da Costa, M., Lim, S. G., & Tan, K. C. (2002). Predicting the decrease of conjugated bilirubin with extracorporeal albumin dialysis MARS using the predialysis molar ratio of conjugated bilirubin to albumin. *Liver Transplantation*, *8*(7), 591−593. Available from https://doi.org/10.1053/jlts.2002.34148.

Ma, Z., Kotaki, M., & Ramakrishna, S. (2005). Electrospun cellulose nanofiber as affinity membrane. *Journal of Membrane Science*, *265*(1−2), 115−123. Available from https://doi.org/10.1016/j.memsci.2005.04.044.

Mahoney, J. R., Hallaway, P. E., Hedlund, B. E., & Eaton, J. W. (1989). Acute iron poisoning. Rescue with macromolecular chelators. *Journal of Clinical Investigation*, 1362−1366. Available from https://doi.org/10.1172/JCI114307.

Müller, J., Wallukat, G., Dandel, M., Bieda, H., Brandes, K., Spiegelsberger, S., Nissen, E., Kunze, R., & Hetzer, R. (2000). Immunoglobulin adsorption in patients with idiopathic dilated cardiomyopathy. *Circulation*, *101*(4), 385−391. Available from https://doi.org/10.1161/01.CIR.101.4.385.

Nakaji, S. (2001). Current topics on immunoadsorption therapy. *Therapeutic Apheresis*, *5*(4), 301−305. Available from https://doi.org/10.1046/j.1526-0968.2001.00360.x.

Nilsson, J. M., Sundquist, S. B., Ahlberg, A., & Bergentz, S. E. (1981). A procedure for removing high titer antibodies by extracorporeal protein A- sepharose adsorption in hemophilia. *Blood*, *58*.

Olbricht, C. J. (1993). Extracorporeal treatment of hypercholesterolaemia. *Nephrology Dialysis Transplantation*, *8*(9), 814−820. Available from https://doi.org/10.1093/oxfordjournals.ndt.a092603.

Onyebuchi, I. I., Ephraim, I., & Afifi, I. N. (2019). Aluminium toxicosis: A review of toxic actions and effects. *Interdisciplinary Toxicology* (70), 45. Available from https://doi.org/10.2478/intox-2019-0007.

Özgür, E., Bereli, N., Türkmen, D., Ünal, S., & Denizli, A. (2011). PHEMA cryogel for in-vitro removal of anti-dsDNA antibodies from SLE plasma. *Materials Science and Engineering C, 31*(5), 915−920. Available from https://doi.org/10.1016/j.msec.2011.02.012.

Palmer, A., Cairns, T., Dische, F., Gluck, G., Gjorstrup, P., Parsons, V., & Welsh Taube, K. D. (1991). Treatment of rapidly progressive glomerulonephritis by extracorporeal immunoadsorption, prednisolone and cyclophosphamide. *Nephrology Dialysis Transplantation, 6*(8), 536−542. Available from https://doi.org/10.1093/ndt/6.8.536.

Perçin, I., Baydemir, G., Ergün, B., & Denizli, A. (2013). Macroporous PHEMA-based cryogel discs for bilirubin removal. *Artificial Cells, Nanomedicine and Biotechnology, 41*(3), 172−177. Available from https://doi.org/10.3109/10731199.2012.712046.

Pitiot, O., Legallais, C., darnige, L., & Vijayalakhsmi, M. A. (2000). A potential set up based on histidine hollow fiber membranesfor the extracorporeal removal of human antibodies. *Journal of Membrane Science*, 166.

Rad, A. Y., Yavuz, H., Kocakulak, M., & Denizli, A. (2003). Bilirubin removal from human plasma with albumin immobilised magnetic poly(2-hydroxyethyl methacrylate) beads. *Macromolecular Bioscience, 3*(9), 471−476. Available from https://doi.org/10.1002/mabi.200350018.

Ray, P. K. (1984). Extracorporeal adsorption of pathologic gammaglobulins and immune complexes in various diseases including cancer. *Plasma Therapy and Transfusion Technology, 4*.

Rech, J., Hueber, A. J., Kallert, S., Requadt, C., Kalden, J. R., & Schulze-Koops, H. (2006). Immunoadsorption and CD20 antibody treatment in a patient with treatment resistant systemic lupus erythematosus and preterminal renal insufficiency [3]. *Annals of the Rheumatic Diseases, 65*(4), 552−553. Available from https://doi.org/10.1136/ard.2005.043026.

Rose, N., & Mackay, I. (Eds.), (2014). *The autoimmune diseases*. Elsevier.

Schettler, V., Neumann, C. L., Hulpke-Wette, M., Hagenah, G. C., Schulz, E. G., & Wieland, E. (2012). Current view: Indications for extracorporeal lipid apheresis treatment. *Clinical Research in Cardiology Supplements, 7*(1), 15−19. Available from https://doi.org/10.1007/s11789-012-0046-6.

Schneider, M., Berning, T., Waldendorf, M., Glaser, J., & Gerlach, U. (1990). Immunoadsorbent plasma perfusion in patients with SLE. *Journal of Rheumatology, 17*.

Şenel, S., Denizli, F., Yavuz, H., & Denizli, A. (2002). Bilirubin removal from human plasma by dye affinity microporous hollow fibers. *Separation Science and Technology, 37*(8), 1989−2006. Available from https://doi.org/10.1081/SS-120003056.

Shi, W., Shen, Y., Jiang, H., Song, C., Ma, Y., Mu, J., Yang, B., & Ge, D. (2010). Lysine-attached anodic aluminum oxide (AAO)-silica affinity membrane for bilirubin removal. *Journal of Membrane Science, 349*(1−2), 333−340. Available from https://doi.org/10.1016/j.memsci.2009.11.066.

Shi, W., Zhang, F., & Zhang, G. (2005). Adsorption of bilirubin with polylysine carrying chitosan-coated nylon affinity membranes. *Journal of Chromatography B: Analytical Technologies in the Biomedical and Life Sciences, 819*(2), 301−306. Available from https://doi.org/10.1016/j.jchromb.2005.02.018.

Shoenfeld, Y., Cervera, R., & Gershwin, M. E. (Eds.), (2008). *Diagnostic criteria in autoimmune diseases*. Humana Press.

Sideman, S., Mor, L., Mordohovich, D., Mihich, M., Zinder, O., & Brandes, J. M. (1981). In vivo hemoperfusion studies of unconjugated bilirubin removal by ion exchange resin. *Transactions - American Society for Artificial Internal Organs, 27*.

Sompayrac L. (2012) How the Immune System Works, 4th Ed., Wiley-Blackwell.

Süfke, S., Lehnert, H., Uhlenbusch-Körwer, I., & Gebauer, F. (2017). Safety aspects of immunoadsorption in IgG removal using a single-use, multiple-pass Protein A immunoadsorber (LIGASORB): Clinical investigation in healthy volunteers. *Therapeutic Apheresis and Dialysis, 21*(4), 405−413. Available from https://doi.org/10.1111/1744-9987.12532.

Sugimoto, K., Yamaji, K., Yang, K. S., Kanai, Y., Tsuda, H., & Hashimoto, H. (2006). Immunoadsorption plasmapheresis using a phenylalanine column as an effective treatment for lupus nephritis. *Therapeutic Apheresis and Dialysis*, *10*(2), 187−192. Available from https://doi.org/10.1111/j.1744-9987.2006.00362.x.

Suzuki, K. (2002). The role of immunoadsorption using dextran-sulfate cellulose columns in the treatment of SLE.

Terman, D. S., Buffaloe, G., Cook, G., Sullivan, M., Mattioli, C., Tillquist, R., & Carlos Ayus, J. (1979). Extracorporeal immunoadsorption: Initial experience in human systemic lupus erythematosus. *The Lancet*, *314*(8147), 824−827. Available from https://doi.org/10.1016/S0140-6736(79)92177-9.

Terman, D. S., Stewart, I., Robinette, J., Carr, R., & Harbeck, R. (1976). Specific removal of DNA antibodies in vivo with an extracorporeal immuno adsorbent. *Clinical and Experimental Immunology*, *24*(2), 231−237.

Tiribelli, C., & Ostrow, J. D. (2005). The molecular basis of bilirubin encephalopathy and toxicity. *Journal of Hepatology*.

Türkmen, D., Yavuz, H., & Denizli, A. (2006). Synthesis ıf tentacle type magnetic beads as immobilized metal chelate affinity support for cytochrome c adsorption. *International Journal of Biological Macromolecules*, 38.

Uzun, L., & Denizli, A. (2006). Bilirubin removal performance of immobilized albumin in a magnetically stabilized fluidized bed. *Journal of Biomaterials Science, Polymer Edition*, *17*(7), 791−806. Available from https://doi.org/10.1163/156856206777656481.

Uzun, L., Yavuz, H., Osman, B., Çelik, H., & Denizli, A. (2010). PHEMA based affinity membranes for in-vitro removal of anti-dsDNA antibodies from SLE plasma. *International Journal of Biological Macromolecules*, 47.

Ventura, R. C. A., Zollner, R. D. L., Legallis, C., Vijayalakshmi, M., & Bueno, S. M. A. (2001). In vitro removal of human IgG autoantibodies by affinity filrtation using immobilized L-histidine onto PEVA hollow fiber membranes. *Biomolecular Engineering*, 17.

Wang, H., Ma, J., Zhang, Y., & He, B. (1997). Adsorption of bilirubin on the polymeric β-cyclodextrin supported by partially aminated polyacrylamide gel. *Reactive and Functional Polymers*, *32*(1), 1−7. Available from https://doi.org/10.1016/S1381-5148(96)00061-2.

Xia, B., Zhang, G., & Zhang, F. (2003). Bilirubin removal by Cibacron Blue F3GA attached nylon-based hydrophilic affinity membrane. *Journal of Membrane Science*, *226*(1−2), 9−20. Available from https://doi.org/10.1016/j.memsci.2003.08.007.

Yan, Y., Ting, Y. Y., Chang, S. J., Cheng, Q. S., Lie, Q. W., Hua, S. X., Hai, C. R., & Zhi, W. J. (1988). A new DNA immune adsorbent for hemoperfusion in SLE therapy: A clinical trial. *Artificial Organs*, 12.

Yavuz, H., Arıca, Y. M., & Denizli, A. (2001). Therapeutic affinity adsorption of iron(III) with dye and ferritin-immobilized pHEMA adsorbent. *Journal of Applied Polymer Science*, *82*, 186−194.

Yavuz, H., Bereli, N., Baydemir, G., Andaç, M., Türkmen, D., & Denizli, A. (2016). *Cryogels: Applications in extracorporeal affinity therapy. Supermacroporous cryogels: Biomedical and biotechnological applications* (pp. 389−418). CRC Press. Available from https://www.crcpress.com/Supermacroporous-Cryogels-Biomedical-and-Biotechnological-Applications/Kumar/p/book/9781482228816.

Yu, Y. H., & He, B. L. (1999). Preparation of immunoadsorbents and their adsorption properties for anti-DNA antibodies in SLE serum. *Reactive and Functional Polymers*, *41*(1), 191−195. Available from https://doi.org/10.1016/S1381-5148(99)00028-0.

Yılmaz, E., Uzun, L., Rad, A. Y., Kalyoncu, U., Ünal, S., & Denizli, A. (2008). Specific adsorbents of the autoantibodies from rheumatoid arthritis patient plasma using histidine-containing affinity beads. *Journal of Bomaterials Science, Polymer Edition*, 19.

Zeitler, H., Ulrich-Merzenich, G., Panek, D., Goldmann, G., Vidovic, N., Brackmann, H. H., & Oldenburg, J. (2012). Extracorporeal treatment for the acute und long-term outcome of patients with life-threatening acquired hemophilia. *Transfusion Medicine and Hemotherapy*, *39*(4), 264−270. Available from https://doi.org/10.1159/000341913.

Zhang, L., & Jin, G. (2005a). Bilirubin removal from human plasma by Cibacron Blue F3GA using immobilized microporous affinity membranous capillary method. *Journal of Chromatography B: Analytical Technologies in the Biomedical and Life Sciences*, *821*(1), 112−121. Available from https://doi.org/10.1016/j.jchromb.2005.04.022.

Zhang, L., & Jin, G. (2005b). New sorbent for bilirubin removal from human plasma: Albumin immobilized microporous membranous PTFE capillaries. *Chinese Chemical Letters*, *16*(11), 1495−1498.

Zhu, B., Iwata, H., Kong, D., Yu, Y., Kato, K., & Ikada, Y. (1999). Preparation of DNA- immobilized immunoadsorbent for treatment of SLE. *Journal of Biomaterials Science. Polymer Edition*, *3*.

Zhu, X. X., Brown, G. R., & St-Pierre, L. E. (1990). Adsorption of bilirubin with polypeptide-coated resins. *Biomaterials, Artificial Cells, and Artificial Organs*, *18*(1), 75−93. Available from https://doi.org/10.3109/10731199009117290.

Physical chemistry of dispersed nanostructures in blood

3

Christian Mayer

Department of Chemistry, Physical Chemistry, University of Duisburg-Essen, Essen, Germany

3.1 Consequences of structural surface

A very central characteristic of nanostructures is their large interface with the environment (Azmi & Shad, 2017). The significance of this effect may be shown on a simple example: a cube with 1 cm side length which consists of 1 mL of bulk material has a surface of 6 cm^2. If this cube is cut into smaller cubes of 1 mm side lengths, its surface increases to 60 cm^2. The same cube cut into cubes of 1 μm side length yields 60,000 cm^2. In the case of nanometer-sized cubes from the same amount of material, we actually obtain the surface of a soccer field: 6×10^7 cm^2 or 6000 m^2! Such an increase in surface has important consequences in terms of the physicochemical properties of the system.

The distribution of the material to smaller particles and the connected increase in surface dA means an increase dE in potential energy. This increase is measured in terms of surface energy determined by the surface tension γ between the dispersed material and the surrounding medium given by $dE = \gamma\, dA$. If 1 mL of water is dispersed into nanosized droplets following the described procedure, this corresponds to an energy gain of 432 J or about 20% of the energy necessary for its evaporation.

Phase transition temperatures will shift toward lower values, as a lower percentage of molecules or atoms are actually bound inside of a lattice or inside a fluid phase. Instead, a growing percentage will reside at the interface where the contact with other molecules or atoms of the same kind is limited to one hemisphere.

For the same reason, the solubility as well as the potential chemical reactivity of the dispersed material is strongly increasing. This affects the thermodynamic properties of the reaction (their driving force) as well as the rate of these processes.

Due to the increased interface, adsorption and desorption processes as well as diffusion of components across the interface become dominant phenomena which are, in many cases, central to the function of the nanostructure.

All these issues affect the preparation, the handling, and the potential use of nanostructures in hematological applications. Therefore they must be considered in all the following topics as the key problems but, at the same time, as the most important functional advantages of nanostructures as compared to more macroscopic alternatives.

Nanotechnology for Hematology, Blood Transfusion, and Artificial Blood. DOI: https://doi.org/10.1016/B978-0-12-823971-1.00003-9

3.2 Size distribution

Thinking of functional nanostructures in blood, we necessarily have to consider droplets or particles in the submicrometer size range. The preparation of these structures may either start from the molecular level (bottom-up) or from macroscopic bulk material (top-down). In both cases, we cannot expect the product to be monodisperse. In practice, we will have to deal with polydispersity, that is, a more or less wide distribution of structure dimensions. In its most simple version, the size distribution function may be approximated by a Gaussian curve in a plot of particle numbers versus particle dimension. In practice, however, most particle size distribution functions are more or less asymmetric. The main source for this asymmetry is based on the higher surface energy for smaller particles or droplets. Due to their higher potential energy, smaller particles are less likely to form in a top-down approach and are more likely to grow in a bottom-up approach. So in most cases, the size distribution functions have a steep slope toward smaller sizes and a shallow tailing toward larger ones.

For hematological applications, the size distribution is crucial as large particles may lead to blocked capillaries. On the other hand, the size (as well as the shape) of the particles influences the distribution of particles within the blood vessels (Tan et al., 2012). Therefore, preparation processes are generally optimized for narrow size distributions and/or smaller nanostructures. Regarding the aspect of surface energy, this will either require high energy dispersal (e.g., by use of ultrasonic devices) or bottom-up processes. In order to exclude large particles, the dispersions may be filtered or centrifuged. The final determination of the size distribution can be achieved either by microscopic imaging [e.g., by transmission electron microscopy or by observation of Brownian motion e.g., using dynamic light scattering (DLS), or video microscopy].

3.3 Dispersion stability

The increased surface energy of dispersed nanostructures creates a driving force toward the agglomeration of particles and other processes of particle growth (like, e.g., Ostwald ripening; Kahlweit, 1975). Also, deviations of the particle density from the density of the liquid environment can lead to their sedimentation or flotation, a process which may be either reversible or irreversible. All these processes may be stopped (or at least made reversible) by a suitable stabilization of the dispersion. In general, a suitable stability (this kind of stability should not be confused with the mechanical stability of the nanostructure itself) can be induced by introducing an energy barrier which keeps the particles from coming into contact with each other. In detail, this approach is described by the DLVO (B. Derjaguin, L. Landau, E. Verwey, T. Overbeek) theory (Russel et al., 1989). In practice, that means that repulsive forces have to be introduced between particles. In the simplest case, this may be a specific charge which is brought onto the particle surface which, together with counterions in the surrounding, forms an interfacial double layer around the particle (Tadros, 2012). If all particles carry a positive or negative charge, the electrostatic repulsion efficiently prevents agglomeration and, in parts, even sedimentation or flotation. At least, all these processes become reversible and the dispersion can be reestablished by shaking or stirring. However, electrostatic stabilization suffers from the presence of electrolytes, a condition which is of course relevant for blood plasma. Dissolved ions reduce the thickness of the interfacial double layer, hereby weakening the interparticle repulsion and destabilizing the dispersion. In addition, multivalent counterions present in blood plasma such as Ca^{2+} even tend to form bridges between negatively charged particles and may actually cause precipitation.

A possible solution to this problem is steric stabilization (Doroszkowski, 1999). In this case, the particle surface is covered by polymer or oligomer chains with high conformative disorder. These chains stick out into the fluid phase and form a dense layer around the structure. When two particles approach each other, these layers are being compressed, leading to a forced loss of disorder (and a corresponding negative entropy change) which then causes an increase in free energy. This results in a repulsive driving force which keeps the structures apart. In contrast to electrostatic stabilization, this effect is insensitive to electrolyte concentration.

The preparative approach to stabilization can consist either of the preparation of nanostructures with an integrated stabilizing layer or by addition of stabilizing surfactants. In the first case, the nanoparticles are synthesized in a way that they carry a uniform charge (for electrostatic stabilization) or an interfacial layer of chains with conformational disorder (such as polyethylene glycol or hydrophilic oligopeptides, for steric stabilization). In the second case, amphiphilic compounds (surfactants) are added which form a stabilizing layer around existing structures. These surfactants consist of a hydrophobic part which interacts with the (generally hydrophobic) nanostructure and a hydrophilic part which interacts with water. The hydrophilic part then either carries a charge for electrostatic stabilization or a hydrophilic chain with high conformational disorder (e.g., based on polyethylene glycol) for steric stabilization.

The performance of stabilization is measured in long-term experiments under optical observation of flocculation or sedimentation phenomena. In addition, the successful application of charge in the case of electrostatic stabilization may be assessed by a determination of the zeta potential. This value is the actual charge of the particle after the more mobile counterions are separated from the surface by shear and can be determined by optical observation of the particle migration in an electrical field. Generally, a large zeta potential (positive or negative) will lead to an efficient electrostatic stabilization.

3.4 **Mechanical stability**

When circulating through blood vessels, dispersed nanostructures are subject to certain mechanical stress. This mainly involves shear forces under laminar or turbulent flow, especially in small capillaries. Red blood cells (erythrocytes) are made to survive these conditions, mainly due to their mechanical strength and flexibility. Hence, their mechanical properties can be seen as the target for artificial particles which are meant to achieve long lifetimes during blood circulation. A direct determination of mechanical properties on nanostructures is possible using atomic force microscopy (Voigtländer, 2020). A simple indentation test allows for the determination of the structural flexibility and the type of irreversible damage caused by punctual load (Altinbas et al., 2006; Bauer, Zähres, et al., 2010). As an example, Fig. 3.1 shows an individual poly(lactide-co-glycolide) nanocapsule before and after indentation measurements and corresponding plots of the force versus the degree of indentation.

The curves reveal information on the spring constant of the capsule during the initial, stiff part of the deformation which is quite similar for all measurements (dotted line in Fig. 3.1). Also, the

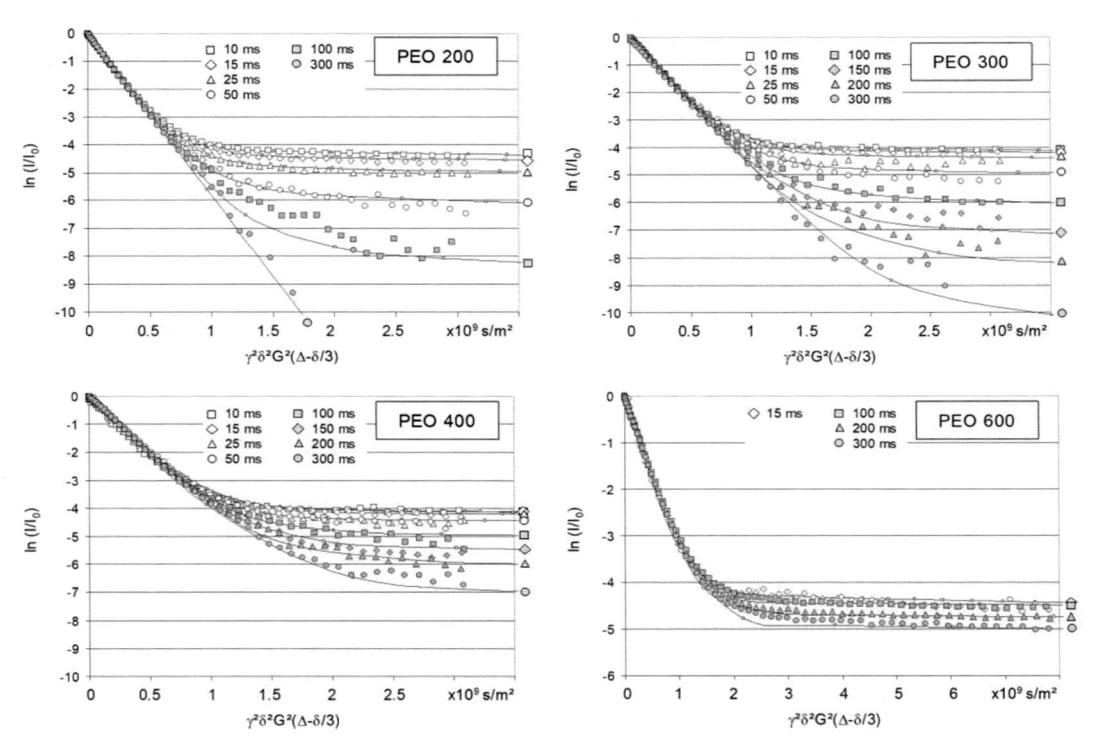

FIGURE 3.1

Atomic force microscopy images of a single poly(lactide-co-glycolide) nanocapsule before (*left*) as well as after one and two indentation tests (*center*). The height profile (bottom left) gives access to the dimensions of the capsule, which is slightly flattened by the contact to the surface. During the indentation experiments, the force is plotted versus the tip position (*right*, all curves are shifted to a starting position at 3 μm).

From Bauer, J., Zähres, M., Zellermann, A., Kirsch, M., Petrat, F., de Groot, H., & Mayer, C., (2010). Perfluorocarbon-filled poly (lactide-co-glycolide) nano- and microcapsules as artificial oxygen carriers for blood substitutes: A physico-chemical assessment. Journal of Microencapsulation, 27(2), 122–132.

deformation is largely reversible. In the given case, the spring constant as determined from the slope of the dotted line amounts to 65 N/m. The data can be compared to those obtained on red blood cells where values around 110 N/m are common. Given that the nanocapsules are much smaller than erythrocytes, their flexibility and toughness is sufficient under the conditions of circulation. The folding behavior of the capsule wall delivers additional information on its toughness and flexibility (Fig. 3.1, *left*).

3.4.1 Brownian motion, diffusion, and more: the essential steps for the carrier function

As an inherent property of dispersed particles, Brownian motion (Freedman, 1983) of nanoparticles in blood has to be considered as the main dynamic process. Its efficiency mainly depends on the size of the particles, but also on the temperature and on the viscosity of the fluid medium (which, by itself, again depends on the temperature). In the absence of concentration or temperature gradients, it is completely isotropic when averaged over time and space. In practice, Brownian motion can be observed indirectly by DLS (Berne & Pecora, 2000) or directly using optical microscopy under dark field illumination (Gooch, 2011). The application of the latter allows for following the track of individual particles. Finally, a computer analysis of a video sequence leads to a particle size distribution function (Finder et al., 2004).

Next to the natural flow process of the fluid medium, Brownian motion is the key to the function of nanoparticles as carrier systems. Long distances of course are more efficiently overcome by pressure gradient-induced flow. On the other hand, laminar flow by itself is practically zero near the boundaries of a tube. Therefore the actual exchange of carrier particles in contact with the surrounding immobile tissue is dominated by Brownian motion. For particles dispersed in blood with its more or less fixed temperature and viscosity, the main system parameter which determines the degree of Brownian mobility is the particle size (not the particle density!): the self-diffusion constant for spherical particles caused by Brownian motion is inversely proportional to the particle diameter (Yang, 1949). In consequence, smaller particles make more efficient carriers than larger particles.

In the case of carrier systems, the migration of the particles to the target tissue is one thing, the release or the exchange of the active component is yet another (Fig. 3.2).

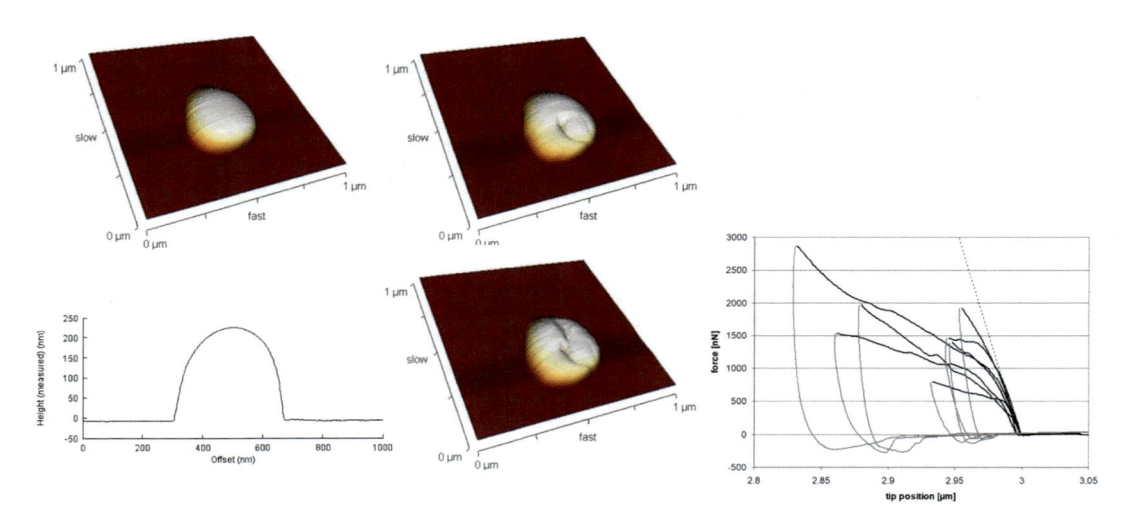

FIGURE 3.2

Physical processes connected to the function of dispersed carrier particles (active ingredient in red). The initial transport steps are flow and Brownian motion. Depending on the type of particles (capsules, solid spheres, nanocrystals), the following release is either caused by permeation, effusion, dissolution, desorption, or a combination thereof. The final step of transport consists of diffusion of the active ingredient. In addition, processes like adsorption, desorption, and exchange (involving an inverted release process) have to be accounted for.

The initial step of the release depends on the type of the particles. In the case of a nanocapsule (a thin spherical membrane encapsulating a liquid core), the active ingredient has to permeate through the membrane (Fig. 3.2, *left*). In the case of a nanosphere (a solid particle, e.g., from a gel, an inorganic matrix, or a solid polymer), the active ingredient has to migrate through a molecular framework, a process called effusion (Fig. 3.2, *center*). In the case of nanocrystals of the active ingredient itself, the initial step consists of its dissolution (Fig. 3.2, *right*). Gel particles may swell and deswell, phenomena which may be connected to release as well. The efficiency of all these processes largely depends on the particle surface and, hence, on the particle's size. Again, the advantage is to the smaller particles due to their larger interface with the fluid environment.

The second step is the same for all cases: diffusion. It is driven by the local concentration gradient: high level of active ingredient at the particle surface, lower in the surrounding. Just like Brownian motion, diffusion depends on the medium viscosity, the temperature, and the size (here of the diffusing molecule). In principal, Brownian motion and diffusion are very much alike if the latter is regarded for the limiting case of a disappearing concentration gradient, a condition where molecules undergo random motion (a condition called self-diffusion).

3.5 Applications of nuclear magnetic resonance spectroscopy

A direct observation of all these processes is possible using nuclear magnetic resonance spectroscopy on particle dispersions (Mayer, 2002, 2005, 2017). A simple NMR spectrum reflects all system constituents including the active component (on this level, the signal's linewidths may serve to estimate the particle size; Mayer & Lukowski, 2000). Combined with pulsed field gradients, a stimulated echo sequence yields data on the diffusive behavior of each component over a variable time period Δ. This information is visualized in a so-called Stejskal−Tanner plot (Stejskal & Tanner, 1965) where the signal intensities are plotted versus a term which depends on the gradient strength G and the observation period Δ (Fig. 3.3). In such a plot, any free diffusion process corresponds to a straight line with a negative slope equal to the self-diffusion constant.

Using this method, every single process depicted in Fig. 3.2 becomes experimentally observable (Bauer et al., 2006; Mayer & Bauer, 2006; Rumplecker et al., 2004). This will be shown on the example of nanocapsules formed by self-association of an amphiphilic copolymer which contains polyethylene glycol of different molar masses as a model active component (Fig. 3.3). All measurements are obtained in the thermal equilibrium without concentration gradients.

First we deal with the Brownian motion of the capsules as a whole (blue arrows in Fig. 3.2). In the Stejskal−Tanner plots in Fig. 3.3, Brownian mobility shows up in the final slope of the curves, especially for short observation periods Δ. In the given system, all slopes for values above 2×10^{-9} second/m^2 on the horizontal axis and $10 < \Delta < 50$ Ms are very shallow and hard to quantify. Still, they can be assigned to a Brownian motion of particles with 200 nm diameter which—in water at room temperature—corresponds to a self-diffusion coefficient of $D = 2 \times 10^{-12}$ m^2/second (Rumplecker et al., 2004). Of course, this value reflects a volume average of the individual values, the distribution of which will vary with the given size distribution of the capsules.

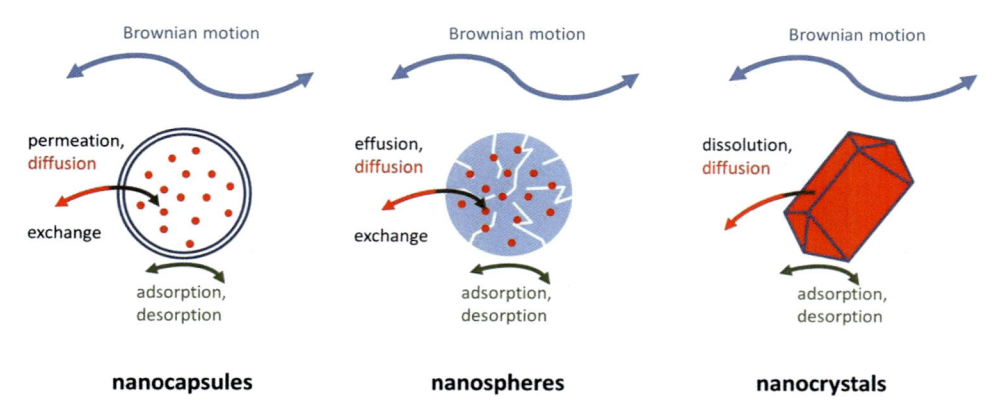

FIGURE 3.3

NMR-measurements using pulsed field gradients on a carrier system formed by an amphiphilic block-copolymer. The plots (so-called Stejskal–Tanner plots) show the relative signal intensity of model active components (polyethylene oxide with average molar masses of 200, 300, 400, and 600 g/mol) as functions of the gradient strength G (horizontal axis) and of the observation period Δ (10-300 Ms, symbols). Solid lines are best fit simulations on the experimental data. The initial negative slopes of the curves are identical to the self-diffusion coefficients of the "active components" in water. The final slopes correspond to the self-diffusion coefficient for the Brownian motion of the whole capsules. The spread of the curves for different observation periods Δ describe the permeation of the model active component through the capsule membrane. As expected, the velocity of permeation (i.e., the spread of the lines for different values of Δ) decreases from PEO 200 to PEO 600.

From Rumplecker, A., Förster, S., Zähres, M., & Mayer, C. (2004). Molecular exchange through vesicle membranes: A pulsed field gradient nuclear magnetic resonance study. Journal of Chemical Physics, 120(18), 8740–8747.

Second, we observe the self-diffusion of the active ingredient (red arrows in Fig. 3.2). It will show up in the initial slopes of the Stejskal–Tanner plots between the values 0 and 0.5 second/m^2 on the horizontal axis. In this case, polyethylene oxide samples of different molar masses (from 200 g/mol for PEO 200 up to 600 g/mol for PEO 600) have been used as model active ingredients. Depending on the molar mass of the PEO, we expect the self-diffusion coefficient to vary accordingly. In fact, the initial (negative) slopes gradually decrease from PEO 200 to PEO 600 corresponding to self-diffusion coefficients of $D_{200} = 6.0 \times 10^{-10}$ m^2/second, $D_{300} = 4.9 \times 10^{-10}$ m^2/second, $D_{400} = 4.3 \times 10^{-10}$ m^2/second, and $D_{600} = 3.6 \times 10^{-10}$ m^2/second (Rumplecker et al., 2004).

Third, all processes which connect the immobilized state and the released state of the active ingredient are directly observable. This includes permeation, effusion, dissolution, and exchange (black arrows in Fig. 3.2). In all these cases, we have a relatively slow process which connects the encapsulated state of the active ingredient characterized by very slow (Brownian) diffusion with the "free" state of the active ingredient with rapid self-diffusion. In the Stejskal–Tanner plots in Fig. 3.3, this shows up as a spread of the final level of the decay curves. Since this level depends on the fraction of molecules which remain immobilized over the whole observation period, and since this observation period is varied between 10 and 300 Ms, we observe a constant loss of this level with increasing Δ. The dependence of this level on Δ (in other words: the spread of the curves) directly reflects the time development of the release process. As expected, smaller tracer molecules permeate more quickly than larger ones. From a numerical simulation of the signal decay plots (solid lines in Fig. 3.3), the permeation rates could be determined for PEO 200 up to PEO 600: $k_{200} = 1.6$/second, $k_{300} = 0.85$/second, $k_{400} = 0.38$/second, and $k_{200} = 0.075$/second (Rumplecker et al., 2004). Smaller permeation processes can be studied using a time-resolved variety of this measurement where an actual concentration gradient is applied.

Fourth, adsorption and desorption processes can be followed by taking advantage of the adsorption-induced cross polarization (cp) phenomenon (Hoffmann & Mayer, 2000). Normally, the approach of NMR cross polarization is limited to the immobilized state. If a dissolved component temporarily binds or adsorbs to the particle surface, cross polarization from hydrogen to carbon nuclei occurs, leading to a strongly increased carbon NMR signal. Hence, adsorption and desorption of dissolved components in a dispersion can be studied by observing the ^{13}C signal under cp conditions and under variation of the cp contact time (Mayer et al., 2002).

These methods are versatile tools for the optimization of carrier systems and for the understanding of release measurements. Among others, they revealed that ethanol strongly accelerates the release process through vesicle membranes (Bauer et al., 2006), they reflect the influence of temperature on the release rate (Leson et al. 2007) and gave access to the finding that chemical cross-linking of nanocapsule membranes leads to reduced permeation rates (Groß-Heitfeld et al., 2014). In addition, a full permeability profile for a given type of capsule can be obtained (Erdmann & Mayer, 2016).

A very special type of exchange is in the focus for dispersed oxygen carriers in artificial blood replacement systems. They are meant to take up oxygen in the lungs and release it to the residual tissue (and vice versa for carbon dioxide). Hence, their key function is represented by gas exchange, especially of oxygen. The oxygen has the particular property to be paramagnetic, so its presence can be detected by NMR again. The paramagnetic interaction has a significant influence on the chemical shift and on the relaxation behavior of the surrounding nuclei. Therefore the concentration of oxygen in carrier particles can be directly monitored. The dispersion can be alternatingly treated with oxygen and nitrogen gas in order to follow the uptake and release, proving the full reversibility of the gas exchange process (Bauer, Zähres, Zellermann, et al., 2010; Wrobeln et al., 2017).

3.6 Conclusion

In general, the full understanding of the physicochemical behavior of dispersed particles is crucial for the preparation, the handling, and—foremost—the function of the system in a hematological application. Of course, the physicochemical assessment cannot predict its physiological performance and cannot replace physiological testing. However, it delivers the elementary basis for the essential design of the particle dispersion. Therefore it is a necessary step in the early stage of their development.

References

Altinbas, N., Fehmer, C., Terheiden, A., Shukla, A., Rehage, H., & Mayer, C. (2006). Polybutylcyanoacrylate nanocapsules prepared from mini-emulsions—A comparison with the conventional approach. *Journal of Microencapsulation, 23*(5), 567–581.

Azmi, M. A., & Shad, K. F. (2017). Nanotechnology: Some basic concepts. In D. Ficai, & A. Mihai (Eds.), *Nanostructures for novel therapy*. Elsevier.

Bauer, A., Hauschild, S., Stolzenburg, M., Förster, S., & Mayer, C. (2006). Molecular exchange through membranes of P2VP-PEO vesicles. *Chemical Physics Letters, 419*, 430–433.

Bauer, A., Kopschütz, C., Stolzenburg, M., Förster, S., & Mayer, C. (2006). The effect of ethanol on the permeability of block copolymer vesicle membranes. *Journal of Membrane Science, 284*, 1–4.

Bauer, J., Zähres, M., Zellermann, A., Kirsch, M., Petrat, F., de Groot, H., & Mayer, C. (2010). Perfluorocarbon-filled poly(lactide-co-glycolide) nano- and microcapsules as artificial oxygen carriers for blood substitutes: A physico-chemical assessment. *Journal of Microencapsulation, 27*(2), 122–132.

Berne, B. J., & Pecora, R. (2000). *Dynamic light scattering*. Courier Dover Publications.

Doroszkowski, A. (1999). The physical chemistry of dispersion. In R. Lambourne, & T. A. Strivens (Eds.), *Paint and surface coatings*. Woodhead Publishing.

Erdmann, C., & Mayer, C. (2016). Permeability profile of poly(alkyl cyanoacrylate) nanocapsules. *Journal of Colloids and Interface Science, 478*, 394–401.

Finder, C., Wohlgemuth, M., & Mayer, C. (2004). Analysis of particle size distribution by particle tracking. *Particle and Particle Systems Characterization, 21*, 372–378.

Freedman, D. (1983). *Brownian motion and diffusion*. Springer.

Gooch, J. W. (2011). Dark field illumination. In J. W. Gooch (Ed.), *Encyclopedic dictionary of polymers*. Springer.

Groß-Heitfeld, C., Linders, J., Appel, R., Selbach, F., & Mayer, C. (2014). Polyalkylcyanoacrylate Nanocapsules: Variation of membrane permeability by chemical cross-linking. *Journal of Physical Chemistry B, 118*, 4932–4939.

Hoffmann, D., & Mayer, C. (2000). Cross polarization induced by temporary adsorption: NMR investigations on nanocapsule dispersions. *Journal of Chemical Physics, 112*(9), 4242–4250.

Kahlweit, M. (1975). Ostwald ripening of precipitates. *Advances in Colloid and Interface Science, 5*(1), 1–35.

Leson, A., Filiz, V., Förster, S., & Mayer, C. (2007). Molecular exchange through vesicle membranes: Determination of the activation energy. *Chemical Physics Letters, 444*, 268–272.

Mayer, C. (2002). NMR on dispersed nanoparticles. *Progress in NMR Spectroscopy, 40*(4), 307–366.

Mayer, C. (2005). NMR studies on nanoparticles. *Annual Reports on NMR Spectroscopy, 55*, 205–258.

Mayer, C. (2017). NMR spectroscopy of nanoparticles. In J. C. Lindon, G. E. Tranter, & D. W. Koppenaal (Eds.), *Encyclopedia of spectroscopy and spectrometry*. Elsevier.

Mayer, C., & Bauer, A. (2006). Molecular exchange through capsule membranes observed by pulsed field gradient NMR. *Progress in Colloid and Polymer Science, 133*, 22–29.

Mayer, C., Hoffmann, D., & Wohlgemuth, M. (2002). Structural analysis of nanocapsules by nuclear magnetic resonance. *International Journal of Pharmacy, 242*, 37–46.

Mayer, C., & Lukowski, G. (2000). Solid state NMR investigations on nanosized carrier systems. *Pharmaceutical Research, 17*(4), 486–489.

Rumplecker, A., Förster, S., Zähres, M., & Mayer, C. (2004). Molecular exchange through vesicle membranes: A pulsed field gradient nuclear magnetic resonance study. *Journal of Chemical Physics, 120*(18), 8740–8747.

Russel, W. B., Saville, D. A., & Schowalter, W. R. (1989). *Colloidal dispersions*. Cambridge University Press.

Stejskal, E. O., & Tanner, J. E. (1965). Spin diffusion measurements: Spin echoes in the presence of a time-dependent field gradient. *Journal of Chemical Physics*, *42*, 288−292.

Tadros, T. (2012). Electrostatic and steric stabilization. In H. Ohshima (Ed.), *Electrical phenomena at interfaces and biointerfaces*. Wiley VCH.

Tan, J., Shah, S., Thomas, A., Daniel Ou-Yang, H., & Liu, Y. (2012). The influence of size, shape and vessel geometry on nanoparticle distribution. *Microfluid Nanofluid*, *14*, 77−87. Available from https://doi.org/10.1007/s10404-012-1024-5.

Voigtländer, B. (2020). *Atomic force microscopy*. Springer.

Wrobeln, A., Schlüter, K. D., Linders, J., Zähres, M., Mayer, C., Kirsch, M., & Ferenz, K. B. (2017). Functionality of albumin-derived perfluorocarbon-based artificial oxygen carriers in the Langendorff-heart. *Artificial Cells, Nanomedicine, and Biotechnology*, *45*(4), 723−730.

Yang, L. M. (1949). Diffusion and the Brownian motion. *Proceedings of the Royal Society A*, *198*, 94−116.

Electromagnetic Casson blood flow in multistenosed porous artery using Caputo–Fabrizio fractional derivatives

4

Dzuliana Fatin Jamil[1], Salah Uddin[2] and Rozaini Roslan[1]

[1]*Department of Mathematics and Statistics, Faculty of Applied Sciences and Technology, Universiti Tun Hussein Onn Malaysia, Pagoh, Malaysia* [2]*Department of Physical and Numerical Sciences, Qurtuba University of Science and Information Technology, Peshawar, Pakistan*

4.1 Introduction

Atherosclerotic disorder occurs in abnormalities such as curved, tapered arteries and stenotic regions. These abnormalities are believed to be partially responsible for the atherosclerotic appearance that impacts cardiovascular system function (Othman Smadi et al., 2006). The investigation of blood flow in a constricted artery is of great concern due to its importance in human vascular diseases. In recent years, the study of blood flow across constricted arteries has gained significant interest due to its great importance in the human cardiovascular system (Abbas et al., 2018; Agarwal & Varshney, 2016; Sankar & Lee, 2011; Singh & Shah, 2010; Srivastava, 2014). Agarwal and Varshney (2016) studied the flow of Herschel-Bulkley fluid through an inclined tube of non-uniform cross-section with multiple stenoses. Mandal et al. (2007) studied the effect of periodic body acceleration on blood flow, by considering the arterial wall as an elastic cylindrical tube with stenosis on the lumen. Tzirtzilakis (2005) designed the mathematical model of biomagnetic fluid dynamics to analyze the Newtonian blood flow in the presence of magnetic field.

Various experiments have been performed to visualize the blood flow pattern caused by stenosis. The blood is treated as Newtonian or non-Newtonian fluid. The non-Newtonian effect becomes more apparent in medium- and small-sized arteries. Sankar and Hemalatha (2007) investigated the non-Newtonian effect of blood in small arteries, by treating the blood as Herschel−Bulkley fluid subjected to various physiological conditions. Siddiqui et al. (2015) have modeled the blood as non-Newtonian Bingham plastic fluid. The model was used to study the flow through a stenosed artery in the presence of slip velocity and body acceleration. As blood flows at low shear rate into the narrow arteries, it behaves like a Casson fluid (Nagarani et al., 2006). Many researchers (Ali et al., 2017; Ali et al., 2019; Maiti et al., 2020; Nejad et al., 2018) have been working on the Casson fluid model for modeling blood flow through narrow arteries. Gross and Aroesty (1972) used the Casson theory in their mathematical analysis to study the pulsatile flow in blood vessels with application to microcirculation. Under the control of periodic external body acceleration, the pulsatile blood flow in a stenosed artery was studied.

The influence of non-Newtonian blood in narrow blood vessels was taken into account by modeling the blood as Casson fluid (Nagarani & Sarojamma, 2008). Biswas and Paul (2013) studied the non-Newtonian fluid models and stated that Casson and Herschel−Bulkley fluid models are more preferred for studying blood flow.

The effectiveness of a magnetic field in treating stenosis has been extensively studied. Tashtoush and Magableh (2008) analyzed the magnetic blood flow in multistenosed arteries. The finite difference method was applied to solve the governing equations. Bose and Banerjee (2015) applied the magnetic drug targeting technique in treating stenosed aortic bifurcation. The principles of FHD and MHD were combined in order to model the blood as a biomagnetic fluid. Mekheimer et al. (2011) analyzed the blood flow through an elastic artery with overlapping stenosis under the effect of magnetic field. Majee et al. (2017) studied the unsteady nonisothermal magnetic blood flow through constricted arteries. Bansi et al. (2018) used the fractional model to study the impacts of magnetic field and heat transfer inside the oscillatory arteries. The investigation of a magnetic field in a porous medium is essential from the theoretical and practical points of view, as most of the natural liquid phenomena are linked with porosity. Hatami et al. (2014) performed the MHD study of non-Newtonian fluid in a porous medium analytically and numerically. Zaman et al. (2017) conducted the numerical study of pulsatile blood flow in an overlapping stenosed porous artery. They found that the elevated permeability parameter would increase the blood velocity, the shear stress, and the flow rate. El-Shahed (2003) studied the effects of pulsatile blood flow in a permeable stenosed artery in the presence of a magnetic field. Ponalagusamy and Priyadharshini (2017) simulated the pulsatile flow of Herschel−Bulkley fluid in a stenosed porous artery. In this study, the impacts of periodic body acceleration and magnetic field on the blood flow were investigated.

Due to the increasing popularity of fractional derivatives, several fractional derivative models have been formulated by inferring the existing fluid models (Abro & Gómez-Aguilar, 2019; Khan et al., 2017; Shah et al., 2019; Uddin et al., 2018). Caputo and Fabrizio (2015) introduced a new derivative technique and applied it in several real-world problems. Ali et al. (2017) has developed a fractional-order model for blood flow (Casson fluid) with the help of Hankel and Laplace transform techniques in order to obtain the exact solutions. Ali Shah et al. (2016) injected magnetic particles into a cylindrical tube numerically under the effects of magnetic field (acting perpendicular to the flow path) and oscillating axial pressure gradient. Considering the significance of fractional derivatives, the researchers (Bakhti et al., 2017) used fractional calculus to study the flow of Oldroyd-B fluid in stenosed arteries. They derived the mathematical model of tapered stenosed artery in the presence of pressure gradient, which might help medical practitioners in treating cardiovascular diseases. Shah and Khan (2016) applied the idea of the Caputo−Fabrizio fractional derivatives to generalize the starting flow of second-grade fluid over a vertical plate and obtained the exact solutions using the Laplace transform.

From the literature review, it is found that the precise solution of the Caputo−Fabrizio fractional derivative in describing the electromagnetic blood flow in multistenosed porous artery has not been described. In the current work, the fractional-order time derivative is applied to model the non-Newtonian Casson fluid. The blood flow is driven by oscillating pressure gradient and periodic body acceleration in the z-direction. The external magnetic and electric fields are considered as well. The fractional derivative model is obtained by converting the first-order derivatives to Caputo−Fabrizio fractional derivatives for the axial blood flow and magnetic particle velocities. The exact solutions are then calculated by means of significant transformations such as Laplace and finite Hankel transforms. The zeroth-order Bessel functions have been used for numerical computations. The graphical results are generated using Mathcad for various fractional and key physical parameters.

4.2 **Methodology**

Let us consider the unsteady blood flow in multistenosed artery aligned in the axial direction (z-axis). The r-axis is lying along the radial direction of the artery. Blood is treated as an incompressible non-Newtonian Casson fluid subjected to oscillating pressure gradient and body acceleration. The blood flow is mainly driven by electric field, upon assuming that the effect of magnetic field is almost negligible. The corresponding momentum equation is therefore a generalization of the preview study conducted by (Ali Shah et al., 2016) with the adding factors of electromagnetic Casson fluid in the multistenosed porous artery. At $t = 0$, the blood and the magnetic particles are treated as stationary (Fig. 4.1).

The governing equations are the Navier–Stokes equations describing the blood flow, the Maxwell's relations describing the magnetic field, and Newton's second law describing the particle motion.

The Maxwell equations are:

$$\nabla \cdot \vec{B} = 0, \nabla \times \vec{B} = \mu_0 \vec{J}, \nabla \times \vec{E} = -\frac{\partial \vec{B}}{\partial t}, \tag{4.1}$$

where $\vec{B}$ is the magnetic flux intensity, μ_0 is the magnetic permeability, $\vec{E}$ is the electric field intensity, and $\vec{J}$ is the current density given by (Hatami et al., 2014; Zaman et al., 2016):

$$\vec{J} = \sigma(\vec{E} + \vec{V} \times \vec{B}), \tag{4.2}$$

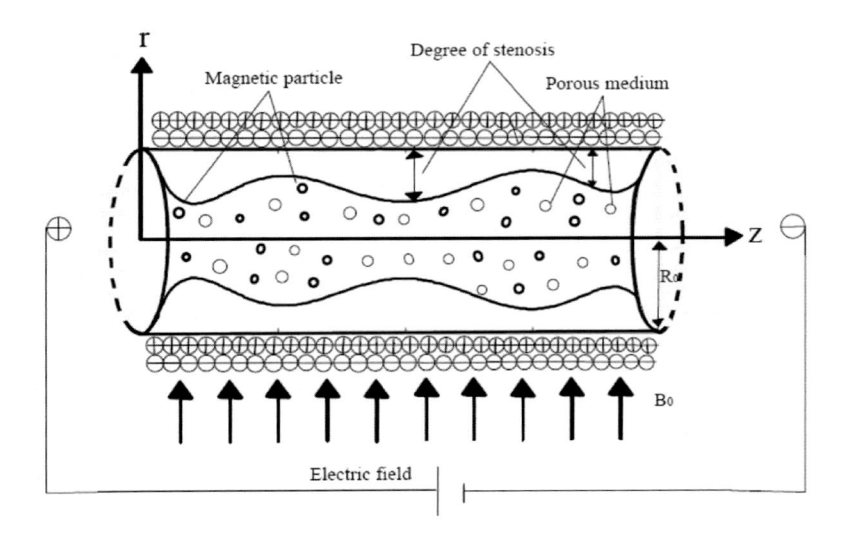

FIG. 4.1

Geometry of an inclined arterial segment with multistenosis.

Here σ is the electrical conductivity and $\vec{V}$ is the velocity field. The electromagnetic force $\vec{F}_{em}$ is defined as:

$$\begin{aligned} \vec{F}_{em} = \vec{J} \times \vec{B} &= \sigma(\vec{E} + \vec{V} \times \vec{B}) \times \vec{B} \\ &= -\sigma B_0^2 u(r,t)\vec{k} - \rho_e E \\ &= -\sigma B_0^2 u(r,t)\vec{k} - \rho_e E_z \vec{k} \end{aligned} \tag{4.3}$$

where $\vec{k}$ is the unit vector of the z-direction and $\vec{V} = u(r,t)\vec{k}$ is the axial velocity of the blood. E_z is the axial component of the applied electric field, while $\rho_e = -\varepsilon\kappa^2\psi(r)$ is the net charge density of the electrolyte solution. The force $\vec{F}_{em}$ will be included in the momentum equations.

The pulsatile characteristics of blood originates from the pumping action of heart and hence the pressure gradient can be expressed as (Agarwal & Varshney, 2016)

$$-\frac{\partial p}{\partial z} = A_0 + A_1\cos(\omega_p t), A_0 > 0. \tag{4.4}$$

where the constants A_0 and A_1 are the amplitudes of the pulsatile magnetic field and the pressure gradient that give rise to systolic or diastolic pressure. ω_p is the pulsatile frequency.

The body acceleration for vibration environment (Shit, 2011) can be written as:

$$F(t) = A_g\cos(\omega_g t + \varphi), \tag{4.5}$$

where A_g denotes the amplitude of the body acceleration, ω_g is the frequency, and φ is the phase angle.

The geometry of the multistenosis in the arterial lumen can be described mathematically (Tashtoush & Magableh, 2008) as:

$$R_z = 1 - \gamma(1.48z - 0.7398z^2 + 0.1485z^3 - 0.013955z^4 + 0.0006145z^5 - 0.000010243z^6) \tag{4.6}$$

where R_z is the radius of the artery in the constricted region, R_0 is the radius of the normal artery, x is the length of stenosis, and γ is the degree of the stenosis.

The momentum equation for fluid flow in an multistenosed artery (Ali et al., 2017; Ali Shah et al., 2016; Sharma et al., 2014) can be written as:

$$\begin{aligned} \frac{\partial u}{\partial t} = &-\frac{1}{\rho}\frac{\partial p}{\partial z} + F(t) + \upsilon\left(1 + \frac{1}{\beta}\right)\left(\frac{\partial^2 u}{\partial r^2} + \frac{1}{r}\frac{\partial u}{\partial r}\right) + \frac{KN}{\rho}(v-u) - \frac{\sigma B_0^2 u}{\rho} \\ &- \frac{\varepsilon\kappa^2\psi(r)E_z}{\rho} - \frac{\mu u}{k_p}, \end{aligned} \tag{4.7}$$

where $u(r,t)$ and $v(r,t)$ denote the blood distribution and magnetic particle velocity in the axial direction, ρ is the fluid density, p is the pressure, and υ is the kinematic viscosity. The material parameter of Casson fluid is $\beta = \frac{\mu_B\sqrt{2\pi_c}}{\tau_r}$, where μ_B is the plastic dynamic viscosity, τ_r is the yield stress of fluid, $2\pi_c$ is the critical value of this product based on the non-Newtonian model, K is the Stokes constant, and N is the number of magnetic particles per unit volume. The term $\frac{KN}{\rho}(v-u)$ is the force due to the relative motion between fluid and magnetic particles. σ and B_0 are the respective electrical conductivity and intensity of the applied magnetic field. It is assumed that the Reynolds number (computed from relative velocity) is small. As such, the force between the magnetic particles and the blood is proportional to the relative velocity.

The motion of magnetic particles is governed by the Newton's second law:

$$m\frac{\partial v}{\partial t} = K(u - v) \tag{4.8}$$

where m is the average mass of the magnetic particles.

The Caputo–Fabrizio derivative operator is:

$$^{CF}D_t^{\alpha\alpha}u(r,t) = \frac{1}{1-\alpha}\int_0^t \exp\left(-\frac{\alpha(t-\tau)}{1-\alpha}\right)\frac{\partial u(r,\tau)}{\partial \tau}d\tau \tag{4.9}$$

The Laplace transform of the Caputo–Fabrizio time derivative can be written as:

$$L\{^{CF}D_t^{\alpha\alpha}u(r,t)\} = \frac{sL\{u(r,t)\} - u(r,0)}{(1-\alpha)s + \alpha} \tag{4.10}$$

The initial boundary conditions of the fluid inside the cylindrical domain of radius R_0 are:

$$\begin{aligned} u(r,0) = 0, \quad v(r,0) = 0, \quad at\,t = 0, \quad r \in [0, R_0], \\ u(R_0, t) = 0, \quad v(R_0, t) = 0, \quad t > 0. \end{aligned} \tag{4.11}$$

In order to perform dimensionless study, the following nondimensional parameters can be introduced

$$r* = \frac{r}{R_0}, \quad t* = \frac{u_0 t}{R_0}, \quad u* = \frac{u}{u_0}, \quad v* = \frac{v}{u_0}, \quad p* = \frac{p}{\rho u_0^2}, \quad z* = \frac{z}{R_0}$$

$$A_g^* = \frac{R_0 A_g}{u_0^2}, \quad \lambda = \frac{k_p u_0}{\mu R_0}, \quad K^2 = \frac{\varepsilon k^2 E_z R_0}{\rho u_0^2} \tag{4.12}$$

where u_0 is the characteristics velocity.

By introducing the above parameters and dropping the * notation, the nondimensional forms of Eqs. (4.7), (4.8), and (4.11) are:

$$\begin{aligned} D_t^\alpha u = A_0 + A_1\cos(\omega_p t) + A_g\cos(\omega_g t + \varphi) + \frac{1}{\text{Re}}\left[1 + \frac{1}{\beta}\right]\left[\frac{\partial^2 u}{\partial r^2} + \frac{1}{r*}\frac{\partial u*}{\partial r*}\right] \\ + R(v - u) - \left(Ha^2 + \frac{1}{\lambda}\right)u + K^2\psi(r), \end{aligned} \tag{4.13}$$

$$G \cdot D_t^\alpha v = u - v \tag{4.14}$$

where $\text{Re} = \frac{R_*^2}{\lambda v}$ is the Reynolds number, $R = \frac{kN\lambda}{\rho}$ is the particle concentration parameter, and $Ha = B_0\sqrt{\lambda}\sqrt{\frac{g}{\rho}}\sin\theta$ is the Hartmann number. The nondimensional boundary conditions are:

$$\begin{aligned} u(r,0) = 0, \quad v(r,0) = 0, \quad at\,t = 0, \quad r \in [0, 1], \\ u(1, t) = 0, \quad v(1, t) = 0, \quad t > 0. \end{aligned} \tag{4.15}$$

Laplace transform is suitable when the temporal variable t is adopted in the blood flow model. After the transformation process, the equation becomes:

$$\begin{aligned} \frac{s\bar{u}(r,s)}{s + \alpha(1-s)} = A_0 + A_1\cos(\omega_p t) + A_g\cos(\omega_g t + \varphi) + \frac{1}{\text{Re}}\left[1 + \frac{1}{\beta}\right]\left[\frac{\partial^2\bar{u}(r,s)}{\partial r^2} + \frac{1}{r}\frac{\partial\bar{u}(r,s)}{\partial r}\right] \\ + R\bar{v}(r,s) - \left(R + Ha^2 + \frac{1}{\lambda}\right)\bar{u}(r,s) + K^2\psi(r), \end{aligned} \tag{4.16}$$

$$G\frac{s\bar{v}(r,s)}{s+\alpha(1-s)} = \bar{u}(r,s) - \bar{v}(r,s),$$ (4.17)

$$\bar{u}(1,s) = 0, \bar{v}(1,s) = 0.$$ (4.18)

From Eq. (4.17), the following equation can be obtained:

$$\bar{v}(r,s) = \frac{s+\alpha(1-s)}{Gs+s+\alpha(1-s)}\bar{u}(r,s)$$ (4.19)

Substituting $\bar{v}(r,s)$ from Eq. (4.19) into Eq. (4.16), the coefficient is:

$$\begin{aligned}
&\left[\frac{s}{s+\alpha(1-s)} - R\left(\frac{s+\alpha(1-s)}{s+sG+\alpha(1-s)}\right) + R + Ha^2 + \frac{1}{\lambda}\right]\bar{u}(r,s) \\
&= \frac{A_0}{s} + \frac{k^2\psi(r)}{s}\frac{A_1s}{s^2+\omega_p{}^2} + \frac{A_g(s\cos\varphi + \omega_g\sin\varphi)}{\omega_g^2+\varphi^2} \\
&\quad + \frac{1}{Re}\left(1+\frac{1}{\beta}\right)\left[\frac{\partial^2\bar{u}(r,s)}{\partial r^2} + \frac{1}{r}\frac{\partial\bar{u}(r,s)}{\partial r}\right],
\end{aligned}$$ (4.20)

By applying finite Hankel transform of zeroth-order [i.e., by applying the boundary condition (4.15) in Eq. (4.20)], the following equation can be obtained:

$$\begin{aligned}
&\left[\frac{s}{s+\alpha(1-s)} - R\left(\frac{s+\alpha(1-s)}{s+sG+\alpha(1-s)}\right) + R + Ha^2 + \frac{1}{\lambda}\right]\bar{u}_H(r_n,s) \\
&= \left[\frac{A_0}{s} + \frac{A_1s}{s^2+\omega_p{}^2} + \frac{A_g(s\cos\varphi + \omega_g\sin\varphi)}{\omega_g^2+\varphi^2}\right]\frac{J_1(r_n)}{r_n} \\
&\quad + \frac{K^2}{s}\frac{r_n}{r_n^2+K^2}J_1(r_n) - \frac{1}{Re}\left(1+\frac{1}{\beta}\right)r_n\bar{u}_H(r_n,s),
\end{aligned}$$ (4.21)

where $\bar{u}_H(r_n,s) = \int_0^1 r\bar{u}(r,s)J_0(r_nr)dr$ represents the finite Hankel transform of the velocity function $\bar{u}(r,s) = LT[u(r,t)]$ and $r_n, n = 1, 2, \dots$ are the positive roots of the equation $J_0(x) = 0$. Here J_0 is the zeroth-order Bessel function and it belongs to the first kind. By simplifying the coefficient of $\bar{u}_H(r_n,s)$ in Eq. (4.21), the following equations can be formulated:

$$\begin{aligned}
\bar{u}_H(r_n,s) &= \left[\frac{s^2x_{5n} + sx_{6n} + \alpha^2}{s^2x_{2n} + sx_{3n} + y_{4n}}\right]\left[\frac{A_0}{s} + \frac{A_1s}{s^2+\omega_p{}^2} + \frac{A_g(s\cos\varphi + \omega_g\sin\varphi)}{\omega_g^2+\varphi^2}\right]\frac{J_1(r_n)}{r_n} \\
&\quad + \frac{K^2}{s}\frac{r_n}{r_n^2+K^2}J_1(r_n)
\end{aligned}$$ (4.22)

$$\begin{aligned}
\bar{u}_H(r_n,s) &= \left[\frac{x_{9n}}{s-x_{7n}} + \frac{x_{10n}}{s-x_{8n}}\right]\left[\frac{A_0}{s} + \frac{A_1s}{s^2+\omega_p{}^2} + \frac{A_g(s\cos\varphi + \omega_g\sin\varphi)}{\omega_g^2+\varphi^2}\right]\frac{J_1(r_n)}{r_n} \\
&\quad + \frac{K^2}{s}\frac{r_n}{r_n^2+K^2}J_1(r_n)
\end{aligned}$$ (4.23)

$$\bar{u}_H(r_n, s) = \frac{J_1(r_n)}{r_n}\left[\frac{s^{-1}}{s - x_{7n}}A_0x_{9n} + \frac{s^{-1}}{s - x_{8n}}A_0x_{10n}\right]$$

$$+ \left(\frac{1}{s - x_{7n}}\right)\left(\frac{s}{s^2 + \omega_p^2}\right)A_1x_{9n} + \left(\frac{1}{s - x_{8n}}\right)\left(\frac{s}{s^2 + \omega_p^2}\right)A_1x_{10n}$$

$$+ \left(\frac{1}{s - x_{7n}}\right)\left(\frac{s}{s^2 + \omega_g^2}\right)A_g\cos\varphi x_{9n} - \left(\frac{1}{s - x_{7n}}\right)\left(\frac{\omega_g}{s^2 + \omega_g^2}\right)A_g\sin\varphi x_{9n} \quad (4.24)$$

$$+ \left(\frac{1}{s - x_{8n}}\right)\left(\frac{s}{s^2 + \omega_g^2}\right)A_g\cos\varphi x_{10n} - \left(\frac{1}{s - x_{7n}}\right)\left(\frac{\omega_g}{s^2 + \omega_g^2}\right)A_g\sin\varphi x_{10n}$$

$$+ \frac{K^2}{s}\frac{r_n}{r_n^2 + K^2}J_1(r_n)$$

Note, the parameters in Eqs. (4.22) and (4.23) introduced for simplifying the coefficient of $\bar{u}_H(r_n, s)$ are:

$$\begin{aligned}
&x_{1n} = Ha^2 + R + \beta_1 r_n^2, \\
&x_{2n} = 1 + G - \alpha - R - R\alpha^2 + 2R\alpha + y_{1n} + \alpha^2 y_{1n} - 2\alpha y_{1n} + Gy_{1n} - G\alpha y_{1n}, \\
&x_{3n} = \alpha + 2R\alpha^2 - 2R\alpha - 2x_{1n}\alpha^2 + 2\alpha x_{1n} + G\alpha x_{1n}, \quad y_{4n} = \alpha^2 y_{1n} - R\alpha^2, \\
&x_{5n} = 1 + \alpha^2 - 2\alpha + G - G\alpha, \quad x_{6n} = -2\alpha^2 + 2\alpha + G\alpha,
\end{aligned} \quad (4.25)$$

$$x_{7n} = \frac{-x_{3n} + \sqrt{x_{3n}^2 - 4x_{2n}x_{4n}}}{2x_{2n}}, \quad x_{8n} = \frac{-x_{3n} - \sqrt{x_{3n}^2 - 4x_{2n}x_{4n}}}{2x_{2n}},$$
$$x_{9n} = \frac{x_{7n}^2 x_{5n} + x_{7n}x_{6n} + \alpha^2}{x_{7n} - x_{8n}}, \quad x_{10n} = \frac{x_{8n}^2 x_{5n} + x_{8n}x_{6n} + \alpha^2}{x_{8n} - x_{7n}}, \quad (4.26)$$

The Laplace transform of the image function $\bar{u}_H(r_n, s)$ discussed in Eq. (4.24) can be obtained by using the Robotnov and Hartley's functions:

$$LT^{-1}\left[\frac{1}{s^w + y}\right] = F_w(-y, t) = \sum_{n=0}^{\infty}\frac{(-y)^n t^{(n+1)w-1}}{\Gamma((n+1)w)}, w > 0 \quad (4.27)$$

$$LT^{-1}\left[\frac{s^z}{s^w + y}\right] = R_{w,z}(-y, t) = \sum_{n=0}^{\infty}\frac{(-y)^n t^{(n+1)w-1-z}}{\Gamma((n+1)w-z)}, \mathrm{Re}(w-z) > 0 \quad (4.28)$$

By inverting the finite Hankel transforms [i.e., Eq. (4.24)], the following equation can be obtained:

$$u(r, t) = \frac{J_1(r_n)}{r_n}[y_{1n}t + y_{2n}t + y_{3n}t + y_{4n}t] \quad (4.29)$$

$$y_{1n}t = \left(\frac{A_0x_{9n}}{x_{7n}} + \frac{r_n^2 K^2 x_{9n}}{x_{7n}(r_n^2 + K^2)}\right)\left(e^{x_{7n}t} - 1\right) + \left(\frac{A_0x_{10n}}{x_{8n}} + \frac{r_n^2 K^2 x_{10n}}{x_{8n}(r_n^2 + K^2)}\right)\left(e^{x_{8n}t} - 1\right) \quad (4.30)$$

$$y_{2n}t = A_1x_{9n}e^{x_{7n}t} * \cos\omega_p t + A_1x_{10n}e^{x_{8n}t} * \cos\omega_p t, \quad (4.31)$$

$$y_{3n}t = A_g\cos\varphi x_{9n}e^{x_{7n}t} * \cos\omega_g t - A_g\sin\varphi x_{10n}e^{x_{8n}t} * \sin\omega_g t, \quad (4.32)$$

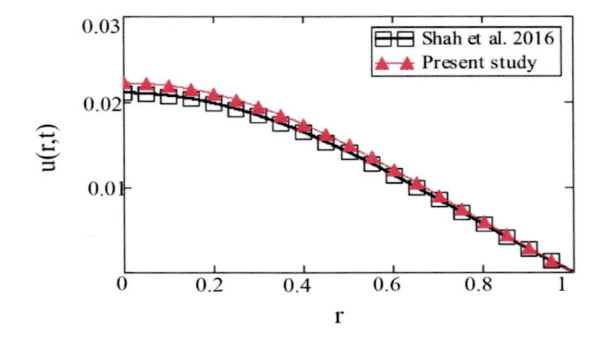

FIGURE 4.2

Comparison of velocity distribution with previous study.

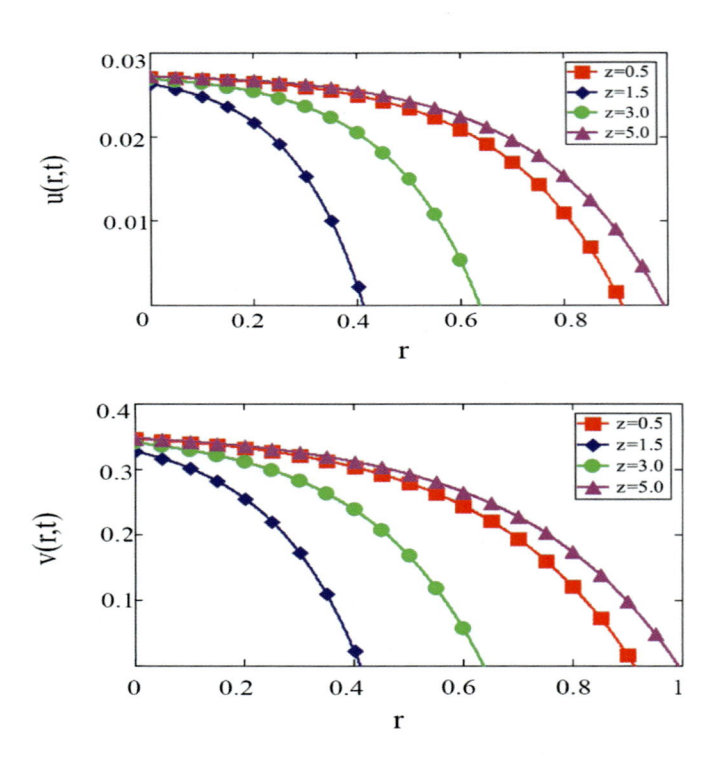

FIGURE 4.3

Axial velocity profiles $u(r,t)$ and $v(r,t)$ at different values of z.

$$y_{4n}t = A_g\cos\varphi x_{10n}e^{x_{8n}t} * \cos\omega_g t - A_g\sin\varphi x_{10n}e^{x_{8n}t} * \sin\omega_g t, \tag{4.33}$$

The magnetic particle velocity can then be obtained from Eq. (4.17):

$$\bar{v}(r,s) = \frac{s + \alpha - \alpha s}{s + Gs + \alpha - \alpha s}\bar{u}(r,s) \tag{4.34}$$

$$v(r,t) = x_{12n}(1 - y_{11n})\left[u(r,t) * e^{y_{12n}t}\right], \quad 0 < \alpha < 1 \tag{4.35}$$

In Eqs. (4.31)–(4.33) and (4.35), $f * g$ represents the convolution product of f and g. The parameters introduced in Eq. (4.35) are:

$$y_{11n} = \frac{1 - \alpha}{G - \alpha + 1}, \quad y_{12n} = \frac{\alpha}{G - \alpha + 1} \tag{4.36}$$

The convolution product of f and g can be calculated as:

$$(f * g)(t) = \int_0^t f(\tau)g(t - \tau)d\tau \tag{4.37}$$

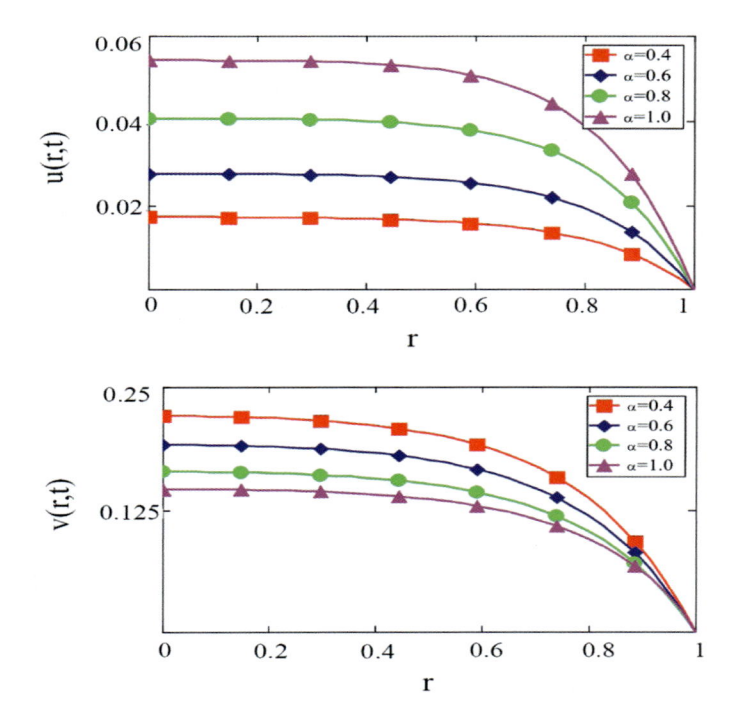

FIGURE 4.4

Axial velocity profiles $u(r,t)$ and $v(r,t)$ for different fractional parameter, α.

4.3 Results and discussions

The effects of magnetic particles and electromagnetic field on the unsteady blood flow through multistenosed porous arteries have been explored in this study. We have provided information on the fractional-order parameter and the other flow properties for blood and magnetic particle distribution by performing mathematical computation via Mathcad. The graphical results are obtained by using the analytical solutions (4.29) and (4.35). The velocity results for several nondimensional parameters, such as fractional parameter α, Casson fluid parameter β, Reynolds number Re, Hartmann number Ha, electric field k, and porosity P, are presented in Figs. (4.2)−(4.9).

Following (Ali et al., 2017; Ali Shah et al., 2016), in the current numerical computation, the following values are set: $A_0 = 0.5, A_1 = 0.6, G = 0.8, R = 0.5, Re = 3, \omega = \frac{\pi}{4}, Ha = 2$ and $\beta = 0.4$. All velocity profiles have been plotted for different fractional parameters and r values. The fractional parameter plays a significant role in controlling the blood velocity. Here, the fractional parameters are set to be $\alpha = 0.4, 0.6, 0.8, 1$. Fractional derivatives explain the memory effects; hence, fractional models can provide adequate information on the complex fluid rheology as compared to ordinary models. The numerical results have been compared against (Ali Shah et al., 2016) as shown in Fig. 4.2. In the current work, the behavior of electromagnetic Casson fluid with magnetic particles passing through multistenosed porous

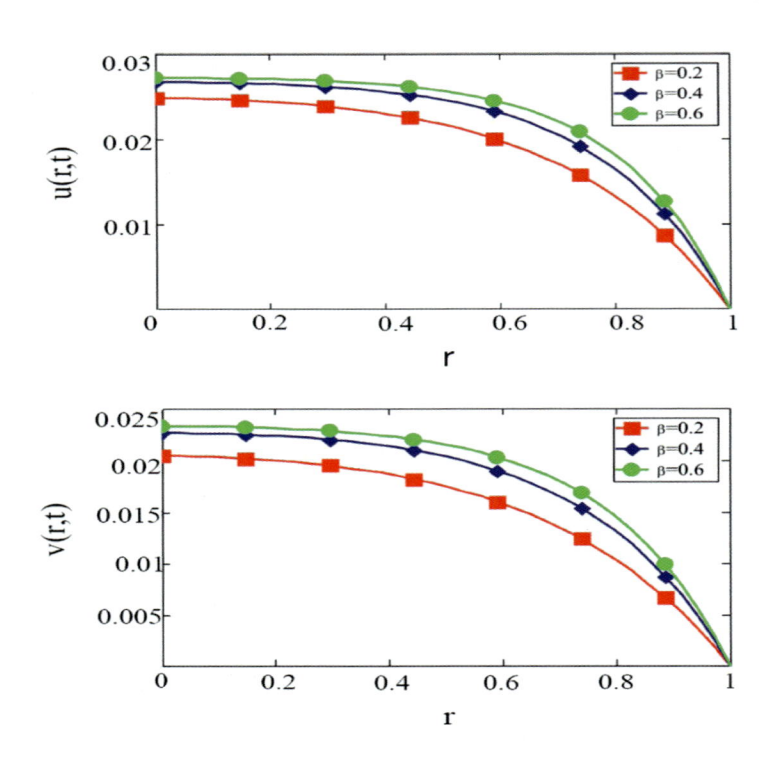

FIGURE 4.5

Axial velocity profiles $u(r,t)$ and $v(r,t)$ for different Casson parameter, β.

arteries is studied analytically using the Caputo–Fabrizio fractional derivative. Meanwhile, Caputo and Fabrizio (2015) simulated the magnetic blood flow in a cylindrical pipe under the influences of magnetic field and oscillating pressure gradient. Their prescribed parameters are: $A_0 = 0.2, A_1 = 0.1, G = 0.8, R = 0.5, \quad Re = 5, \omega = \frac{\pi}{4}, Ha = 2, z = 1,$ and $\beta = 0.25$ for comparison purposes, so they are mutually similar.

Circulatory resistance is a mechanism that regulates the blood flow rate. Fig. 4.3 indicates the effects of severity of the multistenosis artery at different locations: $z = 0.5, 1.5, 3, 5$. It was observed that the flow resistance of blood vessels differs depending on the size of the stenosis. Rising stenotic height will increase the flow resistance as the blood vessel becomes narrower while the blood passes through the area of $r = 0.2$. It is clear that blood circulation is disrupted as the blood flows slower at the narrow stenotic area as compared to the larger part. It is apparent that the fractional parameter alters the velocity profiles. Fig. 4.4 shows the blood and magnetic particle velocities for different fractional parameters, that is, $\alpha = 0.4, 0.6, 0.8$ (for ordinary fluid, $\alpha = 1$). It indicates that the blood represented by the ordinary model flows faster than that by the Casson fluid model with fractional derivatives. This analysis proves the validity of the fractional derivative models. By tuning the fractional parameter, a more physical flow field can be obtained, which is more relevant for a targeted, practical problem.

Fig. 4.5 displays the velocity profiles for various Reynolds numbers (Re). As Re increases, the velocities of blood and magnetic particles increase. The velocity at the central axis changes with

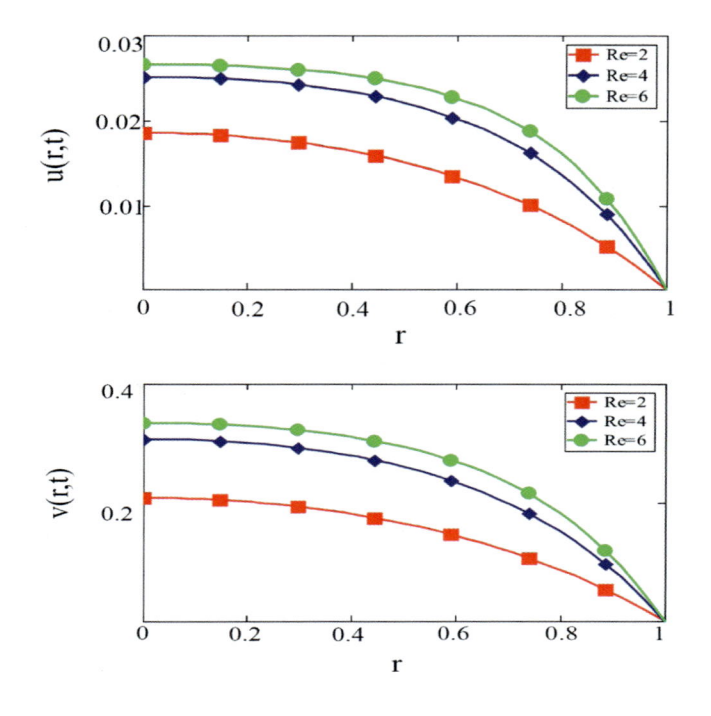

FIGURE 4.6

Axial velocity profiles $u(r,t)$ and $v(r,t)$ for different values of Reynolds number, Re.

respect to the fractional parameter. The non-Newtonian effect (Casson fluid) is more apparent for narrow arteries since higher flow rates are observed. Fig. 4.6 highlights the effect of the Casson fluid parameter on blood flow and magnetic particle motion. The Casson fluid parameter increases with respect to the fluid velocity. The Casson nature is more common in smaller arteries where red blood cells (RBCs) can accumulate near the artery axis (due to rotation), thus forming a cell region. This statement is in perfect agreement with Ali et al. (2017) for a horizontal cylinder. The finding implies that yield stress decreases as β increases and the boundary layer becomes thinner.

The effects of Hartmann number on the blood and magnetic particle velocities are presented in Fig. 4.7. It is apparent that magnetic field decreases the axial flow velocity appreciably. Meanwhile, under the influence of magnetic field, the charged particles would undergo rotational motion. The action of magnetic orientation would induce further the suspension of red blood cells and magnetic particles. This force acts as a resistive drag force that inhibits normal blood flow due to magnetization torque. Hence, the effect of apparent viscosity is more pronounced. The effect of the electric field on the fluid and magnetic particles is reported in Fig. 4.8. It shows that the velocity increases with respect to the strength of electric field. The velocity distributions of blood and particles are significantly affected by the electric field. As observed from the figure, due to the drag force, the flow rate of blood is greater than that of magnetic particles. Increasing the strength of external magnetic field will induce a rapid progression in magnetic particle velocity due to the

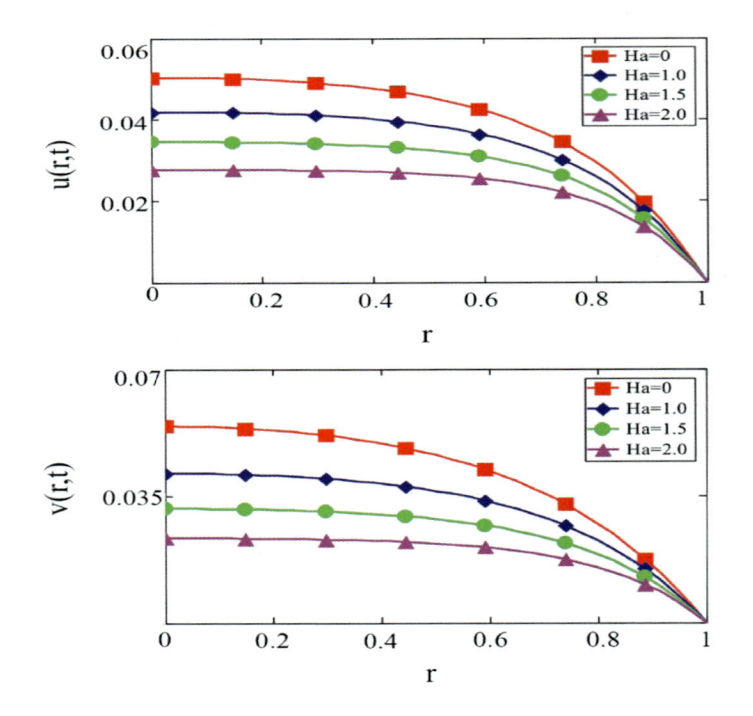

FIGURE 4.7

Axial velocity profiles $u(r,t)$ and $v(r,t)$ for different Hartmann number, Ha.

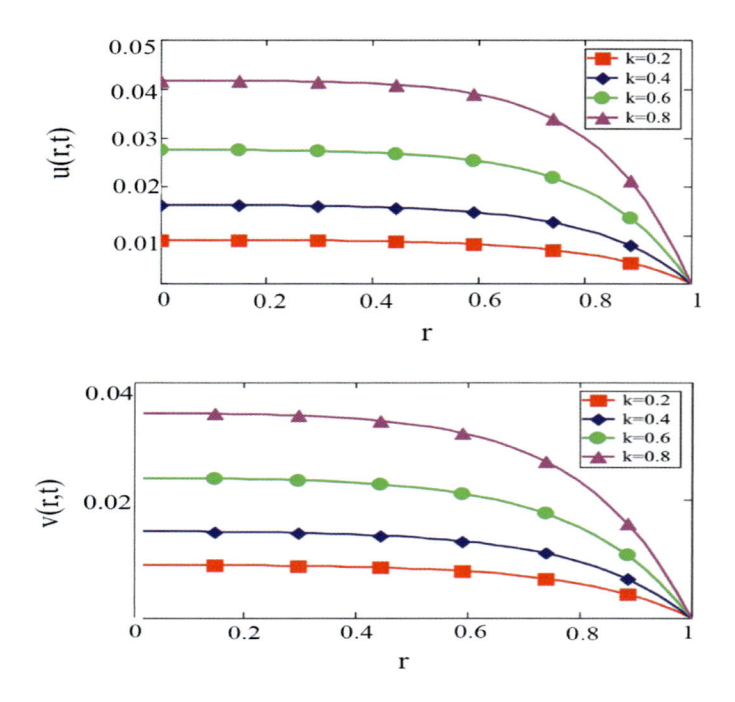

FIGURE 4.8

Axial velocity profiles $u(r,t)$ and $v(r,t)$ for different values of electric field, k.

collisions between the charged particles. Fig. 4.9 illustrates the effects of porosity on blood velocity. It is found that the blood velocity decreases as porosity increases. In this case, the applied magnetic field in the porous artery would affect the blood and magnetic particle distribution.

4.4 Conclusions

The fractional-order blood flow model under the influence of external magnetic and electric fields acting on the non-Newtonian Casson fluid that flows through a multistenosed porous artery has been analyzed. The exact solutions of the Caputo−Fabrizio time fractional derivative have been obtained by using the Laplace and finite Hankel zero-order transformations. Frequently, extracting the ordinary model requires an additional mathematical solution. However, the ordinary model ($\alpha = 1$) for the velocity equation can be obtained directly using the current method since the equation is entirely compliant. Based on the recent changes under the boundary conditions, we have found solutions to address the problem. The blood flow velocity increases with respect to the fractional parameter (blood flow velocity), the Reynolds number, the Casson fluid parameter, and the electric field. However, the blood flow velocity decreases at increasing Hartmann number and porosity. This finding might be useful in the diagnosis and therapeutic treatment of some medical problems.

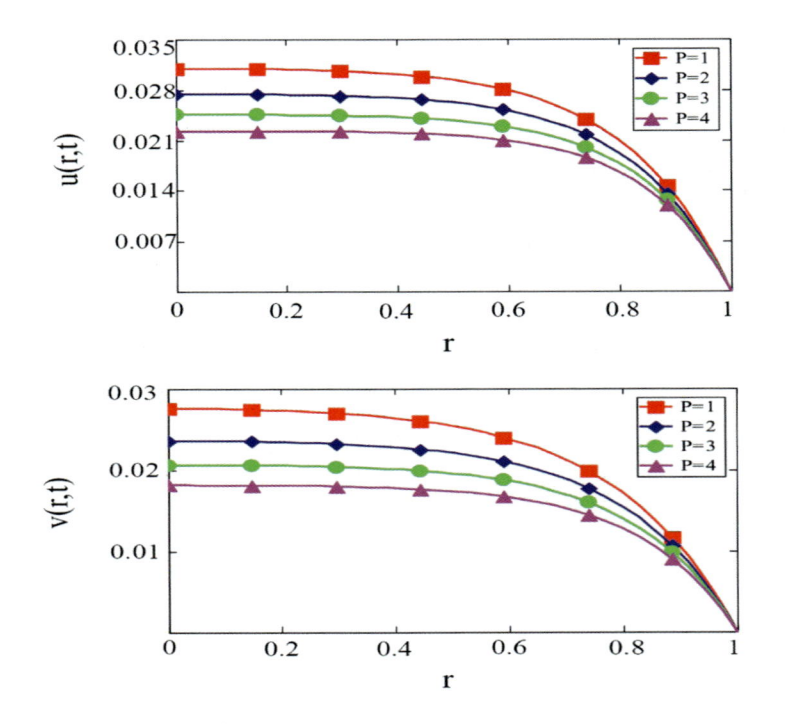

FIGURE 4.9

Axial velocity profiles $u(r,t)$ and $v(r,t)$ for different values of porosity, P.

Acknowledgment

This research was funded by Research Management Centre, Universiti Tun Hussein Onn Malaysia under grant GPPS/H420.

References

Abbas, Z., Shabbir, M. S., & Ali, N. (2018). Numerical study of magnetohydrodynamic pulsatile flow of Sutterby fluid through an inclined overlapping arterial stenosis in the presence of periodic body acceleration. *Results in Physics*, *9*, 753–762.

Abro, K. A., & Gómez-Aguilar, J. F. (2019). A comparison of heat and mass transfer on a Walter's-B fluid via Caputo-Fabrizio versus Atangana-Baleanu fractional derivatives using the Fox-H function. *The European Physical Journal Plus*, *134*(3).

Agarwal, R., & Varshney, N. K. (2016). Pulsatile flow of Herschel-Bulkley fluid through an inclined multiple stenoses artery with periodic body acceleration Raja. *Advanced Applied Science and Research*, *7*(3), 102–113.

Ali, F., Khan,, N., Imtiaz, A., Khan, I., & Sheikh, N. A. (2019). The impact of magnetohydrodynamics and heat transfer on the unsteady flow of Casson fluid in an oscillating cylinder via integral transform: A Caputo—Fabrizio fractional model. *Pramana*, *93*(3), 1—12.

Ali, F., Sheikh, N. A., Khan, I., & Saqib, M. (2017). Magnetic field effect on blood flow of Casson fluid in axisymmetric cylindrical tube: A fractional model. *Journal of Magnetism and Magnetic Materials*, *423*, 327—336.

Ali Shah, N., Vieru, D., & Fetecau, C. (2016). Effects of the fractional order and magnetic field on the blood flow in cylindrical domains. *Journal of Magnetism and Magnetic Materials*, *409*, 10—19.

Bakhti, H., Azrar, L., & Baleanu, D. (2017). Pulsatile blood flow in constricted tapered artery using a variable-order fractional Oldroyd-B model. *Thermal Science*, *21*, 29—40.

Bansi, C. D. K., Tabi, C. B., Motsumi, T. G., & Mohamadou, A. (2018). Fractional blood flow in oscillatory arteries with thermal radiation and magnetic field effects. *Journal of Magnetism and Magnetic Materials*, *456*, 38—45.

Biswas, D., & Paul, M. (2013). Study of blood flow inside an inclined non-uniform stenosed artery. *International Journal of Mathematical Archive*, *2013*(5), 1—10.

Bose, S., & Banerjee, M. (2015). Magnetic particle capture for biomagnetic fluid flow in stenosed aortic bifurcation considering particle-fluid coupling. *Journal of Magnetism and Magnetic Materials*, *385*, 32—46.

Caputo, M., & Fabrizio, M. (2015). A new definition of fractional derivative without singular kernel. *Progress in Fractional Differentiation and Application*, *1*(2), 3—85.

El-Shahed, M. (2003). Pulsatile flow of blood through a stenosed porous medium under periodic body acceleration. *Applied Mathematics and Computation*, *138*(2—3), 479—488.

Gross, J. F., & Aroesty, J. (1972). The mathematics of pulsatile flow in small vessels I. Casson Theory. *Microvascular Research*, *4*, 1—12.

Hatami, M., Hatami, J., & Ganji, D. D. (2014). Computer simulation of MHD blood conveying gold nanoparticles as a third grade non-Newtonian nanofluid in a hollow porous vessel. *Computer Methods and Programs in Biomedicine*, *113*(2), 632—641.

Khan, I., Ali Shah, N., Mahsud, Y., & Vieru, D. (2017). Heat transfer analysis in a Maxwell fluid over an oscillating vertical plate using fractional Caputo-Fabrizio derivatives. *The European Physical Journal Plus*, *132*(4).

Maiti, S., Shaw, S., & Shit, G. C. (2020). Caputo—Fabrizio fractional order model on MHD blood flow with heat and mass transfer through a porous vessel in the presence of thermal radiation. *Physics A: Statistical Mechanics and Its Applications*, *540*.

Majee, S., & Shit, G. C. (2017). Numerical investigation of MHD flow of blood and heat transfer in a stenosed arterial segment. *Journal of Magnetism and Magnetic Materials*, *424*, 137—147.

Mandal, P. K., Chakravarty, S., Mandal, A., & Amin, N. (2007). Effect of body acceleration on unsteady pulsatile flow of non-newtonian fluid through a stenosed artery. *Applied Mathematics and Computation*, *189*(1), 766—779.

Mekheimer, K. S., Haroun, M. H., & Elkot, M. A. (2011). Effects of magnetic field, porosity, and wall properties for anisotropically elastic multi-stenosis arteries on blood flow characteristics. *Applied Mathematics and Mechanics (English (Ed.))*, *32*(8), 1047—1064.

Nagarani, P., & Sarojamma, G. (2008). Effect of body acceleration on pulsatile flow of casson fluid through a mild stenosed artery. *Korea Australia Rheology Journal*, *20*(4), 189—196.

Nagarani, P., Sarojamma, G., & Jayaraman, G. (2006). Exact analysis of unsteady convective diffusion in Casson fluid flow in an annulus—Application to catheterized artery. *Acta Mechanics*, *187*(1—4), 189—202.

Nejad, A. A., Talebi, Z., Cheraghali, D., Shahbani-Zahiri, A., & Norouzi, M. (2018). Pulsatile flow of non-Newtonian blood fluid inside stenosed arteries: Investigating the effects of viscoelastic and elastic walls, arteriosclerosis, and polycythemia diseases. *Computer Methods and Programs in Biomedicine*, *154*, 109—122.

Othman Smadi, S. R., Packirisamy, M., & Stiharu, I. (2006). Modeling of blood flow through multi-stenosis arteries. *IEEE ISIE*, 3400−3403.

Ponalagusamy, R., & Priyadharshini, S. (2017). Nonlinear model on pulsatile flow of blood through a porous bifurcated arterial stenosis in the presence of magnetic field and periodic body acceleration. *Computer Methods and Programs in Biomedicine, 142*, 31−41.

Sankar, D.S., & Hemalatha, K. (2007). A non-Newtonian fluid flow model for blood flow through a catheterized artery—steady flow. *31*, 1847−1864.

Sankar, D. S., & Lee, U. (2011). Nonlinear mathematical analysis for blood flow in a constricted artery under periodic body acceleration. *Communications in Nonlinear Science and Numerical Simulation, 16*(11), 4390−4402.

Shah, N. A., & Khan, I. (2016). Heat transfer analysis in a second grade fluid over and oscillating vertical plate using fractional Caputo−Fabrizio derivatives. *European Physical Journal C, 76*(7).

Shah, N. A., Ahmed, N., Elnaqeeb, T., & Rashidi, M. M. (2019). Magnetohydrodynamic free convection flows with thermal memory over a moving vertical plate in porous medium. *Journal of Applied and Computational Mechanics, 5*(1), 150−161.

Sharma, M. K., Singh, K., & Bansal, S. (2014). Pulsatile MHD flow in an inclined catheterized stenosed artery with slip on the wall. *Journal of Biomedical Science and Engineering, 07*(04), 194−207.

Shit, M. R. G. C. (2011). Pulsatile flow and heat transfer of a magneto-micropolar fluid through a stenosed artery under the influence of body acceleration Introduction. *Journal of Mechanics in Medicine and Biology, 11*(3), 643−661.

Siddiqui, S. U., Shah, S. R., & Geeta. (2015). A biomechanical approach to study the effect of body acceleration and slip velocity through stenotic artery. *Applied Mathematics and Computation, 261*, 148−155.

Singh, S., & Shah, R. (2010). A numerical model for the effect of stenosis shape on blood flow through an artery using power-law fluid. *Advanced Applied Science and Research, 1*(1), 66−73.

Srivastava, N. (2014). The Casson fluid model for blood flow through an inclined tapered artery of an accelerated body in the presence of magnetic field. *International Journal of Biomedical, Engineering and Technology, 15*(3), 198−210.

Tashtoush, B., & Magableh, A. (2008). Magnetic field effect on heat transfer and fluid flow characteristics of blood flow in multi-stenosis arteries. *Heat Mass Transfer, 44*(3), 297−304.

Tzirtzilakis, E. E. (2005). A mathematical model for blood flow in magnetic field. *Physics of Fluids, 17*(7), 1−15.

Uddin, S., Mohamad, M., Sufahani, S., Kamardan, M. G., Mehmood, O. U., Wahid, F., & Roslan, R. (2018). Application of Caputo-Fabrizio fractional order derivative (NFDt) in simulating the MHD flow of the third grade non-Newtonian fluid in the porous artery. *International Journal of Engineering and Technology, 7*(4.30), 527.

Zaman, A., Ali, N., & Sajid, M. (2018). Numerical simulation of pulsatile flow of blood in a porous-saturated overlapping stenosed artery. *Mathematics and Computers in Simulation, 134*, 1−6.

Zaman, A., Ali, N., & Sajid, M. (2016). Slip effects on unsteady non-Newtonian blood flow through an inclined catheterized overlapping stenotic artery. *AIP Advances, 6*(1), 015118.

Mathematical modeling to the motion control of magnetic nano/ microrobotic tools performing in bodily fluids, especially blood/plasma

Ahmet Fatih Tabak

Mechatronics Engineering Department, Kadir Has University, Istanbul, Turkey

5.1 Introduction

A myriad of scientific investigations has been presented in the literature focusing on the use of biomedical microrobots, especially untethered swimmers, for therapeutic purposes for the last six decades. Microrobotic systems have come a long way since the conception of the idea (Andhari et al., 2020; Campo-Deaño, 2016; Ceylan et al., 2017; Ceylan, Yasa, Kilic, et al., 2019; Field et al., 2019; Ghosh et al., 2020; Hu et al., 2020; Kim & Tung, 2015; Leaman et al., 2020; Palagi & Fischer, 2018; Schwarz, Medina-Sánchez, et al., 2020; Ullrich et al., 2014; Wang & Dong, Wu, et al., 2020; Wang & Zhang, 2020; Xu et al., 2015b). However, we are yet to see demonstration of a successful clinical scenario. Success being (1) completion of the given therapeutic task within an acceptable limit of statistical failure; (2) removal of the robots from the tissue one way or the other, for example, retrieval from the tissue or biodegradation via biochemical effects; and (3) complex organisms, that is, mammals, being still alive well after the procedure without showing signs of toxicity, that is, inflammation, allergy, or poisoning (Park & Park, 1996; Remes & Williams, 1992). The aforementioned set of condition, although sounds very strict, is the inescapable prerequisites for robotic devices—particularly to avoid further complications when in contact with blood due to biocompatibility issues (Park & Park, 1996; Remes & Williams, 1992). The task at hand is quite ambitious and will be equally rewarding when a successful microrobotic system is put to clinical use.

One of the main issues with robotic systems, regarding the biomedical applications, is their expected time-dependent behavior. Time-averaged modeling or analysis will fall short as a robotic system should be controllable in real-time, whether fully- or semiautonomous. The models in the literature concerning the microswimmers, for example, bacteria, artificial swimmers, and biohybrid robots, often tend towards the time-averaged behavior assumptions but this does not always address the expectations of the said control effort. A similar problem arises with Stokes flow assumptions with which inertial forces within the flow are omitted, and this has been demonstrated to be a very tricky statement (Tabak, 2018; Tabak, 2019a; Wang & Ardekani, 2012) that needs to be checked to begin with. In other words, a mobile robot must maneuver with an acceptable limit of error regarding the accuracy required by the task and surroundings. To that effect, modeling for other applications of microbiology and modeling for robotics are inherently different as the former benefits from time-averaged analysis more often, whereas the latter requires time-dependent predictions for real-time operation. This modeling constraint leads to several difficulties as the microrobotic systems have the crucial disadvantage of lack of space for sensors in contrast to the larger-scale systems. Any robotic system, as we human beings also do, depends on the sensory information to perceive their surroundings. Furthermore, a robotic system has to decide on the next step based on these readings according to the control law and mission parameters. Microrobots, especially artificial ones, depend on an external computer for intelligent decision-making.

However, robotic-sensing is a different problem: An untethered microrobotic tool swimming in the spinal column (Fountain et al., 2010) cannot carry all the sensors nor could any arbitrary sensor technology be employed to measure the states, for example, position, velocity, proximity, etc. Limits on the overall volume, energy supply, and adverse environmental conditions make it imperative to employ apt models to predict the states of the system as accurately as possible, online or offline, to render the overall system as reliable as possible. Furthermore, an online analysis might be possible, although in a limited capacity, as the arterial network can be numerically reconstructed in part after proper medical scans (Anor et al., 2010). Method and/or path of retrieval of the microrobotic tools should also be considered carefully before the in vivo operation (Iacovacci et al., 2019). However, the entire circulatory system in the targeted tissue or along the intended gait cannot be simply reconstructed (Anor et al., 2010) and if the robot is lost in the tissue without the chance of retrieval, this might lead to not just a failure but the burden of possible clinical consequences (Park & Park, 1996; Remes & Williams, 1992; Yasa et al., 2020).

On top of the outlined issues, the physical complexity of the environment, that is, pulsating flow field (Arcese et al., 2012; Belharet et al., 2012; Sadelli et al., 2017; Wen et al., 2015), non-Newtonian and viscoelastic characteristics of the blood (Arcese et al., 2012; Wen et al., 2015), possible turbulence effects (Al-Azawy et al., 2016; Rahman et al., 2018), and immune system reactions as a defense mechanism against foreign agents introduced to live tissue (Park & Park, 1996), require the utmost attention to detail to build better observers for control purposes. To put it mildly, a medical doctor will use such a device only after it is proven to work within acceptable limits of medical risk. Thus, without a sound clinical scenario, the research on therapeutic microrobotics will always be short on one important aspect for the realization of an actual "swallow-able doctor" (Feynman, 1992) in the foreseeable future.

What follows next is a short but comprehensive discussion of the state-of-the-art for micro-/nano-scale artificial and biohybrid robotic systems. It is followed by the mathematical modeling specifically focusing on bacterial microrobotic systems that might come into contact with endothelial tissue (Arcese et al., 2012), blood plasma (Dutta & Tarbell, 1996), red blood cells (RBCs) (Sriram et al., 2014), and blood clots (Khalil et al., 2017; Khalil et al., 2018), along with a discussion on the accurate prediction of environmental conditions; therefore, covering a wide range of physical phenomena related with cruising in the circulatory system and mechanical grinding of plugs when need be. Finally, a brief introduction on a possible motion control scheme is given for the sake of completeness of the subject matter covered in this chapter. The presented model is strictly limited to the helical microrobot, artificial and biohybrid, dynamics, and deals with the environment in detail only when it directly contributes to the equation of motion (5.1), which will be more than once. The reader shall find more comprehensive information and technical details on diverse aspects of biomedical microrobotics in the references provided in this chapter, and in (Tabak, 2019b).

5.2 **State of the art**

There are exclusive examples that will showcase how microrobots are tailored for specific tasks. Most of the systems proposed are magnetically actuated as the method has an apparent advantage over other methods, such as chemical, optic, and acoustic actuation, imaging, and control (Ahmed et al., 2017; Caldag & Yesilyurt, 2020; Khalil et al., 2018; Park et al., 2021; Sitti & Wiersma, 2020; Wang & Zhang, 2020; Xu et al., 2015a; Yan et al., 2020). They, have not been established as an efficient standalone method unlike magnetic actuation. On the other hand, the control capabilities of all the aforementioned methods are mostly limited when it comes to operating deep in living tissue due to either transmittance, permeability, impedance, or biocompatibility. Although, magnetic actuation has a practical advantage compared to optical actuation as it can penetrate much deeper in the tissue (Sitti & Wiersma, 2020). Furthermore, acoustic fields make it possible to control and limit the dispersion of a magnetic microrobot swarm in a certain volume of interests, which cannot be achieved by pure magnetic actuation (Keya et al., 2018; Xu et al., 2015a). Therefore, a single method might not be the answer for the application; however, the reader will find that magnetic actuation and motion control is widely preferred in the literature.

One specific task, envisioned for the magnetic microrobots, is the delivery of live cells as cargo (Medina-Sánchez et al., 2016; Schwarz, Karnaushenko, et al., 2020; Yasa et al., 2019) in bodily fluids. In these systems, the cells are being actively carried rather than being used to harness propulsion. In addition, there are studies using or proposing to use live cells as an integral part of the robotic tool, resulting in cybernetic microsystems (Alapan et al., 2018; Felfoul et al., 2016; Tabak, 2020a) designed for biomedical applications. In such systems, the live cell propels itself as a semiautonomous component while being magnetically coerced to swim in a specific direction via modulated magnetic fields. The selected bacteria in these studies are either of magnetotactic bacteria species, that is, *Magnetococcus marinus* and *Magnetospirillum gryphiswaldense* (Felfoul et al., 2016; Tabak, 2020a), or modified to accommodate magnetic properties later on, that is, *Escherichia coli* (Alapan et al., 2018); all utilizing helical wave propagation to harness forward thrust using the viscous shear (Brennen & Winet, 1977). There are microrobotic agents designed for traversing the bloodstream and these designs are entirely dominated by either bacteria-inspired propulsion or biohybrid systems: a great majority of the systems are focused on controlling the position of nanoparticles, while only a few make use of Janus particles, helical waves, or live cells, that is, different locomotion modes (Abu-Hamdeh et al., 2020; Alapan et al., 2020; Arcese et al., 2012; Belharet et al., 2012; Campo-Deaño, 2016; Haghdel et al., 2017; Iacovacci et al., 2019; Khalil et al., 2017; Khalil et al., 2018; Sadelli et al., 2017; Wu et al., 2020; Xie et al., 2020; Xu et al., 2020). Some of these designs aim to achieve controlled drug delivery (Abu-Hamdeh et al., 2020; Alapan et al., 2020; Arcese et al., 2012; Haghdel et al., 2017) while a few are focused on more elaborate tasks such as thrombolysis and clot grinding (Khalil et al., 2017; Khalil et al., 2018; Xie et al., 2020).

There are also some studies focused on operating in synovial joints (Go et al., 2020; Tabak, 2020a), where blood vessels are not present, thus there is no flow field expected if the patient is stationary and non-Newtonian behavior of the liquid does not depend on the presence of moving cells in the medium (Arzani, 2018; Bessonov et al., 2016).

Electromagnetic fields can be generated and articulated with stationary but rotating permanent magnets, stationary electromagnetic coils, moving electromagnetic coils, or moving and rotating permanent magnets (Akçura et al., 2018; Du et al., 2020; Manamanchaiyaporn et al., 2020; Pham et al., 2020; Pittiglio et al., 2020; Ryan & Diller, 2017; Salmanipour et al., 2021; Tabak, 2020a; Yang et al., 2020). Moving coils or moving permanents magnets provide certain flexibility to the workspace and dexterity to the actuation system as following the microrobot helps keeping it within the volume of interest while cruising relatively long distances. The said mobility is achieved via robotic arms of different kinematic configurations, that is, single, dual, or multiple kinematic chains (Akçura et al., 2018; Du et al., 2020; Pittiglio et al., 2020; Tabak, 2020a), or a delta robot (Yang et al., 2020). Design of magnetic or electromagnetic actuation involves modeling and optimization of the field and field gradients. Although this particular subject is not in the scope of this chapter, the reader may find great detail in the references provided in this short discussion and in the suggested readings of (Abbott et al., 2007; Abu-Hamdeh et al., 2020; Arcese et al., 2012; Belharet et al., 2012; Nguyen et al., 2020; Peyer et al., 2012; Sadelli et al., 2017; Schamel et al., 2014; Tabak et al., 2011; Xie et al., 2020; Xu et al., 2015b; Xu et al., 2020), some of which will be very briefly revisited for the discussion on micromotion control in the next part.

To this date, only a handful of pre-clinical in vivo uses for microrobots have been envisioned with tangible numerical and physical results. There are a few animal tests (Go et al., 2020; Servant et al., 2015) with swarms, along with a few scenarios involving rather accessible locations in a complex organism such as the joints in the musculoskeletal system, that is, synovial joints (Servant et al., 2015; Tabak, 2020a), and the eyeball, that is, intraocular medium (Kummer et al., 2010; Ullrich et al., 2014). Accessibility is a serious problem for visualization, tracking, control, and energy supply (Tabak, 2019b); therefore, it is not surprising that immediate attention is on the more exposed tissue. On the other hand, the holy grail of the biomedical microrobotic research is to perform successful therapeutic operations in the bloodstream (Abu-Hamdeh et al., 2020; Alapan et al., 2020; Arcese et al., 2012; Belharet et al., 2012; Campo-Deaño, 2016; Haghdel et al., 2017; Iacovacci et al., 2019; Khalil et al., 2017; Khalil et al., 2018; Sadelli et al., 2017; Wu et al., 2020; Xie et al., 2020; Xu et al., 2020), specifically in deep live tissue. Two of these studies (Khalil et al., 2017; Khalil et al., 2018), have employed magnetic helical robots to grind actual blood clots obtained from human subjects, and a novel modeling approach was built and demonstrated by the author of this chapter that was used to predict and control the removal rate. Said approach will be discussed to a certain extent later on.

The next part deals with the robotic model and modeling the environmental conditions pertinent to the circulatory system; as in the flow field characteristics and the rheological behavior of the liquid itself for they are affecting the swimming behavior of microrobots greatly. A series of discussions on the crucial points for controlling individual microrobots and swarms alike in the circulatory system will follow. Furthermore, the model will incorporate effects, that is, forces and torques, during specific operations such as unloading cargo or drilling. The reader will find out that although the model and equations here are based on deterministic observations and calculations, stochastic behavior is also not unexpected due to several key environmental factors (Tabak, 2020a). This concludes the brief introduction to the state-of-the-art; however, various additional studies will be referred to in the modeling section to enrich the discussions on the prediction and control of the rigid-body motion.

5.3 **Swimming in the circulatory system**

5.3.1 **Robotic model**

Before proceeding any further, the reader should appreciate that the result of any model representing a physical construct is dependent on the scenario, that is, boundary and initial conditions. The trouble with modeling is that all plausible effects should be incorporated at the beginning. Furthermore, a hypothetical scenario can serve as a convenient test case, however, it could be impractical due to physical limitations. Thus the model should reflect real conditions to the best of the computational abilities. Prediction of controlled experiments with biorobotic systems is quite tough as the state of living tissue is not easy to either control or guess accurately. However, there is always a tradeoff between the practicality and usefulness of a mathematical model; the computational effort must be modest enough to obtain the results without shadowing the benefit of accuracy, but on the other hand, it cannot be simpler than it should be to reflect the robotic performance and execution of the motion control, properly. Therefore, one is not to either dismiss a physical phenomenon lightly or to include everything indiscriminately, without first investigating the relative significance (Webb, 2006).

To further elaborate, a medical microsystem cannot be allowed to induce thrombosis or immunogenic reactions for sake of the well-being of the patient (Park & Park, 1996). It has been shown that the design of the microrobot directly affects its interaction with the immune cells (Yasa et al., 2020). Also, the microrobot should be properly retracted, if it is not going to degrade via natural means, to minimize the exposure and avoid complications (Iacovacci et al., 2019). The problem with the circulatory system is that it has been established that modeling blood flow and rigid-body motion of an active agent in the bloodstream poses a highly patient-specific problem (Anor et al., 2010; Arzani, 2018; Chakravarty & Sannigrahi, 1999; Goubergrits et al., 2008; Perdikaris et al., 2016; Sriram et al., 2014). Thus one can easily argue that a handful of experiments and simulations simply cannot represent the whole picture. From here on, the reader will find a very comprehensive discussion on the equation of motion, Eq. (5.1), for a microrobotic agent; however, one should keep this discussion in mind at all times.

The swimming microrobot is composed of a rigid body constituting the cargo to be towed and at least one helical tail actuated by an externally controlled rotating magnetic field (Khalil et al., 2017) or the bacterial motor of a single-celled organism (Tabak, 2020a). Regardless, the equation of motion, Eq. (5.1) can be given in the following form including the effects of propulsion, Eq. (5.2), Stokesian-drag, Eq. (5.3), contact, Eq. (5.4), friction, Eq. (5.5), electromagnetic stimulus, Eq. (5.6), electrostatic effect, Eq. (5.7), gravity and buoyancy, Eq. (5.8), the transient behavior of the induced flow field in the immediate vicinity, Eq. (5.9), swarm interaction, Eq. (5.10), the steric force of the inner surfaces of the blood vessels, Eq. (5.11), acoustic waves, Eq. (5.12), and the fretting of blood clots, Eq. (5.13), respectively:

$$\begin{bmatrix} m\dfrac{d^2x}{dt^2} \\ J\dfrac{d^2\theta}{dt^2} \end{bmatrix} = \begin{bmatrix} \sum(F_p) + F_d + F_c + F_f + F_m + F_e + F_g + F_t + F_s + F_{st} + F_a + F_{fr} \\ \sum(T_p) + T_d + T_c + T_f + T_m + T_e + T_g + T_t + T_s + T_{st} + T_a + T_{fr} \end{bmatrix} \tag{5.1}$$

for force and torque components all expressed in the frame of reference of the microrobot. Here, m (kg) is the mass of the microrobot, and J (kg·m^2) denotes the mass moment of the inertia matrix.

All the terms on the right-hand side will be treated as time-dependent and they will be briefly described in the given order. Furthermore, the mass moment of inertia can also be time-dependent due to the rotation and deformation of the tails. The net force and torque acting on the microrobot are usually assumed to add up to zero to satisfy acceleration-free swimming conditions under over-whelming viscosity (Sir Taylor, 1951); however, this approach does not cover what happens to the motion when the flow regime becomes transient rather than viscous for the cargo being towed. One can still omit the rigid-body accelerations of the microrobot on the left-hand side, that is, d^2x/dt^2 (m/s^2) and $d^2\theta/dt^2$ (rad/s^2), based on the argument of numerical accuracy since all components will be of approximations to a degree. It will be possible to obtain a higher resolution with more elaborate modeling techniques such as the finite element method (Abu-Hamdeh et al., 2020; Tabak, 2018); however, the computational requirement renders such an approach unsuitable for real-time applications. Since it will be evident that several components of the equation of motion, Eq. (5.1), are not independent of each other, the problem is quite convoluted, and removing the accelerations will reduce the stiffness of the numerical problem at the discretion of the reader.

The first term on the right-hand side of the equation of motion corresponds to the propulsive effect of the rotating helical tails, $F_p(t)$ (N) and $T_p(t)$ (N·m). The net propulsion force and torque are predicted with:

$$\sum \left(\begin{bmatrix} F_p(t) \\ T_p(t) \end{bmatrix} \right) = \sum_n \left(R_t(t)G(t) \begin{bmatrix} [0 \ 0 \ 0]^T \\ \omega_x \\ 0 \\ 0 \end{bmatrix} \right)_i \qquad (5.2)$$

for n number of independent tails. Here, $G(t)$ denotes the 6-by-6 resistance matrix of a single helical tail. The resistance matrix can contain five distinct information about the helical tails: (1) hydrodynamic interaction caused by nearby boundaries (Brennen & Winet, 1977; Khalil et al., 2017; Lauga et al., 2006); (2) the hydrodynamic interaction with other helical structures (Kim & Power, 2004; Reichert & Stark, 2005); (3) geometry of the helical tail at any given moment (Brennen & Winet, 1977; Keller & Rubinow, 1976); (4) progression of deformation if it is elastic (Khalil et al., 2018); and (5) hydrodynamic impedance for steady periodic effects (Tabak & Yesilyurt, 2014; Tabak, 2018). Furthermore, the resistance matrix, $G(t)$, projects each of the six degrees of freedom (6-dof for short) rigid-body motion onto one another (Keller & Rubinow, 1976; Tabak, 2018), resulting in propulsive force and torque components in all directions. The term $R_t(t)$ is the rotation matrix between the frame of reference of a single helical tail and the inertial frame of the microrobot based on the orientation of the tail with respect to the effective center of mass on which the inertial frame resides. Furthermore, the chirality of the helical tail will be either right-handed or left-handed with constant or variable wave properties, that is, amplitude and wavelength, depending on the initial design of the microrobot (Khalil et al., 2018; Tabak & Yesilyurt, 2014). Finally, the term ω_x (rad/s) denotes the angular velocity of a single helical tail along its long axis.

The terms $F_d(t)$ (N) and $T_d(t)$ (N · m) are the drag force and drag torque components on the entire microrobot opposing its 6-dof rigid-body motion. They are given as:

$$\begin{bmatrix} F_d(t) \\ T_d(t) \end{bmatrix} = K(t) \begin{bmatrix} U(t) - U_{bf}(t) \\ \Omega(t) - \Omega_{bf}(t) \end{bmatrix} \qquad (5.3)$$

where $K(t)$ is the 6-by-6 total resistance matrix of the microrobot. Akin to the matrix $G(t)$ in Eq. (5.2), the matrix $K(t)$ might contain six distinct information: (1) hydrodynamic interaction caused by nearby boundaries (Brennen & Winet, 1977; Higdon & Muldowney, 1995; Khalil et al., 2017; Lauga et al., 2006); (2) the hydrodynamic interaction with other structures (Goldfriend et al., 2015; Kim & Power, 2004; Reichert & Stark, 2005); (3) geometry of the helical tail (and helical body (Khalil et al., 2017; Tabak, 2020a) in some cases) at any given moment (Brennen & Winet, 1977; Cortez et al., 2005; Keller & Rubinow, 1976; Samsami et al., 2020); (4) progression of deformation for elastic components (Khalil et al., 2018); (5) hydrodynamic impedance for steady periodic effects (Tabak & Yesilyurt, 2014; Tabak, 2018); and (6) effective resistance due to Brownian noise (Kim, 1985; Marath & Wettlaufer, 2019). The vectors $U(t)$ and $\Omega(t)$ signify the six-degrees-of-freedom rigid-body motion of the microrobot whereas the vectors $U_{bf}(t)$ and $\Omega_{bf}(t)$ signify the local flow field of the bloodstream.

The microrobot will come into contact with red blood cells (hematocrit—erythrocytes), white blood cells (leukocytes), platelets, proteins, and endothelial cells during the time of flight along with the circulatory system (Bessonov et al., 2016; Dutta & Tarbell, 1996; Krüger-Genge et al., 2019; Sriram et al., 2014; Wen et al., 2015). These encounters will be of sporadic bumps or deliberate contacts for cargo delivery (Arcese et al., 2012; Xu et al., 2020). The contact force can be modeled using the penalty method (Spong & Vidyasagar, 1989) for sporadic bumps that can also have a friction component (Tabak, 2019b), $F_f(t)$ (N), along with a loading/unloading force computation for deliberate operations where no friction can be assumed (Arcese et al., 2012). The contact force vector is parallel to the normal vector, $n_c(t)$, at the point of contact. The resultant terms, $F_c(t)$ (N) and $T_c(t)$ (N·m), which are calculated with respect to the center of mass, denote the contact force and contact torque.

$$\begin{bmatrix} F_c(t) \\ T_c(t) \end{bmatrix} = \begin{bmatrix} kl(t)n_c(t) + b(t)(U(t) + \Omega(t) \times p(t)) \cdot n_c(t) \\ p(t) \times F_c(t) \end{bmatrix} \qquad (5.4)$$

where $p(t)$ (m) is the location vector of the point of contact at the boundary of the microrobot, k (N/m) is the spring constant facilitating the bouncing-back effect, $l(t)$ (m) is the effective penetration depth at any given instance, and $b(t)$ (N·s/m) is the conditional damper constant standing for the viscous energy dissipation of the contact, incorporating adhesive surfaces and van der Waal forces (Arcese et al., 2012; Sadelli et al., 2017) into the equation when surface chemistry results in a sticky contact condition. Thus the contact-specific constant $b(t)$ (N·s/m) requires careful modeling in itself to lump all the aforementioned effects. It is also important to note that the microrobot might not always escape the endothelial layer owing to the presence of hydrodynamic interactions (Kim & Power, 2004). However, there will be a repulsive force known as the steric force, $F_{st}(t)$ (N), between functionalized surfaces of the microrobot and the endothelial cells of the blood vessels (Arcese et al., 2012); therefore, the overall contact phenomenon is expected to be quite complex.

The nonabrasive friction force and torque at the point of contact, $F_f(t)$ (N) and $T_f(t)$ (N·m), should occur if there is a motion present perpendicular to the surface normal at the point of contact, that is, along with the local tangent axis $t_c(t)$ (Tabak, 2019b), as follows:

$$\begin{bmatrix} F_f(t) \\ T_f(t) \end{bmatrix} = \begin{bmatrix} v|F_c(t)|t_c(t) \\ p(t) \times F_f(t) \end{bmatrix} \qquad (5.5)$$

where v denotes the friction coefficient at the contact. The friction force should be in the opposite direction to the tangential projection of $U(t)$ (m/s) and this is guaranteed by $t_c(t)$, which can be calculated and updated at each time-step of the numerical study following these conditions once the vectors $U(t)$ (m/s) and $n_c(t)$ are properly determined. The contact and the friction forces, $F_c(t)$ (N) and $F_f(t)$ (N), are not very easy to predict as the sporadic contact with red blood cells is a highly nonlinear phenomenon due to the expectancy of asymmetric (Arzani, 2018) and multiphase local flow profile (Wen et al., 2015), that is, a non-Newtonian core with a Newtonian behavior near the blood vessel boundaries. It might be prudent to lump the effect of red blood cells into the non-Newtonian viscous modeling, that is, the thixotropy of blood (Dutta & Tarbell, 1996), while the resistance matrices are being constructed. This approach will be discussed at length in the next part.

The main driving force for artificial microrobots and the steering control for biohybrid micro-swimmers are based on the electromagnetic field generated and controlled by a decision loop. The gradient of the field results in a force, $F_m(t)$ (N) dragging the center of the magnetic volume of the body in blood, while rotation of the field lines results in magnetic torque, $T_m(t)$ (N·m), coercing the microrobot to align accordingly. Given the magnetic field density, $B(t)$ (T), and the total magnetization vector for the microrobot, M (A·m^2), the aforementioned vectors can be modeled as (Honda et al., 1996; Zhang et al., 2009):

$$\begin{bmatrix} F_m(t) \\ T_m(t) \end{bmatrix} = \begin{bmatrix} (M \cdot \nabla)B(t) \\ M \times B(t) \end{bmatrix} \tag{5.6}$$

and if the center of magnetic volume and center of mass of the microrobot happen to be apart from each other, a further torque correction is needed. It is important to note that, the field, $H(t) = B(t)/\mu_m$ (A/m), must be carefully modeled and tuned for optimal actuation performance, whereas magnetization vector M must be measured to modulate the amplitude of $B(t)$ (T) to achieve required field strength. Here, μ_m (H/m) is the magnetic permeability of the environment in which the microrobot will perform.

Electrostatic effects owing to the local charge difference, q (C), between the surfaces of the microrobot and the to a surface at a distance of $h_c(t)$ (m) are denoted by $F_e(t)$ (N) and $T_e(t)$ (N·m) and given as (Sadelli et al., 2017):

$$\begin{bmatrix} F_e(t) \\ T_e(t) \end{bmatrix} = \begin{bmatrix} \dfrac{q^2}{4\pi\varepsilon h_c^2(t)} n_{\text{wall}}(t) \\ p(t) \times F_e(t) \end{bmatrix} \tag{5.7}$$

In this equation, ε (F/m) is the local permittivity of blood. Here, it is possible to speculate that the presence of local ion concentration is expected to induce a net force on the microrobot. However, the said effect will not be easy to detect long before it causes any difference in the gait, if it is not tailored and induced deliberately.

Another nontrivial stimulus on the microrobots is gravitational attraction. In the equation of motion, it is often omitted with the assumption that the active agent is neutrally buoyant, which is very hard to achieve. There are bacteria species known to exploit their effective mass, m_{eff} (kg), to traverse along the axis of gravity (Li & Cannon, 1998):

$$\begin{bmatrix} F_g(t) \\ T_g(t) \end{bmatrix} = \begin{bmatrix} m_{\text{eff}} g(t) \\ p_{\text{cov}}(t) \times F_g(t) \end{bmatrix} \tag{5.8}$$

Here, $p_{\text{cov}}(t)$ (m) gives the position of the center of volume with respect to the center of mass to account for inhomogeneities. The vector $g(t)$ (m/s^2) is the time-dependent gravitational attraction vector felt in the frame of the microrobot (Tabak, 2019b).

Inertial effects in micro realm are not limited to gravitational attraction. If the motion of the active agent is proved to be oscillatory or the flow field in which the agent is moving has an oscillatory component, it is well described that acceleration will present itself as a nonlinear effect comparable to the Stokes force in magnitude, in the form of Basset history integral and added mass (Tabak, 2019b; Wang & Ardekani, 2012). The added mass, the second term in the upper right-hand side in the following equation has also a different interpretation in some of the studies and is referred to as the virtual mass (Haghdel et al., 2017; Wen et al., 2015).

$$\begin{bmatrix} F_t(t) \\ T_t(t) \end{bmatrix} = \begin{bmatrix} 6R^2\sqrt{\pi\mu\rho} \int\limits_{-\infty}^{t} \dfrac{\partial U(t)}{\partial t} \dfrac{d\tau}{\sqrt{t-\tau}} - \dfrac{2}{3}\pi R^3 \rho \dfrac{\partial U(t)}{\partial t} \\ p_{\text{body}}(t) \times F_t(t) \end{bmatrix} \tag{5.9}$$

with $p_{body}(t)$ (m) is the position of the center of mass of the cargo with respect to the overall center of mass. The term R here denotes the hydrodynamic radius of the cargo with μ (Pa · s) and ρ (kg/m^3) being viscosity and density of the blood, respectively. In the next part, the reader will be introduced to the modeling approaches used to predict the viscosity of blood. It is very important to note that, if the viscous resistance matrix for the cargo or body of the microrobot, $K(t)$, is built in a way to account for the unsteady effect represented by $F_t(t)$ (N) and $T_t(t)$ (N · m) as given in (Tabak, 2018; Tabak, 2019a), these two components should be omitted in the equation of motion, Eq. (5.1).

The swarm interaction, $F_s(t)$ (N) and $T_s(t)$ (N·m), is the repulsive effect between two swimming microrobots due to hydrodynamic interaction, and should be included in the equation of motion assuming that it is not already accounted for in the resistance matrix, $K(t)$. The repulsive force is the result of hydrodynamic interaction between two active agents in motion at close proximity. The repulsive force is only one of the reasons why a swarm disperses over time. The force acts on two microrobots in the shortest distance and is disruptive for swimming in close formation (Ishikawa et al., 2007) due to the three-dimensional flow field around the rotating helical tails (Tabak, 2018). Although the equation provided here is empirically expressed for two bacteria of spherical cell and helical tail (Ishikawa et al., 2007), the said effect can be also modeled for swarms of microrobots of arbitrary shapes (Goldfriend et al., 2015; Kim & Power, 2004; Kim, 1985; Marath & Wettlaufer, 2019):

$$\begin{bmatrix} F_s(t) \\ T_s(t) \end{bmatrix} = \begin{bmatrix} \dfrac{100e^{-100h_p(t)}}{1 - e^{-100h_p(t)}} d_p(t) \\ p_d(t) \times F_s(t) \end{bmatrix} \tag{5.10}$$

with $h_p(t)$ (m) being the proximity between the opposite surfaces, the vector $d_p(t)$ denoting the direction along with the closest points on the opposite surfaces, and $p_d(t)$ (m) giving the location of the closest point on the surface with respect to the center of mass, of the microrobots. It should be noted that the reported coefficients in this equation were given for a nondiluted suspension of a single bacterium species (Ishikawa et al., 2007). It should also be noted that the interaction between the active agents in suspensions results in turbulence (Dunkel et al., 2013).

A similar interaction-based force is the steric force, i.e., $F_{st}(t)$ (N) and $T_{st}(t)$ (N·m). Although the bacteria are known to be trapped on the walls of channels (Kim & Power, 2004), the expected behavior is the opposite for blood vessels. The steric force, per unit area as defined, is the repulsion force between the microrobot and the endothelial surfaces of the blood vessels and it stems from the difference of superficial chemical composition of the two and can be predicted by integration of a function of thermal energy and material properties over the functionalized area of the microrobot, A_f (m^2), as (Arcese et al., 2012):

$$
\begin{bmatrix} F_{st}(t) \\ T_{st}(t) \end{bmatrix} = \begin{bmatrix} -50 \int_{A_f} (k_b T \rho_g^{3/2} e^{-2\pi(h_s(t)/L_0)}) dA \\ p_w(t) \times F_{st}(t) \end{bmatrix} \tag{5.11}
$$

with Boltzmann constant, k_b (m$^2 \cdot$ kg/s$^2 \cdot$ K); absolute temperature, T (K); the density of the polymer chain used for functionalization, ρ_g (kg/m^3); the distance between the surfaces, $h_s(t)$ (m); the equilibrium length for the grafting procedure, L_0 (m); and the vector giving the position of the point on which the equivalent steric force is being applied, $p_w(t)$ (m). It is important to note that the equation given above is valid for a specific interval of L_0 (m) (Arcese et al., 2012).

The acoustic force of standing waves, which is better suited for motion control rather than actuation (Ahmed et al., 2017; Bruus, 2012; Xu et al., 2015a), $F_a(t)$ (N) and $T_a(t)$ (N·m), exerted on the cargo and body of a group of microrobots on the plane of acoustic propagation can be interpreted as:

$$
\begin{bmatrix} F_a(t) \\ T_a(t) \end{bmatrix} = \begin{bmatrix} R_r^{xyz} \begin{bmatrix} \frac{1}{3}\pi \left[\frac{5(\rho_{rs}/\rho)-2}{2(\rho_{rs}/\rho)+1} - \frac{k_{rs}}{k_{bf}} \right] \omega_a r_{rs}^3 \frac{p_a^2}{\rho c_0^3} sin\left(2\frac{\omega_a}{c_0}z\right) \\ 0 \\ 0 \end{bmatrix} \\ p_{body}(t) \times F_a(t) \end{bmatrix} \tag{5.12}
$$

Here, r_{rs} (m) denotes the radial position of a microrobot with respect to the geometric center of the robotic swarm on the plane of acoustic wave propagation, ρ_{rs} (kg/m^3) is the effective density of the swarm of microrobots within the fluidic volume of interest, k_{rs} (1/Pa) and k_{bf} (1/Pa) are effective compressibilities of the swarm of microrobots and the blood, respectively, ω_a (rad/s) is the frequency of the applied acoustic field, and c_0 (m/s) is the speed of sound in the blood. Furthermore, R_r^{xyz} is the matrix responsible for projecting the radial force of the swarm onto the Cartesian system of individual microrobots depending on its position and orientation. It is important to note that if the microrobot is of the biohybrid kind, its compressibility will also be nonzero. On the other hand, the local compressibility of blood is dependent on the hematocrit of the flow field and it will contribute to the overall compressibility.

The next physical stimulus is modeled in a complicated manner based on the energy transferred to the blood clot and energy used for material removal, that is, fretting, via the tip of the rigid helix of the magnetic microrobot as it comes into contact with it in the blood vessel (Khalil et al., 2017; Khalil et al., 2018). Here, the force and torque vectors of fretting phenomenon, that is, $F_{fr}(t)$ (N) and $T_{fr}(t)$ (N · m), are modeled in terms of geometric properties, that is, the radius of the rigid helix, r_h (m), and position of the tip of the helix with respect to the center of mass, that is, $r_c = [r_{c,x} \; r_{c,y} \; r_{c,z}]^T$ (m); the penetration depth, that is, δ_{bc} (m); surface normal and tangents at the contact, that is, $n_t = [n_{t,x} \; n_{t,y} \; n_{t,z}]^T$; rotation rate of the microrobot, that is, Ω_x (rad/s); and the material properties of the blood cloth, that is, the ultimate tensile strength, τ_{bc} (Pa), the spring constant, k_{bc} (N/m), and damper constant, b_{bc} (N · s/m). The equation and the subsequent model are used to predict the material removal rate, as will be discussed in the next part, with an acceptable error (Khalil et al., 2017).

$$\begin{bmatrix} F_{fr} \\ T_{fr} \end{bmatrix} = \begin{bmatrix} F_{fr,x} \\ F_{fr,y} \\ F_{fr,z} \\ T_{fr,x} \\ T_{fr,y} \\ T_{fr,z} \end{bmatrix} = \left[-\left(k_{bc}\delta_{bc} + b_{bc} \left\{ \begin{array}{l} \partial\delta_{bc}/\partial t \Leftarrow (\partial\delta_{bc}/\partial t > 0) \\ 0 \Leftarrow (\partial\delta_{bc}/\partial t < = 0) \end{array} \right\} \right) n_{t,x} - |T_{fr,x}| n_{t,y}(t)/r_h \right.$$

$$\left. - |T_{fr,x}| n_{t,z}(t)/r_h - \mathrm{sgn}(\Omega_x)\pi\tau_{bc}r_h\delta_{bc}^2 r_{c,z}F_{fr,x} - r_{c,x}F_{fr,z}r_{c,x}F_{fr,y} - r_{c,y}F_{fr,x} \right] \tag{5.13}$$

The physical stimuli briefly discussed so far are deterministic and assumed to dominate the behavior of the microrobot in action. The complexity of the equation of motion (5.1) comes from the fact that each of these elements depends on either the position or velocity of the microrobot and they in turn directly contribute to the equation of motion (5.1). Therefore, the numerical analysis is not very trivial, especially since half of the force components need information on proximity to the surroundings, depending on several frame rotations and kinematic calculations to be carried out simultaneously. Furthermore, the reader will find out in the following parts that the flow field might be stochastic due to several other phenomena that would take the entire model to the stochastic realm.

5.3.2 Rheology of blood

The material properties of the blood, as a complex and inhomogeneous fluid, become important when gravitational attraction, Eq. (5.8), transient effects, Eq. (5.9), acoustic force, Eq. (5.12), and fretting phenomenon, Eq. (5.13) are incorporated into the equation of motion (5.1). On the other hand, even without the presence of said stimuli, the rheology of blood must be articulated within the resistance matrix of the microrobot, $K(t)$, given as:

$$K(t) = \sum_{\text{tails}} \left(\int_0^{\ell_{\text{tail}-i}} \begin{bmatrix} R_{fs}(t)C(t)R_{fs}(t)^{\mathrm{T}} & -R_{fs}(t)C(t)R_{fs}(t)^T S_{\text{tail}}(t) \\ S_{\text{tail}}(t)R_{fs}(t)C(t)R_{fs}(t)^{\mathrm{T}} & -S_{\text{tail}}(t)R_{fs}(t)C(t)R_{fs}(t)^T S_{\text{tail}}(t) \end{bmatrix}_i d\ell \right)$$
$$+ \begin{bmatrix} D_{\text{translation}}(t) & -D_{\text{translation}}(t)S_{\text{body}+\text{cargo}}(t) \\ S_{\text{body}+\text{cargo}}(t)D_{\text{translation}}(t) & D_{\text{rotation}}(t) \end{bmatrix} \tag{5.14}$$

Here, the component $R_{fs}(t)$ is the rotation matrix between the local Frenet–Serret frames along each tail and the inertial frame of the microrobot. The 3×3 matrix $C(t)$ holds the local resistance coefficients for the rotating helical tails. These coefficients are functions of geometry, proximity, and fluid viscosity (Tabak, 2018; Tabak, 2019a). Likewise, the resistance matrices $D_{\text{translation}}(t)$ and $D_{\text{rotation}}(t)$ take care of the viscous resistance on the body and cargo of the microrobot, which can be of a blunt or an arbitrary geometry (Alapan et al., 2018; Felfoul et al., 2016; Khalil et al., 2017; Schwarz, Karnaushenko, et al., 2020; Tabak, 2020a; Yasa et al., 2019). Besides, the drag coefficient on the body can also be expressed in terms of Reynolds number, that is, $\mathrm{Re} = 2\rho \cdot \bar{u} \cdot R/\mu$ with $\bar{u}$ being the amplitude of the relative velocity between the object and the fluid, to incorporate different flow regimes (Haghdel et al., 2017; Wen et al., 2015). More details on the resistive force coefficients can be found in (Haghdel et al., 2017; Tabak, 2019b; Wen et al., 2015). Here, the focus will be on viscosity. Finally, the S matrices in Eq. (5.14) are employed to handle local cross products with respect to the center of mass of the microrobot thus projecting linear velocities to angular and vice versa in the calculation of forces and torques.

The fluid resistance is directly related to the dynamic viscosity of the fluid, the blood. Regarding a Newtonian fluid such as water, the viscosity could be represented as a function of temperature and it would have been more than adequate for microrobotic applications. However, with blood, the viscosity is a quantity demonstrating temporal and spatial differences due to the presence of live cells (Arzani, 2018; Dutta & Tarbell, 1996; Sriram et al., 2014; Wen et al., 2015) along with cross-sectional asymmetry of the stream, multiphase characteristics of the flow field (Arzani, 2018; Sriram et al., 2014; Wen et al., 2015), the size of the blood vessels (Perdikaris et al., 2016), and the pulsating flow conditions (Dutta & Tarbell, 1996; Sadelli et al., 2017; Womersley, 1955). Furthermore, the blood clot is a rather composite material (Chernysh et al., 2020). In return, an active agent swimming in the circulatory system and drilling blood clots is under the influence of a highly nonlinear, non-Newtonian, and viscoelastic behavior. Here on, the reader will find the discussions on how to model the behavior of the blood, i.e., flow field with its non-Newtonian behavior.

5.3.2.1 Blood thixotropy

The main concern is the shear-thinning behavior of the blood, which happens to be directly related to the local RBC concentration, i.e., hematocrit (Sriram et al., 2014). The RBC cells are known to form a non-Newtonian core with the protein molecules (Bessonov et al., 2016; Dutta & Tarbell, 1996; Sriram et al., 2014). In practice, modeling the entire circulatory systems via CFD tools will simply be futile (Anor et al., 2010), although it is possible to lump the flow in the circulatory system as an electrical circuit (Shi et al., 2011). However, local flow fields must be resolved by computational methods which lead to the fact that distinct turbulent and laminar regions with Newtonian and non-Newtonian characteristics will present themselves (Doost et al., 2016; Goubergrits et al., 2008; Rahman & Haque, 2012). Furthermore, a symmetric flow profile assumption for any given location in the blood vessels is not realistic due to spatial dependency of non-Newtonian behavior based on the rouleaux formation, i.e., stacking of RBCs (Arzani, 2018) or the RBC core being the reason for the non-Newtonian phase of the flow field whereas the plasma behaves as a Newtonian fluid near the epidermal tissue surface (Sriram et al., 2014). Therefore, experimental and computational analyses presented in the literature demonstrate that variances in the circulatory network and in the flow field demand a case-by-case study for biomedical microrobotic applications (Anor et al., 2010; Arzani, 2018; Goubergrits et al., 2008; Perdikaris et al., 2016).

To further understand these phenomena and to unequivocally emphasize the importance of acknowledging that a single example, experimental or numerical, cannot represent a solution to the current dilemma of how to control microrobots in circulatory system, we should extend the discussion with the focus being on the properties of the circulatory network. The size of the blood vessels dictates that the introduction of a foreign object with a comparable size will alter the flow field and pressure; therefore, altering the overall flow resistance as a result (Sankar & Hemalatha, 2007). In other words, the diameter of the blood veins and arteries will dictate the overall swimming performance. This complex network can be decomposed into subsystems based on the size (diameter); as in macrovascular for diameter larger than 0.5 mm, mesovascular for diameters down to 10 μm, and microvascular for smaller diameters (Anor et al., 2010; Perdikaris et al., 2016). The macrovascular network can be reconstructed using computer tomography scans, 3D angiography, MRI, and 3D and volumetric intravascular ultrasound (Anor et al., 2010; Bessonov et al., 2016; Evans et al., 1996; Goubergrits et al., 2008; Rim et al., 2013). Furthermore, the microvascular network, i.e., capillaries, surrounds the tissue as a mesh, while the mesovascular network bifurcates, that is, branches, which can be explained by fractals (Anor et al., 2010). The size of the blood vessels affects the flow conditions that can be correlated with Reynolds number and shear rates (Anor et al., 2010; Campo-Deaño, 2016; Dutta & Tarbell, 1996), thus affecting the phase of the flow by introducing higher viscosity and non-Newtonian effects (Arzani, 2018). Fig. 5.1 illustrates the circulatory system, the macrovascular and mesovascular network with bifurcations, and a single microrobot carrying out the operation of opening a clogged artery.

Furthermore, the velocity profile in vivo is observed to not be symmetric or parabolic (Sriram et al., 2014) due to the increased volume fraction of RBC for different chemical and physical conditions (Arzani, 2018; Goubergrits et al., 2008). It is known that Newtonian behavior is observed more often with high shear rates (Anor et al., 2010; Dutta & Tarbell, 1996). Also, a larger diameter will let the bulk of the fluid to flow without the cells dominating the overall behavior; therefore, resulting in Newtonian conditions, although pulsating flow conditions are known to impede this assumption (Wen et al., 2015). It is possible to observe multiphase stream, i.e., non-Newtonian and Newtonian regions side by side, when the shear rate and flow conditions allow (Sriram et al., 2014). Also it is known that the Newtonian flow characteristics become dominant as the volume fraction of RBCs in the blood plasma converges to zero (Wen et al., 2015). Furthermore, the shear rate is a regulatory function for the circulatory system (Sriram et al., 2014) for its building blocks

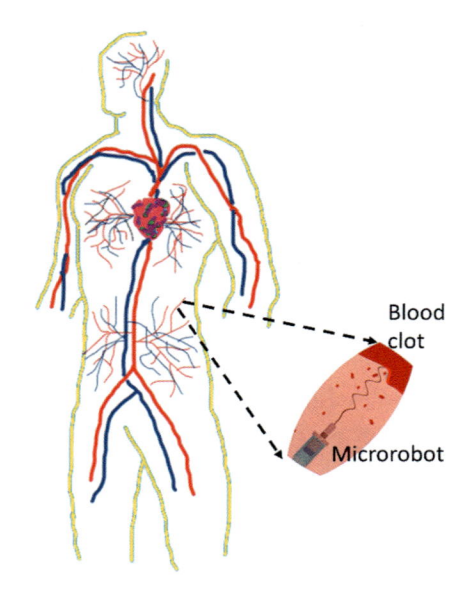

FIGURE 5.1

Illustration of the macrovascular and mesovascular network in the human body, torso, limbs, and head, with the heart at the center, along with a microrobot performing blood clot removal in one of the branches in the circulatory system.

are live cells. For instance, the flow conditions in arteries are known to be responsible for the development of certain arterial diseases (Bessonov et al., 2016; Chakravarty & Sannigrahi, 1999; Goubergrits et al., 2008).

This overall complex chemical and physical relationship between the vascular network, cells, and the flow field is known as the hemodynamics of the circulatory system and it requires utmost attention with special focus on the viscosity (Wen et al., 2015). Even arterial flexibility is known to be a factor in shear stress in the flow field (Anor et al., 2010); as a direct consequence of the local viscosity. The microrobots will be subject to the same phenomena, and an individual micro agent or a swarm might experience the aforementioned multiphase conditions (Anor et al., 2010; Sriram et al., 2014) and various spatial characteristics at the same time. It is known that the surface area in contact with the fluid directly affects the hydrodynamic behavior of the microrobots by altering propulsive and drag forces (Ye et al., 2019). Measuring the flow conditions in vivo online remains unanswered; however, offline scans, measurements, and reconstructions might reveal very important details on what to expect during the time of flight of active agents. There existseveral alternative mathematical models in the literature addressing this issue.

The various analytical approaches used to model the flow field along with the prediction of the shear rate and viscosity of the local bloodstream that will shortly be reviewed here include: the Womersley model (Sadelli et al., 2017; Womersley, 1955), Scott Blair/Casson−Fung model (first proposed by Blair and also applied by Casson, then modified by Fung based on empirical data) (Dutta & Tarbell, 1996; Goubergrits et al., 2008; Sankar & Hemalatha, 2007), Generalized Maxwell model (Dutta & Tarbell, 1996), Herschel−Bulkley fluid model (Haghdel et al., 2017; Sankar & Hemalatha, 2007), Walburn−Schneck model (Goubergrits et al., 2008), (Generalized) Power Law (Abu-Hamdeh et al., 2020; Dutta & Tarbell, 1996; Goubergrits et al., 2008; Haghdel et al., 2017), Quemada rheological model (Sriram et al., 2014), and Carreau−Yasuda model (Sriram et al., 2014); with corrections based on hematocrit levels (Arcese et al., 2012; Goubergrits et al., 2008). It might be also possible to use the Brinkman medium approach (Chen et al., 2020) for very high hematocrit levels. We specifically need the dynamic viscosity viscosity term, μ, to substitute in $C(t)$ in Eq. (5.14); therefore, the reader will find the models regarding the blood viscosity, that is, hemorheology (Bessonov et al., 2016).

The power-law model, which happens to be the most fundamental approach, predicts the viscosity as follows:

$$\mu = m_{pl}\dot{\gamma}^{n_{pl}-1} \tag{5.15}$$

with m_{pl} and n_{pl} known as the consistency index and non-Newtonian index, respectively, and $\dot{\gamma}$ (1/s) denoting the shear rate (Abu-Hamdeh et al., 2020; Goubergrits et al., 2008). Moreover, when the generalized power law is used, the consistency and non-Newtonian indexes are presented by exponential functions of shear rate with empirical fitting coefficients pertaining to the experimental results (Goubergrits et al., 2008).

The Scot Blair/Casson−Fung model incorporates the hematocrit level, H (%), as:

$$\mu = \left(\sqrt{\mu_o(1-H)^{-2.5}} + \sqrt{\frac{(0.625H)^3}{100\dot{\gamma}}}\right)^2 \tag{5.16}$$

where μ_o is equal to 0.0012 (Pa·s) (Goubergrits et al., 2008). The equation is empirical; the constant and the powers are found via curve-fitting over experimental data (Goubergrits et al., 2008). Furthermore, $\mu_o(1 - H)^{-2.5}$ in Eq. (5.16) stands for the Casson viscosity, whereas $(0.625H)^3/(100\dot{\gamma})$ represents the yield stress, on the right-hand side of Eq. (5.16) (Dutta & Tarbell, 1996). The constants are naturally subject to different blood samples from different patients.

The Walburn−Schneck model incorporates the hematocrit, H (%), and the total concentration of fibrinogen and globulin in the stream, TPMA (g/L), to predict the viscosity as:

$$\mu = 0.1 c_1 e^{c_2 H} e^{c_3 TMPA/H^2} \dot{\gamma}^{-Hc_4} \tag{5.17}$$

with the empirical coefficients c_1, c_2, c_3, and c_4 fitted for $H = 0.4$ and TPMA $= 25.9$ (g/L) (Goubergrits et al., 2008).

Goubergrits et al. (2008) further proposed a more definitive model to combine the three approaches with the appropriate empirical corrections for H (%), TPMA (g/L), and temperature, T (K), as:

$$\mu = c_1 e^{c_2 TPMA/H^2} \left(\mu_o(1-H)^{-2.5} + \Delta\mu e^{-(1+\dot{\gamma}/A)e^{(B/\dot{\gamma})}} \right) \tag{5.18}$$

with $\mu_o = 0.0008585$ (Pa·s) and $\Delta\mu = 0.00707$ (Pa·s), c_1, and c_2 being constants to be fitted, A and B being empirical second-degree polynomial functions of H (%) and T (K) as in $A = a_1 H^2 + a_2 H + a_3 T + a_4$ and $B = b_1 H^2 + b_2 H + b_3 T + b_4$ with $a_{\{1,2,3,4\}}$ and $b_{\{1,2,3,4\}}$ being sample-specific coefficients. The values and fits given by the authors are subject to the blood samples.

The Carreau−Yasuda model (Wen et al., 2015) makes use of the volume fraction of the RBCs, υ_{RBC}, in the following form:

$$\mu = m(1 + (\lambda\dot{\gamma})^2)^{(1-n)/2} \tag{5.19}$$

with $m = 122.28(\upsilon_{RBC})^3 - 51.213(\upsilon_{RBC})^2 + 16.305(\upsilon_{RBC})^1 + 1$ and $n = 0.8092(\upsilon_{RBC})^3 - 0.8246(\upsilon_{RBC})^2 - 0.3503(\upsilon_{RBC}) + 1$. Also here, λ (s) is the time−constant that was set as 0.11 for the reported study, and the coefficients are of the curve-fitting results over the experimental results (Wen et al., 2015).

And finally, the Quemada rheological model makes use of the plasma viscosity, μ_p (Pa·s), and hematocrit H (%) as:

$$\mu = \frac{\mu_p}{\left(1 - \left(\frac{k_o + k_\infty \sqrt{\dot{\gamma}/\gamma_c}}{1 + \sqrt{\dot{\gamma}/\gamma_c}}\right) H/2\right)^2} \tag{5.20}$$

with γ_c, k_o, and k_∞ being the coefficients of the modeling approach that again need a proper curve-fitting study (Sriram et al., 2014). This list can further be extended via various empirical equations. The hematology of blood is subject to patient, based on age, gender, medical history, nutrition, geography etc, and the tissue it is passing, such as lungs, liver, kidney, brain, etc. Therefore, it is obvious that the equation and fitting coefficients to be employed could differ from case to case. One can further speculate that all the aforementioned approaches on shear and shear rate modeling could be employed in the same model in the form of a look-up table based on the ever-changing local temporal and spatial conditions.

5.3.2.2 Blood clots

Fig. 5.2 illustrates the grinding and drilling of a blood clot inside a clogged blood vessel. Interacting with the environment in the form of drilling into soft tissue is a problem of conservation of energy, momentum, and mass. The conservation of momentum is the part we dealt with in the previous section, although it will be briefly revisited here. The conservation of energy comes from the fact that the material must be removed from the surface of the soft tissue by breaking the physical bond of the anisotropic material that is the clot. This part is rather tricky as the blood clot is a material interlaced with random fibers of proteins and platelet cells along with other materials (Khalil et al., 2017; Khalil et al., 2018; Tabak, 2019b). Finally, conservation of mass deals with how much material is removed from the clot with the sense of direction instead of calculating the removal rate as an isotropic bulk property. Predicting directional material removal rate is dependent on (1) careful prediction/sensing of the position of the microrobotic tool and (2) accurate prediction of the material removal rate, which is not possible without prior tuning study with the blood clot of the patient (Anor et al., 2010; Arzani, 2018; Chakravarty & Sannigrahi, 1999; Goubergrits et al., 2008; Khalil et al., 2017; Khalil et al., 2018; Perdikaris et al., 2016; Sriram et al., 2014). Nevertheless, a generic approach can be made use of to overcome the subtle variances in material properties given that the initial guess is educated and well thought of.

The fretting force is already given by Eq. (5.13). The force and torque components rely on a series of assumptions and simplifications. For instance, the trailing tip of a rotating rigid helix could be responsible for drilling the blood clot. It is best to be utilized along with chemicals to dissolve, that is, lysis, the fragments peeled off superficially. The potent chemical can also be carried by the microrobot and be released acoustically or in time by just diffusion of sorts (Ceylan, Yasa, Tabak, et al., 2019; Park et al., 2021) to avoid the smaller chunks clogging somewhere else. The drilling operation can be only possible if the tip of the helix penetrates the surface to grind the material until it peels off as it is a composite viscoelastic mixture (Khalil et al., 2017; Khalil et al., 2018; Krasokha et al., 2010). The instantaneous penetration depth of the helix–tip into this

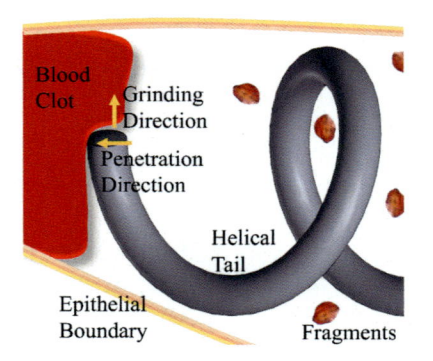

FIGURE 5.2

The tip of the helical microrobot comes into contact with the blood clot and removes small fragments of the composite structure. Due to the rotation of the helix, the fragments disperse in the fluidic environment with a rather uncontrolled manner.

viscoelastic porous structure, δ_{bc} (m), is calculated based on torque equilibrium between the fretting and whatever remains after all the drag and friction effects take the toll out of the magnetic torque applied. The fretting of the said material in volume per unit time will require a certain amount of power transfer from the electromagnetic field, and it can be simplified into the following equation using dimensional analysis (Khalil et al., 2017; Khalil et al., 2018):

$$\left(|T_{m,x}| - \left|\sum_i T_{i,x}\right|\right)\Omega_x \approx \tau_{bc}\pi r_h \Omega_x \delta_{bc}^2/4 \tag{5.21}$$

From here, it is possible to predict the δ_{bc} (m) but one might impose an upper limit per the experimental observations when need be. For instance, the maximum meaningful depth might be simply limited by the diameter of the slender tail. The material properties, that is, the effective spring and damper constants, k_{bc} (N/m) and b_{bc} (N·s/m), respectively, at the point of contact can be determined from here as follows:

$$k_{bc} = (|F_{m,x}| - \left|\sum_i F_{i,x}\right|)/\delta_{bc} \tag{5.22}$$

and

$$b_{bc} = 2\chi(m_{\text{robot}}k_{bc})^{0.5} \tag{5.23}$$

with χ being the unitless damping coefficient of the contact and one might argue that it must be always larger than unity to capture the effect of internal friction of the blood clot and avoid the bouncing off phenomenon. The m_{robot} (kg) is the effective mass of the robot and could also include the added mass in the numerical implementation. Finally, the direction of the contact with respect to the center of mass of the microrobot will affect the force and torque components and requires some trigonometry to handle (Khalil et al., 2017; Khalil et al., 2018). It should be noted that for improved accuracy, the material removal rate should be traced online concerning the 3D position in real-time applications.

The drilling and grinding action can be simulated for different cases, such as the ones reported in (Khalil et al., 2017; Khalil et al., 2018), under different actuation frequencies. As illustrated in Fig. 5.3, the tip of the helical tail will follow a similar trajectory in the blood clot, but first it should get a grip on the surface which does not happen at the first contact but increases gradually in time. It is important to note that the results presented here were obtained without any motion control or numerical tuning in order to match any actual experiments. From here on the reader will find a detailed discussion on the concerns related to motion control of such biomedical microrobotic systems.

5.4 **Motion control**

The success of a microrobotic system is of more importance to pharmaceutics than to biomedical applications. The reason is the cost of drug research. The biggest problem is the pharmacokinetics of the drugs released into the bloodstream. In other words, the adverse reaction of the immune system and local tissue to the drug molecules until they reach their intended target. A drug must be strong enough to reach, let's say, the cancerous area, intact and with enough concentration to make a meaningful impact. This is one of the reasons why developing an effective chemical to eradicate

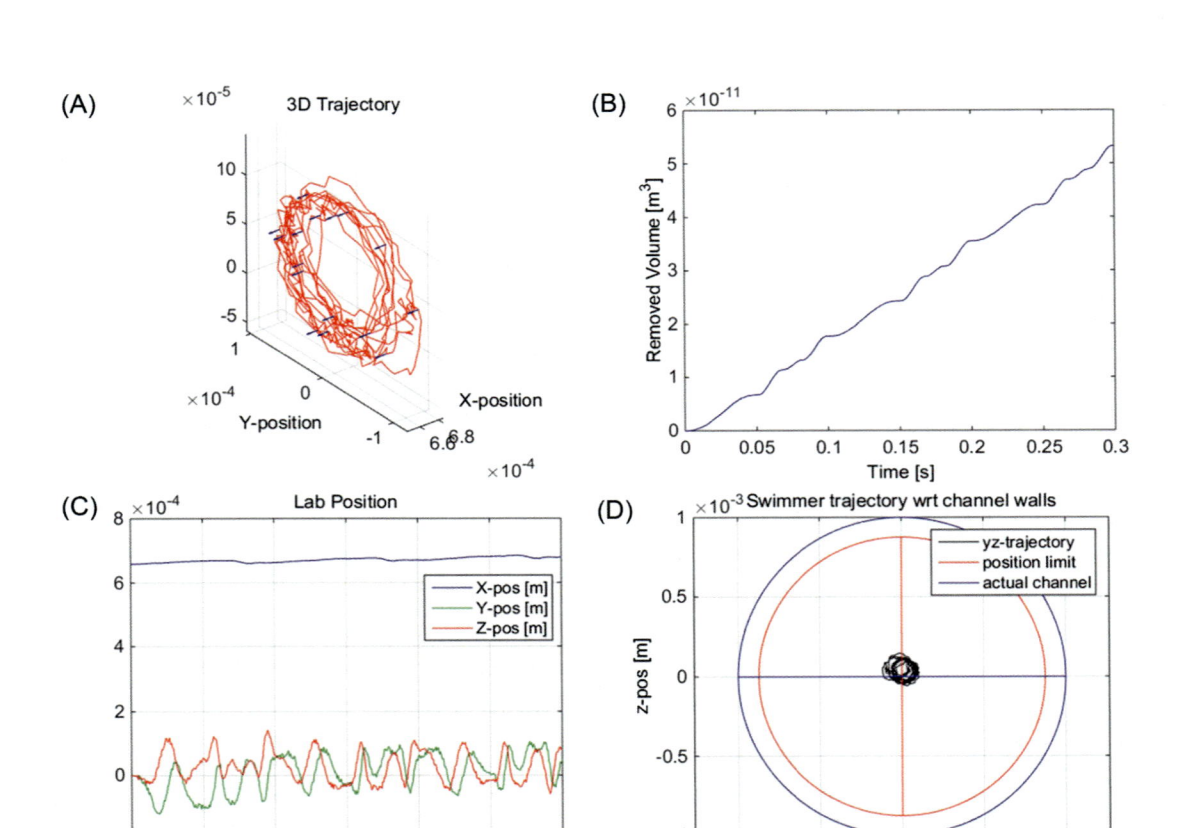

FIGURE 5.3

Simulated blood clot removal based on the reported data for the clot (Khalil et al., 2017; Khalil et al., 2018), with lower actuation frequencies and no specific tuning to match any experimental results: (A) the three-dimensional trajectory of the microrobot in contact with blood clot; (B) the total amount of material removed; (C) the position of the center of mass in the lab frame; (D) 2D view of the gait in the circular cross-section of the blood vessel.

cancerous cells is inherently problematic (Crombag et al., 2016; Felfoul et al., 2016; Garattini, 2007; Jit et al., 2020; Siddiqui & Rajkumar, 2012; Vasan et al., 2019). The only thing that matters, concerning the subject matter covered here, is that a microrobotic device might not catch the attention in medical field unless a successful addressable and controllable cargo towing action is demonstrated with an acceptable statistical failure rate. The said towing action will have to demonstrate carrying and releasing drugs and retrieving the microrobot. Nothing short of this will be deemed worthy if the goal is to use microrobotic systems in the human circulatory system. The problem looks ambitious and the answer remains elusive to this date. It is important to note that since the main actuation and control method present in the current literature is magnetic, the next discussion follows here will consider magnetically actuated and controlled microrobots.

5.4.1 **Controlling a single microrobot**

Although it is quite clear that a single robot will not be enough for most of the cases, it all starts with a single active agent to understand what the practical limitations are. Here, we should talk about the controllability, stability, and what to expect should stochastic effects arise. A single active agent is easier to control and localize in theory but the size poses a real challenge in real applications (Wang & Zhang, 2020), especially in terms of the accuracy of position (Feemster et al., 2020). Furthermore, the physical interaction of a single active agent with immune cells is also directly related to its position and relative velocity (Yasa et al., 2020). Arguably, a hybrid magnetic–acoustic actuation will make a positive impact on the problem at hand (Caldag & Yesilyurt, 2020); however, the type of the tissue will always dictate the outcome of this tug-of-war (Iacovacci et al., 2019), as long as the microrobot cannot leave the environment fast enough.

As discussed before, the microrobot cannot carry necessary sensory arrays to find its way in the labyrinth of the circulatory system. Furthermore, the path to travel is unknown as we can plan the next step only by so much with computer-assisted scans that can help reconstruct mostly down to the mesoscale arterial network, in turn forcing us to guess the condition of the flow field (Anor et al., 2010; Evans et al., 1996; Ludwig, 1950; Perdikaris et al., 2016; Rim et al., 2013). One possible solution is to rely on the sensory capabilities of live bacteria, hence biohybrid microrobots, given their susceptibility to external magnetic fields for assistance under the assumption that they are properly modified for the task at hand (Ghanbari, 2020; Lin et al., 2014; Sun et al., 2020). Besides, they still need external intervention to take the shortest possible route even with no apparent flow field but with a destination to reach (Tabak, 2020a; 2020b).

Artificial or otherwise, the control law imposed on the microrobot should satisfy a set of criteria to be effective, or in other words, the possible detrimental conditions should be anticipated and studied carefully beforehand. One problem is the step-out frequency for magnetic helical microstructures as they cannot follow the magnetic field's rotation due to the viscous torques (Peyer et al., 2012; Tabak et al., 2011). In other words, should the instantaneous viscous torque dominate the rotation, the microrobot will be misaligned with the rotating magnetic field and stall instantaneously. Thus, the maximum frequency applied to the microrobot should be limited. This can further be inspected via stability criterion. Although modeling the applied magnetic field itself is not in the scope of this chapter, it is important to note that the magnetization vector of the microrobot and the relationship between the magnetic force and the applied current on the electromagnetic coils are quite important to apply a stable control law on the system (Abbott et al., 2007; Nguyen et al., 2020; Samsami et al., 2020). In short, the stability criterion tries to explain the limits of magnetic field and rigid-body motion in a more general form.

Secondly, we have to consider the viscoelastic behavior of the fluidic environment that might be triggered by the frequency of the motion. This phenomenon is characterized by the Deborah (De) number defined as the ratio of relaxation time of the fluid, t_r (s), to the period of the oscillations. The optimum velocity for helical structures in micro realm is found to be around De ≈ 1 (Wu et al., 2020). Thus the actuation (or perturbation) frequency of the magnetic field is limited with yet another physical constraint. On the other hand, the field strength has also a lower bound for the criterion of steerability (Schamel et al., 2014): The Péclet (*Pe*) number defined for the perpendicular direction to the axis of rotation should be comparable to the ratio of supplied magnetic energy to the thermal energy freely available (Schamel et al., 2014). Furthermore, the said ratio should be greater than unity to overcome the thermal noise in the environment.

$$Pe = \frac{1}{D_r^{\perp} t_r} \approx \frac{|M \cdot H|}{2k_b T} > 1 \tag{5.24}$$

with $D_r^{\perp}$ (N·m·s/rad) being equal to the diagonal elements of $K(t)$ in the fifth and sixth rows, i.e., $K(4,4)$ and $K(5,5)$, given in full form by Eq. (5.14) assuming there is a sense of symmetry to the overall geometry. Otherwise, equality, Eq. (5.24), should be checked for all lateral directions. This last criterion brings us to the problem of stochastic effects.

Stochastic behavior of an active agent could stem from three main sources: (1) The Brownian noise, that is, the temperature of the environment (Haghdel et al., 2017; Kim, 1985; Marath & Wettlaufer, 2019; Schamel et al., 2014); (2) the upstream, that is, the external flow being in transition or in turbulence (Al-Azawy et al., 2016; Doost et al., 2016; Rahman & Haque, 2012; Rahman et al., 2018; Wen et al., 2015); and (3) the hydrodynamic interactions, that is, bacterial turbulence for swarms (Dunkel et al., 2013; Ishikawa et al., 2007; Keller & Rubinow, 1976; Kim, 1985; Marath & Wettlaufer, 2019). The specific reason for stochastic motion must be singled out to take the necessary action to mitigate the effects if possible. Unfortunately, there is little to no control over the thermal noise and the flow field conditions. The interaction on the other hand could be reduced by putting a greater average distance between the individual active agents of the swarm. For example, the use of radial acoustic forces controlling the volume and dispersion in a relatively larger volume in the flow field (Caldag & Yesilyurt, 2020; Keya et al., 2018; Xu et al., 2015a), a possible method which should be discussed separately, could be feasible.

5.4.2 Control approach for swarms: a brief discussion

There are several swarm studies dealing with the coordinated effort of multiple agents (Belharet et al., 2012; Dong & Sitti, 2020; Keya et al., 2018; Khalil et al., 2018; Servant et al., 2015; Wang & Zhang, 2020; Xie et al., 2020; Xu et al., 2015a). Since individual microrobots cannot be targeted due to various reasons, i.e., resolution, interference, size, impedance mismatch, the composite structure of the tissues, multiphase flow field, etc., the magnetic field is rather focused on the center of volume of the swarm. However, the electromagnetic volume of interest might not cover the entire swarm at all times. This presents a particular problem as unattended active particles will be adrift in the bloodstream and might end up in a much different location than originally intended. There are several risks involved with this situation: (1) the microrobots can end up in a particular microvascular network and clog a vessel; (2) the chemical might be released in a sensitive area followed by a toxic reaction or antibodies being released and attacking the molecules with positive feedback; (3) the microrobot itself being dissolved into the tissue, assuming it was not fully biocompatible, and resulting in a similar chain of reactions; and (4) being stuck in the tissue for a certain time interval, thus generating an interference detrimental for the following attempts, undermining the overall motion control effort. The list can be extended with specific cases; such as if the microrobot ended up in a kidney and acted as a seed for kidney stones over time, just like a pearl in a farmed−scallop.

The outcome of possible problems to specific applications is not easy to predict without careful considerations. One important step toward the successful control of swarms is employing a mobile electromagnetic field that can follow the position and match up with the velocity of the microrobotic swarm. This idea has been recently in focus (Akçura et al., 2018; Dong & Sitti, 2020; Pittiglio et al., 2020; Tabak, 2020a, 2020b; Yang et al., 2020). A robotic arm carrying an electromechanical coil or permanent magnet constitutes a flexible electromagnetic actuation system assuming that necessary sensory feedback is granted. Such control effort should be bilateral: as the robotic arm or arms follow the location of the microrobot or swarms of multiple microrobots, microrobots will be aligned with the external magnetic field; therefore, realizing robotic actuation and/or steering. The control law of such a system might be adaptive (Tabak, 2020a, 2020b, 2020c) to ensure dexterity into the fight time of the microrobots. Fig. 5.4 illustrates multiple robotic arms positioning electromagnetic coils around a certain tissue of interest to coordinate the efforts of multiple microrobots simultaneously. It should also be noted that, regardless of the number of swarms, the steerability and stability criteria should be satisfied to achieve the therapeutic goal. A very simple control law, ζ, example of such an actuation system composed of a single open kinematic chain and a single bacterium (Tabak et al., 2011; Tabak, 2020a, 2020c) could be tailored as given by (5.25) and (5.26) for any given axis, with K_p, K_i, and K_d denoting the proportional, integral, and derivative gains for the PID control scheme. Here, the integral gain, K_i, is being tuned online with the help of yet another constant, K_{i0}, and the instantaneous error, $e(t)$, to improve the set-point tracking performance. Furthermore, the exponential in the denominator of Eq. (5.26), 0.5, can also be tuned to modulate the susceptibility of the integral gain to the instantaneous error. The error can be defined over position (m) or orientation (rad). It is possible to implement more elaborate control law

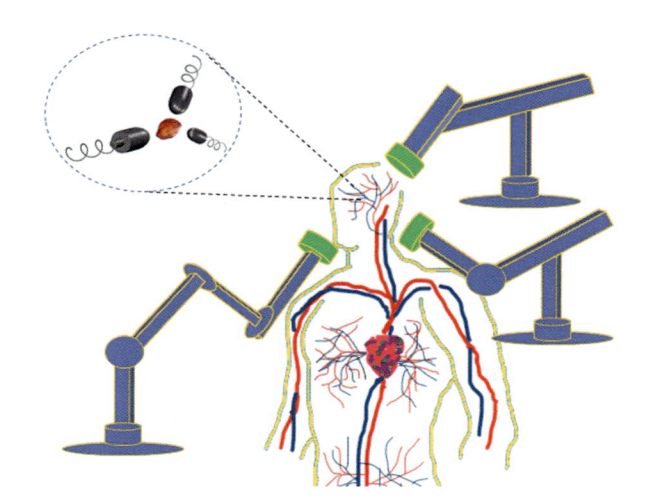

FIGURE 5.4

Robotic arms carrying electromagnetic coils at their end-effector to match the position of the microrobots.

$$\zeta = K_p e(t) + \int K_i e(t)\mathrm{d}t + K_d \dot{e}(t) \tag{5.25}$$

$$K_i = 1/((K_{i0}e(t))^{0.5} + 1) \tag{5.26}$$

strategies depending on the axis, joint, and degrees of freedom (Tabak et al., 2011; Tabak, 2020a, 2020c). Moreover, one can postulate additional adaptive gains for proportional and derivative components to further alter the control law characteristics.

5.5 Conclusion

The future of biomedical microrobotic systems in medicine is full of opportunities as it presents a unique take on the non-invasive medicine. There are numerous promising pre-clinical tests, in vivo and ex vivo, and theoretical studies for motion control. Microrobots performing in the bloodstream and taking on the most ambitious tasks might also revolutionize medicine by reorienting drug research. The microrobotic research efforts are ever-increasing to find a solution to the various problems but were only briefly discussed here. Overall, the journey is perilous, the desired robotic system is highly nonlinear, and the subject matter is extremely interdisciplinary that in return calling for expertise in various fields from science, technology, and medicine. Furthermore, based on the environmental and operational conditions, realistic and inclusive modeling becomes ever more important in biomedical microrobotics as there is little to no room for error.

It has been observed by several studies that using hybrid systems for actuation, sensing, and control is the key to achieve better results, although sensing remains a grand challenge yet to be properly addressed. Building a robust and agile external sensory grid for biomedical needs, although out of the scope of this chapter, should still be mentioned. The control law and the robotic application will inescapably rely on sensory data no matter how well the model represents the dynamics of the system. Some variables must be fed from the real world online so that a model-based observer could be implemented to predict the rest of the states of the system. A sensory array could use systems based on photonics, acoustics, magnetic resonance, and several possible combinations thereof. However, the impedance of composite layers of live tissue will always hinder the accuracy. Furthermore, the sensor components might interfere with each other during operation. Therefore, obtaining clean signals and making use of complex spatial and temporal data will most probably require machine learning and artificial intelligence to recognize patterns and decide fast enough so that the stochastic effects do not prevail over actuation and steering, especially for swarm-oriented applications.

As discussed earlier, the system must perform in real-time; therefore, computationally demanding approaches are often not desirable. Offline studies will be better for extensive statistical analysis, simulation, and rehearsal of the task for contingencies; however, online data acquisition, state estimation, summoning of the control law, and actuation might take up a considerable amount of computational power given that not all of these tasks will have the privilege of abundance in system resources. So, it could be more feasible to reconstruct the tissue digitally following, for instance, computer tomography scans and build a digitial mockup to scale beforehand for the trial-and-error phase cannot be conducted on the actual patient. However, simulating the real conditions with every detail is an overwhelming challenge itself. Similarly, implementation of a physical mock-up with realistic biophysical conditions of live tissue deep in the organism for robotic experiments might be needed to capture the actual conditions. However, this arguably is going to be even more complicated and demanding than modeling the microrobotic tool itself. Although, if successful such a mock-up might offer more reliable results along with the chance of comprehensive system identification. Having said that, we should acknowledge that the mathematical model will reveal key information about the dynamics and behavior of the active agent, performing under adverse conditions, as long as enough attention is paid for the details during its implementation. On the other hand, simulating multiple agents poses yet another challenge of computational nature. To speculate further, one may compare the simulation of a large swarm to the molecular dynamics simulations with motion control if each active agent was going to be actively steered. Either way, mathematical and numerical modeling will be an important part of microrobotic research for years to come.

References

Abbott, J. J., Ergeneman, O., Kummer, M. P., Hirt, A. M., & Nelson, B. J. (2007). Modeling magnetic torque and force for controlled manipulation of soft-magnetic bodies. *IEEE Transactions on Robotics*, *23*(6), 1247−1252. Available from https://doi.org/10.1109/TRO.2007.910775.

Abu-Hamdeh, N. H., Bantan, R. A. R., Aalizadeh, F., & Alimoradi, A. (2020). Controlled drug delivery using the magnetic nanoparticles in non-Newtonian blood vessels. *Alexandria Engineering Journal*, *59*(6), 4049−4062. Available from https://doi.org/10.1016/j.aej.2020.07.010.

Ahmed, D., Dillinger, C., Hong, A., & Nelson, B. J. (2017). Artificial acousto-magnetic soft microswimmers. *Advanced Materials Technology*, *2*(7), 1700050. Available from https://doi.org/10.1002/admt.201700050.

Akçura, N., Çetin, L., Kahveci, A., Alasli, A.K., Can, F.C., & Tamer, Ö. (2018). Guided motion control methodology for microrobots. In: *Proceedings of the 6th international conference on control engineering & information technology*, October 25−27, Istanbul, Turkey. Available from https://doi.org/10.1109/CEIT.2018.8751803.

Alapan, Y., Bozuyuk, U., Erkoc, P., Karacakol, A. C., & Sitti, M. (2020). Multifunctional surface microrollers for targeted cargo delivery in physiological blood flow. *Science Robotics*, *5*(42), eaba5726. Available from https://doi.org/10.1126/scirobotics.aba5726.

Alapan, Y., Yasa, O., Schauer, O., Giltinan, J., Tabak, A. F., Sourjik, V., & Sitti, M. (2018). Soft erythrocyte-based bacterial microswimmers for cargo delivery. *Science Robotics*, *3*(17), aa4423. Available from https://doi.org/10.1126/scirobotics.aar4423.

Al-Azawy, M. G., Turan, A., & Revell, A. (2016). Investigating the impact of non-Newtonian blood models within a heart pump. *International Journal of Numerical Methods in Biomedical Engineering*, e02780. Available from https://doi.org/10.1002/cnm.2780.

Andhari, S. S., Wavhale, R. D., Dhobale, K. D., Tawade, B. V., Chate, G. P., Patil, Y. N., Khandare, J. J., & Banerjee, S. S. (2020). Self-propelling targeted magneto-nanobots for deep tumor penetration and pH-responsive intracellular drug delivery. *Scientific Reports*, *10*, 4703. Available from https://doi.org/10.1038/s41598-020-61586-y.

Anor, T., Grinberg, L., Baek, H., Madsen, J. R., Jayaraman, M. V., & Karniadakis, G. E. (2010). Modeling of blood flow in arterial trees. *Wiley Interdisciplinary Reviews: System Biology and Medicine*, *2*(5), 612−623. Available from https://doi.org/10.1002/wsbm.90.

Arcese, L., Fruchard, M., & Ferreira, A. (2012). Endovascular magnetically guided robots: Navigation modeling and optimization. *IEEE Transactions on Bio-Medical Engineering*, *59*(4), 977−987. Available from https://doi.org/10.1109/TBME.2011.2181508.

Arzani, A. (2018). Accounting for residence-time in blood rheology models: Do we really need non-Newtonian blood flow modelling in large arteries? *Journal of the Royal Society, Interface/The Royal Society*, *15*(146), 20180486. Available from https://doi.org/10.1098/rsif.2018.0486.

Belharet, K., Folio, D., & Ferreira, A. (2012). Control of a magnetic microrobot navigating in microfluidic arterial bifurcations through pulsatile and viscous flow. In: *Proceedings of the IEEE/RSJ international conference on intelligent robots and systems*, October 7–12, Vilamoura, Algarve, Portugal, pp. 2559–2564. Available from https://doi.org/10.1109/IROS.2012.6386030.

Bessonov, N., Sequeira, A., Simakov, S., Vassilevskii, Y., & Volpert, V. (2016). Methods of blood flow modeling. *Mathematical Modelling of Natural Phenomena*, *11*(1), 1–25. Available from https://doi.org/10.1051/mmnp/201611101.

Brennen, C., & Winet, H. (1977). Fluid mechanics of propulsion by cilia and flagella. *Annual Review of Fluid Mechanics*, *9*, 339–398. Available from https://doi.org/10.1146/annurev.fl.09.010177.002011.

Bruus, H. (2012). Acoustofluidics 7: The acoustic radiation force on small particles. *Lab on a Chip*, *12*, 1014–1021. Available from https://doi.org/10.1039/C2LC21068A.

Caldag, H. O., & Yesilyurt, S. (2020). Acoustic radiation forces on magnetically actuated helical swimmers. *Phys. Fluids*, *32*(9), 092012. Available from https://doi.org/10.1063/5.0020930.

Campo-Deaño, L. (2016). Assessing the dynamic performance of microbots in complex fluid flows. *Applied Science*, *6*(12), 410. Available from https://doi.org/10.3390/app6120410.

Ceylan, H., Giltinan, J., Kozielski, K., & Sitti, M. (2017). Mobile microrobots for bioengineering applications. *Lab on a Chip*, *17*, 1705–1724. Available from https://doi.org/10.1039/C7LC00064B.

Ceylan, H., Yasa, I. C., Kilic, U., Hu, W., & Sitti, M. (2019). Translational prospects of untethered medical microrobots. *Progress in Biomedical Engineering*, *1*, 012002. Available from https://doi.org/10.1088/2516-1091/ab22d5.

Ceylan, H., Yasa, I. C., Tabak, A. F., Giltinan, J., & Sitti, M. (2019). 3D-printed biodegradable microswimmer for theranostic cargo delivery and release. *ACS Nano*, *13*(3), 3353–3362. Available from https://doi.org/10.1021/acsnano.8b09233.

Chakravarty, S., & Sannigrahi, A. K. (1999). A nonlinear mathematical model of blood flow in a constricted artery experiencing body acceleration. *Mathematical and Computer Modelling*, *29*(8), 9–25. Available from https://doi.org/10.1016/S0895-7177(99)00067-9.

Chen, Y., Lordi, N., Taylor, M., & Pak, O. S. (2020). Helical locomotion in porous medium. *Physical Review E*, *102*, 043111. Available from https://doi.org/10.1103/PhysRevE.102.043111.

Chernysh, I. N., Nagaswami, C., Kosolapova, S., Peshkova, A. D., Cuker, A., Cines, D. B., Cambor, C. L., Litvinov, R. I., & Weisel, J. W. (2020). The distinctive structure and composition of arterial and venous thrombi and pulmonary emboli. *Scientific Reports*, *10*, 5112. Available from https://doi.org/10.1038/s41598-020-59526-x.

Cortez, R., Fauci, L., & Medovikov, A. (2005). The method of regularized Stokeslets in three dimensions: Analysis, validation, and application to helical swimming. *Physics Fluids*, *17*(3), 031504. Available from https://doi.org/10.1063/1.1830486.

Crombag, M.-R. B. S., Joerger, M., Thürlimann, B., Schellens, J. H. M., Bejinen, J. H., & Huitema, D. R. (2016). Pharmacokinetics of selected anticancer drugs in elderly cancer patients: Focus on breast cancer. *Cancer*, *8*(1), 6. Available from https://doi.org/10.3390/cancers8010006, Basel.

Dong, X., & Sitti, M. (2020). Controlling two-dimensional and cooperative behavior of magnetic microrobot swarms. *International Journal of Robotic Research*, *39*(5), 617–638. Available from https://doi.org/10.1177/0278364920903107.

Doost, S. N., Ghista, D., Su, B., Zhong, L., & Morsi, Y. S. (2016). Heart blood flow simulation: A perspective review. *Biomedical Engineering Online*, *15*, 101. Available from https://doi.org/10.1186/s12938-016-0224-8.

Du., X., Lidong, Y., Yu, J., Chan, K.F., Chiu, P.W.Y., & Zhang, L., 2020. RoboMag: A magnetic actuation system based on mobile electromagnetic coils with tunable working space. In: *Proceedings of the 6th international conference on advanced robotics and mechatronics*, December 18–21, Shenzhen, China, pp. 125–131. https://doi.org/10.1109/ICARM49381.2020.9195280.

Dunkel, J., Heidenreich, S., Drescher, K., Wensink, H. H., Bär, M., & Goldstein, R. E. (2013). Fluid dynamics of bacterial turbulence. *Physical Review Letters*, *110*(22), 228102. Available from https://doi.org/10.1103/PhysRevLett.110.228102.

Dutta, A., & Tarbell, J. M. (1996). Influence on non-Newtonian behavior of blood on flow in an elastic artery model. *Journal of Biomechanical Engineering*, *118*(1), 111−119. Available from https://doi.org/10.1115/1.2795936.

Evans, J. L., Ng, K. H., Wiet, S. G., Vonesh, M. J., Burns, W. B., Radvany, M. G., Kane, B. J., Davidson, C. J., Roth, S. I., Kramer, B. L., Meyers, S. N., & McPherson, D. D. (1996). Accurate three-dimensional reconstruction of intravascular ultrasound data. *Circulation*, *93*(3), 567−576. Available from https://doi.org/10.1161/01.CIR.93.3.567.

Feemster, M., Piepmeier, J. A., Biggs, H., Yee, S., ElBidweihy, H., & Firebaugh, S. L. (2020). Autonomous microrobotic manipulation using visual servo control. *Micromachines*, *11*(2), 132. Available from https://doi.org/10.3390/mi11020132.

Felfoul, O., Mohammadi, M., Taherkhani, S., de Lanauze, D., Xu, Y. Z., Loghin, D., Essa, S., Jancik, S., Houle, D., Lafleur, M., Gaboury, L., Tabrizian, M., Kaou, N., Atkin, M., Vuong, T., Batist, G., Beauchemin, N., Radzioch, D., & Martel, S. (2016). Magneto-aerotactic bacteria deliver drug-containing nanoliposomes to tumour hypoxic regions. *Nature Nanotechnology*, *11*, 941−947. Available from https://doi.org/10.1038/nnano.2016.137.

Feynman, R. P. (1992). There is plenty of room at the bottom. *Journal of Microelectromechanical Systems*, *1*(1), 60−66. Available from https://doi.org/10.1109/84.128057.

Field, R. D., Anandakumaran, P. N., & Sia, S. K. (2019). Soft microrobots: Design components and system integration. *Applied Physics Review*, *6*, 041305. Available from https://doi.org/10.1063/1.5124007.

Fountain, T.W.R., Kailat, P.V., & Abbott, J.J. (2010). Wireless control of magnetic helical microrobots using a rotating-permanent-magnet manipulator. In: *Proceedings of the IEEE international conference on robotics and automation*, May 3−8, 2010, Anchorage, Alaska, USA, pp. 576−581. Available from https://doi.org/10.1109/ROBOT.2010.5509245.

Garattini, S. (2007). Pharmacokinetics in cancer chemotherapy. *European Journal of Cancer*, *43*(2), 271−282. Available from https://doi.org/10.1016/j.ejca.2006.10.015.

Ghanbari, A. (2020). Bioinspired reorientation strategies for application in micro/nanorobotic control. *Journal of Micro-Bio Robotics*, *16*(2), 173−197. Available from https://doi.org/10.1007/s12213-020-00130-7.

Ghosh, A., Xu, W., Gupta, N., & Gracias, D. H. (2020). Active matter therapeutics. *Nano Today*, *31*, 100836. Available from https://doi.org/10.1016/j.nantod.2019.100836.

Go, G., Jeong, S.-G., Yoo, A., Han, J., Kang, B., Kim, S., Nguyen, K. T., Jin, Z., Kim, C.-S., Seo, Y. R., Kang, J. Y., Na, J. Y., Song, E. K., Jeong, Y., Seon, J. K., Park, J.-O., & Choi, E. (2020). Human adipose-derived mesenchymal stem cell-based medical microrobot system for knee cartilage regeneration in vivo. *Science Robotics*, *5*(38), eaay6626. Available from https://doi.org/10.1126/scirobotics.aay6626.

Goldfriend, T., Diamant, H., & Witten, T. A. (2015). Hydrodynamic interactions between two forced objects of arbitrary shape. I. Effect on alignment. *Physics Fluids*, *27*(12), 123303. Available from https://doi.org/10.1063/1.4936894.

Goubergrits, L., Wellnhofer, E., & Kertzscher, U. (2008). Choice and Impact of a non-Newtonian model for wall shear stress profiling of coronary arteries. *IFMBE Proceedings*, *20*, 111−114. Available from https://doi.org/10.1007/978-3-540-69367-3_30.

Haghdel, M., Kamali, R., Haghdel, A., & Mansoori, Z. (2017). Effects of non-Newtonian properties of blood flow on magnetic nanoparticle targeted drug delivery. *Nanomedicine Journal*, *4*(2), 89−97. Available from https://doi.org/10.22038/NMJ.2017.8410.

Higdon, J. J. L., & Muldowney, G. P. (1995). Resistance functions for spherical particles, droplets and bubbles in cylindrical tubes. *Journal of Fluid Mechanics*, *298*, 193−210. Available from https://doi.org/10.1017/S0022112095003272.

Honda, T., Arai, K. I., & Ishiyama, K. (1996). Micro swimming mechanisms propelled by external magnetic fields. *IEEE Transactions on Magnetics*, *32*(5), 5085–5087. Available from https://doi.org/10.1109/20.539498.

Hu, M., Ge, X., Chen, X., Mao, W., Qian, X., & Yuan, W.-E. (2020). Micro/nanorobot: A promising targeted drug delivery system. *Pharmaceutics*, *12*(7), 665. Available from https://doi.org/10.3390/pharmaceutics12070665.

Iacovacci, V., Ricotti, L., Signore, G., Vistoli, F., Sinibaldi, E., & Menciassi, A. (2019). Retrieval of magnetic medical microrobots from the bloodstream. In: *Proceedings of the international conference on robotics and automation*, May 20–24, Montreal, Canada, pp. 2495–2501. Available from https://doi.org/10.1109/ICRA.2019.8794322.

Ishikawa, T., Sekiya, G., Imai, Y., & Yamaguchi, T. (2007). Hydrodynamic interactions between two swimming bacteria. *Biophysical Journal*, *93*(6), 2217–2225. Available from https://doi.org/10.1529/biophysj.107.110254.

Jit, M., Ng, D. H. L., Luangasanatip, N., Sandmann, F., Atkins, K. E., Robotham, J. V., & Pouwels, K. B. (2020). Quantifying the economic cost of antibiotic resistance and the impact of related interventions: Rapid methodological review, conceptual framework and recommendations for future studies. *BMC Medicine*, *18*, 38. Available from https://doi.org/10.1186/s12916-020-1507-2.

Keller, J. B., & Rubinow, S. I. (1976). Swimming of flagellated microorganisms. *Biophysical Journal*, *16*(2), 151–170. Available from https://doi.org/10.1016/s0006-3495(76)85672-x, pt. 1.

Keya, J. J., Kabir, A. M. R., Inoue, D., Sada, K., Hess, H., Kuzuya, A., & Kakugo, A. (2018). Control of swarming of molecular robots. *Scientific Reports*, *8*, 11756. Available from https://doi.org/10.1038/s41598-018-30187-1.

Khalil, I. S. M., Mahdy, D., El Sharkawy, A., Moustafa, R. R., Tabak, A. F., Mitwally, M. E., Hesham, S., Hamdi, N., Klingner, A., Mohamed, A., & Sitti, M. (2018). Mechanical rubbing of blood clots using helical robots under ultrasound guidance. *IEEE Robotics and Automation Letters*, *3*(2), 1112–1119. Available from https://doi.org/10.1109/LRA.2018.2792156.

Khalil, I.S.M., Tabak, A.F., Hageman, T., Ewis, M., Picheli, M., Mitwally, M.E., El-Din, N.S., Abelmann, L., & Sitti, M. (2017). Near-surface effects on the controlled motion of magnetotactic bacteria. In: *Proceedings of the IEEE international conference on robotics and automation*, May 29–June 3, Singapore, pp. 5976–5982. Available from https://doi.org/10.1109/ICRA.2017.7989705.

Khalil, I. S. M., Tabak, A. F., Hamed, Y., Tawakol, M., Klingner, A., El Gohary, N. A., Mizaikoff, B., & Sitti, M. (2018). Independent actuation of two-tailed microrobots. *IEEE Robotics and Automation Letters*, *3*(3), 1703–1710. Available from https://doi.org/10.1109/LRA.2018.2801793.

Khalil, I. S. M., Tabak, A. F., Sadek, K., Mahdy, D., Hamdi, N., & Sitti, M. (2017). Rubbing against blood clots using helical robots: Modeling and in vitro experimental validation. *IEEE Robotics and Automation Letters*, *2*(2), 927–934. Available from https://doi.org/10.1109/LRA.2017.2654546.

Khalil, I. S. M., Tabak, A. F., Seif, M. A., Klingner, A., & Sitti, M. (2018). Controllable switching between planar and helical flagellar swimming of a soft robotic sperm. *PLoS One*, *13*(11), e0206456. Available from https://doi.org/10.1371/journal.pone.0206456.

Kim, J.-W., & Tung, S. (2015). Bio-hybrid micro/nanodevices powered by flagellar motor: Challenges and strategies. *Frontiers in Bioengineering and Biotechnology*, *3*, 100. Available from https://doi.org/10.3389/fbioe.2015.00100.

Kim, M.-J., & Power, T. R. (2004). Hydrodynamic interactions between rotating helices. *Physical Review E*, *69*, 061910. Available from https://doi.org/10.1103/PhysRevE.69.061910.

Kim, S. (1985). Sedimentation of two arbitrary oriented spheroids in a viscous fluid. *International Journal of Multiphase Flow*, *11*(5), 699–712. Available from https://doi.org/10.1016/0301-9322(85)90087-4.

Krasokha, N., Theisen, W., Reese, S., Mordasini, P., Brekenfeld, C., Gralla, J., Slotboom, J., Schrott, G., & Monstadt, H. (2010). Mechanical properties of blood clots — a new test method. *Materialwissenschaft und Werkstofftechnik*, *41*(12), 1019–1024. Available from https://doi.org/10.1002/mawe.201000703.

Krüger-Genge, A., Blocki, A., Franke, R.-P., & Jung, F. (2019). Vascular endothelial cell biology: An update. *International Journal of Molecular Science*, 20(18), 4411. Available from https://doi.org/10.3390/ijms20184411.

Kummer, M. P., Abbott, J. J., Kratochvil, B. E., Borer, R., Sengul, A., & Nelson, B. J. (2010). OctoMag: An electromagnetic system for 5-dof wireless micromanipulation. *IEEE Transactions on Robotics*, 26(6), 1006−1017. Available from https://doi.org/10.1109/TRO.2010.2073030.

Lauga, E., DiLuzio, W. R., Whitesides, G. M., & Stone, H. A. (2006). Swimming in circles: Motion of bacteria near solid boundaries. *Biophysical Journal*, 90(2), 400−412. Available from https://doi.org/10.1529/biophysj.105.069401.

Leaman, E. J., Sahari, A., Traore, M. A., Geuther, B. Q., Morrow, C. M., & Behkam, B. (2020). Data-driven statistical modeling of the emergent behavior of biohybrid microrobots. *APL Bioengineering*, 4(1), 016104. Available from https://doi.org/10.1063/1.5134926.

Li, N., & Cannon, M. C. (1998). Gas vesicle genes identified in *Bacillus* megaterium and functional expression in *Escherichia* coli. *Journal of Bacteriology*, 180(9), 2450−2458. Available from https://doi.org/10.1128/JB.180.9.2450-2458.1998.

Lin, W., Bazylinski, D. A., Xiao, T., Wu, L.-F., & Pan, Y. (2014). Life with compass: Diversity and biogeography of magnetotactic bacteria. *Environmental Microbiology*, 16(9), 2646−2658. Available from https://doi.org/10.1111/1462-2920.12313, special issue on metagenomics, and biomes in health and disease.

Ludwig, G. D. (1950). The velocity of sound through tissues and the acoustic impedance of tissues. *The Journal of the Acoustical Society of America*, 22(6), 862−866. Available from https://doi.org/10.1121/1.1906706.

Manamanchaiyaporn, L., Xu, T., & Wu, X. (2020). An optimal design of an electromagnetic actuation system towards a large homogeneous magnetic field and accessible workspace for magnetic manipulation. *Energies*, 13(14), 911. Available from https://doi.org/10.3390/en13040911.

Marath, N. K., & Wettlaufer, J. S. (2019). Hydrodynamic interactions and the diffusivity of spheroidal particles. *The Journal of Chemical Physics*, 151, 024107. Available from https://doi.org/10.1063/1.5096764.

Medina-Sánchez, M., Schwarz, L., Meyer, A. K., Hebenstreit, F., & Schmidt, O. G. (2016). Cellular cargo delivery: Towards assisted fertilization by sperm-carrying micromotors. *Nano Letters*, 16(1), 555−561. Available from https://doi.org/10.1021/acs.nanolett.5b04221.

Nguyen, K. T., Hoang, M. C., Go, G., Kang, B., Choi, E., Park, J.-O., & Kim, C.-S. (2020). Regularization-based independent control of an external electromagnetic actuator to avoid singularity in the spatial manipulation of a microrobot. *Control Engineering Practice*, 97, 104340. Available from https://doi.org/10.1016/j.conengprac.2020.104340.

Palagi, S., & Fischer, P. (2018). Bioinspired microrobots. *Nature Review Materials*, 3, 113−124. Available from https://doi.org/10.1038/s41578-018-0016-9.

Park, H., & Park, K. (1996). Biocompatibility issues of implantable drug delivery. *Pharmaceutical Research*, 13(12), 1770−1776. Available from https://doi.org/10.1023/A:1016012520276.

Park, J., Kim, J.-y, Pané, S., Nelson, B. J., & Choi, H. (2021). Acoustically mediated controlled drug release and targeted therapy with degradable 3D porous magnetic microrobots. *Advanced Healthcare Materials*, 10(2), 2001096. Available from https://doi.org/10.1002/adhm.202001096.

Perdikaris, P., Grinberg, L., & Karniadakis, G. E. (2016). Multiscale modeling and simulation of brain blood flow. *Physics Fluids*, 28, 021304. Available from https://doi.org/10.1063/1.4941315.

Peyer, K.E., Qiu, F., Zhang, L., & Nelson, B.J. (2012). Movement of artificial bacterial flagella in heterogeneous viscous environments at the microscale. In: *Proceedings of the IEEE/RSJ international conference on intelligent robotics and systems*, October 7−12, Vilamoura, Algarve, Portugal, pp. 2553−2558. https://doi.org/10.1109/IROS.2012.6386096.

Pham, L. N., Steiner, J. A., Leang, K. K., & Abbott, J. J. (2020). Soft endoluminal robots propelled by rotating magnetic dipole fields. *IEEE Transactions on Medical Robotics and Bionics*, 2(4), 598−607. Available from https://doi.org/10.1109/TMRB.2020.3027871.

Pittiglio, G., Chandler, J.H., Richter, M., Venkiteswaran, K., Misra, S., & Valdastri, P. (2020). Dual-arm control for enhanced magnetic manipulation. In: *Proceedings of the international conference on intelligent robots and systems*, October 25−29, Las Vegas, NV, USA, pp. 7211−7218. Available from https://doi.org/10.1109/IROS45743.2020.9341250.

Rahman, M. M., Md., Hossain, A., Mamun, K., Most., & Akhter, N. (2018). Comparative study of Newtonian and non-Newtonian blood flow through a stenosed carotid artery. *AIP Conference Proceedingds*, *1980*, 040017. Available from https://doi.org/10.1063/1.5044327.

Rahman, M.S., & Haque, M.A., 2012. Mathematical modeling of blood flow. In: *Proceedings of the IEEE/ OSA/IPAR international conference on informatics, electronics & vision*, May 18−19, Dhaka, Bangladesh, pp. 672−676. Available from https://doi.org/10.1109/ICIEV.2012.6317446.

Reichert, M., & Stark, H. (2005). Synchronization of rotating helices by hydrodynamic interactions. *European Physical Journal E*, *17*, 493−500. Available from https://doi.org/10.1140/epje/i2004-10152-7.

Remes, A., & Williams, D. F. (1992). Immune response in biocompatibility. *Biomaterials*, *13*(11), 731−743. Available from https://doi.org/10.1016/0142-9612(92)90010-l.

Rim, Y., McPherson, D. D., & Kim, H. (2013). Volumetric three-dimensional intravascular ultrasound visualization using shape-based nonlinear interpolation. *Biomedical Engineering Online*, *12*, 39. Available from https://doi.org/10.1186/1475-925X-12-39.

Ryan, P., & Diller, E. (2017). Magnetic actuation for full dexterity microrobotic control using rotating permanent magnets. *IEEE Transactions on Robotics*, *33*(6), 1398−1409. Available from https://doi.org/10.1109/TRO.2017.2719687.

Sadelli, L., Fruchard, M., & Ferreira, A. (2017). 2D observer-based control of a vascular microrobot. *IEEE Transactions on Automatic Control*, *62*(5), 2194−2206. Available from https://doi.org/10.1109/TAC.2016.2604045.

Salmanipour, S., Yousefi, O., & Diller, E. D. (2021). Design of multi-degrees-of-freedom microrobot driven by homogeneous quasi-static magnetic fields. *IEEE Transactions on Robotics 37*(1), 246−256. Available from https://doi.org/10.1109/TRO.2020.3016511.

Samsami, K., Mirbagheri, S. A., Meshkati, F., & Fu, H. C. (2020). Stability of soft magnetic helical microrobots. *Fluids*, *5*(1), 19. Available from https://doi.org/10.3390/fluids5010019.

Sankar, D. S., & Hemalatha, K. (2007). A non-Newtonian fluid model for blood flow through a catheterized artery − steady flow. *Applied Mathematical Modelling*, *31*(9), 1847−1864. Available from https://doi.org/10.1016/j.apm.2006.06.009.

Schamel, D., Mark, A. G., Gibbs, J. G., Miksch, C., Morozov, K. I., Leshansky, A. M., & Fischer, P. (2014). Nanopropellers and their actuation in complex viscoelastic media. *ACS Nano*, *8*(9), 8794−8801. Available from https://doi.org/10.1021/nn502360t.

Schwarz, L., Karnaushenko, D. D., Hebenstreit, F., Naumann, R., Schmidt, O. G., & Medina-Sánchez, M. (2020). A rotating spiral micromotor for noninvasive zygote transfer. *Advancement of Science*, *7*(18), 2000843. Available from https://doi.org/10.1002/advs.202000843.

Schwarz, L., Medina-Sánchez, M., & Schmidt, O. G. (2020). Sperm-hybrid micromotors: On-board assistance for nature's bustling swimmers. *Reproduction (Cambridge, England)*, *159*(2), R83−R96. Available from https://doi.org/10.1530/REP-19-0096.

Servant, A., Qui, F., Mazza, M., Kostarelos, K., & Nelson, B. J. (2015). Controlled in vivo swimming of a swarm of bacteria-like microrobotic flagella. *Advanced Materials*, *27*(19), 2981−2988. Available from https://doi.org/10.1002/adma.201404444.

Shi, Y., Lawford, P., & Hose, R. (2011). Review of zero-D and 1-D models of blood flow in the cardiovascular system. *Biomedical Engineering Online*, *10*, 33. Available from https://doi.org/10.1186/1475-925X-10-33.

Siddiqui, M., & Rajkumar, S. V. (2012). The high cost of cancer drugs and what we can do about it. *Mayo Clinic Proceedings. Mayo Clinic Proceedings*, *87*(10), 935−943. Available from https://doi.org/10.1016/j.mayocp.2012.07.007.

Sir Taylor, G. (1951). Analysis of the swimming of microscopic organisms. *Proceedings of the Royal Society London A*, *209*(1099), 447−461. Available from https://doi.org/10.1098/rspa.1951.0218.

Sitti, M., & Wiersma, D. S. (2020). Pros and cons: Magnetic vs optical microrobots. *Advanced Materials*, *32*(20), 1906766. Available from https://doi.org/10.1002/adma.201906766, special issue on 'from responsive materials to interactive materials'.

Spong, M. W., & Vidyasagar, M. (1989). *Robot dynamics and control*. N.Y.: John Wiley & Sons, 1989, (Chapter 9).

Sriram, K., Intaglietta, M., & Tartakovsky, D. M. (2014). Non-Newtonian flow of blood in arterioles: Consequences for wall shear stress measurements. *Microcirculation*, *21*(7), 628−639. Available from https://doi.org/10.1111/micc.12141.

Sun, Z., Popp, P. E., Loderer, C., & Revilla-Guarinos, A. (2020). Genetically engineered bacterial biohybrid microswimmers for sensing applications. *Sensors*, *20*(1), 180. Available from https://doi.org/10.3390/s20010180.

Tabak, A. F. (2018). Hydrodynamic impedance of bacteria and bacteria-inspired microswimmers: A new strategy to predict power consumption of swimming micro-robots for real-time applications. *Advanced Theory and Simulations*, *1*(4), 1700013. Available from https://doi.org/10.1002/adts.201700013.

Tabak, A. F. (2019a). Hydrodynamic impedance correction for reduced-order modeling of spermatozoa-like soft micro-robots. *Advanced Theory and Simulations*, *2*(2), 1800130. Available from https://doi.org/10.1002/adts.201800130.

Tabak, A. F. (2019b). Bioinspired and biomimetic micro-robotics for therapeutic applications. In J. Segil (Ed.), *Handbook of biomechatronics*. UK: Elsevier Academic Press. Chapter 12). Available from https://doi.org/10.1016/B978-0-12−812539-7.00010-6.

Tabak, A. F. (2020a). Bilateral control simulations for a pair of magnetically-coupled robotic arm and bacterium for in vivo applications. *Journal of Micro-Bio Robotics*, *16*(2), 199−214. Available from https://doi.org/10.1007/s12213-020-00138-z.

Tabak, A.F. (2020b). Simulated bilateral motion control of a magneto-tactic bacterium via an open kinematic chain. In: *Proceedings of the 2020 IEEE international conference on ubiquitous robots*, June 22−26, Kyoto, Japan, pp. 1−6. Available from https://doi.org/10.1109/UR49135.2020.9144834.

Tabak, A.F. (2020c). Adaptive motion control of modified *E. coli*. In: *Proceedings of the international congress on human-computer interaction, optimization and robotic applications*, June 26−28, Ankara, Turkey, pp. 1−5. Available from https://doi.org/10.1109/HORA49412.2020.9152603.

Tabak, A.F., Temel, F.Z., & Yesilyurt, S. (2011). Comparison on experimental and numerical results for helical swimmers inside channels. In: *Proceedings of the 2011 IEEE/RSJ international conference on intelligent robots and systems*, September 25−30, San Francisco, California, pp. 463−468. Available from https://doi.org/10.1109/IROS.2011.6094620.

Tabak, A. F., & Yesilyurt, S. (2014). Improved kinematic models for two-link helical micro/nanoswimmers. *IEEE Transactions on Robotics*, *30*(1), 14−25. Available from https://doi.org/10.1109/TRO.2013.2281551, special issue on nanorobotics.

Ullrich, F., Fusco, S., Chatzipirpiridis, G., Pané, S., & Nelson, B. J. (2014). Recent progress in magnetically actuated microrobotics for ophthalmic therapies. *European Ophthalmic Review*, *8*(2), 120−126. Available from https://doi.org/10.17925/EOR.2014.08.02.120.

Vasan, N., Baselga, J., & Hyman, D. M. (2019). A review on drug resistance in cancer. *Nature*, *575*, 299−309. Available from https://doi.org/10.1038/s41586-019-1730-1.

Wang, J., Dong, R., Wu, H., Cai, Y., & Ren, B. (2020). A review on artificial micro/nanomotors for cancer-targeted delivery, diagnosis, and therapy. *Nano-Micro Letters*, *12*, 11. Available from https://doi.org/10.1007/s40820-019-0350-5.

Wang, Q., & Zhang, L. (2020). Ultrasound imaging and tracking of micro/nanorobots: From individual to collectives. *IEEE Open Journal of Nanotechnology*, *1*, 6−17. Available from https://doi.org/10.1109/OJNANO.2020.2981824.

Wang, S., & Ardekani, A. M. (2012). Unsteady swimming of small organisms. *Journal of Fluid Mechanics*, *702*, 286−297. Available from https://doi.org/10.1017/jfm.2012.177.

Webb, B. (2006). Validating biorobotic models. *Journal of Neural Engineering*, *3*, R25−R35. Available from https://doi.org/10.1088/1741-2560/3/3/R01.

Wen, J., Liu, K., Khoshmanesh, K., Jiang, W., & Zheng, T. (2015). Numerical investigation of haemodynamics in a helical-type artery bypass graft using non-Newtonian multiphase model. *Computer Methods in Biomechanics and Biomedical Engineering*, *18*(7), 760−768. Available from https://doi.org/10.1080/10255842.2013.845880.

Womersley, J. R. (1955). Method for the calculation of velocity, rate of flow and viscous drag in arteries when the pressure is known. *The Journal of Physiology*, *127*(3), 553−563. Available from https://doi.org/10.1113/jphysiol.1955.sp005276.

Wu, Z., Chen, Y., Mukasa, D., Pak, O. S., & Gao, W. (2020). Medical micro/nanorobots in complex media. *Chemical Society Reviews*, *49*, 8088−8112. Available from https://doi.org/10.1039/D0CS00309C.

Xie, M., Zhang, W., Fan, C., Wu, C., Feng, Q., Wu, J., Li, Y., Gao, R., Li, Z., Wang, Q., Cheng, Y., & He, B. (2020). Bioinspired soft microrobots with precise magneto-collective control for microvascular thrombolysis. *Advanced Materials*, *32*(26), 2000366. Available from https://doi.org/10.1002/adma.202000366.

Xu, H., Medina-Sánchez, M., Maitz, M. F., Werner, C., & Schmidt, O. G. (2020). Sperm-micromotors for cargo delivery through flowing blood. *ACS Nano*, *14*(3), 2982−2993. Available from https://doi.org/10.1021/acsnano.9b07851.

Xu, T., Soto, F., Gao, W., Dong, R., Garcia-Gradilla, V., Magaña, E., Zhang, X., & Wang, J. (2015a). Reversible swarming and separation of self-propelled chemically powered nanomotors under acoustic fields. *Journal of the American Chemical Society*, *137*(6), 2163−2166. Available from https://doi.org/10.1021/ja511012v.

Xu, T., Yu, J., Yan, X., Choi, H., & Zhang, L. (2015b). Magnetic actuation based motion control for microrobots: An overview. *Micromachines*, *6*(9), 1346−1364. Available from https://doi.org/10.3390/mi6091346.

Yan, Y., Jing, W., & Mehrmohammadi, M. (2020). Photoacoustic imaging to tract magnetic-manipulated micro-robots in deep tissue. *Sensors*, *20*(10), 2816. Available from https://doi.org/10.3390/s20102816.

Yang, Z., Yang, L., & Zhang, L. (2020). 3-D visual servoing of magnetic miniature swimmers using parallel mobile coils. *IEEE Transactions on Medical Robotics and Bionics*, *2*(4), 608−618. Available from https://doi.org/10.1109/TMRB.2020.3033020.

Yasa, I. C., Ceylan, H., Bozuyuk, U., Wild, A.-M., & Sitti, M. (2020). Elucidating the interaction dynamics between microswimmer body and immune system for medical microrobots. *Science Robotics*, *5*(43), eaaz3867. Available from https://doi.org/10.1126/scirobotics.aaz3867.

Yasa, I. C., Tabak, A. F., Yasa, O., Ceylan, H., & Sitti, M. (2019). 3D-printed microrobotic transporters with recapitulated stem cell niche for programmable and active cell delivery. *Advanced Functional Materials*, *29*(17), 1808992. Available from https://doi.org/10.1002/adfm.201808992.

Ye, C., Liu, J., Wu, X., Wang, B., Zhang, L., Zheng, Y., & Xu, T. (2019). Hydrophobicity influence on swimming performance of magnetically driven miniature helical swimmers. *Micromachines*, *10*(3), 175. Available from https://doi.org/10.3390/mi10030175.

Zhang, L., Abbott, J. J., Dong, L., Kratochvil, B. E., Bell, D., & Nelson, B. J. (2009). Artificial bacterial flagella: Fabrication and magnetic control. *Applied Physics Letters*, *94*(6), 064107. Available from https://doi.org/10.1063/1.3079655.

Effects of nanoparticles on the blood coagulation system (nanoparticle interface with the blood coagulation system)

6

Huong D.N. Tran[1,2], Fahima Akther[1,2], Zhi Ping Xu[2] and Hang T. Ta[1,2,3]

[1]*Queensland Micro- and Nanotechnology, Griffith University, Nathan, QLD, Australia* [2]*Australian Institute for Bioengineering and Nanotechnology, University of Queensland, St Lucia, QLD, Australia* [3]*School of Environment and Science, Griffith University, Nathan, QLD, Australia*

6.1 Introduction

The blood coagulation system, which comprises cells and plasma coagulation factors that mediate hemostasis at the injury sites, is considered to be critical to the human body (Ilinskaya & Dobrovolskaia, 2013; Martin et al., 2018). Upon injury, damaged endothelial cells expose subendothelial collagens for the initiation of primary hemostasis where platelets aggregate and form a temporary platelet plug. It is followed by secondary hemostasis where a coagulation cascade with the involvement of clotting factors occurs. This results in a fibrin mesh that entraps the platelet plug and red blood cells (RBCs) to form a blood clot and stops the bleeding (Gaston et al., 2018; Rumbaut & Thiagarajan, 2010). Interaction between the normal coagulation system and the fibrinolytic system maintains the delicate thrombohemorrhagic balance in the body. Any interruption in the blood coagulation system might lead to the consequences of either abnormal thrombosis or hemorrhage (de la Harpe et al., 2019; Hante et al., 2019).

The pivotal role of nanoparticles in biomedicine has been affirmed with the continuously increasing number of their applications (Arndt et al., 2020; Bao et al., 2018; Do et al., 2020; Gu et al., 2018; Le, Bach, et al., 2019; Le, Pham, et al., 2019; Nguyen, Nguyen-Tran, et al., 2019; Reimhult, 2019). Nanoparticles can potentially be employed for diagnosis, prevention, and treatment of various diseases (e.g., cancer, cardiovascular disease, degenerative disease, infectious diseases) or for regenerative medicine (de la Harpe et al., 2019; Gu et al., 2018; Hoang Thi et al., 2019; Matus et al., 2018; Nguyen, Bach, et al., 2019; Nguyen, Nguyen-Tran, et al., 2019; Thi et al., 2020; Tran et al., 2019; Tran et al., 2020). Despite the extensive modification and development of nanoparticles with versatile designs, very few can be translated into the clinic due to the lack of the full assessment of their health risks as there are always cell−nanoparticle or blood−nanoparticle interactions in the blood (de la Harpe et al., 2019; Setyawati et al., 2015). Regardless of the administration route and the intended target, nanoparticles can reach the circulatory system due to the ability to permeate epithelium after dermal penetration, oral ingestion, or inhalation (de la Harpe et al., 2019; Fröhlich, 2016). Once inside the bloodstream, they encounter components of the complex physiological environments, including the coagulation system (Hante et al., 2019; Ilinskaya & Dobrovolskaia, 2016; Sobot et al., 2014) (Fig. 6.1). Interaction between nanoparticles and the coagulation system components can potentially interfere with the hemostatic balance in unintended ways, causing lethal coagulation disorders such as disseminated intravascular coagulation and deep vein thrombosis, and thus raise concerns about the clinical safety of these nanoparticles (Fröhlich, 2016; Ilinskaya & Dobrovolskaia, 2016). Therefore making efforts to understand the effect of nanoparticles on the blood coagulation system is highly essential.

The purpose of this chapter is to give a complete depiction of the interface between nanoparticles and the blood coagulation system, which is beneficial for the engineering of nanoparticles and their successful translation to the clinic and market. We present how the nanoparticles interact and affect each component of the coagulation system, and then discuss common in vitro methods to

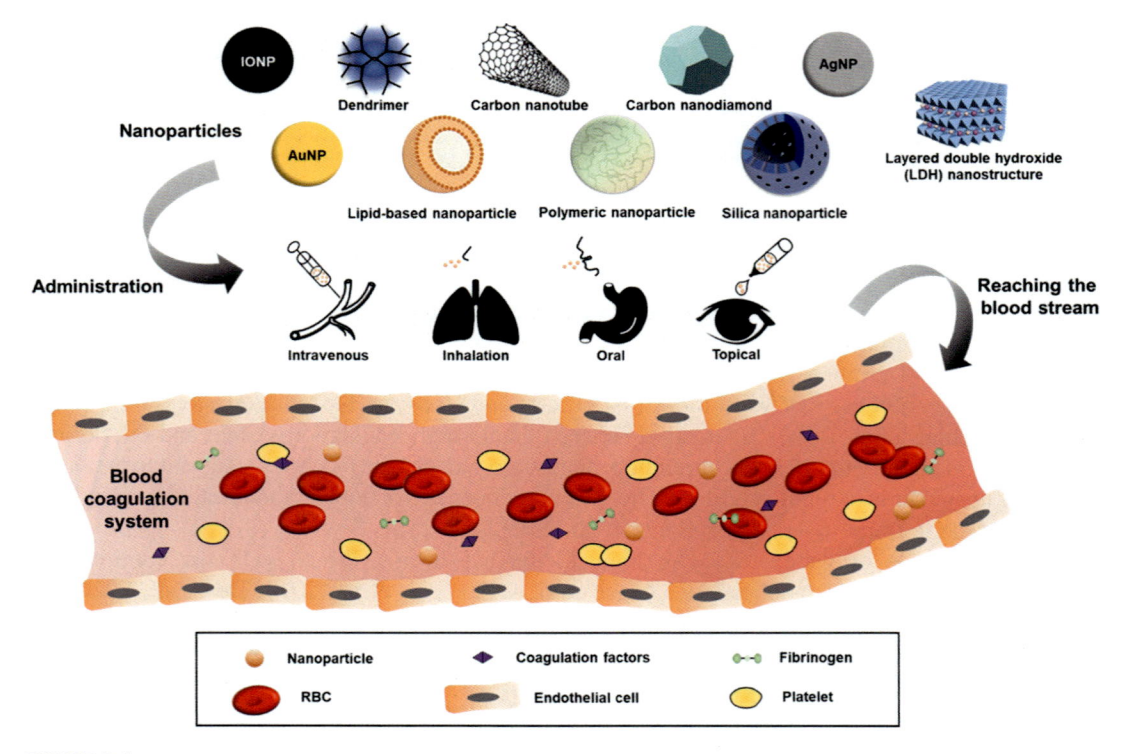

FIGURE 6.1

Nanoparticles encounter the blood coagulation system. Owing to their ability to permeate epithelium, nanoparticles can reach the circulatory system regardless of the administration routes. Once inside the bloodstream, nanoparticles encounter and interact with one or more components of the coagulation system, namely platelets, red blood cells (RBCs), endothelial cells, and plasma coagulation factors. RBCs tend to move to the center of vessels and push platelets toward the periphery, facilitating the collision of platelets and nanoparticles with the vascular endothelium for hemostatic events.

evaluate their effects. Various nanoparticles' parameters that influence the nanoparticle–coagulation system interface will also be discussed. The final section of this chapter addresses two-side effects of engineered nanoparticles for the desirable alteration of hemostasis.

6.2 Interaction of nanoparticles with the blood coagulation system—the underlying mechanisms

Being inside the bloodstream, nanoparticles encounter many blood components and biological systems, including the blood coagulation system. Unintended interactions of nanoparticles with the coagulation system can result in a dysregulation of the hemostatic balance (de la Harpe et al., 2019; Hante et al., 2019). The roles of each component in the coagulation system and possible influences of nanoparticles on them along with the underlying molecular mechanisms will be discussed in the following subsections.

6.2.1 Nanoparticles and vascular endothelium

The studies of nanoparticles associated with the coagulation system usually focus on blood cells and coagulant factors. However, vascular endothelial cells play an important role in the regulation of platelet adhesion, thrombosis, and fibrinolysis (Fröhlich, 2016). Healthy endothelial cells are protected by a glycocalyx layer consisting of heparan sulfate that has the affinity for plasma inhibitory proteins such as antithrombin III and tissue factor pathway inhibitor (TFPI). These proteins, anticoagulant mediators (heparin cofactor II, endothelial protein C receptor (EPCR), and thrombomodulin) expressed on the endothelium surface together with platelet adhesion and aggregation inhibitors [nitric oxide (NO), prostacyclin (PGI_2), and CD39/NTPDase1] secreted by endothelium, help to maintain the thrombo-resistant nature of intact vascular endothelial cells (Ekdahl et al., 2019; Hangge et al., 2017; Reitsma et al., 2007; Yau et al., 2015).

Damage to endothelial cells not only leads to the exposure of tissue factor (TF) (CD142 or FIII), which activates the extrinsic pathway of hemostasis, but also exposes subendothelial collagens that bind FXII to initiate the intrinsic pathway. Moreover, von Willebrand factor (vWF), thromboxane A2 (TXA2), P-selectin (CD62P/GMP-140/PADGEM), and platelet-activating factor (PAF) released by injured endothelial cells along with the exposed collagens are associated with platelets recruitment, adhesion, and activation (Fröhlich, 2016; Ilinskaya & Dobrovolskaia, 2013). Gelderman et al. reported that fullerenol $C_{60}(OH)_{24}$ nanoparticles at $100\,\mu g/mL$ significantly triggered the expression of TF (CD142) on human umbilical vein endothlial cells (HUVECs) after 24 hours of in vitro culture (4% ± 2% CD142[+] cells in control versus 54% ± 20% CD142[+] cells in treatment group) (Gelderman et al., 2008). As also reported, silica nanoparticles (58 nm), especially at high concentration of 50 and $100\,\mu g/mL$, interrupted the NO balance, leading to HUVECs dysfunction (Guo et al., 2015). The interactions between silver nanoparticles (AgNPs) and endothelial cells' membrane, the induction of vWF release, and the reduction of tissue plasminogen activator (tPA) expression at high nanoparticles concentration are the reasons behind endothelium dysfunction caused by Ag NPs and are probably associated with thromboembolic complications (Ragaseema et al., 2012; Sun et al., 2016). After 24 hours of incubation with HUVECs, Ag NPs

(<20 nm) induced cytotoxicity at the concentration of 64 μg/mL (Danielsen et al., 2015), while the toxicity threshold for ZnO NPs (70 nm) toward human aortic endothelial cells (HAECs) was ≥ 15 μg/mL (Liang et al., 2016). Similar to Ag NPs, cationic dendrimer nanoparticles interacted with HUVECs' membrane, and poly(amidoamine) (PAMAM) dendrimer generation 4 and 7 (G4 and G7) (administration doses >10 mg/kg) caused disseminated intravascular coagulation in mice (Greish et al., 2012). Silica nanoparticles, as demonstrated by Feng et al., caused hypercoagulation through inducing vascular endothelial cells dysfunction (Feng et al., 2019). The increased expression of TF and platelet endothelial cell adhesion molecule-1 (PECAM-1 or CD31), as well as the imbalance of the NO/NOS (nitric oxide synthase) system, were detected after the exposure to silica nanoparticles (starting from 1.8 mg/kg of rat).

Organic nanoparticles, by contrast, have little toxicity to vascular endothelial cells. For instance, Liu et al. demonstrated that exposure of HUVECs to mPEG-PLA nanoparticles (around 20 nm) showed no significant effect on the cell viability at the concentration up to 200 μg/mL (Liu et al., 2017). Liposomes (109 and 139 nm) and lipid NPs (50−120 nm) had no cytotoxicity to HUVECs at the concentration up to 100 μg/mL (Matuszak et al., 2016).

6.2.2 Nanoparticles and platelets

Platelet (thrombocyte) is a crucial cellular component in the coagulation system. It is originated from megakaryocytes, anucleate, discoid in shape, and around 2−4 μm in diameter. Around 33% of all platelets is stored in the spleen while the rest circulates in the circulatory system ($\sim$150,000−450,000 platelets/mm^3) without adhering to the intact vascular endothelium (Matus et al., 2018; Nabeshi et al., 2012). Upon injury, the damaged endothelium exposes TF, collagen, and other thrombogenic factors, such as vWF, TXA2, and PAF, for the initiation of primary hemostasis. Once platelets come in contact with vWF and subendothelial collagens for the adhesion to the injured endothelium and vessel wall, they become activated (Broos et al., 2011; Demir et al., 2013; Palta et al., 2014; Rumbaut & Thiagarajan, 2010). The platelet activation process is portrayed by a drastic increase in cytosolic Ca^{2+}, which elicits the reorganization of platelet cytoskeleton, resulting in the shape change (from disk to sphere shape), pseudopodia formation, aggregation, and exocytosis of contents stored inside platelet's granules (De La Cruz et al., 2018) (Table 6.1). Adhesive glycoproteins (vWF, fibrinogen, P-selectin, thrombospondin, and vitronectin), coagulation factors (plasminogen, kininogen, factor V, XI, XIII), plasminogen activator inhibitor-1 (PAI-1), TXA2, PAF, adenosine diphosphate (ADP), and serotonin secreted by activated platelets mediate vasoconstriction and platelet aggregation, activate more platelets and attract them to come to form a weak platelet plug that temporarily seals the injured area (De La Cruz et al., 2018; Hangge et al., 2017; Nabeshi et al., 2012; Rumbaut & Thiagarajan, 2010). There are a certain number of glycoprotein IIb/IIIa (GpIIb/IIIa) receptors presented on the surface of resting platelets (approximately 50,000/platelet) (Sims et al., 1991). Upon activation, GpIIb/IIIa stored in the internal pool of platelets will move to their surface, increasing the number of expressed GpIIb/IIIa. These receptors, both the newly expressed and the already presented ones, undergo a conformation change process, which is related to extracellular ionized calcium and the expression of ligand-induced binding sites to become high-affinity for fibrinogen (Kleiman et al., 1995; Matzdorff & Voss, 2006; Sims et al., 1991). Fibrin forms the bridge between platelets and entraps the platelet plug and other surrounding

Table 6.1 Platelet storage granules and their contents.

Granules	Content class	Factors released
Alpha granules	Adhesive glycoproteins	vWF, thrombospondin, P-selectin, fibrinogen, fibronectin, vitronectin
	Coagulation factors	Plasminogen, kininogens, protein S, factor V, factor XI, factor XIII
	Growth factors	IGF, EGF, PDGF, TGF-β
	Angiogenic factors	PF4 inhibitor, VEGF
	Protease inhibitors	C1-inhibitor, PAI-1, TFPI, α2-antiplasmin, α2-antitripsin, α2-macroglobulin
	Immunoglobulins−chemokines	IL8, IL1β, CD40, CXCL4 (platelet basic protein/NAP-2), CXCL (PF4), CXCL1, CXCL5, CCL5 (RANTES), CCL (MIP-1α)
	Proteases	MMP2, MMP9
Dense granules (or delta granules)	Amines	Serotonin, histamine
	Bivalent cations	Ca^{2+}, Mg^{2+}
	Polyphosphates	ADP, ATP, GDP, GTP
Lysosome granules	Enzymes	Acid proteases, glycohydrolases
Other soluble mediators	NO, TXA2, defensins, PAF	

ADP, adenosine diphosphate; ATP, adenosine triphosphate; EGF, epidermal growth factor; GDP, guanosine diphosphate; GTP, guanosine triphosphate; IGF, insulin-like growth factor; MMP, matrix metalloproteinase; NO, nitric oxide; PAF, platelet-activating factor; PAI-1, plasminogen activator inhibitor-1; PDGF, platelet-derived growth factor; PF4, platelet factor 4; TFPI, tissue factor pathway inhibitor; TGF-β, transforming growth factor β; TXA2, thromboxane A2; VEGF, vascular endothelial growth factor; vWF, von Willebrand factor.
Reproduced with permission from de la Harpe, K. M., Kondiah P. P., Choonara Y. E., Marimuthu T., du Toit L. C., & Pillay V. (2019). The hemocompatibility of nanoparticles: A review of cell−nanoparticle interactions and hemostasis. Cells, 8(10), 1209.

blood cells to form a stable clot (De La Cruz et al., 2018; Hangge et al., 2017; Nabeshi et al., 2012; Rumbaut & Thiagarajan, 2010).

Generally, the interactions of nanoparticles with platelets can affect platelet functions. Different types of nanoparticles with varied size, charge, coating materials, and composition may lead to different outcomes, including activating effect, inhibitory effect, or no effect on platelets.

6.2.2.1 Inorganic nanoparticles

6.2.2.1.1 Carbon-based nanoparticles

The tendency to stimulate platelet aggregation of various types of carbon nanoparticles, including mixed carbon nanoparticles (MCN), single-wall carbon nanotubes (SWCNT), and multiwall carbon nanotubes (MWCNT) (0.2−300 μg/mL), was evaluated and compared with standard urban particulate matter (SRM1648, 1.4 μm) in a study by Radomski et al. Radomski et al. (2005). The results showed that all tested materials induced the platelet aggregation and increased the vascular thrombosis rate in rat carotid arteries model in the order from highest to lowest: MCN ≥ SWCNT > MWCNT > SRM1648. The platelet aggregation induced by these carbon nanoparticles corresponded to the activation of the GpIIb/IIIa receptors and correlated with platelet degranulation, the translocation of P-selectin to the platelet surface, and the tendency to mimic molecular bridges in platelet–platelet interaction. The prothrombotic effect of carbon nanotubes regarding platelet activation and aggregation was further explored in studies by Simak group (Semberova et al., 2009). The results were consistent with the previous study in which SWCNTs (outer diameter <2 nm, $5-15$ μm in length for S1 SWCNT and $1-2$ nm of outer diameter, $5-30$ μm in length for S2 SWCNT) had higher platelet aggregation ($34\% \pm 5\%$ for S1 and $32\% \pm 6\%$ for S2) than MWCNTs (outer diameter was $60-100$ nm, $1-2$ μm in length for M60 and 30 ± 15 nm of outer diameter, $1-5$ μm in length for M30) with platelet aggregation of $27\% \pm 3\%$ (M60) and $38\% \pm 9\%$ (M30). Amorphous carbon nanopowder (outer diameter was ~ 30 nm) showed a weak effect on platelet aggregation ($15\% \pm 2\%$) while fullerenol C60 (~ 1.3 nm), fullerene C60 (~ 0.7 nm), and polystyrene nanobeads (PBs) (20 and 200 nm) had no effect. They reported that the effects of carbon nanotubes on platelet activation, degranulation, and aggregation were accompanied by elevated intracellular $[Ca^{2+}]$ in platelets which is the second key messenger mediating platelet activation. Platelets raised intracellular $[Ca^{2+}]$ by either releasing it from intracellular stores or entering of extracellular Ca^{2+} through plasma membrane channels including SOCE, second messenger-operated Ca^{2+} entry (SMOC), and receptor-operated Ca^{2+} entry (Semberova et al., 2009). As the carbon nanotube-facilitated extracellular Ca^{2+} influx was sensitive to calcium entry blockers 2-APB and SKF 96365, SOCE was proved to be involved in platelet activation induced by carbon nanotubes (De Paoli Lacerda et al., 2011; Semberova et al., 2009). Corbalan et al. proposed that MWCNTs ruptured the dense tubular system—a Ca^{2+} pool—after penetrating the instantly resealed platelet membrane, leading to intracellular Ca^{2+} depletion and activating SOCE (Corbalan et al., 2012).

Carbon nanodiamonds (CNDs) with the size range of $4-10$ nm can evoke platelet activation at low concentration (1 μg/mL). Kumari et al. demonstrated that CNDs elevated the intracellular Ca^{2+} level in platelets and increased the expression of phosphatidylserine on the platelet membrane (Kumari et al., 2014). CND-treated platelets showed reduced viability and altered morphology with developed lamellipodia or filopodia. In vivo results evidenced extensive pulmonary thromboembolism in mice after IV injection of CNDs (Kumari et al., 2014).

6.2.2.1.2 Silver nanoparticles

The accumulation of silver nanoparticles (AgNPs) within platelets can interfere with intraplatelet activities (Jun et al., 2011; Krajewski et al., 2013; Ragaseema et al., 2012; Shrivastava et al., 2009). It has been found that AgNPs, $10-100$ nm in diameter, induced intracellular $[Ca^{2+}]$ (250 μg/mL of AgNPs), which upregulated GpIIb/IIIa (100 μg/mL of AgNPs) and P-selectin expression (100 μg/mL of AgNPs), and serotonin secretion (250 μg/mL of AgNPs) (Jun et al., 2011). Enhanced thrombin and phosphatidylserine generation (250 μg/mL of AgNPs) were observed in fresh human platelets as evidence for platelet aggregation induced by AgNPs. Both AgNPs-induced platelet activation and aggregation were concentration-dependent. Exposure to AgNPs ($0.05-0.1$ mg/kg I.V. or $5-10$ mg/kg intratracheal instillation) in vivo enhanced venous thrombus formation, platelet aggregation, and PS externalization in rat. Accumulated AgNPs (stabilized with sodium polyacrylate, 30 mg/L, $10-15$ nm) triggered α-granule secretion and induced kallikrein-like, FXIIa-like, and thrombin–antithrombin III complex (Krajewski et al., 2013). Further exposure of AgNPs in rat ($0.05-0.1$ mg/kg intravenous or $5-10$ mg/kg intratracheal instillation) induced platelet aggregation, phosphatidylserine externalization, and vascular thrombus formation ex vivo (Jun et al., 2011; Krajewski et al., 2013). In another study, AgNPs (16 nm) only promoted platelet adhesion but not platelet aggregation at the concentration of 50 μg/mL (Laloy et al., 2014). At concentrations up to ~40 μg/mL, neither AgNPs (20 nm) with polyvinyl pyrrolidone coating nor with citrate coating exert any effect on platelet aggregation and coagulation (Huang et al., 2016). However, some other studies reported the antiplatelet properties of AgNPs (stabilized with either citrate or D-glucose) (Bandyopadhyay et al., 2012; Shrivastava et al., 2009). Accumulative AgNPs within platelet granules impeded integrin-mediated platelet responses such as adhesion to immobilized fibrinogen and platelet conformation change, namely retraction of a fibrin clot, in a concentration-dependent manner in vitro and in vivo, regardless of agonists used (Bandyopadhyay et al., 2012; Shrivastava et al., 2009). AgNPs also inhibited platelet aggregation induced by either ADP, thrombin, or collagen in vitro and in mouse whole blood in a dose-dependent manner (Bandyopadhyay et al., 2012). Different dispersing media/coatings/stabilizers used for synthesized AgNPs may be one of the possible reasons for the variability between these studies (Strojan et al., 2017).

6.2.2.1.3 Gold nanoparticles

The effect of gold nanoparticles (AuNPs) on platelets was first demonstrated in rat by Berry et al. (1977). The presence of a high amount of AuNPs in platelets of alveolar capillaries affected platelet aggregation, leading to microthombus and atheromatous plaques formation. Deb et al. presented that the molecular mechanism of platelet aggregation induced by AuNPs (stabilized with citrate) is linked to degranulation and the increased expression level of P-selectin and tyrosine phosphorylation (Deb et al., 2011). The study revealed that platelet response constantly decreased with the increment in the size of AuNPs where AuNPs greater than 60 nm (>40 μM) were inert to platelet as compared to the maximal platelet activation effect of smaller ones (~20 nm) at 40 μM. This might be attributed to the higher accumulation of small AuNPs in platelets (Deb et al., 2011). By contrast, Love et al. found that either AuNPs, Au(+) nanoparticles, or Au(−) nanoparticles (respectively stabilized with either citrate, 11-mercaptoundecanoic acid, or 11-mercaptoundecylamine) of around 30 nm and up to 50 μg/mL did not induce platelet aggregation after short-term exposure probably because of protein corona formation on the surface of examined AuNPs (Love et al., 2012).

6.2.2.1.4 Iron oxide nanoparticles

The effect of iron oxide nanoparticles (IONPs) on platelets is somehow contradictory as they can have either induced (Bircher et al., 2014), inhibitory (Deb et al., 2012; Villegas et al., 2019), or neutral effect (Bircher et al., 2014; Deb et al., 2012; Easo & Mohanan, 2015) on platelet, highly depending on the stabilizing agents coated on the nanoparticle surface. According to the report by Bircher et al., iron carbide nanoparticles coated with carbon (~30 nm) increased the expression of GpIIb/IIIa and P-selectin by platelets, which led to reduced blood clotting time by 25% at the concentration of 1 mg/mL (Bircher et al., 2014). By contrast, PEGylation of iron carbide nanoparticles attenuated the influence of the nanomagnets on coagulation parameters. At the concentration of 0.5 mg/mL, there was no significant effect observed. In other comparable studies, starch-coated IONPs (45 nm, 128−256 μM) (Deb et al., 2012) and dextran-stabilized IONPs (25.3 ± 0.97 nm, 1 mg/mL) (Easo & Mohanan, 2015) did not show any effect on platelet function. However, Deb et al. indicated that citric acid-stabilized iron oxide nanoparticles (FeNP(C)) (35 nm, tested concentration range was 64, 128, 192, and 256 μM) had an antiplatelet property, which was higher than that citric acid has by itself, as reflected in various molecular events including ATP release of dense granules, the level of tyrosine phosphorylation, and the expression of GpIIb/IIIa and CD62P (P-selectin) (Deb et al., 2012). In addition, poly(acrylic acid)-coated IONPs (PAC-IONs) presented the antagonistic effect on platelet aggregation and no effect on platelet activation even up to 62 μg/mL of the nanoparticles (Villegas et al., 2019).

6.2.2.1.5 Silica nanoparticles

As explored in a study by Tavano et al., synthetic amorphous silica nanoparticles (SAS-NPs), bare, and PEGylated organically modified silica nanoparticles (ORMOSIL) had no appreciable effect on platelet activation and aggregation (Tavano et al., 2010). On the contrary, anionic amorphous silica nanoparticles (SiNPs) (10−500 nm) with the concentration varied from 10 to 200 μg/mL were reported to induce platelet activation and aggregation, accompanied by GpIIb/IIIa and CD62P upregulation (Corbalan et al., 2012). Since the thrombotic activity of SiNPs was hindered by inhibitors of ADP and the matrix metalloproteinase-2 (MMP2) pathway, the author discussed that the nanoparticles interact with the Ca^{2+} ion channel and result in extracellular Ca^{2+} influx into platelets cytoplasm, leading to the activation of eNOS for NO generation. After the substrate (L-arginine) is used up, eNOS is uncoupled, and superoxide is produced to interact with NO to form ONOO- (peroxynitrite anion). Low ratio of NO/ONOO- is a marker of oxidative stress and diminished NO-availability, which promotes platelet activation. In other studies, silica nanoparticles (around 58 and 245 nm) were reported to enhance the expression of PECAM-1 (starting from 1.8 mg/kg·bw of rat), result in NO/NOS system imbalance (> 1.8 mg/kg · bw of rat), and an increase in platelet number on endothelial cells (both 10 and 250 μg/mL), promoting platelet adhesion and prethrombotic state (Feng et al., 2019; Saikia et al., 2018). Such a phenomenon is in contrast to another study where silica nanoparticles at 20−200 μg/mL led to the decrease in adhered platelet number as compared to the control and the treatment group at higher concentrations of the nanoparticles (~ 1000 μg/mL) (Nishikawa et al., 2009). The differences between the two studies might be attributed to the porosity, the size (50 vs 250 nm), and the fabrication method (Stöber vs mesoporous silica nanoparticles) of the particles.

6.2.2.1.6 Other inorganic nanoparticles

An in vivo study carried out by Singh et al. depicted an extremely thrombotic effect in mice after intravenous (IV) injection of atomically thin graphene oxide sheets (GO) (Singh et al., 2011). As explored in in vitro tests, GO sheets triggered platelet aggregation through the intracellular release of Ca^{2+} and the activation of Src kinases. At the concentration of 2 μg/mL, this effect of GO sheets was higher than that induced by 1 U/mL of thrombin. Continuing this study, Singh et al. found that amine-modified GO sheets (GO-NH$_2$) (2 and 10 μg/mL) did not show any induced or inhibitory effect on platelets, without noticeable change in the ROS level (Singh et al., 2012). There was no in vivo pulmonary thromboembolism after GO-NH$_2$ exposure.

Rutile titanium (TiO_2) nanorods (0.4−10 μg/mL, 4−6 nm) were reported to cause significant platelet aggregation in rat blood in a concentration-dependent manner (Nemmar et al., 2008). After intratracheal instillation of TiO_2 nanorods in rats, the platelet count was significantly decreased, indicating the platelet aggregation in vivo. The molecular mechanism for platelet response to TiO_2

nanorods is still ambiguous but could be associated with the shape and/or surface feature of the material. However, rutile TiO_2 nanoparticles (67 nm in size) showed no effect on murine platelets with the injection dose of 1 mg/kg in other studies (Bihari et al., 2010; Haberl et al., 2015). As described in some studies, nickel nanoparticles (0.05 mg/mL) or zinc oxide nanoparticles (3:1 v/v ratio of platelet-rich plasma) also cause changes in the platelet shape (Guildford et al., 2009) and promote platelet activation (Šimundić et al., 2013).

6.2.2.2 Organic nanoparticles

6.2.2.2.1 Dendrimers

Several studies have demonstrated that large, cationic poly(amidoamine) (PAMAM) dendrimers (above G4) induced platelet aggregation. By evaluating 12 PAMAM dendrimers of different generations (G3 to G6) functionalized with succinamic acid (anionic), amidoethanol (neutral), and amine (cationic), Marrink et al. revealed that only large and cationic PAMAM dendrimers (amine-G4, amine-G5, and amine-G6) induced platelet aggregation (Dobrovolskaia et al., 2012). Moreover, the aggregation effect was proportional to the number of amine groups on the surface. Since the observed platelet aggregation was neither accompanied with the release of membrane microparticle nor sensitive to inhibitors interfering with platelet activation's pathway, the proposed mechanism is supposed to involve the capability of cationic PAMAM to disrupt platelet membrane integrity and thus induce the aggregation. Computational simulations also supported this proposal (Marrink et al., 2004). In study by Jones et al., large and cationic PAMAM G7 dendrimer nanoparticles (100 µg/mL) showed their effect in altering platelet morphology, which substantially interfered with platelet function, and induced platelet adhesion and aggregation (Jones et al., 2012). Greish et al. also reported that G4 and G7 PAMAM dendrimer nanoparticles caused DIC-like manifestations in mice at a dose >10 mg/kg (Greish et al., 2012). As compared to PAMAM dendrimers, triazine dendrimers (0.01−1 µM) evoked less aggressive platelet aggregation due to differences in the assembly of supramolecular structure and/or cationic charge (Enciso et al., 2016).

6.2.2.2.2 Lipid-based nanoparticles

The effect of lipid-based nanoparticles on platelet is also correlated with surface charge. It has been presented that anionic lipid-based (cetyl alcohol/polysorbate) nanoparticles (50−150 µg/mL) (Koziara et al., 2005) and both anionic and cationic liposome prepared from a photopolymerizable phosphatidylcholine derivative (100−360 µg/0.5 mL platelet) (Juliano et al., 1983) inhibited platelet activation and aggregation in a concentration-dependent manner. However, Reinish et al. reported the reduction in the platelet number after IV injection of anionic liposomes (dose level of 25 mg/kg) in rats in the first 5 minutes (Reinish et al., 1988). The platelet count was recovered 60 minutes postinjection. Similarly, anionic liposomes (phosphatidylcholine: phosphatidic acid = 8:1), not cationic and neutral liposomes, provoked platelet aggregation in vitro and in vivo after IV injection in guinea pigs (Zbinden et al., 1989). The effect was probably due to the interaction between anionic liposomes and FXIII/XI. Moreover, Constantinescu et al. suggested that the interaction of liposomes with platelet was independent of opsonization but dependent on the liposome concentration (Constantinescu et al., 2003). The discrepancies between studies might be attributed to not only the surface charge but also the composition of the lipid-based nanoparticles.

6.2.2.2.3 Other polymeric nanoparticles

Unmodified, carboxyl-modified, and amine-modified polystyrene latex nanoparticles from 50 to 100 nm caused aggregation of platelet in a dose-dependent manner (15−60 µg/mL) except the 50 nm amine-modified ones (Smyth et al., 2015). This aggregation was mediated by secondary agonists released from platelet granules and induced GpIIb/IIIa expression depending on Ca^{2+} influx and protein kinase C signaling pathway. The author also described that these effects were associated with both size and surface modification. In another study, carboxyl-modified polystyrene nanoparticles ($\sim$80 nm, 260 µg/mL) induced platelet aggregation by disrupting platelet membrane and upregulating platelet-activating markers P-selectin and PAC-1, respectively (McGuinnes et al., 2011).

Ramtoola et al. (2011) and Li et al. (2009) found that chitosan (CS), poly(lactic-co-glycolic acid) (PLGA), PLGA-macrogel, and PLGA-CS nanoparticles did not exert any effect on platelet activation in the concentration range of 0.1−500 µg/mL. In Li 's study, PLGA, CS, and PLGA-CS nanoparticles (0.01−100 µg/mL) had slight inhibitory effect toward platelet aggregation induced by collagen (Li et al., 2009). This could be due to the reduced platelet−platelet interaction and/or reduced adsorption of platelets onto collagen fibers.

6.2.3 Nanoparticles and red blood cells

Another significant cellular component for blood coagulation is RBCs (erythrocytes), which has been underestimated in the past. Detailed mechanisms of how RBCs play their roles in hemostasis has been reviewed in-depth previously (Weisel & Litvinov, 2019). Briefly, RBCs attribute to hemostasis through hemorheological properties owning to their abundance and large size (Du et al., 2014). The influence of hemorheology—flow property of blood and its elements—on hemostasis and thrombosis depends on the blood shear rates and viscosity, of which RBC is a main contributor (Mehri et al., 2018; Sriram et al., 2014). The blood viscosity affects platelet distribution within vessels based on the axial margination phenomenon in which RBCs tend to move to the center of vessels and push platelets toward the periphery, facilitating their collision with the vasculature for hemostatic events (Walton et al., 2017).

Interactions of nanoparticles with RBCs can cause RBC aggregation (Wu et al., 2018). For instance, nanodiamonds (100 nm) were found to greatly increase attraction forces between RBCs' membrane, leading to the formation of large and abnormal RBCs aggregates (Avsievich et al., 2019). Meanwhile, no anomalous aggregate was depicted when RBCs were treated with polymeric nanoparticles (600 nm) with platelet-free blood plasma. The concentration of tested nanoparticles was kept at 0.01%. Also, Fe_3O_4 magnetic nanoparticles ($\sim$73 nm, 200 µg/mL) (Ran et al., 2015) induced the aggregation of RBCs. Aggregation of RBCs, especially in small vessels, normally increases the blood viscosity in the center of vessels and platelet margination, resulting in induced endothelium activation and platelet aggregation (Mehri et al., 2018).

Furthermore, nanoparticle interactions with RBCs can also alter the deformability of RBCs, which is the ability of RBCs to change their shape in response to applied stress without resulting in hemolysis (Yedgar et al., 2002). Decrease in RBCs deformability is related to higher risk of thrombosis, since rigid RBCs can block small vessels easily, change the blood flow, and provoke platelet activation (Kwaan & Samama, 2019). Pan et al. demonstrated that the absorption of polystyrene nanoparticles (PSNPs) on murine RBCs significantly reduced RBCs' deformability as a function of elongation index (EI) value at both sizes (200 and 300 nm) as well as both low and high nanoparticles:RBCs ratio (200:1 and 1000:1) (Pan et al., 2018). In contrast, lysozyme-dextran nanogels (LDNGs) did not affect deformability of RBCs even at the nanoparticles:RBCs ratio of 1000:1.

In addition to hemorheology, RBCs also contribute to blood coagulation via the exposure of phosphatidylserine (PS) on the cell surface (Guo et al., 2018). PS is a key phospholipid localized within the plasma membrane. Upon the high shear stress, oxidative stress, or complement attack, damaged RBCs expose PS on the membrane surface, providing a procoagulant surface for the accumulation of coagulation complexes such as prothrombinase and intrinsic tenase that facilitate thrombus formation (Du et al., 2014). Ran et al. reported that IONPs (72.6 ± 0.57 nm, 200 µg/mL) dramatically altered RBCs' rigidity by externalizing PS on the cell surface (the PS-expressed cells reached 40% after 48 hours), which ultimately changed the thrombotic potential of blood (Ran et al., 2015). Moreover, PSNPs exerted mechanical, oxidative, and osmotic stresses on murine RBCs (Pan et al., 2018). As a result, the proportion of RBCs expressing PS increased up to 87% and 92% respectively for low and high nanoparticles: RBCs loading ratios, in comparison to only 0.1% of RBCs only and 0.3% of LDNGs loading for the control.

6.2.4 Nanoparticles and plasma coagulation factors

The majority of circulating plasma coagulation factors are zymogens, precursors of enzymes, which will be converted into the active form once the coagulation cascade is initiated. The other plasma coagulation factors are nonenzymatic that act as either cofactor (e.g., TF (or FIII), FV and FVIII, high-molecular-weight kininogen (HMWK or HK), and protein S) or substrate (e.g., fibrinogen). These factors form a coagulation cascade in secondary hemostasis, which can be divided into extrinsic and intrinsic pathways. Both pathways lead to thrombin generation and ultimately fibrin formation to create a stable blood clot at the injury site. The extrinsic pathway is activated by TF exposed on damaged endothelial cells, initializing the coagulation cascade. In a parallel manner, the intrinsic pathway begins with FXII, prekallikrein (PK), and HMWK (Palta et al., 2014). FXII can be activated via the contact with negatively charged molecules and nanoparticle surfaces such as dextran sulfate, glass, kaolin, celite, and silica (Simak & De Paoli, 2017; Tankersley et al., 1983; van der Graaf et al., 1982; Wiggins & Cochrane, 1979). It can also be autoactivated by the membrane of activated platelets (Bendapudi et al., 2016), resulting in the activation of the kallikrein−kinin system and FXI as well as other downstream zymogens in the intrinsic pathway (Ilinskaya & Dobrovolskaia, 2016). It is important to note that apart from activated FXII (FXIIa), a small amount of thrombin generated by the extrinsic pathway can in turn activate FXI and thus facilitate the activation of the intrinsic pathway and amplification of thrombin generation.

Since nanoparticle surfaces can activate coagulation factor XII to initiate the intrinsic pathway of coagulation, it is reasonable to anticipate that nanoparticles might unintendedly interfere with the coagulation cascade and overall hemostasis. For example, Baker et al. reported that mesocellular foams (MCFs) with the window size >11 nm and the total pore volume at 0.0006 cm^3 facilitated clotting in an FXII-dependent mechanism (Baker et al., 2008). The authors stated that FXII, with a hydrodynamic size of 7.5 nm, can diffuse into and adhere to MCFs' cells, thus they can be activated and initiate a coagulation cascade. Silica nanoparticles (70−1000 nm) at 0.02 mg/mL were reported to activate the intrinsic pathway via their interaction with FXII (Nabeshi et al., 2012).

Decreasing the silica particle size from micrometer to nanometer (30 and 70 nm), that is, increasing the particle surface, resulted in a higher degree of FXII activation after intranasal exposure in mice for 7 days at 500 µg/mouse (Yoshida et al., 2013). Kushida et al. reported that silica nanoparticles at varied concentrations of 0.01−100 nM with the size of 12−85 nm had significant coagulation activity, while those with very small sizes (4−7 nm) did not (Kushida et al., 2014). The reason may be that very small nanoparticles (4−7 nm) have a higher surface curvature, which does not distort the configuration of FXII after its adsorption on the surface of the nanoparticles and affects the activation of other factors such as kallikrein, leading to a coagulant "silent" surface. Besides, coagulant factors such as FXa and vWF were induced whilst anticoagulant factors were reduced after 30 days of exposure to silica nanoparticles (58.11 ± 7.30 nm)in rats (Feng et al., 2019). The tested concentration of the nanoparticles was 1.8−16.2 mg/kg. By evaluating activated partial thromboplastin time (APTT) or partial thromboplastin time (PTT), Burke et al. concluded that MWCNTs (range of diameter was 26−31 nm, median length was 490−580 nm) at the concentration of 100 µg/mL triggered the intrinsic pathway by preferentially interacting with FIXa and acting as a platform to promote its enzyme activity (Burke et al., 2011). Oslakovic et al. reported that amine-modified polystyrene nanoparticles (57.1 and 284 nm in size, 0.5 mg/mL) bound to FVII and IX, which inhibited thrombin formation due to the depletion of these coagulation factors in solution (Oslakovic et al., 2012). By contrast, carboxyl-modified polystyrene nanoparticles (27.8 or 223.9 nm in size, 0.5 mg/mL) act as an active surface to trigger the intrinsic pathway.

Fibrinogen can strongly bind to gold nanoparticles (stabilized with citrate) thanks to cysteine residues presented in alpha, beta, and gamma chains of fibrinogen for Au-S bond formation, which induced the blood clot (Chen et al., 2011). However, another study reported that fibrinogen bound on the gold nanoparticle surface, which was stabilized with citrate, only increased the nanoparticle size but did not usually cause blood coagulation as in the previous study (Dobrovolskaia et al., 2009). Several studies described that interactions between silver nanoparticles and fibrin caused conformation change of fibrin (Ragaseema et al., 2012; Shrivastava et al., 2009, 2011), leading to the inhibition of fibrin polymerization and thrombus formation in vitro (Shrivastava et al., 2011). Nevertheless, it is worth noting that this effect is less pronounced in plasma than in a purified system due to nonspecific interactions of silver nanoparticles with other plasma proteins such as globulin and albumin.

6.3 Common in vitro methods to evaluate the effect of nanoparticles on blood coagulation

There are several methods based on different working principles that can be employed to access the effect of nanoparticles on components of the coagulation system and hemostatic process. A combination of more than one method is usually required to reach any solid conclusion about the nanoparticle effect. Methods that are commonly used are summarized below.

Standard laboratory coagulation tests or standard plasma coagulation tests (SLTs): (Activated) partial thromboplastin time (aPTT, PTT), prothrombin time (PT), and thrombin time (TT) are measured using an automated analyzer to access coagulation. Respectively, PT and aPTT/PTT reflect the activities of coagulation factors during blood coagulation. TT reflects the activity of fibrinogen (Fröhlich, 2016; Simak & De Paoli, 2017; Sperling et al., 2018). Nanoparticles are incubated with citrate-anticoagulated plasma for the test. Interpretation of changes in obtained values is useful to determine the procoagulant/coagulant or proanticoagulant/anticoagulant effects of tested nanoparticles (Burke et al., 2011; Cenni et al., 2008; Dobrovolskaia et al., 2009; Martínez-Gutierrez et al., 2012).

Viscoelastic test: In comparison to standard plasma-based coagulation test, viscoelastic test is performed in citrated whole blood to measure viscoelastic properties during thrombus formation until fibrinolysis (Paniccia et al., 2015; Peng, 2010). Hence, this test provides a global assessment of the entire hemostatic process (da Luz et al., 2013), and thus can be used to evaluate the effect of nanoparticles on clot development, stabilization, and dissolution (Ajdari et al., 2017; Bircher et al., 2014; He et al., 2020; Meng et al., 2012; Zhang et al., 2016). Currently, three commercial systems for the viscoelastic test are available, namely thromboelastography, rotational thromboelastometry, and Sonoclot analysis (da Luz et al., 2013).

Immunoassays: Nanoparticles' effect on blood coagulation can be identified by immunoassay, such as enzyme-linked immune assay (ELISA)—one of the most frequently used. Based on the specificity of the antigen—antibody interaction, ELISA can be used to detect and quantify platelet activation markers such as β-TG, PF4, serotonin, P-selectin (Ferraz et al., 2008; Jones et al., 2012; Mayer et al., 2009; Stevens et al., 2009), prothrombin activation fragments D-dimer (Burke et al., 2011; Feng et al., 2019; Mayer et al., 2009), TF, vWF exposed by damaged endothelial cells (Burke et al., 2011; Feng et al., 2019; Yoshida et al., 2013), eNOS, FXa in serum (Feng et al., 2019), and plasmatic markers indicating the activation of the coagulation cascade (Krajewski et al., 2013).

Synthetic substrate assay: A variety of commercial assay kits have been developed to detect the activity of coagulation factors such as thrombin generation assay (Jones et al., 2012; Kushida et al., 2014; Stevens et al., 2009), FXII activity assay (Yoshida et al., 2013), lactate dehydrogenase (LDH) assay to detect platelet membrane integrity (Guo et al., 2015; Liang et al., 2016; Liu et al., 2017; Mayer et al., 2009; Shrivastava et al., 2009; Singh et al., 2011; Smyth et al., 2015), and NO measurement (Guo et al., 2015; Semberova et al., 2009). These assays are useful to investigate the effect of nanoparticles—coagulation system interactions. The assay is normally based on the detection of chromogenic or fluorogenic substrates, such as intracellular free Ca^{2+} assay (De Paoli Lacerda et al., 2011; Gelderman et al., 2008; Kumari et al., 2014; Shrivastava et al., 2009; Singh et al., 2011, 2012), or intracellular ROS measurement (Guo et al., 2015; Kumari et al., 2014; Liang et al., 2016; Liu et al., 2017; Ran et al., 2015; Singh et al., 2011, 2012; Sun et al., 2016).

Aggregometry: Light transmission aggregometry (LTA) and multiple electrode aggregometry (MEA) are used to assess platelet reactivity and measure platelet aggregation in response to agonists. LTA uses citrated platelet-rich plasma (PRP) and the change in light transmission to detect aggregation while MEA uses whole blood and works by detecting electrical impedance between electrodes (Sun et al., 2019). These tests have been used to study platelet—nanoparticles interactions (Bihari et al., 2010; Corbalan et al., 2012; Deb et al., 2011, 2012; Dobrovolskaia et al., 2012; Haberl et al., 2015; Jones et al., 2012; Jun et al., 2011; Koziara et al., 2005; Kumari et al., 2014; Laloy et al., 2014; Li et al., 2009; Love et al., 2012; Ragaseema et al., 2012; Ramtoola et al., 2011; Santos-Martinez et al., 2012, 2015; Shrivastava et al., 2009; Singh et al., 2011, 2012; Smyth et al., 2015; Tavano et al., 2010; Villegas et al., 2019; Zbinden et al., 1989).

Quartz crystal microbalance with dissipation (QCM-D): QCM-D is a new potential method to investigate the effect of nanoparticles on platelets by detecting platelet aggregation under flow conditions (Santos-Martinez et al., 2012; Santos-Martinez et al., 2015). It has been demonstrated that QCM-D is more sensitive than LTA (Santos-Martinez et al., 2012, 2015).

Flow cytometry: Flow cytometry is a powerful method and provides statistical data related to platelet activation by accessing physical interactions between platelet and nanoparticles (Constantinescu et al., 2003; Singh et al., 2012), platelet surface activation markers such as P-selectin, CD63, PS, GpIIb/IIa, GpIbα, thrombin-mediated PAR-1 (Bihari et al., 2010; Bircher et al., 2014; Corbalan et al., 2012; Deb et al., 2011, 2012; Dobrovolskaia et al., 2012; Jones et al., 2012; Koziara et al., 2005; Kumari et al., 2014; Li et al., 2009; Radomski et al., 2005; Ragaseema et al., 2012; Santos-Martinez et al., 2012; Shrivastava et al., 2009; Singh et al., 2011; Smyth et al., 2015; Tavano et al., 2010; Villegas et al., 2019), change in intracellular Ca^{2+} level (Jun et al., 2011; Shrivastava et al., 2009), or the release of platelet microparticles (Ferraz et al., 2010). The exposure of markers, such as TF, PS, and ICAM-1 (CD54), promotes adhesion and the coagulation cascade on the surface of endothelial cells and RBCs after nanoparticle treatment can be measured as well (Gelderman et al., 2008; Pan et al., 2018). Flow cytometry can also be employed to detect the binding of nanoparticles to platelets and endothelial cells (Semberova et al., 2009; Tavano et al., 2010).

Western blot: This method can detect the presence of specific proteins. Several studies have used western blot to quantify the expression of cellular factors (Guo et al., 2015), expression of VE-cadherin on endothelial cells after exposure to nanoparticles (Sun et al., 2016), expression of other types of marker indicating platelet activation (Ragaseema et al., 2012), or the activation of SOCE (De Paoli Lacerda et al., 2011) induced by the interaction of the coagulation system with nanoparticles.

Real-time polymerase chain reaction (PCR): PCR is another quantitative method used to detect activation markers and coagulation factors indicating the effect of nanoparticles on the coagulation system. Real-time PCR can check mRNA expression of iNOS, eNOS (Guo et al., 2015), expression of VE-cadherin on endothelial cells after exposure to nanoparticles (Sun et al., 2016), the presence of coagulation factors such as TF, vWF, P-selectin (Feng et al., 2019; Semberova et al., 2009), and adhesion markers, namely ICAM-1, PCAM-1 (Semberova et al., 2009).

Microscopy: Microscopy is usually utilized to visualize the effect of nanoparticles on cellular components of the coagulation system (nanoparticle−cell interface). Direct visualization of platelet or RBC aggregates caused by nanoparticles can be easily obtained by optical microscopy (Cenni et al., 2008; Simak, 2016; Zbinden et al., 1989). However, transmission electron microscope (TEM) (Corbalan et al., 2012; De Paoli Lacerda et al., 2011; Kumari et al., 2014; Liang et al., 2016; Radomski et al., 2005; Shrivastava et al., 2009; Singh et al., 2011; Smyth et al., 2015) and scanning electron microscope (SEM) can be also used to access subcellular localization of nanoparticles and the ultrastructure of the nanoparticle surface, and the nanoparticle−blood cell interface with high resolution (Avsievich et al., 2019; Deb et al., 2011; Dobrovolskaia et al., 2012; Ferraz et al., 2008, 2010; Guildford et al., 2009; Jones et al., 2012; Laloy et al., 2014; Ragaseema et al., 2012; Ran et al., 2015; Santos-Martinez et al., 2012, 2015; Šimundić et al., 2013; Singh et al., 2011; Stevens et al., 2009; Sun et al., 2016).

6.4 Factors affecting nanoparticle−blood coagulation system interactions

Different nanoparticles have different effects on blood coagulation components. Changes in any nanoparticle parameters such as size, shape, surface charge, and coating materials might correlate with other ways of interaction and lead to an alternative effect.

6.4.1 **Size**

Silica nanoparticles of smaller size exert more profound impact on the coagulation system (Bauer et al., 2011; Corbalan et al., 2012; Jun et al., 2011; Nabeshi et al., 2012; Nemmar et al., 2014). Among silica nanoparticles with various size of 16, 41, 80, 212, and 304 nm, nanoparticles with the largest size resulted in the release of Weibel—Palade bodies and vWF from endothelial cells after 24 hours of incubation, while this effect only takes a few hours with nanoparticles of smaller size (Bauer et al., 2011). It has been reported that silica nanoparticles with the diameter of 10 nm triggered stronger platelet activation than those larger than 50 nm (Corbalan et al., 2012). Similarly, silica nanoparticles of 30, 50, or 70 nm exhibited increased procoagulant activity as compared to those of 100, 300, 500, or even 1000 nm, due to the increased specific surface area exposed to the coagulation system (Jun et al., 2011; Nabeshi et al., 2012; Nemmar et al., 2014). It is worthy of note that very small size silica nanoparticles (4—7 nm) did not have coagulation activity due to higher surface curvatures (Kushida et al., 2014). However, other studies reported that smaller silica nanoparticles (10—15 nm) inhibited platelet activation in vitro (Shrivastava et al., 2009), while larger nanoparticles (200 nm) caused more pronounced hemostasis in vivo (Kim et al., 2008).

Similar to silica nanoparticles, gold nanoparticles (≤ 50 nm) were easily internalized and accumulated in platelets, inducing platelet activation. In contrast, gold nanoparticles larger than 60 nm were basically inert (Deb et al., 2011; Hecold et al., 2017; Santos-Martinez et al., 2012). In contrast, other studies reported that small gold nanoparticles (5—30 nm) had no effect on platelets while 60 nm-ones prevented platelet aggregation (Aseychev et al., 2013). All of these tested gold nanoparticles were used at 5—40 μM in PRP which corresponded to 0.94—7.5 μg/mL blood.

The effect of particle size was also demonstrated for carbon-based nanoparticles (Meng et al., 2012; Radomski et al., 2005) and silver nanoparticles (Guo et al., 2015). For example, shorter MWCNTs had less effect on platelet activation than the longer ones (Meng et al., 2012). Silver nanoparticles with the diameter of 110 nm exhibited the most effective toxicity to endothelial cells in comparison to 10 and 75 nm ones (Guo et al., 2015). The discrepancies in the size-dependent effect of nanoparticles on coagulation need to be carefully taken because the characterization of nanoparticle size might not be carried out in similar media and technique (e.g., water vs buffer solution or TEM/SEM vs dynamic light scattering). In addition, the impact of a specific nanoparticle on the coagulation system may be also associated with other factors such as surface charge and the concentration of nanoparticles.

6.4.2 **Shape**

The effect of nanoparticle shape on the interaction with the coagulation system has also been presented in some studies. Regarding carbon-based nanoparticles, carbon nanotubes (both multiwalled and single-walled) promoted platelet activation and aggregation while spherical C60 fullerenes did not (Radomski et al., 2005). Cuboidal γ-cyclodextrin nanoscale frameworks showed induced platelet aggregation in comparison to the spherical shape counterparts (He et al., 2019). On the contrary, there were studies showing that carbon-based nanoparticles can cause thrombus formation regardless of their shape (Holzer et al., 2014). Gold nanoparticles with either spherical, hollow sphere, or rod shape did not affect endothelial cells (Bartczak et al., 2012).

6.4.3 Surface charge

Many studies indicated that the surface charge of the nanoparticles is also a key factor orientating their interaction with coagulation system. Positively charged groups on nanoparticles' surface can neutralize and form cross-bridges with negatively charged ionizable sialic acid groups on the platelets' surface, facilitating platelet—platelet interaction and aggregation (Gobbo et al., 2015; Hante et al., 2019). Besides, positively charged nanoparticles can alter platelet morphology (Jones et al., 2012) and disrupt platelet membrane integrity (Dobrovolskaia et al., 2012), inducing the changes in the size and number of platelet aggregates. Large and cationic PAMAM ($\geq$ G4) and triazine dendrimer (G5 and G7) provoked platelet aggregation, in which the aggregation degree was proportional to the number of amine groups on the nanoparticles' surface (Dobrovolskaia et al., 2012).

Other studies showed that the coagulation cascade can also be initiated through the contact with negatively charge nanoparticles' surfaces (Simak & De Paoli, 2017; Tankersley et al., 1983; van der Graaf et al., 1982; Wiggins & Cochrane, 1979). For example, anionic polystyrene (carboxyl-modification) led to the upregulation of activation markers (P-selectin or PAC-1) of platelets whilst cationic polystyrene (amine-modification) led to the interruption of the platelet membrane (McGuinnes et al., 2011). Both positively and negatively charged polystyrene nanoparticles (McGuinnes et al., 2011) can eventually lead to thrombotic events. This is in contradiction with liposomes where both anionic and cationic nanoparticles inhibited platelet activation and aggregation (Juliano et al., 1983), and anionic liposomes reduced the platelet number in rats (Reinish et al., 1988). Nevertheless, there are still contradictory studies that reported the platelet aggregation effect of anionic liposomes (Zbinden et al., 1989), or the independence of the surface charge of polystyrene nanoparticles toward platelet activation (Smyth et al., 2015). Apparently, the charge-dependent effect of nanoparticles on the coagulation system is unpredictable. In physiological conditions, the influence of nanoparticle charge is even more difficult to clarify due to the absorption of plasma proteins on the surface of nanoparticles.

6.4.4 Coating materials

A layer of coating material on the nanoparticle surface can alter its reactivity to the blood coagulation system. Among all, polyethylene glycol (PEG) is the most commonly used polymeric material. Several studies have demonstrated that the presence of PEG on the nanoparticle surface reduced their interference with endothelial cells and platelets, probably due to the capability to prevent protein binding. As a result, unattended hemostasis is reduced and the compatibility of nanoparticles is improved (Koziara et al., 2005; Ragaseema et al., 2012; Santos-Martinez et al., 2014; Su et al., 2017; Tavano et al., 2010; Yu et al., 2012). However, PEGylation of nanoparticles is not successful for all nanoparticles (Burke et al., 2011; Vakhrusheva et al., 2013). In addition to PEG, other polymers, namely dextran (Chowdhury et al., 2013), albumin (Vakhrusheva et al., 2013), starch (Deb et al., 2012), and poly(acrylic acid) (PAA) (Villegas et al., 2019) did not cause any effect on endothelial cells and platelets, or reduce platelet aggregation. It was reported that PAA conjugated on the surface of gold nanoparticles reduced platelet aggregation by binding to fibrinogen and promoting the changes in its conformation (Deng et al., 2011). Gold nanoparticles coated with polyethylenimine and polyvinylpyrrolidone, however, induced platelet aggregation (Hecold et al., 2017).

All the findings above have demonstrated that specific coating material is worth investigating for these commonly used nanoparticles during their interactions with the coagulation system.

6.4.5 Other factors

The surface charges of the nanoparticle can be altered in the physiological fluids due to the binding of plasma proteins or simply pH value, which could come along with the alteration of nanoparticle hydrophobicity (Fröhlich, 2016; Setyawati et al., 2015). A study clarifying the influence of latex polystyrene nanoparticles' hydrophobicity on the blood coagulation carried out by Miyamoto et al. (1990) revealed that hydrophobic latex nanoparticles provoked platelet aggregation to a higher extent than the hydrophilic ones. This could be due to their ability to interact more closely with the cell membrane and activate platelets (Kou et al., 2013). However, further investigation relating to the relationship between hydrophobicity of nanoparticles and the blood coagulation system is rarely found.

Interestingly, the concentration of nanoparticle metal cores (such as gold) had an impact on coagulation. Hsu et al. incorporated polyurethane nanocomposites with gold and revealed that incorporation of a lower gold concentration (43.5 ppm) resulted in less platelet adhesion and activation compared to a higher amount of gold (174 ppm) (Hsu et al., 2006).

6.5 Two-side effect of engineered nanoparticles on the blood coagulation system

Nanoparticles can be purposely engineered for specific interactions with the blood coagulation system to either facilitate or prevent coagulation in order to avoid bleeding or prevent thrombosis, respectively. Usually, nanoparticles can be loaded with drugs and/or decorated with peptides, recombinant factors, or markers on the surface to obtain the desirable effect. Some reviews have discussed nanoparticles that are intended to promote coagulation (Gaston et al., 2018; Ilinskaya & Dobrovolskaia, 2013). For instance, "synthetic platelets" comprising poly(lactic-co-glycolic acid)-poly-L-lysine (PLGA-PLL) nanospheres decorated with PEG terminated RGD peptide were developed by Bertram et al. (2009). This system induced platelet aggregation to halt bleeding at the injury site owning to the interaction between RGD peptide and GpIIb/IIIa on the surface of activated platelets for cross-linking. In the study by Shafir et al., maghemite nanoparticles with recombinant coagulant factor VII (rVII) physically bound on the surface showed comparable activity to free rVII (Shafir et al., 2009). On the other hand, nanoparticles designed to prevent coagulation have also been reviewed elsewhere (Ilinskaya & Dobrovolskaia, 2013). Regarding the target and working mechanism, these nanoparticles can be engineered to have the antithrombotic, antiplatelet, and fibrinolytic effects. To exert the antithrombotic effect, nanoparticles can incorporate anticoagulant drugs inside (e.g., heparin, rutin, dipeptide IleTrp, and adenosine) (Argyo et al., 2012; Jiao et al., 2001, 2002; Nguyen, Nguyen et al., 2019; Wu et al., 2020; Zhao et al., 2018) or are conjugated with a ligand (e.g., thrombin-specific aptamer) (Shiang et al., 2011) that inhibits or delays thrombus formation. Nanoparticles with the antiplatelet effect prohibit platelet activation and aggregation. Liposomes with CD39 incorporated inside (Haller et al., 2006), PAMAM dendrimers conjugated with $P2Y_1$ receptor antagonist MSR2500 (de Castro et al., 2010) or A2A receptor agonist CGS21680 (Kim et al., 2008; Kim et al., 2009), and cubosomes loaded with antiplatelet drug clopidogrel bisulfate (El-Laithy et al., 2018) are representative examples of antiplatelet engineered nanoparticles. Besides, some nanoparticles with PEG functionalization possess an antiplatelet property as well (Koziara et al., 2005; Ragaseema et al., 2012; Santos-Martinez et al., 2014; Su et al., 2017; Tavano et al., 2010; Yu et al., 2012).

Moreover, nanoparticles can also be designed as a carrier of fibrinolytic agents to dissolve existing thrombi (fibrinolytic/thrombolysis effect). Several nanoparticles (e.g., liposomes and polymeric nanoparticles) have been successfully engineered to improve the efficacy of fibrinolytic agents such as urokinase, streptokinase, and tissue plasminogen activator (t-PA) with reduced side effects (Chapurina et al., 2016; Chung et al., 2008; Elbayoumi & Torchilin, 2008; Heeremans et al., 1995; Leach et al., 2003; Nguyen et al., 1990; Su et al., 2020; Zamanlu et al., 2019).

6.6 Conclusion and prospects

Nanoparticles in the bloodstream come into contact and interact with one or more components of the blood coagulation system. This chapter presents a thorough review of possible interactions and influences of various nanoparticles on coagulation system components such as platelets, RBCs, endothelial cells, and plasma coagulation factors. However, there is still plenty of room for more studies in the future as not all commonly examined nanoparticles are investigated and fully understood with regard to the underlying mechanisms. Further research investigating the effects of common nanoparticles on RBCs and specific coagulation factors is going to be of high interest as most of current studies have focused more on the interaction of platelets and endothelial cells with nanoparticles. Several in vitro methods that are often used to assess the effects of nanoparticles on hemostasis have also been briefly mentioned. These methods are usually used in combination. Alterable interferences in the blood coagulation correlating to changes in nanoparticle physiochemical parameters have been examined in many studies but not in a systematic way. The discrepancies in results need to be treated with caution since the characterizations might not be carried out in the comparable methods, setting, and media.. More importantly, the effects of a specific nanoparticle on the coagulation system could be associated with more than one factor. As discussed, coating material is an important factor that can alter nanoparticles' reactivity to the blood coagulation system. However, there is a limited variety of investigated polymeric materials. The effect of metal coating of core—shell nanoparticles, regarding types of metal, thickness of metal coating, and coating method, on the coagulation system has not been explored yet. Therefore more studies are needed to give future insight into the influences of coating materials in hemostasis. It is worth examining specific coating material for commonly used nanoparticles. Moreover, the interface between the blood coagulation system and other factors of nanoparticles such as hydrophobicity, porosity, lipid composition of lipid-based nanoparticles, surface topography may attract much interest in the future. Apparently, in vivo studies are encouraged since the behavior of nanoparticles is not always predictable in physiological conditions due to the absorption of plasma proteins on their surface. Taken together, our chapter is beneficial for the establishment of nanoparticles that can avoid unintended interferences with the hemostatic balance, or purposely increase the interaction with a specific blood coagulation component.

References

Ajdari, N., Vyas, C., Bogan, S. L., Lwaleed, B. A., & Cousins, B. G. (2017). Gold nanoparticle interactions in human blood: A model evaluation. *Nanomedicine: Nanotechnology, Biology and Medicine, 13*(4), 1531−1542.

Argyo, C., Cauda, V., Engelke, H., Rädler, J., Bein, G., & Bein, T. (2012). Heparin-coated colloidal mesoporous silica nanoparticles efficiently bind to antithrombin as an anticoagulant drug-delivery system. *Chemistry−A European Journal, 18*(2), 428−432.

Arndt, N., Tran, H. D., Zhang, R., Xu, Z. P., & Ta, H. T. (2020). Different approaches to develop nanosensors for diagnosis of diseases. *Advanced Science, 7*(24), 2001476.

Aseychev, A., Azizova, O., Beckman, E., Dudnik, L., & Sergienko, V. (2013). Effect of gold nanoparticles coated with plasma components on ADP-induced platelet aggregation. *Bulletin of Experimental Biology and Medicine, 155*(5), 685−688.

Avsievich, T., Popov, A., Bykov, A., & Meglinski, I. (2019). Mutual interaction of red blood cells influenced by nanoparticles. *Scientific Reports, 9*(1), 1−6.

Baker, S. E., Sawvel, A. M., Fan, J., Shi, Q., Strandwitz, N., & Stucky, G. D. (2008). Blood clot initiation by mesocellular foams: Dependence on nanopore size and enzyme immobilization. *Langmuir., 24*(24), 14254−14260.

Bandyopadhyay, D., Baruah, H., Gupta, B., & Sharma, S. (2012). Silver nano particles prevent platelet adhesion on immobilized fibrinogen. *Indian Journal of Clinical Biochemistry, 27*(2), 164−170.

Bao, B. Q., Le, N. H., Nguyen, D. H. T., Tran, T. V., Pham, L. P. T., Bach, L. G., et al. (2018). Evolution and present scenario of multifunctionalized mesoporous nanosilica platform: A mini review. *Materials Science and Engineering: C., 91*, 912−928.

Bartczak, D., Muskens, O. L., Nitti, S., Sanchez-Elsner, T., Millar, T. M., & Kanaras, A. G. (2012). Interactions of human endothelial cells with gold nanoparticles of different morphologies. *Small., 8*(1), 122−130.

Bauer, A. T., Strozyk, E. A., Gorzelanny, C., Westerhausen, C., Desch, A., Schneider, M. F., et al. (2011). Cytotoxicity of silica nanoparticles through exocytosis of von Willebrand factor and necrotic cell death in primary human endothelial cells. *Biomaterials., 32*(33), 8385−8393.

Bendapudi, P. K., Deceunynck, K., Koseoglu, S., Bekendam, R. H., Mason, S. D., Kenniston, J., et al. (2016). *Stimulated platelets but not endothelium generate thrombin via a factor XIIa-dependent mechanism requiring phosphatidylserine exposure.* Washington, DC: American Society of Hematology.

Berry, J., Arnoux, B., Stanislas, G., Galle, P., & Chretien, J. (1977). A microanalytic study of particles transport across the alveoli: Role of blood platelets. *Biomedicine/[publiée pour l'AAICIG], 27*(9−10), 354−357.

Bertram, J. P., Williams, C. A., Robinson, R., Segal, S. S., Flynn, N. T., & Lavik, E. B. (2009). Intravenous hemostat: Nanotechnology to halt bleeding. *Science Translational Medicine, 1*(11), 11ra22−11ra22.

Bihari, P., Holzer, M., Praetner, M., Fent, J., Lerchenberger, M., Reichel, C. A., et al. (2010). Single-walled carbon nanotubes activate platelets and accelerate thrombus formation in the microcirculation. *Toxicology., 269*(2−3), 148−154.

Bircher, L., Theusinger, O. M., Locher, S., Eugster, P., Roth-Z'graggen, B., Schumacher, C. M., et al. (2014). Characterization of carbon-coated magnetic nanoparticles using clinical blood coagulation assays: Effect of PEG-functionalization and comparison to silica nanoparticles. *Journal of Materials Chemistry B., 2*(24), 3753−3758.

Broos, K., Feys, H. B., De Meyer, S. F., Vanhoorelbeke, K., & Deckmyn, H. (2011). Platelets at work in primary hemostasis. *Blood Reviews, 25*(4), 155−167.

Burke, A. R., Singh, R. N., Carroll, D. L., Owen, J. D., Kock, N. D., D'Agostino, R., Jr, et al. (2011). Determinants of the thrombogenic potential of multiwalled carbon nanotubes. *Biomaterials., 32*(26), 5970−5978.

Cenni, E., Granchi, D., Avnet, S., Fotia, C., Salerno, M., Micieli, D., et al. (2008). Biocompatibility of poly (D, L-lactide-co-glycolide) nanoparticles conjugated with alendronate. *Biomaterials.*, *29*(10), 1400−1411.

Chapurina, Y. E., Drozdov, A. S., Popov, I., Vinogradov, V. V., Dudanov, I. P., & Vinogradov, V. V. (2016). Streptokinase@ alumina nanoparticles as a promising thrombolytic colloid with prolonged action. *Journal of Materials Chemistry B.*, *4*(35), 5921−5928.

Chen, G., Ni, N., Zhou, J., Chuang, Y.-J., Wang, B., Pan, Z., et al. (2011). Fibrinogen clot induced by gold-nanoparticle in vitro. *Journal of Nanoscience and Nanotechnology*, *11*(1), 74−81.

Chowdhury, S. M., Kanakia, S., Toussaint, J. D., Frame, M. D., Dewar, A. M., Shroyer, K. R., et al. (2013). In vitro hematological and in vivo vasoactivity assessment of dextran functionalized graphene. *Scientific Reports*, *3*, 2584.

Chung, T.-W., Wang, S.-S., & Tsai, W.-J. (2008). Accelerating thrombolysis with chitosan-coated plasminogen activators encapsulated in poly-(lactide-co-glycolide)(PLGA) nanoparticles. *Biomaterials.*, *29*(2), 228−237.

Constantinescu, I., Levin, E., & Gyongyossy-Issa, M. (2003). Liposomes and blood cells: A flow cytometric study. *Artificial Cells, Blood Substitutes, and Biotechnology*, *31*(4), 395−424.

Corbalan, J. J., Medina, C., Jacoby, A., Malinski, T., & Radomski, M. W. (2012). Amorphous silica nanoparticles aggregate human platelets: Potential implications for vascular homeostasis. *International Journal of Nanomedicine*, *7*, 631.

da Luz, L. T., Nascimento, B., & Rizoli, S. (2013). Thrombelastography (TEG®): Practical considerations on its clinical use in trauma resuscitation. *Scandinavian Journal of Trauma, Resuscitation and Emergency Medicine*, *21*(1), 1−8.

Danielsen, P. H., Cao, Y., Roursgaard, M., Møller, P., & Loft, S. (2015). Endothelial cell activation, oxidative stress and inflammation induced by a panel of metal-based nanomaterials. *Nanotoxicology.*, *9*(7), 813−824.

de Castro, S., Maruoka, H., Hong, K., Kilbey, S. M., Costanzi, S., Hechler, B., et al. (2010). Functionalized congeners of P2Y1 receptor antagonists: 2-Alkynyl (N)-methanocarba 2′-deoxyadenosine 3′, 5′-bisphosphate analogues and conjugation to a polyamidoamine (PAMAM) dendrimer carrier. *Bioconjugate Chemistry*, *21*(7), 1190−1205.

De La Cruz, G. G., Rodríguez-Fragoso, P., Reyes-Esparza, J., Rodríguez-López, A., Gómez-Cansino, R., & Rodriguez-Fragoso, L. (2018). Interaction of nanoparticles with blood components and associated pathophysiological effects. *Unraveling the Safety Profile of Nanoscale Particles and Materials-From Biomedical to Environmental Applications.*

de la Harpe, K. M., Kondiah, P. P., Choonara, Y. E., Marimuthu, T., du Toit, L. C., & Pillay, V. (2019). The hemocompatibility of nanoparticles: A review of cell−nanoparticle interactions and hemostasis. *Cells.*, *8* (10), 1209.

De Paoli Lacerda, S. H., Semberova, J., Holada, K., Simakova, O., Hudson, S. D., & Simak, J. (2011). Carbon nanotubes activate store-operated calcium entry in human blood platelets. *ACS Nano*, *5*(7), 5808−5813.

Deb, S., Patra, H. K., Lahiri, P., Dasgupta, A. K., Chakrabarti, K., & Chaudhuri, U. (2011). Multistability in platelets and their response to gold nanoparticles. *Nanomedicine: Nanotechnology, Biology and Medicine*, *7*(4), 376−384.

Deb, S., Raja, S., Dasgupta, A. K., Sarkar, R., Chattopadhyay, A. P., Chaudhuri, U., et al. (2012). Surface tunability of nanoparticles in modulating platelet functions. *Blood Cells, Molecules, and Diseases*, *48*(1), 36−44.

Demir, E., Burgucu, D., Turna, F., Aksakal, S., & Kaya, B. (2013). Determination of TiO_2, ZrO_2, and Al_2O_3 nanoparticles on genotoxic responses in human peripheral blood lymphocytes and cultured embryonic kidney cells. *Journal of Toxicology and Environmental Health, Part A*, *76*(16), 990−1002.

Deng, Z. J., Liang, M., Monteiro, M., Toth, I., & Minchin, R. F. (2011). Nanoparticle-induced unfolding of fibrinogen promotes Mac-1 receptor activation and inflammation. *Nature Nanotechnology*, *6*(1), 39−44.

Do, V. M. H., Bach, L. G., Tran, D.-H. N., Nguyen, T. N. Q., Hoang, D. T., Nguyen, D. H., et al. (2020). Effective elimination of charge-associated toxicity of low generation polyamidoamine dendrimer eases drug delivery of oxaliplatin. *Biotechnology and Bioprocess Engineering*, 1−11.

Dobrovolskaia, M. A., Patri, A. K., Simak, J., Hall, J. B., Semberova, J., De Paoli Lacerda, S. H. ,, et al. (2012). Nanoparticle size and surface charge determine effects of PAMAM dendrimers on human platelets in vitro. *Molecular Pharmaceutics*, *9*(3), 382−393.

Dobrovolskaia, M. A., Patri, A. K., Zheng, J., Clogston, J. D., Ayub, N., Aggarwal, P., et al. (2009). Interaction of colloidal gold nanoparticles with human blood: Effects on particle size and analysis of plasma protein binding profiles. *Nanomedicine: Nanotechnology, Biology and Medicine*, *5*(2), 106−117.

Du, V. X., Huskens, D., Maas, C., Al Dieri, R., de Groot, P. G., & de Laat, B. (Eds.), (2014). *New insights into the role of erythrocytes in thrombus formation. Seminars in thrombosis and hemostasis.* Thieme Medical Publishers.

Easo, S. L., & Mohanan, P. (2015). In vitro hematological and in vivo immunotoxicity assessment of dextran stabilized iron oxide nanoparticles. *Colloids and Surfaces B: Biointerfaces*, *134*, 122−130.

Ekdahl, K. N., Fromell, K., Mohlin, C., Teramura, Y., & Nilsson, B. (2019). A human whole-blood model to study the activation of innate immunity system triggered by nanoparticles as a demonstrator for toxicity. *Science and Technology of Advanced Materials*, *20*(1), 688−698.

Elbayoumi, T. A., & Torchilin, V. P. (2008). Liposomes for targeted delivery of antithrombotic drugs. *Expert Opinion on Drug Delivery*, *5*(11), 1185−1198.

El-Laithy, H. M., Badawi, A., Abdelmalak, N. S., & El-Sayyad, N. (2018). Cubosomes as oral drug delivery systems: A promising approach for enhancing the release of clopidogrel bisulphate in the intestine. *Chemical and Pharmaceutical Bulletin*, c18−00615.

Enciso, A. E., Neun, B., Rodriguez, J., Ranjan, A. P., Dobrovolskaia, M. A., & Simanek, E. E. (2016). Nanoparticle effects on human platelets in vitro: A comparison between PAMAM and triazine dendrimers. *Molecules.*, *21*(4), 428.

Feng, L., Yang, X., Liang, S., Xu, Q., Miller, M. R., Duan, J., et al. (2019). Silica nanoparticles trigger the vascular endothelial dysfunction and prethrombotic state via miR-451 directly regulating the IL6R signaling pathway. *Particle and Fibre Toxicology*, *16*(1), 1−13.

Ferraz, N., Carlsson, J., Hong, J., & Ott, M. K. (2008). Influence of nanoporesize on platelet adhesion and activation. *Journal of Materials Science: Materials in Medicine*, *19*(9), 3115−3121.

Ferraz, N., Hong, J., & Karlsson Ott, M. (2010). Procoagulant behavior and platelet microparticle generation on nanoporous alumina. *Journal of Biomaterials Applications*, *24*(8), 675−692.

Fröhlich, E. (2016). Action of nanoparticles on platelet activation and plasmatic coagulation. *Current Medicinal Chemistry*, *23*(5), 408−430.

Gaston, E., Fraser, J. F., Xu, Z. P., & Ta, H. T. (2018). Nano-and micro-materials in the treatment of internal bleeding and uncontrolled hemorrhage. *Nanomedicine: Nanotechnology, Biology and Medicine*, *14*(2), 507−519.

Gelderman, M. P., Simakova, O., Clogston, J. D., Patri, A. K., Siddiqui, S. F., Vostal, A. C., et al. (2008). Adverse effects of fullerenes on endothelial cells: Fullerenol C60 (OH) 24 induced tissue factor and ICAM-1 membrane expression and apoptosis in vitro. *International Journal of Nanomedicine*, *3*(1), 59.

Gobbo, O. L., Sjaastad, K., Radomski, M. W., Volkov, Y., & Prina-Mello, A. (2015). Magnetic nanoparticles in cancer theranostics. *Theranostics.*, *5*(11), 1249.

Greish, K., Thiagarajan, G., Herd, H., Price, R., Bauer, H., Hubbard, D., et al. (2012). Size and surface charge significantly influence the toxicity of silica and dendritic nanoparticles. *Nanotoxicology.*, *6*(7), 713−723.

Gu, Z., Yan, S., Cheong, S., Cao, Z., Zuo, H., Thomas, A. C., et al. (2018). Layered double hydroxide nanoparticles: Impact on vascular cells, blood cells and the complement system. *Journal of Colloid and Interface Science*, *512*, 404−410.

Guildford, A., Poletti, T., Osbourne, L., Di Cerbo, A., Gatti, A., & Santin, M. (2009). Nanoparticles of a different source induce different patterns of activation in key biochemical and cellular components of the host response. *Journal of the Royal Society Interface, 6*(41), 1213−1221.

Guo, C., Xia, Y., Niu, P., Jiang, L., Duan, J., Yu, Y., et al. (2015). Silica nanoparticles induce oxidative stress, inflammation, and endothelial dysfunction in vitro via activation of the MAPK/Nrf2 pathway and nuclear factor-κB signaling. *International Journal of Nanomedicine, 10,* 1463.

Guo, H., Zhang, J., Boudreau, M., Meng, J., Yin, J.-j, Liu, J., et al. (2015). Intravenous administration of silver nanoparticles causes organ toxicity through intracellular ROS-related loss of inter-endothelial junction. *Particle and Fibre Toxicology, 13*(1), 1−13.

Guo, L., Tong, D., Yu, M., Zhang, Y., Li, T., Wang, C., et al. (2018). Phosphatidylserine-exposing cells contribute to the hypercoagulable state in patients with multiple myeloma. *International Journal of Oncology, 52*(6), 1981−1990.

Haberl, N., Hirn, S., Holzer, M., Zuchtriegel, G., Rehberg, M., & Krombach, F. (2015). Effects of acute systemic administration of TiO_2, ZnO, SiO_2, and Ag nanoparticles on hemodynamics, hemostasis and leukocyte recruitment. *Nanotoxicology., 9*(8), 963−971.

Haller, C. A., Cui, W., Wen, J., Robson, S. C., & Chaikof, E. L. (2006). Reconstitution of CD39 in liposomes amplifies nucleoside triphosphate diphosphohydrolase activity and restores thromboregulatory properties. *Journal of Vascular Surgery, 43*(4), 816−823.

Hangge, P., Stone, J., Albadawi, H., Zhang, Y. S., Khademhosseini, A., & Oklu, R. (2017). Hemostasis and nanotechnology. *Cardiovascular Diagnosis and Therapy, 7*(Suppl. 3), S267.

Hante, N. K., Medina, C., & Santos-Martinez, M. J. (2019). Effect on platelet function of metal-based nanoparticles developed for medical applications. *Frontiers in Cardiovascular Medicine, 6.*

He, H., Adili, R., Liu, L., Hong, K., Holinstat, M., & Schwendeman, A. (2020). Synthetic high-density lipoproteins loaded with an antiplatelet drug for efficient inhibition of thrombosis in mice. *Science Advances, 6*(49), eabd0130.

He, Y., Xu, J., Sun, X., Ren, X., Maharjan, A., York, P., et al. (2019). Cuboidal tethered cyclodextrin frameworks tailored for hemostasis and injured vessel targeting. *Theranostics., 9*(9), 2489.

Hecold, M., Buczkowska, R., Mucha, A., Grzesiak, J., Rac-Rumijowska, O., Teterycz, H., et al. (2017). The effect of PEI and PVP-stabilized gold nanoparticles on equine platelets activation: Potential application in equine regenerative medicine. *Journal of Nanomaterials, 2017.*

Heeremans, J., Prevost, R., Bekkers, M., Los, P., EMEIS, L., Kluft, C., et al. (1995). Thrombolytic treatment with tissue-type plasminogen activator (t-PA) containing liposomes in rabbits: A comparison with free t-PA. *Liposomes in Thrombolytic Therapy, 102.*

Hoang Thi, T. T., Nguyen Tran, D.-H., Bach, L. G., Vu-Quang, H., Nguyen, D. C., Park, K. D., et al. (2019). Functional magnetic core-shell system-based iron oxide nanoparticle coated with biocompatible copolymer for anticancer drug delivery. *Pharmaceutics., 11*(3), 120.

Holzer, M., Bihari, P., Praetner, M., Uhl, B., Reichel, C., Fent, J., et al. (2014). Carbon-based nanomaterials accelerate arteriolar thrombus formation in the murine microcirculation independently of their shape. *Journal of Applied Toxicology, 34*(11), 1167−1176.

Hsu, Sh, Tang, C. M., & Tseng, H. J. (2006). Biocompatibility of poly (ether) urethane-gold nanocomposites. *Journal of Biomedical Materials Research Part A, 79*(4), 759−770.

Huang, H., Lai, W., Cui, M., Liang, L., Lin, Y., Fang, Q., et al. (2016). An evaluation of blood compatibility of silver nanoparticles. *Scientific Reports, 6*(1), 1−15.

Ilinskaya, A. N., & Dobrovolskaia, M. A. (2013). Nanoparticles and the blood coagulation system. Part I: Benefits of nanotechnology. *Nanomedicine, 8*(5), 773−784.

Ilinskaya, A. N., & Dobrovolskaia, M. A. (2016). *Nanoparticles and the blood coagulation system. Handbook of immunological properties of engineered nanomaterials: Volume 2: Haematocompatibility of engineered nanomaterials* (pp. 261−302). World Scientific.

Jiao, Y., Ubrich, N., Marchand-Arvier, M., Vigneron, C., Hoffman, M., Lecompte, T., et al. (2002). In vitro and in vivo evaluation of oral heparin−loaded polymeric nanoparticles in rabbits. *Circulation.*, *105*(2), 230−235.

Jiao, Y., Ubrich, N., Marchand-Arvier, M., Vigneron, C., Hoffman, M., & Maincent, P. (2001). Preparation and in vitro evaluation of heparin-loaded polymeric nanoparticles. *Drug Delivery*, *8*(3), 135−141.

Jones, C. F., Campbell, R. A., Franks, Z., Gibson, C. C., Thiagarajan, G., Vieira-de-Abreu, A., et al. (2012). Cationic PAMAM dendrimers disrupt key platelet functions. *Molecular Pharmaceutics*, *9*(6), 1599−1611.

Juliano, R., Hsu, M., Peterson, D., Regen, S., & Singh, A. (1983). Interactions of conventional or photopolymerized liposomes with platelets in vitro. *Experimental Cell Research*, *146*(2), 422−427.

Jun, E.-A., Lim, K.-M., Kim, K., Bae, O.-N., Noh, J.-Y., Chung, K.-H., et al. (2011). Silver nanoparticles enhance thrombus formation through increased platelet aggregation and procoagulant activity. *Nanotoxicology.*, *5*(2), 157−167.

Kim, Y., Hechler, B., Gao, Z.-G., Gachet, C., & Jacobson, K. A. (2009). PEGylated dendritic unimolecular micelles as versatile carriers for ligands of G protein-coupled receptors. *Bioconjugate Chemistry*, *20*(10), 1888−1898.

Kim, Y., Hechler, B., Klutz, A. M., Gachet, C., & Jacobson, K. A. (2008). Toward multivalent signaling across G protein-coupled receptors from poly (amidoamine) dendrimers. *Bioconjugate Chemistry*, *19*(2), 406−411.

Kleiman, N. S., Raizner, A. E., Jordan, R., Wang, A. L., Norton, D., Mace, K. F., et al. (1995). Differential inhibition of platelet aggregation induced by adenosine diphosphate or a thrombin receptor-activating peptide in patients treated with bolus chimeric 7E3 Fab: Implications for inhibition of the internal pool of GPIIb/IIIa receptors. *Journal of the American College of Cardiology*, *26*(7), 1665−1671.

Kou, L., Sun, J., Zhai, Y., & He, Z. (2013). The endocytosis and intracellular fate of nanomedicines: Implication for rational design. *Asian Journal of Pharmaceutical Sciences*, *8*(1), 1−10.

Koziara, J., Oh, J., Akers, W., Ferraris, S., & Mumper, R. (2005). Blood compatibility of cetyl alcohol/polysorbate-based nanoparticles. *Pharmaceutical Research*, *22*(11), 1821−1828.

Krajewski, S., Prucek, R., Panacek, A., Avci-Adali, M., Nolte, A., Straub, A., et al. (2013). Hemocompatibility evaluation of different silver nanoparticle concentrations employing a modified Chandler-loop in vitro assay on human blood. *Acta Biomaterialia*, *9*(7), 7460−7468.

Kumari, S., Singh, M. K., Singh, S. K., Grácio, J. J., & Dash, D. (2014). Nanodiamonds activate blood platelets and induce thromboembolism. *Nanomedicine.*, *9*(3), 427−440.

Kushida, T., Saha, K., Subramani, C., Nandwana, V., & Rotello, V. M. (2014). Effect of nano-scale curvature on the intrinsic blood coagulation system. *Nanoscale.*, *6*(23), 14484−14487.

Kwaan, H. C., & Samama, M. (2019). *Clinical thrombosis.* CRC Press.

Laloy, J., Minet, V., Alpan, L., Mullier, F., Beken, S., Toussaint, O., et al. (2014). Impact of silver nanoparticles on haemolysis, platelet function and coagulation. *Nanobiomedicine*, *1*, 4.

Le, N. T. T., Pham, L. P. T., Nguyen, D. H. T., Le, N. H., Tran, T. V., Nguyen, C. K., et al. (2019). Liposome-based nanocarrier system for phytoconstituents. *Novel Drug Delivery Systems for Phytoconstituents*, 45.

Le, V. T., Bach, L. G., Pham, T. T., Le, N. T. T., Ngoc, U. T. P., Tran, D.-H. N., et al. (2019). Synthesis and antifungal activity of chitosan-silver nanocomposite synergize fungicide against Phytophthora capsici. *Journal of Macromolecular Science, Part A*, *56*(6), 522−528.

Leach, J. K., Edgar, A., Patterson, E., Miao, Y., & Johnson, A. E. (2003). Accelerated thrombolysis in a rabbit model of carotid artery thrombosis with liposome-encapsulated and microencapsulated streptokinase. *Thrombosis and Haemostasis*, *90*(07), 64−70.

Li, X., Radomski, A., Corrigan, O. I., Tajber, L., De Sousa Menezes, F., Endter, S., et al. (2009). Platelet compatibility of PLGA, chitosan and PLGA−chitosan nanoparticles. *Nanomedicine.*, *4*(7), 735−746.

Liang, S., Sun, K., Wang, Y., Dong, S., Wang, C., Liu, L., et al. (2016). Role of Cyt-C/caspases-9, 3, Bax/Bcl-2 and the FAS death receptor pathway in apoptosis induced by zinc oxide nanoparticles in human aortic endothelial cells and the protective effect by alpha-lipoic acid. *Chemico-Biological Interactions*, *258*, 40−51.

Liu, F., Huang, H., Gong, Y., Li, J., Zhang, X., & Cao, Y. (2017). Evaluation of in vitro toxicity of polymeric micelles to human endothelial cells under different conditions. *Chemico-Biological Interactions*, *263*, 46−54.

Love, S. A., Thompson, J. W., & Haynes, C. L. (2012). Development of screening assays for nanoparticle toxicity assessment in human blood: Preliminary studies with charged Au nanoparticles. *Nanomedicine.*, *7*(9), 1355−1364.

Marrink, S. J., De Vries, A. H., & Mark, A. E. (2004). Coarse grained model for semiquantitative lipid simulations. *The Journal of Physical Chemistry B*, *108*(2), 750−760.

Martin, K., Ma, A. D. ,, & Key, N. S. (2018). *Molecular basis of hemostatic and thrombotic diseases. Molecular pathology* (pp. 277−297). Elsevier.

Martínez-Gutierrez, F., Thi, E. P., Silverman, J. M., de Oliveira, C. C., Svensson, S. L., Hoek, A. V., et al. (2012). Antibacterial activity, inflammatory response, coagulation and cytotoxicity effects of silver nanoparticles. *Nanomedicine: Nanotechnology, Biology and Medicine*, *8*(3), 328−336.

Matus, M. F., Vilos, C., Cisterna, B. A., Fuentes, E., & Palomo, I. (2018). Nanotechnology and primary hemostasis: Differential effects of nanoparticles on platelet responses. *Vascular Pharmacology*, *101*, 1−8.

Matuszak, J., Baumgartner, J., Zaloga, J., Juenet, M., Da Silva, A. E., Franke, D., et al. (2016). Nanoparticles for intravascular applications: Physicochemical characterization and cytotoxicity testing. *Nanomedicine.*, *11*(6), 597−616.

Matzdorff, A., & Voss, R. (2006). Upregulation of GP IIb/IIIa receptors during platelet activation: Influence on efficacy of receptor blockade. *Thrombosis Research*, *117*(3), 307−314.

Mayer, A., Vadon, M., Rinner, B., Novak, A., Wintersteiger, R., & Fröhlich, E. (2009). The role of nanoparticle size in hemocompatibility. *Toxicology.*, *258*(2−3), 139−147.

McGuinnes, C., Duffin, R., Brown, S. ,L., Mills, N., Megson, I. L., MacNee, W., et al. (2011). Surface derivatization state of polystyrene latex nanoparticles determines both their potency and their mechanism of causing human platelet aggregation in vitro. *Toxicological Sciences*, *119*(2), 359−368.

Mehri, R., Mavriplis, C., & Fenech, M. (2018). Red blood cell aggregates and their effect on non-Newtonian blood viscosity at low hematocrit in a two-fluid low shear rate microfluidic system. *Plos One*, *13*(7), e0199911.

Meng, J., Cheng, X., Liu, J., Zhang, W., Li, X., Kong, H., et al. (2012). Effects of long and short carboxylated or aminated multiwalled carbon nanotubes on blood coagulation. *PLoS One*, *7*(7), e38995.

Miyamoto, M., Sasakawa, S., Ozawa, T., Kawaguchi, H., & Ohtsuka, Y. (1990). Mechanisms of blood coagulation induced by latex particles and the roles of blood cells. *Biomaterials*, *11*(6), 385−388.

Nabeshi, H., Yoshikawa, T., Matsuyama, K., Nakazato, Y., Arimori, A., Isobe, M., et al. (2012). Amorphous nanosilicas induce consumptive coagulopathy after systemic exposure. *Nanotechnology*, *23*(4), 045101.

Nemmar, A., Albarwani, S., Beegam, S., Yuvaraju, P., Yasin, J., Attoub, S., et al. (2014). Amorphous silica nanoparticles impair vascular homeostasis and induce systemic inflammation. *International Journal of Nanomedicine*, *9*, 2779.

Nemmar, A., Melghit, K., & Ali, B. H. (2008). The acute proinflammatory and prothrombotic effects of pulmonary exposure to rutile TiO_2 nanorods in rats. *Experimental Biology and Medicine*, *233*(5), 610−619.

Nguyen, D. H., Bach, L. G., Nguyen Tran, D.-H., Cao, V. D., Nguyen, T. N. Q., Le, T. T. H., et al. (2019). Partial surface modification of low generation polyamidoamine dendrimers: Gaining insight into their potential for improved carboplatin delivery. *Biomolecules*, *9*(6), 214.

Nguyen, P. D., O'rear, E., Johnson, A. E., Patterson, E., Whitsett, T. L., & Bhakta, R. (1990). Accelerated thrombolysis and reperfusion in a canine model of myocardial infarction by liposomal encapsulation of streptokinase. *Circulation Research*, *66*(3), 875−878.

Nguyen, T. D., Nguyen, T. T. T., Ivanov, I. A., Nguyen, K. C., Tran, Q. N., Hoang, A. N., et al. (2019). Nanoencapsulation enhances anticoagulant activity of adenosine and dipeptide IleTrp. *Nanomaterials*, *9*(9), 1191.

Nguyen, T. N. T., Nguyen-Tran, D.-H., Bach, L. G., Du Truong, T. H., Le, N. T. T., & Nguyen, D. H. (2019). Surface PEGylation of hollow mesoporous silica nanoparticles via aminated intermediate. *Progress in Natural Science: Materials International*, *29*(6), 612−616.

Nishikawa, T., Iwakiri, N., Kaneko, Y., Taguchi, A., Fukushima, K., Mori, H., et al. (2009). Nitric oxide release in human aortic endothelial cells mediated by delivery of amphiphilic polysiloxane nanoparticles to caveolae. *Biomacromolecules*, *10*(8), 2074−2085.

Oslakovic, C., Cedervall, T., Linse, S., & Dahlbäck, B. (2012). Polystyrene nanoparticles affecting blood coagulation. *Nanomedicine: Nanotechnology, Biology and Medicine*, *8*(6), 981−986.

Palta, S., Saroa, R., & Palta, A. (2014). Overview of the coagulation system. *Indian Journal of Anaesthesia*, *58*(5), 515.

Pan, D. C., Myerson, J. W., Brenner, J. S., Patel, P. N., Anselmo, A. C., Mitragotri, S., et al. (2018). Nanoparticle properties modulate their attachment and effect on carrier red blood cells. *Scientific Reports*, *8*(1), 1−12.

Paniccia, R., Priora, R., Liotta, A. A., & Abbate, R. (2015). Platelet function tests: A comparative review. *Vascular Health and Risk Management*, *11*, 133.

Peng, H. T. (2010). Thromboelastographic study of biomaterials. *Journal of Biomedical Materials Research Part B: Applied Biomaterials*, *94*(2), 469−485.

Radomski, A., Jurasz, P., Alonso-Escolano, D., Drews, M., Morandi, M., Malinski, T., et al. (2005). Nanoparticle-induced platelet aggregation and vascular thrombosis. *British Journal of Pharmacology*, *146* (6), 882−893.

Ragaseema, V., Unnikrishnan, S., Krishnan, V. K., & Krishnan, L. K. (2012). The antithrombotic and antimicrobial properties of PEG-protected silver nanoparticle coated surfaces. *Biomaterials*, *33*(11), 3083−3092.

Ramtoola, Z., Lyons, P., Keohane, K., Kerrigan, S. W., Kirby, B. P., & Kelly, J. G. (2011). Investigation of the interaction of biodegradable micro-and nanoparticulate drug delivery systems with platelets. *Journal of Pharmacy and Pharmacology*, *63*(1), 26−32.

Ran, Q., Xiang, Y., Liu, Y., Xiang, L., Li, F., Deng, X., et al. (2015). Eryptosis indices as a novel predictive parameter for biocompatibility of Fe_3O_4 magnetic nanoparticles on erythrocytes. *Scientific Reports*, *5*, 16209.

Reimhult, E. (2019). Nanoparticle interactions with blood proteins and what it means: A tutorial review. *Blood and Genomics*, *3*(2), 73−87.

Reinish, L., Bally, M., Loughrey, H., & Cullis, P. (1988). Interactions of liposomes and platelets. *Thrombosis and Haemostasis*, *59*(03), 518−523.

Reitsma, S., Slaaf, D. W., Vink, H., Van Zandvoort, M. A., & Oude Egbrink, M. G. (2007). The endothelial glycocalyx: Composition, functions, and visualization. *Pflügers Archiv: European Journal of Physiology*, *454*(3), 345−359.

Rumbaut, R. E., & Thiagarajan, P. (2010). Platelet-vessel wall interactions in hemostasis and thrombosis. *Synthesis Lectures on Integrated Systems Physiology: From Molecule to Function*, *2*(1), 1−75.

Saikia, J., Mohammadpour, R., Yazdimamaghani, M., Northrup, H., Hlady, V., & Ghandehari, H. (2018). Silica nanoparticle−endothelial *I*nteraction: Uptake and effect on platelet adhesion under flow conditions. *ACS Applied Bio Materials*, *1*(5), 1620−1627.

Santos-Martinez, M. J., Inkielewicz-Stepniak, I., Medina, C., Rahme, K., D'Arcy, D. M., Fox, D., et al. (2012). The use of quartz crystal microbalance with dissipation (QCM-D) for studying nanoparticle-induced platelet aggregation. *International Journal of Nanomedicine*, *7*, 243.

Santos-Martinez, M. J., Rahme, K., Corbalan, J. J., Faulkner, C., Holmes, J. D., Tajber, L., et al. (2014). Pegylation increases platelet biocompatibility of gold nanoparticles. *Journal of Biomedical Nanotechnology*, *10*(6), 1004−1015.

Santos-Martinez, M. J., Tomaszewski, K. A., Medina, C., Bazou, D., Gilmer, J. F., & Radomski, M. W. (2015). Pharmacological characterization of nanoparticle-induced platelet microaggregation using quartz crystal microbalance with dissipation: Comparison with light aggregometry. *International Journal of Nanomedicine, 10*, 5107.

Semberova, J., De Paoli Lacerda, S. H., Simakova, O., Holada, K., Gelderman, M. P., & Simak, J. (2009). Carbon nanotubes activate blood platelets by inducing extracellular Ca^{2+} influx sensitive to calcium entry inhibitors. *Nano Letters, 9*(9), 3312−3317.

Setyawati, M. I., Tay, C. Y., Docter, D., Stauber, R. H., & Leong, D. T. (2015). Understanding and exploiting nanoparticles' intimacy with the blood vessel and blood. *Chemical Society Reviews, 44*(22), 8174−8199.

Shafir, G., Galperin, A., & Margel, S. (2009). Synthesis and characterization of recombinant factor VIIa-conjugated magnetic iron oxide nanoparticles for hemophilia treatment. *Journal of Biomedical Materials Research Part A: An Official Journal of the Society for Biomaterials, the Japanese Society for Biomaterials, and the Australian Society for Biomaterials and the Korean Society for Biomaterials, 91*(4), 1056−1064.

Shiang, Y. C., Hsu, C. L., Huang, C. C., & Chang, H. T. (2011). Gold nanoparticles presenting hybridized self-assembled aptamers that exhibit enhanced inhibition of thrombin. *Angewandte Chemie International Edition, 50*(33), 7660−7665.

Shrivastava, S., Bera, T., Singh, S. K., Singh, G., Ramachandrarao, P., & Dash, D. (2009). Characterization of antiplatelet properties of silver nanoparticles. *ACS Nano, 3*(6), 1357−1364.

Shrivastava, S., Singh, S. K., Mukhopadhyay, A., Sinha, A. S., Mandal, R. K., & Dash, D. (2011). Negative regulation of fibrin polymerization and clot formation by nanoparticles of silver. *Colloids and Surfaces B: Biointerfaces, 82*(1), 241−246.

Simak, J. (2016). *The effects of engineered nanomaterials on the plasma coagulation system. Handbook of immunological properties of engineered nanomaterials: Volume 2: Haematocompatibility of engineered nanomaterials* (pp. 163−192). World Scientific.

Simak, J., & De Paoli, S. (2017). The effects of nanomaterials on blood coagulation in hemostasis and thrombosis. *Wiley Interdisciplinary Reviews: Nanomedicine and Nanobiotechnology, 9*(5), e1448.

Sims, P. J., Ginsberg, M., Plow, E., & Shattil, S. (1991). Effect of platelet activation on the conformation of the plasma membrane glycoprotein IIb-IIIa complex. *Journal of Biological Chemistry, 266*(12), 7345−7352.

Šimundić, M., Drašler, B., Šuštar, V., Zupanc, J., Štukelj, R., Makovec, D., et al. (2013). Effect of engineered TiO_2 and ZnO nanoparticles on erythrocytes, platelet-rich plasma and giant unilamelar phospholipid vesicles. *BMC Veterinary Research, 9*(1), 7.

Singh, S. K., Singh, M. K., Kulkarni, P. P., Sonkar, V. K., Grácio, J. J., & Dash, D. (2012). Amine-modified graphene: Thrombo-protective safer alternative to graphene oxide for biomedical applications. *ACS Nano, 6*(3), 2731−2740.

Singh, S. K., Singh, M. K., Nayak, M. K., Kumari, S., Shrivastava, S., Grácio, J. J., et al. (2011). Thrombus inducing property of atomically thin graphene oxide sheets. *ACS Nano, 5*(6), 4987−4996.

Smyth, E., Solomon, A., Vydyanath, A., Luther, P. K., Pitchford, S., Tetley, T. D., et al. (2015). Induction and enhancement of platelet aggregation in vitro and in vivo by model polystyrene nanoparticles. *Nanotoxicology., 9*(3), 356−364.

Sobot, D., Mura, S., Couvreur, P., Kobayashi, S., & Müllen, K. (2014). *Nanoparticles: Blood components interactions. Encyclopedia of polymeric nanomaterials* (pp. 1−10). Berlin, Heidelberg: Springer.

Sperling, C., Maitz, M., & Werner, C. (2018). *Test methods for hemocompatibility of biomaterials. Hemocompatibility of biomaterials for clinical applications* (pp. 77−104). Elsevier.

Sriram, K., Intaglietta, M., & Tartakovsky, D. M. (2014). Non-Newtonian flow of blood in arterioles: Consequences for wall shear stress measurements. *Microcirculation., 21*(7), 628−639.

Stevens, K. N., Knetsch, M. L., Sen, A., Sambhy, V., & Koole, L. H. (2009). Disruption and activation of blood platelets in contact with an antimicrobial composite coating consisting of a pyridinium polymer and AgBr nanoparticles. *ACS Applied Materials & Interfaces, 1*(9), 2049−2054.

Strojan, K., Leonardi, A., Bregar, V. B., Križaj, I., Svete, J., & Pavlin, M. (2017). Dispersion of nanoparticles in different media importantly determines the composition of their protein corona. *PLoS One*, *12*(1), e0169552.

Su, M., Dai, Q., Chen, C., Zeng, Y., Chu, C., & Liu, G. (2020). Nano-medicine for thrombosis: A precise diagnosis and treatment strategy. *Nano-Micro Letters*, *12*, 1−21.

Su, Y., Zhao, L., Meng, F., Wang, Q., Yao, Y., & Luo, J. (2017). Silver nanoparticles decorated lipase-sensitive polyurethane micelles for on-demand release of silver nanoparticles. *Colloids and Surfaces B: Biointerfaces*, *152*, 238−244.

Sun, P., McMillan-Ward, E., Mian, R., & Israels, S. J. (2019). Comparison of light transmission aggregometry and multiple electrode aggregometry for the evaluation of patients with mucocutaneous bleeding. *International Journal of Laboratory Hematology*, *41*(1), 133−140.

Sun, X., Shi, J., Zou, X., Wang, C., Yang, Y., & Zhang, H. (2016). Silver nanoparticles interact with the cell membrane and increase endothelial permeability by promoting VE-cadherin internalization. *Journal of Hazardous Materials*, *317*, 570−578.

Tankersley, D. L., Alving, B. M., & Finlayson, J. S. (1983). Activation of factor XII by dextran sulfate: The basis for an assay of factor XII. *Blood.*, *62*(2), 448−456.

Tavano, R., Segat, D., Reddi, E., Kos, J., Rojnik, M., Kocbek, P., et al. (2010). Procoagulant properties of bare and highly PEGylated vinyl-modified silica nanoparticles. *Nanomedicine.*, *5*(6), 881−896.

Thi, T. T. H., Nguyen, D. H. T., Nguyen, D. T. D., Nguyen, D. H., & Truong, M.-D. (2020). Decellularized porcine epiphyseal plate-derived extracellular matrix powder: Synthesis and characterization. *Cells Tissues Organs*, *209*(1), 1−9.

Tran, D. H. N., Nguyen, T. H., Vo, T. N. N., Pham, L. P. T., Vo, D. M. H., Nguyen, C. K., et al. (2019). Self-assembled poly (ethylene glycol) methyl ether-grafted gelatin nanogels for efficient delivery of curcumin in cancer treatment. *Journal of Applied Polymer Science*, *136*(20), 47544.

Tran, H. D., Park, K. D., Ching, Y. C., Huynh, C., & Nguyen, D. H. (2020). A comprehensive review on polymeric hydrogel and its composite: Matrices of choice for bone and cartilage tissue engineering. *Journal of Industrial and Engineering Chemistry*.

Vakhrusheva, T. V., Gusev, A. A., & Gusev, S. A. (2013). Vlasova II. Albumin reduces thrombogenic potential of single-walled carbon nanotubes. *Toxicology Letters*, *221*(2), 137−145.

van der Graaf, F., Keus, F. J., Vlooswijk, R. A., & Bouma, B. N. (1982). The contact activation mechanism in human plasma: Activation induced by dextran sulfate. *Blood*, *59*(6), 1225−1233.

Villegas, M. G., Ceballos, M. T., Urquijo, J., Torres, E. Y., Ortiz-Reyes, B. L., Arnache-Olmos, O. L., et al. (2019). Poly (acrylic acid)-coated iron oxide nanoparticles interact with mononuclear phagocytes and decrease platelet aggregation. *Cellular Immunology*, *338*, 51−62.

Walton, B. L., Lehmann, M., Skorczewski, T., Holle, L. A., Beckman, J. D., Cribb, J. A., et al. (2017). Elevated hematocrit enhances platelet accumulation following vascular injury. *Blood, The Journal of the American Society of Hematology*, *129*(18), 2537−2546.

Weisel, J., & Litvinov, R. (2019). Red blood cells: The forgotten player in hemostasis and thrombosis. *Journal of Thrombosis and Haemostasis*, *17*(2), 271−282.

Wiggins, R. C., & Cochrane, C. C. (1979). The autoactivation of rabbit Hageman factor. *The Journal of Experimental Medicine*, *150*(5), 1122−1133.

Wu, H., Su, M., Jin, H., Li, X., Wang, P., Chen, J., et al. (2020). Rutin-loaded silver nanoparticles with antithrombotic function. *Frontiers in Bioengineering and Biotechnology*, *8*, 1356.

Wu, Y.-F., Hsu, P.-S., Tsai, C.-S., Pan, P.-C., & Chen, Y.-L. (2018). Significantly increased low shear rate viscosity, blood elastic modulus, and RBC aggregation in adults following cardiac surgery. *Scientific Reports*, *8*(1), 1−10.

Yau, J. W., Teoh, H., & Verma, S. (2015). Endothelial cell control of thrombosis. *BMC Cardiovascular Disorders*, *15*(1), 1−11.

Yedgar, S., Koshkaryev, A., & Barshtein, G. (2002). The red blood cell in vascular occlusion. *Pathophysiology of Haemostasis and Thrombosis, 32*(5−6), 263−268.

Yoshida, T., Yoshioka, Y., Tochigi, S., Hirai, T., Uji, M., Ichihashi, K.-I., et al. (2013). Intranasal exposure to amorphous nanosilica particles could activate intrinsic coagulation cascade and platelets in mice. *Particle and Fibre Toxicology, 10*(1), 1−12.

Yu, M., Huang, S., Yu, K. J., & Clyne, A. M. (2012). Dextran and polymer polyethylene glycol (PEG) coating reduce both 5 and 30 nm iron oxide nanoparticle cytotoxicity in 2D and 3D cell culture. *International Journal of Molecular Sciences, 13*(5), 5554−5570.

Zamanlu, M., Eskandani, M., Barar, J., Jaymand, M., Pakchin, P. S., & Farhoudi, M. (2019). Enhanced thrombolysis using tissue plasminogen activator (tPA)-loaded PEGylated PLGA nanoparticles for ischemic stroke. *Journal of Drug Delivery Science and Technology, 53*, 101165.

Zbinden, G., Wunderli-Allenspach, H., & Grimm, L. (1989). Assessment of thrombogenic potential of liposomes. *Toxicology., 54*(3), 273−280.

Zhang, Y., Cai, J., Li, C., Wei, J., Liu, Z., & Xue, W. (2016). Effects of thermosensitive poly (N-isopropylacrylamide) on blood coagulation. *Journal of Materials Chemistry B., 4*(21), 3733−3749.

Zhao, W., Liu, Q., Zhang, X., Su, B., & Zhao, C. (2018). Rationally designed magnetic nanoparticles as anticoagulants for blood purification. *Colloids and Surfaces B: Biointerfaces, 164*, 316−323.

Red blood cells under externally induced stressors probed by micro-Raman spectroscopy

7

Jijo Lukose[1], Shamee Shastry[2], Ganesh Mohan[2] and Santhosh Chidangil[1]

[1]Centre of Excellence for Biophotonics, Department of Atomic and Molecular Physics, Manipal Academy of Higher Education, Manipal, India [2]Department of Immunohematology and Blood Transfusion, Kasturba Medical College, Manipal Academy of Higher Education, Manipal, India

7.1 Red blood cells and hemoglobin

Human blood comprises three major components: red blood cells (RBC), white blood cells (WBC), and platelets, all suspended in a yellow liquid medium called plasma. Plasma accounts for approximately 55% of whole blood, whereas the remaining portion is contributed by the cellular components (Molnar & Gair, 2013). RBCs originate from the stem cells in bone marrow, and are responsible for the transport of both carbon dioxide and oxygen to and fro between the cells and lungs (Jensen, 2009). WBCs form the core part of the body's immune system, whereas platelets are the causative factor for processes like blood clotting and wound healing (Golebiewska & Poole, 2015). RBC (or erythrocytes) do not possess nuclei, ribosomes, mitochondria, and other organelles, which are pivotal in other cell types to conduct essential tasks required for cell survival. This structure of RBCs has evolved to facilitate hemoglobin accumulation inside the cell. Hemoglobin is a globular protein that is responsible for oxygen delivery to the peripheral tissues (Giardina et al., 1995). Since RBCs lack mitochondria and rely on anaerobic respiration, they do not utilize any of the transporting gases and thus can effectively carry out direct oxygen delivery to tissues. Hemoglobin forms approximately 90% of red blood cells, in an average of 270−300 million molecules per cell (Tombak, 2019). The hemoglobin level in men is reported to be ~ 16 g/dL, whereas as it is slightly lower for women (~ 14 g/dL) (Billett, 1990). Human RBCs have a biconcave disk-shaped structure, with an average size of 6−8 μm (diameter) and 1.7−2.2 μm (thickness). Their extraordinarily high degree of flexibility in shape facilitates a high surface-to-volume ratio that enables large reversible elastic deformation of the cells as they travel through smaller capillaries during microcirculation (Uzoigwe, 2006). Deformability of RBCs is a crucial element for microcirculation, which is essential for performing gaseous transport. Alterations in the inherent biconcave shape of the cells due to pathological conditions can limit the deformability of cells, which in turn can adversely affect circulation (Diez-Silva et al., 2010).

Hemoglobin is composed of four-protein subunits, and each subunit connects to its neighboring subunits through intermolecular interactions. The ability of hemoglobin to carry out oxygen binding arises due to the presence of a bound prosthetic group called heme which is present in each subunit and which, in turn, is responsible for the red color of blood. The heme group is composed of a central iron atom in the midst of a protoporphyrin ring (Perutz, 1979). The iron atom present in the heme can bind to an oxygen molecule and thus a single hemoglobin molecule has the ability to bind four oxygen molecules. The central iron atom is linked to the four pyrrole nitrogen atoms of the porphyrin ring and the nitrogen atom of a histidine residue in the hemoglobin protein occupies the fifth coordination site of iron (Fig. 7.1). In oxygenated hemoglobin, the sixth coordination site provides the binding site for the oxygen molecule, whereas it remains unoccupied in the case of deoxygenated hemoglobin. The iron atom which lies ~ 0.4 Å outside the porphyrin plane in the case of deoxygenated hemoglobin moves into the plane upon the binding of an oxygen molecule (Franzen et al., 1994; Strekas & Spiro, 1972). Upon oxygenation, the histidine residue that is linked to the iron atom moves closer to the heme group, which ultimately alters the position of other nearby amino acids. This tends to change the shape of the whole protein, which makes it easier for the remaining three hemes to be oxygenated. Binding of one oxygen molecule elevates the tendency of hemoglobin to bind more oxygen, which is termed as cooperative binding (Stefan & Le Novère, 2013).

7.2 Raman spectroscopy of human blood

Recent years have witnessed a keen interest in the study of the interaction of light with biological matter. Researchers have been employing various spectroscopic technologies for the exploration of RBC as well as blood serum properties to explore various mechanisms, the prime motivation being to improve the way one looks at various hematological and other health disorders (Byrne et al., 2020; Devanesan et al., 2014, 2019; Lawaetz et al., 2012; Mostaço-Guidolin & Bachmann, 2011; Paraskevaidi et al., 2017; Rey-Barroso et al., 2020). Spectroscopic tools, in particular Raman

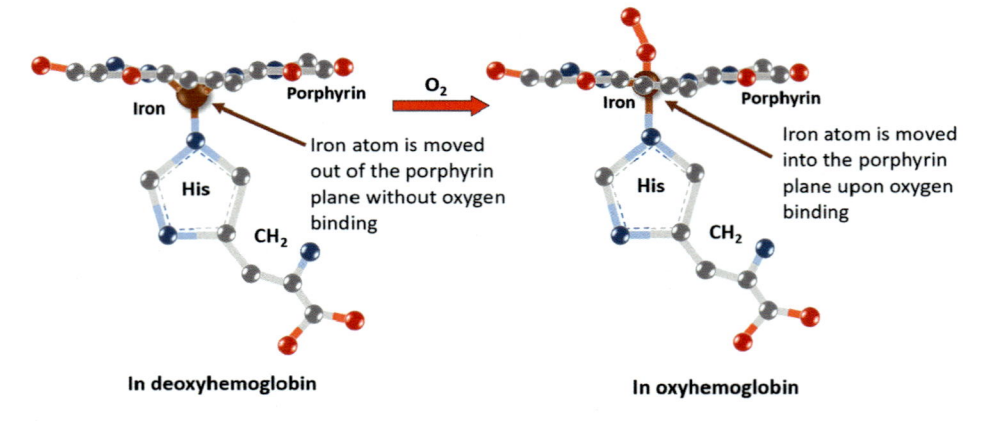

FIGURE 7.1

Schematic of deoxyhemoglobin and oxyhemoglobin (https://commons.wikimedia.org/wiki/File:Heme_deoxy_vs_oxygenated.jpg).

spectroscopy, has been at the forefront for more than four decades as an efficient analytical tool for conducting blood research (Atkins et al., 2017; da Silva et al., 2018; Parlatan et al., 2019). Other conventional analytical techniques explored for the diagnosis and subsequent prognosis of health abnormalities are sometimes tedious and invasive. Having the ability to characterize whole blood and individual blood components effectively, Raman spectroscopy can be a reliable technique for clinical applications, which can facilitate rapid, noninvasive diagnosis of various diseases. Use of Raman spectroscopy for blood-related studies evolved in the early 1970s, with a focus on probing the hemoglobin structure followed by other applications with the gradual improvements and innovations in Raman instrumentation (Brunner et al., 1972; Lippert et al., 1975; Strekas & Spiro, 1972; Yamamoto et al., 1973). This technique is able to provide wide-ranging information regarding the biological analyte of interest and about its environment due to its ability to generate molecular fingerprints of samples. Minimum sample processing steps and the absence of extraneous labeling agents are amongst the most advantageous highlights of this technique (Eberhardt et al., 2015). Minimal interference from water samples, which is the most common ingredient in biological liquid samples, is an added advantage of Raman spectroscopy (Carter & Edwards, 2001).

Raman spectroscopy, in general, is an inelastic light scattering method, which can supply reasonable detailing of the molecular composition, including the specific functional groups existing in biological and chemical samples (Harris & Bertolucci, 1989). In the Raman spectroscopic technique, a monochromatic light beam is allowed to incident on the sample, and the scattered light from the sample is collected. Most of the scattering is an elastic process (Rayleigh scattering), whereas a smaller fraction of scattered light has a frequency less than the incident light (inelastic scattering). During the experiment, Rayleigh scattered light is removed by a filter, and only the inelastic scattered light is recorded. The obtained Raman spectrum can be regarded as a "fingerprint" to identify the chemical species. This spectroscopic tool has gained high interest amongst researchers to discriminate molecules that have a high degree of similarity, due to the unique Raman signatures of each analyte. A schematic diagram illustrating Raman and Rayleigh scattering is given in Fig. 7.2. In Rayleigh/elastic scattering, the excited molecule from the virtual state returns to the ground state by generating a photon with equal energy to that of the incident photon (Das & Agrawal, 2011; Graves & Gardiner, 1989). In the case of inelastic Stokes Raman scattering, the molecule returns to the ground state by emitting a photon with lower energy. In anti-Stokes Raman, the molecule already in the higher vibrational state is raised to the virtual energy level and relaxes back to the lower energy level in the ground state by emitting a photon with energy higher than the incident photon. Since most of the molecules are in the ground state at room temperature, the contribution of anti-Stokes Raman is usually meager.

Raman spectroscopy of blood components has found extensive applications in biomedical research and diagnosis. Health abnormalities such as thalassemia, microcytic anemias, malaria, diabetes, and various cancers can be probed by evaluating the characteristic markers in the whole blood (Bueno Filho et al., 2015; Chen et al., 2016; da Silva et al., 2020; Lin et al., 2014). A recent report has demonstrated the utility of the Raman technique for the detection of colorectal cancer (CRC) from blood samples (Jenkins et al., 2020). This technique has been recommended as a highly cost-effective approach for cancer detection by reducing colonoscopies and the other diagnostic test requirements on patients in the urgent suspected cancer (USC) pathway for CRC. Raman studies conducted with blood samples have also provided better performance compared to the conventional USC pathway for cancer prediction. Barker et al. have recently exploited the capacity of

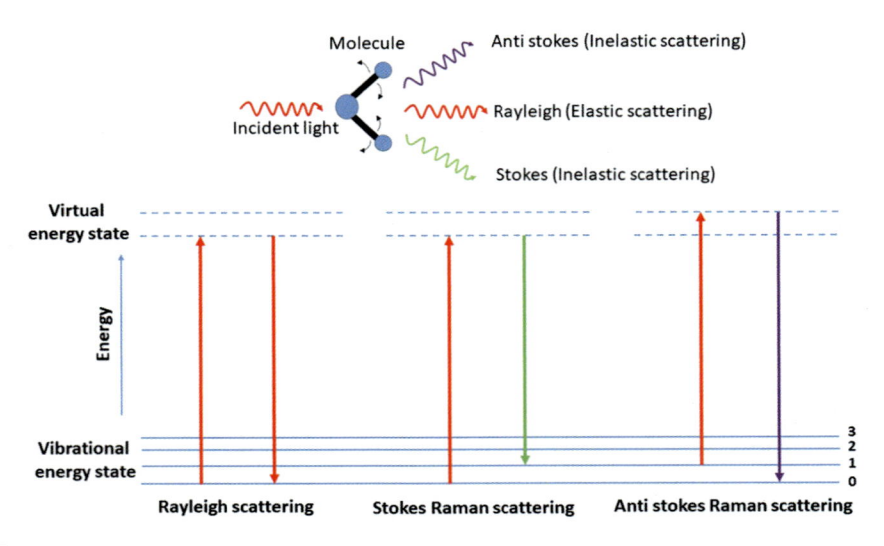

FIGURE 7.2

Schematic diagram of the Raman scattering process.

this analytical technique to investigate the biochemical features of Huntington's disease and its progression by evaluating blood serum (Huefner et al., 2020). The combination of statistical tools with Raman spectroscopy have also enabled researchers to perform the noninvasive detection of endometriosis from serum samples (Parlatan et al., 2019). In a similar manner, the use of multivariate tools has effectively supported Raman spectroscopy for prostate cancer detection by monitoring the prostate specific antigen levels in individuals from serum samples. Differentiation of anemias (iron deficiency anemia and sickle cell anemia) has been reported with 95% accuracy by applying partial least squares discriminant analysis (PLS-DA) on the Raman spectra of blood samples (da Silva et al., 2020). The PLS-DA technique together with Raman spectra of serum have been also effective in providing efficient classification for asthmatic and healthy individuals (Ullah et al., 2019). Detection as well as staging of cancers has been possible by evaluating the molecular signatures in the Raman spectrum of serum samples (Nargis et al., 2019; Pichardo-Molina et al., 2007; Sahu et al., 2013; Wang et al., 2018). In addition, the therapeutic response in breast cancer patients was studied by evaluating the alterations in tryptophan, tyrosine, phenylalanine, proteins, carotenoids, and lipids in blood plasma prior to and post treatment. Raman spectroscopy of blood have also found application in forensic sciences for differentiating human and nonhuman samples, discrimination of samples in accordance with chronological age, gender discrimination from dried blood stains etc. (Boyd et al., 2011; Doty & Lednev, 2018; Fujihara et al., 2017; McLaughlin et al., 2014; Mistek et al., 2016; Virkler & Lednev, 2009).

A schematic diagram of a typical Raman instrument is depicted in Fig. 7.3. The laser light will excite the sample of interest, whereas a filter is used after the laser to remove any exogenous laser output other than the specified wavelength. Sample manipulation for the experiment can be performed using the microscope. An edge filter needs to be kept before the dispersion element in order

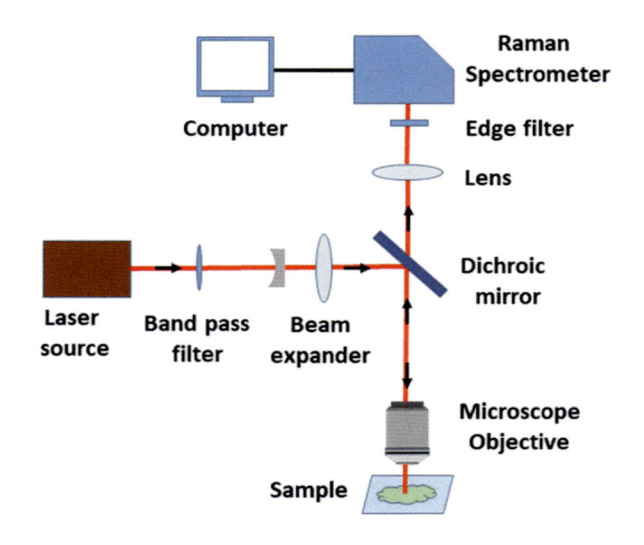

FIGURE 7.3

Schematic diagram of a micro-Raman spectrometer.

to remove the intense Rayleigh scattering light and to extract the comparatively weaker Raman light. Finally, the Raman signatures will be detected using the spectrometer.

7.3 Raman Tweezers—an optical approach to single-cell spectroscopy

Investigation of biological cells has been usually performed on an ensemble of cells by assuming that all the cells in a group behave identically. Later, researchers have recognized the necessity of obtaining measurements at individual cell level than bulk-averaged data, since individual cells in a group can behave differently. Moreover, pivotal information on the behavior of a small number of cells can be hidden in a bulk measurement. Bulk measurements that yield parameters averaged over a large population can also provide misleading information about biological processes, while single-cell measurements can yield information about the heterogeneity in a system. The urge for novel techniques to perform single-cell analysis motivates scientists to consider optical modalities, which can evaluate the biochemistry of cells in their natural state and monitor their dynamics and response to environmental perturbations. However, conventional physical/chemical immobilization techniques to avoid cellular displacement in the sample holder may alter the properties of the cell, which is a major limitation (Avsievich et al., 2020; Xie & Li, 2003). Optical techniques are favorable for these studies since they allow biochemical analysis of living specimens in a minimally invasive and non-destructive manner. Conventional cytometry used for cellular studies often uses external labeling agents (fluorophores) to acquire biochemical details of cell. However, some of these exogenous agents are themselves cytotoxic, which can alter the cellular chemistry and affect cell function. In addition, concerns of photobleaching arises in the case of prolonged cellular studies.

Optical tweezers in combination with Raman spectroscopy (Raman Tweezers) has become a game changer since it can reflect intrinsic biochemical properties of a cell without the addition of exogenous probes. Raman Tweezers is a variant of micro-Raman spectroscopy that is highly advantageous for studying single, live biological cells suspended in an aqueous environment in order to generate their biochemical signatures in response to the microenvironment. Optical tweezers use an optical laser trap created with the aid of a tightly focused near-Gaussian laser beam in order to immobilize the cell in its focal plane, thereby manipulating the live cell (Snook et al., 2009). Optical tweezers are capable of manipulating nanometer- and micron-sized dielectric particles by exerting very small forces through a highly focused laser beam (Snook et al., 2009). Both the scattering force in the direction of light propagation and the gradient force play pivotal roles in the mechanism of optical trapping (as shown in the inset of Fig. 7.4) (Moffitt et al., 2008). If the particle has a size comparatively larger than the incident laser wavelength, then the trapping can be explained using ray optics. Assume a dielectric sphere with a higher refractive index than the surrounding medium. The light ray once it passes through the dielectric sphere bends because of refraction. As per the conservation of momentum, the rate of change of momentum in the case of refracted rays will transfer an equal and opposite fraction of momentum onto the sphere. The gradient force pushes the sphere to the higher intensity region when the dielectric sphere is under a light gradient. Thus a focused laser can create a trap since the light intensity gradient is directed toward

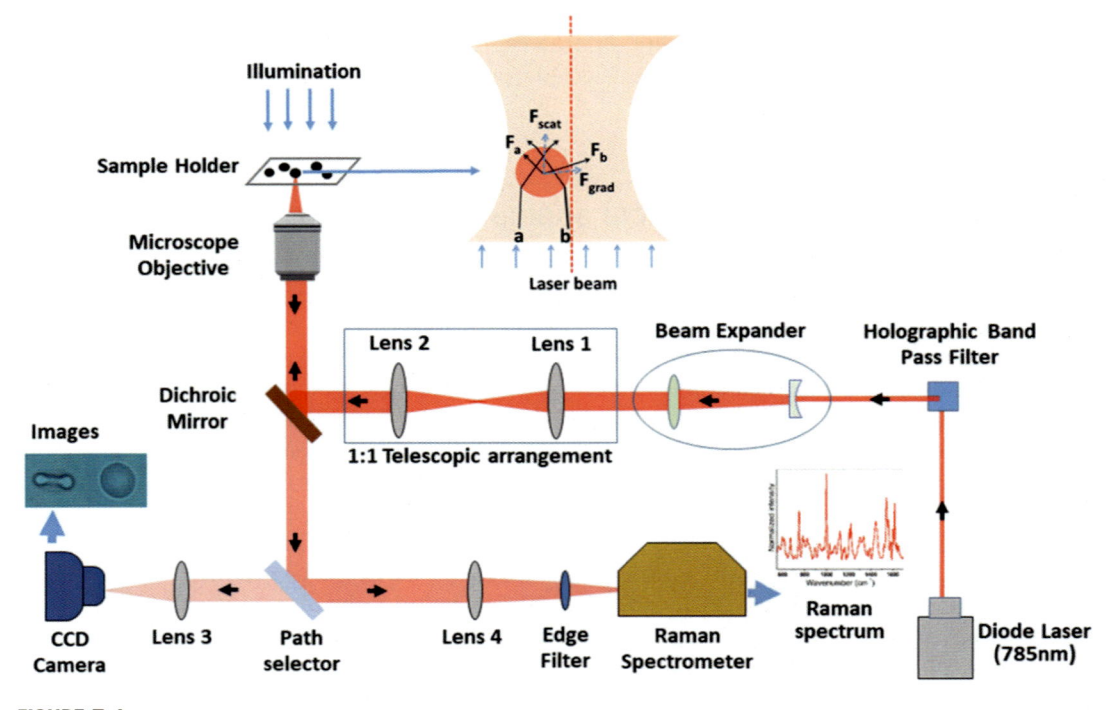

FIGURE 7.4

Schematic of the Raman Tweezers setup

the center (Ashkin & Dziedzic, 1971; Ashkin et al., 1986). The particle will escape along the optical axis if the gradient force is not equal to the scattering force. To obtain a stable trap, the light gradient must be steep, which can be made possible using a high numerical aperture microscope objective (Ashkin, 1992). The combination of this optical trap along with Raman spectroscopy (Raman Tweezers) can provide molecular information of cells without any kinds of chemical fixation that may otherwise harm the cellular structure. In the case of Raman Tweezers, the laser used for trapping the live cell can itself excite the Raman signal simultaneously, which can be acquired using a spectrometer. Thus the Raman tweezers setup can be a valuable tool for the biomedical fraternity to explore detailed information at a single-cell level without any external labeling process. Since water is a weak Raman scatterer, experiments can be easily conducted on cells suspended in aqueous media.

7.4 Raman Tweezers instrumentation

Optical trapping as well as Raman characterization were carried out using the same 785 nm laser source (Starbright Diode Laser, Torsana Laser Tech, Denmark) (Fig. 7.4). A near-IR laser beam can be beneficial for Raman studies on blood samples due to its low absorption, which will minimize the background fluorescence as well as photo damage. Beam expander was used for increasing the beam diameter (nearly 8 mm) to overfill the back aperture of the microscope after which, beam was directed into an inverted microscope (Nikon Eclipse, Ti-U, Japan). The laser beam was later directed onto the microscope objective (100X, Oil immersion) with the use of a dichroic mirror. The red blood cell can be optically trapped with the highly focused laser beam emerging from this microscope objective. Backscattered light from the optically trapped cell was guided toward the spectrometer (Horiba Jobin Yvon iHR320 with 1200 grooves/mm grating blazed at 750 nm) using the same microscope objective. Raman signal was detected via liquid nitrogen cooled charge coupled device (CCD, Symphony CCD-1024 × 256-OPEN-1LS). Initial wavelength calibration of the system usually performed by recording the Raman spectrum of commercially procured polystyrene beads. Visualization of the blood cells suspended in aqueous media was captured using the camera fixed at one of the exit ports of the microscope. Baseline correction is required to remove the background which can be performed using asymmetric least squares smoothing method in MATLAB followed by vector normalization.

7.5 Exogenous stress on red blood cells

7.5.1 Normal saline-induced stress

Infusion of various intravenous fluid is a common practice in medical settings, amongst them normal saline (0.9% sodium chloride) is the most widely used one. In addition to isotonic saline, both hypotonic and hypertonic saline solutions are also being used in clinical practice for the management of acute infections, hyponatremia, and increased intracranial pressure etc. (Duke et al., 2003; Mason et al., 2020). Despite its wide use, literature has implied the adverse effects of these solutions on blood components. Elevated hemolysis of blood cells were reported during blood storage after the washing of cells with normal saline as compared to another crystalloid fluid, Plasmalyte-A (Refaai et al., 2018). The detailed guidelines and regulations on the safe use of optimized concentration, composition, and type of intravenous fluid demanded for various clinical conditions are not yet fully understood. In view of these, Raman Tweezers can be explored to evaluate the impact of stress that normal saline exerts on live RBC. Raman spectra obtained for cells in normal saline and blood plasma are shown in Fig. 7.5 (Lukose et al., 2019a). All the experiments were performed by treating RBCs suspended in blood plasma as control, since it can mimic the physiological condition by keeping the cells intact for Raman measurements. All the experiments were recorded with ~7 mW laser power with an acquisition time of 4 minutes. All the spectra shown here are an average of spectra obtained from 10 different cells of an individual. Hemoglobin is the causative factor for the majority of bands present in the Raman spectra (Fig. 7.5) of an optically trapped RBC. As mentioned in the introduction, the oxygen carrying capacity of heme molecule in the cell is the key entity controlling the gaseous transport by the blood toward the tissues all over the body. As given in Fig. 7.1, the heme adopts a planar configuration upon the binding of oxygen molecule onto the central iron. On the other hand, the iron atom comes out of the porphyrin plane in the absence of oxygen binding, which causes a dome-like structure. These structural variations imparted by the influence of oxygen binding and unbinding can be monitored by evaluating the molecular vibrational fingerprints linked with porphyrin ring via Raman spectroscopy. The bands assigned to methine deformation at 1209 and 1222/cm in heme need to be considered, since these vibrations are highly sensitive to the structural variations occurring due to oxy—deoxy hemoglobin transition. The close proximity of these vibrations toward protein subunits makes them an important point to

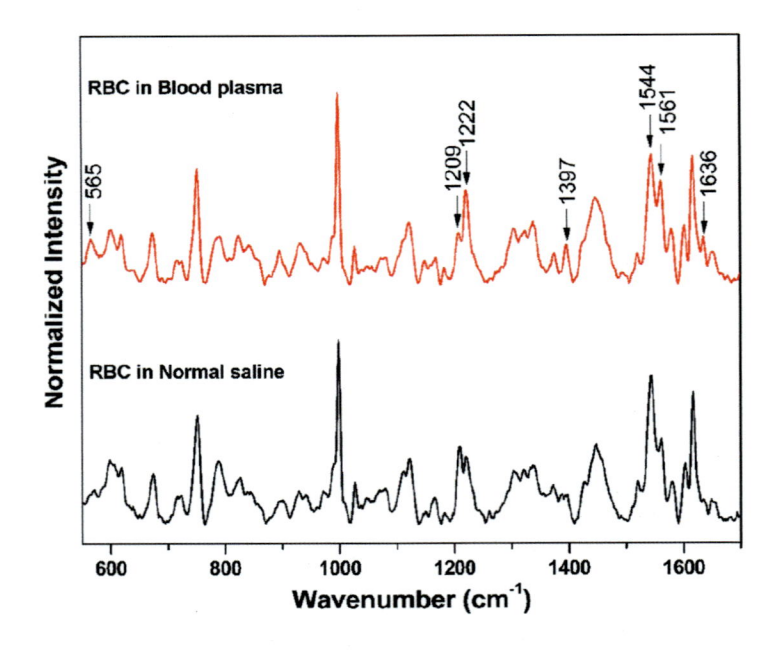

FIGURE 7.5

Average Raman spectra of red blood cells in normal saline and blood plasma (Lukose et al., 2019a).

investigate the alterations in oxyhemoglobin status (Wood & McNaughton, 2002). The spin marker region is also of crucial interest due to its in plane vibrational modes of porphyrin present at 1544 and 1561/cm. These bands evolved from the C-C bonds in the ring are also influenced by the spin state of the iron atom responsible for binding oxygen (Rao et al., 2009). The band at 1544/cm is treated as a marker for the high spin state of iron, which indicates the deoxygenated hemoglobin state (Wood & McNaughton, 2002). The latter band (1561/cm) present in the spin marker is a characteristic feature of low-spin state iron, which can be related to oxygenated hemoglobin. In order to inspect the status of oxygenated hemoglobin, the ratio of oxy/deoxy markers in methine deformation and spin marker region can be evaluated. In addition, the band at 1397/cm due to pyrrole quarter ring stretching is also governed by the spin state and oxidation of the iron atom in the heme. It is clear from the figure that a switching of intensities was observed for the bands 1209 and 1222/cm in the Raman spectra of RBCs suspended in normal saline with respect to plasma. This indicates the tendency of hemoglobin deoxygenation possible once the cells are suspended in normal saline. This assumption is validated by the decrease in the oxyhemoglobin bands 1561 and 1636/cm present in the spin marker region. Moreover, the $Fe-O_2$ stretching band at 565/cm also showed a reduction in its intensity for normal saline treated RBC. Similar tendency is also found in the pyrrole-stretching band at 1397/cm. The bands which undergone transitions have been given in Table 7.1 along with their assignments.

This study has also been extended toward hypotonic and hypertonic normal saline to evaluate the effect of change in extracellular tonicity on RBCs. Both hypertonic and hypotonic saline solutions displayed significant variations in the Raman spectra of red blood cells as compared to the cells in blood plasma (Lukose et al., 2020). The spectra obtained from RBCs in hypertonic solution are given in Fig. 7.6. Overlaid Raman spectra of selected region have shown prominent variations in Raman peak intensities (Fig. 7.7A−D). Increases in band intensities were observed for the peaks around 972, 1244, 1368, and 1384/cm in case of cells in hypertonic solution as compared to the

Table 7.1 Band assignments of red blood cells that underwent major changes in presence of normal saline.

Wavenumber (/cm)	Intensity variation	Band assignment
565	↓	$\nu(Fe\text{-}O_2)$
752	↓	ν_{15}
1209	↑	$\nu_5 + \nu_{18}$
1222	↓	ν_{13} or ν_{42}
1397	↓	ν_{20}
1521	↑	ν_{38}
1544	↑	ν_{11}
1561	↓	ν_2
1603	↑	$\nu(C==C)_{vinyl}$
1617	↓	$\nu(C==C)_{vinyl}$
1636	↓	ν_{10}

Adapted from Lukose, J., Mithun, N., Mohan, G., Shastry, S., & Chidangil S. (2019a). Normal saline-induced deoxygenation of red blood cells probed by optical tweezers combined with the micro-Raman technique. RSC Advances, 9(14), 7878−7884.

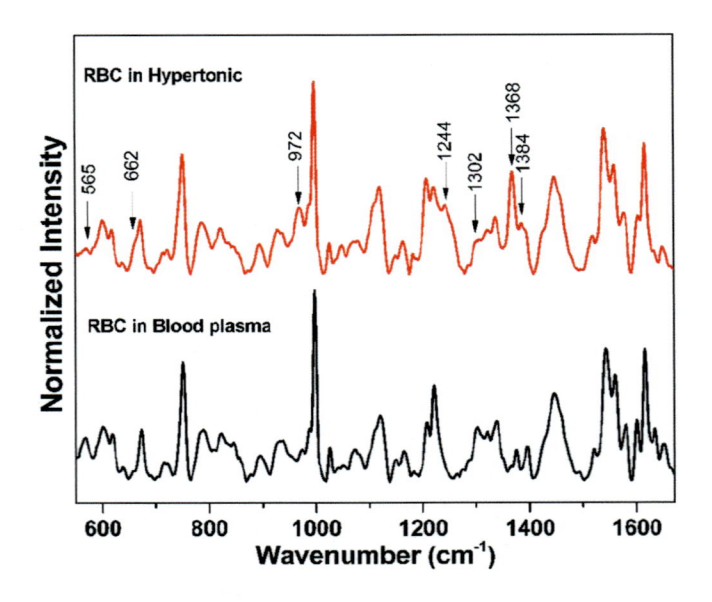

FIGURE 7.6

Raman spectra of trapped red blood cells in blood plasma (control) and hypertonic saline (Lukose et al., 2020).

control. In addition, the presence of a shoulder at 662/cm has been seen also present near the pyrrole deformation band at 674/cm. These changes have been also found in previous literatures dealing with RBCs under heat treatment (Wood et al., 2005). These bands have been reported as the markers of heme aggregate formation arising from the protein denaturation inside the RBCs (Wood et al., 2005). In addition to the effect of heme aggregation, the α-helix of amide III band present at 1302/cm displayed an intensity reduction for cells in hypertonic condition. This mid band is treated as a highly sensitive marker representing conformation of the main chain in the membrane protein (Li et al., 2012). The reduction in this band can be thus attributed as the rupture in hydrogen bond, which sustains the α-helix secondary structure of protein. In addition to these, the conventional markers mentioned above dealing with oxy−deoxy transition have also supported the possibility of deoxygenated hemoglobin formation in hypertonic saline.

The micro-Raman spectra of RBC suspended in hypotonic saline along with the control is shown in Fig. 7.8 (Lukose et al., 2020). All the major bands have experienced an intensity reduction for RBCs treated in hypotonic saline. The porphyrin breathing mode at 752/cm has been considered as an indicator for hemoglobin's intact nature. A decrease in intensity has been observed for this band indicating the probable stress for the cell, which may be due to hemoglobin depletion (Fig. 7.9A). A reduction in intensity has also been observed for the sharp, intense band at 999/cm representing phenylalanine (Fig. 7.9B). The bands assigned to phenylalanine at 897 and 1027/cm also came up with the same trend in hypotonic saline. This can be used to validate the previous assumption of hemoglobin depletion or degradation, since phenylalanine is also present in hemoglobin. On the other hand, previous literature has reported the presence of these bands even in RBC

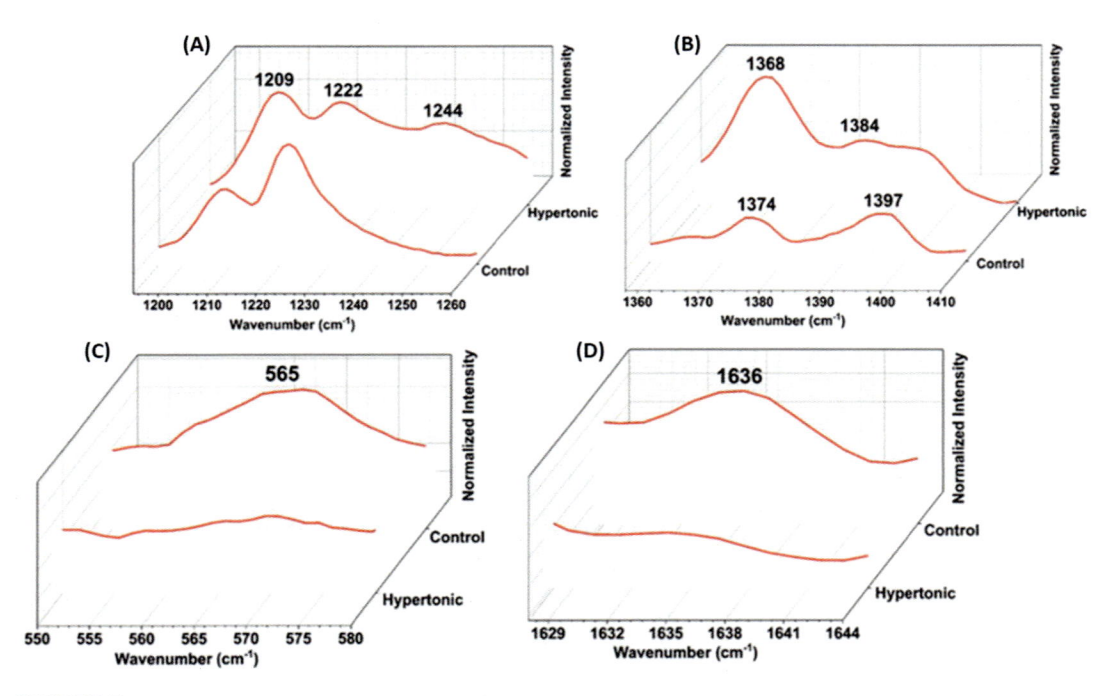

FIGURE 7.7

Raman spectra of red blood cells in blood plasma and hypertonic saline: (A) methine deformation region (B) pyrrole deformation region (C) Fe-O$_2$ stretch, and (D) oxygenation marker (Lukose et al., 2020).

ghosts (Lippert et al., 1975). This indicates the possibility of membrane proteins damage, once the phenylalanine bands are affected. The chance of membrane damage can also be assumed from the decrease in 1447/cm band, which corresponds to the amide I band of protein. Similarly, the Raman band at 1650/cm originating from the Amide I band of protein exhibited a lower signal as compared to blood plasma. (Fig. 7.9D). Moreover, the membrane markers evolve from the lipids at 1069 and 1122/cm have suffered an intensity lowering in hypotonic saline. Hemoglobin deoxygenation is also observed from the variations in methine deformation (Fig. 7.9C), spin marker region, and pyrrole in phase breathing modes.

7.5.1.1 Hydroxyethyl starch induced stress

Hydroxyethyl starch, an intravenous fluid, is often used as an alternative to plasma volume replacement in acute loss of blood. Its efficacy as a volume expander made it the most common colloidal intravenous fluid used in intensive care unit (ICU) settings. Still, concerns regarding the safe use of hydroxyethyl starch (HES) is under debate (Roberts et al., 2018). Adverse effects have been reported in the clinical trials, which experimented with the use of HES in sepsis and critically ill patients (Myburgh et al., 2012). It has been also mentioned that use of this colloidal solution can elevate the bleeding risk factor of the patients in ICU (Haase et al., 2013). In a recent work, Raman Tweezers investigation was performed on single RBCs under the influence of hydroxyethyl starch

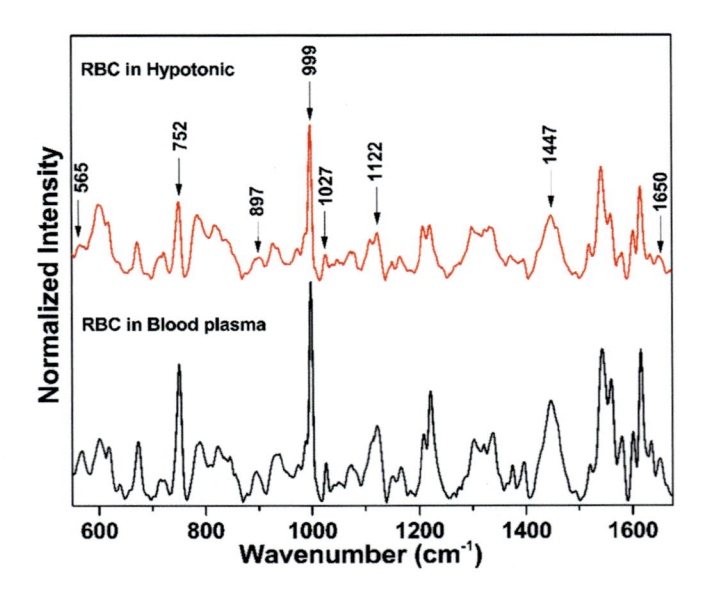

FIGURE 7.8

Raman spectra of red blood cells in blood plasma and hypotonic saline (Lukose et al., 2020).

(Mithun et al., 2020). Experiments were performed at two different laser powers (~ 3 mW and ~ 11 mW) in order to probe the response of cells in HES to laser-induced external stress as compared to that in plasma. High laser power used for trapping RBCs can itself be treated as a mechanical stress and thus the corresponding Raman spectra from the cells can be explored to identify the simultaneous biochemical response of the cell under applied stress. The Raman spectra of cells suspended in control plasma and HES at ~ 3 and ~ 7 mW laser power is given in Fig. 7.10. The status of oxy−deoxy markers in HES and plasma at these laser powers are plotted as bar diagram as shown in Fig. 7.11A−C. A significant reduction in oxygenated hemoglobin level has been observed in HES-treated RBCs. From the Figs. 7.10 and 7.11, the probability of heme aggregation and degradation in HES-treated cells can't be neglected under the influence of a high stress. The shift in pyrrole deformation band from 1375 to 1368/cm^{-1} has been observed in HES at ~ 11 mW laser power, which represents the tendency of heme aggregation resulting from protein denaturation in RBC. A shoulder initiated at 1244/cm^{-1} in case of cells suspended in HES also supports the heme aggregation. In brief, RBCs in HES were found to be more vulnerable to exogenous laser-induced stress than in blood plasma.

7.5.1.2 Bisphenol A and alcohol-induced stress

Bisphenol A (BPA) is widely used in the manufacturing of many consumer goods like plastic water bottles, baby bottles, food containers, home appliances, thermal papers used in billing and tickets, etc. Zhang et al. (2011). Studies have reported the presence of this endocrine disrupting chemical in urine samples collected in India, China, the United States etc. (Lehmler et al., 2018; Zhang et al., 2011). The presence of BPA has been correlated with the elevated chances of diabetes,

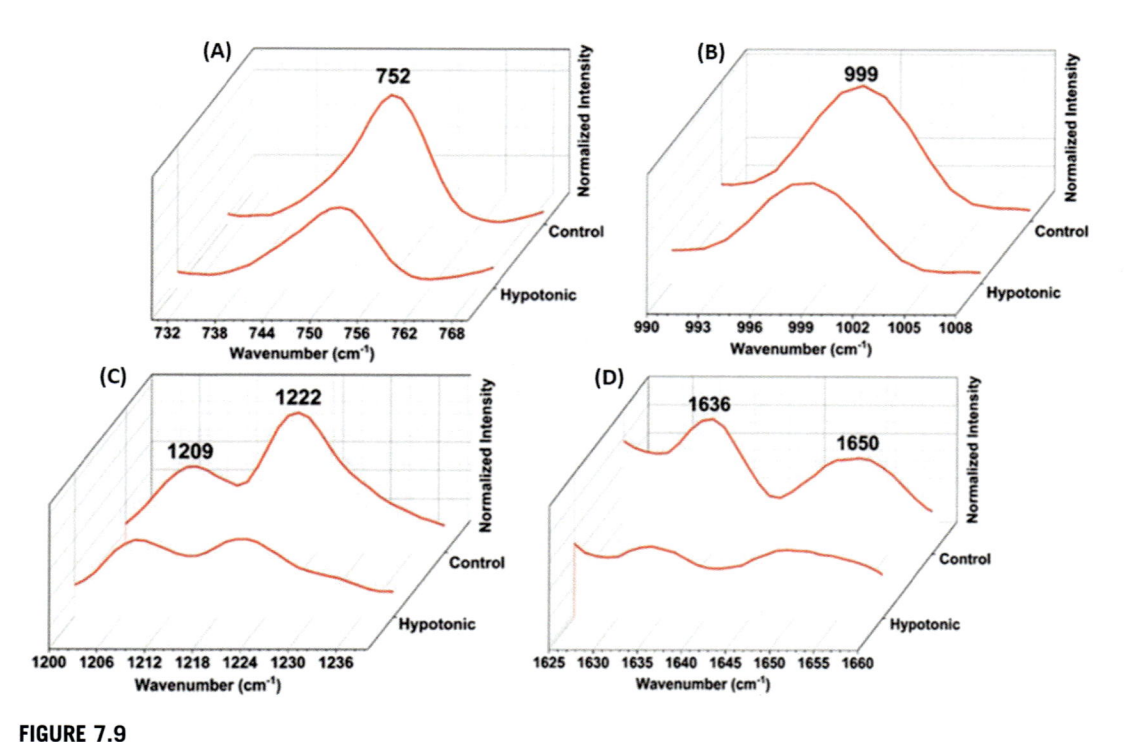

FIGURE 7.9

Overlaid Raman spectra of red blood cells in blood plasma and hypotonic saline: (A) porphyrin breathing mode; (B) phenylalanine major band; (C) methine deformation region; and (D) oxygenation marker and Amide I band (Lukose et al., 2020).

cardiovascular disorders, obesity, etc. (LaKind et al., 2014). Studies have also implicated the risk of immunity disorders with the occurrence of BPA at high dose in blood serum (Ratajczak-Wrona et al., 2020). In view of this, BPA interaction with single, live RBC has been monitored using Raman spectroscopy (Lukose et al., 2019b). BPA solution was prepared by immersing routinely used thermal paper (6.5/cm × 5.5/cm) in saline solution for 30 minutes. The BPA presence in thermal papers was confirmed using UV−Visible absorption and FT-IR spectroscopy prior to the Raman spectroscopy measurements of RBCs (Lukose et al., 2019b). Raman experiments were performed on RBCs suspended in BPA solution for 5, 15, and 30 minutes, as shown in Fig. 7.12 (Lukose et al., 2019b). It is evident that all the heme bands in the Raman spectra have shown an intensity reduction in BPA treated cells compared to the control. The porphyrin breathing mode at 752/cm can be explored to evaluate the vital status of the heme groups present in the hemoglobin. In view of this, the variations for the 752/cm at different time intervals (5, 15, and 30 minutes) were evaluated. As time elapses, the porphyrin breathing mode intensity decreases, as shown in Fig. 7.12A and B. The intensity of this crucial band undergoes a decrease of 98.9% upon exposure to RBC for 30 minutes in BPA solution. Raman data imply that BPA can be toxic to RBCs at higher concentrations, causing complete damage to the cells. It has been also found that RBCs do not

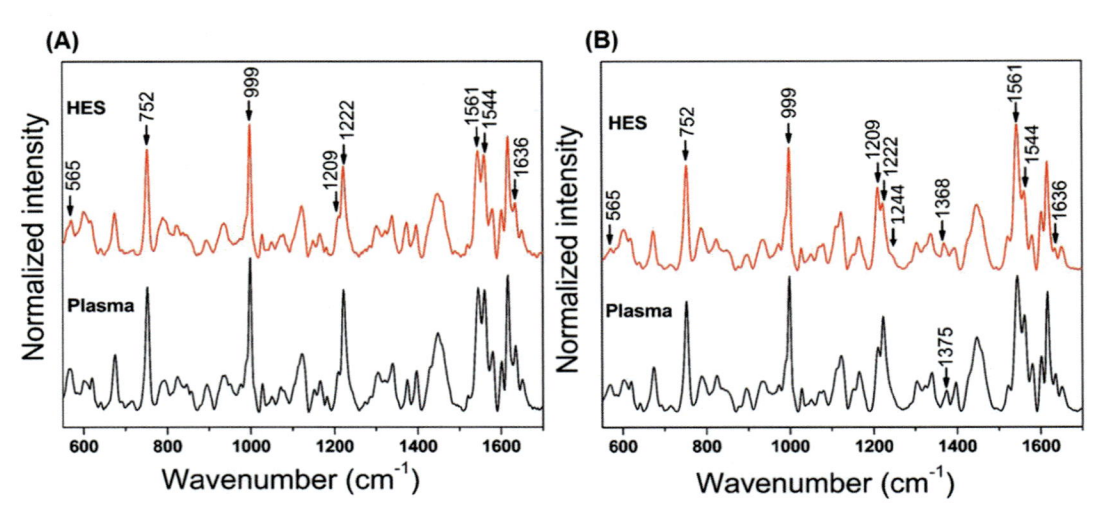

FIGURE 7.10

Raman spectra for red blood cells in blood plasma and hydroxyethyl starch with different laser powers: (A) ~ 3 mW and (B) ~ 11 mW. Mithun et al. (2020)

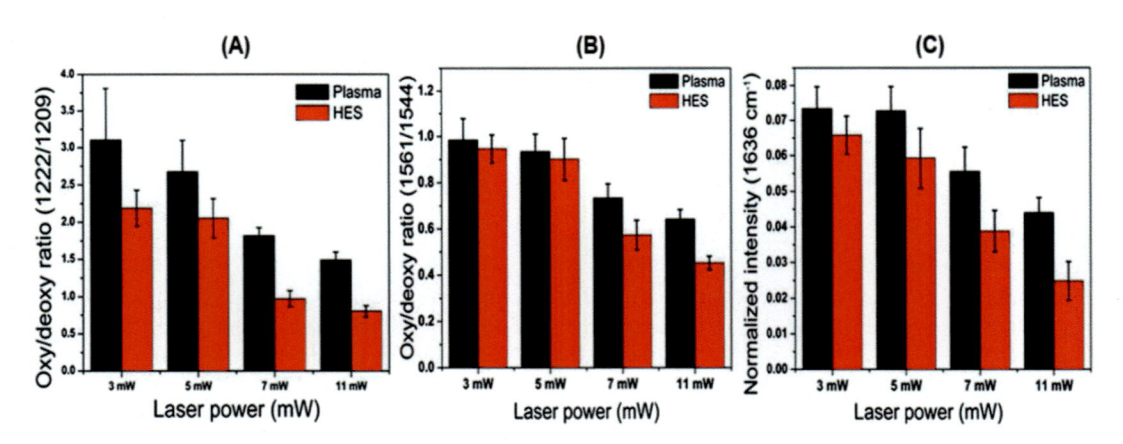

FIGURE 7.11

Bar diagram indicating the oxy–deoxy hemoglobin ratios obtained for red blood cells in control and hydroxyethyl starch at different laser powers for (A) methine deformation region and (B) & (C) spin marker region (Mithun et al., 2020).

face immediate cell burst/death at this BPA concentration, rather it takes $\sim 30-40$ minutes to complete cell burst.

Raman Tweezers have been used to investigate the effect of alcohol on RBCs (Lukose et al., 2019c). The experiments were conducted on RBCs suspended with 5%, 10%, 15%, and 20%

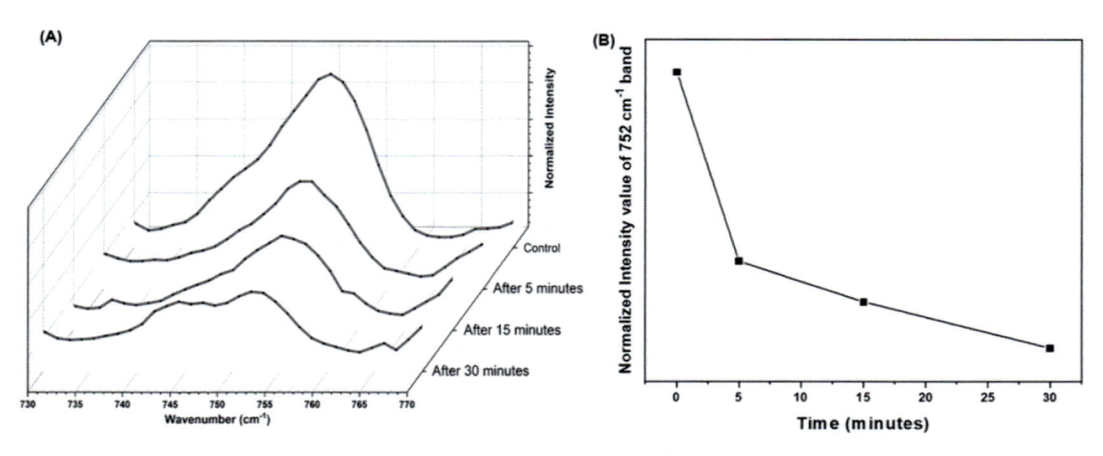

FIGURE 7.12

(A) Overlaid plot of porphyrin breathing mode from Raman spectra of red blood cells treated with ∼390 μg/mL of Bisphenol A for various time intervals. (B) Time versus porphyrin breathing mode intensity plot (Lukose et al., 2019b).

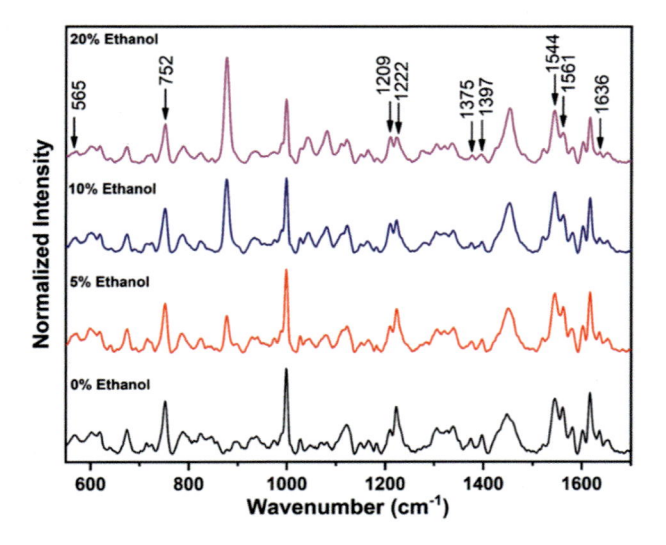

FIGURE 7.13

Raman spectra of red blood cell treated with different ethanol concentrations (Lukose et al., 2019c).

ethanol in blood plasma, as given in Fig. 7.13. The porphyrin breathing mode observed at 752/cm has shown a clear decrease in its intensity in response to increasing alcohol concentration. This may be due to hemoglobin depletion, since this band is treated as a marker indicating the integrity of hemoglobin, as mentioned earlier. A similar reduction is also obtained for phenylalanine band at

999/cm, which is a common component present in both membrane proteins and hemoglobin of RBC. The degradation in the phenylalanine band can be related to the probable hemoglobin damage accompanied with some contribution from the depletion in membrane proteins. The ratio of deoxy/oxy markers in methine deformation and spin marker region is given in Fig. 7.14. The higher ratio (1209/1222/cm) obtained with ethanol concentration represents the presence of more deoxy hemes present in the RBC. The ratios of Raman peak intensities (1544/1561/cm) in the spin marker region have also displayed the similar trend, which confirms the possibility of hemoglobin deoxygenation due to ethanol treatment of RBCs. The Raman band at 1636/cm, which is a characteristic marker

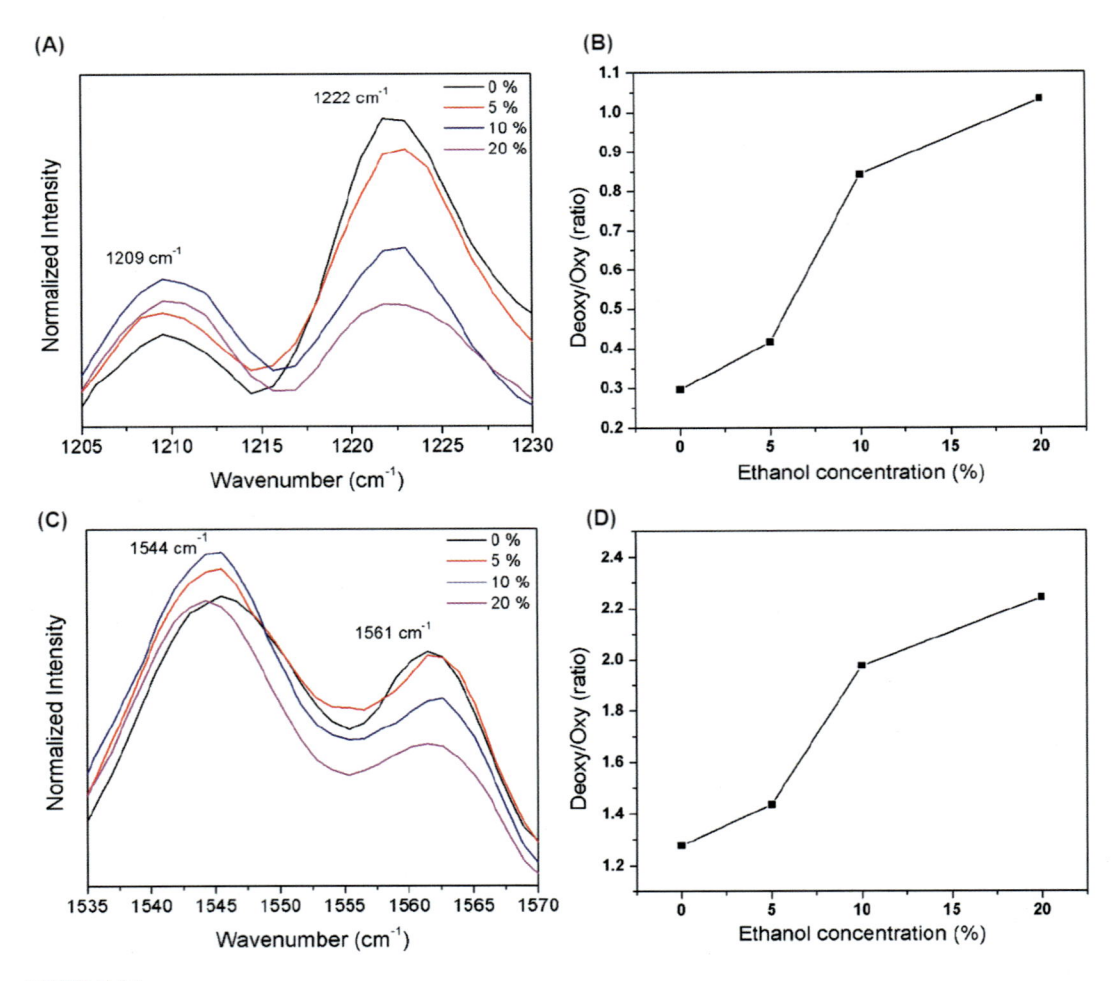

FIGURE 7.14

Raman spectra of red blood cell treated with different ethanol concentrations. Spectral region showing (A) methine deformation region and (B) ratio of peak intensities versus concentration plot; (C) spin marker region and (D) ratio of peak intensities versus ethanol concentration plot (Lukose et al., 2019c).

band for oxygen concentration, also suffered a decrease in intensity. The pyrrole in-phase breathing modes present at 1375 and 1397/cm have also experienced an intensity decrease in cells treated with ethanol. This occurs since this region is also governed by the changes in the iron oxidation state due to oxy−deoxy transition. Moreover, a decrease in intensity has been also found for the Fe-O$_2$ stretch at 565/cm, which again confirms the fact that oxygen ligation to central iron in the porphyrin ring has been also affected due to ethanol.

7.5.1.3 Nanoparticle induced stress

The probable toxicity of nanoparticles to biomolecules is still a debatable topic despite being their ubiquitous use in diverse applications. Metallic nanoparticles such as gold and silver are the most widely employed ones in biomedical scenarios. These nanoparticles have found wide applications in drug transport, photodynamic therapy, molecular imaging, and as antibacterial agents, etc. (Dykman & Khlebtsov, 2011; Elahi et al., 2018). Both in vitro and in vivo toxicity studies have already reported cases of cell membrane injury, deoxyribonucleic acid (DNA) damage, embryo development malformations, inhalation and cardiovascular system diseases upon exposure to nanoparticles (Chen et al., 2015). Thiolate-capped gold nanoparticles have reported to be toxic for nonembryonic stem cells by affecting DNA methylation (He et al., 2018). Pan et.al have found necrosis and mitochondrial damage on cell lines initiated by gold nanospheres (Pan et al., 2009). There are multiple possible routes, which can lead to accidental exposure of these nanoparticles via injection or inhalation, which results in the interaction of blood components with these nanoparticles. These scenarios have highlighted the necessity of evaluating the effect of metallic nanoparticles on RBCs using spectroscopic techniques.

The impact of stress generated on RBC in the presence of both silver and gold nanoparticles has been investigated using Raman Tweezers in our recent work (Barkur et al., 2020). RBCs were treated with both silver and gold nanoparticles with size 10 and 30 nm and Raman spectra were obtained from the optically trapped cells, as given in Figs. 7.15 and 7.16. Nanoparticle treatment stimulated the conversion of oxygenated to deoxygenated hemoglobin in RBCs and silver nanoparticle-treated cells showed higher deoxygenation propensity than gold. One of the probable explanations for deoxygenation may be due to adhesion of the nanoparticle on the cell membrane (Doty & Lednev, 2018). The RBC membrane is made up of a lipid bilayer which is composed of proteins and sialic acid-rich glycoproteins which are of negative charge. This may tend to attract the oppositely charged metallic nanoparticles and the nanoparticles get adsorbed to the cell membrane. The nanoparticle adherence can influence the cell membrane functioning and cause pH alteration. Obstruction in the Na$^+$/H$^+$ channels can reduce the pH of the RBC cytoplasm, which can alter hemoglobin−oxygen binding affinity. In addition, the production of reactive oxygen species may also harm cells by means of lipid peroxidation during nanoparticle treatment. The probability of oxidative stress generated during nanoparticle interaction is evaluated by measuring the thiol values. The decrease in thiol content observed for cells treated with silver nanoparticles indicates the higher oxidative stress production (Fig. 7.17). Oxidative stress can also stimulate the oxygenated to deoxygenated hemoglobin conversion in RBC; the mechanism is detailed elsewhere (Barkur et al., 2020). This can be a validation to the higher deoxygenation tendency observed in Ag nanoparticle-treated cells. Both the suggested mechanisms reasonable for hemoglobin deoxygenation during nanoparticle interaction are shown in the Fig. 7.18.

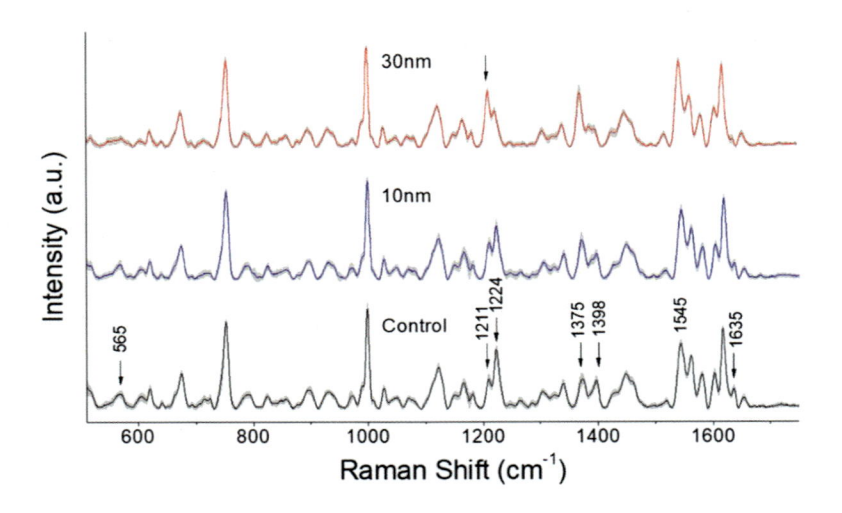

FIGURE 7.15

Raman spectra (shown with standard deviation) from normal red blood cells (RBCs) and RBCs treated with silver nanoparticles of average sizes 10 and 30 nm (Barkur et al., 2020).

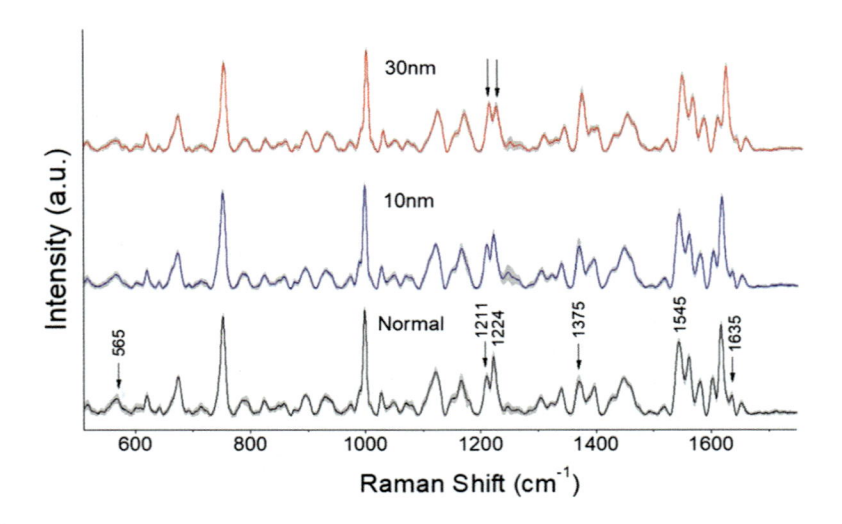

FIGURE 7.16

Mean Raman spectra (shown with standard deviation) from normal red blood cells (RBCs) and RBCs treated with gold nanoparticles of sizes 10 and 30 nm (Barkur et al., 2020).

7.5.1.4 Other red blood cells stress studies using Raman Tweezers technique

The oxygenation response characteristics of normal red blood cells to an applied mechanical stress has been compared with sickle and cord blood red blood cells using Raman Tweezers spectroscopy

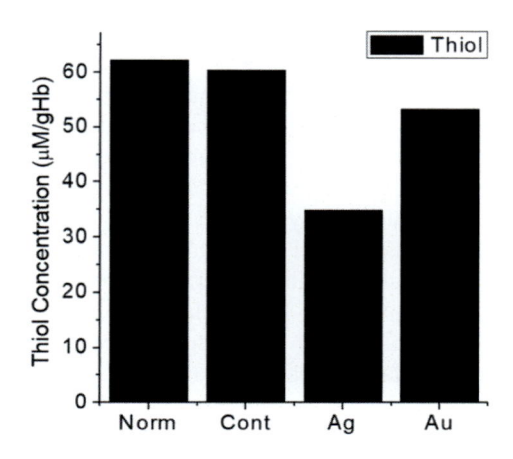

FIGURE 7.17

Thiol measurement for normal red blood cells (RBCs) and RBCs treated with silver and gold nanoparticles (*Ag*, Silver nanoparticles; *Au*, gold nanoparticles; *Cont*, control; *Norm*, normal) (Barkur et al., 2020).

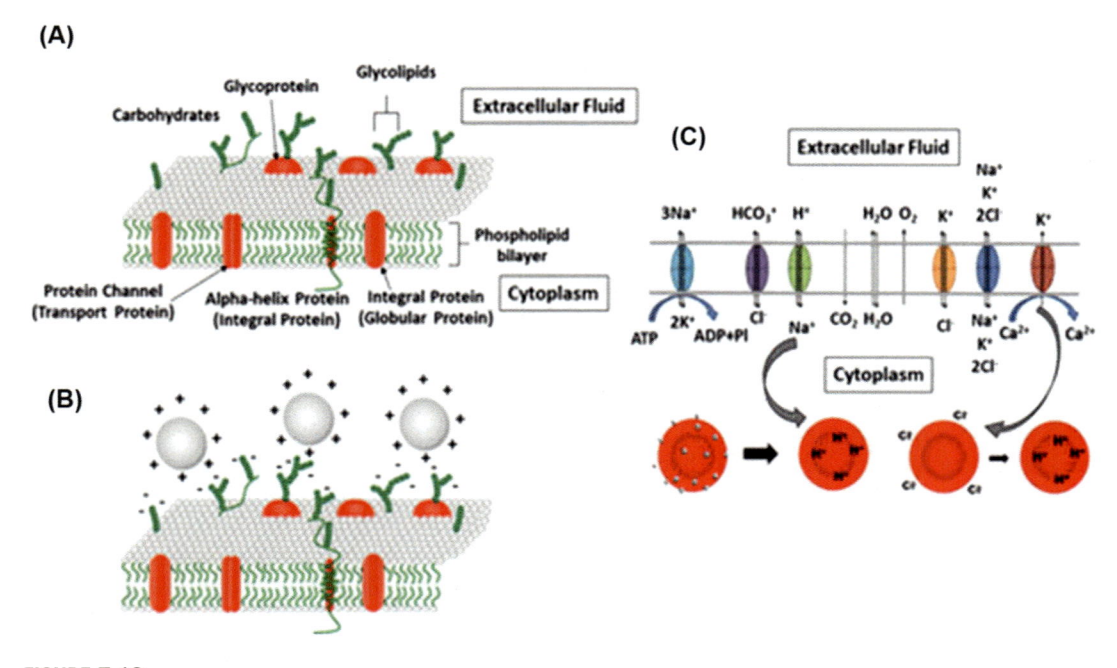

FIGURE 7.18

(A) Schematic diagram depicting the red blood cell membrane structure. (B) Positively charged nanoparticles can get attracted to the negatively charged cell membrane. (C) Mechanism of metal nanoparticle toxicity on erythrocytes (because of nanoparticle adhesion on cell membrane and oxidative stress triggering K^+ channels). Barkur et al. (2020).

(Liu et al., 2013). Raman spectra were obtained from all the three categories of cells trapped at different laser powers. All cells showed a decrease in oxygenation level upon increase in laser power. In comparison to normal cells, sickle cells have been readily deoxygenated under the mechanical stress generated by laser power. This has been attributed to the weak binding affinity of hemoglobin toward oxygen in these cells, which makes them release oxygen easily upon an externally induced mechanical stress. On the other hand, these mechanical perturbations were not able to perturb the cord RBCs to induce deoxygenation as easily as normal adult RBCs. This tendency of cord blood cells has been correlated to the higher binding affinity of fetal hemoglobin toward oxygen as compared to normal cells. In another work, the hemoglobin oxygenation ability in red blood cells collected from control and beta thalassemia patients has been evaluated using this spectroscopic tool (De Luca et al., 2008). The study has reported a decrease in oxygenation capability for cells obtained from thalassemic individuals. In addition, the mechanical properties of the thalassemic and normal cells have been also investigated using a double-trap system (De Luca et al., 2008). An increase in membrane shear modulus has been found for thalassemia cells compared to normal, which can be corroborated with the fact that thalassemia cells have a higher membrane rigidity. The study indicates that the membrane properties are also influenced by this hematological disorder, even though this defect is mainly about the alterations in the chain of globin molecule. Gupta et al. have compared the hemoglobin−oxygen affinity in the cells collected from malarial patients (infected with *Plasmodium vivax*) with healthy cells, where a lesser hemoglobin−oxygen affinity has been observed for malarial RBCs (Dasgupta et al., 2011). They have suggested an increase in intracellular concentration of 2, 3-diphosphoglycerate (2,3-DPG) due to dyserythropoeisis condition occurring in patients infected with *P. vivax* as the probable mechanism behind the decrease in hemoglobin−oxygen affinity. Chowdhury et.al have studied the hemoglobin−oxygen affinity in accordance with the shape variations occurring in red blood cells (discocytes, echinocytes, and sphereocytes) under oxygenated and deoxygenated conditions (Chowdhury et al., 2017). Discocytes were found to be comparatively efficient in releasing oxygen from intracellular hemoglobin under hypoxia conditions generated by the exposure of nitrogen flow. An increase in oxygen affinity has been observed for echinocytes in the Raman investigation, which has been attributed to the ATP decrease in these kinds of blood cells. Raman spectroscopy has also been used to evaluate the effect of organophosphate exposure on blood cells (Singh et al., 2019). The damage induced by Chlorpyrifos, an organophosphate pesticide on RBCS has been observed even at ~ 10 ppb concentration using this technique. The study reported membrane alterations in addition to hemichrome formation in RBCs in response to chlorpyrifos treatment.

In addition to RBCs, researchers have been using Raman Tweezers technique to investigate the biochemistry of other kinds of cells. Fang et al. integrated Raman Tweezers technology with a polydimethylsiloxane chip in order to perform the isolation of a single cell along with its identification in a nondestructive manner within 3 minutes (Fang et al., 2019). The study conducted on BGC823 gastric cancer cells, erythrocytes, lymphocytes, and *Escherichia coli* cells have shown 90% success rate for single-cell isolation. The above study has demonstrated the potential of Raman Tweezers technology to combine with other single-cell analysis techniques, which can pave the way toward multiperspective characterizations of single cells for biological investigations.

7.6 Conclusions

Technological advancements in the field of biomedical optics and photonics have generated more possibilities for exploring cellular dynamics in the past two decades. Optical tweezers has been an indispensable tool for probing and manipulating live, single cells for biochemical investigations using Raman spectroscopy. This combination enables the extraction of molecular fingerprints from biological cells in the presence of external stress agents with minimal sample preparation, without the necessity of any labeling fluorophore tags. Here, we have presented the use of Raman Tweezers technique to study the impact of intravenous fluids, chemicals, and nanoparticles on RBCs at the individual cell level. A comparison of spectral features from normal RBCs and RBCs treated with external factors has revealed that the hemoglobin in red blood cells undergoes deoxygenation, degradation, aggregation, and membrane damage under stress, which can be easily monitored using the Raman Tweezers technique.

Acknowledgment

Authors are thankful to Department of Biotechnology, Govt. of. India for the Raman Tweezers facility, Manipal Academy of Higher Education for other infrastructure and fellowships to Mithun. N. We gratefully acknowledge Professor Deepak Mathur for useful discussions and encouragement in our work with Raman Tweezers.

References

Ashkin, A. (1992). Forces of a single-beam gradient laser trap on a dielectric sphere in the ray optics regime. *Biophysical Journal*, *61*(2), 569−582.

Ashkin, A., & Dziedzic, J. (1971). Optical levitation by radiation pressure. *Applied Physics Letters*, *19*(8), 283−285.

Ashkin, A., Dziedzic, J. M., & Bjorkholm, J. E. (1986). Observation of a single-beam gradient force optical trap for dielectric particles. *Optical Letters*, *11*(5), 288−290.

Atkins, C. G., Buckley, K., Blades, M. W., & Turner, R. F. (2017). Raman spectroscopy of blood and blood components. *Applied Spectroscopy*, *71*(5), 767−793.

Avsievich, T., Zhu, R., Popov, A., Bykov, A., & Meglinski, I. (2020). The advancement of blood cell research by optical tweezers. *Reviews in Physics*, 100043.

Barkur, S., Lukose, J., & Chidangil, S. J. (2020). Probing nanoparticle−cell interaction using micro-Raman spectroscopy: Silver and gold nanoparticle-induced stress effects on optically trapped live red blood cells. *ACS Omega*, *5*(3), 1439−1447.

Billett, H. H. (1990). *Hemoglobin and hematocrit. Clinical methods: The history, physical, and laboratory examinations* (3rd ed.). Butterworths.

Boyd, S., Bertino, M. F., & Seashols, S. J. (2011). Raman spectroscopy of blood samples for forensic applications. *Forensic Science International*, *208*(1−3), 124−128.

Brunner, H., Mayer, A., & Sussner, H. (1972). Resonance Raman scattering on the haem group of oxy-and deoxyhaemoglobin. *Journal of Molecular Biology*, *70*(1), 153−156.

Bueno Filho, A. C., Silveira, L., Yanai, A. L. S. A., & Fernandes, A. B. (2015). Raman spectroscopy for a rapid diagnosis of sickle cell disease in human blood samples: A preliminary study. *Lasers Medical Science, 30*(1), 247−253.

Byrne, H. J., Bonnier, F., McIntyre, J., & Parachalil, D. R. (2020). Quantitative analysis of human blood serum using vibrational spectroscopy. *Clinical Spectroscopy, 2,* 100004.

Carter, E. A., & Edwards, H. G. (2001). *Biological applications of Raman spectroscopy.* New York, NY: Marcel Dekker, Inc.

Chen, F., Flaherty, B. R., Cohen, C. E., Peterson, D. S., & Zhao, Y. (2016). Direct detection of malaria infected red blood cells by surface enhanced Raman spectroscopy. *Nanomedicine: NBM, 12*(6), 1445−1451.

Chen, L. Q., Fang, L., Ling, J., Ding, C. Z., Kang, B., & Huang, C. Z. (2015). Nanotoxicity of silver nanoparticles to red blood cells: Size dependent adsorption, uptake, and hemolytic activity. *Chemical Research in Toxicology, 28*(3), 501−509.

Chowdhury, A., Dasgupta, R., & Majumder, S. K. (2017). Changes in hemoglobin−oxygen affinity with shape variations of red blood cells. *Journal of Biomedical Optics, 22*(10), 105006.

da Silva, A. M., de Brito, P. L., & Silveira, L. (2018). Spectral model for diagnosis of acute leukemias in whole blood and plasma through Raman spectroscopy. *Journal of Biomedical Optics, 23*(10), 107002.

da Silva, W. R., Silveira, L., & Fernandes, A. B. (2020). Diagnosing sickle cell disease and iron deficiency anemia in human blood by Raman spectroscopy. *Lasers Medical Science, 35*(5), 1065−1074.

Das, R. S., & Agrawal, Y. (2011). Raman spectroscopy: Recent advancements, techniques and applications. *Vibrational Spectroscopy, 57*(2), 163−176.

Dasgupta, R., Verma, R. S., Ahlawat, S., Uppal, A., & Gupta, P. K. (2011). Studies on erythrocytes in malaria infected blood sample with Raman optical tweezers. *Journal of Biomedical Optics., 16*(7), 077009.

De Luca, A. C., Rusciano, G., Ciancia, R., Martinelli, V., Pesce, G., Rotoli, B., Selvaggi, L., & Sasso, A. (2008). Spectroscopical and mechanical characterization of normal and thalassemic red blood cells by Raman tweezers. *Optical Express, 16*(11), 7943−7957.

Devanesan, S., AlQahtani, F., AlSalhi, M. S., Jeyaprakash, K., & Masilamani, V. (2019). Diagnosis of thalassemia using fluorescence spectroscopy, auto-analyzer, and hemoglobin electrophoresis—A prospective study. *Journal of Infection and Public Health, 12*(4), 585−590.

Devanesan, S., Saleh, A. M., Ravikumar, M., Perinbam, K., Prasad, S., Abbas, H. A.-S., et al. (2014). Fluorescence spectral classification of iron deficiency anemia and thalassemia. *Journal of Biomedical Optics, 19*(2), 027008.

Diez-Silva, M., Dao, M., Han, J., Lim, C.-T., & Suresh, S. (2010). Shape and biomechanical characteristics of human red blood cells in health and disease. *MRS Bulletin, 35*(5), 382.

Doty, K. C., & Lednev, I. K. (2018). Differentiation of human blood from animal blood using Raman spectroscopy: A survey of forensically relevant species. *Forensic Science International, 282,* 204−210.

Duke, T., Mathur, A., Kukuruzovic, R. H., & McGuigan, M. (2003). Hypotonic versus isotonic saline solutions for intravenous fluid management of acute infections. *Cochrane Database System Review* (3).

Dykman, L., & Khlebtsov, N. G. (2011). Gold nanoparticles in biology and medicine: Recent advances and prospects. *Acta Naturae, 3.* (2 (9)).

Eberhardt, K., Stiebing, C., Matthäus, C., & Schmitt, M. (2015). Advantages and limitations of Raman spectroscopy for molecular diagnostics: An update. *Expert Review on Molecular Diagnosis, 15*(6), 773−787.

Elahi, N., Kamali, M., & Baghersad, M. H. (2018). Recent biomedical applications of gold nanoparticles: A review. *Talanta, 184,* 537−556.

Fang, T., Shang, W., Liu, C., Xu, J., Zhao, D., Liu, Y., & Ye, A. (2019). Nondestructive identification and accurate isolation of single cells through a chip with Raman optical tweezers. *Analytical Chemistry, 91* (15), 9932−9939.

Franzen, S., Lambry, J., Bohn, B., Poyart, C., & Martin, J. (1994). Direct evidence for the role of haem doming as the primary event in the cooperative transition of haemoglobin. *Nature Structutal Biology, 1*(4), 230−233.

Fujihara, J., Fujita, Y., Yamamoto, T., Nishimoto, N., Kimura-Kataoka, K., Kurata, S., Takinami, Y., Yasuda, T., & Takeshita, H. (2017). Blood identification and discrimination between human and nonhuman blood using portable Raman spectroscopy. *International Journal of Legal Medicine, 131*(2), 319−322.

Giardina, B., Messana, I., Scatena, R., & Castagnola, M. (1995). The multiple functions of hemoglobin. *Critical Reviews in Biochemistry and Molecular Biology, 30*(3), 165−196.

Golebiewska, E. M., & Poole, A. W. (2015). Platelet secretion: From haemostasis to wound healing and beyond. *Blood Review, 29*(3), 153−162.

Graves, P. R., & Gardiner, D. J. (1989). *Practical Raman spectroscopy.* Springer-Verlag Berlin Heidelberg.

Haase, N., Perner, A., Hennings, L. I., Siegemund, M., Lauridsen, B., Wetterslev, M., & Wetterslev, J. (2013). Hydroxyethyl starch 130/0.38−0.45 vs crystalloid or albumin in patients with sepsis: Systematic review with *meta*-analysis and trial sequential analysis. *BMJ, 346.*

Harris, D. C., & Bertolucci, M. D. (1989). *An introduction to vibrational and electronic spectroscopy: Courier corporation, symmetry and spectroscopy.*

He, Z., Li, C., Zhang, X., Zhong, R., Wang, H., Liu, J., & Du, L. (2018). The effects of gold nanoparticles on the human blood functions. *Artificial Cells, Nanomedicine, and Biotechnology, 46*(Suppl. 2), 720−726.

Huefner, A., Kuan, W.-L., Mason, S. L., Mahajan, S., & Barker, R. A. (2020). Serum Raman spectroscopy as a diagnostic tool in patients with Huntington's disease. *Chemical Science., 11*(2), 525−533.

Jenkins, C. A., Chandler, S., Jenkins, R., Thorne, K., Woods, F., Cunningham, A., et al. (2020). A new method to triage colorectal cancer referrals in the UK using serum Raman spectroscopy and machine learning. *medRxiv.*

Jensen, F. B. (2009). The dual roles of red blood cells in tissue oxygen delivery: Oxygen carriers and regulators of local blood flow. *Journal of Experimental Biology, 212*(21), 3387−3393.

LaKind, J. S., Goodman, M., & Mattison, D. R. (2014). Bisphenol A and indicators of obesity, glucose metabolism/type 2 diabetes and cardiovascular disease: A systematic review of epidemiologic research. *Critical Review Toxicology, 44*(2), 121−150.

Lawaetz, A. J., Bro, R., Kamstrup-Nielsen, M., Christensen, I. J., Jørgensen, L. N., & Nielsen, H. (2012). Fluorescence spectroscopy as a potential metabonomic tool for early detection of colorectal cancer. *Metabolomics., 8*(1), 111−121.

Lehmler, H.-J., Liu, B., Gadogbe, M., & Bao, W. (2018). Exposure to bisphenol A, bisphenol F, and bisphenol S in United States adults and children: The national health and nutrition examination survey 2013−2014. *ACS Omega, 3*(6), 6523−6532.

Li, N., Li, S., Guo, Z., Zhuang, Z., Li, R., Xiong, K., Chen, S. J., & Liu, S. H. (2012). Micro-Raman spectroscopy study of the effect of Mid-Ultraviolet radiation on erythrocyte membrane. *Journal of Photochemistry and Photobiology B, 112*, 37−42.

Lin, J., Zeng, Y., Lin, J., Wang, J., Li, L., Huang, Z., et al. (2014). Erythrocyte membrane analysis for type II diabetes detection using Raman spectroscopy in high-wavenumber region. *Applied Physics Letters, 104*(10), 104102.

Lippert, J., Gorczyca, L., & Meiklejohn, G. (1975). A laser Raman spectroscopic investigation of phospholipid and protein configurations in hemoglobin-free erythrocyte ghosts. *Biochimica et Biophysics Acta, 382*(1), 51−57.

Liu, R., Mao, Z., Matthews, D. L., Li, C.-S., Chan, J. W., & Satake, N. (2013). Novel single-cell functional analysis of red blood cells using laser tweezers Raman spectroscopy: Application for sickle cell disease. *Experimental Hematology, 41*(7), 656−661.

Lukose, J., Mithun, N., Mohan, G., Shastry, S., & Chidangil, S. (2019a). Normal saline-induced deoxygenation of red blood cells probed by optical tweezers combined with the micro-Raman technique. *RSC Advances, 9*(14), 7878−7884.

Lukose, J., Mithun, N., Priyanka, M., Mohan, G., Shastry, S., & Chidangil, S. (2019b). Laser Raman tweezer spectroscopy to explore the bisphenol A-induced changes in human erythrocytes. *RSC Advances*, *9*(28), 15933−15940.

Lukose, J., Mohan, G., Shastry, S., & Chidangil, S. (2019c). Optical tweezers combined with micro-Raman investigation of alcohol-induced changes on single, live red blood cells in blood plasma. *Journal of Raman Spectroscopy*, *50*(10), 1367−1374.

Lukose, J., Shastry, S., Mithun, N., Mohan, G., Ahmed, A., & Chidangil, S. (2020). Red blood cells under varying extracellular tonicity conditions: An optical tweezers combined with micro-Raman study. *Biomedical Physics and Engineering Express*, *6*(1), 015036.

Mason, A. K., Malik, A., & Ginglen, J. G. (2020). *Hypertonic fluids*. Treasure Island, FL: StatPearls Publishing.

McLaughlin, G., Doty, K. C., & Lednev, I. K. (2014). Raman spectroscopy of blood for species identification. *Analytical Chemistry*, *86*(23), 11628−11633.

Mistek, E., Halámková, L., Doty, K. C., Muro, C. K., & Lednev, I. K. (2016). Race differentiation by Raman spectroscopy of a bloodstain for forensic purposes. *Analytical Chemistry*, *88*(15), 7453−7456.

Mithun, N., Lukose, J., Shastry, S., Mohan, G., & Chidangil, S. (2020). Human red blood cell behaviour in hydroxyethyl starch: Probed by single cell spectroscopy. *RSC Advances*, *10*(52), 31453−31462.

Moffitt, J. R., Chemla, Y. R., Smith, S. B., & Bustamante, C. (2008). Recent advances in optical tweezers. *Annual Review of Biochemistry*, 77.

Molnar, C., & Gair, J. (2013). *Components of the blood. Concepts of biology* (1st Canadian Edition).

Mostaço-Guidolin, L. B., & Bachmann, L. (2011). Application of FTIR spectroscopy for identification of blood and leukemia biomarkers: A review over the past 15 years. *Applied Spectroscopy Reviews*, *46*(5), 388−404.

Myburgh, J. A., Finfer, S., Bellomo, R., Billot, L., Cass, A., Gattas, D., et al. (2012). Hydroxyethyl starch or saline for fluid resuscitation in intensive care. *New England Journal of Medicine*, *367*(20), 1901−1911.

Nargis, H., Nawaz, H., Ditta, A., Mahmood, T., Majeed, M., Rashid, N., et al. (2019). Raman spectroscopy of blood plasma samples from breast cancer patients at different stages. *Spectrochimica Acta, Part A*, *222*, 117210.

Pan, Y., Leifert, A., Ruau, D., Neuss, S., Bornemann, J., Schmid, G., Brandau, W., Simon, U., & Jahnen-Dechent, W. (2009). Gold nanoparticles of diameter 1.4 nm trigger necrosis by oxidative stress and mito-chondrial damage. *Small*, *5*(18), 2067−2076.

Paraskevaidi, M., Morais, C. L., Raglan, O., Lima, K. M., Martin-Hirsch, P. L., Paraskevaidis, E., et al. (2017). Spectroscopy of blood samples for the diagnosis of endometrial cancer and classification of its different subtypes. American Society of Clinical Oncology. *Journal of Clinical Oncology*.

Parlatan, U., Inanc, M. T., Ozgor, B. Y., Oral, E., Bastu, E., Unlu, M. B., et al. (2019). Raman spectroscopy as a non-invasive diagnostic technique for endometriosis. *Science Reports*, 9.

Perutz, M. F. (1979). Regulation of oxygen affinity of hemoglobin: Influence of structure of the globin on the heme iron. *Annual Review of Biochemistry*, *48*(1), 327−386.

Pichardo-Molina, J., Frausto-Reyes, C., Barbosa-García, O., Huerta-Franco, R., González-Trujillo, J., Ramírez-Alvarado, C., et al. (2007). Raman spectroscopy and multivariate analysis of serum samples from breast cancer patients. *Lasers Medical Science*, *22*(4), 229−236.

Rao, S., Bálint, Š., Cossins, B., Guallar, V., & Petrov, D. (2009). Raman study of mechanically induced oxygenation state transition of red blood cells using optical tweezers. *Biophysics Journal*, *96*(1), 209−216.

Ratajczak-Wrona, W., Rusak, M., Nowak, K., Dabrowska, M., Radziwon, P., & Jablonska, E. (2020). Effect of bisphenol A on human neutrophils immunophenotype. *Science Reports*, *10*(1), 1−10.

Refaai, M. A., Conley, G. W., Henrichs, K. F., McRae, H., Schmidt, A. E., Phipps, R. P., et al. (2018). Decreased hemolysis and improved platelet function in blood components washed with plasma-lyte A compared to 0.9% sodium chloride. *American Journal of Clinical Pathology, 150*(2), 146−153.

Rey-Barroso, L., Roldán, M., Burgos-Fernández, F. J., Gassiot, S., Ruiz Llobet, A., Isola, I., & Vilaseca, M. (2020). Spectroscopic evaluation of red blood cells of thalassemia patients with confocal microscopy: A pilot study. *Sensors, 20*(14), 4039.

Roberts, I., Shakur, H., Bellomo, R., Bion, J., Finfer, S., Hunt, B., Myburg, J., Perner, A., & Reinhert, K. (2018). Hydroxyethyl starch solutions and patient harm. *Lancet, 391*(10122), 736.

Sahu, A., Sawant, S., Mamgain, H., & Krishna, C. M. (2013). Raman spectroscopy of serum: An exploratory study for detection of oral cancers. *Analyst., 138*(14), 4161−4174.

Singh, Y., Chowdhury, A., Mukherjee, C., Dasgupta, R., & Majumder, S. K. (2019). Simultaneous photoreduction and Raman spectroscopy of red blood cells to investigate the effects of organophosphate exposure. *Journal of Biophotonics, 12*(5), e201800246.

Snook, R. D., Harvey, T. J., Correia Faria, E., & Gardner, P. (2009). Raman Tweezers and their application to the study of singly trapped eukaryotic cells. *Integrative Biology, 1*(1), 43−52.

Stefan, M. I., & Le Novère, N. (2013). Cooperative binding. *PLoS Computational Biology, 9*(6), e1003106.

Strekas, T. C., & Spiro, T. G. (1972). Hemoglobin: Resonance Raman spectra. *Biochimica et Biophysica Acta, 263*(3), 830−833.

Tombak, A. (2019). *Introductory chapter: Erythrocytes-basis of life*. Erythrocyte: IntechOpen.

Ullah, R., Khan, S., Farman, F., Bilal, M., Krafft, C., & Shahzad, S. (2019). Demonstrating the application of Raman spectroscopy together with chemometric technique for screening of asthma disease. *Biomedical Optical Express, 10*(2), 600−609.

Uzoigwe, C. (2006). The human erythrocyte has developed the biconcave disc shape to optimise the flow properties of the blood in the large vessels. *Medical Hypotheses, 67*(5), 1159−1163.

Virkler, K., & Lednev, I. K. (2009). Blood species identification for forensic purposes using Raman spectroscopy combined with advanced statistical analysis. *Analytical Chemistry, 81*(18), 7773−7777.

Wang, H., Zhang, S., Wan, L., Sun, H., Tan, J., & Su, Q. (2018). Screening and staging for non-small cell lung cancer by serum laser Raman spectroscopy. *Spectrochimica Acta, Part A, 201*, 34−38.

Wood, B. R., Hammer, L., Davis, L., & McNaughton, D. J. (2005). Raman microspectroscopy and imaging provides insights into heme aggregation and denaturation within human erythrocytes. *Journal of Biomedical Optics, 10*(1), 014005.

Wood, B. R., & McNaughton, D. (2002). Micro-Raman characterization of high-and low-spin heme moieties within single living erythrocytes. *Biopolymers., 67*(4−5), 259−262.

Xie, C., & Li, Y.-q (2003). Confocal micro-Raman spectroscopy of single biological cells using optical trapping and shifted excitation difference techniques. *Journal of Applied Physics, 93*(5), 2982−2986.

Yamamoto, T., Palmer, G., Gill, D., Salmeen, I. T., & Rimai, L. (1973). The valence and spin state of iron in oxyhemoglobin as inferred from resonance Raman spectroscopy. *Journal of Biological Chemistry, 248*(14), 5211−5213.

Zhang, Z., Alomirah, H., Cho, H.-S., Li, Y.-F., Liao, C., Minh, T. B., Mohd, M. A., Nakata, H., Ren, N., & Kannan, K. (2011). Urinary bisphenol A concentrations and their implications for human exposure in several Asian countries. *Environmental Science & Technology, 45*(16), 7044−7050.

Drug delivery systems based on blood cells

Aqsa Shahid[1], Aimen Zulfiqar[2], Saima Muzammil[1], Sumreen Hayat[1], Maryam Zain[3], Muhammad Bilal[4] and Mohsin Khurshid[1]

[1]*Department of Microbiology, Government College University, Faisalabad, Faisalabad, Pakistan*
[2]*Department of Chemistry, Government College University, Faisalabad, Pakistan* [3]*Department of Biochemistry and Biotechnology, The Women University Multan, Multan, Pakistan* [4]*School of Life Science and Food Engineering, Huaiyin Institute of Technology, Huaian, P.R. China*

8.1 Introduction

The major therapeutic agents lack the organs and or tissue specificity, therefore they are rapidly cleared from the body. Further, there are many side effects associated with these agents, especially the chemotherapeutic agents that are highly toxic (Ayer & Klok, 2017). The role of drug delivery systems (DDS) was appraised during the past decades as the most promising approaches to counter these specificity and toxicity issues. It is generally believed that the use of such carrier systems can help to enhance the safety and specificity of the therapeutic agents and can further expand their effectiveness (Chi et al., 2020; Ma et al., 2020).

The properties expected from these carriers may include the targeted delivery of drugs, increasing half-life with impact on the nontarget tissues. However, the traditional DDS are unable to achieve the targeted therapies, therefore they are unable to meet the growing demand of modern medicine. Thus a new type of DDS, cell-mediated DDS, has emerged as a promising approach in the past few years to take up the challenges (Yu et al., 2020). This strategy is advantageous due to the unique cellular properties, such as their presence in the circulation, flexible morphology, and lesser or no side effects. Rapid progress in the use of blood cell-based delivery systems is accelerating toward the clinically useful systems for drug delivery leading to significant contributions and breakthroughs in this field as certain red blood cell-based drug delivery systems are in clinical trials (Glassman et al., 2020) as shown in Fig. 8.1.

In this chapter, various blood cell-based approaches for different therapeutic agents are discussed that include the design and drug binding techniques for blood cell-mediated DDSs, along with the prospects and perils, and the chapter provides the future perspectives of living cells-based drug delivery systems.

8.2 Conventional drug delivery systems and their challenges

To decrease the toxic effects of drugs, regulate of drug release, enhance the solubility of drugs, decrease immunogenicity, control the biodistribution, and find specific target sites, drug delivery

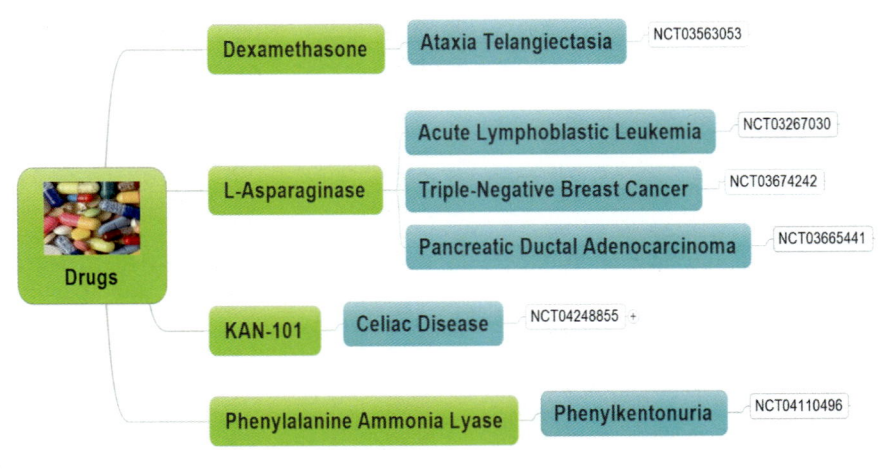

FIGURE 8.1

The clinical trials of red blood cells carrying drugs.

systems provide cargos for treatment. A good drug delivery cargo should have some properties such as being target-specific, safe for the environment, biocompatible, and easy to change toward specific diseases. A controlled drug delivery system has many cargos to carry therapeutics, such as hydrogels (Ashley et al., 2013), films (Prausnitz & Langer, 2008), nanoparticles (NPs) (Parveen et al., 2012), fibers (Cui et al., 2010), micelles (Kedar et al., 2010), and liposomes (Allen & Cullis, 2013). In targeted DDS, molecules that are receptor-specific are highly present on atypical and surrounding cells that are commonly studied. Micromaterials and nanoparticles are specifically designed with target molecules (e.g., proteins, peptides, aptamers, and antibodies) (Sawant & Torchilin, 2012).

Current targeting therapeutics can be divided into two types, that is, active targeting and passive targeting. Passive targeting depends upon the chance EPR (enhanced permeability and retention) effect. This EPR effect may be due to defective lymphatic drainage and leaky vasculature. A process in which NPs (nanoparticles) and larger molecules are released from blood circulation and accumulate more in abnormal tissues than in normal tissues is known as enhanced permeability and retention effect. This phenomenon takes place in defective blood vessels of dense tumorous tissues. Furthermore, due to poor lymphatic drainage the infiltrating molecules cannot be eliminated. On the other hand, active targeting has gained attention in DDS for the last few years. Different biomarkers have been identified on the surface of the tumors and the surrounding cells. Several substances such as cytokines, folate, integrins, and growth factors help in the recognition and targeting of tissue (Brannon-Peppas & Blanchette, 2004; Muro, 2012). Drug vehicles specifically attach to the surface biomarkers and are capable of binding and accumulating in abnormal tissues and the surrounding microenvironment. To increase the efficacy of DDS several CPPs (cell-penetrating peptides) have been invented (Snyder & Dowdy, 2004). Furthermore, to improve the efficiency, both strategies, that is active and passive targeting, are used in combination. Despite advancements in the discovery of certain surface markers and other substances for the targeted drug delivery system, some of them are specific after IV (intravenous) administration, which shows that targeting still depends on chance. To identify the markers on the cell surface and diffuse in the blood circulation for a prolonged period, passive and active targeting requires exogenic drug cargos to reach the targeted binding site. On the other hand, it is difficult to reach the targeted site for drug cargos having a relatively shorter time in blood circulation (Greenwald et al., 2003; Torchilin & Trubetskoy, 1995). Furthermore, humans have a specific protective mechanism for invasion, such as RES (reticuloendothelial system), helping to identify and destroy foreign substances through different physiological functions. Introducing the targeting site-specific drug delivery system is still a challenge for biotechnology.

8.3 Cell-mediated drug delivery systems

To avoid the difficulties in the conventional drug delivery system, a cell-mediated drug delivery system has gained attention in recent years. This innovative approach has shown better therapeutic results with minimal harmful effects due to the unique cell properties, for example, cell structure, metabolism, prolonged circulation time, signaling of the cell, and the abundance of surface ligands.

8.4 Erythrocyte-based delivery systems
8.4.1 Properties of erythrocytes

Red blood cells are a major component of blood, making up more than 99% of total blood cells. The human body produces more than two million red blood cells per second. Erythrocytes are anuclear cells with a thickness of approximately 2.5 μm, a volume of $185-191/\mu m^3$, and a diameter of 7 μm. Red blood cells use hemoglobin (a protein that contains iron) to carry oxygen from the lungs to the other parts of the body. Red blood cells have gained attention due to some unique characteristics which are explained below. Red blood cells have a longer life duration (120 days) in blood circulation than other synthetic carriers and thus can act as a reservoir for the sustained release of drugs (Muzykantov, 2010). For the last few decades, erythrocytes-based DDS have been widely used for numerous characteristics. Erythrocytes help in encapsulating a wide range of drugs due to their unique structures, that is, anuclear and biconcave disks. Red blood cells are isolated by simple methods and can be stored in a refrigerator for a long duration at least for 10 years. Older erythrocytes are identified and eliminated by the reticuloendothelial system, which prevents the formation of toxic by-products, showing that erythrocytes are biodegradable. To target the reticuloendothelial system of bone marrow, spleen, and liver, the routes of red blood cell clearance are widely used. Furthermore, when red blood cells are attached to carriers the activity of vehicles is increased (Ganguly et al., 2006; Murciano et al., 2009). When tPA (tissue-type plasminogen activator) binds with carrier RBCs, the efficacy of tPA as well as intravascular life span is increased. RBCs have proteoglycans, sialic acids, and glycocalyx on the cell surface which help in the uptake of gold nanoparticles (Atukorale et al., 2015). Red blood cells are ideally used for the treatment of endothelium or blood-related disorders by intravascular delivery (Oldenborg et al., 2000; Yoo et al., 2011). Due to these unique properties, the DDS of red blood cells has gained attention in the last few years. The invention of erythrocytes as ideal DDS has also revealed that different substances bind to RBCs and encapsulate the drugs for therapeutics (Harisa Gel et al., 2011; Millán et al., 2004).

8.4.2 The erythrocyte-based carriers

Drugs can be easily transported by using RBCs as cargos by several physical (such as electroporation and hypotonic hemolysis) and chemical (such as biotinylation and covalent conjugation onto surface markers) methods. The strategies for the coupling of drugs to the red blood cell surface is shown in Fig. 8.2. Physical methods are more widely used than chemical methods because these are easier to handle. Erythrocytes can retain their normal structure which is helpful in the transportation of drugs to their targeted sites. The therapeutic molecules loaded by erythrocytes are summarized in Table 8.1.

8.4.3 Electroporation

In this method, an electric current is used to form holes on the cell surfaces (Weaver, 2003). A high-voltage electric current is passed after suspension of the cells in conductive solution which results in the disruption of the cell membrane which results in the formation of abrupt functioning of the cell membrane and causes the encapsulation of the exogenous compounds/molecules in the cells. In this method, the structure of RBCs is retained in case the engulfed molecules are larger in size than that of electropores.

If the size of electropores is larger than the compounds, it causes the swelling of cells which bursts the cell membrane (Tsong, 1991). Many different molecules such as genes, enzymes, and drugs are being encapsulated within the RBCs for a sustained release (Flynn et al., 1994). This method has several disadvantages, such as electrical force, which destroys the morphology of red blood cells and it is difficult to retain the recovery (Lizano et al., 2001; Mangal & Kaur, 1991).

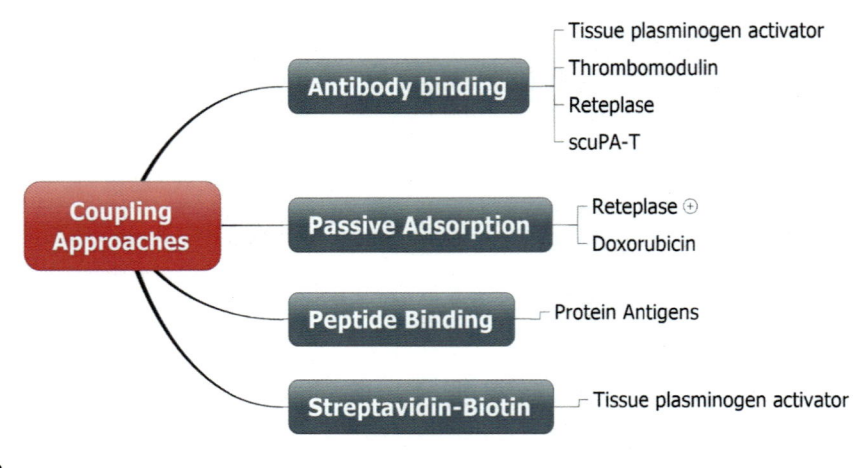

FIGURE 8.2

Strategies for the coupling of drugs to the red blood cell surface.

Table 8.1 Therapeutic molecules loaded by erythrocytes and platelets.

Cells	Therapeutic molecules	Condition	Study type/ model	References
Erythrocytes	Daunorubicin	Acute leukemia	Human	Skorokhod et al. (2004)
	Dexamethasone	Ataxia teleangiectasia	Clinical trial	Chessa et al. (2014)
	Dexamethasone	Chronic obstructive pulmonary disease	Human	Rossi et al. (2001)
	Dexamethasone	Inflammatory bowel disease	Human	Annese et al. (2005)
	Dexamethasone	Ulcerative colitis	Human	Bossa et al. (2013)
	Doxorubicin	Lymphoma	Human	Skorokhod et al. (2007)
	Factor IX	Hemophilia B	In vitro	Sinauridze et al. (2010)
	5-Fluorouracil	Malignant ascites	Mice	Wang et al. (2010)
	Inositol hexaphosphate	Sickle cell anemia	In vitro	Bourgeaux et al. (2010)
	Interferon-α and Ribavirin	Hepatitis C	In vitro	Godfrin et al. (2012)
Platelets	Doxorubicin	Lung cancer	Mice	Sarkar et al. (2013)
	Factor VIII	Hemophilia A	Mice	Shi et al. (2008)
	Factor IX	Hemophilia B	Mice	Shi and Montgomery (2010)
	Ferucarbotran nanoparticles	Transfusion medicine	In vitro	Aurich et al. (2012)

So, to maintain the integrity of the cell, an optimum electric current should be applied during electroporation.

8.4.4 Hypotonic hemolysis

Another technique for the encapsulation of red blood cells is hypotonic hemolysis. It has been observed that erythrocytes act as osmometers in which the size of cells increases in hypotonic solution, whereas the size of cells decreases in a hypertonic solution. In hypotonic conditions, when cells swell the permeability of the cell membrane increases, which helps in the entry of different substances in erythrocytes by passive diffusions, such as nanoparticles, drugs, and biomacromolecules (Sanz et al., 1999). Different methods of hypotonic hemolysis, for example, hypotonic preswelling, hypotonic dialysis, and hypotonic dilution are currently used (DeLoach et al., 1991).

Among these techniques, to encapsulate proteins and enzymes hypotonic dialysis has been widely used since the 1970s because of certain characteristics, that is, it gives a high encapsulation rate, helps in the preservation of cell morphology, and is easy to handle. In this technique, RBCs having HCT values ranging from 70% to 80% are prepared using the dialysis tube, and then these cells are suspended in the buffer (hypotonic) for several hours with continuous shaking (Foroozesh et al., 2011; Muzykantov, 2010). Furthermore, other therapeutic agents, predominantly PNAs (peptide nucleic acids), were delivered in erythrocytes through this technique; more than 80% of encapsulation was achieved after incubation for 18 hours (Chiarantini et al., 2002). When erythrocytes loaded with PNAs were treated with bis-sulfosuccynimidil-suberate and zinc chloride, macrophages were targeted through the process of opsonization. It was observed when PNAs were loaded, it resulted in the inhibition of the protein expression of enzyme NO synthase and decreased synthesis of NO (Chiarantini et al., 2002).

8.4.5 Loading/conjugating therapeutics into/onto cells by chemical modifications

To prevent the degradation of drugs within the carrier cells (such as RBCs, leukocytes, and platelets) and load drugs into blood circulation, cells can be modified by chemical treatment. The outer part of the circulating cell is made up of carbohydrates, proteins, and lipids with different functional groups, which helps in the chemical modification of cells (Chen et al., 2005; Ferguson & Williams, 1988). But just these molecules at the cell surface are not enough for chemical modification. Furthermore, to carry out conjugating reactions different functional groups have been investigated.

8.5 Covalent conjugation onto surface markers

It has been observed that different surfaces of the cell have different functional groups that help in bioconjugation with substances. Some functional groups that are covalently conjugated are used due to their lower toxic effects and ease of labeling, such as thiol and primary amine groups. Furthermore, among primary amines, ester-dependent cross-linkers, such as NHS (N-hydroxy succinimide) are widely used. For the production of immune-camouflaged cells, HPG (hyperbranched polyglycerol) was inserted on the surface of erythrocytes (Rossi et al., 2010). Some functional groups that are abundant with OH group are ideally used for surface modification, such as HPG. In the first step, hyperbranched polyglycerol is reacted with succinic anhydride and N-hydroxy succinimide respectively and then functionalized with SS (succinimidyl succinate). This SS-HPG (succinimidyl succinate hyperbranched polyglycerol) is attached to primary amine groups which are present on the surface of erythrocytes (Rossi et al., 2010).

To prevent the separation of drug cargos during the movement of cells in blood circulation, covalent interaction is used because it employs stronger binding than ligand—receptor bindings. Conjugation reactions help in the covalent biding with thiol and amino groups which are naturally present on cell surfaces. Due to this covalent conjugation drugs are attached to cell surfaces which is more advantageous than endocytosis in terms of drug protection as well as the safety of the cell.

8.5.1 Biotinylation

Biotin is an important component that has a great affinity for streptavidin and avidin proteins (Diamandis & Christopoulos, 1991). Biotinylated cells are attached with several nanoparticles, proteins, and enzymes with higher affinity without changing morphology as well as biological functions of the cell. Several biotinylation reagents, such as N-hydroxy succinimide ester-activated biotins, are widely used to discover the presence of biotins on living cell surfaces that have primary amino groups (Parrott et al., 2003). In buffers, N-hydroxy succinimide ester conjugates with primary amino groups and forms a strong bond at 37°C within a few minutes that can be tolerated by many cells. Furthermore, neutravidin-based nanoparticles were conjugated on biotinylated surfaces of hMSCs (human mesenchymal stem cells) for 48 hours. Consequently, after anchoring with NPs human mesenchymal stem cells showed the maintenance of tumor-homing capability (Cheng et al., 2010). However, erythrocytes were conjugated with antiplatelet drugs for the prophylaxis of fibrinolysis through avidin–biotin bridges (Kwon et al., 2009; Murciano et al., 2003). To target several functional groups, including carbohydrates, sulfhydryl, and carboxyl, various new biotinylation reagents have been introduced.

8.5.2 Recent applications

The encapsulation of 1-asparaginase (1-ASNase) by red blood cells results in its protection from degradation along with increased half-life and with fewer side effects, for example, allergic reactions, compared to naive 1-ANSase. Encapsulated 1-ASNase is considered an important drug for the treatment of acute lymphoblastic leukemia (ALL) due to complete safety guidelines for children and adults (Agrawal et al., 2013; Domenech et al., 2011). For example, etoposide (a chemotherapeutic agent) loaded into RBCs is more potent against macrophages compared to the free form of etoposide, which shows the efficiency of erythrocytes for the drug delivery to the targeted macrophages (Lotero et al., 2003). The mutations in the ECGF1 gene encoded for TP (thymidine phosphorylase) result in a disease known as mitochondrial neurogastrointestinal encephalomyopathy (MNGIE) hence causing accumulation of increased levels of deoxythymidine and deoxyuridine in the body which results in inflammatory bowel disease, deafness, progressive external ophthalmoplegia, sensorimotor polyneuropathy, and gastroenteritis. It was reported in a study that when thymidine phosphorylase is loaded into erythrocytes levels of deoxythymidine as well as deoxyuridine decrease noticeably with fewer symptoms of MNGIE in patients (Moran et al., 2008). According to Harisa et al. increased levels of drug loading and encapsulated efficiencies can be achieved when pravastatin is incorporated into human RBCs (Harisa Gel et al., 2011). Furthermore, after the incorporation of pravastatin into RBCs there were no morphological changes which proved RBCs-mediated DDS is a promising strategy for the incorporation of pravastatin.

The incorporation of an antifolate and antimetabolite agent methotrexate (used for the treatment of blood-related diseases and malignancy) into red blood cells can be used to enhance the lifetime of hepatoma cells of rats which is comparatively less when the cells are treated with the free form of MTX. It was demonstrated by Biagiotti et al. that erythrocytes-based carriers play an important role in the delivery of immunosuppressive agents (Biagiotti et al., 2011). It has been observed that red blood cells can be loaded with different drugs to treat different diseases, such as δ-aminolevulinate is used to treat lead poisoning, β-glucocerebrosidase is used to treat enzyme replacement therapy in Gaucher's disease and gentamicin is used to inhibit bacterial infection. Furthermore, the usage of enalaprilat to treat congestive heart failure and hypertension, and adenosine deaminase to treat adenosine deaminase deficiencies also involves erythrocyte-mediated DDS (Hamidi et al., 2001; Murciano et al., 2003).

8.5.3 Erythrocytes as an enzyme carrier

When enzymes are loaded into the erythrocytes, carrier red blood cells can act as bioreactors. These incorporated enzymes play a crucial role in the removal of the specific substrate from the blood circulation and enable this substrate to enter the erythrocytes via the bloodstream. To treat congenital diseases associated with enzyme deficiency, eliminate toxic substances from the blood, and malignancy, EBRs (erythrocytes bioreactors) can be used.

8.5.4 Enzyme replacement therapy

Deficiency in the activity of several enzymes results in many diseases, to treat these types of diseases, the deficient enzymes are administered into the body. After the administration of free enzymes into the bloodstream, the body provokes an immune response and recognizes them as foreign cells, which results in the removal of the enzyme/drug from the blood. This problem can be overcome by loading enzymes into erythrocytes because it was observed that when enzymes are encapsulated into the erythrocytes the half-life of these enzymes increased and the immune response of the body decreased. The absence of certain lysosomal enzymes results in lysosomal storage disorders such as Fabry disease (α-galactosidase deficient), Gaucher disease (β-glucocerebrosidase deficient), and Slay syndrome (β- glucuronidase deficient) (Bax, Bain et al., 1996; Dale et al., 1977; Deloach & Ihler, 1977; Kaplan et al., 2013; Platt et al., 2018; Spatz, 2004). For enzyme replacement therapy, β-glucocerebrosidase was the first enzyme that was loaded into the erythrocytes. It was observed that the circulation time was increased 4−5 times when β-glucocerebrosidase was incorporated into the red blood cells (Beutler et al., 1977; Humphreys & Ihler, 1980). For the diseases associated with lysosomal enzyme deficiencies, different enzyme replacement therapies are now in use, but this method is highly immunogenic and costly. Furthermore, for enzyme replacement therapy of severe combined immunodeficiency syndrome, phenylketonuria, and MNGIE (neurogastrointestinal encephalomyopathy), adenosine deaminase, phenylalanine hydroxylase, and thymidine phosphorylase, respectively, are used (Bax et al., 2000; Filosto et al., 2018; Moran et al., 2008; Sarkissian et al., 2008; Yew et al., 2013). Furthermore, combination therapy of adenosine deaminase with thymidine phosphorylase helps in the maintenance of hematopoietic stem cell transplantation (Bax et al., 2007). The therapeutic enzymes using RBCs for the amelioration of various disease conditions are summarized in Table 8.2.

8.6 Erythrocytes bioreactors for low-molecular-weight metabolites utilization

To eliminate the metabolites with lower molecular weight from the blood circulation, such as ammonium, cyanide, ethanol, glucose, and methanol, erythrocytes bioreactors can be used. To eliminate methanol, acetaldehyde, and ethanol from the bloodstream, alcohol oxidase, acetaldehyde dehydrogenase, and alcohol dehydrogenase are used, respectively (Alexandrovich & Kosenko, 2017; Lizano et al., 1998; Magnani et al., 1990). To eliminate an excessive amount of glucose, monotherapy with glucose oxidase or combination therapy with hexokinase and glucose oxidase have been described. It was observed in in vivo studies, that when these enzymes were used in

Table 8.2 Therapeutic enzymes encapsulated into RBCs for the amelioration of various conditions.

Condition	Enzyme	References
Alcohol intoxication	Alcohol dehydrogenase	Sanz et al. (1995)
Alcohol intoxication	Alcohol oxidase	Magnani et al. (1993)
Alcohol intoxication	Acetaldehyde dehydrogenase	Magnani et al. (1990)
Antitumor therapy	L-Asparaginase	Agrawal et al. (2013), Bachet et al. (2015), Kwon et al. (2009)
Antitumor therapy	Arginine deiminase	Gay et al. (2016)
Antitumor therapy	L-Methioninase	Gay et al. (2017), Machover et al. (2019)
Cyanide intoxication	Cyanide sulfurtransferase	Cannon et al. (1994), Petrikovics et al. (1994)
Diabetes	Glucose oxidase	Rossi et al. (1992)
Diabetes	Hexokinase	Rossi et al. (1992)
Gaucher disease	β-Glucosidase	Beutler et al. (1977), Dale and Beutler (1976)
Gaucher's disease	Glucuronidase	Thorpe et al. (1975)
Hyperammonemia due to arginase deficiency	Arginase	Adriaenssens et al. (1976)
Hyperammonemia	Glutamate dehydrogenase	Sanz et al. (1995), Sanz et al. (1999)
Hyperammonemia	Glutamine synthetase	Kosenko et al. (2008), Venediktova et al. (2008)
Lysosomal storage diseases	β-Glucuronidase	Thorpe et al. (1975)
Mitochondrial neurogastrointestinal encephalomyopathy (MNGIE)	Thymidine phosphorylase	Levene et al. (2013, 2019)
Phenylketonuria	Phenylalanine hydroxylase	Gámez et al. (2004), Yew et al. (2013)
Phenylketonuria	l-Phenylalanine ammonia lyase	Bell et al. (2017), Rossi et al. (2014)
Severe combined immunodeficiency (SCID) caused by deaminase deficiency	Adenosine deaminase	Bax, Fairbanks et al. (1996), Bax et al. (2007)
Uric acid removal	Uricase	Ihler et al. (1975), Magnani, Mancini et al. (1992)
Urea utilization	Urease + urease (PEGylated)	Baysal and Uslan (2000)
Urea utilization	Urease + alanine dehydrogenase PEGylated)	Hamarat Baysal and Uslan (2002)
Urea utilization	Urease (PEGylated) + alanine dehydrogenase PEGylated)	Baysal et al. (2007)

combination, the level of glucose in the blood circulation remained normal for weeks with higher rates of glucose consumption in mice (Magnani et al., 1988). To remove cyanide from the bloodstream a mitochondrial enzyme named rhodonase was used with the incorporation of rhodanese and thiosulfate (sulfur donor) in erythrocytes, which showed a 40% decrease in the level of cyanide in 10−15 minutes (Leung et al., 1991). Antitumor therapy of RBC bioreactors loaded with different enzymes such as arginine deiminase, asparaginase, and methionine gamma-lyase was also observed.

8.6.1 Ammocytes

The use of erythrocytes bioreactors to eliminate the excessive concentration of ammonium from blood is under investigation. Hyperammonemia is responsible for many fatal diseases such as encephalopathy, Parkinson's disease, prion disease, and Alzheimer's disease (Liu et al., 2018). These neurodegenerative disorders may be associated with chronic liver diseases as well as inborn deficiencies of enzymes (uric acid cycle) such as arginase. To avoid the slow progression toward neurodegenerative disorders and the treatment of excessive concentration of ammonium in blood, the maintenance of ammonium level in blood is required. Innovative therapeutic approaches for the effective treatment of hyperammonemia do not provide satisfactory results. Furthermore, it is suggested by some researchers that some erythrocytes bioreactors called ammocytes can be used for higher efficacy and lower side effects. Glutamate dehydrogenase and glutamine synthetase were used to encapsulate erythrocytes through reversible hypoosmotic dialysis (Venediktova et al., 2008). In mice it was seen in vivo that ammonium was eliminated from the bloodstream in 30−60 minutes by using these bioreactors (Kosenko et al., 2008; Sanz et al., 1999). Furthermore, in other studies, it was observed that other than the encapsulation of these enzymes, the level of ammonium in the bloodstream can also be decreased by using dialyzed erythrocytes at the same rate. But the problem associated with this process was that enzymes stopped the consumption of ammonium after 30−60 minutes because the depletion of α-ketoglutarate and l-glutamic acid in the cell takes place. To overcome this problem, researchers designed other EBRs in which erythrocytes were entrapped by glutamate dehydrogenase and alanine aminotransferase to eliminate ammonium from the bloodstream (Protasov et al., 2019). Furthermore, in this process, α-ketoglutarate and l-glutamic acid were synthesized and utilized during cyclic processes and thus they do not deplete in the cell. In in vivo studies, it was observed that these bioreactors can contribute to the removal of ammonium from the blood even after 2 hours of administration of these EBRs (Kosenko et al., 2008; Sanz et al., 1999).

8.6.2 Inositol hexaphosphate in erythrocytes

A congenital disease associated with the disordered morphology of erythrocytes (sickle-shaped) and anemia is termed sickle cell disease. In sickle cell disease, HbA is replaced by an abnormal form of hemoglobin, HbS (Ataullakhanov et al., 2018). Red blood cells loaded with inositol hexaphosphate can be used to treat patients suffering from sickle cell diseases (Franco et al., 1983; Lamarre et al., 2013). These carrier RBCs are called bioreactors because they show an allosteric effect on hemoglobin (the main constituent of RBCs). In in vitro studies, it was explained by Bourgeaux et al. that the number of sickle cells was lowered 7−7.5-fold when RBCs were administered with the incorporation of inositol hexaphosphate compared to the incorporation of normal erythrocytes into the blood of SCD patients (Bourgeaux et al., 2010). Furthermore, in in vivo studies it was observed by the repeated administration of inositol hexaphosphate-loaded erythrocytes in mice resulted in the decreased rate of anemia. Results of both studies showed that RBCs loaded with inositol hexaphosphate are effective in the treatment of sickle cell disease (Bourgeaux et al., 2012).

8.7 Carrier erythrocytes with a gradual release of the pharmacological agent

No red blood cell incorporated with a drug needs to act as a bioreactor because red blood cells function as a system with the ongoing delivery of therapeutic agents in the blood circulation. To decrease the higher level of the drug just after repeated administration and for maintenance of the level of drug into the bloodstream this method can be used. Furthermore, it is reported that when molecules possessing lower molecular weight are incorporated into RBCs this system works efficiently.

8.7.1 Terpene indole alkaloids

In 1960 it was discovered that some alkaloids including vincristine and vinblastine show a hypoglycemic and anticancer effect (Levêque et al., 1996; Minev, 2011). Vincristine and vinblastine bind with tubulin which interferes with the polymerization of microtubule and prevents anticancer activity. These indole alkaloids are widely used in chemotherapy, but their dose is strictly prescribed due to toxic effects. Vincristine toxicity is associated with alopecia, decreased levels of leukocytes, thrombocytes, and sodium in the blood, and peripheral neuropathy (Arora et al., 2010; Moudi et al., 2013). Side effects of vinblastine are strong extravasation with edema, necrosis in tissues and deep ulcers, myelosuppression, and intestinal diseases. It was observed by Halahakoon et al. that the toxic effects of these drugs can be minimized by incorporating vinblastine and vincristine into erythrocytes. In this study, erythrocytes were lysed by hypotonic lysis, and drugs were incorporated into RBCs. It was reported from in vitro studies that vinblastine and vincristine were released from loaded RBCs almost 100 µg in 60 minutes when incubated at room temperature. But this process is unavailable in the in vivo studies (Trineeva & Khalahakun, 2019).

8.7.2 Glucocorticoids

Since the 19th century, glucocorticoids have been significantly clinically used due to their immunosuppressive, antitoxic, antiallergic, and antiinflammatory properties. Glucocorticoids (steroid hormones) are produced by the adrenal glands. Long-term therapy with steroids can cause life-threatening complications including diabetes, osteoporosis, and immunosuppression (Rossi et al., 2007; Umland et al., 2002). Prednisolone, dexamethasone, hydrocortisone, and methylprednisolone are widely used. But the problem associated with the use of glucocorticoids is their rapid excretion (between 3 and 4 hours) through kidneys and liver metabolism from the body (Czock et al., 2005). To solve this problem, researchers designed carrier erythrocytes which prevent the rapid excretion of the drug into the blood circulation and release the drug into blood circulation gradually. D'Ascenzo et al. designed carrier RBCs by loading dexamethasone-21-phosphate and prednisolone-21- phosphate (prodrugs) into erythrocytes which helps in the release of these drugs slowly into the blood circulation (D'Ascenzo et al., 1997). The encapsulation of RBCs was done by hypoosmotic lysis. It was observed that the rate of elimination of the drug from the body was significantly decreased by using loaded RBCs with glucocorticoids, as compared to monotherapy with glucocorticoids.

8.7.3 Insulin in erythrocytes

Insulin can also be incorporated into the erythrocytes but there are only a few research studies available because the activity of insulin is lost in the erythrocytes and only 4.8%−6% of insulin remains in the loaded RBCs (Ito et al., 1989; Pitt et al., 1983). The incorporation of inhibitors of insulin in the erythrocytes directly stabilizes the amount of insulin in the cells by preventing its degradation. But still, further research is required to carry out this system in vivo (Bird et al., 1983).

8.8 Morphine encapsulation into erythrocytes

By using the glucose hypertonic pulse technique, autologous erythrocytes were lysed and loaded with morphine to check the longer postoperative analgesic effects. In this method, after mixing red blood cells with glucose solution, these RBCs were incubated for half an hour at 37°C; after washing of red blood cells, these RBCs were incubated with morphine solution (Ge et al., 2004; Xuan et al., 2003). In in vivo experiments, it was observed that morphine (free form) shows analgesic effects for 3.2 hours, whereas morphine incorporated into erythrocytes showed a prolonged analgesic effect (i.e., 24 hours) (Luo et al., 2005; Xiaohai et al., 2001).

8.9 Platelet-based delivery systems

8.9.1 Unique properties of platelets

Platelets or thrombocytes make up the smallest component of blood and are approximately 2−3 μm in diameter. Megakaryocytes (large fragments of the bone marrow) are precursors of platelets. Megakaryocytes synthesize hundreds of cells per day. The normal concentration of platelets ranges from 150,000 to 450,000/μL. After 8−10 days, thrombocytes are eliminated from the spleen and liver through RES (Walsh et al., 2015; Wong, 2013). Inside the cytoplasmic granules, platelets have several proteins that help in biological functions (Blair & Flaumenhaft, 2009; Koseoglu & Flaumenhaft, 2013). Thrombocytes are involved in various pathological and biological functions such as inflammation, healing of wounds, atherosclerosis, and thrombosis (Münzer et al., 2014). Platelets also play a role in the maintenance of hemostasis in which damaged capillaries and blood vessels cause the constriction of vessels to inhibit blood loss. Furthermore, platelets interact with collagen and VWF (Von Willebrand factor) for adhesion at the injury site (Shi & Montgomery, 2010).

8.9.2 **The platelet-based carriers**

In blood, platelets are present in small quantities, so it is difficult to separate platelets only by centrifugation. To separate platelets, blood is centrifuged at different speeds, as a result, PPR (platelet-poor plasma) and PRP (platelet-rich plasma) are obtained (Boswell et al., 2012; Sarkar et al., 2013). For aggregometry analysis of platelets, poor plasma acts as a blank. From incubated platelet-rich plasma washed platelets are isolated by a Sepharose column. These steps take place in a sterile environment to prevent the unwanted activation of thrombocytes. Platelets can bind or take up smaller molecular drugs through incubation at room temperature. To encapsulate the drugs within platelets a high-voltage current is applied through the process of electroporation and by incubation at room temperature (Edwards & Biriell, 1994). To encapsulate the antithrombotic drugs these techniques are used in the routine. The therapeutic molecules loaded by platelets are shown in Table 8.1.

8.9.3 **Recent applications**

It has been observed that thrombocytes have a higher rate of loading of drugs into cells, with an elevated level of stimulating cytotoxicity in vivo as well as in vitro, which manifests that thrombocytes can act as good drug vehicles when a small amount of drug is used. To cure patients with a higher risk of hemorrhage and thrombosis antiplatelet drugs are widely used. Antithrombotic agents are used efficiently to manage patients with arterial thrombosis (Chen et al., 2007; Greineder et al., 2013). For four decades, thrombocytes have been labeled with radioactive compounds to measure their survival and recovery rate. But this technique of radiolabeling is prohibited for employment in clinical studies. Instead, to measure the pathogenesis of thrombosis and survival of transfused thrombocytes, the technique of nonradioactive labeling is effectively used in clinical studies (Aurich et al., 2012). When doxorubicin (a chemotherapeutic agent) is delivered to thrombocytes, a higher rate of encapsulation and drug loading concentration is observed. Patients that are treated with platelet-based carriers have higher rates of inhibition as well as apoptosis compared to untreated patients. Furthermore, a study revealed that to manage patients with metastasized cancer platelets a mimetics technique is effectively used (Modery-Pawlowski et al., 2013). The researchers have used the polymeric nanoparticles that were enclosed in the plasma membrane using the human platelet and have found that these platelet membranes-coated nanoparticles have shown less uptake by the phagocytic cells compared to the uncoated nanoparticles. In a clinical approach, when a rat with systemic bacterial infection and a mouse model with coronary restenosis were treated with vancomycin and docetaxel, respectively, through platelet-mimetic NPs, the efficiency of treatment was increased (Hu et al., 2015).

8.10 **Leukocyte-based delivery systems**

8.10.1 **Unique properties of leukocytes**

WBCs help in the elimination of cellular debris and defend the body against infections and foreign bodies through the immune system. The normal concentration of leukocytes in blood ranges from

4×10^9 to $11 \times 10^9/\mu L$ (Hollowell et al., 2005). WBCs are classified into two types based on cytoplasmic granules, that is, (1) granulocytes including basophils, neutrophils, and eosinophils; and (2) agranulocytes including lymphocytes and monocytes. Among these types, lymphocytes, neutrophils, and monocytes are used for the delivery of drugs to cells. Unlike RBCs, white blood cells have a shorter life duration, ranging between 18−20 days. Leukocyte-based DDS gained attention due to its significant role in the immune system, cellular adhesion, penetration in the nonvascular areas via physical and chemical barriers, and cellular interactions.

Neutrophils or PMNs (polymorphonuclear granulocytes) are present in abundance among other types, that is, approximately 40%−75%. Neutrophils reach infection site first and many cytokines are released to eliminate the foreign bodies within a few days (Amulic et al., 2012). Monocytes have a bean-shaped nucleus with no cytoplasmic granules, produced by hematopoiesis. Monocytes are generally present in blood circulation and move toward different organs and are named according to the location, such as lung (alveolar macrophages), liver (Kupffer cells), and kidney (mesangial cells) (Adams & Lloyd, 1997). Lymphocytes can be differentiated by the presence of a larger nucleus covered by a cytoplasmic layer. Furthermore, lymphocytes are classified into two types, that is, B cells and T cells. All types of leukocytes play a significant role in defense against inflammation, tumors, and infections.

8.11 The leukocyte-based carriers

WBCs or leukocytes can be easily isolated as well as purified from the blood through a process known as centrifugation. Furthermore, using a discontinuous sucrose gradient membrane leukocytes can also be separated (Parodi et al., 2013). The processes for drug delivery by leukocytes are well known and are discussed below.

8.11.1 Endocytosis

Eukaryotes engulf food, cells, and some molecules by a cellular process known as endocytosis. Polar and larger substances are not easily internalized via hydrophobic plasma membranes. In endocytosis, distortion of the cell membrane helps to pass these substances. In endocytosis, through the fusion of membrane, the cell membrane expands around the substance and forms large vesicles. Several factors are affecting the process of endocytosis such as shapes, biological properties, charges, and sizes of foreign molecules (Cantón & Battaglia, 2012). Therefore this process is widely used in encapsulating several drug cargos.

Pinocytosis is a type of endocytosis in which cells uptake fluid from the surrounding medium. This process is widely conducted in eukaryotes to internalize the smaller molecules. In early attempts, pinocytosis was widely used to encapsulate drugs into RBCs. Several drugs including chlorpromazine, hydrocortisone, vinblastine, and primaquine can cause internalization of RBCs membrane by pinocytosis (Gopal et al., 2007). Several factors are associated with the process of pinocytosis including pH, drug concentration, electrolyte balance in the cell membrane, and temperature; furthermore, it is an ATP-dependent process (Schrier et al., 1992). These drug delivery systems are absorption dependent and can produce significant cytotoxicity.

Occasionally, phagocytosis occurs in lymphocytes, neutrophils, dendritic cells, and macrophages instead of pinocytosis. The actin-dependent process which is associated with the reorganizations of receptors including mannoses, Fc receptors, and complements is used to internalize larger WBCs into smaller sizes (Aderem & Underhill, 1999). The size of several drug cargos such as liposomes, nanoparticles, and micelles are internalized by the actin-dependent process, then phagocytosed cells fuse with lysosomes which are further broken down by lysosomal enzymes. Phagocytosis has gained attention in the cell-mediated drug delivery system because any foreign invaders, for example, drug-encapsulated particles and drugs are recognized by phagocytosis. In vivo, the elimination of drug-loaded particles occurs rapidly by the reticuloendothelial system, which shows that immune cells are the primary cells to recognize any foreign particle in the body. Phagocytic cells can be directly used as drug delivery carriers rather than designing particles that overcome the clearance of the reticuloendothelial system. Furthermore, the rate of phagocytosis can also be increased by a process termed opsonization. For instance, to promote monocyte−macrophage uptake, albumin and antibodies are coated on SPIO NPs (superparamagnetic iron oxide nanoparticles) (Muzykantov, 2010; Owens & Peppas, 2006). To enhance the therapeutic effects and prevent damage of cell carriers, it is necessary to stop the drug release before reaching targeted cells. Endocytosis is considered an important process to load nanoparticles and drugs. But this method is still challenging due to fission and fusion reactions, cell internalization, and a wide range of receptors. New techniques to develop endocytic cell-mediated DDS are needed. Furthermore, another promising strategy to deliver particles to cells is ligand−receptor binding. Different studies revealed the presence of new surface markers that are helpful in ligand−receptor binding strategies. However, the binding efficiency with the particular cells is not specific, therefore, it is still challenging.

8.11.2 Applications

Leukocyte-based DDS is widely used for delivering drugs across the blood−brain barrier. In the blood−brain barrier, astrocytes and endothelial cells have tight junctions (Knutson & Wessling-Resnick, 2003). These make the barrier between the brain and other cells of the circulatory system, which blocks the entry of harmful substances but allows the passage of essential molecules including glucose, water, lipids, and amino acids. Sometimes, it also blocks the passage of drugs to the brain. But it has been suggested that macrophages can cross the blood−brain barrier and researchers have designed macrophages-based carriers that can deliver drugs to the brain. For example, in in vivo experiments, it was observed in the treatment of Parkinson's disease that nanozymes were delivered by bone marrow-based macrophages (Batrakova et al., 2007; Valable et al., 2008).

The cytotoxic effects of liposomal doxorubicin (a chemotherapeutic agent) are reduced when it is loaded into the human leukemic cell line (THP-1). WBC-based carriers can also act as specific cancer-targeted carriers because they can interact with malignant cells at the tumor site as well as in blood circulation. In an in vivo experiment when liposomal doxorubicin loaded into macrophages was administered into the mice, the volume of the tumor was significantly decreased compared to the administration of free form of liposomal doxorubicin (Choi et al., 2012; Valable et al., 2007). Furthermore, to introduce a novel DDS, efforts are made to incorporate the RNA, small interfering RNA, and small circular DNA into host cells. To incorporate the therapeutic genes, RNA, and plasmid DNA into living cells transduction aand transfection processes are widely used (Naldini et al., 1996). Furthermore, antioxidants such as superoxide dismutase, thioredoxin, and catalase are incorporated into the monocytes to remove the reactive oxygen species (ROS), because excessive production of ROS provokes inflammatory diseases including atherosclerosis, coronary artery disease, and myocardial infarction. Antioxidants are loaded into the macrophages for the clearance of reactive oxygen species which showed better results as compared to merely antioxidants because antioxidants are nonspecific and biodegradable in nature (Circu & Aw, 2010; Hood et al., 2011; Ray et al., 2012; Taniyama & Griendling, 2003).

8.12 Advantages of blood cell-based drug delivery systems

The red blood cells-based DDS is widely used due to numerous factors. For example, RBCs are present in larger concentration and their properties and morphology are known. The surface of erythrocytes allows loading of a higher concentration of drugs with slow molecular release. Furthermore, erythrocytes show good properties of biocompatibility and biodegradability without stimulating immune responses or the production of toxic by-products. Erythrocytes can be engulfed by macrophages in the liver and spleen so RBCs can deliver drug vehicles into the reticuloendothelial system of cells which can be important to manage patients with macrophage-related liver disorders (Bhateria et al., 2014; Gutiérrez Millán et al., 2012; Hu et al., 2012).

The essential feature of a carrier is the manifestation of the benignity of the relation between a material and its biological environment, which is called biocompatibility. As platelets are the constituents of human blood, so they have a higher degree of biocompatibility as compared to the other drug delivery systems. The actual amount of loading of molecules is far less than 500 molecules on 1×10^{-15}/L nanoparticles as reported in the literature. Approximately, 50,000−70,000 molecules can be incorporated into the platelets which represents the higher encapsulation potential of the platelets (Buckley et al., 2000; White, 2005). The targeted sites of platelets are specifically the positions where most proliferation or wound occurs hence decreasing the damage caused by the random targeting. Moreover, the drugs encapsulated by the platelets are not much affected by the immune reactions and the physical stress as the similar systemic clearance of drugs encapsulated in platelets was observed as that of a platelet alone but the circulation time was increased. Controlled drug release has immense importance and a platelets-based drug delivery system can help in the process of controlled drug delivery by mimicry of drugs. It was reported by Sarkar et al. that platelets act synergistically with drugs to increase the adverse effects because the rate of apoptosis is increased when cytotoxic drugs are incorporated into platelets, which means lower drug concentration can be used for treatment with fewer lethal effects. Platelet-based DDS has gained much attention in recent years because in this method the patient's own platelets can act as a carrier to deliver drugs (Sarkar et al., 2013).

8.13 Disadvantages of blood cell-based drug delivery systems

As the most promising DDS, this system still faces a few limitations, such as during encapsulation osmotic pressure may cause the rupture of erythrocytes. The physical structure of erythrocytes can be destroyed during the coupling of drug cargos with RBCs. The structural and biological changes in RBCs may induce the elimination of erythrocytes by the reticuloendothelial system, which results in the decreased circulation time in the blood. Substances coupled with erythrocytes may cause the leakage of RBCs which may induce adverse effects. Furthermore, there is a high risk of blood contamination during the steps of drug delivery and the proper temperature is required for storage (Magnani, Chiarantini et al., 1992; Millán et al., 2004).

Platelet-based DDS has the same disadvantages as erythrocytes-based DDS, such as difficulties of inappropriate storage as well as contamination. When platelets are used as drug delivery cargos for a long time, it causes thrombosis.

8.13.1 Current challenges

The red blood cell-based drug delivery system is considered the most successful approach, but it still faces a few challenges which can have harmful effects. For example, we have discussed the cross-linking of biotin and avidin (for membrane-binding), which can cause hemolysis to biotinylated RBCs through the antigen-presenting cell. For different drug delivery approaches appropriate disposition of biotinylated RBCs is required to avoid serious harmful effects. Despite this, red blood cells-based carriers are the most promising strategy to target the reticuloendothelial system. Further research is needed to target different systems other than RES.

Among all the blood components, thrombocytes are the most sensitive and reactive components, which shows that platelet-based DDS can affect the physiological functions of platelets. Unwanted activation of platelets is associated with both bleeding and thrombosis. Further research is required to discover new methods for maintenance of drugs within platelets before loading into patients because at first the drug is delivered in vitro and the quantity of the drug is reduced (Kelton et al., 1981).

White blood cells are divided into subclasses based on morphology and are present in the lowest concentration in the blood, which has made leukocyte-based DDS more complicated than the other than two systems. Furthermore, this system may lead toward the inhibition of host defense due to overloading in the immune system and reticuloendothelial cells (RES), which is harmful to cells. When different components (such as phagocytes) of the immune system are activated several chemical mediators such as reactive oxygen species (ROS), histamine, and cytokines are released inducing inflammation.

8.14 Conclusions and future prospects

The cell-based drug delivery systems are presenting new outlooks on the prospect of using our cells for the effective delivery of therapeutic drugs and enzymes. Compared to the conventional targeted drug of drugs and even the use of nanodrug delivery systems, these cell-based systems possess additional advantages. It is emphasized that the toxicity of drugs should be a key factor as certain drugs can damage the cell carriers and can change the cellular chemistry after the loading of drugs (Liang et al., 2018; Yu et al., 2020). Therefore most approaches have emphasized the anchoring of therapeutic agents on the cellular surface. Moreover, a rational choice of suitable cells is essential to achieve the targeting efficiency for the developed blood cell-based systems. In this chapter, the commonly used cellular-based delivery systems derived from different circulating cells such as RBCs, platelets, and leukocytes have been described, as these endogenous carriers are advantageous due to being nonbiocompatible, nonimmunogenic, and targeting mechanisms for these carriers are natural. Moreover, the life span of blood cell-based carriers is relatively longer compared to the synthetic carriers and they undergo apoptosis upon aging or damage, hence, they are biodegradable. Also, these blood cell carriers possess higher loading capacities for drugs owing to their large volume.

Despite the blood cell-mediated systems resulting in satisfactory improvement in the therapeutic effects of drugs, still there are several challenges associated with the use of these carriers in clinical practice. One of the important challenges is to maintain the properties of cells during the isolation of cells and loading of therapeutic agents or their reinjection back into the body. Further, the efficiency of drug loading can be lower due to the decreased binding of the drug to the carriers. The leaking of the therapeutic agent from the carrier cells is another important concern. Moreover, the mechanisms of certain diseases are unclear, therefore the drugs and foreign cell complexes may further aggravate the pathological conditions. Even with all these associated challenges the blood cell-based systems have huge potential to transform the therapeutic techniques. The advances in the fields of nanotechnology may open new avenues for the design of these delivery systems. The combinational systems using blood cells and nanotechnology will provide a new path for drug delivery for better therapeutic outcomes. The studies related to the cell-based systems are gaining motion and the advent of new technologies will help to achieve the goal of improving drug delivery.

References

Adams, D. H., & Lloyd, A. R. (1997). Chemokines: Leucocyte recruitment and activation cytokines. *Lancet (London, England).*, 349(9050), 490−495.

Aderem, A., & Underhill, D. M. (1999). Mechanisms of phagocytosis in macrophages. *Annual Review of Immunology*, 17, 593−623.

Adriaenssens, K., Karcher, D., Lowenthal, A., & Terheggen, H. G. (1976). Use of enzyme-loaded erythrocytes in in-vitro correction of arginase-deficient erythrocytes in familial hyperargininemia. *Clinical Chemistry*, 22(3), 323−326.

Agrawal, V., Woo, J. H., Borthakur, G., Kantarjian, H., & Frankel, A. E. (2013). Red blood cell-encapsulated L-asparaginase: Potential therapy of patients with asparagine synthetase deficient acute myeloid leukemia. *Protein and Peptide Letters*, 20(4), 392−402.

Alexandrovich, Y. G., & Kosenko, E. A. (2017). Rapid elimination of blood alcohol using erythrocytes. *Mathematical Modeling and In Vitro Study.*, 2017, 5849593.

Allen, T. M., & Cullis, P. R. (2013). Liposomal drug delivery systems: From concept to clinical applications. *Advanced Drug Delivery Reviews*, 65(1), 36−48.

Amulic, B., Cazalet, C., Hayes, G. L., Metzler, K. D., & Zychlinsky, A. (2012). Neutrophil function: From mechanisms to disease. *Annual Review of Immunology*, 30, 459−489.

Annese, V., Latiano, A., Rossi, L., Lombardi, G., Dallapiccola, B., Serafini, S., et al. (2005). Erythrocytes-mediated delivery of dexamethasone in steroid-dependent IBD patients-a pilot uncontrolled study. *The American Journal of Gastroenterology*, 100(6), 1370−1375.

Arora, R., Malhotra, P., Mathur, A. K., Mathur, A., Govil, C., & Ahuja, P. (2010). *Anticancer alkaloids of* Catharanthus roseus: *Transition from traditional to modern medicine. Herbal medicine: A cancer chemopreventive and therapeutic perspective* (pp. 292−310). New Delhi: Jaypee Brothers Medical Publishers Pvt Ltd.

Ashley, G. W., Henise, J., Reid, R., & Santi, D. V. (2013). Hydrogel drug delivery system with predictable and tunable drug release and degradation rates. *Proceedings of the National Academy of Sciences of the United States of America, 110*(6), 2318−2323.

Ataullakhanov, F. I., Borsakova, D. V., Protasov, E. S., Sinauridze, E., & Zeynalov, A. M. (2018). Erythrocyte: A bag with hemoglobin, or a living active cell? *Voprosy gematologii/onkologii i immunopatologii v pediatrii., 17*, 108−116.

Atukorale, P. U., Yang, Y. S., Bekdemir, A., Carney, R. P., Silva, P. J., Watson, N., et al. (2015). Influence of the glycocalyx and plasma membrane composition on amphiphilic gold nanoparticle association with erythrocytes. *Nanoscale., 7*(26), 11420−11432.

Aurich, K., Spoerl, M. C., Fürll, B., Sietmann, R., Greinacher, A., Hosten, N., et al. (2012). Development of a method for magnetic labeling of platelets. *Nanomedicine: Nanotechnology, Biology, and Medicine., 8*(5), 537−544.

Ayer, M., & Klok, H. A. (2017). Cell-mediated delivery of synthetic nano- and microparticles. *Journal of Controlled Release: Official Journal of the Controlled Release Society., 259*, 92−104.

Bachet, J. B., Gay, F., Maréchal, R., Galais, M. P., Adenis, A., Ms, C. D., et al. (2015). Asparagine synthetase expression and phase I study with L-asparaginase encapsulated in red blood cells in patients with pancreatic adenocarcinoma. *Pancreas, 44*(7), 1141−1147.

Batrakova, E. V., Li, S., Reynolds, A. D., Mosley, R. L., Bronich, T. K., Kabanov, A. V., et al. (2007). A macrophage-nanozyme delivery system for Parkinson's disease. *Bioconjugate Chemistry, 18*(5), 1498−1506.

Bax, B. E., Bain, M. D., Fairbanks, L. D., Simmonds, H. A., Webster, A. D., & Chalmers, R. A. (2000). Carrier erythrocyte entrapped adenosine deaminase therapy in adenosine deaminase deficiency. *Advances in Experimental Medicine and Biology, 486*, 47−50.

Bax, B. E., Bain, M. D., Fairbanks, L. D., Webster, A. D., Ind, P. W., Hershfield, M. S., et al. (2007). A 9-yr evaluation of carrier erythrocyte encapsulated adenosine deaminase (ADA) therapy in a patient with adult-type ADA deficiency. *European Journal of Haematology, 79*(4), 338−348.

Bax, B. E., Bain, M. D., Ward, C. P., Fensom, A. H., & Chalmers, R. A. (1996). The entrapment of mannose-terminated glucocerebrosidase (Alglucerase) in human carrier erythrocytes. *Biochemical Society Transactions, 24*(3), 441s.

Bax, B. E., Fairbanks, L. D., Bain, M. D., Simmonds, H. A., & Chalmers, R. A. (1996). The entrapment of polyethylene glycol-bound adenosine deaminase (Pegademase) in human carrier erythrocytes. *Biochemical Society Transactions, 24*(3), 442s.

Baysal, S. H., & Uslan, A. H. (2000). Encapsulation of urease and PEG-urease in erythrocyte. *Artificial Cells, Blood Substitutes, and Immobilization Biotechnology, 28*(3), 263−271.

Baysal, S. H., Uslan, A. H., Pala, H. H., & Tunçoku, O. (2007). Encapsulation of PEG-urease/PEG-AlaDH within sheep erythrocytes and determination of the system's activity in lowering blood levels of urea in animal models. *Artificial Cells, Blood Substitutes, and Immobilization Biotechnology, 35*(4), 391−403.

Bell, S. M., Wendt, D. J., Zhang, Y., Taylor, T. W., Long, S., Tsuruda, L., et al. (2017). Formulation and PEGylation optimization of the therapeutic PEGylated phenylalanine ammonia lyase for the treatment of phenylketonuria. *PLoS One, 12*(3), e0173269.

Beutler, E., Dale, G. L., Guinto, D. E., & Kuhl, W. (1977). Enzyme replacement therapy in Gaucher's disease: Preliminary clinical trial of a new enzyme preparation. *Proceedings of the National Academy of Sciences of the United States of America, 74*(10), 4620−4623.

Bhateria, M., Rachumallu, R., Singh, R., & Bhatta, R. S. (2014). Erythrocytes-based synthetic delivery systems: Transition from conventional to novel engineering strategies. *Expert Opinion on Drug Delivery, 11*(8), 1219−1236.

Biagiotti, S., Rossi, L., Bianchi, M., Giacomini, E., Pierigè, F., Serafini, G., et al. (2011). Immunophilin-loaded erythrocytes as a new delivery strategy for immunosuppressive drugs. *Journal of Controlled Release.*, *154*(3), 306−313.

Bird, J., Best, R., & Lewis, D. A. (1983). The encapsulation of insulin in erythrocytes. *The Journal of Pharmacy and Pharmacology*, *35*(4), 246−247.

Blair, P., & Flaumenhaft, R. (2009). Platelet alpha-granules: Basic biology and clinical correlates. *Blood Reviews*, *23*(4), 177−189.

Bossa, F., Annese, V., Valvano, M. R., Latiano, A., Martino, G., Rossi, L., et al. (2013). Erythrocytes-mediated delivery of dexamethasone 21-phosphate in steroid-dependent ulcerative colitis: A randomized, double-blind Sham-controlled study. *Inflammatory Bowel Diseases*, *19*(9), 1872−1879.

Boswell, S. G., Cole, B. J., Sundman, E. A., Karas, V., & Fortier, L. A. (2012). Platelet-rich plasma: A milieu of bioactive factors. *Arthroscopy: The Journal of Arthroscopic & Related Surgery: Official Publication of the Arthroscopy Association of North America and the International Arthroscopy Association.*, *28*(3), 429−439.

Bourgeaux, V., Aufradet, E., Campion, Y., De Souza, G., Horand, F., Bessaad, A., et al. (2012). Efficacy of homologous inositol hexaphosphate-loaded red blood cells in sickle transgenic mice. *British Journal of Haematology*, *157*(3), 357−369.

Bourgeaux, V., Hequet, O., Campion, Y., Delcambre, G., Chevrier, A. M., Rigal, D., et al. (2010). Inositol hexaphosphate-loaded red blood cells prevent in vitro sickling. *Transfusion*, *50*(10), 2176−2184.

Brannon-Peppas, L., & Blanchette, J. O. (2004). Nanoparticle and targeted systems for cancer therapy. *Advanced Drug Delivery Reviews*, *56*(11), 1649−1659.

Buckley, M. F., James, J. W., Brown, D. E., Whyte, G. S., Dean, M. G., Chesterman, C. N., et al. (2000). A novel approach to the assessment of variations in the human platelet count. *Thrombosis and Haemostasis*, *83*(3), 480−484.

Cannon, E. P., Leung, P., Hawkins, A., Petrikovics, I., DeLoach, J., & Way, J. L. (1994). Antagonism of cyanide intoxication with murine carrier erythrocytes containing bovine rhodanese and sodium thiosulfate. *Journal of Toxicology and Environmental Health*, *41*(3), 267−274.

Cantón, I., & Battaglia, G. (2012). Endocytosis at the nanoscale. *Chemical Society Reviews*, *41*, 2718−2739.

Chen, H., Mo, W., Su, H., Zhang, Y., & Song, H. (2007). Characterization of a novel bifunctional mutant of staphylokinase with platelet-targeted thrombolysis and antiplatelet aggregation activities. *BMC Molecular Biology*, *8*, 88.

Chen, I., Howarth, M., Lin, W., & Ting, A. Y. (2005). Site-specific labeling of cell surface proteins with biophysical probes using biotin ligase. *Nature Methods*, *2*(2), 99−104.

Cheng, H., Kastrup, C. J., Ramanathan, R., Siegwart, D. J., Ma, M., Bogatyrev, S. R., et al. (2010). Nanoparticulate cellular patches for cell-mediated tumoritropic delivery. *ACS Nano.*, *4*(2), 625−631.

Chessa, L., Leuzzi, V., Plebani, A., Soresina, A., Micheli, R., D'Agnano, D., et al. (2014). Intra-erythrocyte infusion of dexamethasone reduces neurological symptoms in ataxia teleangiectasia patients: Results of a phase 2 trial. *Orphanet Journal of Rare Diseases*, *9*, 5.

Chi, J., Ma, Q., Shen, Z., Ma, C., Zhu, W., Han, S., et al. (2020). Targeted nanocarriers based on iodinated-cyanine dyes as immunomodulators for synergistic phototherapy. *Nanoscale.*, *12*(20), 11008−11025.

Chiarantini, L., Cerasi, A., Fraternale, A., Andreoni, F., Scarí, S., Giovine, M., et al. (2002). Inhibition of macrophage iNOS by selective targeting of antisense PNA. *Biochemistry*, *41*(26), 8471−8477.

Choi, J., Kim, H. Y., Ju, E. J., Jung, J., Park, J., Chung, H. K., et al. (2012). Use of macrophages to deliver therapeutic and imaging contrast agents to tumors. *Biomaterials*, *33*(16), 4195−4203.

Circu, M. L., & Aw, T. Y. (2010). Reactive oxygen species, cellular redox systems, and apoptosis. *Free Radical Biology & Medicine*, *48*(6), 749−762.

Cui, W., Zhou, Y., & Chang, J. (2010). Electrospun nanofibrous materials for tissue engineering and drug delivery. *Science and Technology of Advanced Materials.*, *11*(1), 014108.

Czock, D., Keller, F., Rasche, F. M., & Häussler, U. (2005). Pharmacokinetics and pharmacodynamics of systemically administered glucocorticoids. *Clinical Pharmacokinetics*, *44*(1), 61−98.

D'Ascenzo, M., Antonelli, A., Chiarantini, L., Mancini, U., & Magnani, M. (1997). *Red blood cells as a gluco-corticoids delivery system. Erythrocytes as drug carriers in medicine* (pp. 81−88). Springer.

Dale, G. L., & Beutler, E. (1976). Enzyme replacement therapy in Gaucher's disease: A rapid, high-yield method for purification of glucocerebrosidase. *Proceedings of the National Academy of Sciences of the United States of America, 73*(12), 4672−4674.

Dale, G. L., Villacorte, D. G., & Beutler, E. (1977). High-yield entrapment of proteins into erythrocytes. *Biochemical Medicine, 18*(2), 220−225.

DeLoach, J., Droleskey, R., & Andrews, K. (1991). Encapsulation by hypotonic dialysis in human erythrocytes: A diffusion or endocytosis process. *Biotechnology and Applied Biochemistry, 13*(1), 72−82.

Deloach, J., & Ihler, G. (1977). A dialysis procedure for loading erythrocytes with enzymes and lipids. *Biochimica et Biophysica Acta, 496*(1), 136−145.

Diamandis, E. P., & Christopoulos, T. K. (1991). The biotin-(strept)avidin system: Principles and applications in biotechnology. *Clinical Chemistry, 37*(5), 625−636.

Domenech, C., Thomas, X., Chabaud, S., Baruchel, A., Gueyffier, F., Mazingue, F., et al. (2011). l-asparaginase loaded red blood cells in refractory or relapsing acute lymphoblastic leukaemia in children and adults: Results of the GRASPALL 2005−01 randomized trial. *British Journal of Haematology, 153*(1), 58−65.

Edwards, I. R., & Biriell, C. (1994). Harmonisation in pharmacovigilance. *Drug Safety., 10*(2), 93−102.

Ferguson, M. A., & Williams, A. F. (1988). Cell-surface anchoring of proteins via glycosyl-phosphatidylinositol structures. *Annual Review of Biochemistry, 57*, 285−320.

Filosto, M., Cotti Piccinelli, S., Caria, F., Gallo Cassarino, S., Baldelli, E., Galvagni, A., et al. (2018). Mitochondrial neurogastrointestinal encephalomyopathy (MNGIE-MTDPS1). *Journal of Clinical Medicine., 7*(11).

Flynn, G., Hackett, T. J., McHale, L., & McHale, A. P. (1994). Encapsulation of the thrombolytic enzyme, brinase, in photosensitized erythrocytes: A novel thrombolytic system based on photodynamic activation. *Journal of Photochemistry and Photobiology B: Biology., 26*(2), 193−196.

Foroozesh, M., Hamidi, M., Zarrin, A., Mohammadi-Samani, S., & Montaseri, H. (2011). Preparation and in-vitro characterization of tramadol-loaded carrier erythrocytes for long-term intravenous delivery. *Journal of Pharmacy and Pharmacology., 63*(3), 322−332.

Franco, R. S., Weiner, M., Wagner, K., & Martelo, O. J. (1983). Incorporation of inositol hexaphosphate into red blood cells mediated by dimethyl sulfoxide. *Life Sciences, 32*(24), 2763−2768.

Gámez, A., Wang, L., Straub, M., Patch, M. G., & Stevens, R. C. (2004). Toward PKU enzyme replacement therapy: PEGylation with activity retention for three forms of recombinant phenylalanine hydroxylase. *Molecular Therapy: The Journal of the American Society of Gene Therapy., 9*(1), 124−129.

Ganguly, K., Goel, M. S., Krasik, T., Bdeir, K., Diamond, S. L., Cines, D. B., et al. (2006). Fibrin affinity of erythrocyte-coupled tissue-type plasminogen activators endures hemodynamic forces and enhances fibrinolysis in vivo. *The Journal of Pharmacology and Experimental Therapeutics, 316*(3), 1130−1136.

Gay, F., Aguera, K., Senechal, K., Bes, J., Chevrier, A.-M., Gallix, F., et al. (2016). Arginine deiminase loaded in erythrocytes: A promising formulation for l-arginine deprivation therapy in cancers. *AACR.*

Gay, F., Aguera, K., Sénéchal, K., Tainturier, A., Berlier, W., Maucort-Boulch, D., et al. (2017). Methionine tumor starvation by erythrocyte-encapsulated methionine gamma-lyase activity controlled with per os vitamin B6. *Cancer Medicine., 6*(6), 1437−B52.

Ge, W.-H., Lian, Y.-S., Wang, X.-H., Luo, X., & Xie, P.-H. (2004). Morphological observation of erythrocyte during the preparation of morphine carrier by a hyperosmotic method. *Chinese Pharmaceutical Journal-Beijing, 39*(4), 270−272.

Glassman, P. M., Villa, C. H., Ukidve, A., Zhao, Z., Smith, P., Mitragotri, S., et al. (2020). Vascular drug delivery using carrier red blood cells: Focus on RBC surface loading and pharmacokinetics. *Pharmaceutics., 12*(5).

Godfrin, Y., Horand, F., Franco, R., Dufour, E., Kosenko, E., Bax, B. E., et al. (2012). International seminar on the red blood cells as vehicles for drugs. *Expert Opinion on Biological Therapy, 12*(1), 127−133.

Gopal, V., Kumar, A. R., Usha, A., Karthik, A., & Udupa, N. (2007). Effective drug targeting by erythrocytes as carrier systems. *Current Trends in Biotechnology and Pharmacy.*, *1*(1), 18−33.

Greenwald, R. B., Choe, Y. H., McGuire, J., & Conover, C. D. (2003). Effective drug delivery by PEGylated drug conjugates. *Advanced Drug Delivery Reviews*, *55*(2), 217−250.

Greineder, C. F., Howard, M. D., Carnemolla, R., Cines, D. B., & Muzykantov, V. R. (2013). Advanced drug delivery systems for antithrombotic agents. *Blood*, *122*(9), 1565−1575.

Gutiérrez Millán, C., Colino Gandarillas, C. I., Sayalero Marinero, M. L., & Lanao, J. M. (2012). Cell-based drug-delivery platforms. *Therapeutic Delivery.*, *3*(1), 25−41.

Hamarat Baysal, S., & Uslan, A. H. (2002). In vitro study of urease/AlaDH enzyme system encapsulated into human erythrocytes and research into its medical applications. *Artificial Cells, Blood Substitutes, and Immobilization Biotechnology*, *30*(1), 71−77.

Hamidi, M., Tajerzadeh, H., Dehpour, A. R., & Ejtemaee-Mehr, S. (2001). Inhibition of serum angiotensin-converting enzyme in rabbits after intravenous administration of enalaprilat-loaded intact erythrocytes. *The Journal of Pharmacy and Pharmacology*, *53*(9), 1281−1286.

Harisa Gel, D., Ibrahim, M. F., & Alanazi, F. K. (2011). Characterization of human erythrocytes as potential carrier for pravastatin: An in vitro study. *International Journal of Medical Sciences*, *8*(3), 222−230.

Hollowell, J. G., van Assendelft, O. W., Gunter, E. W., Lewis, B. G., Najjar, M., & Pfeiffer, C. (2005). Hematological and iron-related analytes−reference data for persons aged 1 year and over: United States, 1988−94. *Vital and Health Statistics. Series 11, Data from the National Health Survey*, *247*, 1−156.

Hood, E., Simone, E., Wattamwar, P., Dziubla, T., & Muzykantov, V. (2011). Nanocarriers for vascular delivery of antioxidants. *Nanomedicine (London).*, *6*(7), 1257−1272.

Hu, C. M., Fang, R. H., & Zhang, L. (2012). Erythrocyte-inspired delivery systems. *Advanced Healthcare Materials.*, *1*(5), 537−547.

Hu, C.-M. J., Fang, R. H., Wang, K.-C., Luk, B. T., Thamphiwatana, S., Dehaini, D., et al. (2015). Nanoparticle biointerfacing by platelet membrane cloaking. *Nature*, *526*(7571), 118−121.

Humphreys, J. D., & Ihler, G. (1980). Enhanced stability of erythrocyte-entrapped glucocerebrosidase activity. *The Journal of Laboratory and Clinical Medicine*, *96*(4), 682−692.

Ihler, G., Lantzy, A., Purpura, J., & Glew, R. H. (1975). Enzymatic degradation of uric acid by uricase-loaded human erythrocytes. *The Journal of Clinical Investigation*, *56*(3), 595−602.

Ito, Y., Ogiso, T., Iwaki, M., & Kitaike, M. (1989). Encapsulation of porcine insulin in rabbit erythrocytes and its disposition in the circulation system in normal and diabetic rabbits. *Journal of Pharmacobio-dynamics*, *12*(4), 193−200.

Kaplan, P., Baris, H., De Meirleir, L., Di Rocco, M., El-Beshlawy, A., Huemer, M., et al. (2013). Revised recommendations for the management of Gaucher disease in children. *European Journal of Pediatrics*, *172*(4), 447−458.

Kedar, U., Phutane, P., Shidhaye, S., & Kadam, V. (2010). Advances in polymeric micelles for drug delivery and tumor targeting. *Nanomedicine: Nanotechnology, Biology, and Medicine.*, *6*(6), 714−729.

Kelton, J. G., McDonald, J. W., Barr, R. M., Walker, I., Nicholson, W., Neame, P. B., et al. (1981). The reversible binding of vinblastine to platelets: Implications for therapy. *Blood*, *57*(3), 431−438.

Knutson, M., & Wessling-Resnick, M. (2003). Iron metabolism in the reticuloendothelial system. *Critical Reviews in Biochemistry and Molecular Biology*, *38*(1), 61−88.

Kosenko, E. A., Venediktova, N. I., Kudryavtsev, A. A., Ataullakhanov, F. I., Kaminsky, Y. G., Felipo, V., et al. (2008). Encapsulation of glutamine synthetase in mouse erythrocytes: A new procedure for ammonia detoxification. *Biochemistry and Cell Biology = Biochimie et biologie cellulaire.*, *86*(6), 469−476.

Koseoglu, S., & Flaumenhaft, R. (2013). Advances in platelet granule biology. *Current Opinion in Hematology*, *20*(5), 464−471.

Kwon, Y. M., Chung, H. S., Moon, C., Yockman, J., Park, Y. J., Gitlin, S. D., et al. (2009). L-Asparaginase encapsulated intact erythrocytes for treatment of acute lymphoblastic leukemia (ALL). *Journal of Controlled Release: Official Journal of the Controlled Release Society., 139*(3), 182−189.

Lamarre, Y., Bourgeaux, V., Pichon, A., Hardeman, M. R., Campion, Y., Hardeman-Zijp, M., et al. (2013). Effect of inositol hexaphosphate-loaded red blood cells (RBCs) on the rheology of sickle RBCs. *Transfusion, 53*(3), 627−636.

Leung, P., Cannon, E. P., Petrikovics, I., Hawkins, A., & Way, J. L. (1991). In vivo studies on rhodanese encapsulation in mouse carrier erythrocytes. *Toxicology and Applied Pharmacology, 110*(2), 268−274.

Levene, M., Bain, M. D., Moran, N. F., Nirmalananthan, N., Poulton, J., Scarpelli, M., et al. (2019). Safety and efficacy of erythrocyte encapsulated thymidine phosphorylase in mitochondrial neurogastrointestinal encephalomyopathy. *Journal of Clinical Medicine., 8*(4).

Levene, M., Coleman, D. G., Kilpatrick, H. C., Fairbanks, L. D., Gangadharan, B., Gasson, C., et al. (2013). Preclinical toxicity evaluation of erythrocyte-encapsulated thymidine phosphorylase in BALB/c mice and beagle dogs: An enzyme-replacement therapy for mitochondrial neurogastrointestinal encephalomyopathy. *Toxicological Sciences: An Official Journal of the Society of Toxicology., 131*(1), 311−324.

Levêque, D., Wihlm, J., & Jehl, F. (1996). Pharmacology of Catharanthus alkaloids. *Bulletin du Cancer, 83*(3), 176−186.

Liang, Y., Huo, Q., Lu, W., Jiang, L., Gao, W., Xu, L., et al. (2018). Fluorescence resonance energy transfer visualization of molecular delivery from polymeric micelles. *Journal of Biomedical Nanotechnology., 14*(7), 1308−1316.

Liu, J., Lkhagva, E., Chung, H. J., Kim, H. J., & Hong, S. T. (2018). The pharmabiotic approach to treat hyperammonemia. *Nutrients., 10*(2).

Lizano, C., Pérez, M. T., & Pinilla, M. (2001). Mouse erythrocytes as carriers for coencapsulated alcohol and aldehyde dehydrogenase obtained by electroporation: In vivo survival rate in circulation, organ distribution and ethanol degradation. *Life Sciences, 68*(17), 2001−2016.

Lizano, C., Sanz, S., Luque, J., & Pinilla, M. (1998). In vitro study of alcohol dehydrogenase and acetaldehyde dehydrogenase encapsulated into human erythrocytes by an electroporation procedure. *Biochimica et Biophysica Acta, 1425*(2), 328−336.

Lotero, L. A., Olmos, G., & Diez, J. C. (2003). Delivery to macrophages and toxic action of etoposide carried in mouse red blood cells. *Biochimica et Biophysica Acta (BBA)—General Subjects., 1620*(1), 160−166.

Luo, X., Wang, X., Xu, Z., Cui, S., & Xu, F. (2005). Feasibility of using erythrocytes as morphine carrier for postoperative analgesia after coronary artery bypass grafting. *Chinese Journal of Anesthesiology, 6*, 410−413.

Ma, Q., Cao, J., Gao, Y., Han, S., Liang, Y., Zhang, T., et al. (2020). Microfluidic-mediated nano-drug delivery systems: From fundamentals to fabrication for advanced therapeutic applications. *Nanoscale., 12*(29), 15512−15527.

Machover, D., Rossi, L., Hamelin, J., Desterke, C., Goldschmidt, E., Chadefaux-Vekemans, B., et al. (2019). Effects in cancer cells of the recombinant l-methionine gamma-lyase from brevibacterium aurantiacum. Encapsulation in human erythrocytes for sustained l-methionine elimination. *The Journal of Pharmacology and Experimental Therapeutics, 369*(3), 489−502.

Magnani, M., Chiarantini, L., Vittoria, E., Mancini, U., Rossi, L., & Fazi, A. (1992). Red blood cells as an antigen-delivery system. *Biotechnology and Applied Biochemistry, 16*(2), 188−194.

Magnani, M., Fazi, A., Mangani, F., Rossi, L., & Mancini, U. (1993). Methanol detoxification by enzyme-loaded erythrocytes. *Biotechnology and Applied Biochemistry, 18*(3), 217−226.

Magnani, M., Laguerre, M., Rossi, L., Bianchi, M., Ninfali, P., Mangani, F., et al. (1990). In vivo accelerated acetaldehyde metabolism using acetaldehyde dehydrogenase-loaded erythrocytes. *Alcohol and Alcoholism (Oxford, Oxfordshire), 25*(6), 627−637.

Magnani, M., Mancini, U., Bianchi, M., & Fazi, A. (1992). Comparison of uricase-bound and uricase-loaded erythrocytes as bioreactors for uric acid degradation. *Advances in Experimental Medicine and Biology, 326*, 189−194.

Magnani, M., Rossi, L., Bianchi, M., Giorgio, F., Benatti, U., Guida, L., et al. (1988). Improved metabolic properties of hexokinase-overloaded human erythrocytes. *Biochimica et Biophysica Acta (BBA)— Bioenergetics., 972*(1), 1−8.

Mangal, P. C., & Kaur, A. (1991). Electroporation of red blood cell membrane and its use as a drug carrier system. *Indian Journal of Biochemistry & Biophysics, 28*(3), 219−221.

Millán, C. G., Marinero, M. L., Castañeda, A. Z., & Lanao, J. M. (2004). Drug, enzyme and peptide delivery using erythrocytes as carriers. *Journal of Controlled Release: Official Journal of the Controlled Release Society., 95*(1), 27−49.

Minev, B. (2011). *Cancer management in man: Chemotherapy, biological therapy, hyperthermia and supporting measures.* Springer Science & Business Media.

Modery-Pawlowski, C. L., Master, A. M., Pan, V., Howard, G. P., Sen., & Gupta, A. (2013). A platelet-mimetic paradigm for metastasis-targeted nanomedicine platforms. *Biomacromolecules, 14*(3), 910−919.

Moran, N. F., Bain, M. D., Muqit, M. M. K., & Bax, B. E. (2008). Carrier erythrocyte entrapped thymidine phosphorylase therapy for Mngie. *Neurology, 71*(9), 686−688.

Moudi, M., Go, R., Yien, C. Y., & Nazre, M. (2013). Vinca alkaloids. *International Journal of Preventive Medicine., 4*(11), 1231−1235.

Münzer, P., Borst, O., Walker, B., Schmid, E., Feijge, M. A., Cosemans, J. M., et al. (2014). Acid sphingomyelinase regulates platelet cell membrane scrambling, secretion, and thrombus formation. *Arteriosclerosis, Thrombosis, and Vascular Biology, 34*(1), 61−71.

Murciano, J.-C., Higazi, A. A.-R., Cines, D. B., & Muzykantov, V. R. (2009). Soluble urokinase receptor conjugated to carrier red blood cells binds latent pro-urokinase and alters its functional profile. *Journal of Controlled Release: Official Journal of the Controlled Release Society., 139*(3), 190−196.

Murciano, J.-C., Medinilla, S., Eslin, D., Atochina, E., Cines, D. B., & Muzykantov, V. R. (2003). Prophylactic fibrinolysis through selective dissolution of nascent clots by tPA-carrying erythrocytes. *Nature Biotechnology, 21*(8), 891−896.

Muro, S. (2012). Challenges in design and characterization of ligand-targeted drug delivery systems. *Journal of Controlled Release: Official Journal of the Controlled Release Society., 164*(2), 125−137.

Muzykantov, V. R. (2010). Drug delivery by red blood cells: Vascular carriers designed by mother nature. *Expert Opinion on Drug Delivery, 7*(4), 403−427.

Naldini, L., Blömer, U., Gallay, P., Ory, D., Mulligan, R., Gage, F. H., et al. (1996). In vivo gene delivery and stable transduction of nondividing cells by a lentiviral vector. *Science (New York, NY)., 272*(5259), 263−267.

Oldenborg, P. A., Zheleznyak, A., Fang, Y. F., Lagenaur, C. F., Gresham, H. D., & Lindberg, F. P. (2000). Role of CD47 as a marker of self on red blood cells. *Science (New York, NY)., 288*(5473), 2051−2054.

Owens, D. E., 3rd, & Peppas, N. A. (2006). Opsonization, biodistribution, and pharmacokinetics of polymeric nanoparticles. *International Journal of Pharmaceutics, 307*(1), 93−102.

Parodi, A., Quattrocchi, N., van de Ven, A. L., Chiappini, C., Evangelopoulos, M., Martinez, J. O., et al. (2013). Synthetic nanoparticles functionalized with biomimetic leukocyte membranes possess cell-like functions. *Nature Nanotechnology., 8*(1), 61−68.

Parrott, M., Adams, K., Mercier, G., Mok, H., Campos, S., & Barry, M. (2003). Metabolically biotinylated adenovirus for cell targeting, Ligand screening, and vector purification. *Molecular Therapy: The Journal of the American Society of Gene Therapy., 8*, 688−700.

Parveen, S., Misra, R., & Sahoo, S. K. (2012). Nanoparticles: A boon to drug delivery, therapeutics, diagnostics and imaging. *Nanomedicine: Nanotechnology, Biology, and Medicine., 8*(2), 147−166.

Petrikovics, I., Pei, L., McGuinn, W. D., Cannon, E. P., & Way, J. L. (1994). Encapsulation of rhodanese and organic thiosulfonates by mouse erythrocytes. *Fundamental and Applied Toxicology: Official Journal of the Society of Toxicology.*, *23*(1), 70–75.

Pitt, E., Johnson, C. M., Lewis, D. A., Jenner, D. A., & Offord, R. E. (1983). Encapsulation of drugs in intact erythrocytes: An intravenous delivery system. *Biochemical Pharmacology*, *32*(22), 3359–3368.

Platt, F. M., d'Azzo, A., Davidson, B. L., Neufeld, E. F., & Tifft, C. J. (2018). Lysosomal storage diseases. *Nature Reviews Disease Primers.*, *4*(1), 27.

Prausnitz, M. R., & Langer, R. (2008). Transdermal drug delivery. *Nature Biotechnology*, *26*(11), 1261–1268.

Protasov, E. S., Borsakova, D. V., Alexandrovich, Y. G., Korotkov, A. V., Kosenko, E. A., Butylin, A. A., et al. (2019). Erythrocytes as bioreactors to decrease excess ammonium concentration in blood. *Scientific Reports.*, *9*(1), 1455.

Ray, P. D., Huang, B. W., & Tsuji, Y. (2012). Reactive oxygen species (ROS) homeostasis and redox regulation in cellular signaling. *Cellular Signalling*, *24*(5), 981–990.

Rossi, G. A., Cerasoli, F., & Cazzola, M. (2007). Safety of inhaled corticosteroids: Room for improvement. *Pulmonary Pharmacology & Therapeutics*, *20*(1), 23–35.

Rossi, L., Bianchi, M., & Magnani, M. (1992). Increased glucose metabolism by enzyme-loaded erythrocytes in vitro and in vivo normalization of hyperglycemia in diabetic mice. *Biotechnology and Applied Biochemistry*, *15*(2), 207–216.

Rossi, L., Pierigè, F., Carducci, C., Gabucci, C., Pascucci, T., Canonico, B., et al. (2014). Erythrocyte-mediated delivery of phenylalanine ammonia lyase for the treatment of phenylketonuria in BTBR-Pah(enu2) mice. *Journal of Controlled Release: Official Journal of the Controlled Release Society.*, *194*, 37–44.

Rossi, L., Serafini, S., Cenerini, L., Picardi, F., Bigi, L., Panzani, I., et al. (2001). Erythrocyte-mediated delivery of dexamethasone in patients with chronic obstructive pulmonary disease. *Biotechnology and Applied Biochemistry*, *33*(2), 85–89.

Rossi, N. A., Constantinescu, I., Kainthan, R. K., Brooks, D. E., Scott, M. D., & Kizhakkedathu, J. N. (2010). Red blood cell membrane grafting of multi-functional hyperbranched polyglycerols. *Biomaterials*, *31*(14), 4167–4178.

Sanz, S., Lizano, C., Luque, J., & Pinilla, M. (1999). In vitro and in vivo study of glutamate dehydrogenase encapsulated into mouse erythrocytes by a hypotonic dialysis procedure. *Life Sciences*, *65*(26), 2781–2789.

Sanz, S., Pinilla, M., Garín, M., Tipton, K. F., & Luque, J. (1995). The influence of enzyme concentration on the encapsulation of glutamate dehydrogenase and alcohol dehydrogenase in red blood cells. *Biotechnology and Applied Biochemistry*, *22*(2), 223–231.

Sarkar, S., Alam, M. A., Shaw, J., & Dasgupta, A. K. (2013). Drug delivery using platelet cancer cell interaction. *Pharmaceutical Research*, *30*(11), 2785–2794.

Sarkissian, C. N., Gámez, A., Wang, L., Charbonneau, M., Fitzpatrick, P., Lemontt, J. F., et al. (2008). Preclinical evaluation of multiple species of PEGylated recombinant phenylalanine ammonia lyase for the treatment of phenylketonuria. *Proceedings of the National Academy of Sciences*, *105*(52), 20894–20899.

Sawant, R. R., & Torchilin, V. P. (2012). Multifunctional nanocarriers and intracellular drug delivery. *Current Opinion in Solid State and Materials Science.*, *16*(6), 269–275.

Schrier, S. L., Zachowski, A., & Devaux, P. F. (1992). Mechanisms of amphipath-induced stomatocytosis in human erythrocytes. *Blood*, *79*(3), 782–786.

Shi, Q., Fahs, S. A., Wilcox, D. A., Kuether, E. L., Morateck, P. A., Mareno, N., et al. (2008). Syngeneic transplantation of hematopoietic stem cells that are genetically modified to express factor VIII in platelets restores hemostasis to hemophilia A mice with preexisting FVIII immunity. *Blood*, *112*(7), 2713–2721.

Shi, Q., & Montgomery, R. R. (2010). Platelets as delivery systems for disease treatments. *Advanced Drug Delivery Reviews*, *62*(12), 1196–1203.

Sinauridze, E. I., Vuimo, T. A., Kulikova, E. V., Shmyrev, I. I., & Ataullakhanov, F. I. (2010). A new drug form of blood coagulation factor IX: Red blood cell-entrapped factor IX. *Medical Science Monitor: International Medical Journal of Experimental and Clinical Research, 16*(10), Pi19−Pi26.

Skorokhod, O., Kulikova, E. V., Galkina, N. M., Medvedev, P. V., Zybunova, E. E., Vitvitsky, V. M., et al. (2007). Doxorubicin pharmacokinetics in lymphoma patients treated with doxorubicin-loaded eythrocytes. *Haematologic, 92*(4), 570−571.

Skorokhod, O. A., Garmaeva, T., Vitvitsky, V. M., Isaev, V. G., Parovichnikova, E. N., Savchenko, V. G., et al. (2004). Pharmacokinetics of erythrocyte-bound daunorubicin in patients with acute leukemia. *Medical Science Monitor: International Medical Journal of Experimental and Clinical Research., 10*(4), Pi55−Pi64.

Snyder, E. L., & Dowdy, S. F. (2004). Cell penetrating peptides in drug delivery. *Pharmaceutical Research, 21*(3), 389−393.

Spatz, M. A. (2004). Genetics home reference. *Journal of the Medical Library Association: JMLA, 92*(2), 282−283.

Taniyama, Y., & Griendling, K. K. (2003). Reactive oxygen species in the vasculature: Molecular and cellular mechanisms. *Hypertension, 42*(6), 1075−1081.

Thorpe, S. R., Fiddler, M. B., & Desnick, R. J. (1975). Enzyme therapy. V. In vivo fate of erythrocyte-entrapped beta-glucuronidase in beta-glucuronidase-deficient mice. *Pediatric Research, 9*(12), 918−923.

Torchilin, V. P., & Trubetskoy, V. S. (1995). Which polymers can make nanoparticulate drug carriers long-circulating? *Advanced Drug Delivery Reviews, 16*(2), 141−155.

Trineeva, O., & Khalahakun, A. (2019). Study of desorbtion and exemption of terpeno-indole alkaloids of vinkristin and vinblastin from erythrocitary cell carriers. *Drug Development & Registration, 8*, 16−21.

Tsong, T. Y. (1991). Electroporation of cell membranes. *Biophysical Journal, 60*(2), 297−306.

Umland, S. P., Schleimer, R. P., & Johnston, S. L. (2002). Review of the molecular and cellular mechanisms of action of glucocorticoids for use in asthma. *Pulmonary Pharmacology & Therapeutics, 15*(1), 35−50.

Valable, S., Barbier, E. L., Bernaudin, M., Roussel, S., Segebarth, C., Petit, E., et al. (2007). In vivo MRI tracking of exogenous monocytes/macrophages targeting brain tumors in a rat model of glioma. *Neuroimage, 37*(Suppl. 1), S47−S58.

Valable, S., Barbier, E. L., Bernaudin, M., Roussel, S., Segebarth, C., Petit, E., et al. (2008). In vivo MRI tracking of exogenous monocytes/macrophages targeting brain tumors in a rat model of glioma. *Neuroimage, 40*(2), 973−983.

Venediktova, N. I., Kosenko, E. A., & Kaminsky, Y. G. (2008). Studies on ammocytes: Development, metabolic characteristics, and detoxication of ammonium. *Bulletin of Experimental Biology and Medicine, 146* (6), 730−732.

Walsh, T. G., Metharom, P., & Berndt, M. C. (2015). The functional role of platelets in the regulation of angiogenesis. *Platelets, 26*(3), 199−211.

Wang, G. P., Guan, Y. S., Jin, X. R., Jiang, S. S., Lu, Z. J., Wu, Y., et al. (2010). Development of novel 5-fluorouracil carrier erythrocyte with pharmacokinetics and potent antitumor activity in mice bearing malignant ascites. *Journal of Gastroenterology and Hepatology, 25*(5), 985−990.

Weaver, J. C. (2003). Electroporation of biological membranes from multicellular to nano scales. *IEEE Transactions on Dielectrics and Electrical Insulation., 10*(5), 754−768.

White, J. G. (2005). Platelets are covercytes, not phagocytes: Uptake of bacteria involves channels of the open canalicular system. *Platelets, 16*(2), 121−131.

Wong, A. K. (2013). Platelet biology: The role of shear. *Expert Review of Hematology., 6*(2), 205−212.

Xiaohai, W., Xuan, L., & Shijun, Z. (2001). A clinical study on morphine encapsulated in erythrocytes for postoperative analgesia. *The Journal of Clinical Anesthesilolgy., 2*, 11.

Xuan, L., Xin, X., & Xiao-Hai, W. (2003). Study of erythrocyte as carrier to prolong action duration of morphine. *Journal-Nanjing University Natural Sciences Edition.*, *39*(5), 547−553.

Yew, N. S., Dufour, E., Przybylska, M., Putelat, J., Crawley, C., Foster, M., et al. (2013). Erythrocytes encapsulated with phenylalanine hydroxylase exhibit improved pharmacokinetics and lowered plasma phenylalanine levels in normal mice. *Molecular Genetics and Metabolism*, *109*(4), 339−344.

Yoo, J.-W., Irvine, D. J., Discher, D. E., & Mitragotri, S. (2011). Bio-inspired, bioengineered and biomimetic drug delivery carriers. *Nature Reviews. Drug Discovery*, *10*(7), 521−535.

Yu, H., Yang, Z., Li, F., Xu, L., & Sun, Y. (2020). Cell-mediated targeting drugs delivery systems. *Drug Delivery*, *27*(1), 1425−1437.

Nanosensors for medical diagnosis

Yeşeren Saylan[1], Semra Akgönüllü[1] and Adil Denizli[2]

[1]*Department of Chemistry, Hacettepe University, Ankara, Turkey* [2]*Biochemistry Division, Department of Chemistry, Hacettepe University, Ankara, Turkey*

9.1 Introduction

Tremendous growth has been witnessed owing to the relation between basic sciences and technology for the detection of different analytes in nanosensors in real time for various applications such as food supplies (Chen, Lin, Hong, Yao, & Huang, 2020; Heravizadeh et al., 2020; Rahtuvanoğlu et al., 2020; Wang et al., 2019; Xiao et al., 2020), biomarkers (Esentürk et al., 2019; Pacheco et al., 2018; Rebelo et al., 2019), genetically modified organisms (Chen, Liu, Li, & Wang, 2019; Landete, Langa, Escudero, Peirotén, & Arqués, 2020; Wang et al., 2020; Xiao et al., 2020; Zhang, Wei, He, Huang, & Liu, 2020), proteins (Inci et al., 2020; Khan, Maddaus, & Song, 2018; Saylan & Denizli, 2018; Zhang, Yu, Guo, Lin, & Zhang, 2020), microorganisms (Erdem et al., 2020; Hao et al., 2019; Li et al., 2020), contaminants (Revsbech et al. 2020; Tucci et al., 2019; Wang et al., 2020), drugs (Akgönüllü et al., 2020; Alvau et al., 2018; Battal et al., 2018; Veeralingam & Badhulika, 2020), toxins (Akgönüllü et al. 2020; Hossain, McCormick, & Maragos, 2018; Lvova, 2020; Santana Oliveira, da Silva, de Andrade, Lima Oliveira, 2019), and chemical and biological threat agents (Lafuente et al., 2018; Öztürk & Şehitoğlu, 2019; Saylan et al., 2020; Sharma et al., 2020). Especially, the growing requirement for selective and sensitive biomolecules detection needs the development of accurate, rapid, cost- and user-friendly innovative platforms for the medical diagnosis applications (Poschenrieder et al., 2019). Quantitative measurement of samples is usually performed by conventional methods (Pirzada & Altintas, 2019; Rico-Yuste & Carrasco, 2019). These methods are sensitive but they need expensive and complicated equipment and laboratories and also require professional researchers to work and operate the system (Idil et al., 2017; Saylan et al., 2019). Besides, it is difficult to perform monitoring of biomolecules in real time by these time-consuming and labor-intensive methods (He et al., 2019; Katsarou et al., 2019; Zhao et al., 2020). More platforms with superior detection abilities are needed urgently in medical applications. The progress of powerful, automated, cost-effective, and also highly sensitive and selective biomolecules detection platforms has been a challenge for scientists from different disciplines (He et al., 2020; Kim, Kim, Ahn, Lee, & Nam, 2020a; Kim, Lee, Min, Lee, & Kim, 2020b; Regasa, Lee, Min, Lee, & Kim, 2020). They combine various basic sciences including biology, chemistry, physics, bioengineering, and nanotechnology to maintain growth in risk management due to their distinctive characteristics sensing capabilities (Luo et al., 2019, 2020; Nawaz et al., 2020; Su et al., 2020; Tran et al., 2020). Nanosensors are analytical platforms that combine a sensing element with a transducer that can be optical (Dibekkaya et al., 2016), electrochemical (Rebelo, Pacheco, Cordeiro, Melo, & Delerue-Matos, 2020), piezoelectric (Bakhshpour et al., 2019), thermal (Wang et al., 2019), or magnetic (Huber et al., 2019). The interaction is formed between an analyte and a recognition molecule that translates into a measurable signal (Çalışır et al., 2020). Nanosensors have significant applications in different fields including environmental (Supraja et al., 2019), food analysis (Jia et al., 2020), drug delivery (Zaidi, 2020), medical (Mansuriya & Altintas, 2020), and diagnosis (Cui et al., 2020). Thus, sensitive surfaces are needed for the detection of an analyte in practical applications (Çimen et al., 2020). In this regard, nanosensors are suitable platforms that have various unique properties, such as selectivity, sensitivity, real-time, label-free, low-cost, and quantitative detection (Andryukov, Besednova, Romashko, Zaporozhets, & Efimov 2020; Asghar et al., 2016; Inci et al., 2020; Saylan, Yılmaz, Derazshamshir, Yılmaz, & Denizli, 2017; Sharifi et al., 2019).

This chapter aims to explain the most recent advances in nanosensors. Following the explanation of medical diagnosis needs (Section 9.2), a basic introduction to the nanosensors in diagnosis (Section 9.3) is explained briefly and a broad overview of electrochemical nanosensors (Section 9.3.1), optical nanosensors (Section 9.3.2), piezoelectric nanosensors (Section 9.3.3), and other types of nanosensors (Section 9.3.4) performances and characteristics are discussed for medical diagnosis applications. A discussion about future remarks is added to the end of the chapter (Section 9.4).

9.2 Medical diagnosis needs

Industrialization and modern technologies have generated serious health problems although they have created a more convenient life for people (Iravani, 2020). Unfortunately, significant amounts of toxic, carcinogenic, and mutagenic chemicals are annually produced and released into the environment (Saylan, Erdem, Ünal, & Denizli, 2019b), and then serious diseases can be generated by these materials (Ganapathe et al., 2020). In parallel with the development in conventional methods for the detection of target molecules, there is a still requirement for novel platforms that not only detect biomolecules in real time but additionally report their bioavailability and judge their impacts on humans (Özgür et al., 2020). As a forward step to fulfill this necessity, nanosensors that are based on the design and synthesis of the polymeric materials have been rapidly expanded (Sharifi et al., 2021; Wang et al., 2020). Especially, the COVID-19 outbreak highlights the real power of rapid medical diagnosis to manage the spread of disease and begin timely treatment, and the rapid transformative power of on-site results is evident in broad diagnostic applications this year (Adamson & Jeuken, 2020; Cassedy, Mullins, & O'Kennedy, 2020; Ehtesabi, 2020; Ravi, Cortade, Ng, & Wang, 2020; Chen et al., 2019). However, the need for quality and accurate results have been strengthened and robust support technology is needed to achieve sufficient precision, selectivity, and repeatability with fast result time and ease of use in all industries (Döhla et al., 2020; Garciá-Miranda Ferrari et al., 2020).

9.3 **Nanosensors for diagnosis**

Several diseases are important global common health issues in this century. A well-timed and precise medical diagnosis of a person is the main step for the treatment of the diseases (Saylan & Denizli, 2019). Moreover, the success of medical cures is also based on the primary detection and various studies have been directed to improve rapid, effective, and accurate methods (Koyun et al., 2019). They can be designed using the superiorities of nanosized materials (Büyüktiryaki et al., 2017). Thus the detecting technology has been employed for specific, sensitive, and fast diagnosis and also obtained huge attention within the interdisciplinary communities. The nanosensors can be engineered with a wide diversity relating to the operating principles of electrochemical, optical, or piezoelectric transducers (Arndt et al., 2020; Deng et al., 2020; Fu & Ma, 2020).

9.3.1 **Electrochemical nanosensors**

The fastest growing nanosensor type is the electrochemical nanosensor. Three electrodes—a counter, a reference, and a working electrode—are used in electrochemical nanosensors and these nanosensors measure the electrochemical change of the electrode interfaces (Assavapanumat et al., 2019; Beitollahi et al., 2020; Mohan et al., 2020). Electrochemical nanosensors have several benefits such as simplicity, being cost-friendly, and having extraordinary detectability compared to other nanosensors. The subclasses of electrochemical nanosensors, including potentiometric, conductometric, and voltammetric, are used in medical diagnosis (Pacheco et al., 2018; Dinu & Apetrei, 2020; Ou, Pan, Jow, Chen, & Ling, 2018; Ozcelikay et al., 2019; Semenova, Pinto, Koch, Gernaey, & Junicke, 2020; Zhang et al., 2020). For instance, Rashed et al. demonstrated an electrochemical nanosensor to detect SARS-CoV-2 antibodies employing a commercial impedance platform (Fig. 9.1). They first coated the receptor-binding domain of SARS-CoV-2 spike protein to a well plate that contained sensing electrodes and tested with anti-SARS-CoV-2 antibody CR3022 in different concentrations in less than 5 minutes. They also performed blind testing on six serum specimens received from COVID-19 and non-COVID-19 patients and observed that the nanosensor was capable of diversification in impedance experiments from a negative control for all samples. They compared the impedance values to the standard ELISA test and the results were consistent with a high correlation coefficient ($R^2 = 0.9$) (Rashed et al., 2021). Another virus detection example is Arshad et al.'s study. They prepared a molecularly imprinted polymer-based impedimetric nanosensor to detect dengue virus. They modified screen-printed carbon electrode with nanofibers and coated with dopamine while employing nonstructural protein 1 (NS1—a biomarker for dengue virus) at the polymerization step. They studied the electrochemical properties of the nanosensor by electrochemical impedance spectroscopy and cyclic voltammetry at each modification step. They obtained the results in the wide range (1−200 ng/mL) of NS1 concentration with a 0.3 ng/mL of a limit of detection value. Furthermore, they used the nanosensor for NS1 detection in serum samples and obtained satisfying recovery values (95%−97.14%) with less than 5% standard deviations (Arshad et al., 2020). Nyein et al. introduced microfluidic-based electrochemical sensing patches for real-time detection of glucose, Na^+, K^+, sweat rate, and also chemically induced sweat. They used this nanosensor

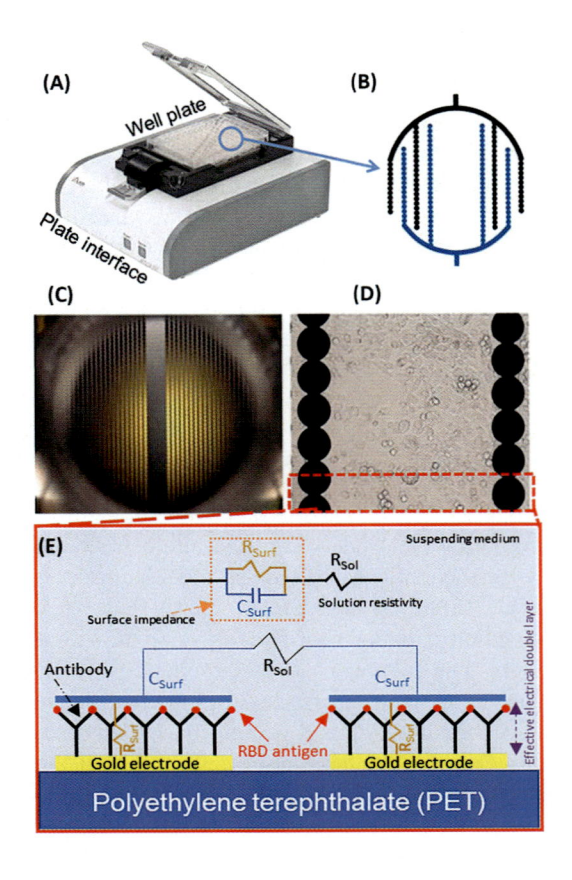

FIGURE 9.1

(A) Image of a commercial well-plate platform with (B) an electrode layout, (C) optic and (D) magnified images of the electrodes, (E) scheme of impedance equivalent circuit model (Rashed et al., 2021).

for examining the regional sweat composition, foreshowing whole-body fluid and loss of electrolytes while exercising, exposing relations of sweat metrics, and following glucose dynamics to discover correlations of sweat-to-blood in healthy and diabetic people. Following a sweat analysis, they concluded that this nanosensor is a critical platform for progressing sweat testing beyond the research for medical diagnosis (Nyein et al., 2019). Besides, Wei et al. developed a metal—organic frame-modified carbon hybrid button-based electrochemical nanosensor for the detection of glucose (Fig. 9.2). They performed kinetic studies in multiple complex mediums including serum, saliva, and urine to show stability, selectivity, and durability of nanosensors. They claimed that the nanosensor has several properties such as reliable nonenzymatic electrocatalysis, high environment tolerance, ease of production, and low cost which also has huge potential for on-site analysis in personalized diagnostic and disease prevention (Wei et al., 2020).

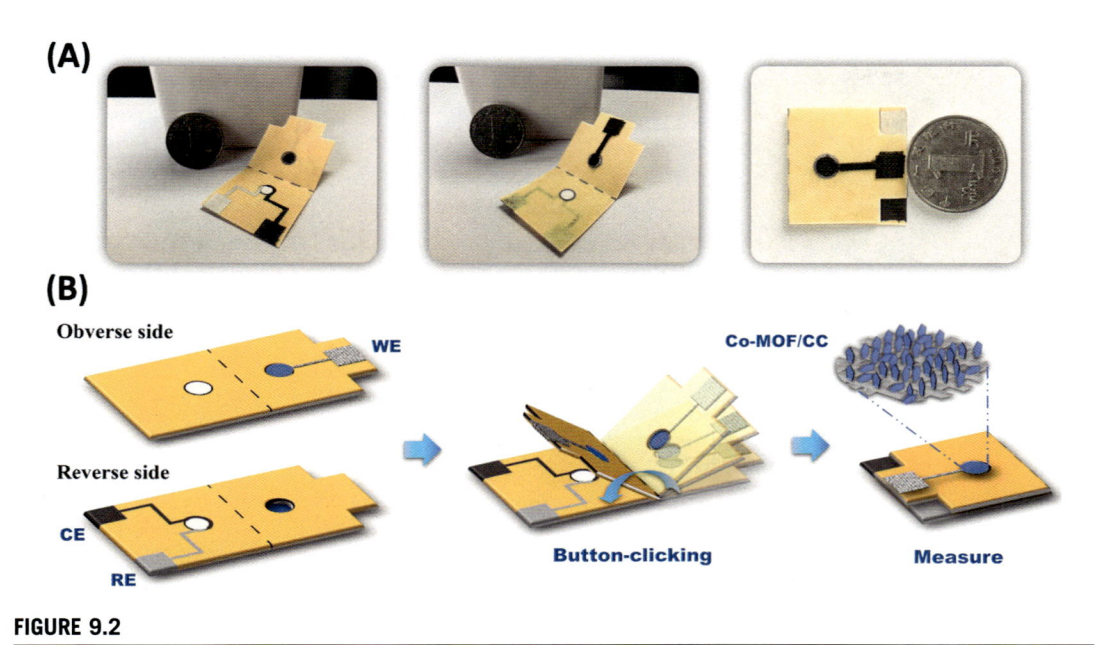

FIGURE 9.2

(A) Button-based and (B) three-dimensional scheme of the nanosensor (Wei et al., 2020).

9.3.2 Optical nanosensors

Optical nanosensors focalize the change of the optical features of the transducer measurements when the reaction occurs. There are several optical nanosensors such as evanescent wave fiber, optrode fiber, time-resolved fluorescence, surface plasmon resonance, resonant, and interferometric in literature and market (Ferreira et al., 2020; Li et al., 2020; Lin et al., 2019; Okazaki, Watanabe, & Kuramitz, 2020; Saylan, Erdem, Cihangir, & Denizli, 2019a; Zhao et al., 2020). These nanosensors use a diversity of sensing modules such as fluorescence, chemiluminescence, light absorption, reflectance, and scattering (Saylan et al., 2017; Sousa, Figueira, Costa, & Raposo, 2020; Van Sau, Ngo, Phan, Tran, & Nguyen, 2020; Yarman, Kurbanoglu, Zebger, & Scheller, 2021). For example, Mocenigo et al. demonstrated an optical nanosensor for antibody detection in serum samples. This nanosensor combined the synthetic peptide nucleic acids and strands related to recognition molecules and labels. They also characterized the nanosensor to detect trastuzumab (an antibody drug for breast cancer) and tumors overexpressing the protein in bodily fluids (Fig. 9.3). They compared the experimental results obtained from patients and ELISA results, which showed a good agreement, and concluded that the optical nanosensors can be used for a drug monitoring system for therapeutic drug management (Mocenigo et al., 2020).

Melentiev et al. showed the new approach for the determination of troponin-T with several advantages including being more rapid (1000-fold) than common approaches and single troponin-T imaging and detecting in real-time prepared in blood serum with a clinical sensitivity (1 pg/mL). They also claimed that this nanosensor can also be employed for monitoring and imaging other biomolecules such as viruses and bacteria in real time. Malinick et al. reported a detection method for the determination of antibodies for multiple sclerosis against gangliosides via surface plasmon resonance imaging

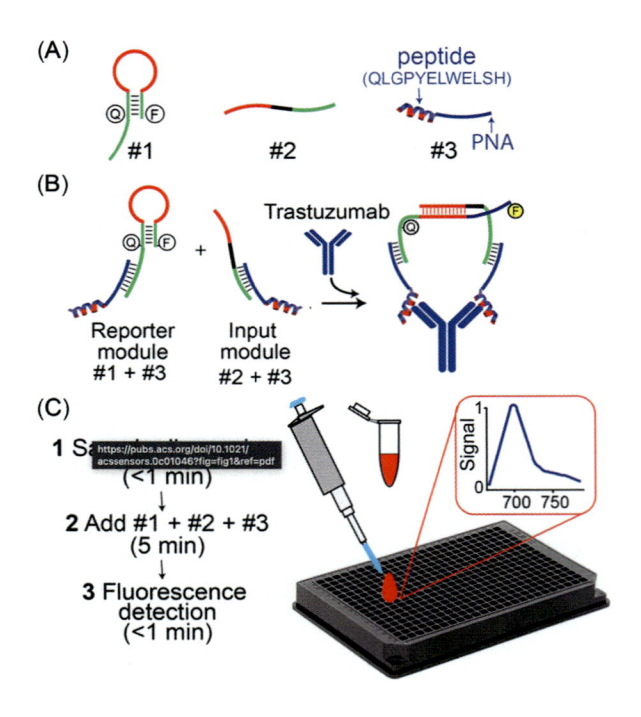

FIGURE 9.3

(A) Detection parts, (B) antibody binds to reporter and input module induces the colocalization, and (C) fluorescent signal of the antibody detection (Mocenigo et al., 2020).

with a microarray combination. They fabricated the ganglioside array and then coated it with a chemical layer for antigen attachment to obtain pseudo-myelin sheath. They performed the detection of antibodies (anti-GT1b, anti-GM1, and anti-GA1) in serum in different concentrations (1−100 ng/ mL). This nanosensor was used for surface functionalization and enabled direct multiple sclerosis biomarker detection that offers an important alternative for improved patient care and assessment (Malinick et al., 2020). Wang et al. designed a nanosensor for rhodamine 6 G detection using the surface-enhanced Raman scattering method by combining a molecularly imprinted polymer. As depicted in Fig. 9.4, they fabricated the gold array at an oil−water interface and then fixed between imprinted polymers and a support layer to obtain a uniformly distributed hot spot. They demonstrated the high specificity of a nanosensor for rhodamine 6G detection via comparing it with other structural analogs and also the high reproducibility of the nanosensor with a stability of the structure to acquire a recyclable surface-enhanced Raman scattering nanosensor (Wang et al., 2020).

9.3.3 Piezoelectric nanosensor

Piezoelectric nanomaterials produce electricity and have been used widely in different devices such as sensors, actuators, and transducers. Piezoelectric nanosensors are also one of the most attractive physical transducers owing to their basicity, low costs, label-free and real-time detection with high sensitivity. Moreover, quartz crystal microbalance nanosensors are the most well-known in the

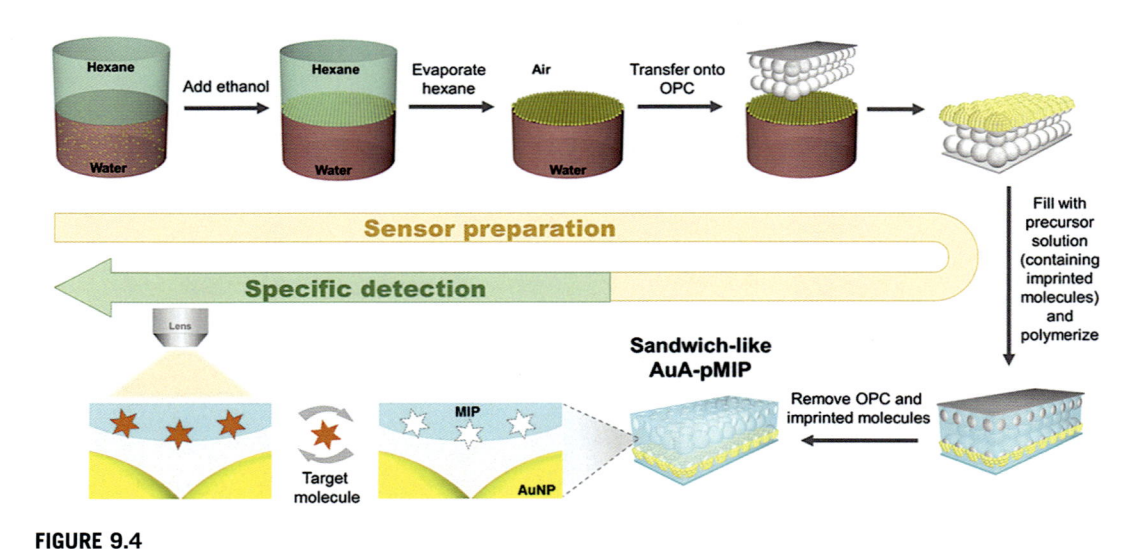

FIGURE 9.4

Scheme of the imprinted nanosensor preparation and specific detection (Wang et al., 2020).

piezoelectric class as practical and convenient sensing tools to measure changes of small mass and frequency on the nanosensor surface. They have been employed for various analytes detection in different matrix (Dong et al., 2020; Inci et al., 2015; Moghadam et al., 2020; Sönmezler et al., 2019; Sriplai et al., 2020). For instance, Hampitak et al. presented a piezoelectric nanosensor that was modified with a graphene layer to measure antibodies in undiluted patient samples. They demonstrated the nanosensor's impact for an antibody toward the phospholipase A2 receptor (a biomarker in primary membranous nephropathy). They constructed the chemical layers via adsorbing denatured bovine serum albumin even in a low concentration on the reduced nanosensor surface and obtained a high specificity and sensitivity that is a better detection limit (100 ng/mL) parallel to the commercial system. They also compared the results with enzyme-linked immunosorbent assay to validate the applicability of usage in medical applications (Hampitak et al., 2020). Chunta et al. generated a quartz crystal microbalance nanosensor using molecularly imprinted polymer to detect actual very-low-density lipoprotein (Fig. 9.5). They used this imprinted nanosensor in a range from 2.5 to 100 mg/dL of very-low-density lipoprotein with a low limit of detection (1.5 mg/dL). They also achieved high recovery values (96%−103%) when the nanosensor was used for very-low-density lipoprotein evaluation at 38−71 mg/dL of concentrations. Furthermore, the nanosensor exhibited a high cross-selectivity with other analogs in which 6%−7% of low-density lipoprotein, 2%−4% high-density lipoprotein, and 1% chylomicrons compared to very-low-density lipoprotein signals (Chunta et al., 2021).

Scarpa et al. suggested a flexible piezoelectric nanosensor to sense sweat pH. They fabricated the nanosensor by using the membranes on a polyimide integrated with the pH-responsive hydrogel and recorded the frequency shift owing to the swelling and shrinking properties of the hydrogel at different pH. The nanosensor showed a responsivity of around 12 kHz/pH in artificial sweat at pH 3−8. They reported that this hydrogel-based nanosensor was the first one that sensed via a flexible resonator, fostering the growth of wearable devices (Scarpa et al., 2020). Curry et al. presented a

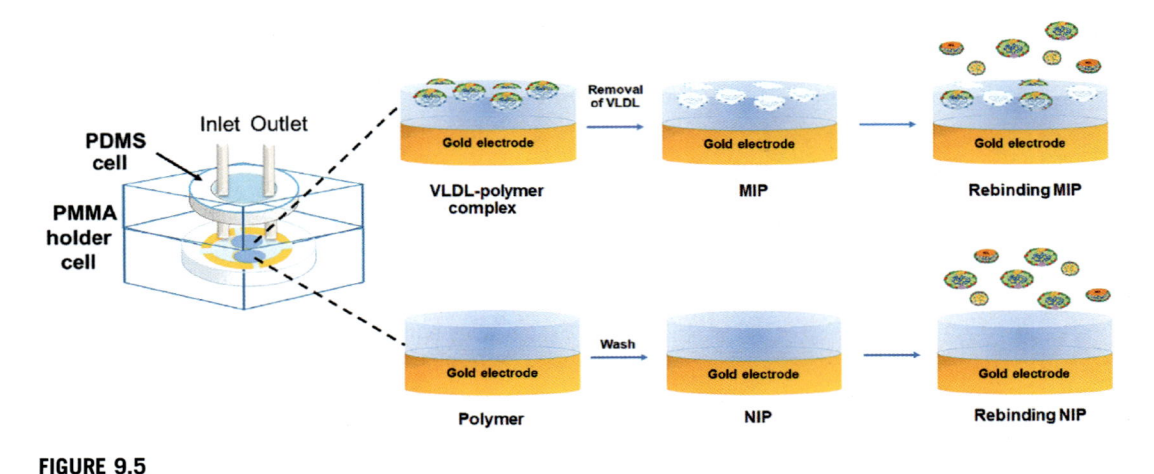

FIGURE 9.5

Scheme of the imprinted and nonimprinted piezoelectric nanosensors preparation (Chunta et al., 2021).

novel method for nanomaterial preparation, device mounting, and electronic combination to produce biocompatible and biodegradable nanofibers to obtain highly stable and efficient performance, and also demonstrated different nanosensor applications such as a sensitive biodegradable pressure for monitoring and ultrasonic transducer for the opening of the blood—brain barrier to simplify the delivery of drugs into the brain (Curry et al., 2020).

9.3.4 Other nanosensors

Nanosensors have been used due to their portability, automated data obtaining, advanced sensitivity, and high-throughput capabilities in the last two decades. The research in the field of nanosensor development has exploded for allergens, bacteria, enzymes, proteins, toxins, and virus detection. Nanosensors are progressively being improved for clinically important biomolecule detection and also diagnosis for medical applications. Many recognition molecules including antibody, aptamer, enzymes, bacteriophages, molecularly imprinted polymers, nucleic acid sequence, and peptide nucleic acids are used for this aim (Menard-Moyon, Bianco, & Kalantar-Zadeh, 2020; Demeke Teklemariam et al., 2020). For instance, Ng et al. evaluated the potential detection platform for biomarkers of cirrhotic protein (intercellular adhesion molecule-1 and mac-2 binding protein glycosylation isomer) using a magnetoresistive nanosensor. The capture probes images of WFA, sICAM-1 and CRP, TNF-α, NF-κB, IL-6 for high- and low-abundance are shown on the top left and right in Fig. 9.6, respectively. They also spotted the positive control of biotin-bovine serum albumin and negative control of bovine serum albumin at the bottom. In a single-use, they discarded the nanosensor following step 3. They also washed the nanosensor after step 3 before continuing onto steps 4 and 5 in a double-use. They employed the combination of biomarkers to show high performance for diagnosis in logistic regression and random forest models (Ng et al., 2020).

Another example is the wearable nanosensors. Lee et al. designed an ultrathin and highly elastic wearable cardiac nanosensor with low cost, capable of wireless energy harvesting, water-resistant, and lightweight (Fig. 9.7). The advantages of the elastic electronics of the wearable cardiac

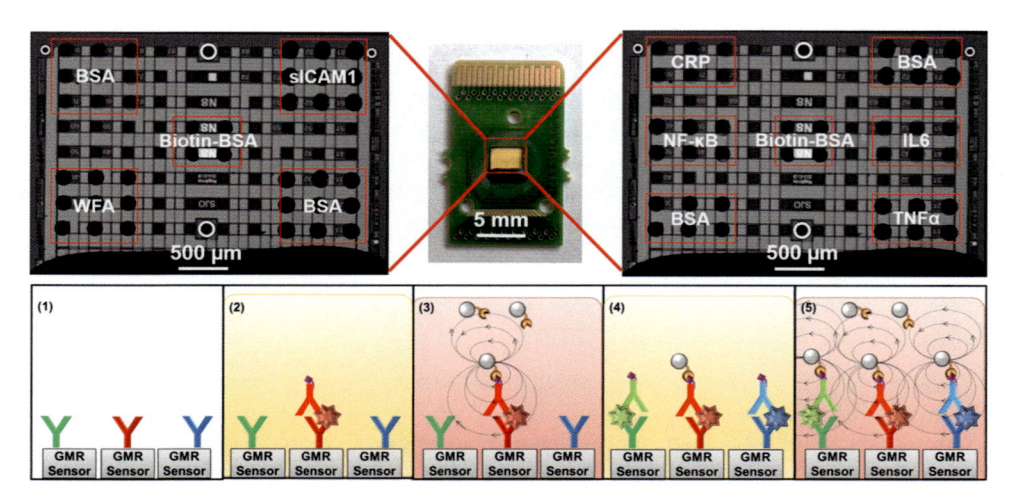

FIGURE 9.6

Magnetoresistive nanosensor for the cirrhosis protein biomarker detection (Ng et al., 2020).

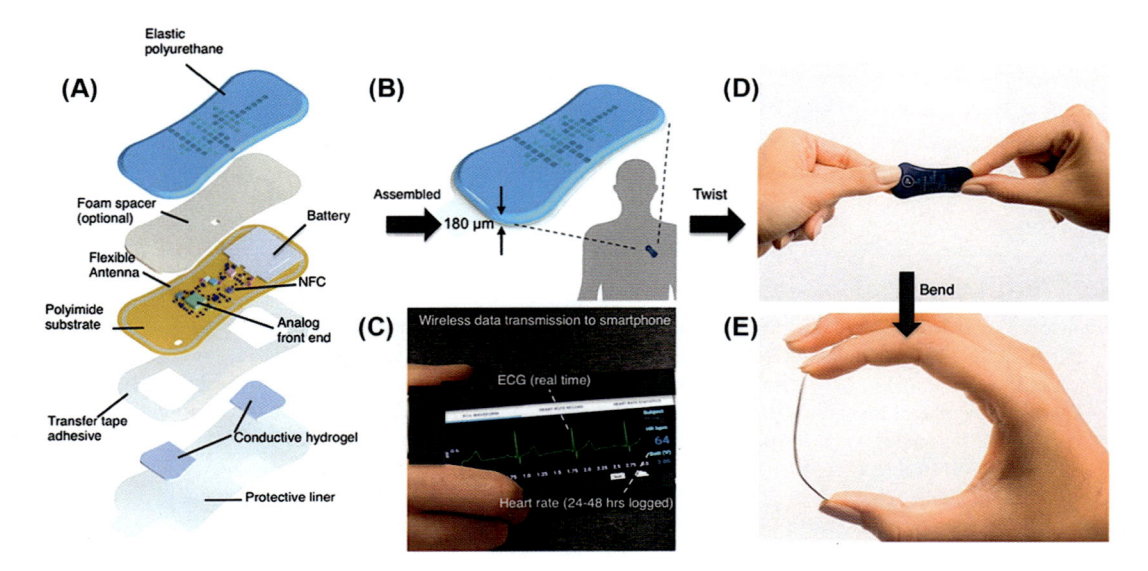

FIGURE 9.7

(A) Exploded image of nanosensor with multiple adhesive, electronic, hydrogel, and polymeric layers; (B) scheme of assembled nanosensor consisting of five distinct layers attached to the torso; (C) results are wirelessly transmitted to a smartphone for scanning of logged heart rate and ECG waveforms, and transmitted from smartphone to a cloud; and (D, E) nanosensor in deformations of mechanical twist and bend (Lee et al., 2018).

nanosensor's, soft encapsulation layers, and adhesives have been demonstrated by theoretical analysis of system-level bending mechanics. It also shows enabling intimate skin coupling. They conducted a clinical feasibility work to apply it to atrial fibrillation patients. They also reported that the wearable cardiac nanosensor's measured cardiac signals matched to the Holter monitor and it was more comfortable. It was concluded that the physical properties and performance of the wearable cardiac nanosensor's show its benefit for monitoring cardiac signals during sleep, daily activity, and exercise with advice for home-based care (Lee et al., 2018).

9.4 Conclusion and vision for future

Recent advances in nanotechnology are supplying nanofabricated tools that are sensitive, small, and cheap for biological molecule analysis. Nanosensor technology has made a great progress in the last years. Nanosensors are phenomenon devices that use biological or chemical interactions for the detection of biomolecules. Nanosensors contain a recognition element that interacts with the target molecule and a physical transducer that transforms the recognition interaction into a practical electrical signal. Transducer elements, including optical, piezoelectric, or electrochemical tools, generate frequency signals, light, or current, respectively. The nanosensors have several key properties such as direct analysis, real-time and label-free electrical signal transduction, high sensitivity, perfect selectivity, and potential for the combination of addressable arrays on a massive scale. The widespread use of nanosensors has been published in the sensing of various biomolecules. The examples described in this chapter illustrate simple and novel nanosensors for the treatment and diagnosis in biomedical applications. The nanosensor surface can be fabricated with polymer thin film or nanoparticles, which present perfect sensitivity, selectivity, reusability, and reliability. It appears probable that there will be numerous applications of nanosensors for the detection of biomarkers, proteins, cells, viruses, and bacteria. As a result, the expansion of smart nanosensors will have a promising future.

References

Adamson, H., & Jeuken, L. J. C. (2020). Engineering protein switches for rapid diagnostic tests. *ACS Sensors*, *5*(10), 3001−3012. Available from https://doi.org/10.1021/acssensors.0c01831.

Akgönüllü, S., Battal, D., Yalcin, M. S., Yavuz, H., & Denizli, A. (2020). Rapid and sensitive detection of synthetic cannabinoids JWH-018, JWH-073 and their metabolites using molecularly imprinted polymer-coated QCM nanosensor in artificial saliva. *Microchemical Journal*, *153*, 104454. Available from https://doi.org/10.1016/j.microc.2019.104454.

Akgönüllü, S., Yavuz, H., & Denizli, A. (2020). SPR nanosensor based on molecularly imprinted polymer film with gold nanoparticles for sensitive detection of aflatoxin B1. *Talanta*, *219*, 1212219. Available from https://doi.org/10.1016/j.talanta.2020.121219.

Alvau, M. D., Tartaggia, S., Meneghello, A., Casetta, B., Calia, G., Serra, P. A., Polo, F., & Toffoli, G. (2018). Enzyme-based electrochemical biosensor for therapeutic drug monitoring of anticancer drug irinotecan. *Analytical Chemistry*, *90*(10), 6012−6019. Available from https://doi.org/10.1021/acs.analchem.7b04357.

Andryukov, B. G., Besednova, N. N., Romashko, R. V., Zaporozhets, T. S., & Efimov, T. A. (2020). Label-free biosensors for laboratory-based diagnostics of infections: Current achievements and new trends. *Biosensors*, *10*(2). Available from https://doi.org/10.3390/bios10020011.

Arndt, N., Tran, H. D. N., Zhang, R., Xu, Z. P., & Ta, H. T. (2020). Different approaches to develop nanosensors for diagnosis of diseases. *Advanced Science*, *7*(24), 2001476. Available from https://doi.org/10.1002/advs.202001476.

Arshad, R., Rhouati, A., Hayat, A., Nawaz, M. H., Yameen, M. A., Mujahid, A., & Latif, U. (2020). MIP-based impedimetric sensor for detecting dengue fever biomarker. *Applied Biochemistry and Biotechnology*, *191*(4), 1384−1394. Available from https://doi.org/10.1007/s12010-020-03285-y.

Asghar, W., Yuksekkaya, M., Shafiee, H., Zhang, M., Ozen, M. O., Inci, F., . . . Demirci, U. (2016). Engineering long shelf life multi-layer biologically active surfaces on microfluidic devices for point of care applications. *Scientific Reports*, 6. Available from https://doi.org/10.1038/srep21163.

Assavapanumat, S., Ketkaew, M., Kuhn, A., & Wattanakit, C. (2019). Synthesis, characterization, and electrochemical applications of chiral imprinted mesoporous Ni surfaces. *Journal of the American Chemical Society*, *141*(47), 18870−18876. Available from https://doi.org/10.1021/jacs.9b10507.

Bakhshpour, M., Piskin, A. K., Yavuz, H., & Denizli, A. (2019). Quartz crystal microbalance biosensor for label-free MDA MB 231 cancer cell detection via notch-4 receptor. *Talanta*, *204*, 840−845. Available from https://doi.org/10.1016/j.talanta.2019.06.060.

Battal, D., Akgönüllü, S., Yalcin, M. S., Yavuz, H., & Denizli, A. (2018). Molecularly imprinted polymer based quartz crystal microbalance sensor system for sensitive and label-free detection of synthetic cannabinoids in urine. *Biosensors and Bioelectronics*, *111*, 10−17. Available from https://doi.org/10.1016/j.bios.2018.03.055.

Beitollahi, H., Tajik, S., Garkani Nejad, F., & Safaei, M. (2020). Recent advances in ZnO nanostructure-based electrochemical sensors and biosensors. *Journal of Materials Chemistry B*, *8*(27), 5826−5844. Available from https://doi.org/10.1039/d0tb00569j.

Büyüktiryaki, S., Say, R., Denizli, A., & Ersöz, A. (2017). Phosphoserine imprinted nanosensor for detection of cancer antigen 125. *Talanta*, *167*, 172−180. Available from https://doi.org/10.1016/j.talanta.2017.01.093.

Cassedy, A., Mullins, E., & O'Kennedy, R. (2020). Sowing seeds for the future: The need for on-site plant diagnostics. *Biotechnology Advances*, *39*. Available from https://doi.org/10.1016/j.biotechadv.2019.02.014.

Çalışır, M., Bakhshpour, M., Yavuz, H., & Denizli, A. (2020). HbA1c detection via high-sensitive boronate based surface plasmon resonance sensor. *Sensors and Actuators, B: Chemical*, *306*, 125712. Available from https://doi.org/10.1016/j.snb.2019.127561.

Menard-Moyon, C., Bianco, A., & Kalantar-Zadeh, K. (2020). Two-dimensional material-based biosensors for virus detection. *ACS Sensors*, *5*(12), 3739−3769. Available from https://doi.org/10.1021/acssensors.0c01961.

Chen, H., Liu, K., Li, Z., & Wang, P. (2019). Point of care testing for infectious diseases. *Clinica Chimica Acta*, *493*, 138−147. Available from https://doi.org/10.1016/j.cc.2019.03.008.

Chen, X. X., Lin, Z. Z., Hong, C. Y., Yao, Q. H., & Huang, Z. Y. (2020). A dichromatic label-free aptasensor for sulfadimethoxine detection in fish and water based on AuNPs color and fluorescent dyeing of double-stranded DNA with SYBR Green I. *Food Chemistry*, *309*, 125712. Available from https://doi.org/10.1016/j.foodchem.2019.125712.

Chunta, S., Boonsriwong, W., Wattanasin, P., Naklua, W., & Lieberzeit, P. A. (2021). Direct assessment of very-low-density lipoprotein by mass sensitive sensor with molecularly imprinted polymers. *Talanta*, *221*, 121549. Available from https://doi.org/10.1016/j.talanta.2020.121549.

Çimen, D., Bereli, N., Günaydın, S., & Denizli, A. (2020). Detection of cardiac troponin-1 by optic biosensors with immobilized anti-cardiac troponin-I monoclonal antibody. *Talanta*, *219*, 121259. Available from https://doi.org/10.1016/j.talanta.2020.121259.

Cui, F., Zhou, Z., & Zhou, H. S. (2020). Molecularly imprinted polymers and surface imprinted polymers based electrochemical biosensor for infectious diseases. *Sensors (Switzerland)*, *20*(4), 996. Available from https://doi.org/10.3390/s20040996.

Curry, E. J., Le, T. T., Das, R., Ke, K., Santorella, E. M., Paul, D., Chorsi, M. T., Tran, K. T. M., Baroody, J., Borges, E. R., Ko, B., Golabchi, A., Xin, X., Rowe, D., Yue, L., Feng, J., Daniela Morales-Acosta, M., Wu, Q., Chen, I. P., . . . Nguyen, T. D. (2020). Biodegradable nanofiber-based piezoelectric transducer. *Proceedings of the National Academy of Sciences of the United States of America*, *117*(1), 214−220. Available from https://doi.org/10.1073/pnas.1910343117.

Demeke Teklemariam, A., Samaddar, M., Alharbi, M. G., Al-Hindi, R. R., & Bhunia, A. K. (2020). Biosensor and molecular-based methods for the detection of human coronaviruses: A review. *Molecular and Cellular Probes*, *54*, 101662. Available from https://doi.org/10.1016/j.mcp.2020.101662.

Deng, J., Zhao, S., Liu, Y., Liu, C., & Sun, J. (2020). Nanosensors for diagnosis of infectious diseases. *ACS Applied Bio Materials*. Available from https://doi.org/10.1021/acsabm.0c01247.

Dibekkaya, H., Saylan, Y., Yılmaz, F., Derazshamshir, A., & Denizli, A. (2016). Surface plasmon resonance sensors for real-time detection of cyclic citrullinated peptide antibodies. *Journal of Macromolecular Science, Part A: Pure and Applied Chemistry*, *53*(9), 585−594. Available from https://doi.org/10.1080/10601325.2016.1201756.

Dinu, A., & Apetrei, C. (2020). Voltammetric determination of phenylalanine using chemically modified screen-printed based sensors. *Chemosensors*, *8*(4), 1−18. Available from https://doi.org/10.3390/chemosensors8040113.

Döhla, M., Boesecke, C., Schulte, B., Diegmann, C., Sib, E., Richter, E., Eschbach-Bludau, M., Aldabbagh, S., Marx, B., Eis-Hübinger, A.-M., Schmithausen, R. M., & Streeck, H. (2020). Rapid point-of-care testing for SARS-CoV-2 in a community screening setting shows low sensitivity. *Public Health*, *182*, 170−172. Available from https://doi.org/10.1016/j.puhe.2020.04.009.

Dong, L., Jin, C., Closson, A. B., Trase, I., Richards, H. R., Chen, Z., & Zhang, J. X. J. (2020). Cardiac energy harvesting and sensing based on piezoelectric and triboelectric designs. *Nano Energy*, *76*. Available from https://doi.org/10.1016/j.nanoen.2020.105076.

Ehtesabi, H. (2020). Application of carbon nanomaterials in human virus detection. *Journal of Science: Advanced Materials and Devices*, *5*(4), 436−450. Available from https://doi.org/10.1016/j.jsamd.2020.09.005.

Erdem, Ö., Cihangir, N., Saylan, Y., & Denizli, A. (2020). Comparison of molecularly imprinted plasmonic nanosensor performances for bacteriophage detection. *New Journal of Chemistry*, *44*(41), 17654−17663. Available from https://doi.org/10.1039/d0nj04053c.

Esentürk, M. K., Akgönüllü, S., Yılmaz, F., & Denizli, A. (2019). Molecularly imprinted based surface plasmon resonance nanosensors for microalbumin detection. *Journal of Biomaterials Science, Polymer Edition*, *30*(8), 646−661. Available from https://doi.org/10.1080/09205063.2019.1600181.

Fatma, Y., Yeşeren, S., Semra, A., Duygu, Ç., Ali, D., Nilay, B., & Adil, D. (2017). Surface plasmon resonance based nanosensors for detection of triazinic pesticides in agricultural foods. In *New pesticides and soil sensors nanotechnology in the agri-food industry Vol. 10* (pp. 679−718). Elsevier BV. Available from https://doi.org/10.1016/b978-0-12-804299-1.00019-9

Ferreira, F., Luxardi, G., Reid, B., Ma, L., Raghunathan, V. K., & Zhao, M. (2020). Real-time physiological measurements of oxygen using a non-invasive self-referencing optical fiber microsensor. *Nature Protocols*, *15*(2), 207−235. Available from https://doi.org/10.1038/s41596-019-0231-x.

Fu, Y., & Ma, Q. (2020). Recent developments in electrochemiluminescence nanosensors for cancer diagnosis applications. *Nanoscale*, *12*(26), 13879−13898. Available from https://doi.org/10.1039/d0nr02844d.

Ganapathe, L. S., Mohamed, M. A., Yunus, R. M., & Berhanuddin, D. D. (2020). Magnetite (Fe_3O_4) nanoparticles in biomedical application: From synthesis to surface functionalisation. *Magnetochemistry*, *6*(4), 1−35. Available from https://doi.org/10.3390/magnetochemistry6040068.

Garciá-Miranda Ferrari, A., Carrington, P., Rowley-Neale, S. J., & Banks, C. E. (2020). Recent advances in portable heavy metal electrochemical sensing platforms. *Environmental Science: Water Research and Technology*, *6*(10), 2676−2690. Available from https://doi.org/10.1039/d0ew00407c.

Hampitak, P., Jowitt, T. A., Melendrez, D., Fresquet, M., Hamilton, P., Iliut, M., Nie, K., Spencer, B., Lennon, R., & Vijayaraghavan, A. (2020). A point-of-care immunosensor based on a quartz crystal microbalance with graphene biointerface for antibody assay. *ACS Sensors*, *5*(11), 3520−3532. Available from https://doi.org/10.1021/acssensors.0c01641.

Hao, J. J., Xie, X., Gu, K. D., Du, W. C., Liu, Y. J., & Yang, H. W. (2019). Research on photonic crystal−based biosensor for detection of *Escherichia coli* colony. *Plasmonics*, *14*(6), 1919−1928. Available from https://doi.org/10.1007/s11468-019-00987-w.

He, C., Ledezma, U. H., Gurnani, P., Albelha, T., Thurecht, K. J., Correia, R., Morgan, S. P., Patel, P., Alexander, C., & Korposh, S. (2020). Surface polymer imprinted optical fibre sensor for dose detection of dabrafenib. *Analyst*, *145*(13), 4504−4511. Available from https://doi.org/10.1039/d0an00434k.

He, C., Ledezma, U. H., Gurnani, P., Albelha, T., Thurecht, K. J., Correia, R., . . . Korposh, S. (2020). Surface polymer imprinted optical fibre sensor for dose detection of dabrafenib. *Analyst*, *145*(13), 4504−4511. Available from https://doi.org/10.1039/d0an00434k.

He, Y., Yang, T., Mo, H., Chen, T., Feng, J., & Zhang, W. (2019). Low-cost potentiometric sensor based on a molecularly imprinted polymer for the rapid determination of matrine in herbal medicines. *Instrumentation Science and Technology*, *47*(6), 581−596. Available from https://doi.org/10.1080/10739149.2019.1609027.

Heravizadeh, O. R., Khadem, M., Dehghani, F., & Shahtaheri, S. J. (2020). Determination of fenthion in urine samples using molecularly imprinted nanoparticles: Modelling and optimisation by response surface methodology. *International Journal of Environmental Analytical Chemistry*. Available from https://doi.org/10.1080/03067319.2020.1808630.

Hossain, M. Z., McCormick, S. P., & Maragos, C. M. (2018). An imaging surface plasmon resonance biosensor assay for the detection of t-2 toxin and masked t-2 toxin-3-glucoside in wheat. *Toxins*, *10*(3). Available from https://doi.org/10.3390/toxins10030119.

Huber, S., Min, C., Staat, C., Oh, J., Castro, C. M., Haase, A., Weissleder, R., Gleich, B., & Lee, H. (2019). Multichannel digital heteronuclear magnetic resonance biosensor. *Biosensors and Bioelectronics*, *126*, 240−248. Available from https://doi.org/10.1016/j.bios.2018.10.052.

Idil, N., Hedström, M., Denizli, A., & Mattiasson, B. (2017). Whole cell based microcontact imprinted capacitive biosensor for the detection of *Escherichia coli*. *Biosensors and Bioelectronics*, *87*, 807−815. Available from https://doi.org/10.1016/j.bios.2016.08.096.

Inci, F., Celik, U., Turken, B., Özer, H. Ö., & Kok, F. N. (2015). Construction of P-glycoprotein incorporated tethered lipid bilayer membranes. *Biochemistry and Biophysics Reports*, *2*, 115−122. Available from https://doi.org/10.1016/j.bbrep.2015.05.012.

Inci, F., Saylan, Y., Kojouri, A. M., Ogut, M. G., Denizli, A., & Demirci, U. (2020). A disposable microfluidic-integrated hand-held plasmonic platform for protein detection. *Applied Materials Today*, *18*, 100478. Available from https://doi.org/10.1016/j.apmt.2019.100478.

Inci, F., Karaaslan, M. G., Mataji-Kojouri, A., Shah, P. A., Saylan, Y., Zeng, Y., . . . Demirci, U. (2020). Enhancing the nanoplasmonic signal by a nanoparticle sandwiching strategy to detect viruses. *Applied Materials Today*, *20*. Available from https://doi.org/10.1016/j.apmt.2020.100709.

Jia, M., Zhongbo, E., Zhai, F., & Bing, X. (2020). Rapid multi-residue detection methods for pesticides and veterinary drugs. *Molecules*, *25*(16), 3590. Available from https://doi.org/10.3390/molecules25163590.

Inci, F., Saylan, Y., Kojouri, A. M., Ogut, M. G., Denizli, A., & Demirci, U. (2020). A disposable microfluidic-integrated hand-held plasmonic platform for protein detection. *Applied Materials Today*, *18*. Available from https://doi.org/10.1016/j.apmt.2019.100478.

Irivani, S. (2020). Nano- and biosensors for the detection of SARS-CoV-2: Challenges and opportunities. *Materials Advances*, 3092−3103. Available from https://doi.org/10.1039/d0ma00702a.

Katsarou, K., Bardani, E., Kallemi, P., & Kalantidis, K. (2019). Viral detection: Past, present, and future. *BioEssays*, *41*(10), 1900049. Available from https://doi.org/10.1002/bies.201900049.

Khan, N. I., Maddaus, A. G., & Song, E. (2018). A low-cost inkjet-printed aptamer-based electrochemical biosensor for the selective detection of lysozyme. *Biosensors*, *8*(1), 7. Available from https://doi.org/10.3390/bios8010007.

Kim, J., Kim, S., Ahn, J., Lee, J. J., & Nam, J. M. (2020a). A lipid-nanopillar-array-based immunosorbent assay. *Advanced Materials*, *32*(26), 2001360. Available from https://doi.org/10.1002/adma.202001360.

Kim, J. B., Lee, S. Y., Min, N. G., Lee, S. Y., & Kim, S. H. (2020b). Plasmonic Janus microspheres created from pickering emulsion drops. *Advanced Materials*, *32*(26), 2001384. Available from https://doi.org/10.1002/adma.202001384.

Koyun, S., Akgönüllü, S., Yavuz, H., Erdem, A., & Denizli, A. (2019). Surface plasmon resonance aptasensor for detection of human activated protein C. *Talanta*, *194*, 528−533. Available from https://doi.org/10.1016/j.talanta.2018.10.007.

Lafuente, M., Pellejero, I., Sebastián, V., Urbiztondo, M. A., Mallada, R., Pina, M. P., & Santamaría, J. (2018). Highly sensitive SERS quantification of organophosphorous chemical warfare agents: A major step towards the real time sensing in the gas phase. *Sensors and Actuators, B: Chemical*, *267*, 457−466. Available from https://doi.org/10.1016/j.snb.2018.04.058.

Landete, J. M., Langa, S., Escudero, C., Peirotén, Á., & Arqués, J. L. (2020). Fluorescent detection of nisin by genetically modified *Lactococcus lactis* strains in milk and a colonic model: Application of whole-cell nisin biosensors. *Journal of Bioscience and Bioengineering*, *129*(4), 435−440. Available from https://doi.org/10.1016/j.jbiosc.2019.10.011.

Lee, S. P., Ha, G., Wright, D. E., Ma, Y., Sen-Gupta, E., Haubrich, N. R., Branche, P. C., Li, W., Huppert, G. L., Johnson, M., Mutlu, H. B., Li, K., Sheth, N., Wright, J. A., Jr., Huang, Y., Mansour, M., Rogers, J. A., & Ghaffari, R. (2018). Highly flexible, wearable, and disposable cardiac biosensors for remote and ambulatory monitoring. *Npj Digital Medicine*, *1*. Available from https://doi.org/10.1038/s41746-017-0009-x.

Li, S., Li, Y., Liu, X., Li, X., Ding, T., & Ouyang, H. (2020). An in-situ electroplating fabricated fabry-perot interferometric sensor and its temperature sensing characteristics. *Coatings*, *10*(12), 1−13. Available from https://doi.org/10.3390/coatings10121174.

Li, S., Li, Y., Liu, X., Li, X., Ding, T., & Ouyang, H. (2020). An in-situ electroplating fabricated fabry-perot interferometric sensor and its temperature sensing characteristics. *Coatings*, *10*(12), 1−13. Available from https://doi.org/10.3390/coatings10121174.

Li, T., Jin, L., Feng, K., Yang, T., Yue, X., Wu, B., Ding, S., Liang, X., Huang, G., & Zhang, J. (2020). A novel low-field NMR biosensor based on dendritic superparamagnetic iron oxide nanoparticles for the rapid detection of Salmonella in milk. *LWT*, *133*, 110149. Available from https://doi.org/10.1016/j.lwt.2020.110149.

Lin, Z. Y., Qu, Z. B., Chen, Z. H., Han, X. Y., Deng, L. X., Luo, Q., ... Zhang, M. (2019). The marriage of protein and lanthanide: Unveiling a time-resolved fluorescence sensor array regulated by pH toward high-throughput assay of metal ions in biofluids. *Analytical Chemistry*, *91*(17), 11170−11177. Available from https://doi.org/10.1021/acs.analchem.9b01879.

Luo, L., Zhang, F., Chen, C., & Cai, C. (2019). Visual simultaneous detection of hepatitis A and B viruses based on a multifunctional molecularly imprinted fluorescence sensor. *Analytical Chemistry*, *91*(24), 15748−15756. Available from https://doi.org/10.1021/acs.analchem.9b04001.

Luo, Y., Liu, F., Li, E., Fang, Y., Zhao, G., Dai, X., Li, J., Wang, B., Xu, M., Liao, B., & Sun, G. (2020). FRET-based fluorescent nanoprobe platform for sorting of active microorganisms by functional properties. *Biosensors and Bioelectronics*, *148*, 111832. Available from https://doi.org/10.1016/j.bios.2019.111832.

Lvova, L. (2020). toxin detection. Chemosensors/toxin detection. *Chemosensors*, *8*(1). Available from https://doi.org/10.3390/chemosensors8010014.

Malinick, A. S., Lambert, A. S., Stuart, D. D., Li, B., Puente, E., & Cheng, Q. (2020). Detection of multiple sclerosis biomarkers in serum by ganglioside microarrays and surface plasmon resonance imaging. *ACS Sensors*, *5*(11), 3617−3626. Available from https://doi.org/10.1021/acssensors.0c01935.

Mansuriya, B. D., & Altintas, Z. (2020). Graphene quantum dot-based electrochemical immunosensors for biomedical applications. *Materials*, *13*(1), 96. Available from https://doi.org/10.3390/ma13010096.

Mocenigo, M., Porchetta, A., Rossetti, M., Brass, E., Tonini, L., Puzzi, L., Tagliabue, E., Triulzi, T., Marini, B., Ricci, F., & Ippodrino, R. (2020). Rapid, cost-effective peptide/nucleic acid-based platform for therapeutic antibody monitoring in clinical samples. *ACS Sensors*, *5*(10), 3109−3115. Available from https://doi.org/10.1021/acssensors.0c01046.

Moghadam, B. H., Hasanzadeh, M., & Simchi, A. (2020). Self-powered wearable piezoelectric sensors based on polymer nanofiber-metal-organic framework nanoparticle composites for arterial pulse monitoring. *ACS Applied Nano Materials*, *3*(9), 8742−8752. Available from https://doi.org/10.1021/acsanm.0c01551.

Mohan, A. M. V., Rajendran, V., Mishra, R. K., & Jayaraman, M. (2020). Recent advances and perspectives in sweat based wearable electrochemical sensors. *TrAC - Trends in Analytical Chemistry*, *131*, 116024. Available from https://doi.org/10.1016/j.trac.2020.116024.

Nawaz, T., Ahmad, M., Yu, J., Wang, S., & Wei, T. (2020). The biomimetic detection of progesterone by novel bifunctional group monomer based molecularly imprinted polymers prepared in UV light. *New Journal of Chemistry*, *44*(17), 6992−7000. Available from https://doi.org/10.1039/c9nj06387k.

Ng, E., Le, A. K., Nguyen, M. H., & Wang, S. X. (2020). Early multiplexed detection of cirrhosis using giant magnetoresistive biosensors with protein biomarkers. *ACS Sensors*, *5*(10), 3049−3057. Available from https://doi.org/10.1021/acssensors.0c00232.

Nyein, H. Y. Y., Bariya, M., Kivimäki, L., Uusitalo, S., Liaw, T. S., Jansson, E., Ahn, C. H., Hangasky, J. A., Zhao, J., Lin, Y., Happonen, T., Chao, M., Liedert, C., Zhao, Y., Tai, L. C., Hiltunen, J., & Javey, A. (2019). Regional and correlative sweat analysis using high-throughput microfluidic sensing patches toward decoding sweat. *Science Advances*, *5*(8). Available from https://doi.org/10.1126/sciadv.aaw9906.

Okazaki, T., Watanabe, T., & Kuramitz, H. (2020). Evanescent-wave fiber optic sensing of the anionic dye uranine based on ion association extraction. *Sensors (Switzerland)*, *20*(10). Available from https://doi.org/10.3390/s20102796.

Ou, S. H., Pan, L. S., Jow, J. J., Chen, H. R., & Ling, T. R. (2018). Molecularly imprinted electrochemical sensor, formed on Ag screen-printed electrodes, for the enantioselective recognition of D and L phenylalanine. *Biosensors and Bioelectronics*, *105*, 143−150. Available from https://doi.org/10.1016/j.bios.2018.01.010.

Ozcelikay, G., Kurbanoglu, S., Zhang, X., Soz, C. K., Wollenberger, U., Ozkan, S. A., . . . Scheller, F. W. (2019). Electrochemical MIP sensor for butyrylcholinesterase. *Polymers*, *11*(12). Available from https://doi.org/10.3390/polym11121970.

Özgür, E., Topçu, A. A., Yılmaz, E., & Denizli, A. (2020). SPR based biomimetic sensor for urinary tract infections. *Talanta*, *212*, 120778.

Öztürk, B. Ö., & Şehitoğlu, S. K. (2019). Pyrene substituted amphiphilic ROMP polymers as nano-sized fluorescence sensors for detection of TNT in water. *Polymer*, *183*, 121868. Available from https://doi.org/10.1016/j.polymer.2019.121868.

Pacheco, J. G., Rebelo, P., Freitas, M., Nouws, H. P. A., & Delerue-Matos, C. (2018). Breast cancer biomarker (HER2-ECD) detection using a molecularly imprinted electrochemical sensor. *Sensors and Actuators, B: Chemical*, *273*, 1008−1014. Available from https://doi.org/10.1016/j.snb.2018.06.113.

Pacheco, J. G., Silva, M. S. V., Freitas, M., Nouws, H. P. A., & Delerue-Matos, C. (2018). Molecularly imprinted electrochemical sensor for the point-of-care detection of a breast cancer biomarker (CA 15−3). *Sensors and Actuators, B: Chemical*, *256*, 905−912. Available from https://doi.org/10.1016/j.snb.2017.10.027.

Pirzada, M., & Altintas, Z. (2019). Nanomaterials for healthcare biosensing applications. *Sensors (Switzerland)*, *19*(23), 5311. Available from https://doi.org/10.3390/s19235311.

Poschenrieder, A., Thaler, M., Junker, R., & Luppa, P. B. (2019). Recent advances in immunodiagnostics based on biosensor technologies—From central laboratory to the point of care. *Analytical and Bioanalytical Chemistry*, *411*(29), 7607−7621. Available from https://doi.org/10.1007/s00216-019-01915-x.

Regasa, M. B., Refera Soreta, T., Femi, O. E., & Ramamurthy, P. C. (2020). Development of molecularly imprinted conducting polymer composite film-based electrochemical sensor for melamine detection in infant formula. *ACS Omega*, *5*(8), 4090−4099. Available from https://doi.org/10.1021/acsomega.9b03747.

Rahtuvanoğlu, A., Akgönüllü, S., Karacan, S., & Denizli, A. (2020). Biomimetic nanoparticles based surface plasmon resonance biosensors for histamine detection in foods. *ChemistrySelect*, *5*(19), 5683−5692. Available from https://doi.org/10.1002/slct.202000440.

Rashed, M. Z., Kopechek, J. A., Priddy, M. C., Hamorsky, K. T., Palmer, K. E., Mittal, N., Valdez, J., Flynn, J., & Williams, S. J. (2021). Rapid detection of SARS-CoV-2 antibodies using electrochemical impedance-based detector. *Biosensors and Bioelectronics*, *171*, 112709. Available from https://doi.org/10.1016/j.bios.2020.112709.

Ravi, N., Cortade, D. L., Ng, E., & Wang, S. X. (2020). Diagnostics for SARS-CoV-2 detection: A comprehensive review of the FDA-EUA COVID-19 testing landscape. *Biosensors and Bioelectronics*, *165*. Available from https://doi.org/10.1016/j.bios.2020.112454.

Rebelo, P., Pacheco, J. G., Cordeiro, M. N. D. S., Melo, A., & Delerue-Matos, C. (2020). Azithromycin electrochemical detection using a molecularly imprinted polymer prepared on a disposable screen-printed electrode. *Analytical Methods*, *12*(11), 1486−1494. Available from https://doi.org/10.1039/c9ay02566a.

Rebelo, T. S. C. R., Costa, R., Brandão, A. T. S. C., Silva, A. F., Sales, M. G. F., & Pereira, C. M. (2019). Molecularly imprinted polymer SPE sensor for analysis of CA-125 on serum. *Analytica Chimica Acta*, *1082*, 126−135. Available from https://doi.org/10.1016/j.ac.2019.07.050.

Regasa, M. B., Refera Soreta, T., Femi, O. E., & Ramamurthy, P. C. (2020). Development of molecularly imprinted conducting polymer composite film-based electrochemical sensor for melamine detection in infant formula. *ACS Omega*, *5*(8), 4090−4099. Available from https://doi.org/10.1021/acsomega.9b03747.

Revsbech, N. P., Nielsen, M., & Fapyane, D. (2020). Ion selective amperometric biosensors for environmental analysis of nitrate, nitrite and sulfate. *Sensors (Switzerland)*, *20*(15), 1−12. Available from https://doi.org/10.3390/s20154326.

Rico-Yuste, A., & Carrasco, S. (2019). Molecularly imprinted polymer-based hybrid materials for the development of optical sensors. *Polymers*, *11*(7), 1173. Available from https://doi.org/10.3390/polym11071173.

Santana Oliveira, I., da Silva Junior, A. G., de Andrade, C. A. S., & Lima Oliveira, M. D. (2019). Biosensors for early detection of fungi spoilage and toxigenic and mycotoxins in food. *Current Opinion in Food Science*, *29*, 64−79. Available from https://doi.org/10.1016/j.cofs.2019.08.004.

Saylan, Y., Akgönüllü, S., & Denizli, A. (2020). Plasmonic sensors for monitoring biological and chemical threat agents. *Biosensors*, *10*(10), 142. Available from https://doi.org/10.3390/bios10100142.

Saylan, Y., Akgönüllü, S., Çimen, D., Derazshamshir, A., Bereli, N., Yılmaz, F., & Denizli, A. (2017). Development of surface plasmon resonance sensors based on molecularly imprinted nanofilms for sensitive and selective detection of pesticides. *Sensors and Actuators, B: Chemical*, *241*, 446−454. Available from https://doi.org/10.1016/j.snb.2016.10.017.

Saylan, Y., Akgönüllü, S., Yavuz, H., Ünal, S., & Denizli, A. (2019). Molecularly imprinted polymer based sensors for medical applications. *Sensors (Switzerland)*, *19*(6). Available from https://doi.org/10.3390/s19061279.

Saylan, Y., & Denizli, A. (2018). Molecular fingerprints of hemoglobin on a nanofilm chip. *Sensors (Switzerland)*, *18*(9), 3016. Available from https://doi.org/10.3390/s18093016.

Saylan, Y., & Denizli, A. (2019). Molecularly imprinted polymer-based microfluidic systems for point-of-care applications. *Micromachines*, *10*(11), 766. Available from https://doi.org/10.3390/mi10110766.

Saylan, Y., Erdem., Cihangir, N., & Denizli, A. (2019a). Detecting fingerprints of waterborne bacteria on a sensor. *Chemosensors*, *7*(3). Available from https://doi.org/10.3390/CHEMOSENSORS7030033.

Saylan, Y., Erdem., Ünal, S., & Denizli, A. (2019b). An alternative medical diagnosis method: Biosensors for virus detection. *Biosensors*, *9*(2). Available from https://doi.org/10.3390/bios9020065.

Saylan, Y., Yılmaz, F., Derazshamshir, A., Yılmaz, E., & Denizli, A. (2017). Synthesis of hydrophobic nanoparticles for real-time lysozyme detection using surface plasmon resonance sensor. *Journal of Molecular Recognition*, *30*(9). Available from https://doi.org/10.1002/jmr.2631.

Scarpa, E., Mastronardi, V. M., Guido, F., Algieri, L., Qualtieri, A., Fiammengo, R., Rizzi, F., & De Vittorio, M. (2020). Wearable piezoelectric mass sensor based on pH sensitive hydrogels for sweat pH monitoring. *Scientific Reports*, *10*(1), 10854. Available from https://doi.org/10.1038/s41598-020-67706-y.

Semenova, D., Pinto, T., Koch, M., Gernaey, K. V., & Junicke, H. (2020). Electrochemical tuning of alcohol oxidase and dehydrogenase catalysis via biosensing towards butanol-1 detection in fermentation media. *Biosensors and Bioelectronics*, *170*. Available from https://doi.org/10.1016/j.bios.2020.112702.

Sharifi, M., Hasan, A., Haghighat, S., Taghizadeh, A., Attar, F., Bloukh, S. H., Edis, Z., Xue, M., Khan, S., & Falahati, M. (2021). Rapid diagnostics of coronavirus disease 2019 in early stages using nanobiosensors: Challenges and opportunities. *Talanta*, *223*, 121704. Available from https://doi.org/10.1016/j.talanta.2020.121704.

Sharifi, M., Attar, F., Saboury, A. A., Akhtari, K., Hooshmand, N., Hasan, A., . . . Falahati, M. (2019). Plasmonic gold nanoparticles: Optical manipulation, imaging, drug delivery and therapy. *Journal of Controlled Release*, *311−312*, 170−189. Available from https://doi.org/10.1016/j.jconrel.2019.08.032.

Sharma, P. K., Kumar, J. S., Singh, V. V., Biswas, U., Sarkar, S. S., Alam, S. I., Dash, P. K., Boopathi, M., Ganesan, K., & Jain, R. (2020). Surface plasmon resonance sensing of Ebola virus: A biological threat. *Analytical and Bioanalytical Chemistry*, *412*(17), 4101−4112. Available from https://doi.org/10.1007/s00216-020-02641-5.

Sousa, R. P. C. L., Figueira, R. B., Costa, S. P. G., & Raposo, M. M. M. (2020). Optical fiber sensors for biocide monitoring: Examples, transduction materials, and prospects. *ACS Sensors*. Available from https://doi.org/10.1021/acssensors.0c01615.

Sönmezler, M., Özgür, E., Yavuz, H., & Denizli, A. (2019). Quartz crystal microbalance based histidine sensor. Artificial cells. *Nanomedicine and Biotechnology*, *47*(1), 221−227. Available from https://doi.org/10.1080/21691401.2018.1548474.

Sriplai, N., Mangayil, R., Pammo, A., Santala, V., Tuukkanen, S., & Pinitsoontorn, S. (2020). Enhancing piezoelectric properties of bacterial cellulose films by incorporation of $MnFe_2O_4$ nanoparticles. *Carbohydrate Polymers*, *231*, 115730. Available from https://doi.org/10.1016/j.carbpol.2019.115730.

Su, C., Li, Z., Zhang, D., Wang, Z., Zhou, X., Liao, L., & Xiao, X. (2020). A highly sensitive sensor based on a computer-designed magnetic molecularly imprinted membrane for the determination of acetaminophen. *Biosensors and Bioelectronics*, *148*, 111819. Available from https://doi.org/10.1016/j.bios.2019.111819.

Supraja, P., Tripathy, S., Krishna Vanjari, S. R., Singh, V., & Singh, S. G. (2019). Electrospun tin (IV) oxide nanofiber based electrochemical sensor for ultra-sensitive and selective detection of atrazine in water at trace levels. *Biosensors and Bioelectronics*, *141*, 111441. Available from https://doi.org/10.1016/j.bios.2019.111441.

Tran, T. T., Clark, K., Ma, W., & Mulchandani, A. (2020). Detection of a secreted protein biomarker for citrus Huanglongbing using a single-walled carbon nanotubes-based chemiresistive biosensor. *Biosensors and Bioelectronics*, *147*, 111766. Available from https://doi.org/10.1016/j.bios.2019.111766.

Tucci, M., Grattieri, M., Schievano, A., Cristiani, P., & Minteer, S. D. (2019). Microbial amperometric biosensor for online herbicide detection: Photocurrent inhibition of *Anabaena variabilis*. *Electrochimica Acta*, *302*, 102−108. Available from https://doi.org/10.1016/j.electacta.2019.02.007.

Van Sau, N., Ngo, Q. M., Phan, T. B., Tran, N. Q., & Nguyen, T. T. (2020). Optical biosensor using near infrared laser for enhancement of detection accuracy. *Journal of Electronic Materials*, *49*(12), 7420−7426. Available from https://doi.org/10.1007/s11664-020-08384-4.

Veeralingam, S., & Badhulika, S. (2020). X (metal: Al, Cu, Sn, Ti)-functionalized tunable 2D-MoS2nanostructure assembled biosensor arrays for qualitative and quantitative analysis of vital neurological drugs. *Nanoscale*, *12*(28), 15336−15347. Available from https://doi.org/10.1039/d0nr03427d.

Wang, D., Zheng, Y., Wei, L., Wei, N., Fan, X., Huang, S., & Xiao, Q. (2020). A signal-amplified whole-cell biosensor for sensitive detection of Hg^{2+} based on Hg^{2+}-enhanced reporter module. *Journal of Environmental Sciences (China)*, *96*, 93−98. Available from https://doi.org/10.1016/j.jes.2020.03.020.

Wang, J., Li, J., Zeng, C., Qu, Q., Wang, M., Qi, W., Su, R., & He, Z. (2020). Sandwich-like sensor for the highly specific and reproducible detection of rhodamine 6G on a surface-enhanced Raman scattering platform. *ACS Applied Materials and Interfaces*, *12*(4), 4699−4706. Available from https://doi.org/10.1021/acsami.9b16773.

Wang, P., Li, H., Hassan, M. M., Guo, Z., Zhang, Z. Z., & Chen, Q. (2019). Fabricating an acetylcholinesterase modulated UCNPs-Cu^{2+} fluorescence biosensor for ultrasensitive detection of organophosphorus pesticides-diazinon in food. *Journal of Agricultural and Food Chemistry*, *67*(14), 4071−4079. Available from https://doi.org/10.1021/acs.jafc.8b07201.

Wang, X., Chen, Y., Chen, X., Peng, C., Wang, L., Xu, X., . . . Xu, J. (2020). A highly integrated system with rapid DNA extraction, recombinase polymerase amplification, and lateral flow biosensor for on-site detection of genetically modified crops. *Analytica Chimica Acta*, *1109*, 158−168. Available from https://doi.org/10.1016/j.ac.2020.02.044.

Wang, Yf, Pan, Mm, Yu, X., & Xu, L. (2020). The recent advances of fluorescent sensors based on molecularly imprinted fluorescent nanoparticles for pharmaceutical analysis. *Current Medical Science*, *40*(3), 407−421. Available from https://doi.org/10.1007/s11596-020-2195-z.

Wang, Z., Jinlong, L., An, Z., Kimura, M., & Ono, T. (2019). Enzyme immobilization in completely packaged freestanding SU-8 microfluidic channel by electro click chemistry for compact thermal biosensor. *Process Biochemistry*, *79*, 57−64. Available from https://doi.org/10.1016/j.procbio.2018.12.007.

Wei, X., Guo, J., Lian, H., Sun, X., & Liu, B. (2020). Cobalt metal-organic framework modified carbon cloth/paper hybrid electrochemical button-sensor for nonenzymatic glucose diagnostics. *Sensors and Actuators, B: Chemical*, *329*, 129205. Available from https://doi.org/10.1016/j.snb.2020.129205.

Xiao, B., Niu, C., Shang, Y., Xu, Y., Huang, K., Zhang, X., & Xu, W. (2020). A 'turn-on' ultra-sensitive multiplex real-time fluorescent quantitative biosensor mediated by a universal primer and probe for the detection of genetically modified organisms. *Food Chemistry*, *330*, 127247. Available from https://doi.org/10.1016/j.foodchem.2020.127247.

Xiao, F., Li, H., Yan, X., Yan, L., Zhang, X., Wang, M., Qian, C., & Wang, Y. (2020). Graphitic carbon nitride/graphene oxide(g-C_3N_4/GO) nanocomposites covalently linked with ferrocene containing dendrimer for ultrasensitive detection of pesticide. *Analytica Chimica Acta*, *1103*, 84−96. Available from https://doi.org/10.1016/j.ac.2019.12.066.

Yarman, A., Kurbanoglu, S., Zebger, I., & Scheller, F. W. (2021). Simple and robust: The claims of protein sensing by molecularly imprinted polymers. *Sensors and Actuators, B: Chemical*, *330*. Available from https://doi.org/10.1016/j.snb.2020.129369.

Zaidi, S. A. (2020). Molecular imprinting: A useful approach for drug delivery. *Materials Science for Energy Technologies*, *3*, 72−77. Available from https://doi.org/10.1016/j.mset.2019.10.012.

Zhang, G., Yu, Y., Guo, M., Lin, B., & Zhang, L. (2019). A sensitive determination of albumin in urine by molecularly imprinted electrochemical biosensor based on dual-signal strategy. *Sensors and Actuators, B: Chemical*, *288*, 564−570. Available from https://doi.org/10.1016/j.snb.2019.03.042.

Zhang, L. P., Wei, Z. H., He, S. N., Huang, Y. P., & Liu, Z. S. (2020). Preparation, characterization, and application of soluble liquid crystalline molecularly imprinted polymer in electrochemical sensor. *Analytical and Bioanalytical Chemistry*, *412*(26), 7321−7332. Available from https://doi.org/10.1007/s00216-020-02866-4.

Zhang, Ym, Zhang, Y., & Xie, K. (2020). Evaluation of Cas12a-based DNA detectio/Cas12a-based DNA detection for fast pathogen diagnosis and GMO test in rice. *Molecular Breeding*, *40*(1). Available from https://doi.org/10.1007/s11032-019-1092-2.

Zhao, L., Huang, L., Luo, G., Wang, J., Wang, H., Wu, Y., . . . Jiang, Z. (2020). An immersive resonant sensor with microcantilever for pressure measurement. *Sensors and Actuators, A: Physical*, 303. Available from https://doi.org/10.1016/j.sna.2019.111686.

Zhao, W. R., Kang, T. F., Xu, Y. H., Zhang, X., Liu, H., Ming, A. J., Lu, L. P., Cheng, S. Y., & Wei, F. (2020). Electrochemiluminescence solid-state imprinted sensor based on graphene/CdTe@ZnS quantum dots as luminescent probes for low-cost ultrasensing of diethylstilbestrol. *Sensors and Actuators, B: Chemical*, *306*, 127563. Available from https://doi.org/10.1016/j.snb.2019.127563.

Applications of nanotechnology in biological systems and medicine

10

Maryam Zain[1], Humaira Yasmeen[2], Sunishtha S. Yadav[3], Sidra Amir[4], Muhammad Bilal[5], Aqsa Shahid[6] and Mohsin Khurshid[6]

[1]*Department of Biochemistry and Biotechnology, The Women University Multan, Multan, Pakistan* [2]*Department of Microbiology and Molecular Genetics, The Women University Multan, Multan, Pakistan* [3]*Center for Medical Biotechnology, Amity Institute of Biotechnology, Amity University, Noida, India* [4]*Department of Biological Sciences, ILM Group of Colleges, Sargodha, Pakistan* [5]*School of Life Science and Food Engineering, Huaiyin Institute of Technology, Huaian, P.R. China* [6]*Department of Microbiology, Government College University Faisalabad, Faisalabad, Pakistan*

10.1 Introduction to nanomedicine and nanoparticles

The 20th century has been revolutionized by progress in nanotechnology as well as its pharmacological activities. Nanotechnology is an important field of science that plays an important role by providing advancements within a short period. In 1959 an American physicist named Richard P. Feynman at CalTech introduced nanotechnology (Kubinova & Sykova, 2010).

Nanotechnology is a branch of science that deals with extremely small-sized particles. The word "nano" has been derived from the Greek term "dwarf" which means extremely minute size. This technology involves characterization, changes, or fabrication of materials, structures, or devices that are unidimensional with lengths 1−100 nm (Duncan, 2011). In nanotechnology individual compounds, molecules, or atoms are used to synthesize devices and materials with special characteristics. Nanotechnology involves two approaches, bottom-up and top-down: in the former approach individual particles are assembled which give rise to nanostructures, while in the latter the size of large particles is reduced which gives rise to the nanostructures like in nanoengineering and nanoelectronics (Duncan, 2011; Limongi et al., 2019).

In recent years nanotechnology has revolutionized the spectrum of technology and its application in our day-to-day life. This technology with multiple and integrated approaches has been applied to almost all the fields and spheres starting from chemistry to biology and pharmaceuticals to medicine. In nanotechnology, newly synthesized nanoparticles (NPs) have characteristics distinct from those of bulk materials (Mody et al., 2010).

Despite the fact, NPs are semisynthetic, some NPs occur naturally. Naturally occurring NPs include organic as well as inorganic compounds produced by volcano eruptions, microbial processes, weathering, or wildfires. It is not necessary to synthesize these particles only in laboratories, but they also have clear existence in nature that has been found since ancient times. They can be incidentally produced as a by-product in any reaction. The main difference between engineered and supplementary nanomaterials is that the former nanomaterial has better-controlled morphology than the latter, moreover, engineered NPs can be deliberately synthesized with innovative characteristics along with their extremely small size (Heiligtag & Niederberger, 2013).

Photochemical reactions, forest fires, and volcanic eruptions are some sources which synthesize nanoparticle naturally in larger quantities which affect air quality significantly throughout the world. However, hair and skin shedding of animals and plants is recurrent in nature and produces NPs naturally. NPs can also be synthesized by human activities such as charcoal burning, industrial operations, and transportation. Only 10% of NPs are synthesized by human activities while naturally 90% of NPs are made. The nanoparticle has a lot of definitions but there was a big challenge to define it internationally in a single definition (Hough et al., 2011).

Based on their sizes, shapes, and properties NPs or materials can be divided into various categories. The different groups comprise polymeric, ceramic, fullerenes, and metal NPs, etc. NPs show several physiochemical properties. They can be used in several applications due to their extremely small size and large surface area. Moreover, optical properties become dominant at this extremely small size which significantly enhances their photocatalytic applications. Magnetic properties, size, and specific morphology of NPs are controlled usefully by synthetic techniques. Besides their useful applications, there are also some disadvantages associated with NPs, their direct discharge in the atmosphere, and their uncontrolled use.

Cell uptake and drug delivery are critically affected by parameters like surface coating, shape, and size of NPs. Their shapes play a significant role in IR absorption. Anisotropic NPs like triangular and rod-shaped AuNPs show strong absorption in the near-infrared (NIR) and visible regions (Ankamwar, 2012). In composition, NPs consist of three layers, first is the topmost layer on the surface having various small polymers, surfactants, molecules, and metal ions, the second one is the shell layer that differs chemically from the core, and the third one is the core that is fundamentally its central portion (Mauricio et al., 2018).

Thus nanotechnology is a highly significant branch of science that is concerned with the preparation of NPs with different sizes, shapes, and chemical compositions with different pharmacological applications (Bleeker et al., 2013; Tinkle et al., 2014; Vineet, 2009). Further, such NPs have a high surface-to-volume ratio which shows many different characteristics as well as enables them to be utilized in special applications. Controlled synthesis is used to obtain their desirable size (Fratoddi et al., 2017). Applications of NPs depend on their functionality and they may be hydrophobic or hydrophilic (Shamsadin et al., 2016). Due to their distinctive properties, NPs are used in various fields like energy storage, agriculture, medicine, electronics, and environmental remediation, etc.

Over the past few years, the application of NPs in the synthesis of novel medicines has increased, because NPs are utilized for better treatment outcomes in different diseases. Due to the implementation of nanotechnology and NPs in the field of the medicine, a more promising field has evolved which is called nanomedicine. Nanomedicine is the consumption of nanomaterials and NPs for the diagnosis, prognosis, prevention, treatment, and control of several diseases (Bleeker et al., 2013; Pita et al., 2016; Tinkle et al., 2014).

The specific properties of nanomaterials and NPs have broadened the application of these in the field of medicines and pharmaceuticals. Nanomedicine is an effective and promising field and will surely introduce certain modifications and advancements in the field of medicine, clinical sciences, and pharmaceuticals by introducing novel medicines, by integrating more effective NPs, by exploiting multiple mechanisms of action, and by increasing bioavailability, thus ensuring maximum efficacy of drugs (Chan, 2006; Méndez-Rojas et al., 2009; Zhang et al., 2012).

Biomedicine involving NPs is known as nanomedicine. Nanomedicine is used for early diagnosis and innovative drug application. Nanotechnology has gained much attention in a relatively short time due to its innovative strategies. The use of nanotechnology in medicine is termed nanomedicine. According to the American Society for Testing and Materials NPs are mostly used in the field of nanomedicines (Ventola, 2012).

Nanomedicine is a comparatively novel field of technology and science. The field of application and research and has been broadened by nanotechnology due to interactions at the nanoscale. We can comprehend the interactions of biomolecules with nanodevices both in the intracellular and extracellular medium. At the nanolevel, the operation allows the manipulation of physical characteristics unlike those detected at the microlevel, like surface-to-volume ratio. In in vitro experiments it was observed that NPs play a crucial role in the delivery of drugs at the targeted site. In vivo, further research is needed for the treatment of cancer and other diseases. Likewise, another application for nanomedicine is drug detoxification that is effectively used in rats. Smaller devices are used in medical technologies that are least hostile and can be implanted inside the body because they quickly respond to biochemical reactions. In comparison with conventional drug delivery, nanomaterials are more sensitive and faster (Ventola, 2012). In medical technology, nanomaterials are used for drug delivery and innovative medicine. Although the study of nanotechnology started properly 15 years ago the knowledge about their toxicity is still limited. There are no such regulatory measures properly designed for such type of research (Rodríguez-López et al.). Nanotechnology is expanding and studies on interaction with various fields have been performed but due to less attention their regulatory measures still cause some problems. Newly synthesized nanomaterials are mostly used without knowing their potential risks and even safety measures. These risky behaviors may provoke harmful effects to the organism at the cellular or tissue level. Scientists should be aware of the potential risks that they trigger in humans. Safety and protective measures should be known. The medical applications of nanomaterials are used in a wide range of medical procedures including therapy and nanodiagnostics (Keles et al., 2016; Padmanabhan et al., 2016). Nanomaterial might trigger reactions that include damage to the arteries and inflammation. They might induce chronic damage to the organs of the body. However, the body has a mechanism of the immune system that can defend the body from foreign interactions of the harmful microorganisms that might include bacteria and viruses or any other foreign agents which are attacking the body.

The inability of such defense mechanisms might result in disease conditions. They are potentially used in the fields of therapy, nanodiagnostics, and imaging (Keles et al., 2016; Padmanabhan et al., 2016). Besides their medical applications, they interact with human cells and tissues, triggering reactions via the interaction of nanomaterial with blood, which results in chronic and acute inflammation. Evolutionary defensive mechanisms have been developed against foreign particles and microorganisms. Whenever these mechanisms become effective, they give immunity to resist invasive agents. Their failure causes illness. The human immune system plays a crucial role in controlling various protective biological processes (Kulkarni et al., 2016; Walls et al., 2016; Weidenbusch & Anders, 2012). These processes coordinate with a variety of tissues and cells to defend the body from the pathogenic agents and various types of cancers (Guillerey et al., 2016; Plitas & Rudensky, 2016). NPs can enter into the body via blood-based carriers, such as erythrocytes, platelets, plasma, and various complement proteins. So what would happen when blood components interact with NPs is a very important thing to be discussed (Anderson, 2001; Gardner et al., 2013). The interaction between humans triggers a tissue reaction when they interact with different person's risks or healthy (Anderson, 2001; Gardner et al., 2013). After entering into the body, NPs act as foreign substances and provoke an immune response. The application of nano-material in medical sciences includes diagnostic imaging and the treatment of cancers.

Blood and vascular endothelium always show an early response to the nanomaterials which first enter into our body. Once nanomaterials enter the bloodstream, they meet the other blood components. Knowledge of the interactive potential between these NPs is needed to understand the toxic effects of these newly developed particles in blood and remote areas of the body.

Nanodiagnostics is the application of nanotechnology to diagnose a disease in the human single cell at the primary stage. There has been a rise in elementary research through the development of spectroscopic and microscopic techniques with ultrahigh sensitivity, three-dimensional and molecular resolution, and the construction of diagnostics tools that are innovative. More precisely, it offers various prospects like

- Rapid diagnostic testing, possibly in the doctor's office or on the bedside, for the preliminary treatment and diagnostic selection and one-to-one dealing at the hospital or even at home.
- There is a need for earlier diagnosis as compare to the efficiency of the current techniques because earlier diagnosis leads to therapeutic opportunities (Bobo et al., 2016).

Various types of NPs work via interaction with the blood components.

10.2 Types of nanoparticles

Various types of NPs are involved in biomedical applications. On these bases, they are divided into the following types

10.2.1 Carbon-based nanoparticles

Carbon nanotubes (CNTs), fullerenes, graphene oxide (GO), nanodiamonds, quantum dots, graphene, derivatives of graphene are included in carbon-based nanomaterials, as shown in Fig. 10.1. They differ in structural dimensions, optical, intense mechanical, thermal, chemical, and mechanical properties due to which they have gained much attention in biomedical applications. Therapeutic molecules are delivered to treat the diseases and to repair the tissue and imaging of cells/tissues have been an area of focus. They have a broad-range one-photon property with biocompatibility and ease in functionality which have enabled them to be good chemotherapeutic agents. Especially, the two-photon inherent fluorescence characteristic of carbon-based NPs in the long-wavelength region allows deep optical imaging of tissues. Carbon-based nanostructures have promising therapeutic and diagnostic applications to treat various diseases.

There are two main types of carbon-based NPs, that is, nanotubes and fullerenes (Gupta et al., 2018; MP, 2009). Fullerenes are the water-insoluble sphere and are made up of 60 carbon atoms. In comparison with diamond and graphite, fullerenes are an allotropic form of carbon. This

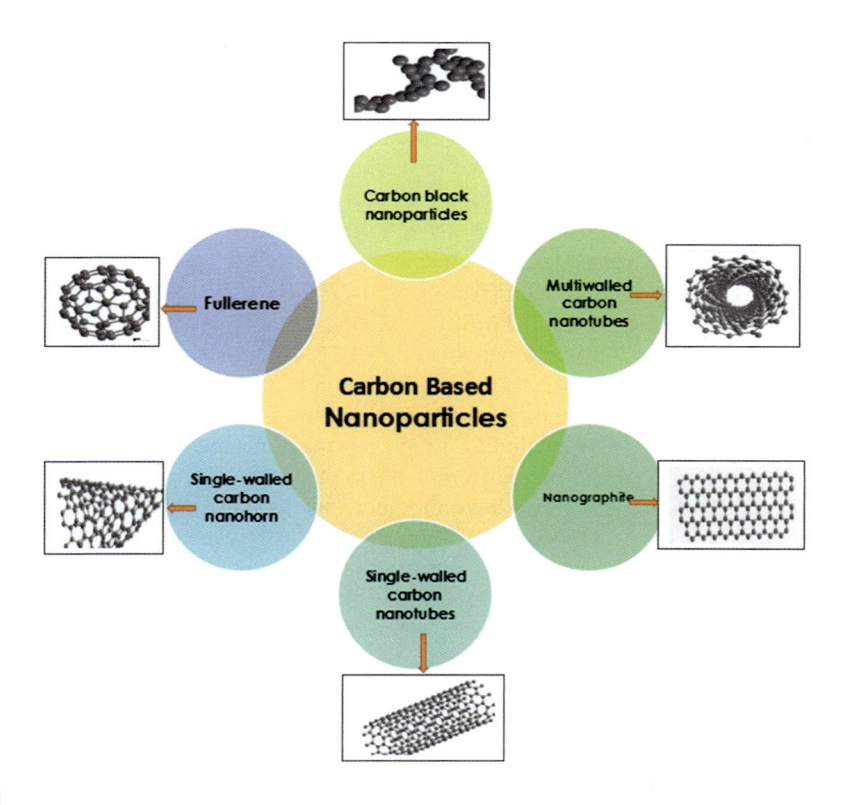

FIGURE 10.1

Carbon-based nanoparticles.

allotropic form of carbon has different diagnostics applications. The fullerenes are also used in various medical imaging procedures. They are also used in drug delivery procedures in combination with other procedures such as photodynamic, acoustics wave therapy, hyperthermia processes, etc. (Augustine et al., 2017; Castro et al., 2017). Fullerenes are water-insoluble spheres containing 60 carbon atoms. The derivatives are effectively used for different therapeutics and diagnostics applications, many researchers have described the use of allotropic forms of carbon for drug delivery and medical imaging along with acoustic wave assisted therapies, hyperthermia, and photodynamic.

CNTs have been categorized into two major categories: single-walled carbon tubes (SWCNT) and multiwalled CNTs (MWCNT). MWCNTs are formed by joining and rolling one or more graphite sheets. The CNTs are utilized in the labeling of biosensors, drug delivery, tissue engineering, and therapeutic applications (Limongi et al., 2017; Maiti et al., 2018).

Carbon is special among the elements as it can articulate steadily secured chains enclosed via hydrogen atoms. Such hydrocarbons are typically used as sources of nonrenewable energy, for example, coal, gaseous petrol, and oil. A small yet significant part is exploited in petrochemical industries as a source of feedstock conveying solvents, paints, plastics, and polymers, etc. The later discovery of CNTs, atoms, different fullerenes, and thin sheets of graphene enhanced the improvement of equipment in the devices business and nanotechnology. Whereas the use of nanoonions, nanorings, nanodiamonds, nanofibers, peapods, etc. is notable, further utilization of CNTs and fullerenes is operationally centered.

CNTs and fullerenes are the most significant classes of NPs that are carbon centered. They have a great advantage in business and industries due to the properties like high electron affinity, quality, structure, electrical conductivity, and flexibility (Galceran AAONnMT, 2015). In their original form they interact with nanocomposites, they show many important applications like commercial filters, due to their physical, mechanical, and chemical characteristics and they can also be used as potent adsorbents of gas in bioremediation (Ngoya et al., 2014). They are also used as organic and inorganic catalysts as supportive media (LFMSSR, 2011).

In the past two decades, carbon-based nanomaterials (CBNs) shows great and more reliable biomedical applications because of several unique physical and chemical properties which include optical diversity, electrical, mechanical, thermal, and structural characteristics. With the help of these specific characteristics, CBNs, including graphene quantum dots, GO, and CNT, have been widely used in biomedical applications. Different biomedical applications include biosensors for drug delivery and cancer therapy. As they have inorganic semiconducting characteristics as well as organic $\pi-\pi$ stacking properties, CBNs are auspicious NPs. They can respond to light at once, by interacting effectively with biomolecules. In the future, they may be used in several biomedical applications. To decrease their toxicity, various chemically based analyses are used that can be successfully applied in biomedical applications like cancer therapy, tissue engineering, drug delivery, and detection of biomolecules (Maiti et al., 2018; Yamashita et al., 2012). To reduce its toxicity and adverse effects and to take more advantage of its various carbon-based NPs that have been synthesized that correctly deliver the therapeutic agent to the tumor while curing osteosarcoma (Wang et al., 2020).

CNTs can be prepared by the procedure of rolling up one or more graphitic sheets, so they are grouped into two different categories as SWCNT and MWCNT. They are practically used in a wide range of medical applications, such as biosensors, tissue engineering, drug delivery, therapeutics, imaging, and labeling (Moya et al., 2020).

The right choice of biomaterials is considered to be the best strategy in the development of nanomedicines, which especially determines the successive biological events. The usage of carbon nanomaterials (CNMs) is increasing due to their medical applications including the targeting of the drug to various sites, imaging of biomolecules, engineering of tissue, as well as sensing of biomolecules. Owing to their ability of high surface area, surface chemistry having multifunctional ability, and excellent optical activity, the CNMs have efficient drug-delivery ability and compatibility of biomolecules. In the last decades, advances were made for the improvement in functionalization of CNMs to reduce health-related problems and increase their biosafety. Recent advances have proved that CNMs can be used with proteins, bioactive peptides, nucleic acids, and drugs to achieve their target sites with very low toxicity, enhancing pharmaceutical ability.

10.2.2 Organic nanoparticles

Organic NPs are prepared from different types of materials, such as lipids and polymers, which have important applications in imaging and therapeutic delivery. Keeping in view of several advantages of organic NPs, their use is increasing for drug delivery and imaging purposes. The state-of-the-art guides the design of future NPs that can be used in therapies. For designing NPs for future generations, the researchers must consider the benefits and harmful effects. Several materials have been used for synthesizing NPs, liposomes, and polymers. Extracellular vesicles are also referred to as organic NPs.

Polymer NPs can be made from two basic types of polymers: natural or synthetic polymers which allow biodegradability and biocompatibility. These are used as an organic approach to solving nanomedicine manufacturing problems (El-Say & El-Sawy, 2017). The NPs can be synthesized by different techniques like fluid technology, dialysis, and emulsifying procedure. For the synthesis of NPs, size, and solubility can be adjusted depending on the type of procedure adopted (Crucho & Barros, 2017).

The organic NPs or precise nanocarriers for drug delivery are CBNs that are characterized based on improved and effective drug-loading capacity and their biocompatibility. The biocompatibility and effectivity both define the spectrum of usage of these nanomaterials in the field of medicine. These are a very important and significant class of nanomaterials since they allow versatile control of chemical composition and morphology. Furthermore, their relatively large size and colloidal stability allow the incorporation of a different combination of several hydrophilic and hydrophobic medicines (Ferreira Soares et al., 2020; Jiang & Pu, 2017; Li et al., 2020; Patra et al., 2018). The categorization of the NPs depends on the way in which they have been synthesized and depending upon the same there are two broad categories. One of them has nanostructures that are based on self-assembly processes like amphiphilic systems. While the other one is through specific synthesis methods like hyperbranched polymers, dendrimers, and CNTs. The most interesting research these days is focusing on the generation of these nanocarriers by the appropriate combination of two methods with the help of a supramolecular approach (Ferreira Soares et al., 2020; Jiang & Pu, 2017; Li et al., 2020).

10.2.3 Inorganic nanoparticles

The availability of various types of inorganic NPs and various chemical methods have shown that inorganic NPs can act as a novel system of drug delivery. Several important issues should be considered before using these inorganic NPs in clinical trials. The first important aspect that needs to be considered is the biocompatibility of the selected inorganic nanosystems. Unlike organic nanomaterials, the clinical trials of inorganic NPs are difficult to do due to a lack of data regarding biosafety and toxicity while carrying out these experiments in vivo, and the excretion routes and assessment of toxicity are not well defined. The conventional inorganic-based nanosystems gave an opportunity to be used in personalized medicine for different types of polygenic and monogenic diseases (Pandey & Dahiya, 2016).

The NPs of the inorganic category have three major types of NPs: metallic (MNPs), quantum dots (QDs), and metal oxide NPs (MONPs). Quantum dots are a type of nanoparticle considered to be semiconductors; they consist of a core that is coated with a shell made up of peptide proteins and polysaccharide molecules so that the toxic materials are not able to leak, which thus results in the increased stability of the NPs.

There are various categories of quantum dots that are used in biosensing and bioimaging, such as cadmium—selenium, gold, and cadmium—tellurium; these all can be used in cell tagging and apoptosis analysis (Mauricio et al., 2018). Magnetic inorganic NPs also include gold, copper, silver, and palladium types, which have high stability that has been proved in various tumor conditions, and they are used as biosensing agents (Vallabani et al., 2018). Oxides of silica, zinc, titania, and zirconia have high stability and antioxidant catalytic actions that make them useful for medical preparation of implants, bioimaging, and delivery of drugs (Limongi et al., 2019; Marino et al., 2018; Mathieu et al., 2019).

10.2.4 Metallic nanoparticles

Metal NPs are prepared from metal precursors. Metallic NPs possess a somewhat different property to alkali and noble metals, such as Cu and silver NPs. They have an absorption band that is broad and in the visible zone of the electromagnetic spectrum (Evanoff & Chumanov, 2005). The different types of metal-based nanomaterials are shown in Fig. 10.2. Moreover, the preparation strategies of metal-based nanomaterials are summarized in Fig. 10.3.

Over time magnetic NPs in the form of nanoshells, cages, and particles have been used as diagnostic and therapeutic agents. Metallic NPs are used in various important medical applications, such as delivering drugs to the areas that are difficult to reach. These particles are also used in the treatment of cancer and other lethal diseases. For such types of medical applications silver as well as gold NPs are significantly used (Nikalje et al., 2015).

10.2.5 Gold nanoparticles

Gold NPs have been used for medicinal purposes since their discovery over 5000 years ago. These particles are used in research studies in various fields. Gold NPs have some unique properties that

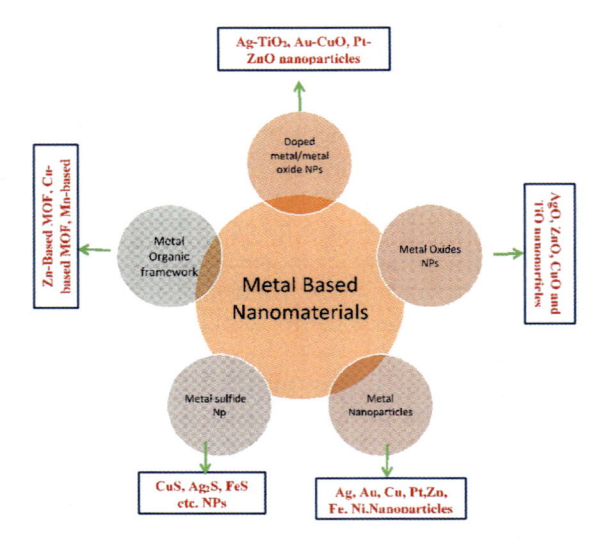

FIGURE 10.2

Different types of metal-based nanomaterials.

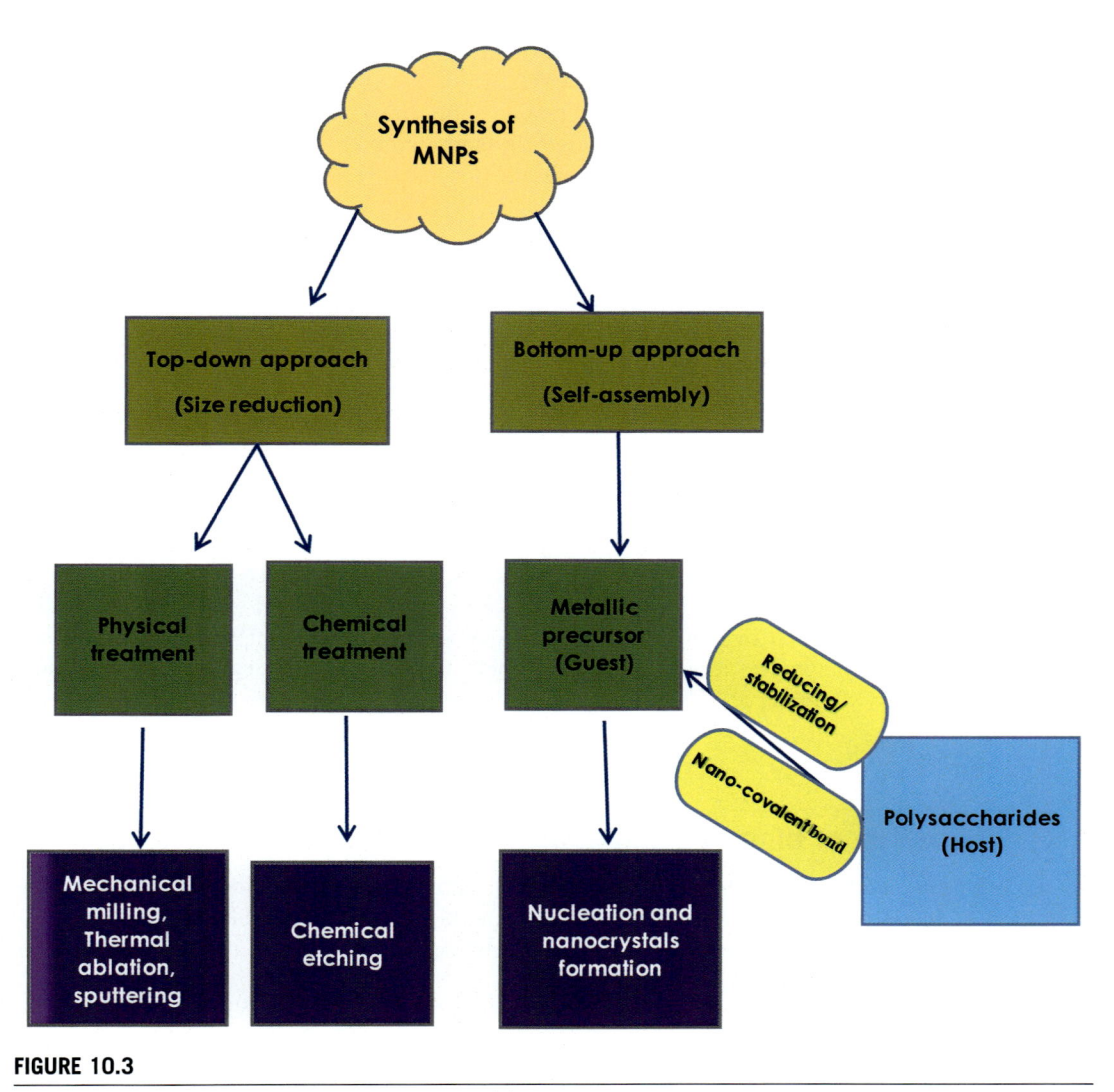

FIGURE 10.3

Preparation strategies of metal-based nanomaterials.

make them different from other particles, such as chemical inertness, optical and electronic properties, and the surface functioning ability, which is due to the overall net negative charge on its outer surface (Dykman & Khlebtsov, 2011). The different properties of gold NPs make them useful as biosensors and for imaging of live cells and tissues (Deb et al., 2011; Laloy et al., 2014).

Gold has some properties that make it unique as it has the property of ease of functionalization with different types of organic molecules, conjugation with ligands and with drug molecules for the delivery of drugs both in the active and passive forms (Sobol et al., 2021).

Gold's chemical inertness allows the ability to experiment in vivo and in vitro. Silver NPs have many biomedical applications because of their very different properties like greater thermal stability, electrical conductivity, catalytic ability, very unique antibacterial, and enhanced optical properties (Zhang et al., 2012). Silver NPs are used in antimicrobial, disinfectant, and photonic applications. These particles are also used in wound dressings, textiles, and device coatings (Zhang et al., 2012).

Plasmonic NPs are a more novel group of NPs that have many advantages in biomedical applications due to their surface plasmon resonance bands in the UV−visible to near-IR spectral range. Their frequency is very sensitive to the physicochemical environment and the distance of the NPs (Dykman, 2010). Plasmon shifts cause changes of color that can be seen by the naked eye and they do not require special instrumentation. The NPs of gold and silver have plasmonic properties. They are also available in different sizes and shapes. The surface properties of plasmonic particles are very advanced and the biofunctionalization can be performed effectively (Dykman, 2010). LSPR of gold and silver NPs can be used for enhancing light-based events and sensing them. One property that made them a most relevant building block is their optoelectrical property. The characterization properties can be changed according to the requirements of biomedicine (Alexis Loiseau et al., 2019).

Iron oxide NPs are widely used in the biomedical field. These particles have greater biocompatibility and high magnetic susceptibility (Marino et al., 2018; Perrone Donnorso et al., 2012). Based on different oxidation states they exist in different forms as maghemite (Fe_2O_3) and magnetite (Fe_3O_4). Both forms are used in biomedical applications (Garino et al., 2019). Iron oxide has a property to oxidize so it requires a biocompatibility shell. Examples of coatings include ceramics, metal, and polymers (Marino et al., 2018; Evanoff & Chumanov, 2005; Garino et al., 2019).

10.3 Interaction of blood components with nanoparticles

For the intravenous administration of drug-loaded NPs, compatibility with blood is required. The interaction of blood components with NPs requires some physicochemical and pathophysiological events and they induce some properties like pharmacokinetics, usefulness, and toxicity (Valetti et al., 2014).

For intravenously administered NPs blood is the first contact medium, but NPs are also administered through other routes, to enter their target areas in tissues and organs. NPs can readily penetrate cells by traversing the biological barriers due to their characteristic size. The particle size also allows them to disrupt the normal biochemical environment of the cell because of increased biological activity. One of the important things that should be considered while designing the NPs is to check the hemocompatibility and safety for therapeutic uses (Weisel et al., 2019).

Blood contains various components, such as white blood cells, red blood cells, and platelets (known as thrombocytes). These blood cells have specified physical structure and chemical properties based on the various important functions, such as in the maintenance of normal homeostasis (Rezaei et al., 2019). NPs' interactions with blood components are therefore not only unavoidable but also possibly unsafe, and hemocompatibility can be a disadvantage associated with NPs usage. Hemocompatibility can be checked at first when NPs are used as therapeutic agents (Weisel et al., 2019).

Red blood cells play an important role in the various hemorheological mechanisms. When NPs interact with blood cells they can either induce erythrocyte accumulation, hemolysis, or decrease deformability that prevents them from performing their basic functions that are required for homeostasis (Rezaei et al., 2019). Red blood cells play their role via axial movement and flow of cells. With the use of NPs, the risk of thrombosis is enhanced by inducing phosphatidylserine which can be observed on the red blood cell membrane, creating a prothrombin surface for the adherence of red blood cells (Scanlon & Sanders, 2018).

After a vascular injury, a hemostatic plug is formed by platelets to prevent blood loss which is the major role of platelets. Nevertheless, during hemorrhaging the platelets play a role in normal hemostasis and are a continuous physiological process that causes the vascular system to balance (Dave, 2020). Vascular damage is a medical problem that requires daily follow-up and treatment with the help of platelets; this is evident from the routine examination of bruising and there is a greater risk of bleeding when there is a low platelet number (Fröhlich, 2016). Platelets not only have the sole responsibility of regulation of homeostasis but are also now known for their role in vascular inflammation, innate immunity, combating tumor growth, and angiogenesis (Fröhlich, 2016; Laloy et al., 2014). As platelets are one of the important constituents of blood cells in circulation so the interaction of NPs with any of their distinguished functions can have devastating effects. During primary hemostasis procedures, platelet normalization is the critical stage to secure hemostasis.

White blood cells play a vital role in the maintenance of homeostasis. These cells are divided into further categories as monocytes, lymphocytes, and granulocytes and they have an important role in immunity and body defense systems. But now, current studies have also found that these immune cells dynamically result in normal hemostasis and thrombosis (Dave, 2020). Various authors agree that the innate immune system and coagulation are linked to each other. It is important to understand the interaction of blood especially the blood cells and the immune pathways with the NP for an effective synthesis of various blood cells, the artificial production of blood cells, and their modifications for better transport of oxygen to the tissues (De La Cruz et al., 2017).

All NPs or biomaterials meant for human use initiate a tissue response with the sick or normal tissue. This mechanism is activated by the contact of the material with the body tissues when implanted in the tissues. On entering the bloodstream, they meet various blood cells (white cells, red cells, and platelets) and plasma proteins. It is important to understand how they interact with these components to find out their effective toxic potential.

10.4 Emerging applications of nanoparticles in biological systems and medicine

10.4.1 Functionalization of nanoparticles

The use of metal-based NPs dates to the discovery of Faraday (1857) that NPs can exist in solution. Further studies have characterized their color and morphology (Kumar et al., 2018; Yaqoob et al., 2020). Metal NPs are inorganic NPs that include metal (MNPs) and metal oxide NPs (MONPs) (Vemuri et al., 2019). Their characteristics, such as size, morphology, diameter, molar volume, and physiochemical properties, determine their applicability in various fields. However, their application depends on the synthesis process. Functionalization of different functional groups such as DNA, RNA, peptides, and antibodies makes noble MNPs (Ag, Au, and Pt) an ideal choice in cancer diagnosis and treatment (Yaqoob et al., 2020).

The characterization of MNP is largely dependent on the preparation method (Nair & Laurencin, 2007). Therefore the choice of preparation method is critical for achieving desired properties of NPs. They are either prepared by physical or chemical methods. However, the use and accumulation of toxic agents and the formation of hazardous by-products make them of least interest. Green synthesis of NPs is a useful method that is more advantageous than traditional methods (Ahmed et al., 2016; Kaliaraj, 2017). Among the top-down and bottom-up approaches, the bottom-up technique is widely used (Wang et al., 2017; Yip et al., 2014).

MNPs of copper, gold, silver, and palladium have been widely used in various diagnostic applications due to their high stability in hypoxic tumor conditions. They have been also used in imagining techniques, especially contrast imaging and biosensing agents. Metallic oxide NPs have a greater chemical stability as well as catalytic properties which make them the right choice to be used in medical applications, drug delivery, and imaging techniques for living systems. In addition to metal NPs, magnetic NPs are highly biocompatible and stable. They have a high surface-to-volume ratio. They are highly specific in the diagnosis and isolation of lymphoma from biological fluids. Furthermore, functionalized MNPs bind specifically to CD44 receptors, thus they can specifically isolate cancer cells (Vemuri et al., 2019).

Besides their synthetic synthesis, they are synthesized naturally as well. Magnetosomes and magnetotactic bacteria produce iron oxide NPs which have been tested against tumors. For this purpose, heat along with a magnetic field was applied repeatedly in lab-controlled mice to destroy tumors. They have been used to treat cancer via localized heat production, precise drug release, and enriched radiation effect. MNPs are efficient in cancer treatment because of the controlled targeting of cancer cells (Alphandéry, 2020). AuNPs are used to recognize cancer cells and precisely target the delivery of NPs into the cell nucleus (Dreaden et al., 2012). Moreover, silver-coated nanowires are compatible with adenocarcinoma epithelial cells. Likewise, lead NPs have anticancer properties (Fan et al., 2018). For breast cancer treatment, the use of iron NPs is also suggested (Gao et al., 2018). Studies have shown that MnO_2 NPs have higher hemocompatibility and histocompatibility than their side effects (Chen et al., 2019; Yaqoob et al., 2020). Lysis of erythrocytes determines the success rate of any treatment. If the lysis rate is high, the potential role of the drug to treat disease becomes less. MNPs are efficient in diagnosis, even zinc oxide NPs that are cytotoxic to other cell types. Functionalization of MNPs directly affects hemocompatibility and thus lysis.

Interestingly it has been observed that NPs with positive charge have a higher rate of hemolysis than negatively charged NPs. Positively charged NPs interact with the negatively charged plasma membrane, therefore resulting in lysis. However, the hemolytic activity of negatively charged NPs is attributed to several accessible groups to erythrocyte membranes. Therefore the hemolytic potential of each NPs is accessed differently. Generally, positive NPs are nonhemolytic, therefore their use would not affect the morphology of erythrocytes (Lozano-Fernández et al., 2019). The hemocompatibility of NPs provides initial insights into the interaction of nanomedicine with blood. They can activate platelets and subpopulations of blood cells, which stimulate other unintentional responses leading to the formation of tumor cells (Lozano-Fernández et al., 2019).

Fig. 10.4 shows a generalized diagram of the types of NPs and their biomedical applications. Table 10.1 presents various applications of MNPs.

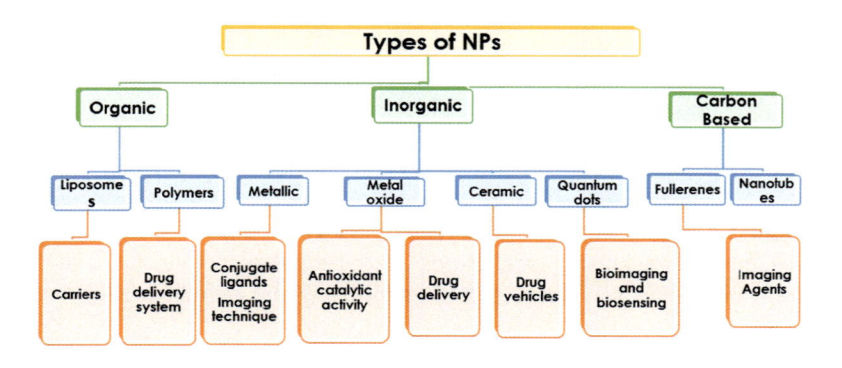

FIGURE 10.4

Generalized diagram of the types of nanoparticles and their biomedical applications.

Table 10.1 Applications of metallic nanoparticles (MNPs).		
Nanomaterials	**Targeted sites**	**Applications**
Platinum nanoparticles	Cancer cell	Toxicity evaluation
Gold/Ag nanoparticles	Cancer cell	Imaging therapy, photothermal therapy
Gold nanoparticles	Cancer cell	Radiosensitizer applications
Platinum nanoparticles	Cancer cell	Toxicity evaluation
Gold nanoparticles	Cancer cell	Radiosensitizer applications
Gold, iron oxide particles linked with glutathione	Cancer cell	Radiosensitizer applications
Platinum nanoparticles	Cancer cell	Therapeutic evaluation
Gold nanoparticles linked with glucose	Cancer cell	Radiosensitizer applications
Gold nanoparticles	Cancer: glioblastoma-based multiform	Radiosensitizer applications
Silver nanoshells	Cancer cell	Photothermal ablation
Silver nanoparticles linked with polyvinylpyrrolidone	Brain cancer	Therapeutic evaluation
Silica−gold nanoshells	Cancer	Photothermal-based therapies
Gold-branched shell nanostructures	Breast cancer	Imaging therapy, and chemotherapeutic therapy
FeO nanoparticle with Si/Au nanoshells	Head and neck cancer	Photothermal therapies
Silica−gold nanoshells	Brain tumor	Photothermal-based therapies
Magnetic nanoparticles	Cancer cell	Chemotherapies, biosensors, and imaging applications
Silica−gold nanoshells	Cancer	Photothermal-based therapies

10.4.2 Nanomedicines

NPs have been serving as an alternative approach to overcome the problem of entering the blood system. However, NPs have some limitations, like being unable to reach the brain via blood. For this reason, NPs bind with the surface of endothelial cells by adsorptive transcytosis or receptor-mediated transcytosis. In adsorptive transcytosis, NPs are attached and delivered into the body. The properties of NPs allow their binding and delivery into the target cell. While in receptor-mediated transcytosis, NPs are fabricated with different types of ligands to bind to specific receptors and thus this promotes the process of endocytosis. Based on types of receptors, there are two types of receptor-mediated transcytosis: clathrin-mediated endocytosis and caveolin endocytosis (Ahn et al., 2020; Lee & Leong, 2020).

The interaction of the NPs with these types of endocytosis plays an important role in the interaction of NPs with blood. The properties of NPs themselves determine their fate. These properties are size, charge, surface modification, and transportation. The entrance of NPs into endothelial cells via endocytosis depends on the size of NPs. Estimated size of less than 200 nm is considered as the optimal size for entrance into endothelial cells efficiently.

The charge on NPs determines their route of administration. Positively charged NPs transport more easily than negative or neutral NPs due to the negative charge of membranes. While negative or neutral NPs display a reduced rate of adsorption, and thus longer circulation time. PEGylation-coating of NPs with PEG also increases circulation time by producing the steric barrier around NPs. Thus avoiding opsonization and their removal by the mononuclear phagocytotic mechanism. This low cell uptake can be improved using zwitterionic molecules. Surface modification of NPs with different types of ligands is carried out, whose aim is to increase diffusion across cell membranes. The attachment of many ligands on the surface keeps the NPs attached to the cell membrane surface. However, the attachment of ligands with increased affinity for receptors may resolve the low diffusion rate. Alternatively, combination therapy of NPs with peptides can penetrate the cell which avoids the endocytosis pathway and delivers NPs directly to the cell cytoplasm (Ahn et al., 2020; Chung et al., 2020).

10.4.3 Nanoprobes for cell tracking

Stem cell therapy is regenerative medicine. Stem cells have the property of multipotent self-renewal to treat many incurable diseases such as cancers, autoimmune disorders, Alzheimer's and Parkinson's disease, bone defects, liver diseases, and heart failure (Bianco et al., 2013; Granero-Moltó et al., 2009; Hayrapetyan et al., 2015; Mao & Mooney, 2015; Mimeault et al., 2007; Ramdasi et al., 2015; Robinton & Daley, 2012; Segers & Lee, 2008). It is imperative to track and understand the fate of stem cells. It is also important to observe the regeneration ability of transplanted stem cells to improve the efficiency and safety of the treatment. Apart from biodistribution, it is crucial to check the stem cell viability for potential stem cells. For this purpose, the in vivo monitoring of transplanted viable stem cells via external fluorescent probes is still a big challenge (Du et al., 2010; Zhang et al., 2019).

During the last two decades, the technology of cell tracking has gained the attention of scientists working in the field of biological sciences. They are interested in studying developmental biology, disease prognosis, cancer biology, differentiation of cells, molecular pathways, and different areas involving cell tracking. Nanotechnology is an emerging field in which various functional nanoprobes have been made for the visualization of several cellular or molecular behaviors in vivo as well as in vitro. In comparison to traditional methods that use dyes, the use of nanoprobes has proved to have great advantages, such as strong signal strength, narrow absorption with emission bands, and greater photostability (Cheng et al., 2013; Das et al., 2016; Wang et al., 2013).

The in vitro experiments of cell tracking have played an important role in identifying several unanswered questions, for example, the influence of cytoskeletal proteins on differentiation, the pattern of differentiation of hematopoietic stem cells and neural cells (Wang et al., 2013; Yi et al., 2017), signaling of cytokines during differentiation can be checked in cell tracking system (Yi et al., 2017), and the development of hemogenic endothelium (Bevington et al., 2017; Yuan et al., 2017). For a complete understanding of these complex processes at cellular levels, it is imperative to work at the scale of an individual molecule. Several progressively advanced optical techniques are under consideration to study the dynamics of biomolecules in vivo and in vitro. This technique is noninvasive yet highly sensitive and has wide applications for the detection of evolving molecules in living systems. It requires optimized probes to detect a single molecule (Leduc et al., 2013; Liang et al., 2016).

The general property of an ideal probe includes high specificity, smaller size (no interference with the targeted biological molecules), monovalent, longer photostability, and the ability to produce strong signals for localization. The ideal probes are still under development and many challenges are still to be addressed, such as poor photostability and crowded environments inside the living systems (Leduc et al., 2013; Liang et al., 2016).

For in vivo tracking, fluorescent NPs are widely used. NIR NPs are more sensitive and have greater tissue penetration. They have improved temporal and spatial resolution of fluorescence imaging-based stem cell tracking for the reason that they exhibit reduced scattering, absorption, and autofluorescence of NIR fluorescence in tissues (Chen et al., 2018; Leduc et al., 2013).

Quantum dots (semiconducting nanocrystals) have also shown longer photostability and they have been used to study live cells for several minutes (Chen et al., 2018). However, the larger diameter of functionalized quantum dots has not proven to be suitable for molecular tracking in some specific areas inside the cells (Liang et al., 2016). In this regard, self-illuminated quantum dots are produced by using bioluminescence resonance energy transfer which can be used to sense the viable cells endogenously (Dreaden et al., 2012; Zhang et al., 2019). Another alternative approach is to use gold NPs having a diameter of 5 nm. They present greater absorption cross-sections and produce strong and optically stable signals in the cellular environments (Sheung et al., 2018).

10.5 Conclusion

Nanotechnology has revolutionized several fields as it has applications in many fields such as electronics, energy, cosmetics, agriculture, food, and health. The greater surface area per unit volume of them provides a greater surface activity that accelerates the chemical reactions and increases catalysis which ultimately improves their efficiency. At the cellular level, several organelles lie in the range of the nanoscale in living organisms. This fact predicts the use of nanotechnology in biological systems and medicine. Based on chemical and physical properties they are divided into several categories. NPs have different applications depending upon their composition and properties. Moreover, major applications of nanoprobes include the fields of developmental biology, disease prognosis, cancer biology, differentiation of cells, molecular pathways, and various areas involving cell tracking.

References

Ahmed, S., Ahmad, M., Swami, B. L., & Ikram, S. (2016). A review on plants extract mediated synthesis of silver nanoparticles for antimicrobial applications: A green expertise. *Journal of Advanced Research*, 7(1), 17−28.

Ahn, S. I., Sei, Y. J., Park, H. J., Kim, J., Ryu, Y., Choi, J. J., Sung, H. J., MacDonald, T. J., Levey, A. I., & Kim, Y. (2020). Microengineered human blood−brain barrier platform for understanding nanoparticle transport mechanisms. *Nature Communications*, 11(1), 1−12.

Alexis Loiseau, V. A., Boitel-Aullen, G., Lam, M., Salmain, M., & Boujday, S. (2019). Silver-based plasmonic nanoparticles for and their use in biosensing. *Biosensors (Basel)*, 9(2), 78.

Alphandéry, E. (2020). Natural metallic nanoparticles for application in nano-oncology. *International Journal of Molecular Sciences*, 21(12), 4412.

Anderson, J. M. (2001). Biological responses to materials. *Annual Review of Materials Research* (31), 81−100.

Ankamwar, B. (2012). Size and shape effect on biomedical applications of nanomaterials. *Biomedical Engineering-Technical Applications in Medicine*, 93−114.

Augustine, S., Singh, J., Srivastava, M., Sharma, M., Das, A., & Malhotra, B. D. (2017). Recent advances in carbon based nanosystems for cancer theranostics. *Biomaterials Science*, 5(5), 901−952.

Astefaneia, A., Núñez, O., & Galceran, M. T. (2015). Characterisation and determination of fullerenes: A critical review. *Analytica Chimica Acta*, 882, 1−21.

Bevington, S. L., Cauchy, P., Withers, D. R., Lane, P. J., & Cockerill, P. N. (2017). T cell receptor and cytokine signaling can function at different stages to establish and maintain transcriptional memory and enable T helper cell differentiation. *Frontiers in Immunology*, 8, 204.

Bianco, P., Cao, X., Frenette, P. S., Mao, J. J., Robey, P. G., Simmons, P. J., et al. (2013). The meaning, the sense and the significance: Translating the science of mesenchymal stem cells into medicine. *Nature Medicine*, 19(1), 35−42.

Bleeker, E. A., de Jong, W. H., Geertsma, R. E., Groenewold, M., Heugens, E. H., Koers-Jacquemijns, M., et al. (2013). Considerations on the EU definition of a nanomaterial: Science to support policy making. *Regulatory Toxicology and Pharmacology: RTP.*, 65(1), 119−125.

Bobo, D., Robinson, K. J., Islam, J., Thurecht, K. J., & Corrie, S. R. (2016). Nanoparticle-based medicines: A review of FDA-approved materials and clinical trials to date. *Pharmaceutical Research*, 33(10), 2373−2387.

Castro, E., Hernandez Garcia, A., Zavala, G., & Echegoyen, L. (2017). Fullerenes in biology and medicine. *Journal of Materials Chemistry B.*, 5(32), 6523−6535.

Chan, V. S. (2006). Nanomedicine: An unresolved regulatory issue. *Regulatory Toxicology and Pharmacology: RTP.*, 46(3), 218−224.

Chen, G., Zhang, Y., Li, C., Huang, D., Wang, Q., & Wang, Q. (2018). Recent advances in tracking the transplanted stem cells using near-infrared fluorescent nanoprobes: Turning from the first to the second near-infrared window. *Advanced Healthcare Materials*, 7(20)1800497.

Chen, J., Meng, H., Tian, Y., Yang, R., Du, D., & Li, Z. (2019). Recent advances in functionalized MnO2 nanosheets for biosensing and biomedicine applications. *Nanoscale Horizons* (4), 321−338.

Cheng, L., Wang, C., & Liu, Z. (2013). Upconversion nanoparticles and their composite nanostructures for biomedical imaging and cancer therapy. *Nanoscale, 5*(1), 23−37.

Chung, B., Kim, J., Nam, J., Kim, H., Jeong, Y., Liu, H. W., . . . Chung, S. (2020). Evaluation of cell-penetrating peptides using microfluidic in vitro 3D brain endothelial barrier. *Macromolecular Bioscience*1900425.

Crucho, C. I. C., & Barros, M. T. (2017). Polymeric nanoparticles: A study on the preparation variables and characterization methods. *Materials Science & Engineering C, Materials for Biological Applications, 80,* 771−784.

Das, B. D. P., Pal, P., & Dhara, S. (2016). Single step synthesized sulfur and nitrogen doped carbon nanodots from whey protein: Nanoprobes for longterm cell tracking crossing the barrier of photo-toxicity. *RSC Advances, 6*(65), 60794−60805.

Dave, A. L. H. D. (2020). *Physiology, hemostasis*. StatPearls Publishing LLC.

Deb, S., Patra, H. K., Lahiri, P., Dasgupta, A. K., Chakrabarti, K., & Chaudhuri, U. (2011). Multistability in platelets and their response to gold nanoparticles. *Nanomedicine: Nanotechnology, Biology, and Medicine, 7*(4), 376−384.

Dreaden, E. C., Alkilany, A. M., Huang, X., Murphy, C. J., & El-Sayed, M. A. (2012). The golden age: Gold nanoparticles for biomedicine. *Chemical Society Reviews, 41,* 2740−2779.

Du, J., Yu, C., Pan, D., Li, J., Chen, W., Yan, M., et al. (2010). Quantum-dot-decorated robust transductable bioluminescent nanocapsules. *Journal of the American Chemical Society, 132*(37), 12780−12781.

Duncan, T. V. (2011). Applications of nanotechnology in food packaging and food safety: Barrier materials, antimicrobials and sensors. *Journal of Colloid and Interface Science, 363*(1), 1−24.

Dykman, L. A., & Khlebtsov, N. G. (2011). Gold nanoparticles in biology and medicine: Recent advances and prospects. *Acta Naturae, 3*(2), 34−55.

Dykman, N. G. K. D. (2010). Plasmonic nanoparticles: Fabrication, optical properties, and biomedical applications, Chapter: 2 In V. V. Tuchin (Ed.), *Handbook of photonics for biomedical science*. Boca Raton: CRC Press.

El-Say, K. M., & El-Sawy, H. S. (2017). Polymeric nanoparticles: Promising platform for drug delivery. *International Journal of Pharmaceutics, 528*(1-2), 675−691.

Evanoff, D. D., Jr., & Chumanov, G. (2005). Synthesis and optical properties of silver nanoparticles and arrays. *Chemphyschem: A European Journal of Chemical Physics and Physical Chemistry, 6*(7), 1221−1231.

Fan, G., Dundas, C. M., Zhang, C., Lynd, N. A., & Keitz, B. K. (2018). Sequence-dependent peptide surface functionalization of metal-organic frameworks. *ACS Applied Materials & Interfaces, 10*(22), 18601−18609.

Ferreira Soares, D. C., Domingues, S. C., Viana, D. B., & Tebaldi, M. L. (2020). Polymer-hybrid nanoparticles: Current advances in biomedical applications. *Biomedicine & Pharmacotherapy = Biomedecine & Pharmacotherapie, 131,* 110695.

Fratoddi, I. M. R., Fontana, L., Venditti, I., Familiari, G., Battocchio, C., Magnono, E., Nappini, S., Leahu, G., Belardini, A., et al. (2017). Electronic properties of a functionalized noble metal nanoparticles covalent network. *Journal of Physical Chemistry C* (121), 18110−18119.

Fröhlich, F. (2016). Action of nanoparticles on platelet activation and plasmatic coagulation. *Current Medicinal Chemistry* (23), 408−430.

Gerardo González De La Cruz, Patricia Rodríguez-Fragoso, Jorge Reyes-Esparza, Anahí Rodríguez-López, Rocío Gómez-Cansino and Lourdes Rodriguez-Fragoso (2017). Interaction of nanoparticles with blood components and associated pathophysiological effects. Unraveling the safety profile of nanoscale particles and materials—from biomedical to environmental applications. In: Andreia C. Gomes and Marisa P. Sarria, (Eds.), IntechOpen, https://doi.org/10.5772/intechopen.69386.

Gao, Q., Xie, W., Wang, Y., Wang, D., Guo, Z., Gao, F., et al. (2018). A theranostic nanocomposite system based on radial mesoporous silica hybridized with Fe_3O_4 nanoparticles for targeted magnetic field responsive chemotherapy of breast cancer. *RSC Advances, 8*(8), 4321−4328.

Gardner, A. B., Lee, S. K., Woods, E. C., & Acharya, A. P. (2013). Biomaterials-based modulation of the immune system. *BioMed Research International*, 732182.

Garino, N., Limongi, T., Dumontel, B., Canta, M., Racca, L., Laurenti, M., et al. (2019). A microwave-assisted synthesis of zinc oxide nanocrystals finely tuned for biological applications. *Nanomaterials*, 9(2).

Granero-Moltó, F., Weis, J. A., Miga, M. I., Landis, B., Myers, T. J., O'Rear, L., et al. (2009). Regenerative effects of transplanted mesenchymal stem cells in fracture healing. *Stem Cells*, 27(8), 1887−1898.

Guillerey, C., Huntington, N. D., & Smyth, M. J. (2016). Targeting natural killer cells in cancer immunotherapy. *Nature Immunology*, 17(9), 1025−1036.

Grossi, V., Urbani, A., Giugni, A., Cantalini, C., Santucci, S., & Passacantando, M. (2009). Simultaneous growth of MWCNTs at different temperatures in a variable gradient furnace. In: *Solid State Phenomena* 2009, 154, (pp. 77−82). Trans Tech Publications Ltd.

Gupta, T. K. B. P., Chappidi, S. R., Sastry, S. Y., Paggi, M., et al. (2018). Advances in carbon based nanomaterials for bio-medical applications. *Current Medicinal Chemistry*.

Hayrapetyan, A., Jansen, J. A., & van den Beucken, J. J. J. P. (2015). Signaling pathways involved in osteogenesis and their application for bone regenerative medicine. *Tissue Engineering Part B: Reviews*, 21(1), 75−87.

Heiligtag, F. J., & Niederberger, M. (2013). The fascinating world of nanoparticle research. *Materials Today*, 16(7−8), 282.

Hough, R. M., Noble, R. R. P., & Reich, M. (2011). Natural gold nanoparticles. *Ore Geology Reviews*, 42(1), 55−61.

Jiang, Y., & Pu, K. (2017). Advanced photoacoustic imaging applications of near-infrared absorbing organic nanoparticles. *Small (Weinheim an der Bergstrasse, Germany)*, 13(30).

Kaliaraj, G.S.S., & Manivasagan, P. (2017). Green synthesis of metal nanoparticles using seaweed polysaccharides. In: *Seaweed polysaccharides: Isolation, Biological and Biomedical Applications*, Elsevier.

Keles, E., Song, Y., Du, D., Dong, W. J., & Lin, Y. (2016). Recent progress in nanomaterials for gene delivery applications. *Biomaterials Science*, 4(9), 1291−1309.

Kubinova, S., & Sykova, E. (2010). Nanotechnologies in regenerative medicine. *Minimally Invasive Therapy & Allied Technologies: MITAT: Official Journal of the Society for Minimally Invasive Therapy*, 19(3), 144−156.

Kulkarni, O. P., Lichtnekert, J., Anders, H. J., & Mulay, S. R. (2016). The immune system in tissue environments regaining homeostasis after injury: Is "Inflammation" always inflammation? *Mediators of Inflammation*, 2016, 2856213.

Kumar, H., Venkatesh, N., Bhowmik, H., & Kuila, A. (2018). Metallic nanoparticle: A review. *Biomedical Journal of Scientific & Technical Research* (4), 3765−3775.

Laloy, J., Minet, V., Alpan, L., Mullier, F., Beken, S., Toussaint, O., et al. (2014). Impact of silver nanoparticles on haemolysis, platelet function and coagulation. *Nanobiomedicine*, 1, 4.

Leduc, C., Si, S., Gautier, J., Soto-Ribeiro, M., Wehrle-Haller, B., Gautreau, A., et al. (2013). A highly specific gold nanoprobe for live-cell single-molecule imaging. *Nano Letters*, 13(4), 1489−1494.

Lee, C. S., & Leong, K. W. (2020). Advances in microphysiological blood-brain barrier (BBB) models towards drug delivery. *Current Opinion in Biotechnology*, 66, 78−87.

LFMSSR., & Coville, S. D. M. N. J. (2011). Nitrogen-doped carbon nanotubes as a metal catalyst support. *Applied Nanoscience*, 1(67), 67−77.

Li, X., Porcel, E., Menendez-Miranda, M., Qiu, J., Yang, X., Serre, C., et al. (2020). Highly porous hybrid metal-organic nanoparticles loaded with gemcitabine monophosphate: A multimodal approach to improve chemo- and radiotherapy. *ChemMedChem*, 15(3), 274−283.

Liang, F., Zhang, Y., Hong, W., Dong, Y., Xie, Z., & Quan, Q. (2016). Direct tracking of amyloid and tu dynamics in neuroblastoma cells using nanoplasmonic fiber tip probes. *Nano Letters*, 16(7), 3989−3994.

Limongi, T., Canta, M., Racca, L., Ancona, A., Tritta, S., Vighetto, V., et al. (2019). Improving dispersal of therapeutic nanoparticles in the human body. *Nanomedicine: Nanotechnology, Biology, and Medicine, 14* (7), 797–801.

Limongi, T., Tirinato, L., Pagliari, F., Giugni, A., Allione, M., Perozziello, G., et al. (2017). Fabrication and applications of micro/nanostructured devices for tissue engineering. *Nano-Micro Letters, 9*(1), 1.

Lozano-Fernández, T., Dobrovolskaia, M., Camacho, T., González-Fernández, Á., & Simón-Vázquez, R. (2019). Interference of metal oxide nanoparticles with coagulation cascade and interaction with blood components. *Particle and Particle Systems Characterization* (36)).

Maiti, D., Tong, X., Mou, X., & Yang, K. (2018). Carbon-based nanomaterials for biomedical applications: A recent study. *Frontiers in Pharmacology, 9*, 1401.

Mao, A. S., & Mooney, D. J. (2015). Regenerative medicine: Current therapies and future directions. *Proceedings of the National Academy of Sciences, 112*(47), 14452–14459.

Marino, A., Battaglini, M., De Pasquale, D., Degl'Innocenti, A., & Ciofani, G. (2018). Ultrasound-activated piezoelectric nanoparticles inhibit proliferation of breast cancer cells. *Scientific Reports, 8*(1), 6257.

Mathieu, M., Martin-Jaular, L., Lavieu, G., & Thery, C. (2019). Specificities of secretion and uptake of exosomes and other extracellular vesicles for cell-to-cell communication. *Nature Cell Biology, 21*(1), 9–17.

Mauricio, M. D., Guerra-Ojeda, S., Marchio, P., Valles, S. L., Aldasoro, M., Escribano-Lopez, I., et al. (2018). Nanoparticles in medicine: A focus on vascular oxidative stress. *Oxidative Medicine and Cellular Longevity, 2018*, 6231482.

Méndez-Rojas, M. A., Angulo-Molina, A., and Aguilera-Portillo, G. (2009). Chapter 61: Nanomedicine: Small steps, big effects. In: *CRC concise encyclopedia of nanotechnology*, (1st ed.) CRC Press.

Mimeault, M., Hauke, R., & Batra, S. K. (2007). Stem cells: A revolution in therapeutics' recent advances in stem cell biology and their therapeutic applications in regenerative medicine and cancer therapies. *Clinical Pharmacology & Therapeutics, 82*(3), 252–264.

Mody, V. V., Siwale, R., Singh, A., & Mody, H. R. (2010). Introduction to metallic nanoparticles. *Journal of Pharmacy & Bioallied Sciences, 2*(4), 282–289.

Moya, A., Hernando-Perez, M., Perez-Illana, M., San Martin, C., Gomez-Herrero, J., Aleman, J., et al. (2020). Multifunctional carbon nanotubes covalently coated with imine-based covalent organic frameworks: Exploring structure-property relationships through nanomechanics. *Nanoscale, 12*(2), 1128–1137.

Nair, L.S., & Laurencin C.T. (2007). Silver nanoparticles: Synthesis and therapeutic applications. *Journal of Biomedical Nanotechnology 3*(4), 301–316.

Ngoya, J.M., Wagner, N., Riboldi, L., & Bolland, O. (2014). A CO_2 capture technology using multi-walled carbon nanotubes with polyaspartamide surfactant. *Energy Procedia, 63*, 2230–2248.

Nikalje, A. P., Ansari, A., Bari, S., & Ugale, V. (2015). Synthesis, biological activity, and docking study of novel isatin coupled thiazolidin-4-one derivatives as anticonvulsants. *Archiv der Pharmazie, 348*(6), 433–445.

Padmanabhan, P., Kumar, A., Kumar, S., Chaudhary, R. K., & Gulyas, B. (2016). Nanoparticles in practice for molecular-imaging applications: An overview. *Acta Biomaterialia, 41*, 1–16.

Pandey, P., & Dahiya, M. (2016). A brief review on inorganic nanoparticles. *Journal of Critical Reviews, 3*(3), 18–26.

Patra, J. K., Das, G., Fraceto, L. F., Campos, E. V. R., Rodriguez-Torres, M. D. P., Acosta-Torres, L. S., et al. (2018). Nano based drug delivery systems: Recent developments and future prospects. *Journal of Nanobiotechnology, 16*(1), 71.

Perrone Donnorso, M., ME., De Angelis, F., La Rocca, R., Limongi, T., et al. (2012). Nanoporous silicon nanoparticles for drug delivery applications. *Microelectronic Engineering* (98), 626–629.

Pita, R., Ehmann, F., & Papaluca, M. (2016). Nanomedicines in the EU-regulatory overview. *The AAPS Journal, 18*(6), 1576–1582.

Plitas, G., & Rudensky, A. Y. (2016). Regulatory T cells: Differentiation and function. *Cancer Immunology Research*, *4*(9), 721−725.

Ramdasi, S., Sarang, S., & Viswanathan, C. (2015). Potential of mesenchymal stem cell based application in cancer. *International Journal of Hematology-Oncology and Stem Cell Research*, *9*(2), 95.

Rezaei, R., Safaei, M., Mozaffari, H. R., Moradpoor, H., Karami, S., Golshah, A., et al. (2019). The role of nanomaterials in the treatment of diseases and their effects on the immune system. *Open Access Macedonian Journal of Medical Sciences*, *7*(11), 1884−1890.

Robinton, D. A., & Daley, G. Q. (2012). The promise of induced pluripotent stem cells in research and therapy. *Nature*, *481*(7381), 295−305.

Scanlon, V. C., & Sanders, T. (2018). Essentials of anatomy and physiology. In: F. Davis, (Ed.). Philadelphia, Pennsylvania, USA.

Segers, V. F. M., & Lee, R. T. (2008). Stem-cell therapy for cardiac disease. *Nature*, *451*(7181), 937−942.

Shamsadin, S. P. Z., Folkerts, G., Garssen, J., Moin, M., Adcock, I. M., Movassaghi, M., Ardestani, M. S., Moazzeni, S. M., & Mortaz, E. (2016). Conjugated alpha-alumina nanoparticle with vasoactive intestinal peptide as a nono-drug in treatment of allergic asthma in mice. *European Journal of Pharmacology*, *791*, 811−820.

Sheung, J. Y., Ge, P., Lim, S. J., Lee, S. H., Smith, A. M., & Selvin, P. R. (2018). Structural contributions to hydrodynamic diameter for quantum dots optimized for live-cell single-molecule tracking. *The Journal of Physical Chemistry. C, Nanomaterials and Interfaces*, *122*(30), 17406−17412.

Sobol, N. B., Korsen, J. A., Younes, A., Edwards, K. J., & Lewis, J. S. (2021). Immuno PET imaging of pancreatic tumors with (89)Zr-labeled gold nanoparticle-antibody conjugates. *Molecular Imaging and Biology*, *23*(1), 84−94.

Tinkle, S., McNeil, S. E., Muhlebach, S., Bawa, R., Borchard, G., Barenholz, Y. C., et al. (2014). Nanomedicines: Addressing the scientific and regulatory gap. *Annals of the New York Academy of Sciences*, *1313*, 35−56.

Valetti, S., Mura, S., Noiray, M., Arpicco, S., Dosio, F., Vergnaud, J., et al. (2014). Peptide conjugation: Before or after nanoparticle formation? *Bioconjugate Chemistry*, *25*(11), 1971−1983.

Vallabani, N. V. S., SS., & Karakoti, A. (2018). Magnetic nanoparticles: Current trends and future aspects in diagnostics and nanomedicine. *Current Drug Metabolism*, *20*, 457−472.

Vemuri, S. K., Banala, R. R., Mukherjee, S., Uppula, P., Gpv, S., VG, A., et al. (2019). Novel biosynthesized gold nanoparticles as anti-cancer agents against breast cancer: Synthesis, biological evaluation, molecular modelling studies. *Materials Science & Engineering C, Materials for Biological Applications*, *99*, 417−429.

Ventola, C. L. (2012). The nanomedicine revolution: Part 1: Emerging concepts. *P & T: A Peer-Reviewed Journal for Formulary Management*, *37*(9), 512−525.

Ventola, C. L. (2012). The nanomedicine revolution: Part 2: Current and future clinical applications. *P & T: A Peer-reviewed Journal for Formulary Management*, *37*(10), 582−591.

Vineet, K. S. K. (2009). Plant mediated synthesis of silver and gold nanoparticles and their applications. *Journal of Chemical Technology and Biotechnology (Oxford, Oxfordshire: 1986)* (84), 151−157.

Walls, J., Sinclair, L., & Finlay, D. (2016). Nutrient sensing, signal transduction and immune responses. *Seminars in Immunology*, *28*(5), 396−407.

Wang, C., Cheng, L., & Liu, Z. (2013). Upconversion nanoparticles for photodynamic therapy and other cancer therapeutics. *Theranostics*, *3*(5), 317−330.

Wang, C., Gao, X., Chen, Z., Chen, Y., & Chen, H. (2017). Preparation, characterization and application of polysaccharide-based metallic nanoparticles: A review. *Polymers*, *9*(12), 689.

Wang, X., Mao, M., Zhu, S., Xing, S., Song, Y., Zhang, L., et al. (2020). A novel nomogram integrated with inflammation-based factors to predict the prognosis of gastric cancer patients. *Advances in Therapy*, *37*(6), 2902−2915.

Weidenbusch, M., & Anders, H. J. (2012). Tissue microenvironments define and get reinforced by macrophage phenotypes in homeostasis or during inflammation, repair and fibrosis. *Journal of Innate Immunity*, *4*(5-6), 463−477.

Weisel, J. W., Litvinov, R. I., Weisel, J. W., & Litvinov, R. I. (2019). *Journal of Thrombosis and Haemostasis*, *17*(2), 271−282.

Yamashita, T., Yamashita, K., Nabeshi, H., Yoshikawa, T., Yoshioka, Y., Tsunoda, S. I., et al. (2012). Carbon nanomaterials: Efficacy and safety for nanomedicine. *Materials*, *5*(2), 350−363.

Yaqoob, A. A., Ahmad, H., Parveen, T., Ahmad, A., Oves, M., Ismail, I. M., Qari, H. A., Umar, K., & Mohamad Ibrahim, M. N. (2020). Recent advances in metal decorated nanomaterials and their various biological applications: A review. *Frontiers in Chemistry* (8), 341.

Yi, D. K., Nanda, S. S., Kim, K., & Selvan, S. T. (2017). Recent progress in nanotechnology for stem cell differentiation, labeling, tracking and therapy. *Journal of Materials Chemistry B*, *5*(48), 9429−9451.

Yip, J.L., Wong, K.H., Leung, P.H.M., Yuen, C.W.M., Cheung, M.C. (2014). Investigation of antifungal and antibacterial effects of fabric padded with highly stable selenium nanoparticles. *Journal of Applied Polymer Science*, *131*(17), 40728.

Yuan, G. C., Cai, L., Elowitz, M., Enver, T., Fan, G., Guo, G., et al. (2017). Challenges and emerging directions in single-cell analysis. *Genome Biology*, *18*(1), 84.

Zhang, W., Wei, S., Wu, Y., Wang, Y.-L., Zhang, M., Roy, D., et al. (2019). Poly (ionic liquid)-derived graphitic nanoporous carbon membrane enables superior supercapacitive energy storage. *ACS Nano*, *13*(9), 10261−10271.

Zhang, X. Q., Xu, X., Bertrand, N., Pridgen, E., Swami, A., & Farokhzad, O. C. (2012). Interactions of nanomaterials and biological systems: Implications to personalized nanomedicine. *Advanced Drug Delivery Reviews*, *64*(13), 1363−1384.

Lab-on-a-chip (lab-on-a-phone) for analysis of blood and diagnosis of blood diseases

11

Fahima Akther[1,2], Huong D.N. Tran[1,2], Jun Zhang[2], Nam-Trung Nguyen[2] and Hang T. Ta[1,2,3]

[1]*Australian Institute for Bioengineering and Nanotechnology, University of Queensland, St Lucia, QLD, Australia*
[2]*Queensland Micro- and Nanotechnology, Griffith University, Nathan, QLD, Australia* [3]*School of Environment and Science, Griffith University, Nathan, QLD, Australia*

11.1 Introduction

Human whole blood is a major biofluid because it carries the key physiological and pathological information (Kuan & Huang, 2020). Blood mainly consists of blood cells such as red blood cells (RBCs), white blood cells (WBCs), platelets, and plasma. Under normal conditions, each component exhibits distinct physiological properties, including size and the total number of blood cells, and the concentration of different blood proteins (Kuan & Huang, 2020). The quantity of various molecular components present in the blood, including proteins, nucleic acids, and metabolites, can be used as diagnostic parameters for many diseases (Liu et al., 2019). Deviation from the normal parameters can be linked to a specific disease condition. A blood test could provide a rapid and precise diagnostic platform by analyzing different circulating biomarkers. A conventional blood test requires multiple steps such as mixing, lysing, centrifugation, and filtration for accurate and sensitive detecting of the biomarkers (Lee et al., 2019). Furthermore, the process requires bulky and expensive laboratory equipment operated by skilled personnel, which increases the cost of the test. Therefore conventional analyzers usually use a considerably large amount of blood that prolongs the data acquisition time, typically from 1 day to several days (Kuan & Huang, 2020; Mabey et al., 2004), limiting the capability of point-of-care (POC) diagnosis in emergencies. As a result, extensive investigations are conducted to develop easy, efficient, and rapid POC diagnostic tools that can be operated at the patient site with minimum skills (Arshavsky-Graham & Segal, 2020).

To address this issue, miniaturized lab-on-a-chip (LOC) devices have been becoming popular for blood analysis. LOC devices must function without bulky infrastructure, as well as independent of location for direct use in POC treatment (Emde et al., 2020). Recent advancement in microfluidic technology plays a vital role in handling and processing of a small amount of liquid, and the integration of small biosensors in the LOC devices for real-time observation (Mach et al., 2013). More importantly, microfluidic LOC devices significantly reduce the sample and reagent volume, and shorten analyzing time with higher accuracy and reproducibility (Li, 2005). For instance, Kuan et al. (2016) reported a microfluidic device that was able to process whole blood and detect Hgb (Hemoglobin) and HbA1c (glycated hemoglobin) within 30 minutes by using only 5 μL of blood, which corresponds to a 300-fold reduction from the required volume for a standard analyzer. On the other hand, a thermoplastic chip could detect ABO/Rh blood typing within a minute using just 1 μL of blood sample via colorimetric detection (Chen et al., 2015). Huet et al. (2017) designed an automated passive microfluidic biochip that was able to detect the real-time positive and negative agglutination of RBCs in less than 2 minutes with 100% accuracy. Furthermore, the integration of the smartphone with the microfluidic LOC devices gradually shifts the diagnosis platform from centralized laboratories to individual homes driven monitoring (Liu et al., 2019). Smartphone's features such as a high-resolution camera, in-built apps, and computational units can be integrated with other analytical sensing systems to produce highly sensitive medical testing (Arshavsky-Graham & Segal; Liu et al., 2019). Low cost, simple operation technique, and portability make smartphone-based LOC devices a good alternative in resource-limited areas. In addition, smartphone apps allow the storage of data in a database or the wireless transmission of data to remote sites (Lee et al., 2020; Romeo et al., 2016).

This chapter provides a concise overview of recent development of LOC devices for whole blood analysis. We mainly discuss the fabrication techniques and the working principle of the chips for blood cell separation, blood typing, and disease diagnosis in POC platforms. This chapter also discusses the widely applied sensing modalities for analyzing blood diseases in smartphone-based LOC devices. Here, we highlight the detection principles for each sensing technique with several relevant examples. Finally, we discuss the main challenges and promising prospects in this field. A graphical overview of this chapter is presented in Fig. 11.1.

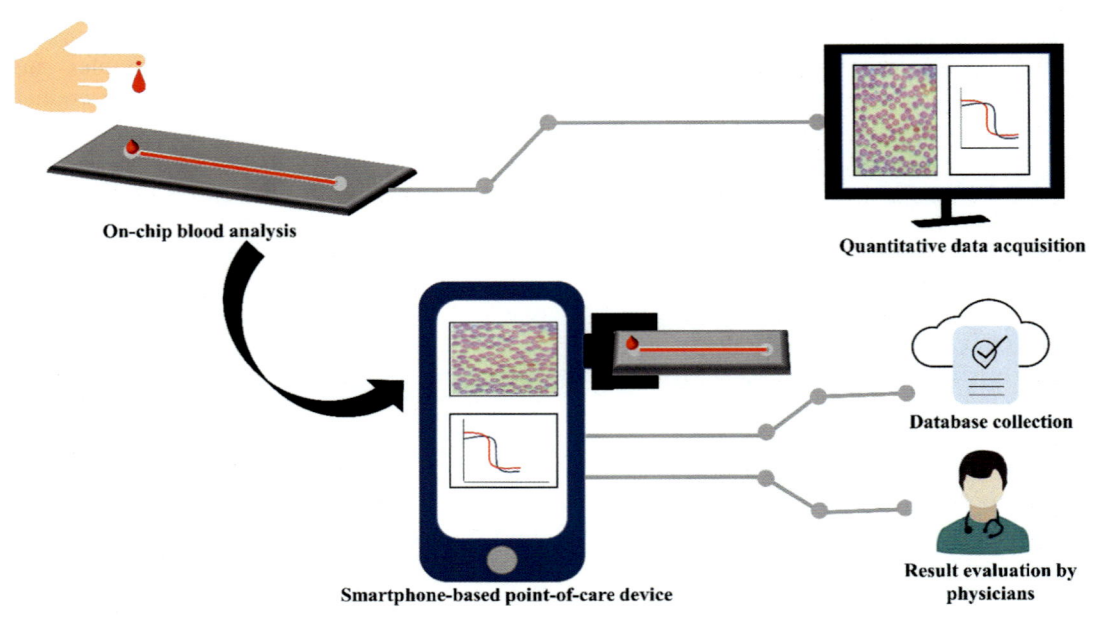

On-chip blood analysis

Quantitative data acquisition

Database collection

Result evaluation by physicians

Smartphone-based point-of-care device

FIGURE 11.1

Schematic representation of the uses of lab-on-a-chip (LOC) devices for point-of-care blood analysis and the modulation of smartphone technologies with LOC devices for on-site or remote data analysis.

11.2 On-chip whole blood analysis and disease diagnosis

Blood is a widely used biological sample for regular clinical analysis. Blood components carry critical physiological and pathological information of the human body. Quantification of cellular components of blood is commonly required for disease identification and management. For example, a complete blood count (CBC) provides an overall health status of a patient by monitoring the shape, size, and the number of each cell type, as well as the concentration of different proteins or metabolites in the blood (Kuan et al., 2018). This sophisticated test requires expensive automated hematology analyzers and flow cytometers for precise data acquisition. Moreover, conventional techniques require a large volume of blood (~ 1 mL) and trained staff for handling, which restrain the blood analysis in POC settings (Kuan et al., 2018). LOC technology is emerging as a miniaturized point-of-care diagnostic platform for disease diagnosis from a single drop of blood. Due to the advanced control over fluid behavior (Nikoleli et al., 2018), microfluidic LOC technology has superior advantages such as the reduction of sample and reagent volume, rapid responses, shorter analysis time, along with the cut down of cost per test (Lee et al., 2019; Nikoleli et al., 2018). This section provides an overview of LOC devices for the analysis of blood components and blood diseases. Table 11.1 lists a summary of the characteristic features of the LOC devices for analyzing human whole blood.

11.2.1 On-chip separation of blood components

Whole blood is a mixture of different components, and each has its unique attributes. Analyzing the blood components is a key diagnostic step where the accurate separation of the blood components plays a critical role in the precision of a specific test. In order to design a successful LOC blood analyzer, some characteristic attributes should be considered. For instance, the LOC-based devices must show sufficient sensitivity to the microscale blood sample. Also, the design should facilitate the incorporation of the on-chip sensor for the separation of blood components, rapid detection and communication of the signals, and translation of the signals into a readable clinical outcome (Lee & Lee, 2013). Recent studies mainly focus on developing simple LOC devices for whole blood processing in a single chip (Dixon et al., 2020; Kim et al., 2019; Kuan et al., 2018; Lee et al., 2019; Li et al., 2013; Madadi et al., 2015; Nguyen et al., 2015; Yang et al., 2019). Furthermore, LOC devices can process whole blood without any external instrument such as a centrifuge (Kuan & Huang, 2020). Kuan et al. (2018) designed a multichannel PDMS (polydimethylsiloxane) device for simultaneous processing of plasma, RBCs, and WBCs in a single device without any interference. The chip consisted of a separated whole blood inlet, a buffer inlet, and a bifurcation zone. The bifurcation zone contained 10 side channels and one main channel. Six bead-packed (10 μm beads) side channels were used for plasma extraction, while four-necked side channels (2 μm neck) and the main channel were used to trap RBCs and WBCs, respectively. All side channels were positioned at 60-degree angles with the main channel to maintain the uniform flow throughout the device. The unique chip design separated the blood components by following the bifurcation law and crossflow method. According to the bifurcation law or Zweifach-Fung effect, at the bifurcation point of the microvascular network, the daughter channels with the high flow velocity tend to pull the higher fraction of the RBCs (Mantegazza et al., 2020). When the blood and the

Table 11.1 A summary of the characteristic features of the lab-on-a-chip devices used in blood analysis.

Application	Chip design	Mechanism	Injected sample	Required time	Required sample volume	Target	Results	References
Analyzing whole blood, including plasma and RBCs (red blood cells) extraction, and WBCs (white blood cells) trapping.	– A PDMS (polydimethylsiloxane) chip consisted of a separated whole blood inlet, a buffer inlet, and a bifurcation zone. – The bifurcation zone contained ten side channels and one main channel. – Six bead-packed (10 μm-bead) side channels were used for plasma extraction, while four-necked side channels (2 μm-necked) and the main channel were used for trapping the RBCs and WBCs, respectively.	Bifurcation law and crossflow method.	Whole blood	20 min	6 μL	– Plasma – RBCs – WBCs	– Plasma extracted by the microfluidic device had a minimum dilution factor (0.76X) and low hemolysis effect. – Extracted RBCs could use for blood typing. – Up to ∼1800 WBCs could be trapped in 20 min.	Kuan et al. (2018)
Measurement of whole blood viscosity, hematocrit, and RBCs deformability	– The chip consisted of two parts: microviscometer and microhemocytometer. – The microviscometer was molded from PDMS and the electronic components for the hemocytometer was made of a printed circuit board. – Microviscometer had 10 arrays, each of which had 100 microchannels.	– Viscosity was measured by the fluid distribution according to the hydraulic resistances from the given equation: $$R = \frac{12\mu L}{wh^3 N}\left(1 - \frac{192h}{\pi^5 w}\sum_{n=1,3,5,\ldots}^{\infty}\frac{1}{n^5}\tanh(n\pi w/2h)\right)^{-1}$$ where, R, μ, L, w, h, and N denote the hydraulic resistance, viscosity, channel length, width, height, and number of channels filled with fluids, respectively. – Hematocrit was estimated from the electrical characteristic of the blood cytoplasmic resistant, and plasma resistant, while RBC deformation was detected from the change of membrane capacitance.	– Whole blood – Chemically hardened RBC with plasma	∼5 min	500 μL	– Whole blood – RBCs	– The normalized difference for the whole blood viscosity measurement by the physiometer was $0.8\% \pm 1.4\%$ less than the values obtained from the rotational cone-and-plate viscometer. – For hematocrit measurement, physiometer showed lower coefficient of variance (0.3%–1.2%) than the centrifuge (2.0%–2.7%). – For RBCs deformability measurement, a strong linear correlation ($R^2 = 0.97$) was observed between the deformability index acquired from physiometer and estimated from the image acquisition.	Kim et al. (2019)

– Blood/plasma separation	– The simple chip design with a main channel (5 mm long and – 240 μm wide) and a side channel (1 mm long and 120 μm) was manufactured by suing lamination-based microfabrication. – Optically transparent indium tin oxide coated glass was used to form electrodes and underlying substrate.	This dielectrophoresis (DEP)-based microfluidic device continually extracted the plasma by generating nonuniform electric field with a higher gradient at the junction area.	– Whole blood	~7 μL	15 min	Plasma	– The purity of the extracted plasma by the DEP plasma extractor was close to 100% with a yield of approximately 31%.	Yang et al. (2019)
Blood/plasma separation	– The digital microfluidic (DMF) device consisted of two plates: inkjet-printed bottom plate (black outline) and ITO (indium tin oxide)-PET top plate – The top plates were formed from ITO-PET substrates coated with FluoroPel PFC 1101 V as – The bottom plate consisted of 77 roughly square interdigitated driving electrodes (2.8 × 2.8 mm), 8 reservoir electrodes (7.6 × 6.4 mm), and 8 dispensing electrodes (5.6 × 2.0 mm). – Three integrated porous membranes with the bottom plate: plasma separation membrane for trapping RBCs, transport membrane for delivery of filtered plasma into electrode array, and impregnated wax plug to limit the flow of plasma in the TM.	Membrane filtration method.	– Whole blood	~50 μL	4 min	Plasma	The device performance was compared with the standard centrifugation technique and no cell was observed in extracted plasma under microscope which indicated almost 100% purity of the plasma.	Dixon et al. (2020)
Blood Typing	– A single-channel microfluidic chip was fabricated by using cyclic olefin polymer (COP). – The channel was embedded with dried anti-An or anti-B dried reagents.	Passive fluid flow	Undiluted or diluted blood with PBS (phosphate buffer saline) (1:5)	6.5 μL	Less than 10 min for undiluted blood and 100 s for diluted blood	Agglutination of RBCs	The embedded reagents triggered the agglutination reaction and exhibited 100% accuracy in identifying ABO blood typing.	Huet et al. (2018)

(*Continued*)

Table 11.1 A summary of the characteristic features of the lab-on-a-chip devices used in blood analysis. *Continued*

Application	Chip design	Mechanism	Injected sample	Required time	Required sample volume	Target	Results	References
— Blood/ plasma separation — Blood typing	The device consisted of a PDMS main channel for the blood transportation, a top plasma collector channels, and a bottom etched glass that has the array of pillars for plasma extraction.	Cell separation from the plasma was done by applying the hydrodynamic forces.	— Whole blood — IgM antibodies.	50 μL	10 min	— Plasma — Agglutination of RBCs	— The device could successfully yield a 12% of plasma with 100% purity. — The ABO/Rh grouping was successfully done by observing the agglutination in both antigens of RBCs (forward) and antibodies of plasma (reverse) via naked eyes and microscopic images. — The device could not detect the agglutination in highly viscous sample containing hematocrit higher than 50% but for hematocrit lower than that the results showed an agreement for all 4 types of blood.	Karimi et al. (2019)
Blood typing	Long-range surface plasmon resonance (LR-SPR) was consisted of high refractive index glass, Cytop film layer, and thin gold film. — The anti-A or anti-B was covalently immobilized on the gold surface.	— This LR-SPR based device worked by observing change in refractive index.	— RBC-An and RBC-B samples — Anti-An and Anti-B monoclonal antibody	The lowest detection limits were 1.58×10^5 cells/mL for RBC-An and 3.83×10^5 cells/mL for RBC-B.	10 min	RBC	— The results of the ABO blood typing produced by the LR-SPR based chip were consistent with those obtained from the gold standard test of agglutination. — The LR-SPR chip showed higher sensitivity compared to the Short-range SPR devices.	Tangkawsakul et al. (2016)
Blood typing	A surface plasmon resonance (SPR) array coupled with multiple anti-RBC antibodies was designed and placed in a flow chamber for multiplex blood typing.	The blood samples ran through the flow chamber and blood typing was determined by monitoring RBC-antibody binding on the SPR array.	— Diluted whole blood in PBS (120-fold dilution). — IgM and IgG antibodies.	200 μL	5 min	RBC	— The chip was reusable and could use to test at least 100 samples. — The ABO/Rh typing results gave 100% match with the results obtained from classical serology with all antibody except anti-E/e monoclonals	Szittner et al. (2019)

				350−500 μL	∼5 min	RBC		Yamamoto et al. (2020)
− Blood typing − Cross matching	− The PDMS device consisted of a dilutor, a homogenizer, and four detectors. − The dilutor was divided into three inlets for whole blood, PBS, and air, a T-shaped and a Y-shaped junction.	− The dilution was performed by applying a microbubble motion. − Preloaded blood typing reagents were remained stand by in the reaction chamber and blood typing was determined by the RBCs-specific antibody reaction. − Crossmatch was done by mixing RBCs with serum.	− Dilution of the blood with PBS was done automatically inside the device. − Anti-A, anti-B, and − anti-D reagents	350−500 μL	∼5 min	RBC	− Nonuniformity of the mixing of blood-PBS was notices at the end of the mixing channel because of the low flow disturbance created from the microbubbles. − The device exhibited high sensitivity to detect weak agglutinations even at level 1 + without any illumination. − No data was available for cross-matching.	Yamamoto et al. (2020)
− Blood/ plasma separation − Blood cross- matching	− The device consisted of three PDMS layers (3 mm thickness), one thin PDMS membrane layer (25 μm thickness), and two plasma separation membranes connected with a pressure chamber.	− Donor and recipients' blood were injected to the chip via two inlet and then separated individual's plasma through the two plasma separation membranes. − The cross-matching results were achieved by cross-reacting the donor's blood plasma with the recipient's whole blood (minor test) or recipient's blood plasma with the donor's whole blood (major test) by pushing and releasing the pressure chamber.	Whole blood	50 μL	10 min	− Plasma − RBC	− The transfusion suitability was interpreted depended on the degree of agglutination. 1. Compatible when agglutination did not occur in both major and minor test. 2. Incompatible when agglutination occurred in the major test but not in the minor test. 3. Incompatible when agglutination occurred in the minor test but not in the major test, but a little amount of blood could be transfused in emergencies. 4. Completely incompatible if agglutination occurred in both major and minor test.	Park and Park (2018)
Disease diagnosis- Anemia	− The paper-based microfluidic colorimetric chip was fabricated on Whatman (Grade-1) cellulose filter paper with the mean pore of 11 μm by inkjet printing. − A hydrophilic reaction pads for colorimetric detection was attached to guide the fluid flow through the porous network.	− Hemoglobin (Hgb) was detected from the redox reaction between 3,3′-Dimethyl- [1,1′-biphenyl]-4,4′-diamine (o-tolidine) and hydrogen peroxide yielding greenish-blue colored oxidized o-tolidine products − The Hgb concentration was estimated from the colorimetric signal intensity.	− Whole blood − Drabkin's solution	2 μL	2.5−3.5 min	Hgb concentration	− The results obtained from the colorimetric device showed strong correlation with the standard hematology analyzer ($r = 0.909$), − The device showed 87.5%, and 100% sensitivity for detecting mild, and severe anemia, respectively with 100% specificity in both cases. − The color-scale yielded quantitative measurement within 1.5 g/dL Hgb	Biswas et al. (2018)

(Continued)

Table 11.1 A summary of the characteristic features of the lab-on-a-chip devices used in blood analysis. *Continued*

Application	Chip design	Mechanism	Injected sample	Required time	Required sample volume	Target	Results	References
							concentration for 91% sample.	
Disease diagnosis- Anemia	– A rectangular microfluidic channel with a length, width, and height of 3 cm × 200 × 50 μm was fabricated by using PDMS. – A miniature microscope was designed to measure the intensity of the transmitted light.	Hemoglobin level was optically detected from whole blood without hemolysis by illuminating the blood in a microfluidic channel at a peak wavelength of 540 nm. – The absorbance of the samples was measured by using a CMOS sensor coupled with a lens to magnify the image onto the detector.	Diluted blood with calcium free Tyrode buffer	–	–	Hgb	– The device could measure Hgb concentration approximately 3.20 g/dL, 4 g/dL, and 5.5 g/dL for severe, moderate, mild cases of anemia, successively. – The device was enabled to detect the optical density for the flowing blood at a shear rate of 500 s^{-1}.	Taparia et al. (2017)
Disease diagnosis- Anemia	On-wafer based device was fabricated by silicon wafer and connected to computer interfaced software controlled electronic circuit.	The electrical parameters of the erythrocyte suspensions with varying levels of hematocrit were obtained by impedance/ capacitance spectroscopy and current-voltage (I-V) measurement.	Diluted blood with PBS	20 μL	<3 s	Hematocrit	– The detection limit for the hematocrit measurement was 3.5%. – The impedance value increased with increase hematocrit level while reduction was observed in capacitance values. – The device was highly sensitive and could detect the change in capacitance and impedance with the variation of the hematocrit level by 1%.	Chakraborty et al. (2020)
Disease diagnosis- Acute promyelocyte leukemia	– The microfluidic chip consisted of eight parallel microfluidic channels with separated inlets and outlets. – A chip holder was used to connect the tubes with the chip. – The quantity of the liquid flow through all channels could	– In bead-based sandwich ELISA, anti-PML-biotin antibody was used as capture antibody and the anti-RAR_-HRP (horseradish peroxidase) antibody as detection antibody. – The detection mechanism was followed by several	– NB-4, HL-60 and MV4–11 cell lines – Five primary patient samples: Three FAB (French–American–British classification system) type M3 samples and one FAB type M4 and one M5 (PML-RAR_-negative samples). – PBS	– The assay solution contained 1 10–0.1 μg total cell lysate and 0.1 μg protein from patient's sample.	~1 h	PML-RAR fusion protein	– The result suggested the fluorescence intensity of the positive signals was at least 20 times higher than the negative signals. – The chip was sensitive at very low concentration of fusion protein and was able to detect	Emde et al. (2020)

			Diluted blood with PBS				from 5 ng/μL concentration.	
	maintain via a sample loop in the three-way valve, where flow could control individually through all channels by using a motor selection valve.	steps: 1. the PML-RARα expression in cells was shown with subsequent cell lysis outside the chip. 2. The cell lysate was then incubated with the sandwich ELISA in the microfluidic chip. 3. The PML region of the fusion protein was bound with the anti-PML antibody, while RAR_ region was bound with the anti-RAR_-HRP antibody. 4. The reduction reaction of the fusion protein was catalyzed by HRP and converted resazurine to resorufine which was determined by measuring fluorescence intensity.					− The device required 98.5% less assay components than the conventional systems.	
Disease diagnosis-Leukemia	− The microfluidic PDMS device was consisted of a prefilter region and a single cell trapping region. − Prefilter section was constructed as pillar with a pitch of 25 μm. − The device had 16 parallel trapping channels.	− Size based cell separation via single-cell trapping array was done by passive hydrodynamic forces. − Identification of leukemia cell was done by analyzing phasor-FLIM (fluorescence lifetime imaging microscopy) imaging of the single cell.	Diluted blood with PBS	A large volume of sample could continuously use to trap the single cell.	3 min	− Leukemia cells- THP-1, Jurkat and K562 cells, − WBCs	− The trapping efficacy of WBC and leukemia cells was 78%. − Different leukemia cell lines were quantitatively differentiated from each other with AUC (area under curve) values higher than 0.95, which indicating high sensitivity and specificity for single cell analysis. − When 2% hematocrit sample was running through the device the rate of 0.2 mL/h, ∼72,000 single leukemia cells and WBCs could be trapped in 6 min.	Lee et al. (2018)

(*Continued*)

Table 11.1 A summary of the characteristic features of the lab-on-a-chip devices used in blood analysis. *Continued*

Application	Chip design	Mechanism	Injected sample	Required time	Required sample volume	Target	Results	References
Disease diagnosis- Acute lymphoblastic Leukemia	A herringbone straight channel was fabricated by using PDMS and coated the surface with capture antibodies.	The device separated peripheral blood lymphoblasts by using affinity separations technique.	Lymphoblasts spiked in blood	—	30 min	Lymphoblast	The device isolated CCRF-CEM lymphoblasts, patient-derived lymphoblast COG-LL332, and COG-LL317 cell lines with 82%−97%, 80%−97%, and 57%−92% purity, respectively with the initial spike concentrations of <1%.	Li et al. (2017)
Disease diagnosis- Multiple myeloma (MM)	The PDMS chip consisted of a sample prefilter region with multiple layers of micropillars and a cell capture region with 8 separate channels.	Cell sorting was based on the difference between clonal plasma cell and blood cell physical properties such as deformability, size, hydrodynamic properties through some microstructural units in the chip.	Human myeloma cancer cell lines spiked in healthy donor blood	— 1 mL for myeloma cell spiked blood. — 0.5 mL for patient's blood.	More than 4 h at the flow rate of 0.5 mL/h for 1 mL.	Clonal plasma cell	— The chip was able to filter out 99.99% of WBCs. — The capture efficacy of the device for clonal plasma cells from the myeloma cell spiked blood was approximately 40%−55%. — The device was enabled to isolate clonal plasma cell from MM relapsing patients and 117 cells were captures from 0.5 mL of blood while no cell was captured from the healthy donor.	Ouyang et al. (2019)
Disease diagnosis- Plasma cell disorders (PCDs) including monoclonal gammopathy of undetermined significance (MGUS), smoldering multiple myeloma (SMM), and multiple myeloma (MM).	— The device consisted of 50 sinusoidal channels with dimensions of 25 μm (width) and 150 μm (depth), fabricated in COC polymer using hot embossing and arranged in a Z-configuration. — The surface of the channels was modified with covalently attached anti-CD138 antibodies for selecting clonal plasma cells that expressed CD138.	— The device was designed to isolate the circulating clonal plasma cells via positive affinity selection from the whole blood. — To check the device efficacy, approximately ~ 500 RPMI-8226 cells were seeded into the whole blood from healthy donors (1 mL).	— Patient's whole blood — RPMI-8226 cell	1 mL of whole blood	1 mL of whole blood can be processed in ~1 h	Clonal plasma cells	— The device showed high sensitivity for isolating clonal plasma cells directly from the patient's blood even those with MUGS, while the higher cell burden was noticed in the patients with symptomatic MM. — Approximately 69% RPMI-8226 cells recovery was achieved by this device.	Kamande et al. (2018)

PBS (phosphate buffer saline) were introduced through the blood and buffer inlets, respectively, a boundary or a crossover layer between the two fluids was formed. The flow rate ratio between the whole blood and the PBS was 1:10, which pushed the cells onto the wall of the blood inlet and directed towards the bifurcation region. Because of the small pore size of 1.55 µm of the bead-packed channels, only blood plasma could flow through these channels. On the other hand, due to the smaller size and higher deformability, RBCs could squeeze through the 2 µm-necked channels. Only the bigger size WBCs traveled to the main channel and retained in the trapping unit. The dilution factor of the acquired plasma was checked by measuring the absorbance to confirm the quality of the plasma for further biomarker detection. The results demonstrated successful plasma extraction with a minimum dilution factor of 0.76 ×. The hemolysis effect and the intensity profile of the extracted plasma from the chip were similar to that from the standard centrifugation. Extracted RBCs were directly checked using blood typing, demonstrating the extraction of fully functional RBCs from the device. Utilizing the geometry of the trapping units, the device was able to trap up to ∼1800 WBCs in 20 minutes from 6 µL of blood.

A cyclic olefin copolymer (COC)-based LOC device was developed for blood/plasma separation via asymmetric capillary force on the POC platform. The main advantage of this design was that the lateral transportation of whole blood by capillary pumping without any external power. The detection was done by a capillary-driven lateral flow colorimetric assay. Another simple dielectrophoresis-based microfluidic device was designed to separate human plasma without any dilution to minimize the potential error of the traditional plasma separator (Yang et al., 2019). The chip design was straightforward with the main channel and a side channel (Fig. 11.2A). At the bottom surface of the chip, two thin electrodes were positioned perpendicularly to the main channel and parallel to each other, which generated the nonuniform electric field with a higher gradient on the junction area. When blood was infused through the main channel, the blood cells experienced a strong negative dielectrophoresis force against the hydrodynamic force that repelled the cells from the flow direction and directed them towards the side channels. The chip was completely optically transparent, allowing for real-time visualization of plasma separation under a microscope. Most importantly, the purity of the extracted plasma was almost 100% with a yield of approximately 31%.

Moreover, a microfluidic-based physiometer, equipped with a temperature controller, was designed to study the blood viscosity, hematocrit level, and RBCs deformability on a chip (Kim et al., 2019). The physiometer consisted of two main parts: a PDMS molded microviscometer and an electronic printed circuit board that worked as a microhemocytometer. The microviscometer was composed of 10 arrays, and each array had 100 microchannels. The results obtained from the device showed a strong correlation with the corresponding gold standard methods. The viscosity measured by the physiometer only showed $0.8\% \pm 1.4\%$ differences with the rotational cone-and-plate viscometer. The coefficient of variation for hematocrit measurement was $0.3\% - 1.2\%$ less for the physiometer than the result obtained from centrifugation. Also, measuring the deformability index using a physiometer and standard image processing exhibited a strong coefficient of determination of 0.97.

However, plasma separation from a finger-prick volume of blood remains an enduring challenge for the POC platform (Kersaudy-Kerhoas & Sollier, 2013). Dixon et al. (2020) proposed a digital-based portable microfluidic device to overcome this limitation. Their design had an integrated porous membrane for automated plasma separation. The chip was able to perform the complex,

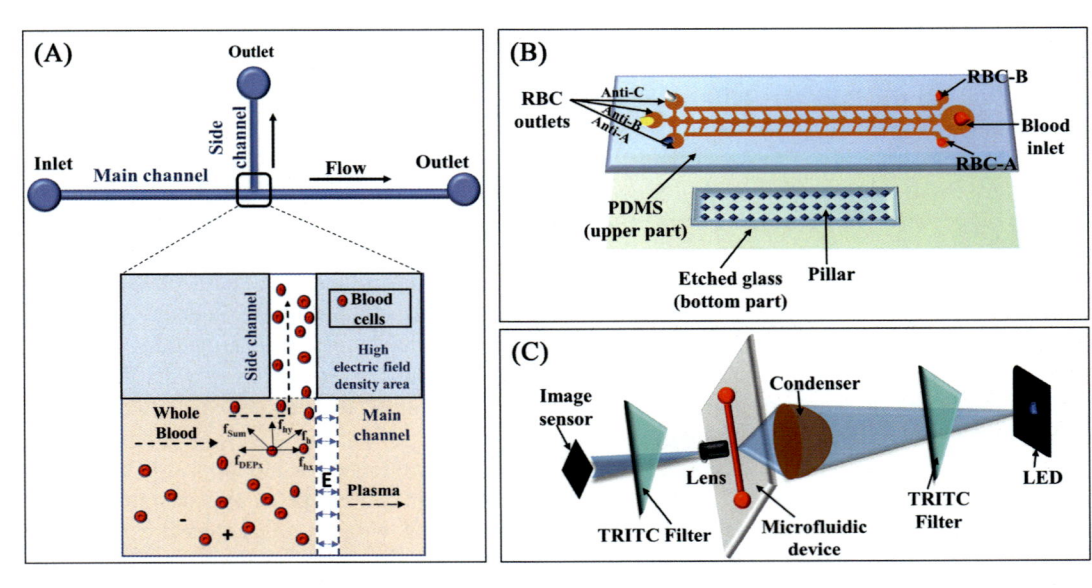

FIGURE 11.2

Schematic presentation of microfluidic chips for on-chip blood analysis and disease diagnosis. (A) A simple dielectrophoresis-based microfluidic device, consisted of a main channel and a side channel with ITO (indium tin oxide) etched two electrodes at the bottom glass surface for blood plasma separation by generating nonuniform electric field, redrawn from Yang, F., Zhang, Y., Cui, X., Fan, Y., Xue, Y., Miao, H. et al. (2019). Extraction of cell-free whole blood plasma using a dielectrophoresis-based microfluidic device. *Biotechnology Journal, 14*(3), Article e1800181. (B) A passive microfluidic blood typing detector, consisted of a top part and a bottom part. The top part was composed of a fish bone-shaped main channel and two plasma collector channels, while the bottom part was consisted of an array of pillars for RBCs filtration.

Adapted from Karimi, S., Mehrdel, P., Farré-Lladós, J., & Casals-Terré, J. (2019). A passive portable microfluidic blood-plasma separator for simultaneous determination of direct and indirect ABO/Rh blood typing. Lab on a Chip, 19*(19), 3249–3260. (C) An optically modified rectangular shaped microfluidic device for determining the haemoglobin level by measuring the optical density of the samples, redrew from Taparia, N., Platten, K. C., Anderson, K. B., & Sniadecki, N. J. (2017). A microfluidic approach for hemoglobin detection in whole blood.* AIP Advances, 7*(10), 105102.*

multistep diagnostic assay from finger-prick blood collection on a single chip. The blood samples could be loaded directly without any prior processing onto the device by finger prick, making their design accessible for POC diagnosis.

11.2.2 On-chip blood typing

Rapid and accurate blood typing is crucial for blood transfusion, organ transplantation, and any critical medical condition such as pregnancy (Karimi et al., 2019; Zhang et al., 2017). Among more than 30 blood typing methods, ABO/Rh (rhesus) systems attract the most attention due to the high mortality from mismatched ABO/Rh reactions during blood transfusion (Klein et al., 2015; Zhang et al., 2017). The ABO blood-type is categorized based on the presence or absence of certain antigens (An and B) on the surface of the RBCs (forward typing) and antibodies (anti-An and anti-B) in the plasma (reverse typing) (Karimi et al., 2019; Mujahid & Dickert, 2015), while the Rh type is identified based on the presence (Rh +) or absence (Rh-) of the D antigen in the RBCs (Yamamoto et al., 2020). The basic principle of blood typing relies on the agglutination of the RBCs because of the specific antigen-antibody reaction (Goodell et al., 2010). Forward blood typing is performed by mixing blood cells with the anti-A reagent and anti-B reagent separately. Agglutination of RBCs with the given reagent indicates a positive test result. Agglutination of RBCs with both reagents denotes type AB, while no agglutination, in either case, indicates type O. In reverse grouping, blood plasma is treated against known RBCs from groups An and B (Mujahid & Dickert, 2015). The plasma containing anti-An antibody reacts with the group B RBCs, while anti-B reacts with the group A RBCs and causes agglutination. Agglutination of RBCs, in either case, denotes a positive result. On the other hand, agglutination of RBCs in both cases indicates group O, while no agglutination means group AB. The reverse grouping is mainly done to determine the accuracy of the forward grouping. Similarly, Rh typing is done by adding RBCs with a reagent containing D antibodies. Agglutination of RBCs indicates Rh-positive, while no reaction means Rh-negative (Mujahid & Dickert, 2015). Conventionally blood group is mainly detected using tube test method, slide method, microplate-based method, gel column agglutination, and affinity column technology (Yamamoto et al., 2020) that require special laboratory equipment operated by skilled personnel with long turnaround times (Zhang et al., 2017). Hence, developing simple and cost-effective blood grouping systems is the key demand for facilitating such tools in emergencies and resource-limited areas.

Researchers are aiming to develop LOC-based POC devices for blood typing that could detect both antigens on RBCs (forward) and antibodies from plasma (reverse) from a single drop of blood (Casals-Terré et al., 2020; Chang et al., 2018; Karimi et al., 2019). Zhang et al. (2017) exploited a simple dye-assisted paper-based assay using immobilized antibodies and bromocresol green dye for rapid and reliable detection of ABO/Rh antigens. The device was capable of on-chip blood/plasma separation by capillary action and simultaneous forward-reverse blood grouping within 2 minutes on a single POC chip. The mechanism relies on the visual readout strategy of the reaction between bromocresol green (BCG) and serum protein. The device exhibited teal blue when the blood-type antigens were present and detected the brown color if the antigens were absent. Alongside, passive microfluidic chips in blood-plasma separation become popular because such devices do not require any external energy for fluid movement (Maria et al., 2017). Huet et al. (2018) prepared a passive microfluidic chip with embedded anti-An or anti-B dried reagents inside the microchannel for enhancing capillary flow. The chip provided a rapid and sensitive ABO forward blood typing platform for real-time observation and quantitative measurement of agglutination from a single blood drop with a microscope and a computational unit. Because of the simple operation techniques, only 5 seconds would require to fill the microchannel with the diluted blood sample. It took approximately 2 minutes to generate the data by dissolving the dried reagents in the blood.

Conversely, the working principle of the passive microfluidic devices relies only on capillary forces and may face clogging problems during cell-plasma separation (Karimi et al., 2019; Maria et al., 2017). Karimi et al. (2019) proposed a design by combining a passive microfluidic blood-plasma separator with a blood typing detector for both forward and reverse groupings. Their design allowed the separation of RBCs from plasma by applying hydrodynamic forces through the generation of stagnation zones, which minimized the clogging of cells and maximized the amount of the extracted plasma in the plasma collector. In fluid dynamics, the stagnation zone is a point in a flow field where the local fluid velocity is zero. The chip consisted of a top part containing a fishbone-shaped main channel with dead-end branches (RBCs outlet) and two plasma collector channels, while the bottom part consisted of an array of diamond-shaped pillars for RBCs filtration and directed the separated plasma towards the plasma collector channel (Fig. 11.2B). The system required only 50 µL of blood without any preanalytical modification. Initially, a droplet of blood was infused into the inlet of the main channel. The plasma was extracted through the arrays of the pillar and collected into the plasma collector channels by capillary forces. Moreover, during this step, the channel geometry allowed the generation of a stagnation zone at the dead-end branches and trapped the RBCs in the RBCs outlets. Specific reagents (anti-A, anti-B, anti-D) for ABO/Rh forward blood typing were added directly into the three RBCs outlets separately for observing agglutination of RBCs. For reverse typing, two known blood samples from Group A and group B were added to the plasma collector separately to observe the possible agglutination of the RBCs. The device achieved a 12% yield of plasma with 100% purity within 10 minutes. The agglutination of RBCs was easily detectable via naked eyes, but the validation was done by analyzing microscopic images to minimize the error. Moreover, to improve the device sensitivity, advanced methods were developed, such as surface plasmon resonance (Szittner et al., 2019; Tangkawsakul et al., 2016) and lab-on-a-disk platform (Chen et al., 2019). In addition, on-chip cross-matching POC tests were carried out by checking the hemocompatibility between the recipient's RBCs and the donor's plasma to prevent the hemolytic transfusion reaction during emergencies (Park & Park, 2018; Yamamoto et al., 2020).

11.2.3 On-chip disease diagnosis

A blood test is considered to be an essential diagnostic tool for monitoring many chronic health conditions. Human blood contains massive information regarding overall health functions. With the rapid technological advancement in LOC, extensive research has been conducted to invent faster and efficient microfluidic POC diagnosis devices for blood diseases. Anemia is perceived as one of the most prevalent blood disorders (Biswas et al., 2018). According to the World Health Organization, approximately a quarter of the world's population is severely affected by anemia (Biswas et al., 2018; Taparia et al., 2017). The causes are related to the fundamental hematologic ailments, iron deficiency, and chronic conditions such as kidney disease, cirrhosis, and infections. This condition is diagnosed by measuring the Hgb concentration in blood. Biswas et al. (2018) reported a simple paper-based colorimetric microfluidic device for POC detection of anemia by measuring the blood Hgb level. This user-friendly microfluidic POC chips comprised 16 circular detection sites, which were separated by a hydrophobic barrier. The colorimetric detection was based on the Hgb catalyzed redox reaction between 3,3′-Dimethyl-[1,1′- biphenyl]-4,4′-diamine (o-tolidine) and hydrogen peroxide, which only cost approximately 0.02$ per test. Heme groups of hemoglobin acts as a chemical catalyst to breakdown the hydrogen peroxide into nascent oxygen and water. Nascent oxygen then oxidizes o-tolidine (Goyal & Basak, 2009). After completing the reaction on the detection side, the color yield was ranging from the lighter shades of bluish-green to deeper-green of oxidized o-tolidine products based on different Hgb levels. The findings were validated by comparing the results from a standard hematology analyzer. This device exhibited

100% specificity with 87.5% and 100% sensitivity for mild and severe anemia successively. Besides, Taparia et al. (2017) developed an optically modified chip to measure the blood Hgb level without cell lysis by measuring the optical density of the samples for different hematocrit levels (Fig. 11.2C). The detection mechanism of Hgb level on their chip was the optical illumination of the samples at 540-nm wavelength and determining the absorbance of the sample by a complementary metal-oxide semiconductor (CMOS) sensor coupled with a lens to magnify the image onto the detector. Another study reported an on-chip quantitative estimation of hematocrit level by measuring the variation of the dielectric/capacitive nature of an erythrocyte cell suspension (Chakraborty et al., 2020). In addition, other groups developed POC diagnostic tool for iron deficiency by using aqueous multiphase systems (Hennek et al., 2016), and molecular fingerprints of Hgb (Saylan & Denizli, 2018) on-chip.

Furthermore, early detection of blood cancer is the prerequisite for successful individualized therapy. Microfluidic LOC techniques become a reliable and affordable platform for isolating different circulating blood cancer cells from the blood of patients in a high-throughput manner (Lee et al., 2018). However, the separation process is still challenging because of the size overlapping between the cancer cells and WBCs. To overcome this challenge, Lee et al. (2018) combined passive hydrodynamic control with phasor-FLIM (fluorescence lifetime imaging microscopy) imaging for label-free size-based cell separation and rapid screening. The device consists of a prefilter region to control the unwanted cell aggregation and a single cell trapping region. Each chip contains 16 identical arrays of highly packed 100 single-cell traps. Because of the height differences between the trapping unit and the main delivery channel, the smaller sized RBCs ran smoothly towards the outlet, but WBCs and leukemia cells were retained into the trapping region. The trapped leukemia cells were easily differentiable from the WBCs in the phasor-FLIM lifetime map based on their significant shift towards the shorter fluorescence lifetime. On-chip lymphoblast isolation was accomplished using affinity separation technique with a purity of approximately 82%−97%, 80%−97%, and 57%−92% for CCRF-CEM lymphoblasts, patient-derived lymphoblast COG-LL332, and COG-LL317 cell lines, respectively (Li et al., 2017). A recent study reported the development of a microfluidic chip for acute promyelocyte leukemia (APL) by the detection of promyelocytic leukemia-retinoic acid receptor α (PML-RARα) fusion protein, which is considered as a diagnostic biomarker in 95% APL cases (Emde et al., 2020). The chip could detect the PML-RARα in cell lysates by a sandwich ELISA (enzyme-linked immunosorbent assay) system on the surface of magnetic streptavidin-coated microparticles within an hour where conventional lab-based techniques require 1−3 days.

Plasma cell disorder, such as multiple myeloma (MM), is responsible for about 20% of death from hematological cancer. The current gold standard for MM detection includes expensive and invasive bone marrow aspiration for detecting clonal plasma cells. However, because of the lower sensitivity and nonuniform distribution of the clonal cells in the bone marrow, this test can produce inaccurate results. On the other hand, studies revealed that some clonal cells leave the bone marrow and entered the peripheral blood of the MM patients (Nowakowski et al., 2005; Ouyang et al., 2019). To isolate and detect clonal cells, microfluidic devices were developed to capture the circulating clonal plasma cells from peripheral blood of the MM patients (Kamande et al., 2018; Ouyang et al., 2019). Ouyang et al. (2019) designed a microfluidic platform to separate circulating clonal plasma cells (cCPC) based on the physical and mechanical properties of the cells. cCPCs possess unique physical and mechanical attributes such as larger size (30−50 μm) (Ouyang et al., 2019) and high cell membrane elastic modulus (~ 540 Pa) (Feng et al., 2010) that differentiate them easily from the ordinary blood cells. The device showed approximately 40%–55% capture efficacy from the myeloma cell spiked blood sample. cCPC. Moreover, an affinity selection-based microfluidic chip showed higher sensitivity in cCPC detection among patients with MGUS (monoclonal gammopathy of undetermined significance), smoldering MM, and symptomatic MM (Kamande et al., 2018). Additionally, incorporation of PCR (Polymerase chain reaction) or RT (reverse transcriptase)-PCR technique in microfluidic chips also allows for the detection of genetic blood disorders like thalassemia (Lien et al., 2009) and sickle cell anemia (Zhu, Palla, et al., 2013).

11.3 Smartphone-based platform for blood analysis and disease diagnosis

Smartphone-based biosensor system provides innovative solutions for rapid and accurate on-site qualitative and quantitative POC diagnosis for many diseases (Alawsi & Al-Bawi, 2019; Xu et al., 2018). This widely accessible portable device reduces the size and cost of the laboratory instruments and the professional operation demands (Romeo et al., 2016; Xu et al., 2018). The main advantage of the smartphone-based devices is that the complex sensing system is achievable by integrating additional features such as high-resolution miniature cameras, optical sensors, dongles, and electrical circuits with the in-built mobile sensors for the detection of a wide range of biological signals (Xu et al., 2015, 2018). Moreover, the powerful processor and the smartphone memories make it possible to store the test result, perform on-site data processing and conversion for real-time feedback, and wireless transmission of data to remote sites (Lee et al., 2020; Romeo et al., 2016). Smartphone-based biosensor systems can be categorized according to the analytical method and sensing modality (Xu et al., 2018). Detection via smartphone is generally based on different optical measurements such as bright-field, fluorescence, and/or colorimetric. In this section, we will discuss the widely applied sensing modalities for analyzing blood in smartphone-based LOC devices. Table 11.2 summarizes different smartphone-based devices used in blood analysis.

11.3.1 Microscopic imaging biosensor

Smartphone-based imaging sensors are one of the widely used sensing modalities because of the consistent advancement in the optical imaging hardware of the mobile phone over the past two decades (Ozcan, 2014). By advancing the integrated circuit design and manufacturing technologies, the resolutions of the embedded CMOS image sensor (CIS) cameras in smartphones reach more than 40 MP recently (Ozcan, 2014). Furthermore, by combining different compatible optomechanical attachments with the in-built phone lens and optoelectronic image sensors, the smartphone can be modulated into a microscope. Most smartphone-based microscopes are optical microscopes. A smartphone-based microscope mainly consists of a visible light source and a system of lenses to magnify sample images (Xu et al., 2015). The main purpose of such modification is to develop portable and cost-effective bright-field and fluorescence microscopy for on-site disease diagnosis (Vashist et al., 2014). Also, microscopic images taken with the smartphone can be processed and analyzed easily using simple image recognition software such as ImageJ (Wicks et al., 2017).

Table 11.2 A summary of smartphone-based devices used in blood analysis with various sensing modalities.

Sensing modality	Detect target	Types of smartphone devices	Accessories	Characterization Time	Characterization Sample volume	Characterization Cost	Application	Accuracy	Reference
Microscopic imaging-Bright-field, Fluorescence	WBCs (white blood cells), RBC (red blood cells), Hemoglobin (Hgb)	Samsung Galaxy SII with 8MPixel color camera and built-in lens with a focal length of f ~4 mm	− A cellphone-based attachment fixed on the top of the cellphone camera unit, worked as battery holder and universal port. − Two AA batteries. − Three separate add-on components for holding the samples. − An inexpensive plano-convex lens for fluorescence imaging. − Light-emitted-diodes (LEDs) for bright-field imaging.	<10 s	~10 μL	−	− Counting of RBC and WBC. − Determining the concentration of Hgb.	In a good agreement with Sysmex KN21 hematology analyzer and the correlation coefficient was ~0.98 and ~0.92 for cell count and Hgb concentration, respectively between the two methods.	Zhu, Sencan, et al. (2013)
Microscopic imaging— Bright-field	RBCs	Nokia N73 camera phones, with a 3.2-megapixel (204861536 pixel) CMOS camera and a 5.664.2 mm sensor, yielding a 2.7 mm pixel spacing.	− A 20× wide field microscope eyepiece (Model NT39−696, Edmunds Optics). − A 0.85 NA 60× Achromat objective (Model NT38−340, Edmunds Optics).	−	A peripheral blood smear was developed by using patient's blood	−	Sickle cell anemia	This system deliberated enough image resolution and contrast for the direct observation of sickled cells in blood smears taken from Hgb SS disease patients with no additional contrast-enhancing techniques.	Breslauer et al. (2009)
Microscopic imaging— Fluorescence	WBCs	Sony-Erickson U10i Aino with ~8 Mpixel color RGB sensor	− A simple lens − A plastic color filter − 3 LEDs − A battery	−	>0.1 mL	Cell phone attachments cost was <14 USD.	Cell density	A decent match was observed between the phone fluorescent images and the conventional fluorescent microscopic images.	Zhu, Yaglidere et al. (2011)
Microscopic imaging— Fluorescence	WBCs	Sony-Erickson U10i Aino with 8 Mpixel color RGB sensor and a built-in lens in front of the CMOS sensor with a focal length of f ~4.65 mm.	− A plano-convex lens − A plastic color filter − Two LEDs − Batteries	−	<10 μL	Cell phone attachment cost was <4 USD	Cell density	~95% accuracy against a reference hematology analyzer with a correlation coefficient of ~0.93.	Zhu, Mavandadi, et al. (2011)

(Continued)

Table 11.2 A summary of smartphone-based devices used in blood analysis with various sensing modalities. *Continued*

Sensing modality	Detect target	Types of smartphone devices	Accessories	Characterization			Application	Accuracy	Reference
				Time	Sample volume	Cost			
Microscopic imaging—Fluorescence	WBCs	iPhone 8	− A three-layer paper-based microfluidic device for cell sorting. − XFox Professional 300X Optical Glass Lenses − a 500 nm long-pass optical filter − A LED illuminator	5 min	1−4 μL	−	− Sorting of WBCs from whole blood − Quantification of WBCs	∼95% accuracy was noticed compared to the hemocytometer-based manual counting.	Bills et al. (2019)
Optical-intrinsic photothermal response	Hgb	Samsung Galaxy S8 smartphone (Android 8.0.0 operating system), containing a Samsung Exynos 8895 system-on-chip (SoC) with f eight CPU cores.	− Photothermal angular scattering (PTAS) sensor. − A capillary tube for loading blood sample. − Laser diodes (LDs) as probe and PT (photo-therma) excitation light sources. − A circular aperture (700 μm in diameter) was used to control the intensity of probe light. − A complimentary metal-oxide semiconductor (CMOS) sensor for data acquisition and transferring the data to the mobile device for processing.	<8 s	<150 nL	<0.20 USD for analyzing each sample	Detection of anemia	∼99% accuracy against a reference hematology analyzer with 1.21% of coefficient of variation.	(Lee et al., 2020)
Colorimetric	Hgb	An android phone with AutoCAD 360 app	− A 3D printed point-of-care chip with a rotating ring channel	The auto-mixing of reagents with blood via capillary force was demonstrated within 1 s	∼ 5 μL	Cost per test ∼ 50 cents	Detection of anemia	− The correlation analysis between 3D device and the hematology analyzer gave the coefficient of variation in the range of ∼ 0.2%−5%. − Sensitivity of the test was 81.2% for detecting severe anemia, and 100% for mild anemia.	Plevniak et al. (2016)

								− The diagnostic specificity is 100% for detecting severe anemia, and 83.3% for mild anemia.	
Colorimetric	Hgb	Lenovo S660 (Android version 4.1.2).	− 24 well plate − In-house Hgb calculator app	−	5 µL	Cost per test ∼$0.15	Detection of anemia	− The value showed the close agreement to the value achieved from the automated hematology Analyzer with the coefficient of determination of 0.976. − Sensitivity of the test was about 94%. − Specificity of the test was about 90%	Ghatpande et al. (2016)
Colorimetric	RBCs	Galaxy S II, Samsung	− A disposable microfluidic device. − (PDMS) light diffuser. − a PMMA [Poly (methyl methacrylate)] white acrylic box for avoiding variable external light.	−	10 µL	−	Quantification of blood hematocrit	The limit of detection (LOD) obtained from the developed platform was 0.1% of hematocrit with a sensitivity of 0.53 GSV (Gray-scale-valuation) (a.u.)/ hematocrit%	Kim et al. (2017)
Colorimetric	RBCs	Galaxy S II, Samsung	− disposable microfluidic device. − In-house histogram app.	−	10 µL	−	Quantification of blood hematocrit	LOD obtained from the developed platform was 10% of hematocrit	Jalal et al. (2017)
Electrochemical	WBCs	HTC	− Batteries. − A portable potentiostat. − HC-05 Bluetooth Serial − Module. − Power regulator. − Microcontroller.	1 min	10 µL	−	− Leukocytosis − Leukopenia	High repeatability as low as 10% of coefficient of variation.	Wang et al. (2017)

11.3.1.1 Bright-field imaging

Bright-field imaging modality is the simplest optical microscopy modality for POC sensing in smartphone-based devices (Liu et al., 2019; Xu et al., 2018). The primary imaging mechanism of smartphone-based microscopy depends on shadow imaging. Generally, the sample is illuminated uniformly by using a white light source. The light is transmitted through the sample and creates contrasts because of the varied attenuation of the transmitted light in different densities (Drey et al., 2013). Then, CSI camera of the phone captures the images. Zhu, Sencan, et al. (2013) developed a cellphone-based imaging cytometry platform where a 3-part optomechanical system measured the RBCs and WBCs densities and Hgb concentrations (Fig. 11.3A). The device consists of three parts: (1) a cell phone (Samsung Galaxy SII), (2) a cell phone base attachment, which worked as a battery holder and universal port, and (3) three separate add-on sample holders. Samples were placed separately in three different add-on compartments for RBC imaging/counting, WBC imaging/counting, and Hgb concentration measurement. Light-emitting-diodes (LEDs) were used as an illuminator for the bright-field images. For detecting the test value, only approximately 10 µL of the sample was required. A custom-developed application enabled the rapid analysis of the captured microscopic images, which provided the information related to the cell number and Hgb concentration within 10 seconds. For both RBCs and WBCs, the RGB (red, blue, green) values of the acquired images were converted into HS (hue, saturation) values and the maximum intensity contrast between the cells and background was calculated to count the cells. RGB model refers to the biological processing of colors in the human visual system, while HSV (hue, saturation, values) model corresponds to the human perception of the color similarity (Kahu et al., 2019; Loesdau et al., 2014). From HSV images, it is possible to separate the chromatic (hue and saturation) and achromatic (value) information, making it advantageous to measure color information independently from value information (Loesdau et al., 2014). Besides, Hgb concentration was measured by the light transmission intensity of the lysed blood samples. The data for cell counting was comparable with the standard bench-top hematology analyzer (Sysmex KN21), and the correlation coefficient was ~ 0.98 between the two methods. The Hgb concentration from the cellphone-based system demonstrated a good agreement with that obtained from the Sysmex KN21 hematology analyzer with a correlation coefficient of ~ 0.92. The data could be stored in the phone memory or transmitted to the central server by a wireless network for remote diagnosis that made this cost-effective blood analyzer advantageous to use in resource-limiting areas.

11.3.1.2 Fluorescence imaging

Smartphone-based fluorescence bioimaging sensor has been widely used because of the simple working principle, high specificity, and strong sensitivity in the detection of the biomarkers (Arts et al., 2016; Priye et al., 2017; Xu et al., 2018). The optomechanical modification of such devices is different from the bright-field imaging technique. Smartphone-based fluorescence microscope generally contains an excitation, and an emission filter, and an excitation source such as laser diodes (Xu et al., 2018). A simple, cost-effective cellphone-based fluorescence microscopic platform was developed for wide-field imaging and rapid quantification of fluorescently labeled biomarkers using a small blood volume (Zhu, Yaglidere, et al., 2011). The existing cellphone camera was modified by compact and light-weighted (28 g) attachments, including 3 LEDs, a simple lens, and a mechanical holder for the plastic color filter for achieving 10 mm resolution across

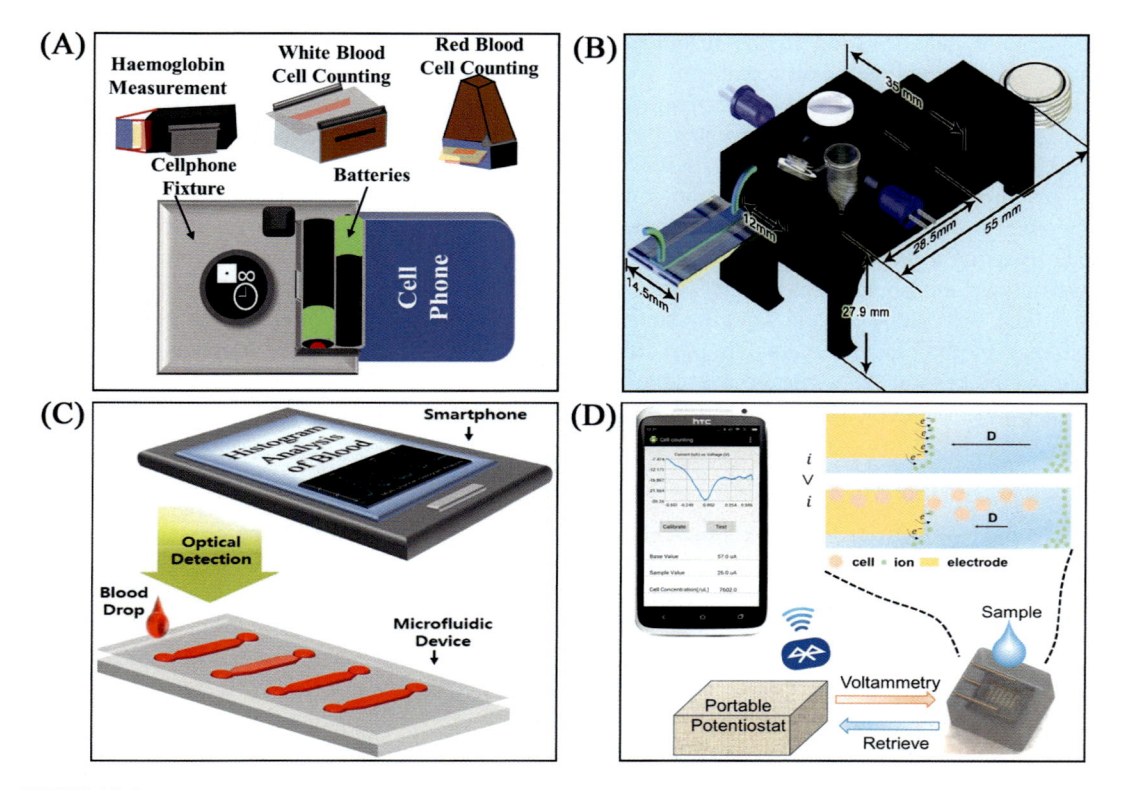

FIGURE 11.3

Examples of smartphone-based bioimaging sensors. (A) A bright-field imaging sensor for counting blood cells and measuring hemoglobin concentration.

Adapted from Zhu, H., Sencan, I., Wong, J., Dimitrov, S., Tseng, D., Nagashima, K. et al. (2013). Cost-effective and rapid blood analysis on a cell-phone. Lab on a Chip, 13(7), 1282–1288. (B) A wide-field fluorescent imaging sensor on the smartphone for blood cells counting, adapted from Zhu, H., Mavandadi, S., Coskun, A. F., Yaglidere, O., & Ozcan, A. (2011). Optofluidic fluorescent imaging cytometry on a cell phone. Analytical Chemistry, 83(17), 6641–6647. (C) A colorimetric optical sensor for determining blood haematocrit, adapted from Jalal, U. M., Kim, S. C., & Shim, J. S. (2017). Histogram analysis for smartphone-based rapid hematocrit determination. Biomedical Optical Express, 8(7), 3317–3328. (D) An electrochemical counter for label-free counting of white blood cells, adapted from Wang, X., Lin, G., Cui, G., Zhou, X., & Liu, G. L. (2017). White blood cell counting on smartphone paper electrochemical sensor. Biosensors and Bioelectronics, 90, 549–557.

a field-of-view of 81 mm^2. The main advantage of this design was the changeable LEDs and the plastic filter for different excitation/emission wavelengths that made the device suitable for a wide range of fluorophores. The device feasibility was investigated by detecting the fluorescently labeled WBCs from whole blood samples. The WBCs were labeled with STYO16 nucleic acid staining and excited with blue LEDs at 470 nm peak wavelength. The images were taken by the phone camera and compared to that taken by the conventional fluorescent microscope. The number of the cells showed a decent match in both methods that suggested the usefulness of the device for wide-field imaging and quantification in the resource-limited settings. The same group developed another optofluidic imaging cytometry platform for detecting fluorescently labeled WBCs (Fig. 11.3B) (Zhu, Mavandadi, et al., 2011). The main difference of this device from the previous design was the use of disposable microfluidic channels for the continuous flow of the samples with a syringe pump at the flow rate of ~ 1 μL/minute for 5−6 minutes. The team also reduced the manufacturing cost by approximately US\$ 10 and the sample volume (~ 10 times lower) than the old design. Moreover, the imaging resolution of the device was improved up to ~ 2 μm in fluorescent imaging mode by using a plano-convex lens with a focal length of ~ 0.6 mm. The microfluidic device was placed above the phone camera and also coupled with the cheap LEDs to use as a multilayered optofluidic waveguide for the sample excitation. The microscopic fluorescence movie of the cells was taken by the phone camera. The rapid digital photo processing was done by using the contour detection algorithm. The system was able to count the labeled cells automatically. The results obtained from the device matched well with the standard hematology analyzer with 95% accuracy.

11.3.2 Colorimetric detection

Colorimetric optical sensing is the most commonly used technique in smartphone-based POC diagnosis, where an ambient light sensor collects the photons from the sample directly and then capture the images by using the phone camera (Dutta, 2019; Liu et al., 2019). A simple smartphone-based colorimetric test mainly requires illumination and image processing with the in-built CMOS image sensor (Liu et al., 2019). This test detects the concentration of biomarkers in the sample with the help of the color reagent, based on the absorbance or reflected intensity differences of the biomarker-reagent complexes (Heo et al., 2014). Plevniak et al. (2016) reported a smartphone integrated 3D microfluidic auto-mixing device that can perform colorimetric quantification of Hgb level. The unique design of the device with a rotating ring channel enabled the rapid mixing of the blood and oxidizing agent inside the device within a second by capillary force. An in-house written color-scale analytical app with the smartphone output app (AutoCAD 360) was used for the colorimetric detection of blood Hgb. The visual readout was depending on the oxidation−reduction reaction in the blood Hgb and oxidizing agent. With increased Hgb concentration, the visual representation of the tested samples presented in a color span blue, green, yellow, orange, and red. The photospectrometer measurements calibrated the visual data for quantitative measurement of Hgb concentrations. The device is suitable for the color-blind patients. Besides the low cost, this colorimetric based microfluidic device showed almost 100% specificity for detecting severe anemia, while 83.3% for mild anemia. Another smartphone-based application was developed for the accurate and sensitive colorimetric measurement of blood Hgb level by Ghatpande et al. (2016). The colorimetric estimation was measured using an in-house app, named Hb calculator, operated by a Lenovo S660 (Android version 4.1.2). The mobile phone app converted the variation of color intensity into digitalized values to estimate the Hgb concentration. This simple device only used 5 μL of blood for analysis and showed approximately 94% specificity with 90% sensitivity in Hgb detection.

Moreover, Kim et al. (2017) directly used the intrinsic programs of the smartphone for analyzing blood hematocrit and imaging. A specialized optical setup, including an optical diffuser and a reflector, was used to avoid the image burning via the smartphone flashlight and the ambient light effect during picture collection. The results showed a linear increase of GSV (Gray-scale-valuation) value with the raising hematocrit concentration of the sample. The same group later proposed another smartphone-based microfluidic chip for rapid detection and quantification of blood hematocrit by an in-house "Histogram" app (Fig. 11.3C) (Jalal et al., 2017). The main advantage of the later design was the integration of a software-based histogram algorithm instead of the hardware modification. The integrated "Histogram" app could automatically detect the blood samples in the microfluidic channels and quantify the blood hematocrit in equal and varying optical conditions with auto-calibration. However, the limit of detection (LOD) obtained from the previous design was 0.1% of hematocrit, while the later one needed at least 10% of hematocrit present in the sample.

11.3.3 Electrochemical biosensing

An electrochemical-based sensing system works by transferring the electrochemical information into an analytically useful signal. In an electrochemical sensing system, two components including a chemical (molecular) recognition system and a physicochemical transducer for signal detection, are required to form a sensing electrode (Faridbod et al., 2011). POC devices with an electrochemical sensor are widely used in medical testing and analytical chemistry because of the high precision, rapid responses, cost-efficacy, and portability (Yu, Zhang, Buscaglia, et al., 2016; Yu, Zhang, Chang, et al., 2016; Zhang et al., 2018). A smartphone-based electrochemical counter was proposed by Wang et al. (2017) for label-free WBC counting (Fig. 11.3D). The device was based on microporous paper with gold microelectrodes. When the cells were trapped on the membrane and covered the microelectrodes, the voltammetry signal was generated from a portable potentiostat. Simultaneously, the received signal was retrieved and transmitted to the smartphone through the Bluetooth module. A smartphone app analyzed the signals and evaluated the cell concentration value within a minute. The system could quantify the cell concentration, covering from physiological to the pathological range, by using only 10 μL of blood.

11.4 Conclusion and perspective

Microfluidic LOC devices are bridging the gap between the laboratory centralized blood test and POC platform for the early detection of diseases. LOC devices reduce the sample volume on a micrometer scale and minimizes the data acquisition time. Combining smartphone-based sensing technology with LOC drastically cuts down the per-test cost to less than a dollar with high accuracy and sensitivity. Though there are significant research efforts to develop simple, low cost, portable POC devices, the commercialization of such devices remains restricted. Scalability is an important requirement for commercialization, where complex device structures create challenges. Introducing 3D printing in device fabrication instead of lithographic techniques proves beneficial in case of scalability. Meanwhile, most LOC devices are not fully automatic and require external equipment for sample handling and preparation. Moreover, data acquisition and processing also need expensive instruments such as microscopes and computational units. Besides, smartphone-based devices use in-house programs that are not available for mass application. Researchers should consider using the in-built smartphone Apps or design user-friendly programs for translating data into a simple visual readout. Smartphone-based devices can cause potential contamination because of poor hygiene control. Automation of all assay steps such as sample and reagent integration, mixing, separation, washing, and data acquisition might minimize the error and contamination. To achieve this aim, researchers can consider installing simple reagent storage and micropump for automatic handling without manual intervention. For disease diagnosis, most studies used spiked samples by adding a known quantity of cells of interest, such as tumor cells with the blood that made it challenging to translate the test results into the clinical setup. Clinical samples directly collected from the patients should be analyzed instead of spiked samples to validate the integrity of the devices. Finally, multiplexing, or simultaneous detection of different blood analytes is still not possible by LOC devices. In the future, scientists should make significant efforts to develop a multiplex POC platform from complete blood analysis to disease diagnosis.

Abbreviations

CMOS	Complementary metal-oxide-semiconductor
cCPC	Circulating clonal plasma cell
CIS	CMOS image sensor
Hgb	Hemoglobin
LEDs	Light-emitting-diodes
LOC	Lab-on-a-chip
MM	Multiple myeloma
PBS	Phosphate buffer saline
PDMS	Polydimethylsiloxane,
POC	Point-of-care
RBCs	Red blood cells
Rh	Rhesus
WBCs	White blood cells

References

Alawsi, T., & Al-Bawi, Z. (2019). A review of smartphone point-of-care adapter design. *Engineering Reports*, *1*(2), e12039.

Arshavsky-Graham, S., & Segal, E. (2020) *Lab-on-a-chip devices for point-of-care medical diagnostics* (pp. 1−19). In: Advances in Biochemical Engineering/Biotechnology. Berlin, Heidelberg: Springer. Available from: https://doi.org/10.1007/10_2020_127

Arts, R., den Hartog, I., Zijlema, S. E., Thijssen, V., van der Beelen, S. H. E., & Merkx, M. (2016). Detection of antibodies in blood plasma using bioluminescent sensor proteins and a smartphone. *Analytical Chemistry*, *88*(8), 4525−4532.

Bills, M. V., Nguyen, B. T., & Yoon, J.-Y. (2019). Simplified white blood cell differential: An inexpensive, smartphone- and paper-based blood cell count. *IEEE Sensors Journal*, *19*(18), 7822−7828.

Biswas, S. K., Bandyopadhyay, S., Kar, S., Som, N. K., & Chakraborty, S. (2018). Anemia diagnosis on a simple paper-based assay. *bioRxiv*, 439224.

Breslauer, D. N., Maamari, R. N., Switz, N. A., Lam, W. A., & Fletcher, D. A. (2009). Mobile phone based clinical microscopy for global health applications. *PLoS One*, *4*(7), e6320.

Casals-Terré, J., Farré-Lladós, J., López, J. A., Vidal, T., & Roncero, M. B. (2020). Enhanced fully cellulose based forward and reverse blood typing assay. *Journal of Biomedical Materials Research. Part B, Applied Biomaterials*, *108*(2), 439–450.

Chakraborty, S., Das, S., Das, C., Chandra, S., Sharma, K. D., Karmakar, A., et al. (2020). On-chip estimation of hematocrit level for diagnosing anemic conditions by Impedimetric techniques. *Biomedical Microdevices*, *22*(2), 38.

Chang, Y. J., Fan, Y. H., Chen, S. C., Lee, K. H., & Lou, L. Y. (2018). An automatic lab-on-disc system for blood typing. *SLAS Technology*, *23*(2), 172–178.

Chen, J.-Y., Huang, Y.-T., Chou, H.-H., Wang, C.-P., & Chen, C.-F. (2015). Rapid and inexpensive blood typing on thermoplastic chips. *Lab on a Chip*, *15*(24), 4533–4541.

Chen, Y.-W., Li, W.-T., Chang, Y., Lee, R.-H., & Hsiue, G.-H. (2019). Blood-typing and irregular antibody screening through multi-channel microfluidic discs with surface antifouling modification. *Biomicrofluidics.*, *13*(3), 034107.

Dixon, C., Lamanna, J., & Wheeler, A. R. (2020). Direct loading of blood for plasma separation and diagnostic assays on a digital microfluidic device. *Lab on a Chip*, *20*(10), 1845–1855.

Drey, L. L., Graber, M. C., & Bieschke, J. (2013). Counting unstained, confluent cells by modified bright-field microscopy. *Biotechniques*, *55*(1), 28–33.

Dutta, S. (2019). Point of care sensing and biosensing using ambient light sensor of smartphone: Critical review. *TrAC Trends in Analytical Chemistry*, *110*, 393–400.

Emde, B., Kreher, H., Bäumer, N., Bäumer, S., Bouwes, D., & Tickenbrock, L. (2020). Microfluidic-based detection of AML-specific biomarkers using the example of promyelocyte leukemia. *International Journal of Molecular Science*, *21*(23).

Faridbod, F., Gupta, V. K., & Zamani, H. A. (2011). Electrochemical sensors and biosensors. *International Journal of Electrochemistry*, *2011*, 352546.

Feng, Y., Ofek, G., Choi, D. S., Wen, J., Hu, J., Zhao, H., et al. (2010). Unique biomechanical interactions between myeloma cells and bone marrow stroma cells. *Progress in Biophysics and Molecular Biology*, *103*(1), 148–156.

Ghatpande, N. S., Apte, P. P., Joshi, B. N., Naik, S. S., Bodas, D., Sande, V., et al. (2016). Development of a novel smartphone-based application for accurate and sensitive on-field hemoglobin measurement. *RSC Advances*, *6*(106), 104067–104072.

Goodell, P. P., Uhl, L., Mohammed, M., & Powers, A. A. (2010). Risk of hemolytic transfusion reactions following emergency-release RBC transfusion. *American Journal of Clinical Pathology*, *134*(2), 202–206.

Goyal, M. M., & Basak, A. (2009). Estimation of plasma haemoglobin by a modified kinetic method using o-tolidine. *Indian Journal of Clinical Biochemistry*, *24*(1), 36–41.

Hennek, J. W., Kumar, A. A., Wiltschko, A. B., Patton, M. R., Lee, S. Y., Brugnara, C., et al. (2016). Diagnosis of iron deficiency anemia using density-based fractionation of red blood cells. *Lab on a Chip*, *16*(20), 3929–3939.

Heo, J. H., Cho, H. H., Lee, J. W., & Lee, J. H. (2014). Achromatic–chromatic colorimetric sensors for on-–off type detection of analytes. *Analyst.*, *139*(24), 6486–6493.

Huet, M., Cubizolles, M., & Buhot, A. (2017). Real time observation and automated measurement of red blood cells agglutination inside a passive microfluidic biochip containing embedded reagents. *Biosensors & Bioelectronics*, *93*, 110–117.

Huet, M., Cubizolles, M., & Buhot, A. (2018). Red blood cell agglutination for blood typing within passive microfluidic biochips. *High Throughput*, *7*(2).

Jalal, U. M., Kim, S. C., & Shim, J. S. (2017). Histogram analysis for smartphone-based rapid hematocrit determination. *Biomedical Optical Express*, *8*(7), 3317–3328.

Kahu, S. Y., Raut, R. B., & Bhurchandi, K. M. (2019). Review and evaluation of color spaces for image/video compression. *Color Research & Application*, *44*(1), 8–33.

Kamande, J. W., Lindell, M. A. M., Witek, M. A., Voorhees, P. M., & Soper, S. A. (2018). Isolation of circulating plasma cells from blood of patients diagnosed with clonal plasma cell disorders using cell selection microfluidics. *Integrative Biology*, *10*(2), 82−91.

Karimi, S., Mehrdel, P., Farré-Lladós, J., & Casals-Terré, J. (2019). A passive portable microfluidic blood-−plasma separator for simultaneous determination of direct and indirect ABO/Rh blood typing. *Lab on a Chip*, *19*(19), 3249−3260.

Kersaudy-Kerhoas, M., & Sollier, E. (2013). Micro-scale blood plasma separation: From acoustophoresis to egg-beaters. *Lab on a Chip*, *13*(17), 3323−3346.

Kim, B. J., Lee, Y. S., Zhbanov, A., & Yang, S. (2019). A physiometer for simultaneous measurement of whole blood viscosity and its determinants: Hematocrit and red blood cell deformability. *Analyst.*, *144*(9), 3144−3157.

Kim, S. C., Jalal, U. M., Im, S. B., Ko, S., & Shim, J. S. (2017). A smartphone-based optical platform for colorimetric analysis of microfluidic device. *Sensors and Actuators B: Chemical*, *239*, 52−59.

Klein, H. G., Cortés-Puch, I., & Natanson, C. (2015). Blood-transfusion decisions not simple. *Nature*, *521* (7552), 289.

Kuan, D.-H., & Huang, N.-T. (2020). Recent advancements in microfluidics that integrate electrical sensors for whole blood analysis. *Analytical Methods*, *12*(26), 3318−3332.

Kuan, D. H., Wang, I. S., Lin, J. R., Yang, C. H., Huang, C. H., Lin, Y. H., et al. (2016). A microfluidic device integrating dual CMOS polysilicon nanowire sensors for on-chip whole blood processing and simultaneous detection of multiple analytes. *Lab on a Chip*, *16*(16), 3105−3113.

Kuan, D. H., Wu, C. C., Su, W. Y., & Huang, N. T. (2018). A microfluidic device for simultaneous extraction of plasma, red blood cells, and on-chip white blood cell trapping. *Science Reports*, *8*(1), 15345.

Lee, D. H., Li, X., Ma, N., Digman, M. A., & Lee, A. P. (2018). Rapid and label-free identification of single leukemia cells from blood in a high-density microfluidic trapping array by fluorescence lifetime imaging microscopy. *Lab on a Chip*, *18*(9), 1349−1358.

Lee, J., & Lee, S.-H. (2013). Lab on a chip for in situ diagnosis: From blood to point of care. *Biomedical Engineering Letters*, *3*(2), 59−66.

Lee, J., Song, J., Choi, J.-H., Kim, S., Kim, U., Nguyen, V.-T., et al. (2020). A portable smartphone-linked device for direct, rapid and chemical-free hemoglobin assay. *Scientific Reports*, *10*(1), 8606.

Lee, K. K., Kim, M. O., & Choi, S. (2019). A whole blood sample-to-answer polymer lab-on-a-chip with superhydrophilic surface toward point-of-care technology. *Journal of Pharmaceutical and Biomedical Analysis*, *162*, 28−33.

Microfluidics in lab-on-a-chip: Models, simulations and experiments. In D. Li (Ed.), *Microscale heat transfer fundamentals and applications*. Dordrecht, Netherlands: Springer.

Li, T., Fan, Y., Cheng, Y., & Yang, J. (2013). An electrochemical Lab-on-a-CD system for parallel whole blood analysis. *Lab on a Chip*, *13*(13), 2634−2640.

Li, W., Zhang, Y., Reynolds, C. P., & Pappas, D. (2017). Microfluidic separation of lymphoblasts for the isolation of acute lymphoblastic leukemia using the human transferrin receptor as a capture target. *Analytical Chemistry*, *89*(14), 7340−7347.

Lien, K. Y., Liu, C. J., Kuo, P. L., & Lee, G. B. (2009). Microfluidic system for detection of alpha-thalassemia-1 deletion using saliva samples. *Analytical Chemistry*, *81*(11), 4502−4509.

Liu, J., Geng, Z., Fan, Z., Liu, J., & Chen, H. (2019). Point-of-care testing based on smartphone: The current state-of-the-art (2017−2018). *Biosensors and Bioelectronics*, *132*, 17−37.

Hue and saturation in the RGB color space. In M. Loesdau, S. Chabrier, & A. Gabillon (Eds.), *Image and signal processing*. Cham: Springer International Publishing.

Mabey, D., Peeling, R. W., Ustianowski, A., & Perkins, M. D. (2004). Diagnostics for the developing world. *Nature Reviews: Microbiology*, *2*(3), 231−240.

Mach, A. J., Adeyiga, O. B., & Di Carlo, D. (2013). Microfluidic sample preparation for diagnostic cytopathology. *Lab on a Chip*, *13*(6), 1011–1026.

Madadi, H., Casals-Terré, J., & Mohammadi, M. (2015). Self-driven filter-based blood plasma separator microfluidic chip for point-of-care testing. *Biofabrication.*, *7*(2), 025007.

Mantegazza, A., Clavica, F., & Obrist, D. (2020). In vitro investigations of red blood cell phase separation in a complex microchannel network. *Biomicrofluidics.*, *14*(1), 014101.

Maria, M. S., Chandra, T. S., & Sen, A. K. (2017). Capillary flow-driven blood plasma separation and on-chip analyte detection in microfluidic devices. *Microfluid Nanofluidics*, *21*(4), 72.

Mujahid, A., & Dickert, F. L. (2015). Blood group typing: From classical strategies to the application of synthetic antibodies generated by molecular imprinting. *Sensors*, *16*(1), 51, Basel.

Nguyen, J., Wei, Y., Zheng, Y., Wang, C., & Sun, Y. (2015). On-chip sample preparation for complete blood count from raw blood. *Lab on a Chip*, *15*(6), 1533–1544.

Nikoleli, G.-P., Siontorou, C. G., Nikolelis, D. P., Bratakou, S., Karapetis, S., & Tzamtzis, N. (2018). Chapter 13—Biosensors based on microfluidic devices lab-on-a-chip and microfluidic technology. In D. P. Nikolelis, & G.-P. Nikoleli (Eds.), *Nanotechnology and biosensors* (pp. 375–394). Elsevier.

Nowakowski, G. S., Witzig, T. E., Dingli, D., Tracz, M. J., Gertz, M. A., Lacy, M. Q., et al. (2005). Circulating plasma cells detected by flow cytometry as a predictor of survival in 302 patients with newly diagnosed multiple myeloma. *Blood*, *106*(7), 2276–2279.

Ouyang, D., Li, Y., He, W., Lin, W., Hu, L., Wang, C., et al. (2019). Mechanical segregation and capturing of clonal circulating plasma cells in multiple myeloma using micropillar-integrated microfluidic device. *Biomicrofluidics.*, *13*(6), 064114.

Ozcan, A. (2014). Mobile phones democratize and cultivate next-generation imaging, diagnostics and measurement tools. *Lab on a Chip*, *14*(17), 3187–3194.

Park, J., & Park, J. K. (2018). Finger-actuated microfluidic device for the blood cross-matching test. *Lab on a Chip*, *18*(8), 1215–1222.

Plevniak, K., Campbell, M., & Mei, H. (2016). 3D printed microfluidic mixer for point-of-care diagnosis of anemia. *Annual International Conference on IEEE Engineering, Medical and Biological Society*, *2016*, 267–270.

Priye, A., Bird, S. W., Light, Y. K., Ball, C. S., Negrete, O. A., & Meagher, R. J. (2017). A smartphone-based diagnostic platform for rapid detection of Zika, chikungunya, and dengue viruses. *Scientific Reports*, *7*, 44778.

Romeo, A., Leung, T. S., & Sánchez, S. (2016). Smart biosensors for multiplexed and fully integrated point-of-care diagnostics. *Lab on a Chip*, *16*(11), 1957–1961.

Saylan, Y., & Denizli, A. (2018). Molecular fingerprints of hemoglobin on a nanofilm chip. *Sensors*, *18*(9), Basel.

Szittner, Z., Bentlage, A. E. H., van der Donk, E., Ligthart, P. C., Lissenberg-Thunnissen, S., van der Schoot, C. E., et al. (2019). Multiplex blood group typing by cellular surface plasmon resonance imaging. *Transfusion*, *59*(2), 754–761.

Tangkawsakul, W., Srikhirin, T., Shinbo, K., Kato, K., Kaneko, F., & Baba, A. (2016). Application of long-range surface plasmon resonance for ABO blood typing. *International Journal of Analytical Chemistry*, *2016*, 1432781.

Taparia, N., Platten, K. C., Anderson, K. B., & Sniadecki, N. J. (2017). A microfluidic approach for hemoglobin detection in whole blood. *AIP Advances*, *7*(10), 105102.

Vashist, S. K., Mudanyali, O., Schneider, E. M., Zengerle, R., & Ozcan, A. (2014). Cellphone-based devices for bioanalytical sciences. *Analytical and Bioanalytical Chemistry*, *406*(14), 3263–3277.

Wang, X., Lin, G., Cui, G., Zhou, X., & Liu, G. L. (2017). White blood cell counting on smartphone paper electrochemical sensor. *Biosensors and Bioelectronics*, *90*, 549–557.

Wicks, L. C., Cairns, G. S., Melnyk, J., Bryce, S., Duncan, R. R., & Dalgarno, P. A. (2017). EnLightenment: High resolution smartphone microscopy as an educational and public engagement platform. *Wellcome Open Research, 2*, 107.

Xu, D., Huang, X., Guo, J., & Ma, X. (2018). Automatic smartphone-based microfluidic biosensor system at the point of care. *Biosensors and Bioelectronics, 110*, 78−88.

Xu, X., Akay, A., Wei, H., Wang, S., Pingguan-Murphy, B., Erlandsson, B., et al. (2015). Advances in smartphone-based point-of-care diagnostics. *Proceedings of the IEEE, 103*(2), 236−247.

Yamamoto, K., Sakurai, R., & Motosuke, M. (2020). Fully-automatic blood-typing chip exploiting bubbles for quick dilution and detection. *Biomicrofluidics., 14*(2), 024111.

Yang, F., Zhang, Y., Cui, X., Fan, Y., Xue, Y., Miao, H., et al. (2019). Extraction of cell-free whole blood plasma using a dielectrophoresis-based microfluidic device. *Biotechnology Journal, 14*(3), e1800181.

Yu, Y., Zhang, Q., Buscaglia, J., Chang, C.-C., Liu, Y., Yang, Z., et al. (2016). Quantitative real-time detection of carcinoembryonic antigen (CEA) from pancreatic cyst fluid using 3-D surface molecular imprinting. *Analyst., 141*(14), 4424−4431.

Yu, Y., Zhang, Q., Chang, C.-C., Liu, Y., Yang, Z., Guo, Y., et al. (2016). Design of a molecular imprinting biosensor with multi-scale roughness for detection across a broad spectrum of biomolecules. *Analyst., 141* (19), 5607−5617.

Zhang, H., Qiu, X., Zou, Y., Ye, Y., Qi, C., Zou, L., et al. (2017). A dye-assisted paper-based point-of-care assay for fast and reliable blood grouping. *Science Translational Medicine, 9*(381).

Zhang, Q., Kaisti, M., Prabhu, A., Yu, Y., Song, Y.-A., Rafailovich, M. H., et al. (2018). Polyaniline-functionalized ion-sensitive floating-gate FETs for the on-chip monitoring of peroxidase-catalyzed redox reactions. *Electrochimica Acta, 261*, 256−264.

Zhu, H., Mavandadi, S., Coskun, A. F., Yaglidere, O., & Ozcan, A. (2011). Optofluidic fluorescent imaging cytometry on a cell phone. *Analytical Chemistry, 83*(17), 6641−6647.

Zhu, H., Sencan, I., Wong, J., Dimitrov, S., Tseng, D., Nagashima, K., et al. (2013). Cost-effective and rapid blood analysis on a cell-phone. *Lab on a Chip, 13*(7), 1282−1288.

Zhu, H., Yaglidere, O., Su, T. W., Tseng, D., & Ozcan, A. (2011). Cost-effective and compact wide-field fluorescent imaging on a cell-phone. *Lab on a Chip, 11*(2), 315−322.

Zhu, J., Palla, M., Ronca, S., Warpner, R., Ju, J., & Lin, Q. (2013). A MEMS-based approach to single nucleotide polymorphism genotyping. *Sensors and Actuators. A, Physical, 195*, 175−182.

Lab-on-a-chip for analysis of blood 12

Hayder A. Abdulbari

Center of Excellence for Advanced Research in Fluid Flow, Faculty of Chemical and Natural Resources Engineering, University Malaysia Pahang, Pahang, Malaysia

12.1 Introduction to microfluidics science

The booming and rapid development of the microelectronic technologies in the 1960s started a new life-changing era to humankind with new living standards. This rapid development introduced unique opportunities to explore endless scientific corners that had not been investigated before and to revisit existing works to enrich the fundamental knowledge and phenomena behind them. This remarkable development in microelectronics technologies was associated with other important developments in microfabrication methods that facilitated the fabrication and testing of new micro-scaled systems (Bragheri et al., 2016). Still, the development of nonelectronic microsystems was not as fast and as huge as electronic systems until the 1970s, where polymer and silicon microfabrication were first introduced (Rapp, 2017). This important and game-changing development provided an excellent opportunity to start a completely new scientific path to investigate closely known physical phenomenon that were not being "observed" or controlled before with the conventional testing methods (Chen et al., 2019; Su et al., 2015; Tay et al., 2016). By then, the "microfluidics" terminology started to be the major definition of a new era in microflow systems and microfabrication methods (MEMS) (Kumar et al., 2019). Microfluidics is a multidisciplinary technology that combines many fields, such as chemistry, biology, biotechnology, and medicine (Bhattacharya et al., 2018; Sackmann et al., 2014; Whitesides, 2006). Microfluidics technology focuses on the design, manufacture, and experimentation of miniaturized fluid flow systems, which have endured tremendous growth over the last 20 years (Abdulbari & Basheer, 2017a, 2017b; Kratz et al., 2020; Leal et al., 2020; Xuan, 2019). Being a multidisciplinary area, this progressively increasing sector of research has a wide variety of uses in biomedical diagnostics (Dimov et al., 2008; Fachin et al., 2017; Karabacak et al., 2014), chemical analysis (Livak-Dahl et al., 2011), automobile, and electronics industries (Ricco, 2006).

Microfluidic mixing and separation techniques are one of the state-of-art technologies that are rapidly evolving and are considered as the heart of any microfluidic chip (Günther & Jensen, 2006). Extensive experiments have shown outstanding mixing effectiveness through simple geometries of microfluidic chips and devices (Gobby et al., 2001; Lee et al., 2011; Paik et al., 2003; Panić et al., 2004). Of the fluids that flow in microchannels, molecular diffusion processes regulate the mixing among deformed fluid elements. Increasing contact area between two species is therefore one of the most effective ways of achieving close-to-perfect matching that cannot be achieved in large-scale mixers. In microfluidic mixing systems, fluids can usually be combined within 55−300 ms, and thus the need for large system throughputs can be easily met with a clear structure. At the macroscale, phase separation is driven by the phase density differences, and thus by the difference in gravity forces. Nevertheless, gravitational forces are marginal in microfluidics, which allows the use of surface forces to completely distinguish two phases in one single step. Castell and coworkers reported 100% separation efficiency with a microfluidic device designed for separating two immiscible liquids (chloroform and water) (Castell et al., 2009). According to their findings, the capillary component used to separate the liquids utilized differences in the wettability of both forms.

12.2 Basic principles

It was known that downscaling alone is not enough to reproduce the role of devices in the macro world after the innovations of microfluidic devices (Song et al., 2018). Nevertheless, there has been the main increase in industrial interest in the area following intensive research dedicated to understanding fluid behavior at the microscale level. When analyzing the fluid responses at the microscale level, the condition may be reduced to wet areas controlled by the influence of internal forces. In other words, the interfacial forces have more significant effects compared to the fluid volume-dependent properties, such as gravitational forces. This behavior was explained by the scaling law principle, which is defined as the geometry ratio (GR) of a microchannel. As the miniaturization process concentrates on decreasing the size of a given component, the geometry ratio becomes an intrinsic parameter and plays an essential role in determining the action of the fluids flow. Eqs. (12.1) and (12.2) define mathematically the geometry ratio where the ratio is extremely large as the surface area approaches infinity. The major changes in GR values indicate that surface forces such as friction, viscosity, and stress are becoming more prominent in microfluidic systems, that is, intermolecular interfaces (Song et al., 2018).

$$GR \equiv \frac{\text{Surface Area}}{\text{System Volume}} \tag{12.1}$$

$$GR \equiv \frac{(A)}{(V)} \propto \frac{L^2}{L^3} \propto \frac{1}{L} \tag{12.2}$$

The Knudsen number (K_n) is an important dimensionless parameter that describes those effects, as in Eq. (12.3) (Hajmohammadi et al., 2018).

$$K_n = \frac{\lambda}{L} \tag{12.3}$$

where "λ" is the mean free path of the molecules owing to decreasing volume (L). Typically, if the magnitude of the Knudsen number is greater than 0.001, the standard gradient approach can be applied with sufficient accuracy. In other words, assuming that the flow has no-slip status, it is safe; that is, when flowing, it will not bind to the channel wall. In liquids that flow in a narrow space, as is the case in microfluidics, due to the tight organization of the liquid molecules, they are commonly observed to behave as a gradient and encounter no-slip and no-temperature hop conditions at boundaries. This behavior is also seen by internal forces because of the regulation at the microscale.

12.3 Microfabrication technologies and methods

The rapid developments in microfluidics technology in the past few years was associated with the introduction of different microfabrication methods that satisfy the need for each application. Until 1998 the microfluidics chips fabricated using specially designed micromachining and microfabrication technologies were expensive, complex, and mostly for fundamental research purposes like laser micromachining and photolithography. In 1998 Xia and Whiteside (Xia & Whitesides, 1998) introduced the soft lithography fabrication method using transparent polymers like polydimethylsiloxane (PDMS). This method provided an excellent platform to fabricate low-cost, transparent microfluidic chips for different kinds of applications. Despite all these advantages, the soft lithography microfabrication method had its own drawbacks. that is, specialized equipment and a dust-controlled cleanroom are needed. In the past few years, several advanced and cost-feasible microfabrication methods were developed and applied to serve the growing demands for low-cost and high-performance microfluidic chips. In this section, we will discuss some of the important microfabrication methods that are widely used in the formation of microfluidic chips.

12.3.1 Machining

Micromachining is a nonlithographic technique that uses laser or milling equipment. Machining does not restrict the substrate material, but the materials must be soft and ductile to process (Ashman & Kandlikar, 2006). The commercial CO_2 laser can be used to create microfluidic devices with a depth of $100-300\,\mu m$, where the total channel width can be up to $85\,\mu m$ (Klank et al., 2002). Huang and Liu (Huang et al., 2010), suggested that microchannel roughness may be minimized by preheating polymethyl methacrylate (PMMA) substrates up to $70°C-90°C$. Certain processing methods other than CO_2 laser machining, such as IR laser (Romoli et al., 2011), foil-assisted CO_2 laser (Chung & Lin, 2011), and direct-write laser (Cheng et al., 2004) were added. The machining itself also involves a combination of a few microchannel manufacturing techniques such as wet etching, photolithography (Muluneh & Issadore, 2013), and others involving cleanroom facilities, thus increasing manufacturing costs.

12.3.2 Embossing

Embossing, also known as hot embossing, is gaining significant attention as a process of producing microchannels; it requires the impression of structures on substrates, typically polymeric material using a preformed master. Once the substrate has been heated up to its glass transition temperature (T_g), an embossing force is used to imprint the substrate using the leader. Thus the master's material is required to endure several cycles of elevated temperatures and cycles of forces embossing. Nickel has been described as a feasible stamp material; however, its cost is high and does not allow for rapid modification of the design (Shinohara et al., 2007). It is known that the SU-8 stamp on silicon, glass, and PDMS could withstand more than 50 embossing cycles in which the stamp manufacturing operation can be completed in less than 4 hours. The embossing process takes a lot of time and is not suited for research purposes for prototyping multiple microchannels. Konstantinou et al. proposed that the approach of embossing would produce 50 instruments a week in which this process meets the need for improvements in the configuration without lengthy procedure (Konstantinou et al., 2016). Roller hot embossing is invented using the traditional hot embossing method, and it can be used as a substratum based on PMMA. This approach applies to the mass production of microfluidic devices, where the printing on transparent sheets and film materials is a continuous process (Friend & Yeo, 2010; Velten et al., 2009).

12.3.3 Injection molding

Injection molding is a nonphotolithographic process, which is also widely used in microfluidic system manufacturing. Metal mold masters such as brass and aluminum have been reported as appropriate materials for microchannel replication. The microchannel profile, such as width and depth, is influenced by the injection molding working conditions, so research into the optimal conditions for the process started to attract interest. This method is suitable for the manufacturing of microfluidic polypropylene devices that can be used in the field of biochemistry and medicine. Apart from polypropylene, thermoplastics such as polycarbonate, PMMA, cycloolefin polymers, and copolymers may also be used as substrates (Fu et al., 2011; Jakeway et al., 2000; McCormick et al., 1997; Mela et al., 2005; Shan et al., 2009).

12.3.4 Etching

Etching is a method where chemicals are used to peel away the wafer's surface for channel processing. Different etching methods have been created in the production of microchannels, such as silicon etching, chemical etching, and laser-induced etching. The etching is done with a layer of silicon with the patterned sheet, and then etching the silicon. Different processes such as wet anisotropic, wet isotropic, dry anisotropic, and dry isotropic can etch silicon. The system of wet anisotropy often involves the use of simple solutions, such as potassium hydroxide (KOH). Acidic solutions such as hydrogen fluoride (HF) and nitric acid (HNO_3) are used in wet isotropic areas where the solution's structure influences the final microchannel form (Kang et al., 1998; Schwartz & Robbins, 2019).

12.3.5 Soft lithography

The introduction of soft lithography was a major changing point for microfabrication science, where affordable, transparent, and effective microfluidic chips became able to be produced. PDMS elastomeric properties allow the elastomer to be readily separated from the wafer and easily attached to another piece of polymer or glass during the sealing process. The traditional way of soft lithography, however, involves the manufacture of a special silicon master consisting of a photoresist with a microchannel pattern on it. This process is much more time-consuming and requires cleanroom equipment, thus increasing costs. Due to the hydrophobic nature of PDMS, when treating functional macromolecules such as protein adsorption, surface molecular property treatments were needed (Becker, 2002; Guo et al., 2015; Martin & Aksay, 2005; Wong & Ho, 2009).

12.3.6 Lab-on-chip

Lab-on-chip (LOC) is a terminology that can summarize hundreds of devices that have been developed and commercially applied as point-of-care (POC) assessment tools in the past few decades using microfluidics technology. Conventional LOC equipment involves multiple functionalities: sample transport, sample preparation (Mach et al., 2013), isolation and detection, and analysis modules. Label-free particle or cell separation is crucially important to most of the analytical and preparatory techniques used in the fields of chemical, biochemical, and clinical analysis, which have led to groundbreaking advances in the speed of analysis, separation resolution, and process automation (Plouffe & Murthy, 2014). Also, microfluidic separation devices can sometimes be a part of portable point-of-care or on-site detection systems. Several variations of microfluidic cell sorters, which implement different sorting mechanisms, have been designed and manufactured. The selected technique of particle sorting is typically determined by the nature of the application, which is strongly dependent on the characteristics of the sample, and the final objective of the analysis also should be taken into consideration (Chen et al., 2009). A few strategies are available for this objective, based on cell/particle characteristics, including the sorting of particles in fluids or the separation of particulate matter from fluids. Individual particles can behave or interfere with the analyte, in which case they must be extracted from the sample.

In conventional cell isolation methods, scale separation is widely used to isolate mixtures of cells or particles prior to subsequent examination or culture (Li et al., 2015). Cells show differences in their hydrodynamic radii owing to shifts in thickness, form, or mechanical properties, based on their state of health (Park et al., 2010). Established cell separation techniques can also be categorized into two main classes to improve selected subpopulations. The first group is based on physical descriptors such as differences in size, shape, and density, and includes filtration and centrifugation techniques commonly used for heterogeneous samples. The second category consists of affinity approaches such as solid matrix capture (beads, tubes, fibers), fluorescence-activated cell sorting, and magnetic cell sorting, which are focused on biochemical cell surface characteristics and biophysical criteria.

Standard fractionation instruments, varying in scale from laptop to room-sized versions, are standard equipment for most clinical laboratories and are used to analyze and isolate cells and other biological particles. The disadvantages to established, conventional flow measurements are that these instruments are costly and require substantial external facilities, such as laboratories, staff, and reagents. They are not suitable for integration with other analysis steps, and the time needed to process signal data limits the rate at which cells can be detected. A great deal of work is therefore being done to develop techniques that will not only be affordable and therefore more easily obtainable but would also be more effective and possibly be able to test particle properties that are not currently available. In the scientific testing context, cell sorting and counting instruments are examples of technical tools planned and programmed for use in consolidated labs, much like the very earliest computers. Biologic samples (blood, urine, other biological fluids) are often obtained from patients in hospitals, at home, or in different environments and sent to these regional testing labs where the examination takes place. That being said, the transport of samples demands money and time, and the quality of samples may decay due to natural biological processes.

12.4 Microfluidic technology for blood and cells testing applications

A blood test is considered an initial step in the diagnosis of many diseases and medical conditions, based on the assumption that irregular blood cells generally possess specific physicochemical and biological characteristics (e.g., scale, deformability, and chemical structure). Empowering the advantages of microfluidics technology, alterations in physicochemical properties of the blood cells may be used for fast and precise medical assessment. In this section, an overview of the typical medical blood and blood cells testing techniques developed with the aid of the microfluidics technology is presented, especially for biomedical applications, as well as a synopsis of some design considerations regarding microfluidic devices.

12.4.1 Complete blood count

A commonly performed complete blood count (CBC) assay is usually performed to evaluate the concentrations of 10 different components of each significant blood cell: white blood cells, red blood cells, and platelets (George-Gay & Parker, 2003). Essential components assessed via this method include the number of red blood cells, hemoglobin, and hematocrit. Traditionally, this searching test is usually conducted in two main approaches, namely, manual count and automated count (Bentley et al., 1993). In both approaches, the blood sample is collected and preprepared by centrifuge separation of the blood cells.

The CBC testing indicates the density of each type of cell, the size of the cell, the percentage of particular blood cells, and the density of different proteins, creatinine, or metabolites (Verbrugge & Huisman, 2015). Computerized hematology analysis tools and flow cytometers are the most standard techniques in health facilities or research labs for highly advanced CBC analysis. Lysed blood cells are sunk into a cuvette for spectrophotometric analysis for molecular level identification (Magnes et al., 2012). Accurate measures of CBC typically involve reliable and high-quality processes for obtaining and processing whole blood samples to prevent any background intervention, like centrifugation, fractionation, lysis, or dilution. The conventional testing methods have their own drawbacks such as hemolysis possibility as in inappropriate centrifugation, long testing time, and large blood sample volumes required ($\sim$ milliliters), besides human errors that are directly related to the training quality. Microfluidics is an optimal strategy for simplifying entire blood collection, focused on the capacity to incorporate different practical elements and possible features of automation (Bi et al., 2016; Chin et al., 2011; Yang et al., 2012). Besides, the needed sample volume can be efficiently minimized due to its miniature channel size and surface area.

The unique mixing and separation efficiencies in the microscaled blood test devices provided a great opportunity to introduce an optimized portable device that addresses all the mentioned drawbacks. With such technology, the separation of blood cells or plasma may generally be divided into active or passive methods of separation (Li et al., 2020; Nivedita & Papautsky, 2013). External forces (acoustic force, dielectrophoretic force, electromagnetic force, or a combination of the above forces) are manipulated by effective separation methods to guide targeted cells to a particular direction or location. Passive techniques of separation mostly utilize the variations in cell properties to distinguish the cells, such as height, form, or stiffness (Janmey et al., 2020).

In general, active separation approaches allow for high-throughput, high-selectivity cell separation chosen for the sorting or isolation of uncommon cells (e.g., circulating tumor cells or bacteria) in vast volumes of blood samples. Usually, conversely, the samples must be either purified cells or blood lysing of RBCs. The explanation behind this is the large number of RBCs in whole blood, which reduces the successful trapping of WBCs or uncommon cells. One example is the use of a magnetic field to separate magnetically marked WBCs from whole, unprocessed blood, specifically. Within a particular area of vigorous magnetic field intensity, magnetically trapped cells aggregate and are not suitable for cell counting or single-cell research. Passive fraction approaches, by comparison, do not involve having any cycle of cell marking or advanced micro/nanomanufacturing, rendering them more cost-effective and sufficient to isolate blood cells from plasma, or vice versa.

For plasma, one basic microfluidic system uses a microsubtrench across the flow path that can capture sedimentation-based blood cells and allow for plasma purification. Specific microchannel structures for plasma filtration include deterministic lateral displacement (LDD) structures, curved microchannel series, and microfluidics laminar microvortices. A further efficient method for plasma extraction is to use the bifurcation rule (or the so-called Zweifach−Fung effect) to direct most RBCs through the main microchannel (higher flow rate), and the plasma can be collected from the side microchannel (lower flow rate). A related configuration uses a somewhat narrower side microchannel that enables both RBCs and plasma to move in, whereas WBCs move downstream from the main channel. The passive method of separation reduces the difficulty of cell labeling. The fundamental problem, though, is the lower performance in contrast with successful separation approaches. Even though specific microfluidic-based blood cell isolation or plasma extraction strategies have been illustrated, several approaches can only remove one form of blood product by extracting or disregarding the other elements, which reduces the observable variables in a single whole blood sample. By refining the configuration of the bifurcation channel and incorporating cellular trapping devices, a microfluidic system for the continuous flow of entire blood distribution to separate different blood components is needed. The device must contain two types of side channels (with and without packed beads), and a series of WBC trapping units based on hydrodynamics.

Fundamentally speaking, there seems to be a shortfall of microfluidic innovations that links the difference between both the desire for high-performance blood count and the specific demand for compact uses. The Coulter method and the flow cytometry method in microfluidics has shown the viability of absolute WBC counts and a three-part WBC differential. Compared to the five-part WBC differential count given by modern laboratory blood counters, it is still restricted. Additionally, the immunoassays used in previous methods struggle from the drawbacks of low-temperature storage (as low as 4°C) and limited life span. In contrast, long shelf life and room temperature storage specifications are essential considerations for portable and easy point-of-care implementations. Besides, a handheld tool that offers the WBC count and WBC differential alone may function as a practical clinical check for point-of-care systems. It may be used, for instance, to track exposure to astronauts to space radiation during space flight. In another scenario, as a rapid diagnostic test for distinguishing bacterial and viral infections, the WBC count with differential may be used in physician offices. For all the above purposes, the production of portable blood cell counting technology with enhanced efficiency, with an emphasized need for the WBC count and the WBC differential count, is also of high significance.

Baratchi et al. and Ossowski et al. established a plastic substrate microfluidic cytometer and used immunological assays for counting WBCs (Baratchi et al., 2014; Nahavandi et al., 2014; Ossowski et al., 2015). The microfluidic system was designed by hot thermoplastic material (polycarbonate) embossing, and the structures obtained were used to have a two-dimensional hydrodynamic orientation. Blood samples were pretreated with fluorescent-conjugated immuno-assays, and optical detections were conducted on a microscope-modified device involving light scattering and fluorescence analysis. They were able to illustrate the differentiation of three-part WBCs (lymphocyte, monocyte, and granulocyte) and the lymphocyte subtype counts.

Meanwhile, through calculating electrical impedances, Holmes et al. recorded counting the blood cells on a microchip (Holmes et al., 2007). The microchip was produced by complete wafer bonding of two glass substrates with polyimide templates to create the microfluidic channels, and electrodes on the substrates were prepatterned by metal electrodes. When the pretreated sample of blood flows through the microfluidic device, changes in the electrical impedance caused by individual cells were measured at multiple AC frequencies. They were able to show the count of the WBCs, RBCs, and platelets, as well as the viability of the three-part WBC differential count, utilizing this approach. There have also been studies of the usage of microfluidic cytometry methods for particular blood cell analyzes, such as the measurement of protein expression on the surface of WBC and the enumeration of two WBC subtypes (CD^{4+} lymphocyte cells and CD^{8+}) in HIV diagnosis.

12.5 Basic and complete metabolic panel

A basic metabolic panel (BMP) evaluates blood concentrations of specific molecules such as electrolytes, calcium, and glucose (Kildow et al., 2018). Usually, the metabolic panel is selected to evaluate the muscles, bones, and internal organs of patients by assessing oxygen, potassium, electrolytes, albumin, total protein, and biomarkers for kidney function (urea nitrogen and creatinine) and liver function (alkaline phosphatase, alanine aminotransferase) (Di Sebastiano & Mourtzakis, 2012). Typically, the lipoprotein test, which contains total cholesterol, high-density, and low-density lipoprotein cholesterol (known as "healthy" and "poor" cholesterol), and triglycerides, is performed to determine and control the likelihood of cardiovascular disease (Lee et al., 2019; Zhai et al., 2020). Generally, the blood coagulation assessment shows partial operated thromboplastin time (aPTT), prothrombin time (PT), and thrombin time (TT) to analyze the intrinsic, extrinsic, and common pathways of the coagulation cascade, as well as the renal function assessment to quantify the platelet phase of hemostasis. Commonly the blood enzyme analysis tests cardiovascular enzymes like troponin T, creatine kinase, and myoglobin to detect heart failure or heart muscle harm.

Microfluidics was introduced to the metabolomics testing systems only lately, but it also has the opportunity to play a major role in this area. Lengthy, computationally expensive sample preparation procedures comprised of various extraction procedures and solvothermal reactions, for example, are typically associated with metabolomic studies (Danoy et al., 2020). These processes could be performed in a microfluidic layout with less time and greater efficiency. To date, the overwhelming number of microfluidic devices primarily concentrated on combining specific separation and detection systems for biological sample assessment Furthermore, a majority of microfluidic systems were designed and fabricated in a manner that incorporates biological systems (cells) with microfluidic devices to carry out a metabolomic evaluation of target samples. The use of commercially available microfluidic devices for metabolic assessments is described throughout many findings. Miyado et al. utilized a Shimadzu quartz glass microfluidic device to separate and diagnose metabolites of nitric oxide in human plasma (Miyado et al., 2008). The study introduced a high-throughput analysis that does not demand deproteinizations as an initial separation step before the analysis. The zwitterionic surfactant N-cyclohexyl-2-aminoethanesulfonic acid (CHES) was used to achieve separations in less than 60 seconds. A Shimadzu MCE 2010 method was used in another study to isolate antimicrobial metabolites from Pseudomonas fluorescens, and linear UV imaging was implemented to estimate the percentage of resorcinol, monoacetylphloroglucinol, and 2,4-diacetylphlorglucinol. The designed quartz microfluidics device used a channel of 30 μm about 30 μm, which was 25 mm long. Holcomb et al. developed a microfluidic electrochemical detection system (ECD) with an electrode detector system to track the metabolites and xenobiotics of small molecules. The designed device had eight gold working electrodes that were individually modified, which were used to selectively detect analytes utilizing potential phase amperometry, and the device can be used to electrochemically overcome comigration species (Guihen & Glennon, 2005).

Many other reports outlined the use of other sample pretreatment procedures on microfluidic systems, such as mixing, sample labeling, and metabolite generation. Such as the study reported by Urbanski et al. in which they described embryo metabolism identification using an incorporated microfluidic device (Microfluidic tools for metabolomics, 2008). The system was manufactured from PDMS using multilayer soft lithography procedures and composed of two layers; an upper flow channel network for charging and blending samples and assays, and a lower core network layer for valving and fluid control. The delivery of the sample and reagent aliquots, mixing, data acquisition and data analysis could be made autonomously after the initial sample and reagent loading onto the chip. Ma et al., in a new different study, outline the integration of sol −gel-based bioreactors into a microfluidic chip for drug metabolite generation (Ma et al., 2009). Metabolite assessment and cytotoxicity assays for cells have also been performed on-chip, which also requires the device to consist of three layers. A follow-up study utilized a quartz middle layer that hosts a human liver microsome sol-gel bioreactors for the production of drug metabolites. An upper PDMS layer is composed of solution reservoirs for the middle layer, whereas a lower PDMS layer contained hepG2 cells for drug and drug metabolite cytotoxicity assays located below the sol-gel bioreactors. The influence of the UDP-glucuronosyltransferase-generated acetaminophen and acetaminophen metabolites throughout the sol-gel bioreactors were examined. The production of metabolites was controlled through capillary electrophoresis with UV−vis spectroscopy, while the cytotoxicity of the hepG2 cells was tracked through live/dead fluorescent stain. Incorporating bioreactors of drug metabolism along with multiple functionalities of analysis on a microfluidic system such as this has opportunities for fast metabolite characterization of drugs.

Through the use of targeted analytical methods such as immunoassays eliminates the need for thorough preprocessing of samples and therefore enables a direct biofluid assessment. In a study conducted by Murphy et al., the authors identified a competitive immunoassay for the simultaneous determination of micromosaic patterned proteins and metabolites (Murphy et al., 2009). For some of these assays, the bovine serum albumin conjugate of the target analyte was coated on the activated silicon nitride surface by sliding through the substrate surface of the microfluidic network. The reference sample containing the targeted analyte and its fluorescently labeled antibody was then structured in a coordinated system, and the competing immunoassay was evaluated using a fluorescent microscope. Using this approach, the metabolic biomarkers thyroxine and 3-nitrotyrosine can be easily observed.

Jo et al. presented the incapacitating of neuropeptide release from the *Aplysia californica* sac cell neurons to different regions of the solid substrate using microfluidic networks (Dolly et al., 2011). The potential of the residuals to restrict an analyte of interest on a solid substrate advocated additional sample cleaning prior to analysis. In this study, the authors were able to rinse off unretained substances and spray a solid matrix substrate with neuropeptide immobilization for direct matrix-assisted laser desorption ionization (MALDI)-Ms testing. The microfluidic systems consisted of a PDMS upper layer with reservoirs for neural stem cells and microfluidic channels for controlling fluid flow, as well as a bottom silicon layer functionalized with C18 for the capture of neuropeptides. With the aid of this system, the team was able to monitor neuropeptide release pre, during, and after neural pathways.

12.5.1 Lipid panel

In clinical analysis/diagnosis, the identification of total lipid levels in the blood is of high significance since it is an essential key measure of the risk of developing cardiovascular disease (Dorfman et al., 2020). There are three major categories of the Total lipid, namely, "Total lipid" includes low-density lipoproteins (LDL), very-low-density lipoproteins, and high-density lipoproteins (HDL) (Maldonado et al., 2001). There is indeed a favorable and unfavorable cause and effect relationship between LDL ("bad" cholesterol) and HDL ("good" cholesterol) rates, respectively, and heart disease and stroke. To help determine the probability of heart attack, serum cholesterol content may be assessed using the lipid panel or lipid profile in whereby the amounts of gross, HDL, LDL, and triglycerides are calculated. The colorimetric, photometric, fluorometric, spectrophotometric, HPLC and electrochemical techniques are popular techniques for serum cholesterol analyses. There is a great need for affordable healthcare for total lipid profiling for point-of-care diagnosis for resource-limited regions such as villages and remote zones. Specific lipase and phospholipase biosensors are evolving as a critical resource in clinical science, as they are used as diagnostic instruments to monitor plasma and blood levels of triglycerides, cholesterol, and phospholipids. Many approaches include fast cholesterol identification by combining enzymatic biomarkers and electrochemical monitoring, as presented by Ruecha et al., where the analysis was conducted in capillary electrophoresis of poly (dimethylsiloxane) microchips (Ruecha et al., 2017). Cholesterol detection approach, also commonly used as illustrated by Wisitsoraat et al. using a functionalized carbon nanotube (CNT) electrode by direct growth of CNTs on a glass plate (Wisitsoraat et al., 2010). The linear measurement range for the cholesterol sensor based on CNTs was between 50 and 400 mg/dL.

Additionally, Okazaki et al. achieved a unique efficiency liquid chromatography (HPLC) quantitative identification of triglycerides (TGL) in serum lipoproteins and serum-free glycerol (FG) (Okazaki et al., 2005). A few other electrochemical methods for triglyceride biosensors have also been found for a specific operating electrode, such as platinum (Pt), gold (Au), carbon (C), and indium tin oxide. Using a Mobile Cholesterol Program for Quick Diagnostics ("smartCARD"),

Oncescu et al. built a system for assessing cholesterol levels in the blood using a smartphone camera for reading cholesterol levels. Although all of these handheld devices are functionally simple, they are only used to detect one variable at a time (Oncescu et al., 2014). The National Cholesterol Education Plan proposed for all adults above the age of 20 to be assessed for coronary heart disease risk requires total cholesterol (Tc), LDL cholesterol, HDL cholesterol, and TGL identification. Checking the total lipid level in resource-poor rural areas is treated as extravagant expenditure unless the emergency occurs. The need for ultra-low-cost, rapid devices that can meet resource-poor people's healthcare demands remains a challenge. Paper-based microfluidic devices in this sector are evolving as an excellent tool for "On Demand-Devices," particularly in clinical research applications (Parween et al., 2019). A few of the biggest challenges when using paper-based microfluidic devices are the user's heterogeneity of color distribution in the detection zones, sensitivity, and poor end-color judgment. Significant attention has been given to the chemical modification of cellulose for biochemical sensing applications, such as weak protein interaction with cellulose may result in protein washing off, and denaturation. Wang et al. used chitosan modified μPADs to covalently immobilize antibodies (Wang et al., 2012). Hassan et al. reported poly(carboxy betaine) (PCB)-modified cellulose paper as a possible paper-based microfluidic diagnostic tool and used it to detect glucose for an extended period (42 days) (Hassan et al., 2019). Substantial improvement of sensitivity in assays has been identified in almost all the above cases. Further development in this area would further help increase sensing specificity and sensitivity for biochemical sensing applications of these clinical diagnostics.

12.6 **Sexually transmitted disease tests**

Reliable and affordable POC testing is required to monitor epidemics of sexually transmitted diseases (STD), as patients can seek prompt diagnosis and treatment. Traditional POC procedures work inadequately for *Chlamydia trachomatis* or *Neisseria gonorrhoeae*, and need improved assays. *Trichomonas vaginalis* diagnostics depends on wet preparation, with some significant advances. Serological syphilis POC assays may influence resource-poor conditions, with multiple assays used with only one accessible in certain countries. Given the inadequacy of POC assays for treatable bacterial infections, the development of technological advancements provides hope across to advance POC diagnoses.

For many STDs, diagnostic tests have improved markedly, despite the limitations associated with applying them. Modern treatment of STD requires the use of many cultures or serological techniques. Though techniques such as Gram stain microscopy, wet preparation, and direct fluorescent antibody can be called POC tests, they are of little utility outside a laboratory, and many are reactive and nonspecific. Due to lower susceptibility, other immunological methods such as ELISA are no longer recommended for chlamydia by the CDC. Nowadays, the gold-standard assays for the identification of CT and NG are considered nucleic acid amplification tests (NAATs), with several licensed commercial assays available. Fortunately, biomedical research has advanced over the last 10 years to produce several exciting new POC or "near-patient" assays. In 2018 Yang et al. presented a low-cost, combined, and pattern-to-answer paper-origami study to diagnose three bovine reproductive diseases in semen samples obtained at a rural India test site (Yang et al., 2018).

The paper-based assay presented by detects pathogen DNA from one viral pathogen, bovine herpes virus-1 (BoHV-1), and two bacteria (*Brucella* and *Leptospira*) and has demonstrated near-perfect sensitivity (97.4%–98.7%) and specificity (99.4%–99.9%) in urogenital specimens. The paper-based microfluidic device is a modular platform for testing samples directly from patients, which requires no hands-on manipulation from specimen loading and results are available in a short time. A statistical model also was documented for the ability of the newest generation of POC tests, including the Cepheid assay or another modern POC study, to minimize the prevalence of CT and NG in a large population environment (O'Sullivan et al., 2019). The researchers determined that if the latest POC test surpassed a sensitivity of 95% and a baseline sampling range of 44% per annum, the occurrence of NG could be minimized from 7.1% to 5.7% and the occurrence of CT could be minimized from 11.9% to 8.9%. The incidence might be as small as 0.6% for NG and 1.5% for CT if test coverage is expanded to 60%–80%. The study estimates that decreasing the period between testing and diagnosis alone will have a limited influence on the decrease in prevalence. However, the use of highly reactive POC tests has immense potential to boost the screening result, as they can accomplish the cumulative effect of lowering the care duration to zero and effectively raising the treatment rate to 100%.

The Cepheid GenXpert assay for NG, similar to the assay for CT, promises to deliver near-patient trials integrating microfluidic technologies with real-time PCR. The cartridge-based test extracts NG DNA from the symptomatic and asymptomatic patients in female endocervical swabs, patient-collected vaginal swabs, and female and male urine specimens (Gaydos & Hardick, 2014).

A plethora of treponemal assessments are available for the use of any diagnostic test. Rapid POC assessments have been established that can be carried out on the blood of the fingerstick and are commonly used in resource-limited settings, for example, Trinity Health Check (Diagnostics Direct, LLC, Stone Harbor, NJ, United States) is FDA cleared for use in the United States. The treponemal POC screening for syphilis has a 95.6% favorable consensus and a 90.5% unfavorable consensus for gold-standard research, with 90.6% agreement outcomes.

Trichomonas infections, triggered by parasite TV, are particularly prevalent STIs, with reports of 3.7 million infections per year in the United States and 180 million worldwide (Zhang et al., 2018). They constitute the most prevalent curable STD in young, sexually active women and are correlated with weak pregnancy outcomes such as low birth weight and premature birth.

A new rapid prototype TV POC assay, in combination with the Atlas Io POC platform, has been developed for use. The assessment incorporates novel electrochemical endpoint identification, with the target being a multicopy section of the TV genome (Cristillo et al., 2017). In a quantitative test procedure, 90 clinical vaginal swab extracts were used to validate the efficiency of the design analysis, showing 95.5% (42/44) and 95.7% (44/46) responsiveness and specificity, respectively.

There seems to be no doubt that reliable and affordable POC tests are strongly required to track the world 'scostly epidemics of STIs so that patients can seek prompt diagnosis and appropriate care for such infections (Moore, 2006). The groundbreaking advancements in microfluidic technologies are the promise of better POC diagnostics for STIs. Outstanding serological POC diagnostic tests were developed for the detection of anti-HIV antibodies and p24 antigen, as well as syphilis, but the production of accurate POC tests for the detection of curable bacterial STIs, such as chlamydia and gonorrhea, is struggling to keep up and has been insufficient to date. The further production of POC diagnostics for such infections needs considerable expenditure of resources, energy, and testing effort. The vulnerability of increasing antimicrobial resistance to gonorrhea poses a significant obstacle, both as a public health concern and demonstrates the immediate need for new POC assays and quick testing to diagnose antibiotic susceptibility or tolerance to currently active antibiotics in order to curb the tolerance to grow and assist with antibiotic-saving steps.

Acknowledgment of the need to assess the efficiency of new POC tests for STIs in real-world settings is crucial. While responsiveness and specificity are frequently used to determine a test's output in trial conditions, it is the PPV and NPV that might be more relevant when analyzing assays, because the prevalence influences those values. Therefore it would be necessary to consider the predictive values of a study that would rely on the infection's population incidence, which may dramatically influence test output in the specific environment, if it is very small.

Given the existing shortage of appropriate POC testing for treatable bacterial infections, with the implementation of emerging technical developments such as low-cost microfluidic paper-based systems and the potential to integrate advanced technology with healthcare services, significant development can be anticipated in the coming years (Akyazi et al., 2018; Boobphahom et al., 2020; Hu et al., 2016; Sharma et al., 2015; Vidic et al., 2019).

The POC diagnostic tests for testicular cancers are promising but are not yet as responsive as NAATs; more improvements are needed but must be anticipated in the immediate future as developments such as isothermal amplification and microfluidic innovations provide the potential to offer NAAT sensitivity and specificity to POC diagnostic tests for TV. Excellent FDA-cleared POC assessments are needed for serological assays, syphilis, and HSV, and expectations are high that new diagnostic tests can satisfy the needs and demand for these assays under current growth.

The prospects for the production of modern affordable and reliable POCs must rely on the engagement of public health authorities and industry to effectively collaborate in getting these assays, such as those focused on chip technology and biosensors, via regulatory criteria and into practical usage in the United States and resource-limited environments. In the next 5 years, the POC research system will be loaded up with new POC assays.

References

Abdulbari, H. A., & Basheer, E. A. (2017a). Investigating the enhancement of microfluidics-based electrochemical biosensor response with different microchannel dimensions. *Current Analytical Chemistry*, *13*(5). Available from https://doi.org/10.2174/1573411012666160920145311.

Abdulbari, H. A., & Basheer, E. A. M. (2017b). Electrochemical biosensors: Electrode development, materials, design, and fabrication. *ChemBioEng Reviews*, *4*(2), 92−105. Available from https://doi.org/10.1002/cben.201600009.

Akyazi, T., Basabe-Desmonts, L., & Benito-Lopez, F. (2018). Review on microfluidic paper-based analytical devices towards commercialisation. *Analytica Chimica Acta*, *1001*, 1−17. Available from https://doi.org/10.1016/j.ac.2017.11.010.

Ashman, S., & Kandlikar, S. G. (2006). A review of manufacturing processes for microchannel heat exchanger fabrication. In *ASME 4th international conference on nanochannels, microchannels, and minichannels, Parts A and B*. ASMEDC. https://doi.org/10.1115/icnmm2006-96121.

Baratchi, S., Khoshmanesh, K., Sacristán, C., Depoil, D., Wlodkowic, D., McIntyre, P., & Mitchell, A. (2014). Immunology on chip: Promises and opportunities. *Biotechnology Advances*, *32*(2), 333−346. Available from https://doi.org/10.1016/j.biotechadv.2013.11.008.

Becker, H. (2002). Polymer microfluidic devices. *Talanta*, *56*(2), 267–287. Available from https://doi.org/10.1016/s0039-9140(01)00594-x.

Bentley, S. A., Johnson, A., & Bishop, C. A. (1993). A parallel evaluation of four automated hematology analyzers. *American Journal of Clinical Pathology*, *100*(6), 626–632. Available from https://doi.org/10.1093/ajcp/100.6.626.

Bi, H., Duarte, C. M., Brito, M., Vilas-Boas, V., Cardoso, S., & Freitas, P. (2016). Performance enhanced UV/vis spectroscopic microfluidic sensor for ascorbic acid quantification in human blood. *Biosensors and Bioelectronics*, *85*, 568–572. Available from https://doi.org/10.1016/j.bios.2016.05.054.

Boobphahom, S., Nguyet Ly, M., Soum, V., Pyun, N., Kwon, O.-S., Rodthongkum, N., & Shin, K. (2020). Recent advances in microfluidic paper-based analytical devices toward high-throughput screening. *Molecules*, *25*(13), 2970. Available from https://doi.org/10.3390/molecules25132970.

Bragheri, F., Martinez Vazquez, R., & Osellame, R. (2016). *Microfluidics*. Three-dimensional microfabrication using two-photon polymerization: Fundamentals, technology, and applications (pp. 310–334). Elsevier Inc. Available from https://doi.org/10.1016/B978-0-323-35321-2.00016-9.

Castell, O. K., Allender, C. J., & Barrow, D. A. (2009). Liquid–liquid phase separation: Characterisation of a novel device capable of separating particle carrying multiphase flows. *Lab on a Chip*, *9*(3), 388–396. Available from https://doi.org/10.1039/b806946h.

Chen, C. H., Cho, S. H., Tsai, F., Erten, A., & Lo, Y.-H. (2009). Microfluidic cell sorter with integrated piezoelectric actuator. *Biomedical Microdevices*, *11*(6), 1223–1231. Available from https://doi.org/10.1007/s10544-009-9341-5.

Chen, H., Yu, Z., Bai, S., Lu, H., Xu, D., Chen, C., Liu, D., & Zhu, Y. (2019). Microfluidic models of physiological or pathological flow shear stress for cell biology, disease modeling and drug development. *TrAC Trends in Analytical Chemistry*, *117*, 186–199. Available from https://doi.org/10.1016/j.trac.2019.06.023.

Cheng, J.-Y., Wei, C.-W., Hsu, K.-H., & Young, T.-H. (2004). Direct-write laser micromachining and universal surface modification of PMMA for device development. *Sensors and Actuators B: Chemical*, *99*(1), 186–196. Available from https://doi.org/10.1016/j.snb.2003.10.022.

Chin, C. D., Laksanasopin, T., Cheung, Y. K., Steinmiller, D., Linder, V., Parsa, H., Wang, J., Moore, H., Rouse, R., Umviligihozo, G., Karita, E., Mwambarangwe, L., Braunstein, S. L., van de Wijgert, J., Sahabo, R., Justman, J. E., El-Sadr, W., & Sia, S. K. (2011). Microfluidics-based diagnostics of infectious diseases in the developing world. *Nature Medicine*, *17*(8), 1015–1019. Available from https://doi.org/10.1038/nm0.2408.

Chung, C. K., & Lin, S. L. (2011). On the fabrication of minimizing bulges and reducing the feature dimensions of microchannels using novel CO_2 laser micromachining. *Journal of Micromechanics and Microengineering*, *21*(6), 065023. Available from https://doi.org/10.1088/0960-1317/21/6/065023.

Cristillo, A. D., Bristow, C. C., Peeling, R., Van Der Pol, B., de Cortina, S. H., Dimov, I. K., Pai, N. P., Jin Shin, D., Chiu, R. Y. T., Klapperich, C., Madhivanan, P., Morris, S. R., & Klausner, J. D. (2017). Point-of-care sexually transmitted infection diagnostics: Proceedings of the STAR sexually transmitted infection-clinical trial group programmatic meeting. *Sexually Transmitted Diseases*, *44*(4), 211–218. Available from https://doi.org/10.1097/OLQ.0000000000000572.

Danoy, M., Poulain, S., Jellali, R., Gilard, F., Kato, S., Plessy, C., Kido, T., Miyajima, A., Sakai, Y., & Leclerc, E. (2020). Integration of metabolomic and transcriptomic profiles of hiPSCs-derived hepatocytes in a microfluidic environment. *Biochemical Engineering Journal*, *155*, 107490. Available from https://doi.org/10.1016/j.bej.2020.107490.

Dimov, I. K., Garcia-Cordero, J. L., O'Grady, J., Poulsen, C. R., Viguier, C., Kent, L., Daly, P., Lincoln, B., Maher, M., O'Kennedy, R., Smith, T. J., Ricco, A. J., & Lee, L. P. (2008). Integrated microfluidic tmRNA purification and real-time NASBA device for molecular diagnostics. *Lab on a Chip*, *8*(12), 2071. Available from https://doi.org/10.1039/b812515e.

Di Sebastiano, K. M., & Mourtzakis, M. (2012). A critical evaluation of body composition modalities used to assess adipose and skeletal muscle tissue in cancer. *Applied Physiology, Nutrition, and Metabolism*, *37*(5), 811−821. Available from https://doi.org/10.1139/h2012-079.

Dolly, J. O., Wang, J., Zurawski, T. H., & Meng, J. (2011). Novel therapeutics based on recombinant botulinum neurotoxins to normalize the release of transmitters and pain mediators. *FEBS Journal*, *278*(23), 4454−4466. Available from https://doi.org/10.1111/j.1742-4658.2011.08205.x.

Dorfman, L., Ghersin, I., Khateeb, N., Daher, S., Shamir, R., & Assa, A. (2020). Cardiovascular risk factors are not present in adolescents with inflammatory bowel disease. *Acta Paediatrica*. https://doi.org/10.1111/apa.15237.

Fachin, F., Spuhler, P., Martel-Foley, J. M., Edd, J. F., Barber, T. A., Walsh, J., Karabacak, M., Pai, V., Yu, M., Smith, K., Hwang, H., Yang, J., Shah, S., Yarmush, R., Sequist, L. V., Stott, S. L., Maheswaran, S., Haber, D. A., Kapur, R., & Toner, M. (2017). Monolithic chip for high-throughput blood cell depletion to sort rare circulating tumor cells. *Scientific Reports*, *7*(1), 10936. Available from https://doi.org/10.1038/s41598-017-11119-x.

Friend, J., & Yeo, L. (2010). Fabrication of microfluidic devices using polydimethylsiloxane. *Biomicrofluidics*, *4*(2), 026502. Available from https://doi.org/10.1063/1.3259624.

Fu, G., Tor, S. B., Hardt, D. E., & Loh, N. H. (2011). Effects of processing parameters on the micro-channels replication in microfluidic devices fabricated by micro injection molding. *Microsystem Technologies*, *17* (12), 1791−1798. Available from https://doi.org/10.1007/s00542-011-1363-2.

Gaydos, C., & Hardick, J. (2014). Point of care diagnostics for sexually transmitted infections: Perspectives and advances. *Expert Review of Anti-Infective Therapy*, *12*(6), 657−672. Available from https://doi.org/10.1586/14787210.2014.880651.

George-Gay, B., & Parker, K. (2003). Understanding the complete blood count with differential. *Journal of PeriAnesthesia Nursing*, *18*(2), 96−117. Available from https://doi.org/10.1053/jpan.2003.50013.

Gobby, D., Angeli, P., & Gavriilidis, A. (2001). Mixing characteristics of T-type microfluidic mixers. *Journal of Micromechanics and Microengineering*, *11*(2), 126−132. Available from https://doi.org/10.1088/0960-1317/11/2/307.

Guihen, E., & Glennon, J. D. (2005). Rapid separation of antimicrobial metabolites by microchip electrophoresis with UV linear imaging detection. *Journal of Chromatography A*, *1071*(1−2), 223−228. Available from https://doi.org/10.1016/j.chroma.2004.12.031.

Günther, A., & Jensen, K. F. (2006). Multiphase microfluidics: From flow characteristics to chemical and materials synthesis. *Lab on a Chip*, *6*(12), 1487−1503. Available from https://doi.org/10.1039/b609851g.

Guo, L., Feng, J., Fang, Z., Xu, J., & Lu, X. (2015). Application of microfluidic "lab-on-a-chip" for the detection of mycotoxins in foods. *Trends in Food Science & Technology*, *46*(2), 252−263. Available from https://doi.org/10.1016/j.tifs.2015.09.005.

Hajmohammadi, M. R., Alipour, P., & Parsa, H. (2018). Microfluidic effects on the heat transfer enhancement and optimal design of microchannels heat sinks. *International Journal of Heat and Mass Transfer*, *126*, 808−815. Available from https://doi.org/10.1016/j.ijheatmasstransfer.2018.06.037.

Hassan, Q., Li, S., Ferrag, C., & Kerman, K. (2019). Electrochemical biosensors for the detection and study of α-synuclein related to Parkinson's disease—A review. *Analytica Chimica Acta*, *1089*, 32−39. Available from https://doi.org/10.1016/j.ac.2019.09.013.

Holmes, D., Sun, T., Morgan, H., Holloway, J., Cakebread, J., & Davies, D. (2007). Label-free differential leukocyte counts using a microfabricated, single-cell impedance spectrometer. In *2007 IEEE Sensors*. IEEE. https://doi.org/10.1109/icsens.2007.4388687.

Hu, J., Cui, X., Gong, Y., Xu, X., Gao, B., Wen, T., Lu, T. J., & Xu, F. (2016). Portable microfluidic and smartphone-based devices for monitoring of cardiovascular diseases at the point of care. *Biotechnology Advances*, *34*(3), 305−320. Available from https://doi.org/10.1016/j.biotechadv.2016.02.008.

Huang, Y., Liu, S., Yang, W., & Yu, C. (2010). Surface roughness analysis and improvement of PMMA-based microfluidic chip chambers by CO_2 laser cutting. *Applied Surface Science*, *256*(6), 1675−1678. Available from https://doi.org/10.1016/j.apsusc.2009.09.092.

Jakeway, S. C., de Mello, A. J., & Russell, E. L. (2000). Miniaturized total analysis systems for biological analysis. *Fresenius' Journal of Analytical Chemistry*, *366*(6−7), 525−539. Available from https://doi.org/10.1007/s002160051548.

Janmey, P. A., Fletcher, D. A., & Reinhart-King, C. A. (2020). Stiffness sensing by cells. *Physiological Reviews*, *100*(2), 695−724. Available from https://doi.org/10.1152/physrev.000130.2019.

Kang, S.-W., Chen, J.-S., & Hung, J.-Y. (1998). Surface roughness of (110) orientation silicon based micro heat exchanger channel. *International Journal of Machine Tools and Manufacture*, *38*(5−6), 663−668. Available from https://doi.org/10.1016/s0890-6955(97)00115-6.

Karabacak, N. M., Spuhler, P. S., Fachin, F., Lim, E. J., Pai, V., Ozkumur, E., Martel, J. M., Kojic, N., Smith, K., Chen, P., Yang, J., Hwang, H., Morgan, B., Trautwein, J., Barber, T. A., Stott, S. L., Maheswaran, S., Kapur, R., Haber, D. A., & Toner, M. (2014). Microfluidic, marker-free isolation of circulating tumor cells from blood samples. *Nature Protocols*, *9*(3), 694−710. Available from https://doi.org/10.1038/nprot.2014.044.

Kildow, B. J., Karas, V., Howell, E., Green, C. L., Baumgartner, W. T., Penrose, C. T., Bolognesi, M. P., & Seyler, T. M. (2018). The utility of basic metabolic panel tests after total joint arthroplasty. *The Journal of Arthroplasty*, *33*(9), 2752−2758. Available from https://doi.org/10.1016/j.arth.2018.05.003.

Klank, H., Kutter, J. P., & Geschke, O. (2002). CO_2-laser micromachining and back-end processing for rapid production of PMMA-based microfluidic systems. *Lab on a Chip*, *2*(4), 242. Available from https://doi.org/10.1039/b206409j.

Konstantinou, D., Shirazi, A., Sadri, A., & Young, E. W. K. (2016). Combined hot embossing and milling for medium volume production of thermoplastic microfluidic devices. *Sensors and Actuators B: Chemical*, *234*, 209−221. Available from https://doi.org/10.1016/j.snb.2016.04.147.

Kratz, C., Furchner, A., Sun, G., Rappich, J., & Hinrichs, K. (2020). Sensing and structure analysis by in situ IR spectroscopy: From mL flow cells to microfluidic applications. *Journal of Physics: Condensed Matter*, *32*(39), 393002. Available from https://doi.org/10.1088/1361-648x/ab8523.

Kumar, M., Yadav, S., Kumar, A., Sharma, N. N., Akhtar, J., & Singh, K. (2019). MEMS impedance flow cytometry designs for effective manipulation of micro entities in health care applications. *Biosensors and Bioelectronics*, *142*, 111526. Available from https://doi.org/10.1016/j.bios.2019.111526.

Leal, S., Cristelo, C., Silvestre, S., Fortunato, E., Sousa, A., Alves, A., Correia, D. M., Lanceros-Mendez, S., & Gama, M. (2020). Hydrophobic modification of bacterial cellulose using oxygen plasma treatment and chemical vapor deposition. *Cellulose*. Available from https://doi.org/10.1007/s10570-020-03005-z.

Lee, C.-Y., Chang, C.-L., Wang, Y.-N., & Fu, L.-M. (2011). Microfluidic mixing: A review. *International Journal of Molecular Sciences*, *12*(5), 3263−3287. Available from https://doi.org/10.3390/ijms12053263.

Lee, T., Kim, J., Uh, Y., & Lee, H. (2019). Deep neural network for estimating low density lipoprotein cholesterol. *Clinica Chimica Acta*, *489*, 35−40. Available from https://doi.org/10.1016/j.cc.2018.11.022.

Li, J., Ni, T., Liu, H., Wu, L., Pan, Y., Zhao, Y., & Zhu, Y. (2020). Functional poly(carboxybetaine methacrylate) coated paper sensor for high efficient and multiple detection of nutrients in fruit. *Chinese Chemical Letters*, *31*(5), 1099−1103. Available from https://doi.org/10.1016/j.cclet.2019.11.005.

Li, P., Mao, Z., Peng, Z., Zhou, L., Chen, Y., Huang, P.-H., Truica, C. I., Drabick, J. J., El-Deiry, W. S., Dao, M., Suresh, S., & Huang, T. J. (2015). Acoustic separation of circulating tumor cells. *Proceedings of the National Academy of Sciences of the United States of America*, *112*(16), 4970−4975. Available from https://doi.org/10.1073/pnas.1504484112.

Livak-Dahl, E., Sinn, I., & Burns, M. (2011). Microfluidic chemical analysis systems. *Annual Review of Chemical and Biomolecular Engineering*, *2*(1), 325−353. Available from https://doi.org/10.1146/annurev-chembioeng-061010-114215.

M, K. R., Bhattacharya, S., DasGupta, S., & Chakraborty, S. (2018). Collective dynamics of red blood cells on an in vitro microfluidic platform. *Lab on a Chip*, *18*(24), 3939−3948. Available from https://doi.org/10.1039/c8lc01198b.

Ma, B., Zhang, G., Qin, J., & Lin, B. (2009). Characterization of drug metabolites and cytotoxicity assay simultaneously using an integrated microfluidic device. *Lab on a Chip*, *9*(2), 232−238. Available from https://doi.org/10.1039/b809117j.

Mach, A. J., Adeyiga, O. B., & Di Carlo, D. (2013). Microfluidic sample preparation for diagnostic cytopathology. *Lab on a Chip*, *13*(6), 1011−1026. Available from https://doi.org/10.1039/c2lc41104k.

Magnes, J., Raley-Susman, K. M., Melikechi, N., Sampson, A., Eells, R., Bello, A., & Lueckheide, M. (2012). Analysis of freely swimming c. elegans using laser diffraction. *Open Journal of Biophysics*, *02*(03), 101−107. Available from https://doi.org/10.4236/ojbiphy.2012.23013.

Maldonado, E. N., Romero, J. R., Ochoa, B., & Aveldaño, M. I. (2001). Lipid and fatty acid composition of canine lipoproteins. *Comparative Biochemistry and Physiology Part B: Biochemistry and Molecular Biology*, *128*(4), 719−729. Available from https://doi.org/10.1016/s1096-4959(00)00366-3.

Martin, C. R., & Aksay, I. A. (2005). Microchannel molding: A soft lithography-inspired approach to micrometer-scale patterning. *Journal of Materials Research*, *20*((8)), 1995−2003. Available from https://doi.org/10.1557/jmr.20050.0251.

McCormick, R. M., Nelson, R. J., Alonso-Amigo, M. G., Benvegnu, D. J., & Hooper, H. H. (1997). Microchannel electrophoretic separations of DNA in injection-molded plastic substrates. *Analytical Chemistry*, *69*(14), 2626−2630. Available from https://doi.org/10.1021/ac9701997.

Mela, P., van den Berg, A., Fintschenko, Y., Cummings, E. B., Simmons, B. A., & Kirby, B. J. (2005). The zeta potential of cyclo-olefin polymer microchannels and its effects on insulative (electrodeless) dielectrophoresis particle trapping devices. *Electrophoresis*, *26*(9), 1792−1799. Available from https://doi.org/10.1002/elps.200410153.

Microfluidic tools for metabolomics. (2008). https://dspace.mit.edu/handle/1721.1/46495.

Miyado, T., Wakida, S., Aizawa, H., Shibutani, Y., Kanie, T., Katayama, M., Nose, K., & Shimouchi, A. (2008). High-throughput assay of nitric oxide metabolites in human plasma without deproteinization by lab-on-a-chip electrophoresis using a zwitterionic additive. *Journal of Chromatography A, 1206*(1), 41-4. Available from https://doi.org/10.1016/j.chroma.2008.07.065. Epub 2008 Jul 26. PMID: 18692851.

Moore, A. (2006). *Trichomoniasis. Sexually transmitted diseases* (pp. 229−241). Humana Press. Available from https://doi.org/10.1007/978-1-59745-040-9_10.

Muluneh, M., & Issadore, D. (2013). Hybrid soft-lithography/laser machined microchips for the parallel generation of droplets. *Lab on a Chip*, *13*(24), 4750−4754. Available from https://doi.org/10.1039/c3lc50979f.

Murphy, B. M., Dandy, D. S., & Henry, C. S. (2009). Analysis of oxidative stress biomarkers using a simultaneous competitive/non-competitive micromosaic immunoassay. *Analytica Chimica Acta*, *640*((1−2)), 1−6. Available from https://doi.org/10.1016/j.ac.2009.03.003.

Nahavandi, S., Baratchi, S., Soffe, R., Tang, S.-Y., Nahavandi, S., Mitchell, A., & Khoshmanesh, K. (2014). Microfluidic platforms for biomarker analysis. *Lab on a Chip*, *14*(9), 1496−1514. Available from https://doi.org/10.1039/c3lc51124c.

Nivedita, N., & Papautsky, I. (2013). Continuous separation of blood cells in spiral microfluidic devices. *Biomicrofluidics*, *7*(5), 54101. Available from https://doi.org/10.1063/1.4819275.

Okazaki, M., Usui, S., Ishigami, M., Sakai, N., Nakamura, T., Matsuzawa, Y., & Yamashita, S. (2005). Identification of unique lipoprotein subclasses for visceral obesity by component analysis of cholesterol profile in high-performance liquid chromatography. *Arteriosclerosis, Thrombosis, and Vascular Biology*, *25*(3), 578−584. Available from https://doi.org/10.1161/01.atv.0000155017.60171.88.

Oncescu, V., Mancuso, M., & Erickson, D. (2014). Cholesterol testing on a smartphone. *Lab on a Chip*, *14*(4), 759−763. Available from https://doi.org/10.1039/c3lc51194d.

Ossowski, P., Raiter-Smiljanic, A., Szkulmowska, A., Bukowska, D., Wiese, M., Derzsi, L., Eljaszewicz, A., Garstecki, P., & Wojtkowski, M. (2015). Differentiation of morphotic elements in human blood using optical coherence tomography and a microfluidic setup. *Optics Express*, *23*(21), 27724. Available from https://doi.org/10.1364/oe.23.027724.

O'Sullivan, S., Ali, Z., Jiang, X., Abdolvand, R., Ünlü, M. S., Silva, H. P., da, Baca, J. T., Kim, B., Scott, S., Sajid, M. I., Moradian, S., Mansoorzare, H., & Holzinger, A. (2019). Developments in transduction, connectivity and AI/machine learning for point-of-care testing. *Sensors*, *19*((8)), 1917. Available from https://doi.org/10.3390/s19081917, Basel, Switzerland.

Paik, P., Pamula, V. K., & Fair, R. B. (2003). Rapid droplet mixers for digital microfluidic systems. *Lab on a Chip*, *3*(4), 253−259. Available from https://doi.org/10.1039/b307628h.

Panić, S., Loebbecke, S., Tuercke, T., Antes, J., & Bošković, D. (2004). Experimental approaches to a better understanding of mixing performance of microfluidic devices. *Chemical Engineering Journal*, *101*(1−3), 409−419. Available from https://doi.org/10.1016/j.cej.2003.10.026.

Park, Y., Best, C. A., Badizadegan, K., Dasari, R. R., Feld, M. S., Kuriabova, T., Henle, M. L., Levine, A. J., & Popescu, G. (2010). Measurement of red blood cell mechanics during morphological changes. *Proceedings of the National Academy of Sciences of the United States of America*, *107*(15), 6731−6736. Available from https://doi.org/10.1073/pnas.0909533107.

Parween, S., Debishree Subudhi, P., & Asthana, A. (2019). An affordable, rapid determination of total lipid profile using paper-based microfluidic device. *Sensors and Actuators B: Chemical*, *285*, 405−412. Available from https://doi.org/10.1016/j.snb.2019.01.064.

Plouffe, B. D., & Murthy, S. K. (2014). Perspective on microfluidic cell separation: A solved problem? *Analytical Chemistry*, *86*(23), 11481−11488. Available from https://doi.org/10.1021/ac5013283.

Rapp, B. E. (2017). *Introduction to maple. Microfluidics: Modelling, mechanics and mathematics* (pp. 9−20). Elsevier. Available from https://doi.org/10.1016/b978-1-4557-3141-1.50002-2.

Ricco, A. J. (2006). *Microfabricated biosensing devices: MEMS, microfluidics, and mass sensors. Biosensing* (pp. 79−106). Netherlands: Springer. Available from https://doi.org/10.1007/1-4020-4058-x_6.

Romoli, L., Tantussi, G., & Dini, G. (2011). Experimental approach to the laser machining of PMMA substrates for the fabrication of microfluidic devices. *Optics and Lasers in Engineering*, *49*(3), 419−427. Available from https://doi.org/10.1016/j.optlaseng.2010.11.013.

Ruecha, N., Lee, J., Chae, H., Cheong, H., Soum, V., Preechakasedkit, P., Chailapakul, O., Tanev, G., Madsen, J., Rodthongkum, N., Kwon, O.-S., & Shin, K. (2017). Paper-based digital microfluidic chip for multiple electrochemical assay operated by a wireless portable control system. *Advanced Materials Technologies*, *2*(3), 1600267. Available from https://doi.org/10.1002/admt.201600267.

Sackmann, E. K., Fulton, A. L., & Beebe, D. J. (2014). The present and future role of microfluidics in biomedical research. *Nature*, *507*(7491), 181−189. Available from https://doi.org/10.1038/nature13118.

Schwartz, B., & Robbins, H. (2019). Chemical etching of silicon: IV. etching technology. *Journal of The Electrochemical Society*, *123*(12), 1903−1909. Available from https://doi.org/10.1149/1.2132721.

Shan, X., Jin, L., Soh, Y. C., & Lu, C. W. (2009). A polymer-metal hybrid flexible mould and application for large area hot roller embossing. *Microsystem Technologies*, *16*(8−9), 1393−1398. Available from https://doi.org/10.1007/s00542-009-0991-2.

Sharma, S., Zapatero-Rodríguez, J., Estrela, P., & O'Kennedy, R. (2015). Point-of-care diagnostics in low resource settings: Present status and future role of microfluidics. *Biosensors*, *5*(3), 577−601. Available from https://doi.org/10.3390/bios5030577.

Shinohara, H., Mizuno, J., & Shoji, S. (2007). Fabrication of a microchannel device by hot embossing and direct bonding of poly(methyl methacrylate). *Japanese Journal of Applied Physics*, *46*(6A), 3661−3664. Available from https://doi.org/10.1143/jjap.460.3661.

Song, P., Fisher, A. C., Meng, L., & Nguyen, H. V. (2018). *Microfluidics for chemical analysis. Microfluidics: Fundamental, devices and applications* (pp. 211−235). Wiley-VCH Verlag GmbH & Co. KGaA. Available from https://doi.org/10.1002/9783527800643.ch6.

Su, W., Gao, X., Jiang, L., & Qin, J. (2015). Microfluidic platform towards point-of-care diagnostics in infectious diseases. *Journal of Chromatography A, 1377*, 13−26. Available from https://doi.org/10.1016/j.chroma.2014.12.041.

Tay, A., Pavesi, A., Yazdi, S. R., Lim, C. T., & Warkiani, M. E. (2016). Advances in microfluidics in combating infectious diseases. *Biotechnology Advances, 34*(4), 404−421. Available from https://doi.org/10.1016/j.biotechadv.2016.02.002.

Velten, T., Schuck, H., Haberer, W., & Bauerfeld, F. (2009). Investigations on reel-to-reel hot embossing. *The International Journal of Advanced Manufacturing Technology, 47*(1−4), 73−80. Available from https://doi.org/10.1007/s00170-009-1975-1.

Verbrugge, S. E., & Huisman, A. (2015). Verification and standardization of blood cell counters for routine clinical laboratory tests. *Clinics in Laboratory Medicine, 35*(1), 183−196. Available from https://doi.org/10.1016/j.cll.2014.10.008.

Vidic, J., Vizzini, P., Manzano, M., Kavanaugh, D., Ramarao, N., Zivkovic, M., Radonic, V., Knezevic, N., Giouroudi, I., & Gadjanski, I. (2019). Point-of-need DNA testing for detection of foodborne pathogenic bacteria. *Sensors, 19*(5), 1100. Available from https://doi.org/10.3390/s19051100, *Basel, Switzerland*.

Wang, S., Ge, L., Song, X., Yu, J., Ge, S., Huang, J., & Zeng, F. (2012). Paper-based chemiluminescence ELISA: Lab-on-paper based on chitosan modified paper device and wax-screen-printing. *Biosensors and Bioelectronics, 31*(1), 212−218. Available from https://doi.org/10.1016/j.bios.2011.10.019.

Whitesides, G. M. (2006). The origins and the future of microfluidics. *Nature, 442*(7101), 368−373. Available from https://doi.org/10.1038/nature05058.

Wisitsoraat, A., Sritongkham, P., Karuwan, C., Phokharatkul, D., Maturos, T., & Tuantranont, A. (2010). Fast cholesterol detection using flow injection microfluidic device with functionalized carbon nanotubes based electrochemical sensor. *Biosensors and Bioelectronics, 26*(4), 1514−1520. Available from https://doi.org/10.1016/j.bios.2010.07.101.

Wong, I., & Ho, C.-M. (2009). Surface molecular property modifications for poly(dimethylsiloxane) (PDMS) based microfluidic devices. *Microfluidics and Nanofluidics, 7*(3), 291−306. Available from https://doi.org/10.1007/s10404-009-0443-4.

Xia, Y., & Whitesides, G. M. (1998). Soft lithography. *Annual Review of Materials Science, 28*(1), 153−184. Available from https://doi.org/10.1146/annurev.matsci.28.1.153.

Xuan, X. (2019). Recent advances in direct current electrokinetic manipulation of particles for microfluidic applications. *Electrophoresis*. Available from https://doi.org/10.1002/elps.201900048.

Yang, X., Forouzan, O., Brown, T. P., & Shevkoplyas, S. S. (2012). Integrated separation of blood plasma from whole blood for microfluidic paper-based analytical devices. *Lab on a Chip, 12*(2), 274−280. Available from https://doi.org/10.1039/c1lc20803a.

Yang, Z., Xu, G., Reboud, J., Ali, S. A., Kaur, G., McGiven, J., Boby, N., Gupta, P. K., Chaudhuri, P., & Cooper, J. M. (2018). Rapid veterinary diagnosis of bovine reproductive infectious diseases from semen using paper-origami DNA microfluidics. *ACS Sensors, 3*(2), 403−409. Available from https://doi.org/10.1021/acssensors.7b00825.

Zhai, J., Li, H., Wong, A. H.-H., Dong, C., Yi, S., Jia, Y., Mak, P.-I., Deng, C.-X., & Martins, R. P. (2020). A digital microfluidic system with 3D microstructures for single-cell culture. *Microsystems & Nanoengineering, 6*(1). Available from https://doi.org/10.1038/s41378-019-0109-7.

Zhang, Y., Ceylan Koydemir, H., Shimogawa, M. M., Yalcin, S., Guziak, A., Liu, T., Oguz, I., Huang, Y., Bai, B., Luo, Y., Luo, Y., Wei, Z., Wang, H., Bianco, V., Zhang, B., Nadkarni, R., Hill, K., & Ozcan, A. (2018). Motility-based label-free detection of parasites in bodily fluids using holographic speckle analysis and deep learning. *Light, Science & Applications, 7*, 108. Available from https://doi.org/10.1038/s41377-018-0110-1.

Nanotechnology for blood test to predict the blood diseases/blood disorders

13

Setti Sudharsan Meenambiga, Punniavan Sakthiselvan, Sowmya Hari and Devasena Umai

Department of Bio-Engineering, School of Engineering, Vels Institute of Science, Technology and Advanced Studies, Chennai, India

13.1 Introduction

Nanotechnology, the art of constructing structures and particles of the size 10^{-9} nm, is finding applications in every corner of science. Owing to the unique size-based properties at the nanoscale, there has been a lot of focus on the use of nanostructures and nanomaterials in the biomedical field in the last few years. This has given rise to a new discipline called nanomedicine where various branches of sciences like engineering, material science, physics, and chemistry amalgamate with biology. Nanomedicine involves the use of nanoparticles, nanostructures, and nanodevices to diagnose, target, and treat diseases. Nanodiagnostics is an emerging field and is exhibiting its great potential to identify diseases and disorders. In this chapter, we will be focussing on various nanodiagnostic methods that are used for detecting blood diseases and disorders.

Blood is a specialized fluid connective tissue which consists of four main components. Plasma, the liquid part of blood contains proteins and salts. There are three types of cells with important functions, which constitute the solid part of blood. Red blood cells (RBCs) play a key role as oxygen carriers. White blood cells defend the body by fighting infections. Platelets are responsible for blood clotting. Any impairment in the function or production of one or more of these components gives rise to blood diseases or disorders which may be acute or chronic. Blood diseases and disorders may be inherited or they may arise from other factors like nutritional deficiency, mutation in genes, and may sometimes be due to infections. The most common types of blood disorders include anemia which is characterized by a low number of RBCs, and platelet disorders like hemophilia and clotting disorders. Blood diseases include the three main types of blood cancers: lymphoma, leukemia, and myeloma. Based on the prevailing symptoms, several diagnostic methods like complete blood count, imaging tests like ultrasound, magnetic resonance imaging (MRI), computed tomography (CT), positron emission tomography (PET) scans, bone marrow biopsy, and genomic testing have been used.

Effective diagnosis of a disease is essential to identify where the problem is and provide the correct medication so that the treatment will be more effective. The methods used for diagnosis should be such that the analyte molecules related to a particular disease or disorder and the sensing molecules interact significantly, resulting in even small amounts of the analyte molecules being

Nanotechnology for Hematology, Blood Transfusion, and Artificial Blood. DOI: https://doi.org/10.1016/B978-0-12-823971-1.00005-2

detected. Consequently, this allows for the early detection of diseases with much precision and high sensitivity. This is where nanodiagnostics serve the purpose. Various nanostructures and nanodevices are being developed which can open new directions for accurate and rapid diagnosis, especially at the molecular level. Nanodiagnostic methods are also less invasive and cost-effective. In addition, point-of-care testing (POC) can also be made possible through nanotechnology to get exact, real-time results in just a few minutes (Pourmand et al., 2012).

Some of those diagnostics, which are being developed for diagnosing blood diseases and disorders will be discussed below.

Nanoparticles, one of the best outcomes of nanotechnology, are basically particles of $1-100$ nm in size. An important feature of nanoparticles is that a nanoparticle has a very large surface area to volume ratio, due to which it differs from its bulk counterpart in several properties. Hence, they are greatly used in biomedical applications. Nanoparticles are basically classified into organic and inorganic. Organic nanoparticles mainly include dendrimers, polymeric nanoparticles, micelles, and liposomes. Inorganic nanoparticles include all nanoparticles of metals and metal oxides, quantum dots, and carbon nanotubes. In this chapter, nanoparticles involved in detecting blood diseases and disorders will be highlighted.

Researchers are showing a keen interest in nanodiagnostics because nanostructures and nanodevices are small in size with a large surface area. They are highly stable and biocompatible. Through nanotechnology, real-time diagnosis is possible, which is time- and cost-effective. As nanodevices are tiny in size, they can be designed to be portable and used for POC diagnosis.

In further sections, blood diseases and disorders like anemia, hemophilia, formation of blood clots, and cancers of blood (including leukemia, lymphoma, and myeloma) and the nanodiagnostic methods used to detect these disorders are discussed.

13.2 Blood disorders

13.2.1 Anemia

Anemia, a global health concern affecting one third of world's population, is a condition with a decrease in RBCs/hemoglobin concentration compared with the normal level. Infants and female adolescents are at high risk due to diverse causes leading to mortality, morbidity, and impaired behavioral development in infants. At the biological level, anemia is defined as an imbalance in erythrocyte count production due to inefficient erythropoiesis and loss of erythrocytes due to hemolysis. Anemia is diagnosed as different types on the basis of RBC morphology and the biological mechanism of its causation. Nutritional disorders, infection/inflammation, and genetic disorders of hemoglobin form the major criteria to cause different types of anemia, especially in low- and middle-income countries (Chaparro & Suchdev., 2019).

Increased blood loss results in acute and chronic anemia, which is due to gastrointestinal blood loss and postpartum hemorrhage. Classification based on excessive hemolysis includes acquired and hereditary anemia, in which acquired anemia is primarily by malarial infection, whereas hereditary anemia includes hemoglobin disorders. When the concentration of hematopoietic nutrients, such as iron, vitamins A, B12, B6, C, D and E, folate, riboflavin, and copper, are insufficient for the normal production of RBC, nutritional anemia arises. Based on the reports of the World

Health Organization, more than 50% of anemia cases are caused by iron deficiency (Hurrell & Zimmermann, 2007).

Iron is an essential element for the metabolism of all living organisms. Oxygen and electron transport, deoxyribonucleic acid synthesis, hormone synthesis, and cell cycle control are all regulated by iron. A broad spectrum of diseases are caused by iron deficiency with a variety of clinical manifestations and in severe cases leading to neurodegenerative disorders. Iron forms a free radical leading to excess iron concentration which must be regulated carefully to prevent tissue damage. The major causes of iron deficiency are mainly as a result of fast metabolism, pregnancy, and excess blood loss caused by pathological infections of whipworm and hookworm. Female adolescents are at high risk of iron deficiency due to menstrual cycles and the low amount of iron paves the way for anemia during pregnancy. The low bioavailability and low solubility of iron is the major factor for iron deficiency anemia (IDA) which could be corrected through iron-supplemented diets.

Many diseases cause chronic anemia, also called anemia of inflammation (AI), which is the second major cause next to IDA. AI is normocytic with a reduced reticulocyte count in which cytokines are released after infection leading to altered iron metabolism. The inflammatory cytokines block iron absorption and mobilization in the circulation, thus reducing the production and life pan of RBCs (Nairz et al., 2016). Soil-transmitted helminths, such as hookworm, feed on human intestinal mucosa, thus causing blood loss leading to IDA, which could be treated with antihelminthic medications such as albandazole. The malarial parasite requires iron for its survival and this in turn causes severe IDA and the anemic condition caused by malaria is multifactorial, resulting in increased hemolysis of RBC. Normocytic anemia with reduced reticulocyte count and unaltered iron stores is prevalent among HIV-infected patients and the degree of anemia increases as the disease progresses.

Around 33,000 children are born every year with genetic hemoglobin disorders globally, leading to sickle cell anemia and thalassemia. Sickle cell anemia causes chronic hemolytic anemia, which is due to the defect in the β-globin chain damaging the blood vessels. Thalassemia on the other hand is classified as either α-thalassemia and β-thalassemia, based on the defect in the α- or β-globin chain which causes an autosomal recessive disorder leading to severe anemia and impaired erythropoiesis.

13.2.2 Bleeding disorders

A group of disorders that share the inability to form a proper blood clot are called bleeding disorders. A bleeding disorder may be a condition that affects the way your blood normally clots. The clotting process is also known as coagulation, in which blood changes from a liquid to solid. In some cases, certain conditions prevent the clotting process, which can result in heavy or prolonged bleeding. They are normally characterized by extended bleeding after injury, trauma, surgery, or menstruation. Improper clotting is often caused by defects in blood components like platelets and/or clotting proteins, also called clotting factors. The body produces 13 clotting factors. If any of them are defective or deficient, blood coagulation is affected; a light, moderate, or severe bleeding disorder may result.

When the blood cannot clot properly it develops into bleeding disorders. Our body needs blood proteins and blood cells such as clotting factor and platelets for the blood to clot. A plug is formed at the site of a damaged or injured blood vessel when the platelets clump together, after which a fibrin clot is formed by the clotting factors. This prevents blood from flowing out of the blood vessel by keeping the platelets in place. The clotting factors or platelets don't work the way they should or are in short supply in people with bleeding disorders. Excessive or prolonged bleeding can occur when the blood doesn't clot which can also lead to spontaneous or sudden bleeding in the muscles, joints, or other parts of the body.

Bleeding disorders may also be caused by (1) vitamin K deficiency, (2) decreased RBC count, and (3) side effects from certain medications. Anticoagulants are medications that can interfere with the clotting of blood.

Bleeding disorders can either be acquired or inherited. Inherited disorders are passed down through genetics, whereas acquired disorders can develop or spontaneously occur later in life. Some bleeding disorders can result in severe bleeding following an accident or injury, while in other disorders, heavy bleeding can happen suddenly and for no reason. There are a number of different bleeding disorders, but the following are the most common ones:

- Hemophilia A and B are conditions that occur when there are low levels of clotting factors in your blood, which can cause heavy or unusual bleeding into the joints. Though hemophilia is not common, it can have life-threatening complications and side effects.
- Factor II, V, VII, X, or XII deficiencies are bleeding disorders related to abnormal bleeding and blood clotting problems.
- von Willebrand's disease (VWD) is the most common inherited bleeding disorder which develops when the blood lacks the von Willebrand factor that helps the blood to clot.

The proteins in the blood that control bleeding are called clotting factors. Problems with factor VIII is called hemophilia A and problems with factor IX is known as hemophilia B. Rare clotting factor deficiencies are bleeding disorders in which any one of the other clotting factors (i.e., factors I, II, V, V + VIII, VII, X, XI, or XIII) are absent or not working properly. Not much is known about these disorders because they are diagnosed so rarely (Zhang, Wei, et al., 2019; Zhang, Li, et al., 2019).

von Willebrand Factor (VWF) is synthesized and secreted by vascular endothelium to form part of the perivascular matrix. By binding with a receptor on the platelet surface membrane (glycoprotein Ib/IX), von Willebrand factor promotes the platelet adhesion phase of hemostasis, thus connecting the platelets to the vessel wall. von Willebrand factor is also necessary to maintain normal plasma factor VIII levels. In response to stress, exercise, pregnancy, inflammation, or infection, levels of VWF can temporarily increase. Although VWD, is a hereditary disorder that may cause factor VIII deficiency, like hemophilia A, the factor VIII deficiency in VWD is usually only moderate.

13.2.3 Blood cancer

Hematologic malignancy involves cancer in blood, lymphocytes, lymph node, lymphatic vessels, lymphatic system, bone marrow, spleen, thymus, and lymphoid tissues and is commonly known as blood cancer. The most common blood cancers are leukemia and myeloma, which start in the bone marrow, and lymphoma, which starts in the lymphatic system. The main pathogenesis in leukemia and myeloma are no proper formation of blood cells like RBCs, white blood cells, and platelets and they mainly affect the immune functioning of the body.

Leukemia is a malignant tumor in blood forming organs due to clonal and distorted proliferation of hematopoietic stem cells that affect the production of precursors of blood cells and development of leukocytes. The cells in blood include RBCs that help in the transport of oxygen, white blood cells, that provide immunity against pathogens, and platelets that help in blood clotting. In normal conditions, RBCs are produced in higher numbers compared to white blood cells and platelets. But in leukemia, white blood cells are produced in more numbers leading to crowding of white blood cells in the blood. Leukemia in the bone marrow leads to improper production of white blood cells. So, the immune cells are highly affected. Fever, tiredness, chills, loss of blood cells leading to paleness, swollen and painful glands, and pins and needles in bone and joints are the symptoms associated with leukemia. This disease is also associated with weak blood vessels, which often rupture leading to bleeding (Manisha, 2012; Shafique & Tehsin, 2018). Based on the type of cell (myeloid or lymphoid) and the extent of cell differentiation, it is classified as acute lymphoblastic leukemia, acute myelogenous leukemia, chronic lymphocytic leukemia, and chronic myelogenous leukemia (Davis et al., 2014). Acute lymphoblastic leukemia is one of the most common leukemias in adults that is developed due to the genetic changes in the lymphoid precursor cells, leading to altered proliferation and differentiation. Many genetic translocations have been identified in the genes like ETV6-RUNX1, TCF3-PBX1, and BCR-ABL1. Abnormal accumulation of myeloblasts cells in bone marrow is the common condition in acute myeloblastic leukemia that leads to bone marrow failure and death. An increase in the large number of resting lymphocytes with less proliferative ability is the main pathogenesis of leukemia. Chronic lymphocytic leukemia is a malignant tumor of the lymphoid precursor which has increased accumulation of $CD5^+$ B cells in the blood, bone marrow, and lymphoid tissues.

A malignant tumor due to improper proliferation and differentiation of lymphoid cells and their precursors is known as a lymphoma. Lymphomas are classified into two types: (1) Hodgkin lymphomas; and (2) non-Hodgkin lymphomas. Based on the cells involved they are further subclassified into B cell lymphoma, T cell lymphoma, and natural killer cell lymphomas. Hodgkin lymphomas are a B cell-associated cancer, with a characteristic increase in Hodgkin and Reed-Sternberg (HRS) cells that are derived from the B cells. These HRS cells are very low in normal conditions, while in Hodgkin's lymphoma they are elevated and also the HRS cells in Hodgkin lymphoma express unusual markers. Many risk factors are associated with Hodgkin lymphoma. Persons affected with Epstein Barr virus-induced glandular fever and hepatitis are highly prone to Hodgkin lymphoma. Obese men are at higher risk of developing Hodgkin lymphoma. Non-Hodgkin lymphoma is a large group of cancers mainly involving white blood cells and T cells. The main pathogenesis involves the suppression of the immune system. Non-Hodgkin lymphomas are common in patients with HIV. Patients with autoimmune diseases like systemic lupus erythrematosus, rheumatoid arthritis, and lymphomagnesis are at increased risk of developing non-Hodgkin lymphoma (Singh et al., 2020).

13.3 Current diagnostic and therapeutic strategies for blood disorders

Effective diagnosis of the disease becomes inevitable to identify where the problem is and provide the correct medication so that the treatment will be less tedious. Blood contains a lot of biomolecules like enzymes and antigens. The onset of a disease or a disorder can be characterized by the excess production of some of these biomolecules. There are some biomolecules which will be produced during the progression of a disease and these will serve as an indication of the disease as well. The methods used for the diagnosis of a disease will be effective if the analyte molecules related to that disease or disorder and the sensing molecules interact significantly such that even small amounts of the analyte molecules could be detected. The therapeutic strategies currently followed for blood disorders in the case of anemia and bleeding disorders are mainly aimed at finding out the deficiency factor and correcting it accordingly.

A typical reduction in the number of RBCs in the circulatory system and subsequent reduction in their oxygen-carrying capacity can be defined as an anemic condition. But in clinical practice, anemia is commonly diagnosed by a reduced hemoglobin count for IDA. Serum ferritin level (<30 ng/mL) is the most clear test to access the iron store in the case of IDA. Perl's iron staining method was also used for the diagnosis of IDA but it has its own limitations due to its invasive procedure. This test stains bone marrow and allows visualization of the absence of iron in both erythroblasts and macrophages for IDA (Camaschella et al., 2016). Low volumes of RBCs and hemoglobin along with an increase in RBC width in the initial stages of iron deficiency are typical markers of IDA. RBC width is unaltered in anemia of chronic disease and it requires a differential diagnosis. Reduced hemoglobin volume is a parameter that is present in the early stages of erythropoiesis restriction stimulated by iron. Also, the absence of iron causes the accumulation of zinc instead of iron in the protoporphyrin ring of hemoglobin which could be a screening test for IDA.

Anemia of inflammation or anemia of chronic disease (ACD) is multifactorial, and is most prevalent in admitted patients eliciting an inflammatory response. Initial diagnosis of ACD involves blood film and bone marrow analysis, reticulocyte concentration, biluribin levels, stool analysis, and assessing the renal function. Blood film morphology should be analyzed for ACD which indicates microcytic red cell morphology with hemoglobin concentration of $8-9.5$ g/dL. Serum ferritin levels are usually increased during chronic inflammatory states and hence serum ferritin levels are increased in ACD when compared to IDA. Iron dextran is the most commercially used intravenous iron and has the advantage of infusing total iron in one single administration with lower cost. But iron dextran has the disadvantage of causing anaphylactic reactions which are life-threatening.

Iron therapy and carcinogenesis are highly related as iron (FeII) on reaction with hydrogen peroxide forms highly reactive free radicals which can cause DNA damage and aid in tumor progression. Treatment for ACD should focus on improving the oxygen-carrying capacity and also on detecting the basic major underlying disease. Anemia leads to poor prognosis for terminally ill patients with cancers and cognitive heart failure. Iron therapy could not be very effective in treating ACD as this anemic condition is not an absolute deficiency and is a relative deficiency. Erythropoietic-stimulating agents help in treating cancers, kidney diseases, and HIV patients as these counteract the antiproliferative effects of cytokines and also increase iron uptake along with heme biosynthesis. The diagnosis and treatment of ACD involves the use of iron supplements along with erythropoietic-stimulating agents, which could benefit from the understanding of the pathogenesis of ACD. This leads to more future targets of treatment (Madu & Ughasoro, 2017).

The three extensive treatments for bleeding disorders are as follows:

- Risk reduction

    Patients with bleeding disorders should avoid medications that thin the blood and may also need to make some changes to lifestyle or activities to reduce their risk of bleeding. Physicians can take steps to reduce the risk of operative and postoperative bleeding.

- Medications
  Several drugs are available that help prevent clots from dissolving and for improving blood coagulation.
- Replacement therapy
  Patients with moderate to severe bleeding disorders may require transfusion of clotting factors or blood platelets. Clotting factors may be lab-synthesized proteins or donated human products. As a preventive measure, patients with severe bleeding disorders may receive clotting factor transfusions. However, repeated administrations of such factors result in increasing tolerance and resistance leading to escalating dosages for efficacy.

One of the main reasons for the increased mortality rate associated with blood cancer is the difficulties associated with its diagnosis. Effective treatment of cancer relies mainly on the efficient diagnosis. Physical signs of leukemia include pale skin, enlargement of liver, spleen, and lymph nodes. Increase in white blood cells or platelets and Beta-2-microglobulin helps in leukemia diagnosis. Further bone marrow biopsy, X-ray, PET–MRI, and CT scans could be used to confirm the presence and extent of the spread of leukemia, lymphoma, and myeloma.

Chemotherapy, radiotheraphy, and bone marrow transplantation are the commonly followed treatment options for blood cancer. Chemotherapy uses multiple anticancer drugs to be given at a time for blood cancer which helps to inhibit the growth of cancer cells. The most commonly used chemo drugs include Vincristine, Daunorubicin, Cytarabine L-asparaginase, 6-mercaptopurine (6-MP), and methotrexate. These chemo drugs affect the bone marrow cells resulting in lower blood cell counts. Tumor lysis syndrome is most common with leukemia patients; it occurs when the contents of the broken leukemia cells enter into the bloodstream causing abnormal functioning of kidneys. This ultimately affects the heart and central nervous system.

Chemotherapy is given before a stem cell transplant which requires a matching healthy donor. Radiation therapy is another treatment option which relieves pain and discomfort. However, these treatments methods lead to adverse effects like immunosuppresion and tolerance to correcting factors. Chemotherapy leads to severe complications like neutropenia (low levels of neutrophils), peripheral neutropathy, nausea, vomiting, fatigue, alopecia (hair loss), and many others.

13.4 Nanodiagnostic approaches to detect blood disorders

Nanotechnology-based techniques have recently attracted popularity as an approach that may overcome the problems of current diagnostic techniques through their unique physical properties (i.e., shape, size, surface charge, and dimension) and specific mode of actions. These techniques may be applied to develop accurate, reliable, rapid, safe, cost-efficient, sensitive, specific, and easily accessible techniques for the detection of bleeding disorders. The diagnostic tests that use nanoscale particles as tags or labels are faster and more sensitive and only small amounts of sample material are needed, which are the certain advantages of applying nanotechnology to molecular diagnostics.

Structures with one or more dimensions having their size in the nanoscale are usually referred to as nanostructures. These nanosized structures can turn out to be phenomenal diagnostic tools owing to their size-dependent properties. In recent years, there has been a lot of interest in nanostructure-based diagnostic assays, both in vitro and in vivo. Nanostructures have excellent optical properties due to which they have the ability to detect the analyte molecules and produce signals which are better than conventional dye molecules (Bellah et al., 2012). Nanoparticles, nanotubes, nanowires, nanocrystals, nanofibres, nanocrystals, and quantum dots fall under the category of nanostructures. Among these, nanoparticles and quantum dots have been widely used for the diagnosis of blood diseases and disorders. Quantum dots are basically semiconductor nanocrystals. Owing to their optical and electrical properties, quantum dots have been used as fluorescent labels in in vivo medical imaging. They have a core surrounded by a shell which can be functionalized with various biomolecules. Quantum dots are advantageous over other labeling markers in the diagnosis of various diseases as they have broad absorption spectra, narrow emission bands, and extended photostability (Nune et al., 2009).

13.4.1 Nanodiagnostic sensors to detect anemia

The fact that biomolecular interactions are highly specific is exploited in the development of biosensors. There are several biological elements like antibodies, aptamers, or enzymes that can be used as bioreceptors. The transducer converts the interaction between the recognition molecule (bioreceptor) and the analyte molecule (target) into a signal which is then processed to give an output of the interaction. Numerous products of nanotechnology, including nanoparticles and nanotubes, can be incorporated into biosensors for enhanced sensing; this is known as a nanosensor. Nanosensors commonly make use of optical-based transducers. They work on the basis of measuring the light absorbance, fluorescence, scattering, reflectance, or refractive index that occurs when the recognition molecule and analyte molecule interact. There are two approaches by which optical sensing can be done, that is, label-based and label-free. Surface plasmon resonance (SPR), colorimetric analysis, fluorescent-based analysis, and biobarcode are some of the methods where optical detection is applied to detect target molecules. The distance between the nanoparticles determines their optical properties. In colorimetric biosensors, the aggregation of the nanoparticles occurs when the analyte binds to the bioreceptor molecule functionalized on a nanoparticle. This leads to a shift in their extinction spectrum which results in a colorimetric response.

13.4.1.1 Nanosensors to detect serum ferritin levels

Iron level in the body is assessed usually by serum ferritin level which is the specific biochemical test and most widely used iron status indicator. The serum ferritin level increases in concentration during infection/inflammation and hence the ferrtin level should be very carefully monitored for detecting the severity of IDA. Noninvasive methods of detection are preferable, especially in children, for the regular monitoring of serum ferritin levels, and hence nanobiosensors to detect saliva ferritin levels were developed. Saliva contains most of the information of the blood but in minute quantities. Hence, highly sensitive biosensors to detect significantly lower levels of saliva ferritin were the focus (Höller et al., 2018).

Yen et al. (2016) developed a silicon nanowire biosensor to detect ferritin level in a phosphate buffer saline which was highly sensitive and a label-free detection method. They developed polycrystalline-silicon horn-like nanowire field effect transistors (poly-Si NW FETs) with good electrical characteristics. The poly-Si NW FETs was able to measure serum ferritin levels below 50 pg/mL in a microfluidic channel (Namdari et al., 2016). Very few studies have been done to detect serum ferritin levels using FETs, which are highly sensitive and could detect lower levels of serum ferritin. Although, silicon nanowires with FETs have a very high sensitivity with lower detection limit, the design and synthesis of silicon nanowires are generally expensive and nontrivial. Graphene when compared to silicon nanowires are less expensive and graphene was used to develop GFET (graphene field effect transistor) for ferritin level detection (Oshin et al., 2020). GFET biosensors were designed using an innovative and cost-effective method whose detection level gives reliable data on the iron deficiencies in the human body. The lowest detection limit of about 5.3 ng/L could be detected using this GFET biosensor. It is simple to fabricate and transfer graphene to the substrate when compared to silicon nanowires.

Recently, a nanosensor was designed on a micelle which is sensitized by porphyrin and this micelle on interacting with iron produces fluorescence which can be detected by a camera. The amount of fluorescence is inversely proportional to the iron concentration as an increase in the concentration of protein-bound iron causes the system to quench less. This causes a reduction in fluorescence, which indicates the increase in iron concentration bound to serum proteins (Halder et al., 2019).

13.4.1.2 Carbon dots to detect iron levels in deep tissues

Carbon dots have been extensively used as nanoparticles in the imaging and sensing of biological samples. Nitrogen-doped carbon dots have been developed by conjugating with fluorescent isothiocyanate, which has high biocompatibility and low cytotoxicity. Two-photon fluorescence imaging serves as a multifunctional probe with excellent deep tissue imaging with high resolution and high specificity toward Fe^{3+} ions, serving as a good diagnosis of iron deficiency in deeper tissues (Lesani et al., 2020). This bionanosensor technology was used by group of scientists in Australia to develop a hypersensitive nanosensor to help in detecting iron disorders leading to the earlier diagnosis of life-threatening diseases. The sensor developed was tested on pig skin and the nanoprobe had excellent characteristics of deep tissue imaging and the ability to measure iron levels in the body and deeper tissues. Further research is aimed at testing the probe on animal models which helps in visualizing the complex biological tissues and to develop lab-on-a-chip systems. This system requires small blood volumes from the patient to gain an insight of ferric iron deficiency in the body for early intervention and treatment.

13.4.1.3 Diamond nanosensor

Nanosensors were developed using tiny diamonds containing nitrogen vacancy defect centers, which serve as excellent nanosensors to detect ferritin molecules (Ermakova et al., 2013). The lattice defects occur when two carbon atoms are replaced by a single nitrogen atom and an empty lattice site. These lattice defects could detect a very weak magnetic field and hence the weak magnetic field created by ferritin bound iron molecules could be measured. Accurate measurement of iron level is done by measuring the ferritin level which is an iron-binding protein. Hence, this nanodiamond sensor helps in detecting iron levels in blood with a noninvasive approach.

13.4.2 Nanotechnology-based methods for diagnosis of bleeding disorders

Nanopharmaceuticals are often used to detect diseases at much earlier stages and therefore the diagnostic applications could rely on conventional procedures using nanoparticles. Nanopharmaceuticals are an emerging field where the therapeutic delivery system and the sizes of the drug particle work on the nanoscale. Delivering the acceptable dose of a particular active agent to specific disease site still remains difficult within the pharmaceutical industry. Therefore by offering site-specific targeting of active agents, nanopharmaceuticals have enormous potential in addressing this failure of traditional therapeutics. Nanopharmaceuticals can reduce toxic systemic side effects, thereby leading to better patient compliance.

13.4.2.1 Nanomaterials for diagnosis of blood coagulation

In recent decades the growing field of nanomedicine has created a need for the investigation of nanomaterials biocompatibility including their anticoagulation properties. Anticoagulation properties of these nanomaterials, such as metal oxide nanoparticles of titanium and zinc oxide, metal nanoparticles such as gold, silver, and platinum; (Dakshayani et al., 2019; Ehmann et al., 2015), carbon nanomaterials of graphene oxide and carbon nanowires, and modified polymeric nanomaterials have been studied and they were proven to be used successfully in the treatment and detection of hemostasis diseases. Nanoparticles can specifically interact with the coagulation system in different ways. Generally, this interaction can be in two different ways: interaction with cells, such as epithelial cells, monocytes, and platelets or contact with plasma coagulation factors (Matus et al., 2018).

Polymeric nanoparticles (Nps), metallic nanoparticles, zinc oxide Nps, nanoceria, and electrospun nanofibers have been used in wound healing and to stop bleeding. For instance, electrospun N-alkylated chitosan (NACS) fibers are studied as an efficient hemostasis agent. NACS are often used to halt bleeding by converting whole liquid blood into a gel immediately. It has been explained that, by adhering and physically sealing the bleeding wound based on mainly the electrostatic interaction between the positive charge of the protonated amine group of Chitosan and negative charge of erythrocytes cell membranes is the main mechanism of clotting by Chitosan-based hemostatic agent. These fibers are shown to be in favor of the activation of platelets and coagulation factors (Wang et al., 2018). Muthiah Pillai et al. (2019) reported chitosan hydrogel incorporated with nano-Whitlockite (WH: $Ca_{18}Mg_2 (HPO_4)_2(PO_4)_{12}$), which includes approximately up to 20 wt.% of the inorganic phase of human bone. WH consists of Ca^{2+}, Mg^{2+}, and PO^{-34} ions which activate different coagulation factors involved in coagulation cascade.

13.4.2.2 Gold nanoparticles to detect blood proteins

Gold nanoparticles are used widely in bioimaging as they are inert, comparatively less toxic, and they can be surface functionalized. Gold nanoparticles have very good optical properties. They have the ability to absorb and scatter light which varies with the shape and size of the nanoparticles. When excited by a light of specific wavelength, the conduction electrons undergo a collective oscillation called SPR (Huang & El-Sayed, 2010). This property of SPR makes gold nanoparticles excellent markers for the optical detection of the label-free interactions with the target analyte using spectroscopic or colorimetric methods. Gold nanoparticles can be conjugated with antibodies, fluorescent dyes, or aptamers (Hu et al., 2011). Gold nanoparticles can also be used as contrast enhancers in CT imaging (Kattumuri et al., 2007). As there are numerous methods by which gold nanoparticles can be used, they are gaining a lot of interest in diagnosing diseases.

Gold nanoparticles conjugated with fibrinogen is one example of such label-free colorimetric biosensor which is used to detect picomolar concentrations of thrombin in blood (Chen et al., 2010). Localized surface plasmon resonance (LSPR) biosensor is another label-free biosensor. It consists of a sensor surface coated with a nanoparticle layer and a bioreceptor. Binding of an analyte causes a shift in the SPR band. In addition, to improve the sensitivity, sandwich LSPR biosensors are used. It consists of a sensor surface immobilized with a bioreceptor like aptamer or antibody. A second aptamer or a secondary antibody is conjugated with a nanoparticle. When the analyte gets sandwiched between the two, there is a wavelength shift in the SPR spectrum at the surface of the film (Razak et al., 2017). For the detection of proteins in blood, there are several immunoassays, such as enzyme linked immunosorbent assay (ELISA), radioimmunoassay, and western blot. But POC detection is not possible in these assays. For rapid, POC detection, lateral flow immunoassays employing nanoparticles like carbon, gold, magnetic nanoparticles, carbon nanotubes, and quantum dots as labels have been developed.

13.4.3 Nanotechnology in diagnosis of blood cancer

Nanotechnology is one major approach in the diagnosis and treatment of blood cancer. Nanotechnology is highly explored for the diagnosis and treatment of various diseases. In cancer diagnosis, nanotechnology is used to detect the cancer biomarkers, cancer-associated proteins, tumor DNA in circulatory system, any circulating tumor and exosomes. Nanomaterials have large surface to volume ratio, hence they can be densely coated with the aptamers, antibodies, and oligonucleotides, which can specifically bind and detect cancer cells. Nantotechnology-based detection techniques are very efficient as they are highly sensitive, time- and cost-effective (Zhang, Wei, et al., 2019; Zhang, Li, et al., 2019). In cancer, the nanomaterials are made to detect specific biomarkers like proteins, nucleic acids, or carbohydrates that are highly expressed in cancer and can be measured in blood, tissues, saliva, or urine.

13.4.3.1 Nanoparticles in cancer diagnosis

Nanoparticles are highly explored for the diagnosis and treatment of various diseases. For cancer diagnosis and treatment many types of organic and inorganic nanoparticles are widely used. Gold nanoparticles have size-tunable optical properties, strong absorption, and scattering, which makes them highly efficient to detect cancer. Polyethylene glycol-coated gold nanoparticles with surface-enhanced Raman scattering conjugated to single-chain variable fragment (ScFv) antibody has been designed to detect epidermal growth factor receptor, a notable biomarker in cancer. An electrochemical immunosensor made of magnetic beads coated with gold nanoparticles conjugated to horseradish peroxidase and secondary antibody was designed to specifically detect cancer. Gold nanoparticles conjugated to ssDNA oligonucleotides were designed to detect a unique chromosomal abnormality in the BCR-ABL gene in chronic myeloid leukemia. Gold nanoparticles with surface-enhanced Raman scattering specifically detected the B cell overexpression of CD45 and CD19 through flow cytometry (Khoshfetrat & Mehrgardi, 2017; Song et al., 2016).

Nanoparticles with supermagnetic properties made from iron oxide, especially magnetite (Fe_3O_4) and maghemite (γ-Fe_2O_3), have been highly explored for cancer treatment and diagnosis. Overexpressed CD20 biomarker in leukemia and lymphoma was detected by magnetic nanoparticles coated with avidin attached to biotinylated anti-CD20 antibody. Fluorescent dye Cy5.5-labeled nanoparticles conjugated to anti-CD20 antibody has been designed to detect B cell malignancies (Sahoo et al., 2017; Tazi et al., 2011).

13.4.3.2 Aptamer-conjugated nanoparticles

Selective detection of surface proteins in cancer leukemia can be identified using specific probes made of synthetic nucleic acids, called aptamers. This ssDNA or RNA can recognize specific proteins and peptides through a process called systematic evolution of ligands by exponential enrichment (SELEX). Cell SELEX technique is an efficient technique to generate aptamers with different targets simultaneously. However, for designing the aptamers, the target is needed to be known first. Aptamers have many advantages like high affinity, easy synthesis, less toxicity, ability to fold in three-dimensional conformations, high sensitivity with a detection limit of 10 cells/mL (Vinhas et al., 2017). Aptamers conjugated to nanoparticles provide high selectivity and sensitivity in cancer diagnosis. Aptamer conjugated to gold nanoparticles can bind specifically to the target cancer cells with a characteristic color change from red to violet (Tan et al., 2019). Also the aptamer changes the conformation when bound to the target, which can be easily detected using a sensor integrated with it (Hubbe et al., 2019).

Aptamer conjugated to DNA has been developed to detect the protein tyrosine kinase 7 (PTK7), a cell surface protein that is overexpressed in acute lymphoblastic leukemia (Poturnayová et al., 2019). Different types of aptamers against platelet-derived growth factor (PDGF), vascular endothelial growth factor (VEGF), tenascin-C, nuclear factor kappa-light-chain-enhancer (NFκB) of activated B cells, and prostate-specific membrane antigen have been designed in cancer diagnosis (Kim et al., 2018). Aptasensors are sensors attached to aptamers that can easily detect the conformational change when aptamers bind to cancer cell.

13.4.3.3 Optical and fluorescent aptasensors

Optical aptasensors are aptamers integrated to fluorescent dye and nanoparticles are used in the diagnosis of cancer. The cancer cells are detected by fluorescent signal from the aptamers. Many different types of optical biosensors have been developed to detect the cancer cell for example, chemiluminescence resonance energy transfer (CRET) aptasensor detected VEGF with a detection limit of 5 ppm. In acute lymphoma leukemia cells, a fluorescence aptasensor based on terbium (III)-aptamer showed promising results in the detection of T cell acute lymphoblastic leukemia cell line (Wu et al., 2019). Grapheme oxide conjugated fluorescent aptasensor worked as a quencher and target agent to selectively target leukiemia cells. On binding to target cells, the fluorescent aptamers, since they are complementary to the cancer cell DNA, get removed from the graphene oxide and bind to the target cancer cells. This results in the emission of fluorescence which can be easily detected (Tan et al., 2018).

An efficient electrochemical biosensor was developed for the detection of mucin 1, a surface protein highly upregulated in cancer. In leukemia cells, electrochemiluminiscence aptasensors were used to detect adenosine triphosphate (Negahdary et al., 2019).

A novel colorimetric aptasensors with sensing element peroxidase DNAzyme detected human leukemic lymphoblasts rapidly by change in the color of the substrate on binding of the sensor to the target cell. It is highly advantageous as it is easy, rapid, and does not require costly apparatus, DNA modification, or labeling (Zhu et al., 2010).

13.4.3.4 Quantum dots

Quantum dots are nanocrystals with a luminescent property that are simulated with UV light. They function as an efficient probe in cancer detection. They are efficient fluorescent probes with good biocompatibility, less toxic, highly stable, and are proved to be an efficient bioimaging technique. Combining aptamers and quantum dots has been proved to be efficient in the diagnosis of leukemia. Quantum dots conjugated with lectin are used to identify leukemia through flow cytometry (Zhelev et al., 2005). Zinc oxide carbon quantum dots conjugated to concanavalin A nanoprobe have been successfully developed for the detection and monitoring of leukemia (Zhang et al., 2013).

Inorganic quantum dots are highly advantageous compared to organic quantum dots due to their high resistance for photobleaching. In cancer detection, the quantum dots are attached to functionalized antibodies or complementary molecules to specifically target the cancer cells. The cadmium-coated quantum dots are toxic when exposed to UV light, hence nontoxic polymer coated quantum dotsa and micelle-encapsulated quantum dots are also developed. Aptamer-conjugated quantum dots attached to doxorubicin (DOX), a fluorescent drug, have been developed to detect cancer cells. When the QD−aptamer−DOX conjugates encounter a cancer cell, DOX will be taken up by the cancer cell through endocytosis and emit fluorescence which can be detected (Choi et al., 2010).

13.4.3.5 Magnetic nanoparticles

Magnetic nanoparticles have exceptional magnetic properties due to which they can be used as contrast agents in MRI. In the detection of leukemia, magnetic and fluorescent nanoparticles functionalized with aptamers have been used. Here, the aptamers are bound to the specific target cells and subsequently the magnetic nanoparticles isolate the target cells and fluorescent nanoparticles enhance detection with high sensitivity (Herr et al., 2006). Magnetic nanoparticles conjugated with antibodies have shown 95% efficiency in isolating lymphoma cells which were then imaged by flow cytometry and confocal microscopy (Sahoo et al., 2017). Likewise, functionalization with hyaluronic acid has helped in the detection of leukemia. Liposomes and chitosan-based nanoparticles have also been used along with magnetic nanoparticles and applied in imaging techniques (Limongi et al., 2019).

13.4.3.6 Nanowires

Nanowires have a length to width ratio of 1000 or more and a nanorange diameter of 10−200 nm and are now used for cancer biosensing. Due to their large surface area, fluorescent nanowires and magnetic nanowires are widely used in fluorescence imaging and MRI (Nana et al., 2019). Silver nanoparticles combined with silicon nanowires have gained attention due to their ability to increase SERS. Mutations in DNA are detected using molecular beacons probes designed using gold nanoparticle−silicon nanowire. These nanowire conjugated nanoparticles are highly sensitive and were able to detect the DNA even at very low concentrations (Abdul Rashid et al., 2013). The tube-like morphology of nanowires makes them easier to penetrate inside and detect the target cells (Lee et al., 2018). Shalek et al. (2012) illustrated the effect of silicon nanowire to detect the dysregulated signaling pathway in chronic lymphocytic leukemia.

13.4.3.7 Cantilevers

Nanocantilevers are highly efficient method in cancer diagnosis, in which the lever is fixed on one end can be made to bind to specific target proteins associated with cancer. The binding of the cancer proteins leads to surface tension changes causing the bending of the cantilevers. From the sensing of cantilevers, the presence of cancer biomarkers can be easily detected. The main working procedure of the cantilever is a monoclonal antibody specific to the particular antigen attached to the cantilevers and a second antibody attached to the nanoparticles is designed. This sandwiches the target protein in-between which is then detected (Hassanzadeh et al., 2011).

13.4.3.8 Carbon nanotubes to detect DNA mutation

Carbon nanotubes are large cylindrical layers of graphite, with atoms arranged in hexagons. They are time-effective compared to the ELISA technique and less expensive compared to immunoassay techniques. Carbon nanotube sensors attached to antibodies specifically detect cancer biomarkers on the surface of the blood cells in the presence of laser light. When the nanotube sensor encounters a cancer cell, a drop in electron transport in the sensor leads to a decrease in light-induced current. Mutations in DNA can also be detected using a record player with a nanotube tip, that traces the DNA and analyzes the mutated regions. In leukemia cells, the overexpression of P-glycoprotein was detected with the help of a single-walled carbon nanotube (SWNT) with high precision (Gulati et al., 2018).

13.4.3.9 Nanopores

Nanopores are tiny nanodevice with pores with nanoscale diameter. The ssDNA is made to pass through the nanopores. Any mutation in the DNA can be identified by the change in shape and electrochemical property of the base (Ganesh et al., 2011). Nanopores are used for rapid sequencing of genes. In acute myeloid leukemia gene mutations in six genes, NPM1, FLT3, CEBPA, TP53, IDH1, and IDH2, were rapidly identified using nanopore sequencing (Cumbo et al., 2019).

13.4.3.10 Nanoshells and nanorods

Nanoshells are nanobeads designed to absorb and scatter light near the infrared region. They are spherical in shape and can get accumulated in the cancer cell due to their high permeability. Nanoshells can be coated with tumor-targeting moieties for enhanced tumor specificity (Morton et al., 2010). For example, to detect cancer cells, nanoshells can be coated with specific antibodies. Initially, the antibodies are attached to polyethylene glycol (PEG) which is then attached to the nanoshell surface through the thiol group (Jaishree & Gupta, 2012). Gold nanoshells are nanoparticles with a dielectric core bound to a gold outer shell. The main advantage of gold nanoshells is their good physical, chemical, and optical properties. They are highly biocompatible and nontoxic (Shanbhag et al., 2017). Capsule-like nanoparticles, called nanorods, of length 10−20 nm are highly advantageous due to their shape and can be synthesized from many elements. Various nanorods made of carbon, zinc oxide, and gold, and magnetic nanorods are widely exploited for the diagnosis of cancer (Ghassan et al., 2019). CD33 antibodies conjugated nanorods specifically detected CD33 expressed human acute and chronicle leukemia, and successfully performed thermolysis of cancer cells (Liopo et al., 2012). Table 13.1 shows the nanotechnology-based methods for the diagnosis of blood cancer.

13.5 Nanomedicines for treatment of blood diseases

Extensive research on nanodrug delivery systems for iron is being carried out to treat anemia. The potential harmful effects of iron supplementation on the gastrointestinal tract makes it essential for a

Table 13.1 Nanotechnology-based diagnosis of blood cancer.

Nanoparticle	Cancer	Target/Biomarker	References
Aptamer conjugated gold coated magnetic nanoparticle	Acute myeloid leukemia	PTK7	Khoshfetrat and Mehrgardi (2017)
Silver nanoparticles	Acute myeloid leukemia	Leukemia cells	Khetani et al. (2015)
Quantam dots conjugated aptamers	Acute lymphocytic leukemia	PTK7	Yu et al. (2016)
Aptamers conjugated fluorescent silica nanoparticles	Acute lymphocytic leukemia	PTK7	Tan et al. (2016)
Superparamagnetic iron oxide nanoparticles	leukemia	CD34	Jaetao et al. (2009)
ssDNA-Gold nanoparticles	Chronic myeloid leukemia	E14a2	Vinhas et al. (2017)
Gold nanobeacons	Chronic myeloid leukemia	E13a2, E14a2	Cordeiro et al. (2016)
Anti-CD20- polymeric nanoparticles	Chronic lymphocytic leukemia	CD20	Capolla et al. (2015)
PEG coated SERS gold nanoparticles	B cell cancer	CD19, CD20, CD45	MacLaughlin et al. (2013)
Anti-CD20-conjugated magnetic nanoparticles	Lymphoma	CD20	Sahoo et al. (2017)
Quantum dots conjugated oligonucleotide probe	Acute leukemia and follicular lymphoma	Bcl2, surviving, myeloperoxidase,	Tholouli et al. (2006)
Quantum dots conjugated oligonucleotide probe	Chronic myeloid leukemia	Leukemia positive DNA	Sharma et al. (2012)
CdSe quantum dots	Leukemia	Lectin	Mashinchian et al. (2014)
Fluorescent semiconductor quantum dots	B cell non-Hodgkin's lymphoma	CD20	Shariatifar et al. (2019)
Single-walled carbon nanotubes	B cell lymphoma	CD20	Chinen et al. (2015)
Carbon nanotube	Myeloma, lymphoma	Arginase 1	Baldo et al. (2016)
Superparamagnetic iron oxide nanoparticles	Acute leukemia	CD34	Samir et al. (2015)
Liposome or lanthanide oxyfluoride nanoparticle	Leukemia	CD33	Simard and Leroux (2009)

better drug delivery system such as nanobased drug delivery. Iron supplementation leads to adverse effects causing inflammation of the intestinal mucosa by generating free radicals. Hence, encapsulation of iron using a nanoparticle aids in increasing the solubility and bioavailability of the molecules.

13.5.1 **Nanomedicines in treating anemia**

Intestinal absorption of nonheme iron was improved by designing anchovy muscle protein hydrosylate (AMPH) to form nanosized products. AMPH was tested in rats which showed enhanced bioavailability and increased hemoglobin regeneration (Wu et al., 2014). Thus the nanoform of iron particles have increased bioavailability. In another study, iron oxo-hydroxide nanoparticles were developed which resembled the ferritin core and were tested on animal models. The designed nanomaterial has much less cytotoxic effect and no undesirable effects which could be effectively used in treating IDA. A safer form of iron supplement was developed by modified tartarate inclusion in the nanodisperse ferrihydrite, which reduces the risk of free radicals (Powell et al., 2014).

13.5.1.1 Lipid nanoparticles

Solid lipid nanoparticles have excellent properties of greater absorption, stability, and biocompatibility when compared to liposomes which are highly unstable. Lipid nanoparticles enhance the transport of particles across the gut lipid membrane and increase the ability of cellular uptake of the loaded molecules. Lipids also act as barriers by being permeable to selective molecules on the gastrointestinal tract (Zariwala et al., 2013). Ferrous sulfate, a form of iron supplementation was loaded with lipid nanomaterial and the rate of iron absorption was 13.42% more than the unloaded free form of ferrous sulfate. In another study by Hosny et al. (2015) iron-loaded lipid nanoparticles were developed by the process of emulsification and ultrasonication. By performing in vivo pharmacokinetic studies on male albino rats, the drug release was found to be efficient and the bioavailability of iron loaded with lipid nanoparticles was higher when compared to commercially available iron tablets. Lipid nanoparticle iron delivery system could be administered through various routes of drug delivery such as oral, tropical, rectal routes, etc.

13.5.1.2 Vitamin-coated nanoparticles

Multivitamins like folic acid, ascorbic acid, and nicotinic acid were capped with iron oxide nanoparticles and tested in anemic albino rats. European Egyptian Pharmaceutical Industries holds patent for iron oxide nanoparticles coated with multivitamins. Among the vitamins, folic acid and nicotinic acid help in cell formation, whereas ascorbic acid is required for iron absorption. This formulation was tested in anemia-induced albino rats and the hemoglobin level was found to increase to 14.6 g/dL which could correct the anemia in albino rats within 4 days (Mahmoud & Helmy, 2014). Similarly, magnetite nanoparticles coated with ascorbic acid were developed as vitamin C (ascorbic acid) aids in increased intestinal iron absorption. Intraperitoneal or oral route of administration effectively cured anemia and resulted in stimulated erythropoiesis observed through bone marrow studies of albino rats. This invention was patented by Salah et al. (2010) and it increased the RBC count and hemoglobin count considerably in anemic rats.

13.5.1.3 Nanomineral water

In 2016 nanomineral water was prepared which contains the elemental form of iron mixed with water. Bound iron forms such as iron sulfate, iron gluconoate, and iron fumarate have their own limitation of poor intestinal absorption. The nanoiron preparation could be directly absorbed in the bloodstream and reaches the cells with 100% bioavailability. Nanoiron was found to be a good option for pregnant women who have a high chance of blood loss and also for women suffering from miscarriage.

13.5.2 Nanomedicines for bleeding disorders

Some new drugs for bleeding disorder diseases treatment had been approved by FDA, including Elzonris for BPDCN, Cablivi (caplacizumab-yhdp) for aTTP, and Jakafi (ruxolitinib), and Inrebic (fedratinib) for myelofibrosis. These had set good examples for researchers, academic and industrial, to get a better understanding of how to move forward the development and clinical trial of nanomedicine for bleeding disorders diseases. A straightforward path will be to integrate the approved drugs for bleeding disorders diseases treatment into a nanocarrier, forming a promising candidate of nanomedicine.

13.5.2.1 Hemostatic mechanisms of nanomaterials

The blood coagulation balance is achieved by the interaction of the blood platelets with the vascular endothelial cells and the plasma coagulation system. These systems prevent thrombosis in healthy organisms and enable blood clotting to stop bleeding in the events of vascular damage. The effect of coagulation factor is essential in the hemostatic procedure and can be categorized into three pathways. Tissue factor (TF) is released when tissue and blood vessels are damaged, in the extrinsic coagulation pathway (ECP). FX and FIX are activated when TF forms a complex with the transformation of the accelerator precursors FVII or activated FVIIa (TF-FVIIa). The fact that ECP was first started in pathological coagulation, is widely believed. The process of coagulation is significantly promoted once TF comes into the blood. In the intrinsic coagulation pathway, which is usually associated with vascular wall injury, subendothelial tissue components (collagen) are exposed and FXII is activated by collagen into FXIIa; high-molecular-weight kininogen (HMWK) combined with a small amount of FXIIa, followed by convertion of prekallikrein to kallikrein, can quickly feedback-activate FXII with HMWK. The ionized calcium (Ca^{2+}) reactivates FXI and the activated factor XII (FXIIa) reactivates FXI and FXIa. FXIa forms a complex with Ca^{2+}, PF3 (phosphatidylserine) and FVIIIa (activated by thrombin), which in turn activates FX as FXa. It is thought that the FXI is directly activated by initial thrombin and converts FXI into FXIa. Besides, activated FXa forms a complex prothrombin with PF3, Ca^{2+}, and FVa (activated by thrombin) in the common coagulation pathway. Prothrombin (Fll) is converted to thrombin (FIIa) by prothrombinase, and thrombin converts fibrinogen (Fg) to soluble fibrin monomer; FXIII is activated into FXIIIa by thrombin which causes FMs molecules to be cross-linked to the insoluble and stable fibrin, which results in blood coagulation.

The surface charge of nanoparticles has been proved to be strongly associated with hemostasis. It is well-known that chitosan and its derivatives form a cell thrombus or thrombus in combination with negatively charged RBCs, white blood cells, and platelets and are positively charged, which promotes the secretion of glycosaminoglycans such as hyaluronic acid to speed up wound healing. Other materials like oxidized regenerated cellulose and surgical hemostatic gauze made of oxidized cellulose are based on the hemostatic mechanism in which the acidic carboxyl groups in the molecule and hemoglobin, form a sticky plastic block when combined with Fe^{3+} ions, closing capillaries and stopping bleeding, providing a matrix for platelets adhesion, aggregation, and leading to the activation of the body coagulation mechanism.

13.5.2.2 TiO₂ nanotubes

TiO_2 nanotubes are often readily prepared by anodization of titanium in an electrolyte comprising dimethyl sulfoxide and HF, after which it is dispersed by sonication if needed. Studies on the interaction between blood and TiO_2 nanotubes demonstrate significantly stronger clot formation at reduced clotting times in comparison with pure blood. TiO_2 nanotubes can also play a role as a bioactive scaffold to facilitate fibrin formation. These understandings suggest that TiO_2 nanotube-functionalized bandage could be used to stop hemorrhage or stimulate fast hemostasis. Sun et al. (2018) indicated that when blood was in contact with nanotube-decorated gauze bandages or with nanotubes directly, the activated clotting time of blood was reduced approximately by 10% (from 285 seconds for pure blood to 260 seconds). The blood directly in contact with nanotubes was indicated to have increased blood clot strength by $\sim 75\%$ (from 2.21 kPa for pure blood to 3.87 kPa for blood containing 1 mg/mL of nanotubes), which suggests that the gauze bandages decorated with TiO_2 nanotubes could improve both the rate of blood clot formation and the strength of blood.

13.5.2.3 Carbon nanotubes

Many studies have been steadfast in the investigation of the interaction between carbon nanotubes and blood proteins and cells in recent years. It has been shown that SWNTs added either to isolated platelets or platelet-rich plasma can stimulate platelets to aggregate. Studies have also shown that the effect of SWNT on blood coagulation is not only limited to inducing platelet aggregation and SWNTs interact simultaneously with many factors of hemostasis. Carboxylated SWNTs reduce the clot formation time by five times as compared with the platelet-poor plasma, which has been shown by further study on modified SWNTs. Interestingly, functionalized SWNTs exhibit procoagulant and proaggregating properties and diminish their negative influence on blood cells when administrated into the bloodstream of animals. These promising effects for hemostasis applications could be attributed to the interaction between nanotubes and cells and significant changes in nanotube surface properties after surface modification or functionalization of SWNTs.

13.5.2.4 Platelet-inspired nanomedicine

Platelets are generated from megakaryocytes in the bone marrow and are the smallest cells in the blood and have no nucleus. Molecular mechanisms of platelet-mediated hemostasis include primary and secondary hemostasis. Platelets bind rapidly to specific proteins (e.g., von Willebrand factor) and collagen at the bleeding site during the primary hemostasis, followed by interplatelet cross-linking via fibrinogen, which is recognized through the active GPIIb-IIIa on the surface of platelet molecules. Fibrin is formed and deposited in the process of coagulation cascade on the phosphoserine-rich surface of active platelets. Such a coagulation process includes activation, adhesion, and aggregation of platelets, as well as deposition and maturation of fibrin, when taken together. Thrombocytopenia can be caused by poor coagulation, and hypercoagulability can lead to thrombosis. In a bleeding complication case, natural platelets or platelet-derived products are offered for transfusion. These products suffer from short shelf life, contamination, and risks of infection/immunoreaction (only if prior serological testing was conducted).

To overcome such issues, artificial platelet-like biomaterials are attracting increasing attention. Platelets are loaded with thousands of biocomponents and are of micrometer size. Scientists have focused on the "platelet-inspired" nanomaterials that could simplify the configuration of the nanomaterials and function well as platelets during the hemostasis process. A recent emerging trend would be to enhance platelet adhesion and hemostatic plug formation to harness more platelet's functions, such as clot contraction, and to build a minimal system. Fig. 13.1 shows the formation of a blood clot and the artificial platelet which will be able to perform some essential functions of

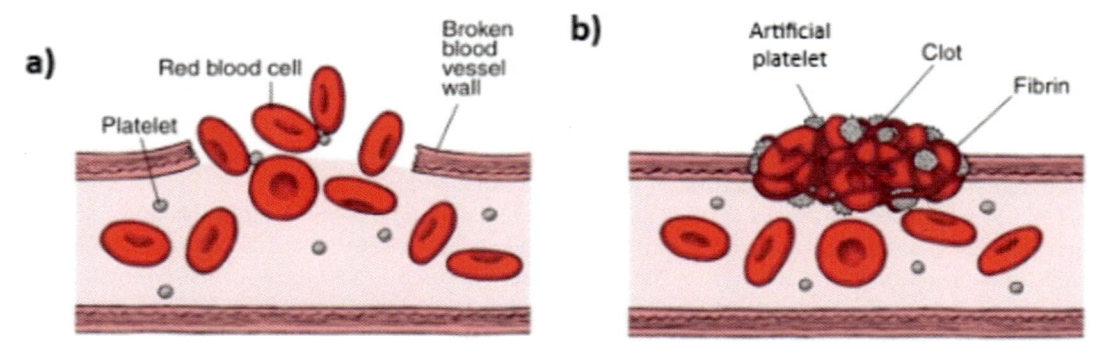

FIGURE 13.1

Formation of blood clot: (A) damaged blood vessel; (B) repaired blood vessel wall.

natural platelets, including vesicle encapsulation, protein production, attachment, and fusion (Majumder & Liu, 2017). Fibrin-binding antibodies that are coupled to platelet-like particles (PLPs) have shown the capability to mimic natural platelet functions to bind the wound site, stabilize clot structure, and enhance clot formation (Brown et al., 2014). A core−shell PLP was incorporated by further improvements to facilitate temporal control over clot retraction and mimic the antimicrobial action of platelets by integrating with gold nanoparticles (Sproul et al., 2019), which might improve healing outcomes after hemostasis.

SynthoPlate is a synthetic liposomal platelet surrogate that has shown bifunctional features, adhesion and aggregation, due to the integration of three peptides: von Willebrand Factor-binding peptide, peptides that bind collagen, and GPIIb-IIIa-binding fibrinogen-mimetic peptide (Shukla et al., 2017). After a femoral artery injury, the SynthoPlate was incorporated in animal testing and the result showed that blood loss was reduced, blood pressure was stabilized, and survival rate was improved (Hickman et al., 2018).

13.5.3 Nanotechnology in treatment of blood cancer

Currently the treatment for blood cancer includes chemotherapy and radiation. However, these treatment methods involve long-term side effects and also can lead to the development of multidrug resistance. Other treatment techniques like stem cell transplant require a matching healthy donor. In recent times, nanotechnology is widely used in the treatment and drug delivery for various diseases due to the selective delivery of drugs to the target cells without damaging the healthy cells.

13.5.3.1 Quantum dots in treatment of cancer

Wogonin, a natural flavonoid with antiproliferative effects, was conjugated to cadmium−telluride quantum dots with diameter 4 nm specifically induced apoptosis in leukemia cells. In myeloma cancer, polyethylene glycol-modified cadmium−telluride quantum dots conjugated to doxorubicin selectively delivered the drug to the cancer cell in a pH-controlled manner and improved the efficacy of the drug (Chen et al., 2018).

13.5.3.2 Nanoparticles in treatment of cancer

Nanoparticles have various advantages like increased retention, permeability of drugs, and selective targeting of targeted cells and are widely used for the treatment of various diseases. Various organic, inorganic, and hybrid nanoparticles are generated from different materials and are used in the diagnosis and treatment of cancers. Wogonin or cytarabine conjugated magnetic nanoparticles successfully destroyed the proliferation of tumor cells in leukemia.

Gold nanoparticles conjugated to oligonucleotides and dasatinib, a potent drug in inhibiting tyrosine kinase inhibitor, showed promising effects in chronic myeloid leukemia treatment. In chronic myeloid leukemia, gold nanoparticles functionalized with fludarabine were used to selectively inhibit tyrosine kinase inhibitor in cancer cells. In chronic myeloid leukemia, another gene that is highly upregulated is BIRC5. In order to control the level of BIRC5 expression, gold nanoparticles functionalized with complementary oligonucleotide conjugated to the drug dasatinib were designed. The oligonucleotide binds to the target mRNA and the dasatinib drug is released that levels the BIRC5 expression. In acute myeloid leukemia, the highly expressed folate receptors were successfully detected using nanoconjugates. Oligonucleotide conjugated nanoparticles with imatinib efficiently decreased the cancer cell survival (Gossai et al., 2014).

Overexpression of *Mcl-1* expression was controlled by lipid nanoparticles in lymphoma cancer cells. Studies in murine model showed that topoisomerase I poison SN-38 bound to lipid nanoparticles reduced tumors and improved survival rate (Knapp et al., 2016). Polymeric nanoparticles, synthesized from copolymers, show promising effects in the treatment, imaging, and diagnosis of cancer cells. Micelles made of PEG—polylactic acid encapsulated with porous silicon (pSi), and coated with E-selectin thioaptamer (ESTA) conjugated to parthenolide were designed to selectively target and reduce acute myeloid leukemia cells. ESTA bound to E-selectin in the bone marrow endothelial cells, where the drug parthenolide was delivered and reduced the severity of cancer (Zong et al., 2016). In acute lymphocytic leukemia, nanoparticles selectively delivered the drug doxorubicin to CD19 positive cancer cells, thus protecting the healthy cell from the drug's cytotoxic effects (Krishnan et al., 2015).

13.5.3.3 Nanoliposomes in treatment of cancer

Nanoliposomes as carriers for contrast agents such as iodine for MRI, and CT scans have been shown to prevent rapid clearance of the contrast agent from the body efficiently, thereby improving the cardiac imaging and effectiveness of total blood pool in animal models. An improved liposomal formulation of iodine indicated a high blood pool iodine concentration that facilitated excellent contrast between the myocardium and blood in the right and left ventricles, aorta, pulmonary trunk, and inferior vena cava. There is significantly lower liver and spleen contrast, as is predicted from the delayed clearance of the PEGylated liposomal iodine formulation via the reticuloendothelial system. Brightness and contrast of image in the CV regions are reduced because normal iodine generally collects in areas such as spleen and liver. This long duration at stable, high opacity makes liposomal iodine a promising effective micro-CT agent for contrast enhancement within submillimeter vessels without significant renal clearance (Godin et al., 2010). Another emerging field is represented by CT tomography where nanoparticles were shown the potential to increase imaging contrast. Liposomes are double-layered vesicles with one aqueous layer attached to one or more layers of phospholipids. They form a structure similar to the phospholipid bilayer of biological membrane. In leukemia, the overregulated CXCR4 was successfully antagonized using BAT1 conjugated liposomes (McCallion et al., 2019).

Anti-CD19 norcantharidin-loaded liposome specifically destroyed CD19 positive acute lymphocytic leukemia. Tyrosine–protein kinase transmembrane receptor, expressed in chronic lymphocytic leukemia, were targeted by miR-29b-loaded immunoliposomes (Houshmand et al., 2020).

13.6 Future perspectives and conclusion

Since the 1990s the list of FDA-approved nanotechnology-based products and clinical trials has increased staggeringly and includes synthetic polymer particles, liposome formulations, micellar nanoparticles, protein nanoparticles, nanotubes, nanocrystals, and many other nanobased products often in combination with drugs or biologics.

Treatment strategy combined to diagnosis is termed as theranostics. Thus the integration of treatment and diagnosis at the nanoscale level is termed as nanotheranostics. This is highly advantageous as they reduce the overall time and cost for the process and also provide precise results. For example, B cell lymphoma was specifically targeted and treated with gold nanoparticles combined with p-mercaptobenzoic acid and drug rituximab. siRNA bound diatomite nanoparticles selectively downregulated the antiapoptotic factor BCL2. This approach can be exploited in targeting antiapoptotic factor in B cell lymphoma and leukemia (Vinhas et al., 2017).

A nanobot or nanorobot is an invention which scientists are trying hard to implement to make a breakthrough in the field of biomedicine. It would be a milestone in this field as its applications and benefits would be immense. Nanobots, made by using DNA origami or carbon nanotubes, can be used to build biosensors of various types. These can also be made to circulate in the body to detect analyte molecules, such as proteins or nucleic acids, related to a particular disease. Early diagnosis of diseases and continuous monitoring of body functions are some of the merits of nanobots. Another novel optical biosensor is the biobarcode sensor. Biobarcodes consist of nanoscale bands of metals that are coated with a targeting molecule to which specific analytes can bind. They are advantageous as they can detect low volumes of the analyte and can be multiplexed (Hu et al., 2011).

Nowadays, nanomedicine has revolutionized the way we discover and administer the drugs in biological systems, in spite of regulatory mechanisms for nanomedicines along with safety/toxicity assessments. Due to the advances in nanomedicine, our ability to diagnose the diseases and combine diagnosis with therapy has also became a reality. Nanoparticles aim to improve pharmaceutical areas ranging from drug delivery to the detection of diseases and to create more effective medical treatments. There is an increase in the demand for new strategies and devices to diagnose and treat diseases accurately, easily, and efficiently. The implementation of new methods in the nanotechnology field will lead to the design and standardization of alternative therapies specific to each patient and disease. Preclinical testing focuses mainly on the characterization of nanomedicine, using original and multifaceted instruments. The nanoparticles are the therapeutics tools for the researchers and physicians to make the diagnosis and treatment techniques for blood diseases.

References

Abdul Rashid, J. I., Abdullah, J., Yusof, N. A., & Hajian, R. (2013). The development of silicon nanowire as sensing material and its applications. *Journal of Nanomaterials, 2013.*

Baldo, S., Buccheri, S., Ballo, A., Camarda, M., La Magna, A., Castagna, M. E., Romano, A., Iannazzo, D., Di Raimondo, F., Neri, G., & Scalese, S. (2016). Carbon nanotube-based sensing devices for human Arginase-1 detection. *Sensing and Bio-sensing Research, 7*, 168−173.

Bellah, M., Christensen, S. M., & Iqbal, S. M. (2012). Nanostructures for medical diagnostics. *Journal of Nanomaterials, 2012*, 22.

Brown, A. C., Stabenfeldt, S. E., Ahn, B., Hannan, R. T., Dhada, K. S., & Herman, E. S. (2014). Ultrasoft microgels displaying emergent platelet-like behaviours. *Nature Materials, 13*, 1108−1114.

Camaschella, C., Hoffbrand, A.V., & Hershko, C. (2016). Iron deficiency and disorders of haem synthesis. PostGraduate Haematology. Ed, 7.

Capolla, S., Garrovo, C., Zorzet, S., Lorenzon, A., Rampazzo, E., Spretz, R., Pozzato, G., Núñez, L., Tripodo, C., Macor, P., & Biffi, S. (2015). Targeted tumor imaging of anti-CD20-polymeric nanoparticles developed for the diagnosis of B-cell malignancies. *International Journal of Nanomedicine, 10*, 4099.

Chaparro, C. M., & Suchdev, P. S. (2019). Anemia epidemiology, pathophysiology, and etiology in low-and middle-income countries. *Annals of the New York Academy of Sciences, 1450*(1), 15.

Chen, C. K., Huang, C. C., & Chang, H. T. (2010). Label-free colorimetric detection of picomolar thrombin in blood plasma using a gold nanoparticle-based assay. *Biosensors and Bioelectronics, 25*(8), 1922−1927.

Chen, D., Chen, B., & Yao, F. (2018). Doxorubicin-loaded PEG-CdTe quantum dots as a smart drug delivery system for extramedullary multiple myeloma treatment. *Nanoscale Research Letters, 13*(1), 1−9.

Chinen, A. B., Guan, C. M., Ferrer, J. R., Barnaby, S. N., Merkel, T. J., & Mirkin, C. A. (2015). Nanoparticle probes for the detection of cancer biomarkers, cells, and tissues by fluorescence. *Chemical Reviews, 115* (19), 10530−10574.

Choi, Y. E., Kwak, J. W., & Park, J. W. (2010). Nanotechnology for early cancer detection. *Sensors (Basel), 10*(1), 428−455.

Cordeiro, M., Giestas, L., Lima, J. C., & Baptista, P. M. V. (2016). BioCode gold-nanobeacon for the detection of fusion transcripts causing chronic myeloid leukemia. *Journal of Nanobiotechnology, 14*(1), 38.

Cumbo, C., Minervini, C. F., Orsini, P., Anelli, L., Zagaria, A., Minervini, A., Coccaro, N., Impera, L., Tota, G., Parciante, E., & Conserva, M. R. (2019). Nanopore targeted sequencing for rapid gene mutations detection in acute myeloid leukemia. *Genes, 10*(12), 1026.

Dakshayani, S. S., Marulasiddeshwara, M. B., Kumar, S., Golla, R., Devaraja, S. R. H. K., & Hosamani, R. (2019). Antimicrobial, anticoagulant and antiplatelet activities of green synthesized silver nanoparticles using *Selaginella* (Sanjeevini) plant extract. *International Journal of Biological Macromolecules, 131*, 787−797.

Davis, A. S., Viera, A. J., & Mead, M. D. (2014). Leukemia: An overview for primary care. *American Family Physician, 89*(9), 731−738.

Ehmann, H. M., Breitwieser, D., Winter, S., Gspan, C., Koraimann, G., Maver, U., Sega, M., Köstler, S., Stana-Kleinschek, K., Spirk, S., & Ribitsch, V. (2015). Gold nanoparticles in the engineering of antibacterial and anticoagulant surfaces. *Carbohydrate Polymers, 117*, 34−42.

Ermakova, A., Pramanik, G., Cai, J. M., Algara-Siller, G., Kaiser, U., Weil, T., Tzeng, Y. K., Chang, H. C., McGuinness, L. P., Plenio, M. B., & Naydenov, B. (2013). Detection of a few metallo-protein molecules using color centers in nanodiamonds. *Nano Letters, 13*(7), 3305−3309.

Ganesh, E. N., Kaushik, R. R., & Krishna, K. M. (2011). Study of nano device for effective detection, diagnosis and treatment of cancer. *Computer Sciences and Telecommunications, 1*, 51−67.

Ghassan, A. A., Mijan, N. A., & Taufiq-Yap, Y. H. (2019). Nanomaterials: An overview of nanorods synthesis and optimization. In Morteza Sasani Ghamsari, & Soumen Dhara (Eds.), *Nanorods and nanocomposites*. IntechOpen. Available from http://doi.org/10.5772/intechopen.84550.

Godin, B., Sakamoto, J. H., Serda, R. E., et al. (2010). Emerging applications of nanomedicine for therapy and diagnosis of cardiovascular diseases. *Trends in Pharmacological Science, 31*, 199−205.

Gossai, N. P., Naumann, J. A., Li, N. S., Zamora, E. A., Gordon, D. J., Piccirilli, J. A., et al. (2014). Drug conjugated nanoparticles activated by cancer cell specific mRNA. *Oncotarget, 7*, 38243−38256.

Gulati, P., Kaur, P., Rajam, M. V., Srivastava, T., Mishra, P., & Islam, S. S. (2018). Single-wall carbon nanotube based electrochemical immunoassay for leukemia detection. *Analytical Biochemistry, 15*(557), 111−119.

Halder, A., Shikha, D., Adhikari, A., Ghosh, R., Singh, S., Adhikari, T., & Pal, S. K. (2019). Development of A nano-sensor (FeNSOR) based device for estimation of iron ions in biological and environmental samples. *IEEE Sensors Journal, 20*(3), 1268−1274.

Hassanzadeh, P., Fullwood, I., Sothi, S., & Aldulaimi, D. (2011). Cancer nanotechnology. *Gastroenterology and Hepatology from Bed to Bench., 4*(2), 63−69.

Herr, J. K., Smith, J. E., Medley, C. D., Shangguan, D., & Tan, W. (2006). Aptamer-conjugated nanoparticles for selective collection and detection of cancer cells. *Analytical Chemistry, 78*(9), 2918−2924.

Hickman, D. A., Pawlowski, C. L., Shevitz, A., Luc, N. F., Kim, A., & Girish, A. (2018). Intravenous synthetic platelet (SynthoPlate) nanoconstructs reduce bleeding and improve 'golden hour' survival in a porcine model of traumatic arterial hemorrhage. *Scientific Report, 8*, 3118.

Höller, U., Bakker, S. J., Düsterloh, A., Frei, B., Köhrle, J., Konz, T., Lietz, G., McCann, A., Michels, A. J., Molloy, A. M., & Murakami, H. (2018). Micronutrient status assessment in humans: Current methods of analysis and future trends. *Trends in Analytical Chemistry, 102*, 110−122.

Hosny, K. M., Banjar, Z. M., Hariri, A. H., & Hassan, A. H. (2015). Solid lipid nanoparticles loaded with iron to overcome barriers for treatment of iron deficiency anemia. *Drug Design, Development and Therapy, 9*, 313.

Houshmand, M., Garello, F., Circosta, P., Stefania, R., Aime, S., Saglio, G., & Giachino, C. (2020). Nanocarriers as magic bullets in the treatment of leukemia. *Nanomaterials, 10*(2), 276.

Hu, Y., Fine, D. H., Tasciotti, E., Bouamrani, A., & Ferrari, M. (2011). Nanodevices in diagnostics. *Wiley Interdiscip Reviews: Nanomed Nanobiotechnol, 3*(1), 11−32.

Huang, X., & El-Sayed, M. A. (2010). Gold nanoparticles: Optical properties and implementations in cancer diagnosis and photothermal therapy. *Journal of Advanced Research, 1*(1), 13−28.

Hubbe, H., Mendes, E., & Boukany, P. E. (2019). Polymeric nanowires for diagnostic applications. *Micromachines, 10*(4), 225.

Hurrell, R. F., & Zimmermann, M. B. (2007). Nutritional iron deficiency. *Lancet, 370*, 511−520.

Jaetao, J. E., Butler, K. S., Adolphi, N. L., Lovato, D. M., Bryant, H. C., Rabinowitz, I., Winter, S. S., Tessier, T. E., Hathaway, H. J., Bergemann, C., & Flynn, E. R. (2009). Enhanced leukemia cell detection using a novel magnetic needle and nanoparticles. *Cancer Research, 69*(21), 8310−8316.

Jaishree, V., & Gupta, P. D. (2012). Nanotechnology: A revolution in cancer diagnosis. *Indian Journal of Clinical Biochemistry, 27*(3), 214−220.

Kattumuri, V., Katti, K., Bhaskaran, S., Boote, E. J., Casteel, S. W., Fent, G. M., Robertson, D. J., Chandrasekhar, M., Kannan, R., & Katti, K. V. (2007). Gum arabic as a phytochemical construct for the stabilization of gold nanoparticles: In vivo pharmacokinetics and X-ray-contrast-imaging studies. *Small (Weinheim an der Bergstrasse, Germany), 3*(2), 333−341.

Khetani, A., Momenpour, A., Alarcon, E. I., & Anis, H. (2015). Hollow core photonic crystal fiber for monitoring leukemia cells using surface enhanced Raman scattering (SERS). *Biomedical Optical Express, 6*(11), 4599−4609.

Khoshfetrat, S. M., & Mehrgardi, M. A. (2017). Amplified detection of leukemia cancer cells using an aptamer-conjugated gold-coated magnetic nanoparticles on a nitrogen-doped graphene modified electrode. *Bioelectrochemistry (Amsterdam, Netherlands), 114*, 24−32.

Kim, M., Kim, D. M., Kim, K. S., Jung, W., & Kim, D. E. (2018). Applications of cancer cell-specific aptamers in targeted delivery of anticancer therapeutic agents. *Molecules (Basel, Switzerland), 23*(4), 830.

Knapp, C. M., He, J., Lister, J., & Whitehead, K. A. (2016). Lipidoid nanoparticle mediated silencing of Mcl-1 induces apoptosis in mantle cell lymphoma. *Experimental Biology and Medicine, 241*, 1−7.

Krishnan, V., Xu, X., Kelly, D., Snook, A., Waldman, S. A., Mason, R. W., et al. (2015). CD19-targeted nano-delivery of doxorubicin enhances therapeutic efficacy in B-cell acute lymphoblastic leukemia. *Molecular Pharmaceutics*, *12*, 2101–2111.

Lee, H., Choi, M., Lim, J., Jo, M., Han, J. Y., Kim, T. M., & Cho, Y. (2018). Magnetic nanowire networks for dual-isolation and detection of tumor-associated circulating biomarkers. *Theranostics*, *8*(2), 505–517.

Lesani, P., Singh, G., Viray, C. M., Ramaswamy, Y., Zhu, D. M., Kingshott, P., Lu, Z., & Zreiqat, H. (2020). Two-photon dual-emissive carbon dot-based probe: Deep-tissue imaging and ultrasensitive sensing of intracellular ferric ions. *ACS Applied Materials & Interfaces*, *12*(16), 18395–18406.

Limongi, T., Susa, F., & Cauda, V. (2019). Nanoparticles for hematologic diseases detection and treatment. *Hematology and Medical Oncology*, *4*(3), 1015761.

Liopo, A. V., Conjusteau, A., Konopleva, M., Andreeff, M., & Oraevsky, A. A. (2012). Laser nanothermolysis of human leukemia cells using functionalized plasmonic nanoparticles. *Nano Biomedicine and Engineering*, *4*(2), 66.

MacLaughlin, C. M., Mullaithilaga, N., Yang, G., Ip, S. Y., Wang, C., & Walker, G. C. (2013). Surface-enhanced Raman scattering dye-labeled Au nanoparticles for triplexed detection of leukemia and lymphoma cells and SERS flow cytometry. *Langmuir: The ACS Journal of Surfaces and Colloids*, *29*(6), 1908–1919.

Madu, A. J., & Ughasoro, M. D. (2017). Anemia of chronic disease: An in-depth review. *Medical Principles and Practice*, *26*(1), 1–9.

Mahmoud, M. B. M., & Helmy, S. H. A. (2014). Novel formula of iron-based nanocomposites for rapid and efficient treatment of iron deficiency anemia. European Egyptian Pharmaceutical Industries, Patent no. WO2014135170A1. <http://www.google.co.in/patents/WO2014135170A1?cl = en> (Accessed: 20 Nov 2020).

Majumder, S., & Liu, A. P. (2017). Bottom-up synthetic biology: Modular design for making artificial platelets. *Physical Biology*, *15*, 013001.

Manisha, P. (2012). Leukemia: A review article. *International Journal of Advanced Research in Pharmaceutical & Bio Sciences*, *1*(4), 397–408.

Mashinchian, O., Johari-Ahar, M., Ghaemi, B., Rashidi, M., Barar, J., & Omidi, Y. (2014). Impacts of quantum dots in molecular detection and bioimaging of cancer. *BioImpacts: BI*, *4*(3), 149.

Matus, M. F., Vilos, C., Cisterna, B. A., Fuentes, E., & Palomo, I. (2018). Nanotechnology and primary hemostasis: Differential effects of nanoparticles on platelet responses. *Vascular Pharmacology*, *101*, 1–8.

McCallion, C., Peters, A. D., Booth, A., Rees-Unwin, K., Adams, J., Rahi, R., Pluen, A., Hutchinson, C. V., Webb, S. J., & Burthem, J. (2019). Dual-action CXCR4-targeting liposomes in leukemia: Function blocking and drug delivery. *Blood Advances*, *3*(14), 2069–2081.

Morton, J. G., Day, E. S., Halas, N. J., & West, J. L. (2010). Nanoshells for photothermal cancer therapy. *Methods in Molecular Biology*, *624*, 101–117.

Muthiah Pillai, N. S., Eswar, K., Amirthalingam, S., Mony, U., Kerala Varma, P., & Jayakumar, R. (2019). Injectable nano whitlockite incorporated chitosan hydrogel for effective hemostasis. *Acs Applied Bio Materials*, *2*, 865–873.

Nairz, M., Theurl, I., Wolf, D., & Weiss, G. (2016). Iron deficiency or anemia of inflammation? *Wiener Medizinische Wochenschrift*, *166*(13–14), 411–423.

Namdari, P., Daraee, H., & Eatemadi, A. (2016). Recent advances in silicon nanowire biosensors: Synthesis methods, properties, and applications. *Nanoscale Research Letters*, *11*(1), 406.

Nana, A. B. A., Marimuthu, T., Kondiah, P. P. D., Choonara, Y. E., Du Toit, L. C., & Pillay, V. (2019). Multifunctional magnetic nanowires: Design, fabrication, and future prospects as cancer therapeutics. *Cancers (Basel)*, *11*(12), 1956.

Negahdary, M., Moradi, A., & Heli, H. (2019). Application of electrochemical aptasensors in detection of cancer biomarkers. *Biomedical Research Therapeutics*, *6*, 3315–3324.

Nune, S. K., Gunda, P., Thallapally, P. K., Lin, Y. Y., Laird Forrest, M., & Berkland, C. J. (2009). Nanoparticles for biomedical imaging. *Expert Opinion on Drug Delivery*, *6*(11), 1175−1194.

Oshin, O., Kireev, D., Hlukhova, H., Idachaba, F., Akinwande, D., & Atayero, A. (2020). Graphene-based biosensor for early detection of iron deficiency. *Sensors*, *20*(13), 3688.

Poturnayová, A., Buríková, M., Bízik, J., & Hianik, T. (2019). DNA aptamers in the detection of leukemia cells by the thickness shear mode acoustics method. *Chemphyschem: A European Journal of Chemical Physics and Physical Chemistry*, *20*(4), 545−554.

Pourmand, A., Pourmand, M. R., Wang, J., & Shesser, R. (2012). Application of nanomedicine in emergency medicine; Point-of-care testing and drug delivery in twenty - first century. *DARU Journal of Pharmaceutical Science*, *20*(26), 3.

Powell, J. J., Bruggraber, S. F., Faria, N., Poots, L. K., Hondow, N., Pennycook, T. J., Latunde-Dada, G. O., Simpson, R. J., Brown, A. P., & Pereira, D. I. (2014). A nano-disperse ferritin-core mimetic that efficiently corrects anemia without luminal iron redox activity. *Nanomedicine: Nanotechnology, Biology and Medicine*, *10*(7), 1529−1538.

Razak, K. A., Makhsin, S. R., Zakaria, N. D., & Nor, N. M. (2017). Gold nanoparticles for diagnostic development. In A. Ismail, N. M. Nor, J. M. Abdullah, A. Acosta, & M. E. Sarmiento (Eds.), *Sustainable diagnostics for low resource areas*. Malaysia: Penerbit Universiti Sains.

Sahoo, S. L., Liu, C. H., & Wu, W. C. (2017). Lymphoma cell isolation using multifunctional magnetic nanoparticles: Antibody conjugation and characterization. *RSC Advances*, *7*(36), 22468−22478.

Salah, E.D.T.A., Bakr, M.M., & Kamel, H.M. (2010). Magnetite nanoparticles as a single dose treatment for iron deficiency anemia. Innovative Research and Development Co. (Inrad), Patent no. WO2010034319A1. <http://www.google.co.in/patents/WO2010034319A1?cl = en> (Accessed: 20 Nov 2020).

Samir, A., Elgamal, B. M., Gabr, H., & Sabaawy, H. E. (2015). Nanotechnology applications in hematological malignancies. *Oncology Reports*, *34*(3), 1097−1105.

Shafique, S., & Tehsin, S. (2018). Computer-aided diagnosis of acute lymphoblastic leukaemia. *Computational and Mathematical Methods in Medicine*, *2018*.

Shalek, A. K., Gaublomme, J. T., Wang, L., Yosef, N., Chevrier, N., Andersen, M. S., Robinson, J. T., Pochet, N., Neuberg, D., Gertner, R. S., & Amit, I. (2012). Nanowire-mediated delivery enables functional interrogation of primary immune cells: Application to the analysis of chronic lymphocytic leukemia. *Nano Letters*, *12*(12), 6498−6504.

Shanbhag, P. P., Iyer, V., & Shetty, T. (2017). Gold nanoshells: A ray of hope in cancer diagnosis and treatment. *Nuclear Medicine and Biology Imaging*, *2*(2), 1−5.

Shariatifar, H., Hakhamaneshi, M. S., Abolhasani, M., Ahmadi, F. H., Roshani, D., Nikkhoo, B., Abdi, M., & Ahmadvand, D. (2019). Immunofluorescent labeling of CD20 tumor marker with quantum dots for rapid and quantitative detection of diffuse large B-cell non-Hodgkin's lymphoma. *Journal of Cellular Biochemistry*, *120*(3), 4564−4572.

Sharma, A., Pandey, C. M., Sumana, G., Soni, U., Sapra, S., Srivastava, A. K., Chatterjee, T., & Malhotra, B. D. (2012). Chitosan encapsulated quantum dots platform for leukemia detection. *Biosensors and Bioelectronics*, *38*(1), 107−113.

Shukla, M., Sekhon, U. D., Betapudi, V., Li, W., Hickman, D. A., & Pawlowski, C. L. (2017). In vitro characterization of SynthoPlate (synthetic platelet) technology and it's in vivo evaluation in severely thrombocytopenic mice. *Journal of Thrombosis and Hemostasis: JTH*, *15*, 375−387.

Simard, P., & Leroux, J. C. (2009). pH-sensitive immunoliposomes specific to the CD33 cell surface antigen of leukemic cells. *International Journal of Pharmaceutics*, *381*(2), 86−96.

Singh, R., Shaik, S., Negi, B. S., Rajguru, J. P., Patil, P. B., Parihar, A. S., & Sharma, U. (2020). Non-Hodgkin's lymphoma: a review. *Journal of Family Medicine and Primary Care*, *9*(4), 1834.

Song, S., Hao, Y., Yang, X., Patra, P., & Chen, J. (2016). Using gold nanoparticles as delivery vehicles for targeted delivery of chemotherapy drug fludarabine phosphate to treat hematological cancers. *Journal of Nanoscience and Nanotechnology, 16*(3), 2582−2586.

Sproul, E. P., Nandi, S., Chee, E., Sivadanam, S., Igo, B. J., Schreck, L., et al. (2019). Development of biomimetic antimicrobial platelet-like particles comprised of microgel nanogold composites. *Regenerative Engineering and Translational Medicine, 6*, 299−309.

Sun, H., Lu, L., Yanjie, B., Huilin, Y., Huan, Z., Chunde, L., & Lei. (2018). Nanotechnology-enabled materials for hemostatic and anti-infection treatments in orthopedic surgery. *International Journal of Nanomedicine, 13*, 8325−8338.

Tan, J., Lai, Z., Zhong, L., Zhang, Z., Zheng, R., Su, J., Huang, Y., Huang, P., Song, H., Yang, N., & Zhou, S. (2018). A graphene oxide-based fluorescent aptasensor for the turn-on detection of CCRF-CEM. *Nanoscale Research Letters, 13*(1), 1−8.

Tan, J., Yang, N., Hu, Z., Su, J., Zhong, J., Yang, Y., Yu, Y., Zhu, J., Xue, D., Huang, Y., & Lai, Z. (2016). Aptamer-functionalized fluorescent silica nanoparticles for highly sensitive detection of leukemia cells. *Nanoscale Research Letters, 11*(1), 1−8.

Tan, Y., Li, Y., & Tang, F. (2019). Nucleic acid aptamer: A novel potential diagnostic and therapeutic tool for leukemia. *Onco Targets Therapy, 12*, 10597−10613.

Tazi, I., Nafil, H., & Mahmal, L. (2011). Monoclonal antibodies in hematological malignancies: Past, present and future. *Journal of Cancer Research and Therapeutics, 7*(4), 399.

Tholouli, E., Hoyland, J. A., Di Vizio, D., O'Connell, F., MacDermott, S. A., Twomey, D., Levenson, R., Yin, J. A. L., Golub, T. R., Loda, M., & Byers, R. (2006). Imaging of multiple mRNA targets using quantum dot based in situ hybridization and spectral deconvolution in clinical biopsies. *Biochemical and Biophysical Research Communications, 348*(2), 628−636.

Vinhas, R., Mendes, R., Fernandes, A. R., & Baptista, P. V. (2017). Nanoparticles—Emerging potential for managing leukemia and lymphoma. *Frontiers in Bioengineering and Biotechnology, 18*(5), 79.

Wang, X., Guan, J., Zhuang, X., Li, Z., Huang, S., Yang, J., et al. (2018). Exploration of blood coagulation of N-Alkyl chitosan nanofiber membrane in vitro. *Biomacromolecules, 19*, 731−739.

Wu, H., Zhu, S., Zeng, M., Liu, Z., Dong, S., Zhao, Y., Huang, H., & Lo, Y. M. (2014). Enhancement of non-heme iron absorption by anchovy (*Engraulis japonicus*) muscle protein hydrolysate involves a nanoparticle-mediated mechanism. *Journal of Agricultural and Food Chemistry, 62*(34), 8632−8639.

Wu, S., Yang, N., Zhong, L., Luo, Y., Wang, H., Gong, W., Zhou, S., Li, Y., He, J., Cao, H., & Huang, Y. (2019). A novel label-free terbium (iii)-aptamer based aptasensor for ultrasensitive and highly specific detection of acute lymphoma leukemia cells. *Analyst, 144*(12), 3843−3852.

Yen, L. C., Pan, T. M., Lee, C. H., & Chao, T. S. (2016). Label-free and real-time detection of ferritin using a horn-like polycrystalline-silicon nanowire field-effect transistor biosensor. *Sensors and Actuators B: Chemical, 230*, 398−404.

Yu, Y., Duan, S., He, J., Liang, W., Su, J., Zhu, J., Hu, N., Zhao, Y., & Lu, X. (2016). Highly sensitive detection of leukemia cells based on aptamer and quantum dots. *Oncology Reports, 36*(2), 886−892.

Zariwala, M. G., Elsaid, N., Jackson, T. L., Lopez, F. C., Farnaud, S., Somavarapu, S., & Renshaw, D. (2013). A novel approach to oral iron delivery using ferrous sulphate loaded solid lipid nanoparticles. *International Journal of Pharmaceutics, 456*(2), 400−407.

Zhang, M., Liu, H., Chen, L., Yan, M., Ge, L., Ge, S., & Yu, J. (2013). A disposable electrochemiluminescence device for ultrasensitive monitoring of K562 leukemia cells based on aptamers and ZnO@ carbon quantum dots. *Biosensors & Bioelectronics, 49*, 79−85.

Zhang, N., Wei, M. Y., & Qiang, Ma (2019). Nanomedicines: A potential treatment for blood disorder diseases. *Frontiers in Bioengineering and Biotechnology, 7*, 1−16.

Zhang, Y., Li, M., Gao, X., Chen, Y., & Liu, T. (2019). Nanotechnology in cancer diagnosis: Progress, challenges and opportunities. *Journal of Hematology and Oncology, 12*(1), 137.

Zhelev, Z., Ohba, H., Bakalova, R., Jose, R., Fukuoka, S., Nagase, T., Ishikawa, M., & Baba, Y. (2005). Fabrication of quantum dot-lectin conjugates as novel fluorescent probes for microscopic and flow cytometric identification of leukemia cells from normal lymphocytes. *Chemical Communications (Cambridge, England), 15*, 1980−1982.

Zhu, X., Cao, Y., Liang, Z., & Li, G. (2010). Aptamer-based and DNAzyme-linked colorimetric detection of cancer cells. *Protein & Cell, 1*(9), 842−846.

Zong, H., Sen, S., Zhang, G., Mu, C., Albayati, Z. F., Gorenstein, D. G., et al. (2016). In vivo targeting of leukemia stem cells by directing parthenolide-loaded nanoparticles to the bone marrow niche. *Leukemia: Official Journal of the Leukemia Society of America, Leukemia Research Fund, U.K, 30*, 1582−1586.

Advanced methods for clearing blood clots using mechanical thrombectomy devices and untethered microrobots

Dalia Samir Ahmed Mahdy

Fraunhofer Institute for Silicate Research ISC, Würzburg, Germany

14.1 Introduction

Thrombosis is the obstruction of blood flow inside a blood vessel due to the formation of a blood clot. The movement of this blood clot through the human circulatory system to the brain or lungs can cause serious cardiovascular diseases such as myocardial infarctions and ischemic strokes. Cardiovascular diseases are among the leading causes of death and long-term disability worldwide (Mozaffarian et al., 2015). Traditional treatment of clogged blood vessels depends on chemical lysis therapy with anticoagulants, either by systemic infusion into the circulatory system or by catheter-directed thrombolysis. Thrombolysis can also be assisted or accelerated by thermal and acoustic energy. Further, mechanical thrombectomy devices have been presented for the capture and removal of blood clots to restore the flow inside occluded blood vessels. A minimally-invasive potential approach for the treatment of occluded blood vessels could be achieved by untethered microrobotic systems designed for intravascular therapy. Traditional treatment for flow restoration inside clogged vessels have been limited to the administration of a thrombolytic drug, infused over an hour into the systemic circulation (Tissue plasminogen activator for acute ischemic stroke, 1995). Research studies have investigated combined techniques to extend the treatment window and efficiency of thrombolytic therapy. In addition to other approaches such as ultrasound-accelerated thrombolysis and laser-assisted thrombolysis.

14.2 Thrombolysis

14.2.1 Systemic administration of tissue plasminogen activators

Systemic administration of thrombolytic agents is performed through intravenous or intraarterial thrombolysis. The strategy of the treatment is decided based on the location and size of occlusion. Many thrombolytic drugs have been developed along with advancements in thrombolytic therapy with anticoagulants. Anticoagulants acting as tissue plasminogen activators for the treatment of occluded blood vessels such as streptokinase and urokinase have been approved by the US Food and Drug Administration for clinical use in the treatment of acute ischemic stroke, pulmonary embolism, and acute myocardial infarction. In 1995 the National Institute of Neurological Disorders and Stroke study group assessed the safety and effectiveness of intravenous thrombolysis, using recombinant tissue plasminogen activator through a randomized clinical study. They noted that patients with ischemic stroke treated with tissue plasminogen activators within 3 hours after the onset of symptoms were at least 30% more likely to have minimal or no disability at 3 months than those who received a placebo (Lee et al., 2010; Tissue plasminogen activator for acute ischemic stroke, 1995).

The main function of tissue plasminogen activators is to accelerate the conversion of plasminogen into plasmin. Plasmin momentarily splits fibrin toward the degradation of the interlaced network of the blood clot. Thrombolysis depends on the generation of sufficient plasmin from plasminogen by a tissue plasminogen activator. Early randomized trials and meta-analysis suggested that fibrinolysis and in specific intraarterial fibrinolysis has a noticeable influence on the recanalization of occluded blood vessels. Systemic administration of anticoagulant drugs has been long associated with a relatively long treatment window and risky side effects, including major bleeding complications. In order to extend the treatment window of conventional therapy, clinical studies investigated further techniques for the acceleration and enhancement of thrombolytic therapy.

14.2.2 Catheter-directed thrombolysis

Catheter interventions with or without localized administered anticoagulants have shown clinical success. These treatments for clogged blood vessels are efficient in dissolving or reducing blood clots and normalizing hemodynamics. Catheter-directed thrombolysis is a minimally-invasive approach designed for localized treatment of occluded blood vessels, by the infusion of a tissue plasminogen activator into or within the sight of the thrombus. In 1983 Zeumer et al. described the successful vascular recanalization of three out of five reported cases with arterial thrombosis, by selective perfusion of streptokinase using a coaxial catheter system (Zeumer et al., 1983). The efficiency and safety of catheter-directed thrombolytic therapy have been demonstrated by Semba et al. with urokinase for the treatment of deep venous thrombosis. Patients with iliofemoral DVT have undergone treatment with a urokinase dose infused over an average of 30 hours; lysis was complete in 18 (72%), partial in five (20%), and not achieved in two (8%) of 25 treated limbs (Semba & Dake, 1994). Further, the Prolyse in Acute Cerebral Thromboembolism (PROACT) II study is the first positive randomized clinical trial for endovascular treatment of patients with occlusion of the middle cerebral artery detected by angiography. Thrombolysis has been achieved by intraarterial infusion of recombinant prourokinase within 6 hours of the onset of acute ischemic stroke (Furlan et al., 1999).

Catheter-directed thrombolysis using streptokinase versus heparin has been reported by Elsharawy et al. for the treatment of iliofemoral venous thrombosis. Patients were randomized to either thrombolysis followed by anticoagulant or anticoagulant alone. Lysis of clots and deep venous reflux were assessed with ultrasound duplex and plethysmography after 6 months. Early findings support catheter-directed thrombolysis over systemic infusion of anticoagulants (Elsharawy & Elzayat, 2002). Further, Manninen et al. have presented selective catheter-directed thrombolysis in the treatment of a group of patients with a mean age of 48 years old with acute deep vein thrombosis. They used venous access using ultrasound guidance into various sites depending on the size and location of the thrombus.

Selective catheter-directed therapy was performed on the patients assisted by injections of urokinase and endovascular stenting (Manninen et al., 2012).

Another approach for clearing occluded blood vessels is by the addition of laser ablation or photoablation to catheter-based devices. The deployment of laser ablation in thrombolytic therapy with catheter-based devices depends on the transfer of laser energy through optic fibers to the distal end of the catheter for the therapy to take place. An example of optoacoustic thrombolysis systems, Celliers et al. have invented a catheter-based device for generating acoustic vibrations in biological environments. The optical energy is deposited in a water-based absorbing fluid and generates an acoustic impulse in the fluid through thermodynamic mechanisms. This approach of producing acoustic vibrations can be used for treating thrombus along with the incorporation of thrombolytic drug treatments through the catheter system (Celliers et al., 2000).

14.2.3 Ultrasound-accelerated thrombolysis

Ultrasound has been deployed to accelerate or enhance thrombolysis based on tissue plasminogen activators. A few decades ago, Trübestein et al. presented thrombolysis by ultrasound as a novel method for clearing clogged blood vessels (Trübestein et al., 1976). Ultrasound speeds up the enzymatic breakdown of the fibrin in the thrombus, which in turn increases the transfer of enzymes and accelerates its dissolution rate (Francis et al., 1995). Sakharov et al. have described two mechanisms explaining the acceleration of fibrinolysis with ultrasound. The first mechanism is acoustic streaming which requires the presence of the thrombus in a sufficient amount of blood flow to act as a medium. The second mechanism is the rise in temperature at the site of the thrombus, which is challenging to be utilized in vivo (Sakharov et al., 2000). Alexandrov et al. have investigated the influence of lysis of thrombus in brain ischemia using ultrasound and systemic infusion of tissue plasminogen activator combined. They used 2 MHz, single-element pulsed-wave ultrasonography to monitor the degradation of the recanalization of the intracranial occlusion in vivo on a group of patients. Results of the randomized clinical trials show that the continuous monitoring of the occlusion using ultrasonography aided the thrombolytic therapy with administered tissue plasminogen activators (Alexandrov et al., 2004).

Parikh et al. have evaluated the clinical outcomes of ultrasound-accelerated thrombolysis in patients with deep vein thrombosis. Results show that ultrasound-accelerated thrombolysis results in better lysis with a relatively lower drug dose and treatment time. The incorporation of ultrasound reduces the total time of infusion and shows a stronger influence of complete lysis and reduced bleeding complications (Parikh et al., 2008). In conclusion, therapeutic ultrasound has been showing great potential for localized therapy. Research studies showed the influence of ultrasound on the augmentation of the effectiveness of thrombolytic drugs. Other studies are still investigating the techniques of optimizing the use of ultrasound to enhance thrombolysis.

14.3 Mechanical thrombectomy

Thrombectomy is a minimally-invasive surgical procedure for removing a blood clot from inside a blood vessel for recanalization of the occluded vessel. Various mechanical thrombectomy devices have been presented and tested. Using a medical imaging modality, the doctor guides an instrument through blood vessels toward the location of the occlusion. Mechanical thrombectomy devices are designed with the aim of extending the treatment window of thrombolysis and catheter-directed thrombolysis, in addition to the treatment of patients with medical conditions that may restrict or prevent treatment with thrombolysis. For example; patients with bleeding diathesis cannot undergo thrombolysis since the probable bleeding complications that will be life-threatening in such a case. Examples of mechanical thrombectomy systems are the MERCI retriever system (concentric medical), Penumbra system (Penumbra Inc.), and the Solitaire X revascularization system as shown in Fig. 14.1.

The MERCI Retriever is a device designed for the treatment of ischemic strokes. The Merci Retrieval System (Concentric Medical) consists of the Merci Retriever, the Merci balloon-guided catheter, and the Merci microcatheter. The balloon-guided catheter is a catheter with a large 2.1 mm lumen and a balloon located at its distal tip. The Merci Retriever is a tapered wire with five helical loops of decreasing diameter (from 2.8 to 1.1 mm) at its distal end. It is advanced through a microcatheter in a straight configuration and proceeds in a helical shape once it is delivered into the occluded artery in order to entrap the thrombus. A selective angiogram is performed where the Merci retriever is used in order to evaluate the size and tortuosity of the vessels. The microcatheter is guided into the occluded vessel and moved past the thrombus using traditional catheterization techniques. Then, the Merci retriever is advanced through the microcatheter helical loops that are deployed beyond the thrombus. Afterward, the Merci retriever is retracted at the

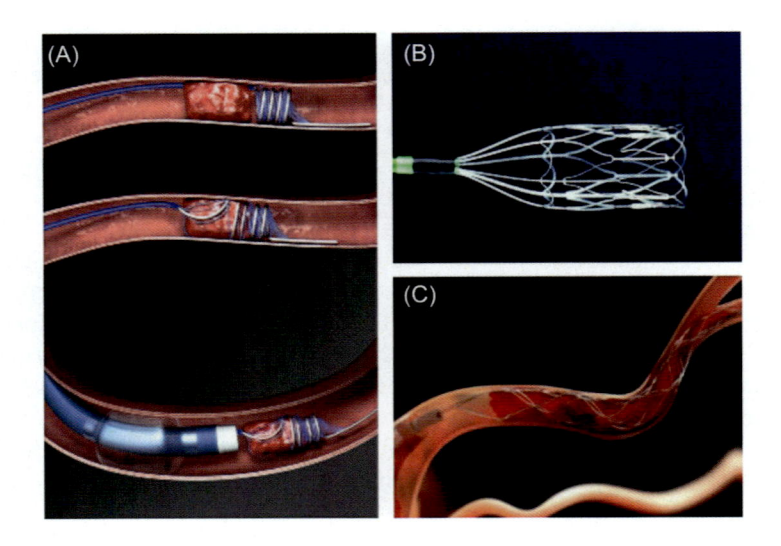

FIGURE 14.1

Mechanical thrombectomy devices used for revascularization of clogged blood vessels. (A) The MERCI retriever system (Concentric medical); (B) a schematic representation of the thrombus removal ring of the Penumbra system (Penumbra Inc.) as presented by Bose et al. (2008); and (C) a schematic representation of the Solitaire X revascularization system.

contact of the thrombus and the coil is reformed to enclose the blood clot and allow it to get withdrawn toward the tip of the balloon-guided catheter and evacuated outside the body.

A clinical research study has been conducted using the Merci retriever for the treatment of 30 patients with moderate to severe stroke. Results suggest that the mechanical removal of thrombus performed within 8 hours from symptom onset in moderate and severe stroke is safe and efficient for half the group of successfully recanalized patients with minimal risk of symptomatic hemorrhage or complications related to the usage of the device. Treatment of occluded blood vessels using mechanical thrombectomy devices shows higher rates of revascularization in comparison to thrombolysis using tissue plasminogen activators (Pierre Gobin et al., 2004).

Further, the Penumbra system is a mechanical thrombectomy system designed for the treatment of patients with acute ischemic stroke secondary to intracranial large vessel occlusions. The main function of the system is the fast and efficient removal of blood clots. The system has two main components. First is a reperfusion microcatheter that is attached to a suction pump and specially designed for the navigation through the intracranial circulation and to aspirate the thrombus causing the occlusion. The second is the separator, which is a microwire controlled by an operator and is inserted through the reperfusion catheter. Its distal end is a cone to help clear ingested clots from the catheter tip and support continuous aspiration. In 2009 the Penumbra Pivotal Stroke Trial Investigators conducted a study to assess the clinical safety and effectiveness of the penumbra system in the revascularization of patients presenting with acute ischemic stroke secondary to intracranial large vessel occlusive disease. A total of 125 target vessels were treated by the Penumbra system. Results suggested the Penumbra system performs safe and effective revascularization in patients with ischemic stroke secondary to large vessel occlusive disease within 8 hours from symptom onset (Penumbra Pivotal Stroke Trial Investigators, 2009).

Furthermore, the Solitaire X thrombectomy device (Medtronic Plc), featuring a unique stent retriever-based technology, restores blood flow and retrieves blood clots from occluded blood vessels in the brain for patients experiencing an acute ischemic stroke due to large vessel occlusion. The overlapping design of the stent allows the device to expand and shrink according to the size of the occluded vessel. The treatment window is 6−16 hours after onset symptoms. The device can be delivered to the site of the occlusion using a microcatheter providing flexibility with a lower clot crossing profile.

14.4 **Using untethered helical robots in clearing clogged blood vessels**

Microrobotic systems hold promise in diverse biomedical applications owing to their small size and maneuverability. These unique features would allow them to swim inside the human body to achieve tasks that cannot be accomplished through common therapeutic interventions. Their potential biomedical applications range from single-cell manipulation and microassembly to targeted drug delivery and localized therapy (Ghosh & Fischer, 2009; Nelson et al., 2010). Nature has always been an inspiration for novel designs of robotic motion from swimming and wiggling motions to walking or flying. Biological organisms have evolved to adapt and survive in the environment. Thus concepts from biologically inspired robots allow scientists to design and produce real systems that can maintain some of the desired properties of biological organisms. These properties include robustness, adaptability, and quick response. Various propulsion mechanisms have been proposed to provide locomotion in low Reynolds number environments. For example; electric fields generated by miniature diodes (Chang et al., 2007), chemical energy by using an internal catalytic chemical engine (Ma et al., 2016) or by ejecting microbubbles via the decomposition of hydrogen peroxide (Solovev et al., 2009) and magnetic fields (Peyer et al., 2013). The actuation of micro/nanorobots using magnetic fields has been widely deployed for potential biomedical applications. Researchers have designed robust systems to generate magnetic fields for the wireless control of micro/nanorobots, or in other words to externally actuate micro/nanorobots with magnetic fields. The various presented designs of the control systems allow the customization of workspace and multiple degrees-of-freedom (DoF), in order to actuate and control microrobots using several motion control techniques such as open-loop control and closed-loop or path-following control. Systems for magnetic actuation can be categorized as electromagnetic actuation systems and rotating permanent magnetic systems.

A potential minimally-invasive biomedical application of microrobots is the clearing of clogged blood vessels via direct interaction with the clot using a microrobot toward its dissolution. The illustration (Fig. 14.2) shows a schematic representation of the interaction of the tail of a helical robot with the three-dimensional fibrin network of a blood clot. Traditional fibrinolytic therapy for the ischemic disease includes catheter-directed localized thrombolysis or systemic administration of thrombolytic agents such as streptokinase. However, medical precautions and tests are necessary to avoid side effects such as excessive bleeding. Mechanical thrombectomy devices provide an alternative approach for the treatment of clogged blood vessels. They depend on the use of microsized

FIGURE 14.2

A schematic representation of the interaction of the tip of the helical robot with the fibrin network of the blood clot.

From Mahdy, D., Hamdi, N., Hesham, S., El Sharkawy, A., & Khalil, I. S. (c.2018). The influence of mechanical rubbing on the dissolution of blood clots. In 2018 40th annual international conference of the IEEE engineering in medicine and biology society (EMBC) (pp. 1660–1663). IEEE.

catheters and wires for the retrieval of blood clots and flow restoration in the vessel. Nonetheless, the use of such devices is associated with the risks and complications of surgical procedures. The deployment of microrobotics in the mechanical rubbing of blood clots could minimize the risks associated with the use of fibrinolytic drugs or mechanical thrombectomy devices.

14.4.1 Helical robot and actuation system

The helical robot (Fig. 14.3) consists of magnetic head and a helical tail. The helical tail is of length L, wavelength λ and helix angle α. Parameters of the helical robot are given in the table below:

The head of the helical robot is a permanent magnet (Neodymium grade N40, Amazing Magnets LLC, California, United States), with an edge length of 500 µm. It is rigidly attached to one end of the helical tail to provide a magnetic dipole moment, with an axial magnetization oriented perpendicular to the spring axis and magnetic flux density (Br) of 1420 mT. The screw-like tail of the robot is designed and fabricated using an aluminum spring with length and diameter of 4 mm and 100 µm, respectively. The pitch of the helical body is 500 µm and the helix angle is approximately 35 degrees.

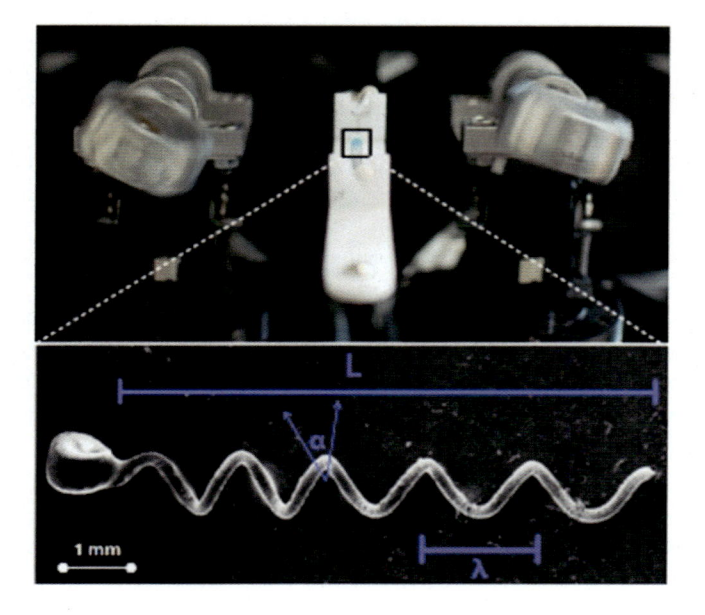

FIGURE 14.3

A permanent magnet-based robotic system is used to actuate the helical robot. The catheter segment accommodating the helical robot, blood clot, and the swimming medium is positioned at equal distance between two rotating dipole fields (Hosney et al., 2015). The helical robot consists of a magnetic head and a helical tail of length L, wave length λ, and helix angle α. Forward motion of the robot is obtained as the rotation of the permanent magnets exerts a magnetic torque on the dipole of the helical robot head.

In the micro- and nanoscales, viscous forces are dominant over inertial forces. The ratio between these forces is called Reynolds number (Re) and is calculated as $Re = \rho \, v \, L/\mu$. Where ρ is the density of the fluid (995 kg/m^3), v is the velocity of the robot before rubbing (20×10^{-3} m/second), and L is its length (4×10^{-3} m), and μ is the dynamic viscosity of the fluid (0.8882 cP). The length of the helical robot is 4 mm. A low Reynolds number is achieved by using a swimming media with relatively high viscosity. For the presented helical robot Reynolds number is calculated as $= 0.089$, which is considered a low Reynolds number as it is less than 1. Parameters of the helical robot, such as length, helix outer diameter, pitch, number of turns etc., contribute to the locomotion speed and behavior of the helical robot. Analysis and optimization of such parameters is required for enhanced performance of the robots of each particular robot design and application.

A permanent magnet-based robotic system with an open configuration is used for the control and steering of the helical robot (Fig. 14.3). The system has been presented for the steering and control of a helical microrobot (Hosney et al., 2015). Open configuration systems have two main advantages over closed configuration systems; they can be scaled up to the size of in vivo devices, and are easier for the incorporation of a medical imaging modality. The system consists of two linear motion stages. Each motion stage carries a robotic base (Cyton Gamma 300, Robai, Cambridge, United States) with three degrees-of-freedom. The end-effector of each robotic base holds a DC motor (Maxon 47.022.022-00.19-189 DC Motor, Maxon Motors, Sachseln, Switzerland) that rotates a permanent magnet (N40 Neodymium, Amazing Magnets LLC, California, United States) with outer diameter and thickness of 38 and 20 mm, respectively. This system is mounted on a tuned damped optical table (M-ST-UT2-58-12, Newport, California, United States) to minimize vibration. The two rotating dipole fields exert magnetic torque on the magnetic dipole of the helical robot (Fig. 14.2). The resultant magnetic field at the position of the helical robot is 5.5 mT. A rotating magnetic field is generated by the two rotating permanent magnets to actuate and navigate the helical robot in a three-dimensional, two-dimensional, or one-dimensional workspace. The robotic base changes the orientation of the permanent magnets to direct the robot toward the reference positions. This control is achieved by maintaining the axis of the permanent magnets parallel to the longitudinal axis of the microrobot.

The two rotating permanent magnets allow the microrobot to rotate and move forward by exerting a magnetic torque on the dipole of the head of the robot. It is assumed that the rotation angle of the permanent magnets and the helical robot are aligned. The workspace accommodating the robot and swimming medium is placed in-between the magnets at equal distance. The utilization of two synchronized rotating permanent magnets mitigates the magnetic field gradient and reduces the magnetic forces along the lateral direction of propulsion. Such forces cause undesirable drifting and instability in the locomotion of the robot. Therefore this strategy allows the locomotion control of the helical robot with almost uniform field in the center of the workspace.

14.4.2 Characterization of blood clots

Blood clots are prepared based on the protocol proposed by Hoffmann and Gill (Hoffmann & Gill, 2012), where 5 mL of fresh venous blood is drawn from two blood donors between the age of 25 and 28, neither having records of blood coagulation disorders or acute illness and without a recent history of oral contraceptive use or anticoagulant therapy. Local Institutional Ethical Board approval is obtained for the preparation protocol of the blood clots, and donors give written informed consent. Blood donors have similar health conditions and daily habits for the aimed consistency of the results. The blood sample is inserted into a vacutainer tube without anticoagulant (Fig. 14.4A) held upright and kept inside a water bath at 37°C for 1 hour to coagulate forming a columnar blood clot (Fig. 14.4B). For the rheological analysis of the blood clot samples, the blood sample is allowed to clot in a beaker with a round base forming a disk-shaped blood clot, which is inserted between the parallel plates of the rheometer for testing (Fig. 14.4C).

The characterization of the morphology of blood clots supports a better understanding of the clotting mechanism and the composition and structure of the clot. Red blood cells (RBCs) and platelets form a big portion of the thrombus mass, especially in venous thrombosis. Cines et al. have discussed the process of clot formation and contraction, in which platelets aggregate to form a temporary sealant and fibrinogen is converted into a network of fibrin polymers, causing contraction of the blood clot by the cytoplasmic proteins inside platelets. Such contractions help the retention of RBCs inside the fibrin network of the clot. They have observed the morphology of entrapped blood cells inside a blood clot (Cines et al., 2014). Fig. 14.5 presents scanning electron microscopy images of dried fragments from the blood clot samples. The image indicates the entrapped blood cells inside the three-dimensional fibrin network of the clot, which in turn validates that it has a similar structure to the blood clots forming inside the human circulatory system. The sample is heated up from 25°C to 37°C in the rheometer, enclosed inside a temperature control unit, and sheer modulus is measured against increasing temperature during the heating process.

Quantification of the mechanical properties of blood clots is fundamental to understanding many aspects of cardiovascular disease and its treatment. In addition, it is essential for the

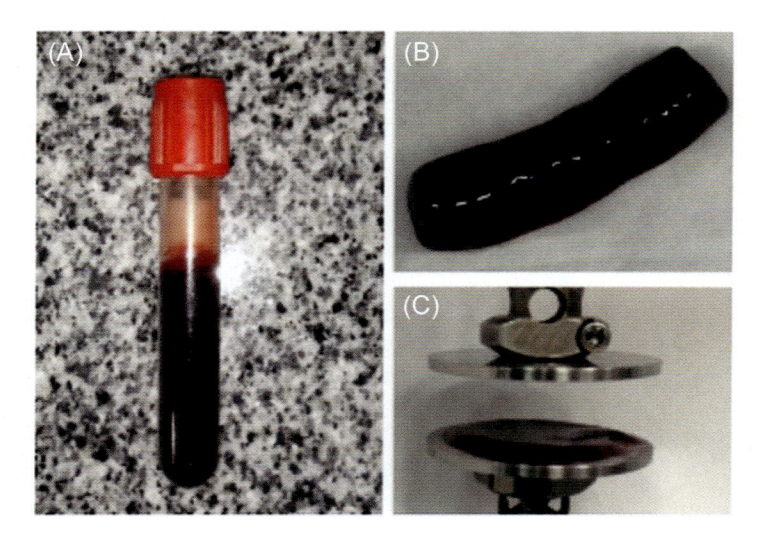

FIGURE 14.4

Samples are withdrawn from blood donors and inserted into a vacutainer tube without anticoagulant (A). The blood sample is evacuated from the tube and a columnar blood clot is obtained after drying (B). For the rheological analysis, the blood sample is allowed to clot inside a beaker with a round base forming a disk-shaped sample(C).

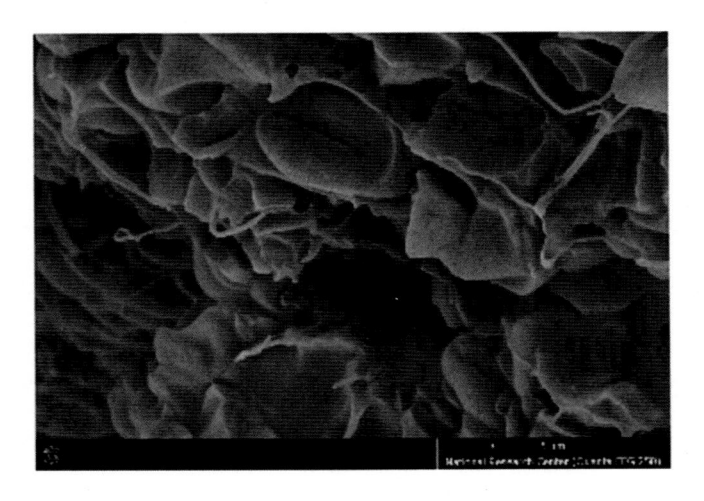

FIGURE 14.5

Morphology and mechanical properties of the blood clots are characterized using scanning electron microscopy imaging.

modeling of mechanical rubbing and its influence on the dissolution of clots. Epidemiological studies have shown a relationship between the mechanical properties of blood clots and myocardial infarction, as the in vitro blood clots formed in patients with myocardial infarction show a rigid fibrin network structure compared to the control (Scrutton et al., 1994). Further, Rozenberg et al. have studied blood coagulation and the mechanical properties of thrombosis by means of viscometric analysis. They have found that the degree of platelet aggregation during the thrombosis process is a function of the measured sheer, and they found that the morphology of the formed clots in vitro are similar to in vivo clots (Rozenberg & Dintenfass, 1964).

Following the discussed contributions, we now move to the characterization of the mechanical properties of blood clots. Shear modulus and ultimate shear strength of blood clot samples are measured. A rheology test is done on 1-hour-old blood clot samples using a Bohlin Gemini instrument (Malvern Instruments, United Kingdom), shown in Fig. 14.6. The instrument is equipped with a parallel plates measurement system. Samples are placed between a lower plate with diameter of 40 mm and an upper plate with diameter of 25 mm. The clots are surrounded by oil with viscosity of 0.06 Pa to avoid drying and denaturation of the samples. The gap between the plates is 1.569 mm and oscillatory shear with maximum shear stress of 0.1 Pa is applied at frequency of 1 Hz. At room temperature (25°C) and body temperature (37°C).

Measurements of shear modulus of a representative trial of shear modulus measurement of blood clots is shown in Fig. 14.7A. At room temperature (25°C) and body temperature (37°C), shear modulus is measured as 40.5 ± 0.6 and 41.4 ± 0.5 Pa, respectively. In addition, the stress–strain relation of the blood clots for two representative trials is shown in Fig. 14.7B. This characterization experiment indicates that the ultimate shear strength of 1-hour-old blood clots is approximately 1 kPa. This value represents the force under which the blood clot breaks or fractures.

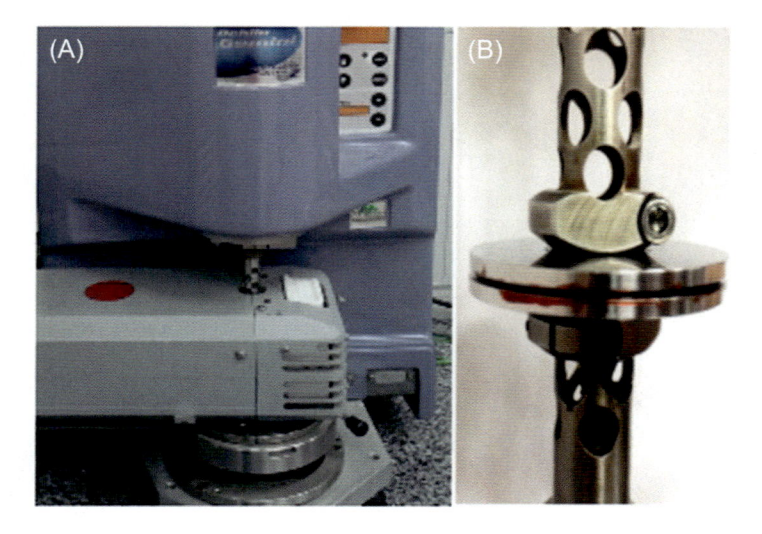

FIGURE 14.6

(A) A Bohlin Gemini instrument (Malvern Instruments, United Kingdom) is used for the characterization of the shear modulus and ultimate shear strength of blood clot samples. (B) The instrument comprises a parallel plates system.

Ultimate shear strength value can be used in the modeling of mechanical rubbing to predict the theoretical removal rate of the blood clot during rubbing.

14.4.3 Removal of blood clots

The dissolution of blood clots is experimentally investigated using two distinct groups, that is, dissolution using full dose of streptokinase (Group 1) and mechanical rubbing (Group 2). In each trial, the clots are prepared and inserted into catheter segments that are mounted between two rotating permanent magnets to exert magnetic torque on the dipole of the helical robot. The volume of the blood clot is calculated from camera feedback (TavA1000−120kc, Basler Area Scan Camera, Basler AG, Ahrensburg, Germany) using a morphological filtering algorithm implemented on Matlab (MathWorks Inc.) during the experiment. The camera provides an orthogonal top view of the blood clot. The volume detection algorithm is based on morphological filtering with the following steps: (1) frames are acquired from camera feedback as a sequence of images, morphological operations are performed on a frame by frame; (2) the acquired frame is manually cropped around the blood clot to reduce the processing time; (3) the cropped RGB image is converted into a gray image; (4) threshold is applied on the obtained gray image resulting in a black and white image where the white pixels represents the blood clot; and (5) finally the dimensionless area of the white pixels is calculated and stored for each frame.

Prior to the volume dissolution detection of blood clots with camera feedback, the volume detection algorithm is tested on a gelatin thrombus model under the same settings for validation. Gelatin is dyed with Methylene blue (dye) and inserted inside the catheter segment, taking an

(A) (B)

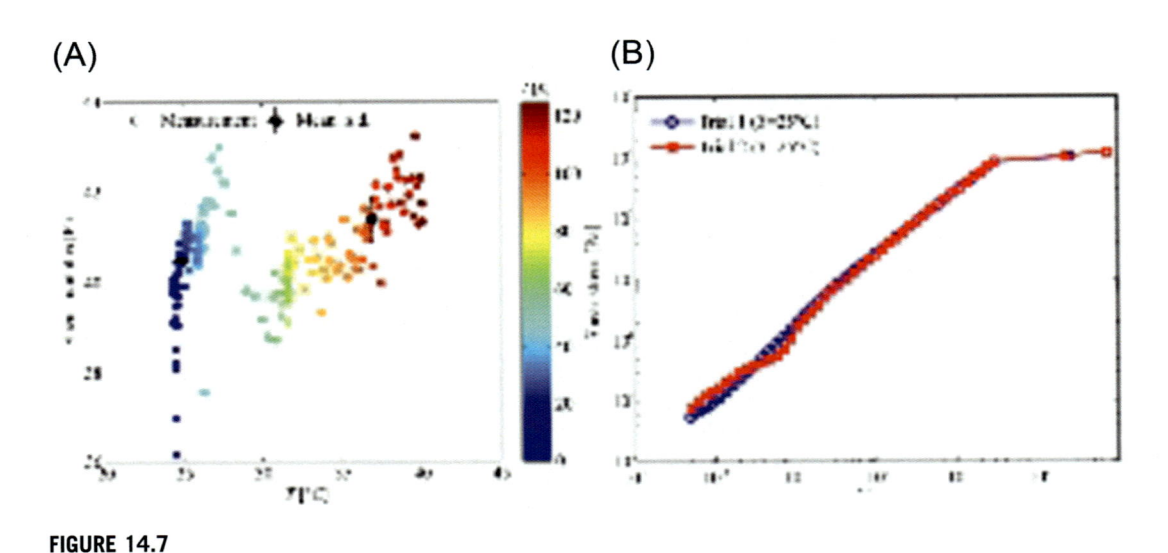

FIGURE 14.7

(A) The shear modulus of 1-h-old samples is 40.5 ± 0.6 and 41.4 ± 0.5 Pa at room temperature (25°C) and body temperature (37°C), respectively. (B) The shear stress—strain diagram of the blood clots indicates that the ultimate shear strength (τ) is 1000 Pa.

From Mechanical rubbing of blood clots using helical robots under ultrasound guidance. (c.2018). IEEE Robotics and Automation Letters, 3*(2). https://doi.org/2377—3766 3.*

almost cylindrical shape, and the catheter segment is filled with phosphate buffered saline with flow rate of 10 mL/hour. The volume of gelatin thrombus model as shown in Fig. 14.8 below was calculated to be almost constant throughout the experiment.

Now we move to the dissolution of blood clots. The dissolution of blood clots is experimentally investigated using two distinct groups, that is, dissolution using a full dose of streptokinase (Group 1) and mechanical rubbing using helical robots (Group 2). Blood clots are prepared (as explained in Section 14.2.4.1) and inserted into a polyvinyl chloride catheter segment. All experiments are done in the presence of a flow rate of 10 mL/hour. This flow rate is devised based on the administration and infusion rates for adult patients (maximum flow rates in small arterioles, capillaries, and venules are approximately 0.25, 0.045, and 0.00324 mL/hour). For instance, 10 mL of streptokinase is usually given intravenously in a 1-hour infusion. The flow rate is provided and controlled using a syringe pump (Genie Plus, GT-4201D-12, Kent Scientific, Connecticut, United States). The pump is fixed to one end of the catheter segment while the other end is inserted into a beaker for the collection of samples when needed.

14.4.3.1 Group 1: dissolution using full dose of streptokinase

The physiology of the fibrin—clot formation is relatively known. A blood clot or thrombus consists of blood cells entrapped in a matrix of the protein fibrin. Enzyme-mediated dissolution of the fibrin—clot is known as fibrinolysis. There are three major classes of fibrinolytic drugs: tissue plasminogen activator (tPA), streptokinase, and urokinase. Drugs in these three classes all

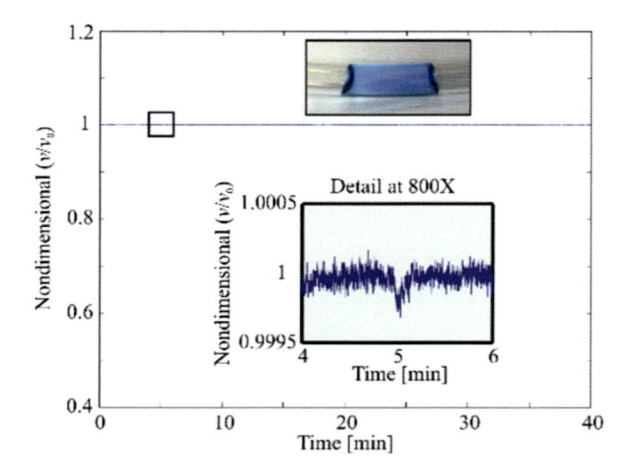

FIGURE 14.8

Volume detection of gelatin thrombus model. Graph shows that the change of the dimensionless volume of gelatin thrombus model is almost zero over 40 min. The viscoelastic properties of gelatin allow the model to fill the cylindrical volume of the catheter segment.

have the ability to effectively dissolve blood clots, but they differ in their precise mechanisms in ways that alter their selectivity for blood clots. Streptokinase has been the choice for treatment of myocardial infarction for more than 40 years (Sherry et al., 1959). It is an enzyme secreted by several species of streptococci bacteria that work by converting fibrin-bound plasminogen to plasmin, a natural fibrinolytic agent which breaks down fibrin contained in a blood clot leading to its lysis. In continuation, streptokinase is commonly used for myocardial infarction, arterial and venous thrombosis, and pulmonary embolism. Dotter et al. have demonstrated the treatment of 17 patients with arterial thromboembolism by selective doses of streptokinase delivered by catheter near or directly into blood clots. They have shown that such a method offers improved lysis with the targeted delivery of the drug, and enables the treatment with lower doses of streptokinase to reduce possible bleeding and other side effects associated with fibrinolytic agents (Dotter et al., 1974).

In this group, the dissolution of blood clots is achieved under the influence of a streptokinase-based thrombolytic agent at flow rate of 10 mL/hour. The medium inside the catheter segment is a mixture of one vial of the drug (Sedonase, SEDICO Co., Egypt). Each drug vial contains: streptokinase 1,500,000 I.U. and phosphate buffered saline; the vial is reconstituted with 5 mL of phosphate buffered saline and rolled gently forming the mixture. The volume of the clot is calculated throughout a lysis period of 40 minutes. The average dissolution rate calculated from 6 trials is -0.17 ± 0.032 mm^3/minutes. In the representative trial shown in Fig. 14.9, the initial and final volumes are calculated to be 94.24 and 91.99 mm^3, respectively. The measured dissolution rate is -0.19 mm^3/minutes. Furthermore, the formation of fluid channels inside the clot is observed after approximately 20 minutes of streptokinase injection.

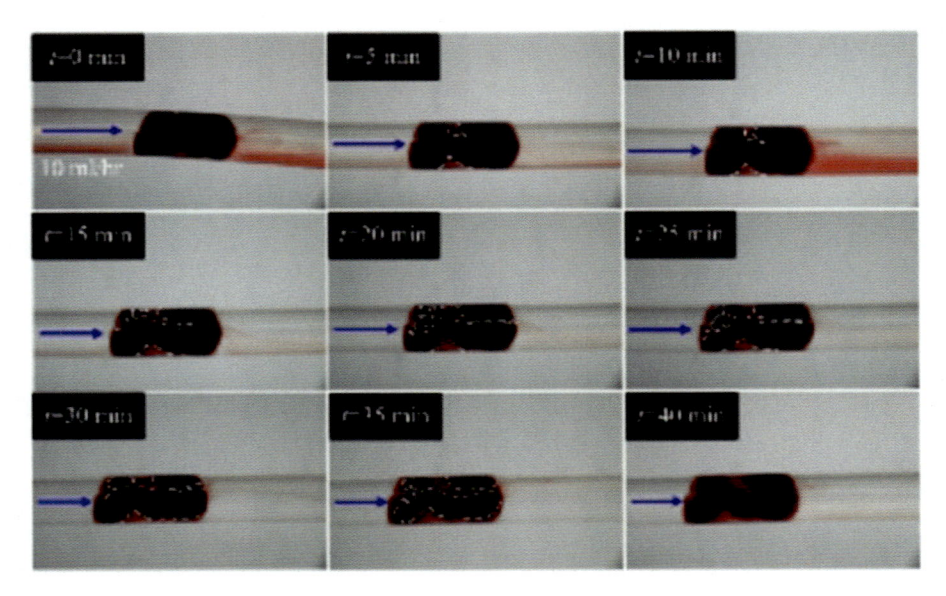

FIGURE 14.9

A representative experiment of dissolution of a blood clot using full dose of streptokinase (1,500,000 I.U.). The blue arrow indicates the direction of the flow and formation of fluid channels (white dashed lines) inside the clot is observed.

From Khalil, I. S., Tabak, A. F., Sadek, K., Mahdy, D., Hamdi, N., & Sitti, M. (c.2017). Rubbing against blood clots using helical robots: Modeling and in vitro experimental validation. IEEE Robotics and Automation Letters, 2(2), 927–934.

Although streptokinase have been long used in thrombolytic therapy, the side effects of the drug include excessive bleeding, chemically driven liver damage, thrombocytopenia (low blood platelets count), and osteoporosis (thinning of the bones, with reduction in bone mass, due to depletion of calcium and bone protein) (Van Beek et al., 2009). Therefore removal of blood clots by mechanical rubbing with helical robots is presented as an alternative potential approach to mitigate such risks.

14.4.3.2 Group 2: mechanical rubbing against blood clots

The helical robot is propelled along the x-axis toward the clot and against a similar flow rate to that used in the chemical lysis experiments (10 mL/hour). The permanent magnet-based robotic system (explained in Section 14.2.1) is used to actuate the robot, as shown in Fig. 14.3. Each permanent magnet generates a magnetic field of 0.552 T on its surface. The two permanent magnets are rotated and the linear speed of the robot is measured to be 15 mm/second, at a frequency of 35 Hz. The step-out frequency of the robot is experimentally measured to be 67.3 Hz, and a linear increase of the swimming speed is observed versus the angular frequency of the rotating permanent magnets within this range. Therefore the particular choice of the actuation frequency affects the swimming

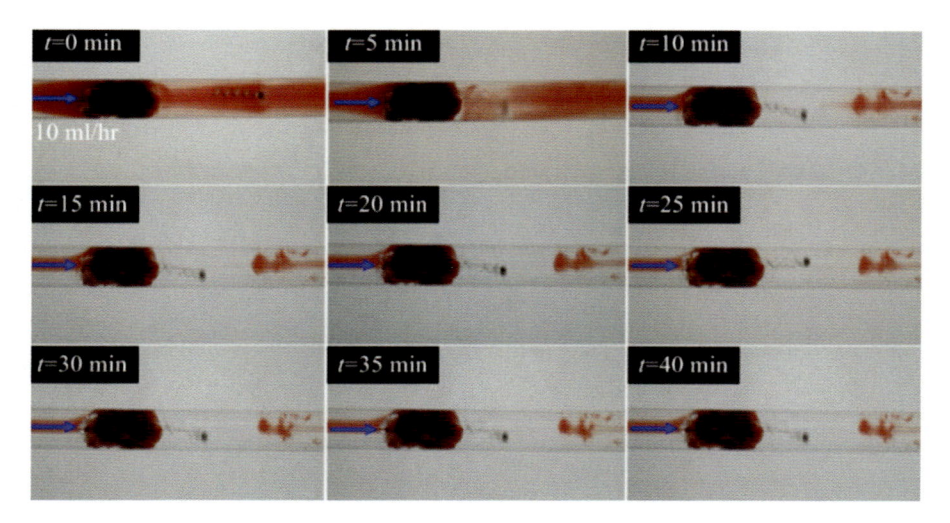

FIGURE 14.10

A representative experiment of mechanical rubbing against a blood clot using a helical robot at a frequency of 35 Hz.

Data from Khalil, I. S., Tabak, A. F., Sadek, K., Mahdy, D., Hamdi, N., & Sitti, M. (c.2017). Rubbing against blood clots using helical robots: Modeling and in vitro experimental validation. IEEE Robotics and Automation Letters, 2(2), 927–934.

speed, that is, the approaching speed to the clot, before the rubbing behavior. The resultant magnetic field at the position of the robot is measured as 5.5 mT using a Tesla meter (3MH3A tesla meter, Senis, Switzerland). The rubbing behavior is observed once the tip of the tail comes into contact with the blood clot. A representative trial of mechanical rubbing of blood clots is shown in Fig. 14.10. The initial and final volumes of the clot are calculated to be 94.24 and 60.65 mm^3, respectively after 40 minutes of rubbing against the clot. The dissolution rate at a frequency of 35 Hz is -0.885 mm^3/minutes.

The efficiency of mechanical rubbing is compared to chemical lysis under similar experimental settings. A representative trial of each along with a control experiment is compared. Mechanical rubbing achieves a higher removal rate than chemical lysis. The ratio between the measured volume and the initial volume of the clot is measured at every time instant during the lysis and rubbing. Further, the volume ratio (v/v_0) is measured in the absence of lysis and rubbing (zero-input response) to evaluate the volume dissolution due to experimental settings that do not include the effect of the robot or the drug. This response shows that the blood clot does not undergo any change in its volume. Nevertheless, lysis and rubbing have constant rates of change. The dissolution rate of the lysis is -0.19 mm^3/minutes in this trial. The rubbing is done at 35 and 40 Hz, and a flow rate of 10 mL/hour. The dissolution rate of the rubbing is -0.885 and -0.315 mm^3/minutes at 35 and 40 Hz, respectively.

In contrast to chemical lysis, the removal rate of the clot can be controlled via rubbing using the frequency of the rotating dipole fields. The influence of the rubbing frequency on the removal

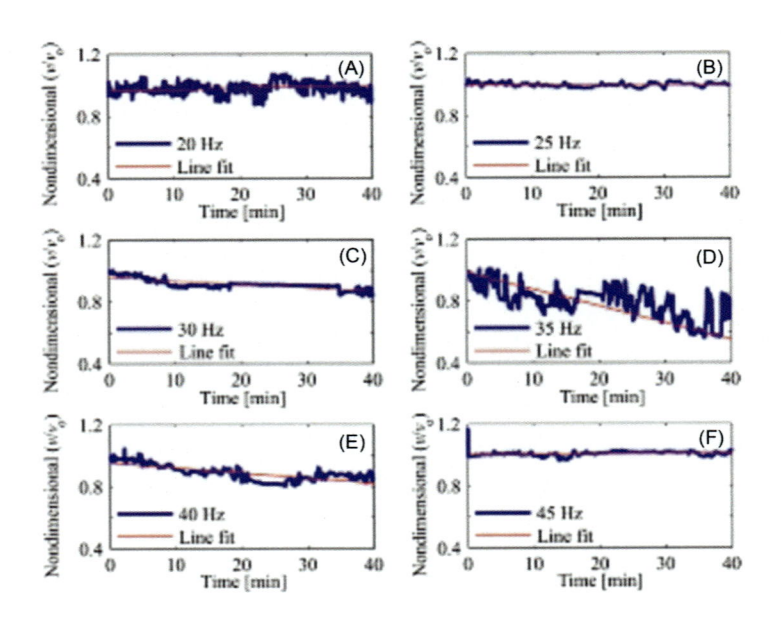

FIGURE 14.11

The influence of rubbing frequency on the removal rate of the blood clot is experimentally investigated between 20 and 45 Hz.

From Khalil, I. S., Tabak, A. F., Sadek, K., Mahdy, D., Hamdi, N., & Sitti, M. (c.2017). Rubbing against blood clots using helical robots: Modeling and in vitro experimental validation. IEEE Robotics and Automation Letters, 2(2), 927–934.

rate of the blood clot between 20 and 45 Hz is investigated. The removal rate by the robot is almost negligible at frequencies below 20 Hz Fig. 14.11A and B. Removal rates of -0.230, -0.885, and -0.315 mm^3/minutes are measured at rubbing frequencies of 30, 35, and 40 Hz, respectively, as shown in Fig. 14.11C−E. The maximum removal rate is observed at 35 Hz. At and above 45 Hz, negligible removal of the clot is observed (Fig. 14.11F).

Mechanical rubbing is not effective below 20 Hz and this could be attributed to the presence of a flow against the helical robot. At a result, the robot does not generate sufficient thrust to come into contact with the clot and achieve effective rubbing at these frequencies. In addition, rubbing is also not effective at and above a frequency of 45 Hz. The material removal (fretting) phenomenon is affected by at least two physical stimuli. First, the flow inside the catheter segment exerts a drag force against the robot. This flow is essential to provide fair comparison between chemical lysis and mechanical rubbing. Therefore the drag force decreases the material removal at low frequencies. Second, the non-Newtonian nature of the blood clot affects the removal rate. Resistive force theory, which can be used to model and describe mechanical rubbing, dictates that the velocity is in a linear relationship with the rotation rate. Power of the penetration increases with the square of the velocity. However, the effective viscosity of the blood clot under sudden impact should increase exponentially (or in an equivalent manner). Therefore, the damping effect becomes dominant as the frequency increases. This means that penetration decreases as spring behavior becomes more negligible. Hence, once again penetration depth and fretting becomes ineffective at relatively higher frequencies.

The experimental results do not show significant difference within the measurement error between chemical lysis and rubbing at 20, 25, 30, and 45 Hz. However, the rubbing results at 35 and 40 Hz show a significant increase in the removal rate of the clot compared to lysis under similar conditions. In conclusion, it is experimentally shown that mechanical rubbing against blood clots using helical robots results in higher removal rates than chemical lysis using a thrombolytic agent (Table 14.1). Streptokinase achieves dissolution rate of -0.17 ± 0.032 mm^3/minutes ($n = 6$), whereas rubbing achieves removal rate of -0.56 ± 0.27 mm^3/minutes ($n = 6$) at frequency of 35 Hz.

14.4.4 The influence of mechanical rubbing on blood clots

In this section, the influence of mechanical rubbing on blood clots is investigated using three different methods: (1) analysis of the blood clot weight; (2) analysis of the red blood cells and platelets count; (3) spectrophotometric analysis of blood clots. These tests are further investigated in order to validate the influence of mechanical rubbing of blood clots. Also for better understanding and observation of the changes in structure occurring to blood clots due to mechanical rubbing.

14.4.5 Blood clot weight analysis

As discussed earlier, mechanical rubbing of blood clots results in their volume dissolution. It is expected that the decrease in volume of blood clots implies a loss in their weight according to the law of conservation of matter. Hence, the weight of blood clots is measured before and after applying mechanical rubbing. Measurements are done via an electronic balance (ABS 220-4 Analytical Balance, KERN & SOHN GmbH, Balingen, Germany). The blood clot is measured before insertion into the catheter segment, and then is extracted from the catheter segment and weighed again after 40 minutes of mechanical rubbing. The percentage of reduction in the weight of the blood clot w_r is calculated by the following equation:

$$w_r = \frac{w(t_f) - w(t_0)}{w(t_0)} \times 100$$

where $w(t_f)$ is the weight of blood clot after 40 minutes of mechanical rubbing, $w(t_0)$ is the initial weight of the blood clot. In the absence of mechanical rubbing (zero-input response), weight reduction is measured as $51.9\% \pm 3.1\%$ ($n = 3$) and is attributed to the effect of the

Table 14.1 Comparison between chemical lysis and mechanical rubbing at 35 Hz.

Experimental group	Dissolution rate (mm^3/min)
Chemical lysis with full dose	-0.17 ± 0.032
Rubbing at 35 Hz	-0.56 ± 0.27

flowing solution in the catheter segment at a rate of 10 mL/hour on the freshly formed blood clot (1 hour old), in addition to the loss due to insertion and extraction of the blood clot from the catheter segment. Weight reduction percentage measurements (w_r) under the influence of rotating magnetic fields with varying frequency in the range of 20–45 Hz are shown in Table 14.2 below.

An increase in weight reduction is observed for all values of ω in comparison to 0 Hz. This increase validates the dissolution of blood clots under the influence of mechanical rubbing. However, the weight decrease does not highlight all the changes related to the effect of the mechanical rubbing on the fibrin network of the blood clot. Hence, the cell count of the samples past the robot and blood clot is calculated, which reflects the breakdown of the fibrin network causing the release of RBCs and platelets previously entrapped within the blood clot.

14.4.6 Cell count analysis

Venous blood clots are typically rich in fibrin and RBCs. Fibrin cross-linking by Factor XIII (fibrin stabilizing factor) and the elastic properties of the blood clot play an important role in the retention of RBCs inside the fibrin network of the clot. After clot formation, active platelets pull on the fibrin fibers and reduce the size of the blood clot in a physiological mechanism called clot retraction. This retraction mechanism increases the retention of RBCs inside the blood clot (Topaz, 2018). The interaction between the rotating tip of the helical robot and the blood clot allows the RBCs and platelets to break free from the fibrin network of the blood clot. In this section, cell count is carried out to estimate the number of released cells from the fibrin network of the blood clot, and to investigate the influence of the mechanical rubbing frequency on the numbers of released cells. A hemocytometer (Neubauer Improved, Germany) is used for the cell count under a microscope (MF Series 176-Measuring Microscope, Mitutoyo America Corporation). Hemocytometer is a counting chamber device originally designed for counting blood cells. A mixture with a volume of 1.5 mL past the robot and the blood clot is collected every 5 minutes into a small tube for cell count analysis. A volume of 10 μL of the sample is added on the hemocytometer under the microscope with magnification of 20 $\times$. Images of each sample are captured using a camera (avA1000–120kc, Basler Area Scan Camera, Basler AG, Germany) mounted on the microscope. Cell count is later carried out using circular objects detection on Matlab.

The counting chamber is shown in Fig. 14.12 counting square (marked as a big square) which consists of 16 small squares. The cell count is calculated from five counting squares each with an area of 1 mm^2 and depth of 0.1 mm. Therefore the following equation is used to calculate the number of cells in each sample:where c is the number of counted cells. This number is multiplied by 10^4 which is the volume of each counting square and divided by 5

Table 14.2 Measurements of weight reduction percentage under the influence of rotating magnetic fields with varying frequency in the range of 20–45 Hz.

ω (Hz)	0	20	25	30	35	40	45
w_r (%)	51.9 ± 3.1	71.4 ± 5.1	74.4 ± 11.1	67.8 ± 4.5	72.6 ± 5	73.5 ± 8.8	65.7 ± 2.6

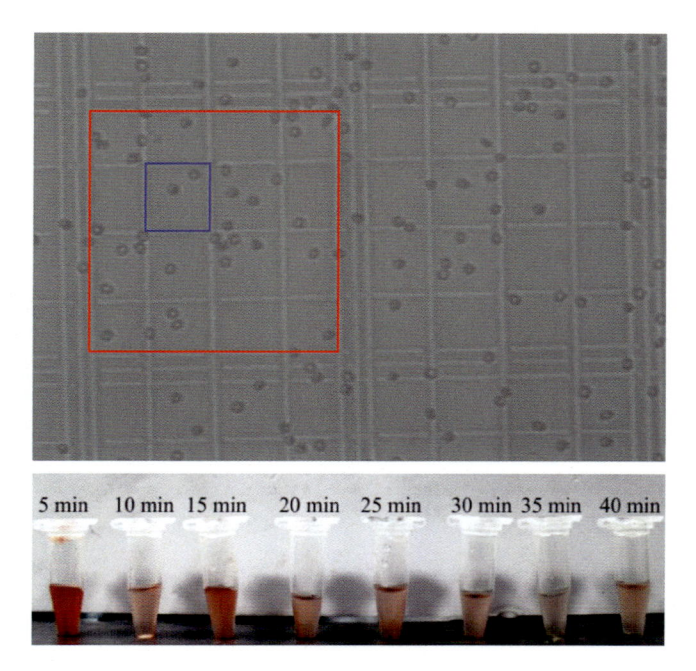

FIGURE 14.12

A hemocytometer (Neubauer Improved, Germany) is used for a cell count under a microscope (MF Series 176-Measuring Microscope, Mitutoyo America Corporation). Mixture past the helical robot and the blood clot is collected every 5 min during the experiments for cell count analysis.

$$\text{Cells/mL} = \frac{c \times 10^4}{5}$$

which is the number of counting squares. The total number of cells after 40 minutes of mechanical rubbing under the influence of varying frequency in the range of 20−45 Hz is shown in Fig. 14.13.

First, cell count for the mixture, which past the robot and blood every 5 minutes during the experiments, is shown as clot. The averaged sum of the cell count after 40 minutes of rubbing provides a maximum value of $654 \pm 108 \times 10^4$ cells/mL as calculated at 40 Hz, compared to $54 \pm 12 \times 10^4$ cells/mL in the absence of mechanical rubbing. Averages and standard deviations are calculated from three trials. Cell count at all frequencies in the range of 20−45 Hz is increased in comparison to the cell count at 0 Hz. The increased cell count indicates the increased number of released RBCs and platelets from the fibrin network of the blood clot, validating the mechanical breakdown of fibrin by the rotating tip of the helical robot.

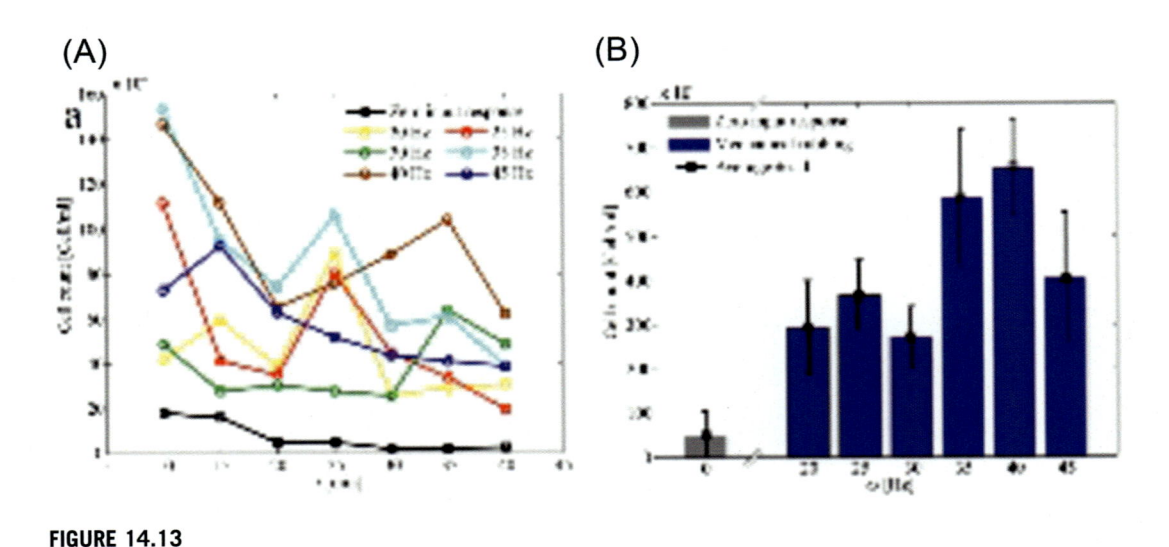

FIGURE 14.13

Count of red blood cells and platelets in samples collected past the robot and blood clot.

From Mahdy, D., Hamdi, N., Hesham, S., El Sharkawy, A., & Khalil, I. S. (c.2018). The influence of mechanical rubbing on the dissolution of blood clots. In 2018 40th annual international conference of the IEEE engineering in medicine and biology society (EMBC) (pp. 1660–1663). IEEE.

14.4.7 Spectrophotometric analysis

Spectrophotometry is the method of measuring the amount of light absorbed by a substance by measuring the attenuation of light after traveling through this substance. It is widely used for quantitative analysis in a variety of applications in biochemistry, chemical engineering, and clinical applications. In this section, spectrophotometric analysis is carried for further validation of the efficiency of mechanical rubbing. Samples past the robot and blood clot are collected every 5 minutes during the experiments. Absorbance of the collected samples is measured using a spectrophotometer (V-730 UV–Visible Spectrophotometer, Oklahoma city, United States). The device generates a beam of light that interacts with the tested sample and measures the amount of light absorbed by the sample over a certain range of wavelength. This allows the calculation of the total amount of chemical substance in the tested sample using the Beer–Lambert law, which relates the attenuation of light to the properties of the material through which the light travels as follows:

$$A = \epsilon bc$$

where A is the absorbance measured by device, ϵ is the wavelength-dependent molar absorptivity coefficient, b is the path length of the cuvette in which the sample is contained, and c is the compound concentration. The Beer–Lambert law states that the proportion of incident light absorbed by a transparent medium is independent of the intensity of light, provided that there are no other physical or chemical changes in the medium. The absorption of light is directly proportional to both the concentration of the absorbing medium and the thickness of the medium in the light path.

The procedure of spectrophotometric analysis is carried out as follows. First, a baseline measurement of the samples is done for the optimum wavelength selection within the range of visible light (400–800 nm). Maximum absorbance was measured at wave length ($\lambda = 416$ nm). Second, absorbance of each sample is measured at the selected wave length (λ). Finally, the Beer–Lambert law is used to calculate the concentration (in this case, the concentration is 521,880 as reported by Prahl, 1999) and $b = 1$ cm. Absorbance is linearly proportional to the concentration of the substance in the same sample. Thus increased absorbance implies increased concentration and correspondingly increased number of blood cells released from the blood clot. The concentration of samples calculated past the robot and blood clot under the influence of varying frequency of rotating magnetic fields in the range of 20–45 Hz is shown in Fig. 14.14. Maximum total concentration was

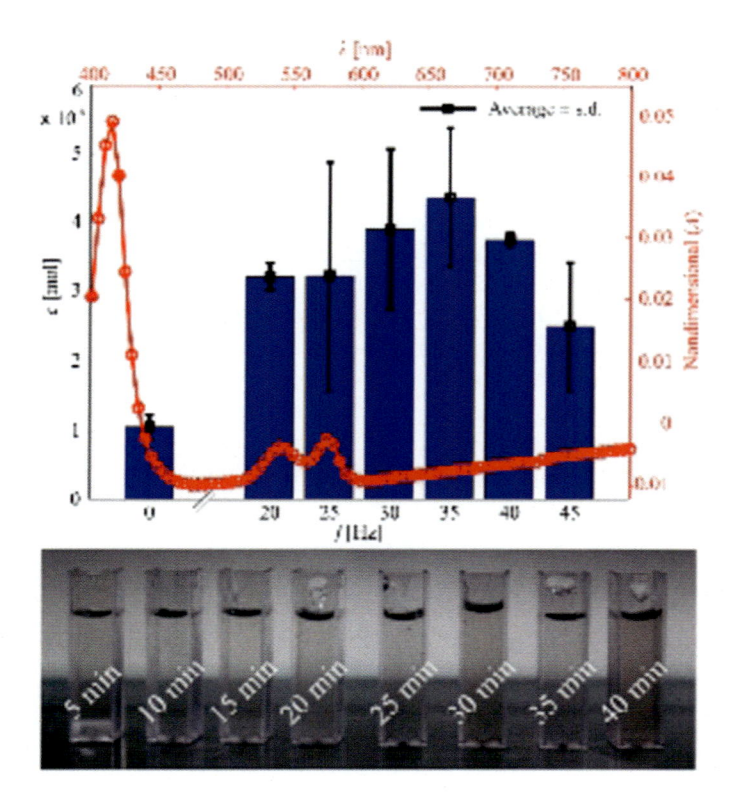

FIGURE 14.14

Spectrophotometric analysis is performed to study the influence of mechanical rubbing on the concentration of blood clots. First, the baseline is selected as $\lambda = 416$ nm. Then, absorbance is measured under the influence of a rotating magnetic field with varying frequency in the range of 20–45 Hz.

From Mahdy, D., Hamdi, N., Hesham, S., El Sharkawy, A., & Khalil, I. S. (c.2018). The influence of mechanical rubbing on the dissolution of blood clots. In 2018 40th annual international conference of the IEEE engineering in medicine and biology society (EMBC) (pp. 1660–1663). IEEE.

measured as 4.35×10^{-6} mol at a frequency of 35 Hz, compared to 1.05×10^{-6} mol in the absence of mechanical rubbing. The values of absorbance measured at all frequencies is higher than absorbance measured at 0 Hz, which shows that the concentration of RBCs and platelets is also increased. The increased concentration of RBCs and platelets in the samples collected postmechanical rubbing validates its efficiency.

14.4.8 **Characterization in rabbit aorta**

In order to investigate the behavior of the helical robot in a real blood vessel, locomotion of the helical robot is characterized in rabbit aorta. The aorta is an elastic artery with an expandable media for circulating blood, it is the main artery that originates in the heart and delivers oxygenated blood to the organs. The use of arteries in experiments is clinically relevant to the potential biomedical application of clearing clogged blood vessels, since the major cause of ischemic diseases such as stroke and myocardial infarction is the obstruction of the corresponding artery by blood clots. The aorta consists of three main layers from inside to outside, that is, tunica intima, tunica media, and tunica adventitia. Tunica media is the muscular layer of arteries and veins, it provides elasticity and controls the diameter of a blood vessel. In arteries, the tunica adventitia is supported by external elastic lamina that increase the elasticity needed for greater expansion in case of relatively higher flow rate.

A rabbit weighting 1.5 kg is dissected and its aorta is isolated. The ends of the aorta are connected to a catheter segment of 3 mm inner diameter for fluid circulation, and to provide a stationary locomotion model for the helical robot the during experiments. The aorta is connected to a syringe pump (Genie Plus, GT4201D-12, Kent Scientific, Connecticut, United States) and flow of 90 mL/hour is induced against the direction of propulsion, which maintains the viability of the aorta during the experiments. The diameter of the aorta is measured as 4 ± 0.3 mm, this variation in the diameter could be attributed to the elasticity of the aorta. The speed of the helical robot is characterized inside the rabbit aorta in Fig. 14.15.

The maximum speed in rabbit aorta is measured as 11.3 ± 0.52 versus 14.8 ± 0.37 mm/second in a catheter segment at actuation frequency of 8 Hz. Averaged speed and standard deviations are calculated from five trials ($n = 5$). A representative trial at actuation frequency of 7 Hz is shown in Fig. 14.15. The speed reduction inside the rabbit aorta in comparison to the catheter segment could be attributed to the elastic properties of the aorta, and the interaction between the robot and the inner layer of the aorta compared to the interaction with the channel wall of the catheter segment. A statistical test is conducted using analysis of variance (ANOVA) to investigate the influence of the actuation frequency and the host model (rabbit aorta and catheter segment) on the swimming speed of the helical robot. Results show statistical significance ($F_0 > F_\alpha$), where F_0 is calculated as 3.25 and 14.94 and F_α is calculated as 2.48 and 4.12 for the actuation frequency and the host model, respectively. Related research studies have investigated the locomotion of microrobots inside rabbit aorta. Jeong et al. have demonstrated the penetration of an induced arterial thromboembolism in the aorta of a live pig using an intravascular therapeutic microrobot system, the path of the robot in vivo was tracked and controlled with X-rays (Jeong et al., 2016). Choi et al. have achieved position control of a microrobot in a pulsating flow of blood vessels using an electromagnetic actuation system. They have performed in vivo experiments in the aorta of a 4-month-old female micropig, to evaluate the locomotion of the microrobot in a real blood vessel under similar conditions to coronary interventions (Choi et al., 2010)

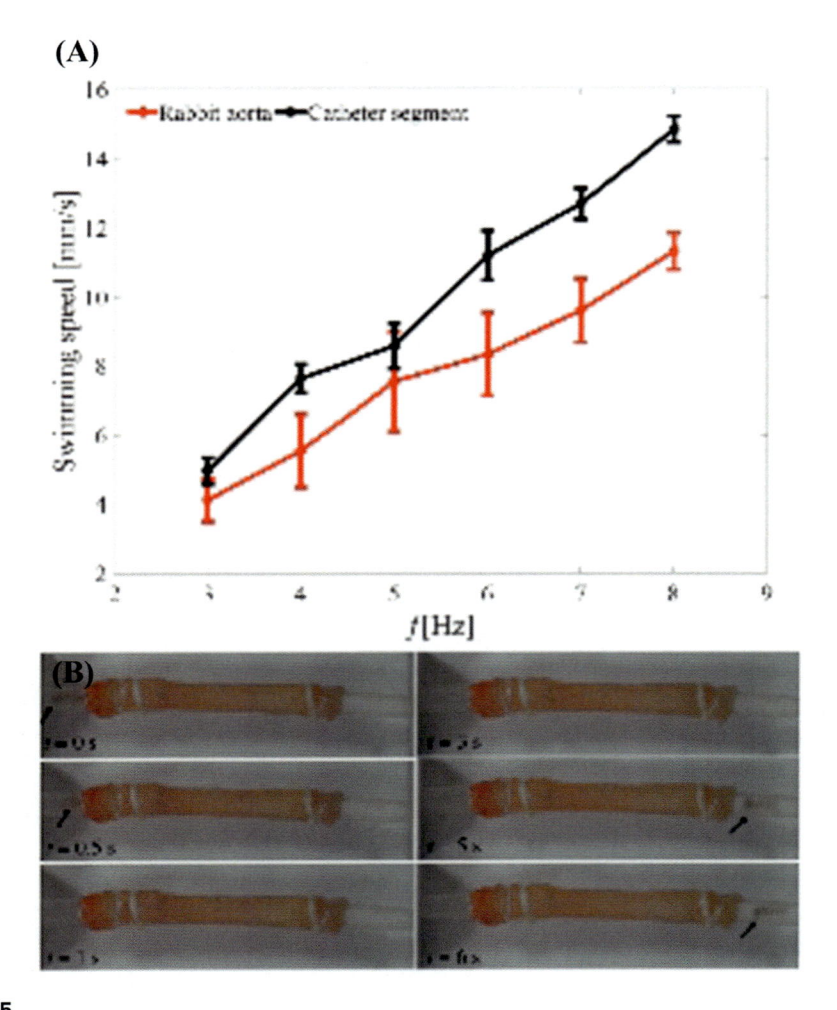

FIGURE 14.15

The speed of the helical robot is characterized in rabbit aorta under the influence of rotating magnetic field with varying frequency (A). A representative trial is shown at frequency of 7 Hz (B).

From Mahdy, D., Hesham, S., Mansour, M., Mohamed, A., Basla, I., Hamdi, N., Khalil, I. S., & Misra, S. (c.2019). Characterization of helical propulsion inside in vitro and Ex vivo models of a rabbit aorta. In 2019 41st annual international conference of the IEEE engineering in medicine and biology society (EMBC) (pp. 5283–8286). IEEE.

14.5 **Conclusions and future work**

This chapter focused on the recent advanced methods for clearing clogged blood vessels. The traditional treatment of thrombosis involves systemic infusion of an anticoagulant drug. Such treatment has side effects such as excessive bleeding and cannot be conducted for patients with certain blood diseases and bleeding disorders. In addition, mechanical thrombectomy devices have been utilized by researchers and clinicians, such as the MERCI retriever system (concentric medical), Penumbra system (Penumbra Inc.), and the Solitaire X revascularization system. These devices usually depend on the insertion of wires and microcatheters for the ablation of blood clots. However, the use of such devices is associated with postsurgery complications. An alternative minimal invasive approach for recanalization of clogged vessels is via untethered microrobots. A helical robot actuated by a permanent magnet-based robotic system is presented for mechanical rubbing of blood clots. Mechanical rubbing results in higher removal rates than chemical lysis using a thrombolytic agent under the same settings. Experimental results show that mechanical rubbing at achieves average removal rate of 0.17 ± 0.032 mm^3/minutes (at frequency of 35 Hz) in comparison to 0.85 ± 0.042 mm^3/minutes with chemical lysis using a streptokinase-based thrombolytic agent.

The influence of mechanical rubbing on the weight of the blood clot has been experimentally studied, and cell count and absorbance has been calculated for the collected samples past the helical robot and the blood clot. During mechanical rubbing, the interaction of the tip of the rotating helical robot with the three-dimensional fibrin network of the blood clot results in tearing of the fibrin and its decomposition. The effect of this interaction is validated by the count of released RBCs and platelets and concentration of the samples collected past the robot and the blood clot. Maximum decreased weight percentage was measured as $74.4\% \pm 11.1\%$ at 25 Hz, cell count as $654 \pm 108 \times 10^4$ cells/mL ($n = 3$) at 40 Hz and concentration as 4.35×10^{-6} mol ($n = 2$) at 35 Hz. Compared to $51.9\% \pm 3.1\%$, $54 \pm 12 \times 104$ cells/mL ($n = 3$) and concentration as 1.05×10^{-6} mol ($n = 2$) in the absence of mechanical rubbing.

The conditions that will be encountered by the helical robot during swimming inside the human body are challenging. To investigate similar conditions, the swimming speed of the helical robot is characterized inside a segment of a dissected rabbit aorta, under the influence of rotating magnetic fields with varying frequency in the range of 3−8 Hz. The robot swims against a flow of PBS at 90 mL/hour with a maximum speed of 11.3 ± 0.52 mm/second inside the aorta compared to 14.8 ± 0.37 mm/second in catheter segment at actuation frequency of 8 Hz. The outer diameter of the aorta is measured as 4 ± 0.3 mm varying according to the expansion by the applied flow rate. We attribute the reduction of swimming speed to the elastic properties of the aorta in comparison to a rigid catheter segment.

The influence of rubbing in combination with chemical lysis at different doses of a fibrinolytic agent could be studied. The comparative study between mechanical rubbing, rubbing in combination with different percentages of fibrinolytic agent, and pure chemical lysis is essential to optimize the integration between mechanical rubbing and the chemical lysis of blood clots. Experimental results in this work are conducted against a flow rate of 10 mL/hour. It is essential to modify our system to enable mechanical rubbing against greater flow rates comparable to medium arteries and veins. Finally, it is required to fabricate the helical robot with a biodegradable material to allow it to be absorbed by the human body after completing its task.

References

Alexandrov, A. V., Molina, C. A., Grotta, J. C., Garami, Z., Ford, S. R., Alvarez-Sabin, J., Montaner, J., Saqqur, M., Demchuk, A. M., & Moyé, L. A. (2004). Ultrasound-enhanced systemic thrombolysis for acute ischemic stroke. *New England Journal of Medicine, 351*(21), 2170−2178.

Bose, A., Henkes, H., Alfke, K., Reith, W., Mayer, T. E., Berlis, T. E., Branca, V., & Po, S. (2008). Sit for the penumbra phase 1 stroke trial investigators. *American Journal of Neuroradiology*, *29*(7), 1409–1413. Available from https://doi.org/10.3174/ajnr.A1110.

Celliers, P., Da Silva, L., Glinsky, M., London, R., Maitland, D., Matthews, D., & Fitch, P. (2000). *Opto-acoustic thrombolysis*. Google Patents.

Chang, S. T., Paunov, V. N., Petsev, D. N., & Velev, O. D. (2007). Remotely powered self-propelling particles and micropumps based on miniature diodes. *Nature Materials*, *6*(3), 235–240.

Choi, J., Jeong, S., Cha, K., Qin, L., Li, J., Park, J., Park, S., & Kim, B. (2010). Position stabilization of micro-robot using pressure signal in pulsating flow of blood vessel. Sensors, 2010 IEEE (. 723–726). IEEE.

Cines, D. B., Lebedeva, T., Nagaswami, C., Hayes, V., Massefski, W., Litvinov, R. I., Rauova, L., Lowery, T. J., & Weisel, J. W. (2014). Clot contraction: Compression of erythrocytes into tightly packed polyhedra and redistribution of platelets and fibrin. *Blood, The Journal of the American Society of Hematology*, *123* (10), 1596–1603.

Dotter, C. T., Rösch, J., & Seaman, A. J. (1974). Selective clot lysis with low-dose streptokinase. *Radiology*, *111*(1), 31–37.

Elsharawy, M., & Elzayat, E. (2002). Early results of thrombolysis vs anticoagulation in iliofemoral venous thrombosis. A randomised clinical trial. *European Journal of Vascular and Endovascular Surgery*, *24*(3), 209–214.

Francis, C. W., Blinc, A., Lee, S., & Cox, C. (1995). Ultrasound accelerates transport of recombinant tissue plasminogen activator into clots. *Ultrasound in Medicine & Biology*, *21*(3), 419–424.

Furlan, A., Higashida, R., Wechsler, L., Gent, M., Rowley, H., Kase, C., Pessin, M., Ahuja, A., Callahan, F., & Clark, W. M. (1999). Intra-arterial prourokinase for acute ischemic stroke: The PROACT II study: A randomized controlled trial. *JAMA: The Journal of the American Medical Association*, *282*(21), 2003–2011.

Ghosh, A., & Fischer, P. (2009). Controlled propulsion of artificial magnetic nanostructured propellers. *Nano Letters*, *9*(6), 2243–2245.

Hoffmann, A., & Gill, H. (2012). Diastolic timed vibro-percussion at 50 Hz delivered across a chest wall sized meat barrier enhances clot dissolution and remotely administered streptokinase effectiveness in an in-vitro model of acute coronary thrombosis. *Thrombosis Journal*, *10*(1), 23.

Hosney, A., Klingner, A., Misra, S., & Khalil, I. S. (2015). Propulsion and steering of helical magnetic micro-robots using two synchronized rotating dipole fields in three-dimensional space. In *2015 IEEE/RSJ international conference on intelligent robots and systems (IROS)* (pp. 1988–1993). IEEE.

Jeong, S., Choi, H., Go, G., Lee, C., Lim, K. S., Sim, D. S., Jeong, M. H., Ko, S. Y., Park, J.-O., & Park, S. (2016). Penetration of an artificial arterial thromboembolism in a live animal using an intravascular thera-peutic microrobot system. *Medical Engineering & Physics*, *38*(4), 403–410.

Lee, M., Hong, K.-S., & Saver, J. L. (2010). Efficacy of intra-arterial fibrinolysis for acute ischemic stroke: Meta-analysis of randomized controlled trials. *Stroke: A Journal of Cerebral Circulation*, *41*(5), 932–937.

Ma, X., Jang, S., Popescu, M. N., Uspal, W. E., Miguel-López, A., Hahn, K., Kim, D.-P., & Sánchez, S. (2016). Reversed janus micro/nanomotors with internal chemical engine. *ACS Nano*, *10*(9), 8751–8759.

Manninen, H., Juutilainen, A., Kaukanen, E., & Lehto, S. (2012). Catheter-directed thrombolysis of proximal lower extremity deep vein thrombosis: A prospective trial with venographic and clinical follow-up. *European Journal of Radiology*, *81*(6), 1197–1202.

Mozaffarian, D., Benjamin, E. J., Go, A. S., Arnett, D. K., Blaha, M. J., Cushman, M., De Ferranti, S., Després, J.-P., Fullerton, H. J., & Howard, V. J. (2015). Executive summary: Heart disease and stroke sta-tistics—2015 update: A report from the American Heart Association. *Circulation*, *131*(4), 434–441.

Nelson, B. J., Kaliakatsos, I. K., & Abbott, J. J. (2010). Microrobots for minimally invasive medicine. *Annual Review of Biomedical Engineering*, *12*, 55–85.

Parikh, S., Motarjeme, A., McNamara, T., Raabe, R., Hagspiel, K., Benenati, J. F., Sterling, K., & Comerota, A. (2008). Ultrasound-accelerated thrombolysis for the treatment of deep vein thrombosis: Initial clinical experience. *Journal of Vascular and Interventional Radiology, 19*(4), 521−528.

Penumbra Pivotal Stroke Trial Investigators. (2009). The penumbra pivotal stroke trial: Safety and effectiveness of a new generation of mechanical devices for clot removal in intracranial large vessel occlusive disease. *Stroke: A Journal of Cerebral Circulation, 40*(8), 2761−2768.

Peyer, K. E., Zhang, L., & Nelson, B. J. (2013). Bio-inspired magnetic swimming microrobots for biomedical applications. *Nanoscale, 5*(4), 1259−1272.

Pierre Gobin, Y., Starkman, S., Duckwiler, G. R., Grobelny, T., Kidwell, C. S., Jahan, R., Pile-Spellman, J., Segal, A., Vinuela, F., & Saver, J. L. (2004). MERCI 1: A phase 1 study of mechanical embolus removal in cerebral ischemia. *Stroke: A Journal of Cerebral Circulation, 35*(12), 2848−2854.

Prahl, S. (1999). Optical absorption of hemoglobin. http://Omlc.Ogi.Edu/Spectra/Hemoglobin.

Rozenberg, M., & Dintenfass, L. (1964). Thrombus formation in vitro: A rheological and morphological study. *Australian Journal of Experimental Biology and Medical Science, 42*(1), 109−115.

Sakharov, D. V., Hekkenberg, R. T., & Rijken, D. C. (2000). Acceleration of fibrinolysis by high-frequency ultrasound: The contribution of acoustic streaming and temperature rise. *Thrombosis Research, 100*(4), 333−340.

Scrutton, M., Ross-Murphy, S., Bennett, G., Stirling, Y., & Meade, T. (1994). Changes in clot deformability—A possible explanation for the epidemiological association between plasma fibrinogen concentration and myocardial infarction. *Blood Coagulation & Fibrinolysis: An International Journal in Haemostasis and Thrombosis, 5*(5), 719−723.

Semba, C. P., & Dake, M. D. (1994). Iliofemoral deep venous thrombosis: Aggressive therapy with catheter-directed thrombolysis. *Radiology, 191*(2), 487−494.

Sherry, S., Fletcher, A. P., & Alkjaersig, N. (1959). Fibrinolysis and fibrinolytic activity in man. *Physiological Reviews, 39*(2), 343−382.

Solovev, A. A., Mei, Y., Bermúdez Ureña, E., Huang, G., & Schmidt, O. G. (2009). Catalytic microtubular jet engines self-propelled by accumulated gas bubbles. *Small, 5*(14), 1688−1692.

Tissue plasminogen activator for acute ischemic stroke. (1995). *New England Journal of Medicine, 333*(24), 1581−1588. Available from https://www.nejm.org/doi/full/10.1056/nejm199512143332401.

Topaz, O. (2018). *Cardiovascular thrombus: From pathology and clinical presentations to imaging, pharmacotherapy and interventions.* Academic Press.

Trübestein, G., Engel, C., Etzel, F., Sobbe, A., Cremer, H., & Stumpff, U. (1976). Thrombolysis by ultrasound. *Clinical Science and Molecular Medicine, 51*(s3), 697s−698s.

Van Beek, E. J., Büller, H. R., & Oudkerk, M. (2009). *Deep vein thrombosis and pulmonary embolism.* John Wiley & Sons.

Zeumer, H., Hacke, W., & Ringelstein, E. (1983). Local intraarterial thrombolysis in vertebrobasilar thromboembolic disease. *American Journal of Neuroradiology, 4*(3), 401−404.

Nanotechnology for stroke treatment 15

Yanjun Yang[1], Yoong Sheng Phang[2] and Yiping Zhao[2]

[1]*School of Electrical and Computer Engineering, College of Engineering, The University of Georgia, Athens, GA, United States* [2]*Department of Physics and Astronomy, The University of Georgia, Athens, GA, United States*

15.1 Introduction

Conventional therapy for dealing with cerebral stroke (Fig. 15.1A) and associated complications is primarily focused on employing chemotherapeutic agents to lyse the blood clot or reduce neuroinflammation. Although a great deal of effort has been spent to investigate potential thrombolytic agents, recombinant tissue plasminogen activator (rt-PA) remains the only FDA (the Food and Drug Administration) approved thrombolytic agent for treating ischemic stroke in clinical setting thus far. The rt-PA is a serine protease with a molecular weight of ~ 70 kDa and is an enzyme that catalyzes the conversion of plasminogen to plasmin, which then binds to fibrin and breaks up the blood clot. The detailed molecular reaction mechanism for rt-PA is shown in Fig. 15.1B, which consists of one activation pathway (solid arrows) and multiple inhibition pathways (dotted arrows) (Cesarman-Maus & Hajjar, 2005; Bhattacharjee et al., 2014). In the activation pathway (solid arrows), the rt-PA molecules can preferentially attach to plasminogen (PLG), which is secreted as a single chain glycoprotein by the liver and circulates in the blood in an inactive form. The formation of the rt-PA-PLG ternary complex can release plasmin (PLM), which binds to the fibrin network of blood clot and initiates cleavage of fibrinogen or soluble fibrin from the C-terminal end of fibrin α-polypeptide chain. Such a process gradually breaks down fibrin into fibrin degradation products, which eventually dissolves blood clots. There are at least two inhibition pathways, as shown in Fig. 15.1B. Endothelial cells can release plasminogen activator inhibitors, which prevent plasminogen from cleaving into plasmin, block fibrinolysis, and confer an overall procoagulation effect when they bind to rt-PAs. In addition, free plasmin in circulating blood could rapidly form a complex with circulating α2-plasmin inhibitor and become inactivated. Thus in circulating blood, the activity of plasmin is tightly regulated to prevent excessive fibrinolysis, which is manifested by a bleeding tendency. The National Institute of Neurological Disorders and Stroke study suggested that eight out of 18 stroke patients who receive rt-PA according to a strict protocol will recover by three months after the event without significant disability. This is compared to six out of 18 stroke patients (one-third) who recover substantially regardless of treatment (Group, 1995).

While rt-PA-based thrombolysis is considered safe, side effects shown on some patients have brought major concerns about the safety and the efficacy of rt-PA treatment. The intravenous use of rt-PA appears to be effective only within the first 4−5 hours after the initial onset of the

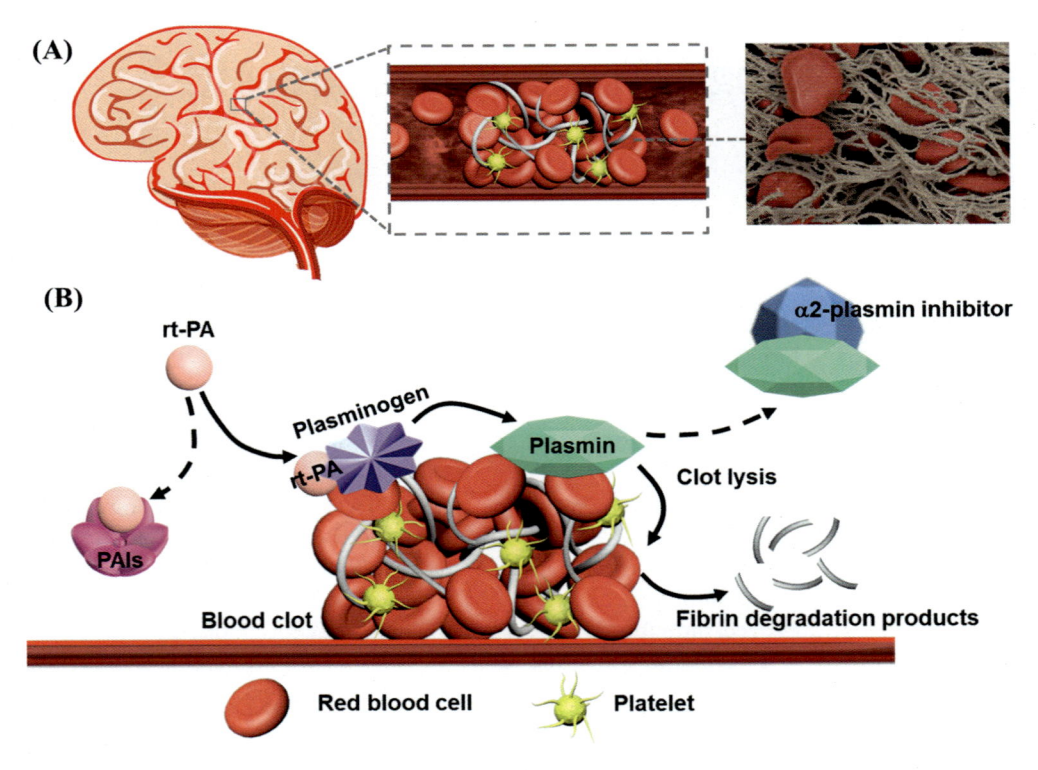

FIGURE 15.1

(A) The structure of blood clot in brain and colorized scanning electron micrograph of a whole blood clot (Hsu et al., 2015). (B) The reaction mechanism of rt-PA-clot.

Reprint with permission from Hsu, B., Conway, W., Tschabrunn, C., Mehta, M., Perez-Cuevas, M., Zhang, S., & Hammond, P. (2015). Clotting mimicry from robust hemostatic bandages based on self-assembling peptides. ACS Nano, 9(9), 9394–9406. *Copyright 2015 American Chemical Society.*

symptoms (Brott & Bogousslavsky, 2000; Harold et al., 2007). Current acute stroke therapy requires infusion of the rt-PA drug through a catheter placed within the blocked vessel. It has been reported that in about 6%−7% cases, the usage of rt-PA can cause symptomatic intracranial hemorrhages (SIH) because rt-PA is free to diffuse throughout the body (Abdalla & Mäder, 2009). The risk of rt-PA-related SIH is significantly increased if rt-PA is administered more than 3 hours after the stroke, which limits its use in therapeutics (Laverdure et al., 2001). SIH may lead to death if it is left unattended. In about half of all cases, the rt-PA fails to lyse the clot and recanalize the middle cerebral artery (Adams Harold et al., 2007). As a result, rt-PA treatment is rarely suitable for ischemic stroke patients and is used only on about 1%−2% of them (Alexandrov et al., 2004). Clearly, in order to improve this therapy, that is, to enhance the blood clot lysis rate and reduce the rt-PA-mediated risks, new strategies with better safety and improved efficiency are needed. To achieve such a goal, there are some challenges that must be overcome. First, the consumption and concentration of rt-PA should be reduced during the blood circulation in order to minimize the side effect of hemorrhage, while the local rt-PA concentration should be increased to boost the blood clot lysis rate. Such a challenge requires that the rt-PA molecules are protected during circulation in the blood, maintain their thrombolytic activity, and concentrate onto the clot site. Collectively, this is called targeted delivery. Second, the reaction rate of rt-PA and clot at the blood clot site should be enhanced. This means that once targeted delivery is achieved, other mechanisms can be introduced to improve the lysis speed. Third, the new method should be biocompatible and non-toxic, and should not induce other side effects, that is, the newly developed method should be safe and reliable.

In the last two decades, many works have demonstrated that the nanoparticle (NP)-based drug carrier method can significantly enhance stroke treatment efficacy and meet the challenges stated above due to the following advantages: first, multiple drugs can be incorporated into NPs so that multiple actions with synergistic effect can be enabled for superior stroke therapy. Second, the surface of the NPs can be modified with targeting ligands that can specifically bind these drug-loaded carriers to the stroke-affected site and reduce side effects. Third, stimuli responsive NPs can be designed to enable accurate and temporal control of drug release as well as sequential release. This chapter will discuss the general principles for improving thrombolytic efficiency using nanotechnology and highlight different enhancement mechanisms for stroke treatment. Nanotechnology-based stroke treatment method can be divided into two general catalogs, passive NP treatment and active NP treatment. The passive NP treatment includes a wide repertoire of organic, inorganic, and hybrid NPs with diverse shapes, sizes, and surface modifications. Stimulated material or smart delivery mechanisms have been developed to control the local rt-PA release. To further improve the treatment efficacy, active NP treatments with increased local temperature or enhanced local mass transportation have been deployed recently using external magnetic field, ultrasound, or light.

15.2 **Principles for improving thrombolytic efficiency**

The details of the rt-PA-based thrombolytic process are very complicated, and many parameters can be used to fine tune the thrombolytic efficiency with different mathematical or numerical models (Bannish et al., 2017; Gu et al., 2019). However, from a fundamental point of view, this complicated reaction process can be simplified as the following (Fig. 15.2A) (Cheng, Huang, et al., 2014): (1) rt-PA and PLG molecules diffuse to the clot surface and bind to lysine sites; (2) PLG molecules on the fibrin surface are activated into PLM by the neighboring rt-PA molecules and PLM molecules start to cleave the local fibrin fibers into soluble products (P); and (3) lysed molecules (P) leave the fibrin surface and expose new lysine sites. The process can be described as,

$$S + T \overset{k_T}{\to} ST(SP) \overset{k_P}{\to} S + P \tag{15.1}$$

with rt-PA absorption rate k_T, product desorption rate k_p, and a fast transition reaction $ST \to SP$. The thrombolytic rate is determined by how fast subprocesses happen sequentially and is dominated by the slowest subprocess. For different concentrations of rt-PA, the overall thrombolysis could either belong to a diffusion-limited (mass transport dominated) reaction or an activation controlled reaction. The rate of diffusion-limited reaction depends closely on the rt-PA bulk concentration,

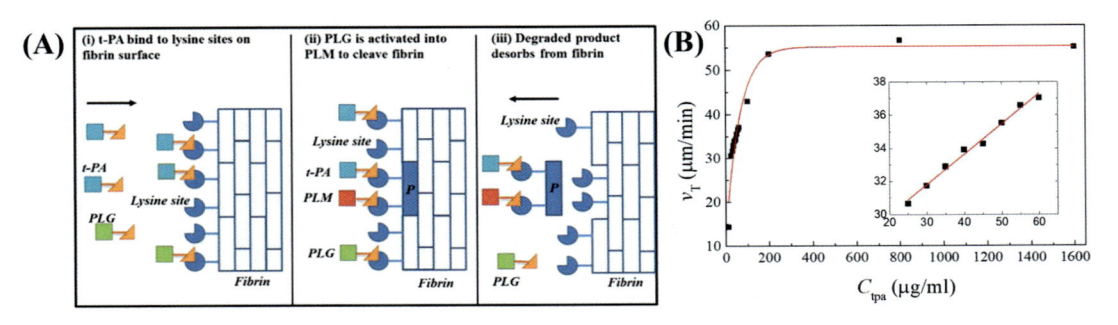

FIGURE 15.2

(A) Simplified schemes of rt-PA (or t-PA)-mediated thrombolytic reactions on the clot surface (Cheng, Huang, et al., 2014). (B) The clot boundary moving speed v_T versus the t-PA concentration C_{tPA} (Cheng, Huang, et al., 2014).

Reprint with permission from Cheng, R., Huang, W., Huang, L., Yang, B., Mao, L., Jin, K., ZhuGe, Q., & Zhao, Y. (2014). Acceleration of tissue plasminogen activator-mediated thrombolysis by magnetically powered nanomotors. ACS Nano, 8(8), 7746–7754. Copyright 2014 American Chemical Society.

while the activation controlled reaction is independent of the rt-PA concentration. Fig. 15.2B shows experimental results of the fibrin clot lysis speed v_T versus the rt-PA concentration C_{tPA}. When $C_{tPA} > 180\ \mu g/mL$, the lysis speed is saturated and is independent of C_{tPA}, and when $C_{tPA} \leq 180\ \mu g/mL$, v_T increases almost linearly with C_{tPA}. Usually, rt-PA-mediated thrombolysis under clinical administration is a diffusion-limited reaction, that is, the transport (diffusion) of rt-PA to the surface of the clot is the slowest step in determining the overall reaction rate. For a diffusion-limited reaction, the reaction rate k is determined by von Smoluchowski's equation,

$$k_T = 4\pi C_{tPA} D_{tPA}(r_{tPA} + r_{site}) \tag{15.2}$$

where D_{tPA} and C_{tPA} are the diffusivity and concentration of rt-PA at reaction site, respectively, and r_{tPA} and r_{site} are the radii of rt-PA molecules and lysine site. Clearly, one needs to increase the reaction rate constant k_T to enhance the thrombolytic process. Note that for thrombolysis, the clot lysis process should only occur at the blood clot site, that is, the enhanced k_T should only be increased locally. Based on Eq. (15.2), two principles can be applied to increase k_T: first, to increase the local concentration C_{tPA} of rt-PA near the blood clot; second, to enhance the mass transport coefficient D_{tPA}. Fig. 15.3 summarizes the current nanotechnology-based strategy for enhanced stroke treatment.

15.2.1 Nanocarriers to increase local concentration C_{tPA}

In practice, the local rt-PA concentration is likely to be lower than its administered concentration due to the loss and inactivation of rt-PA molecules in circulation, its adsorption probability to fibrin, and the inherent limitation of diffusion in a stagnant channel (Collet et al., 2000). The ideal scenario is that the administered and local concentrations are equal. Thus it is critical to

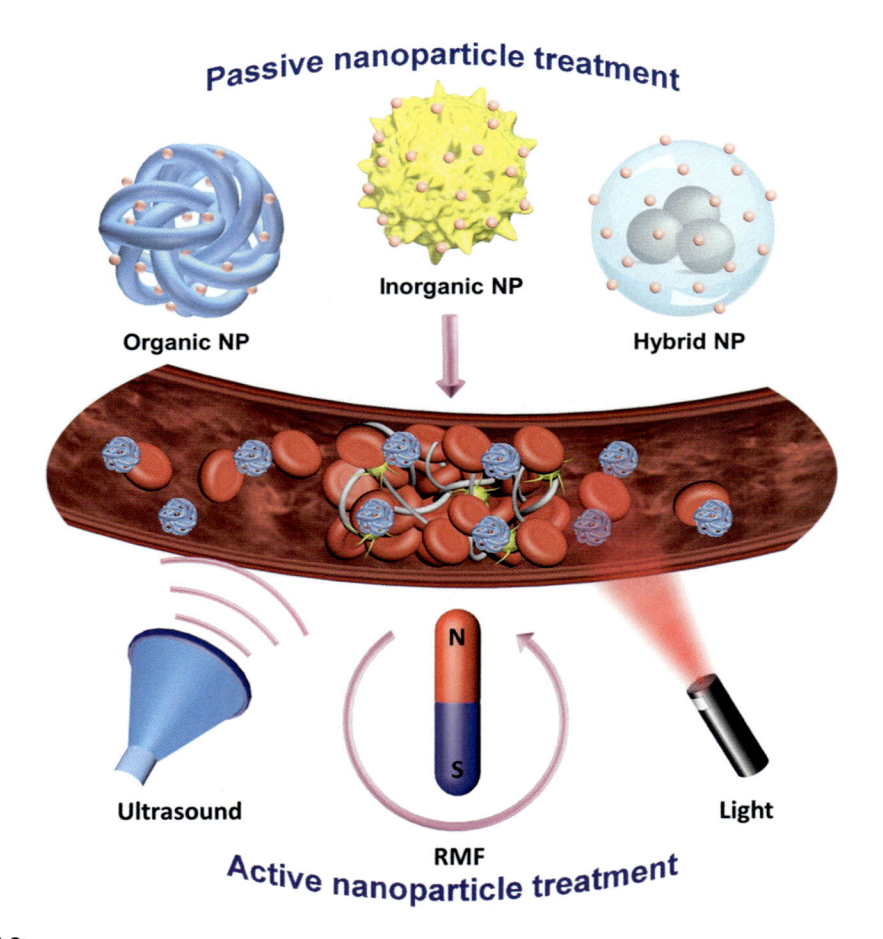

FIGURE 15.3

Scheme illustration of current nanotechnology-based strategy for enhanced stroke treatment. Top: various nanocarriers for targeted delivery and enhanced local lysis agent concertation. Bottom: various external interactions such as ultrasound, magnetic field, and light to enhance thrombolytic efficiency.

enhance the local concentration of rt-PA. In general, the local concentration C_{tPA} can be enhanced via targeted delivery of nanocarriers. The ideal nanocarriers should satisfy the following conditions: (1) the rt-PA molecules should be chemically bound to or encapsulated inside the carriers; (2) the nanocarriers can be concentrated to the location of the blood clot by a certain guided mechanism in the blood circulation system; and (3) after concentration to the blood clot, the rt-PAs in the carriers can be released locally through a specific mechanism. This way, the amount of administered rt-PA in the circulation system can be minimized while their activity can be preserved, so that the potential side effects caused by rt-PA can also be reduced, minimized, or even eliminated.

15.2.2 Enhanced mass transport D_{tPA}

There are several methods one can implement to enhance the diffusivity D_{tPA}. First, D_{tPA}^T is temperature-dependent and is given by the Einstein's equation,

$$D_{tPA}^T = \frac{k_B T}{6\pi\eta r_{tPA}} \tag{15.3}$$

where η is the viscosity of the liquid, k_B is the Boltzmann constant, and T is the temperature. Both Eqs. (15.2) and (15.3) imply that if the local temperature around the blood clot site can be raised, the thrombolytic procedure can be enhanced. The local temperature change can be achieved through hyperthermia either via magnetic NPs (Kumar & Mohammad, 2011) or plasmonic NPs (Qin & Bischof, 2012). Magnetic hyperthermia is based on the fact that magnetic nanoparticles (MNPs) can convert electromagnetic energy from an external high-frequency field to heat. This is due to the magnetic hysteresis of the MNPs when they are subjected to an alternating magnetic field. Photothermal therapy (PTT) is an alternative way to increase the local temperature by leveraging localized photoabsorbing chromophores to treated areas. In particular, metallic NPs with absorption peaks within the near-infrared (NIR) wavelength range can act as effective chromophores. The electrons in the metallic NP experience a coherent, collective oscillation upon the excitation of external electromagnetic waves and causes significant light scattering or absorption, depending on the size, shape, and arrangement of the particles (Eustis & El-Sayed, 2006). Such a property is known as localized surface plasmon resonance and it can generate a light absorption about five times larger than that of photoabsorbing dyes (Huang et al., 2007). Through electron-phonon relaxation, the absorbed photon energy inside the NP can be converted to heat and is conducted to the local environment.

An alternative way to enhance mass transport is to apply hydrodynamic principles to agitate the solution, much like the bar magnets used in general chemistry experiments. If the drug solution is filled with many small magnetic bars (submicrometer scale) and these bars are agitated by external magnetic field, then the diffusivity D_{tPA} can be enhanced by the hydrodynamic flow produced by these actively moving magnetic bars. The hydrodynamically enhanced diffusivity D_{tPA} can be expressed as $D_{tPA} = D_{tPA}^T + D_{tPA}^C$, where D_{tPA}^C is convectional flow enhanced diffusivity by moving magnetic bars. The enhanced diffusivity is not limited to moving magnetic bars. As long as a hydrodynamic flow can be induced locally around the treatment area by a certain mechanism, an enhanced diffusivity can be produced.

15.3 Passive nanoparticle treatment

Drug carrying NPs whose thrombolytic capabilities are not enhanced by external interactions such as magnetic field, ultrasound, etc. are defined as passive NPs (PNPs). PNPs can be classified into inorganic, organic, or hybrid NPs based on their chemical composition, as shown in Fig. 15.4. A brief summary of the properties and performance of these PNPs for stroke treatment are presented in Table 15.1. PNPs often act as vehicles for the delivery of rt-PA (or other thrombolytic agents) to the clot site and are further conjugated with molecules that possess clot homing capabilities to achieve targeted delivery and enhanced therapeutic effects. Such PNP carriers provide two benefits:

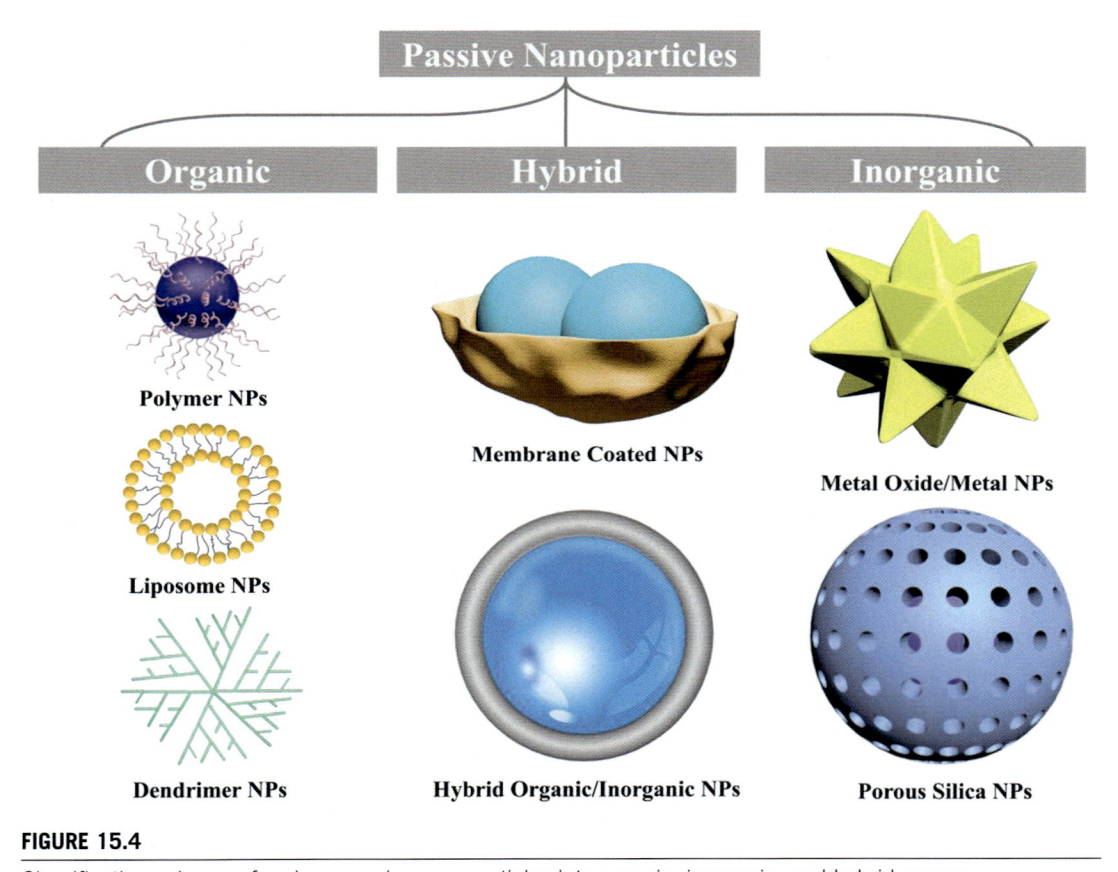

FIGURE 15.4

Classification scheme of various passive nanoparticles into organic, inorganic, and hybrid groups.

to prolong the circulation time for rt-PA and to increase the local concentration of rt-PA via targeted delivery and release. Three primary criteria should be addressed in the design of an efficient PNP for stroke treatment: the NP should be capable of (1) carrying and releasing a sufficient therapeutic payload, (2) circulating through the blood vessels and remaining active in the bloodstream long enough for effective treatment, and (3) targeting the ischemic region.

Usually rt-PA can be loaded onto a NP by surface adsorption or encapsulation (Fig. 15.5A). In either case, NPs may actively release rt-PA as they are transported through the bloodstream. Therefore, the nature of the immobilization of rt-PA on the NP surface or the encapsulation of rt-PA inside the NP are instrumental to ensure that an ample amount of therapeutic drug arrives at the thrombus site. Drug molecules are conjugated onto the surface of a NP by one or both of the following methods: physisorption (Juenet et al., 2018; Li, Zhang, et al., 2020; Wang et al., 2012; Huang, Zhang, et al., 2019), in which molecules are adsorbed on the NP surface by intermolecular forces; and/or chemisorption (Rao et al., 2019; Zhang et al., 2020; Bao et al., 2018), in which molecules are bound to the NP via ionic or covalent bonds. Most studies on PNPs achieve controlled

Table 15.1 Summary of the properties and performance of these passive NPs for stroke treatment.

	Material	Size (nm)	Shape	Drug and dosage	Drug-loading mechanism	Targeting moiety/ activity	Drug release mechanism	Therapeutic mechanisms and efficiency	Other salient features	References
Organic	Dextran and polysaccharide polymer- NP	136	Sphere	rt-PA (2.5 mg/kg)	Physisorption (LE = 82.3%)	Fucoidan	Physical impetus (shear stress)	Reduced thrombus density to 29.5% in 30 min	Full fibrinolytic activity of the rt-PA recovered	Juenet et al. (2018)
	Cationic polymer dendrigraft	65	Sphere	Rapamycin (1 mg/kg)	Encapsulation (EE = 64.9%)	CREKA	Chemical stimulus (ROS)	Reduced infarct area to less than 20%	ROS responsive and fibrin binding polymers. No obvious cytotoxicity	Lu et al. (2019)
	Poly(lactic-co-glycolic acid) PLGA polymer	1200	Discoid	rt-PA (2.5 mg/kg)	Surface adsorption (LE ~ 100%)	N/A	N/A	Reduced thrombus to 58% of initial size in 35 min	Less than 10% of rt-PA lost in bloodstream during transport	Colasuonno et al. (2018)
	Phosphatidylserine Liposome	150–170	Sphere	Streptokinase	Encapsulation (EE = 40%)	RGD and DAEWVDVS peptide	Physical impetus (phospholipase-A_2)	Delayed vessel occlusion with an efficiency close to that of free streptokinase	Render targeted fibrinolysis without affecting systemic hemostasis	Pawlowski et al. (2017)
	Liposome	164.6	Sphere	rt-PA (0.5 mg/mL)	Encapsulation (EE = 33.4%)	cRGD	Chemical stimulus (pH)	Near complete in vitro clot lysis within 75 min	97% of fibrinolytic activity retained after release	Huang, Yu, et al. (2019)
	Liposome	150	Sphere	uPA (100 U/g)	Encapsulation (EE = 29%)	cRGD	Chemical stimulus (pH)	Reduced thrombus size to < 30% in 40 min at a quarter of the dosage of free uPA	Long circulation half-life of 2.5 h compared to 15 min for free uPA	Zhang et al. (2018)
	Liposome	96	Sphere	ZL006	Encapsulation (EE = 79.12%)	Stroke homing peptide	Chemical stimulus (pH)	Significant reduction of infarct volume to < 30% 24 h after	Dual targeting enhanced local concentration of ZL006 to ischemic tissue	Zhao et al. (2016)

	Material	Size (nm)	Shape	Drug	Loading	Targeting	Release	Effect	Effect	Reference
								Middle Cerebral Artery Occlusion (MCAO)	and reduced drug side effects	
	Hydrogel	100–300	Sphere	rt-PA	Encapsulation	N/A	Chemical stimulus (thrombin enzyme)	When used as a biomaterial surface coating, reduced clot weight tenfold compared to uncoated biomaterial surfaces	Rate of rt-PA activity recovery linearly correlated with thrombin concentration	Li et al. (2017)
	PEG/polyglutamic acid peptide dendrimer	105	Dendrite	Nattokinase	Physisorption	N/A	Chemical stimulus (pH)	Decreased thrombus weight by 2.25 times more than free nattokinase in vitro	Dendrimer NP can protect nattokinase from environmental deactivation	Zhang et al. (2018)
	PAMAM dendrimer	>200	Dendrite	Salvianic acid	Physisorption (LE = 13.6%)	COG1410	Chemical stimulus (ROS: Reactive oxygen species)	Decreased infarct size to less than 20%	No toxicity to any bodily organs	Li, Zhang, et al. (2020)
	Magnetite nanoporous matrix	<200	Sphere	uPA (0.5 kU/mg)	Entrapment in pores	Targeted by external magnetic field	No release	Condensation of NPs to clot by magnetic field in vitro accelerated lysis rate by 250% compared to free uPA	Prolonged thrombolytic activity due to entrapment and nonrelease of uPA	Drozdov et al. (2016)
Inorganic	Magnetic mesoporous silica NP	50	Porous sphere	uPA (658.6 U/mg)	Physisorption	Magnetic field	N/A	86 ± 6% clot lysis efficiency in in vitro dynamic flow tests	In vitro targeting by magnetic field allowed for twofold higher lysis efficiency	Wang et al. (2012)

(Continued)

Table 15.1 Summary of the properties and performance of these passive NPs for stroke treatment. *Continued*

	Material	Size (nm)	Shape	Drug and dosage	Drug-loading mechanism	Targeting moiety/ activity	Drug release mechanism	Therapeutic mechanisms and efficiency	Other salient features	References
								compared to free uPA		
	Carbon dot	8	Sphere	uPA	Chemisorption of carbon dots onto uPA	N/A	No release	Achieved thrombolytic rate of uPA − CDs of up to 97.57%, which is higher than free uPA.	Biocompatible, and can act as fluorescence probe for bioimaging	Niu et al. (2020)
	Ceria NP	19.17	Sphere	Edaravone	Chemisorption	Angiopep-2	Chemical stimulus (pH)	ROS elimination properties decreased infarct volume by 66%	Low toxicity and excellent hemo/ histocompatibility.	Bao et al. (2018)
Hybrid	Platelet membrane coated dextran polymer core	167.2	Spherical core−shell structure	rt-PA (0.5 mg/ kg) and ZL006e (4 mg/kg)	rt-PA surface adsorbed, ZL006 encapsulated (LE = 1.3%, EE = 10.2%)	Platelet membrane	Chemical stimulus (thrombin enzyme)	Full recovery of rt-PA thrombolytic activity and higher thrombolytic rate than free rt-PA after 20 min	ZL006e provided neuroprotection after thrombolysis by rt-PA	Xu et al. (2019)
	Platelet membrane coated boronic ester/dextran polymer core	194.6	Spherical core−shell structure	NR2B9C (0.3 mg/kg)	Encapsulation (EE = 62.33%)	Stroke homing peptide	Chemical stimulus (ROS)	After 24 h, reduced infarct size to less than 30% of initial size	Blank NPs could reduce ROS levels	Lv et al. (2018)
	Platelet membrane nanocarrier encapsulating $\gamma - Fe_2O_3$	200	Sphere	L-arginine	Encapsulation	Dual targeting by platelet membrane and magnetic field gradient	N/A	Significant recanalization in ischemic area after 0.5−1 h after administration	In situ generation of NO prevents platelet aggregation to prevent thrombotic plaque formation	Li, Li, et al. (2020)

	Nanoparticle	Size (nm)	Shape	Drug	Loading method	Targeting	Stimulus	In vivo efficacy	Advantages	Reference
	Platelet membrane coated PLGA polymer NPs	200	Sphere	Lumbrokinase (8×10^4 U/kg)	Encapsulation (EE = 4.14%)	Platelet membrane	N/A	Significant thrombus reduction (below 20%) at low dose compared to free lumbrokinase	Reduced hemorrhagic risk and no significant cytotoxicity	Wang, Wang, et al. (2020)
	Platelet membrane coated PLGA polymer NPs	167.0	Sphere	rt-PA	Chemisorption	Platelet membrane	N/A	Reduced thrombus area to less than 10% versus $\sim 25\%$ for free rt-PA	Low bleeding risk by evasion of fibrinolytic state	Xu, Zhang, et al. (2020)
	Fe_3O_4 NPs embedded in multiarm peptide dendrimers	<100	Spherical core−shell structure	Nattokinase	Encapsulation	Dual targeting by RGD and external magnetic field	N/A	threefold higher blood clot dissolution than free nattokinase in 5.5 h in vitro.	Thrombus dissolution could be enhanced by external magnetic field	Zhang et al. (2020)
	Magnetic mesoporous silica/ polyglutamic acid peptide dendrimer	90	Spherical core−shell structure	Nattokinase	Physisorption	Dual targeting by RGD and external magnetic field	N/A	Decreased thrombus weight to 55.3% in 5.5 h.	Avoids aggregation with other NPs and electrostatic interactions with normal cells due to negative charge	Huang, Zhang, et al. (2019)
	Platelet membrane coated gold nanorods	50	Core−shell structure	uPA (60,000 IU/ mg) or Lysozyme	Encapsulation (EE = 50.3%)	Platelet Membrane	Chemical stimulus (light)	Reduced blood clot weight from ~ 75 mg to ~ 30 mg in 3 h	80.9% of lysozyme could be released after NIR irradiation for 3 h	Yang et al. (2018)

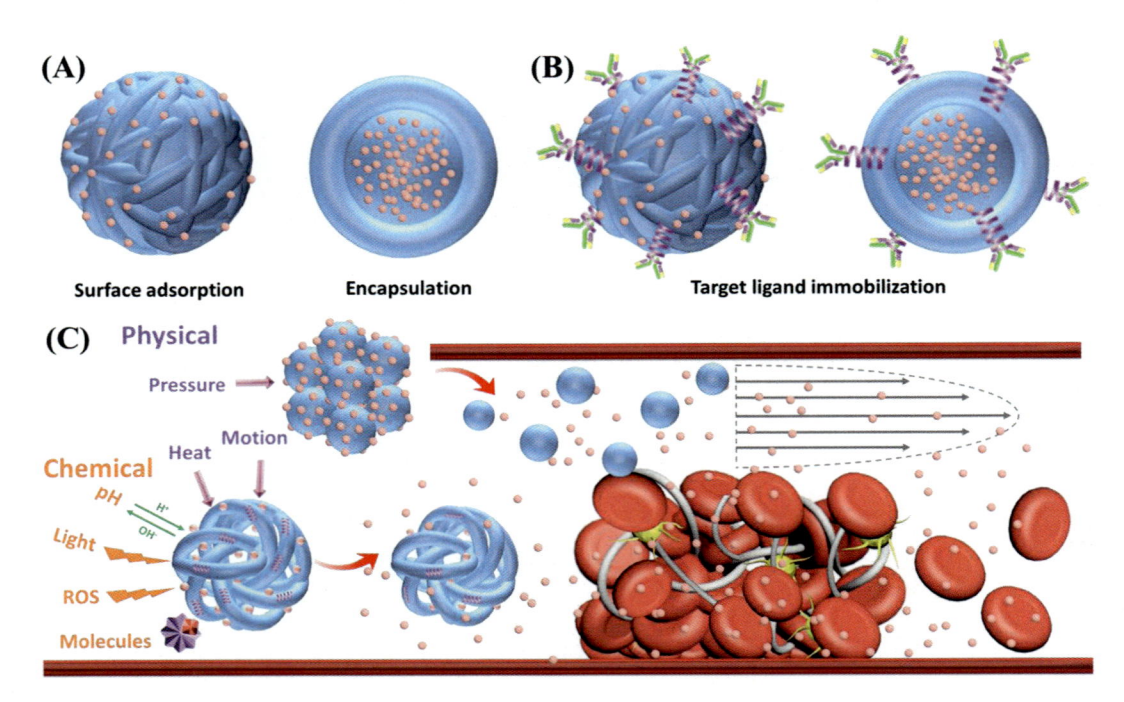

(A)

Surface adsorption   Encapsulation

(B)

Target ligand immobilization

(C) **Physical**

Pressure →

Chemical

Heat   Motion

pH

Light

ROS

Molecules

FIGURE 15.5

Scheme illustration: (A) Loading of rt-PA onto a NP by surface adsorption or encapsulation. (B) Target ligand immobilization. (C) Two main mechanisms to trigger drug release: physical impetus and chemical stimulus.

rt-PA release by using thrombin-cleavable chemical bonds, linking rt-PA to the NP's surface, although some have used both physisorption and chemisorption simultaneously for different drugs and moieties (Li, Zhang, et al., 2020).

For surface adsorption-based drug-loading mechanisms, the loading ratio of the drugs is determined by the surface area-to-volume ratio (SVR) (Biener et al., 2009). The size, shape, topology, and material properties of a NP dictate the type and amount of rt-PA (or other thrombolytic molecules) that can be loaded (Chou et al., 2011). A large SVR is favorable for a higher loading efficiency of surface adsorbed molecules. For example, a spherical NP of radius r has an SVR of $3/r$, that is, the smaller the particle, the larger the SVR. Thus smaller particles possess several significant advantages as drug carriers, including the ability to carry larger amounts of rt-PA, undergo increased surface modifications, spread easily through the bloodstream, and release rt-PA rapidly at the clot sites (Mauricio et al., 2018). Some PNPs, usually inorganic, feature a porous structure for drug-loading which can further increase the SVR. Most PNPs for stroke treatment have sizes between 10 and 200 nm, as this is the optimal size range for minimizing renal filtration and clearance by the phagocytes of the mononuclear phagocyte system. The SVR is also influenced by particle geometry, including spheres, cylinders, disks, and cubes. Notably, spherical NPs are good candidates for drug delivery and are frequently used in PNPs, but in certain cases, anisotropic shapes could provide favorable configurations for binding and cellular uptake for higher efficiencies due to a larger SVR (Sun et al., 2014; Zhang, Li, et al., 2019).

As presented in Table 15.1, surface adsorption-based drug mechanisms have been adapted for inorganic, organic, and hybrid NPs. In organic NPs, rt-PA can be surface adsorbed by linking onto specific functional groups exposed on the NP surface. While covalent binding provides an effective platform for preventing premature drug release, it can decrease the thrombolytic activity on-site. Thus electrostatic linkage has been the mechanism of choice for surface adsorption on organic NPs. For example, Juenet et al. electrostatically linked rt-PA to aminated dextran functionalized with fucoidan (a targeting ligand), and flow cytometry showed that roughly 80% of the NPs were uniformly loaded with rt-PA on their surfaces (Juenet et al., 2018). However, it should be noted the small proportion of NPs that remain unloaded have the potential to compete with loaded NPs, resulting in a lowered thrombolytic activity. Nevertheless, the polymer NPs were able to reduce thrombus density to less than 30% in vivo within 30 minutes (Juenet et al., 2018). A similar approach has also been implemented to attach rt-PA, antioxidants, and other drugs to other organic NPs, such as poly(lactic-co-glycolic acid) (PLGA) polymer NPs, various dendrimer NPs (Li, Zhang, et al., 2020; Zhang et al., 2018), etc. Inorganic NPs often possess a porous structure to efficiently entrap rt-PA and other therapeutics, as porous NPs can further increase the SVR by tuning the size of the NP as well as its pore size and pore volumes (Wang et al., 2012). Drozdov et al. developed a magnetite NP with a high specific surface area (~ 120 m^2/g) and a nanoporous matrix structure (Drozdov et al., 2016). The average matrix pore size of 8 nm was complementary to the size of uPA (Urokinase), the thrombolytic drug used in the study. Because the uPA could fit snugly into these pores, the NPs could form plasmin when interacting with plasminogen without releasing the uPA, allowing the NP to exert thrombolysis over long periods of time. Large clot dissolution rates were achieved without releasing uPA in vitro, corresponding to a uPA concentration of 12.5 wt.% in the NP. Biodegradable porous inorganic NPs, such as porous silica NPs, have also been developed for thrombolytic and postischemic stroke treatment (Wang et al., 2012; Huang et al., 2018; Balasubramanian et al., 2020). For hybrid NPs featuring the membrane cloaking technique, including platelet membranes (Lv et al., 2018; Li, Li, et al., 2020; Wang, Wang, et al., 2020; Xu, Zhang, et al., 2020; Yang et al., 2018), red blood cells (RBCs) (Chen et al., 2020) and exosomes (Zhang, Wu, et al., 2019), rt-PA is often conjugated to the membrane surfaces by using various peptides as linker molecules for triggered drug release, which will be further discussed later in this section (Rao et al., 2019). For example, Xu et al. conjugated rt-PA to the surface of a platelet membrane coating a PLGA polymer core using bifunctional maleimide linkers (Xu, Zhang, et al., 2020). These NPs could accumulate selectively at the thrombus site and prolong the rt-PA circulation half-life, leading to a significantly enhanced thrombolytic activity in comparison to free rt-PA (Xu, Zhang, et al., 2020).

If rt-PA is loaded by encapsulation, the NP acts as a shell to protect the drug from degradation and leaves the surface available for further functionalization. In some cases, encapsulation of rt-PA can be advantageous over surface adsorption, because exposure of the blood to the antithrombotic agents on the material surface may cause hemostatic disorders and side effects under normal conditions (Li et al., 2017). In contrast to surface adsorption, the encapsulation requires a larger void to surface ratio (VSR). For example, for a spherical shell particle with an inner radius r and a shell thickness d, the VSR $\sim \frac{r^3}{3(r+d)^2}$. For $r \gg d$, VSR $\sim \frac{r}{3}$. Thus the rt-PA-encapsulating NPs tend to be larger in size (between 100 and 200 nm), as shown in Table 15.1. In drug-encapsulating NPs, avoiding drug leakage during transport and sufficient recovery of the drug's thrombolytic activity after release at the clot site are critical. Much like surface adsorption, the structure of the NP determines different modes of encapsulation and the encapsulation efficiencies.

Liposome NPs are the most efficient drug-encapsulating NPs due to their lipid bilayer shell composition, which allows hydrophilic drugs to be encapsulated within the hollow core and hydrophobic drugs to be entrapped between the hydrophobic lipid tails (Gao et al., 2013). They have also proven to be among the most efficient organic NPs for preventing rt-PA leakage. Using liposome NPs, Zhang et al. showed that less than 25% of initially encapsulated uPA leaked from the NP before reaching the thrombus site (Zhang et al., 2018). The efficacy of encapsulation by liposomes was further corroborated by Huang et al., who found that the increased repulsion between the negative surface charge of liposome NPs and the shielding effect of surface modification with polyethylene glycol (PEG) could reduce rt-PA leakage by over 40% compared to nonmodified groups (Huang, Yu, et al., 2019). Hybrid NPs usually use the core–shell configuration to encapsulate drugs, particularly with dual or multifunctionalities. For example, Lv et al. encapsulated a neuroprotectant called NR2B9C into an NP made of a boronic ester modified dextran polymer, coated with a RBC membrane (Lv et al., 2018). It was shown that large amounts of encapsulated NR2B9C were released only when triggered by the high intracellular reactive oxygen species (ROS) microenvironment of ischemic brain tissues. In addition to the neuroprotection provided by the encapsulated NR2B9C, blank NPs were found to have ROS scavenging capabilities which further contributed to the reduction of infarct volume below 40% of its initial size after 24 hours intravenous administration of the NPs in a concentration of 0.3 mg/kg.

Another important issue is to have an efficient targeting mechanism to accumulate PNPs, release the drug at the clot site, and avoid nonspecific distribution. Targeting can be achieved by using NPs with homing capabilities to the ischemic region, that is, by conjugating the NPs with targeting ligands (see Fig. 15.5B), which include small organic particles, peptides, antibodies, designed proteins, and nucleic acid aptamers due to their affinity for specific biochemical targets abundant in pathological tissues in the clot site (Rizvi & Saleh, 2018). Targeting ligands have also been selected based on their specificity to molecules expressed by platelets or other molecules accumulating in the thrombus site. Alternatively, NPs can be externally guided and condensed onto the clot by an external magnetic field, as is the case with many inorganic and hybrid NPs. Regardless of the drug-loading mechanisms, the targeting ligands are always immobilized on the outer surface of the NPs, as shown in Fig. 15.5B.

The most widely used targeting ligands for PNPs are arginine-glycine-aspartic peptide (RGD) and cyclic arginine-glycic-aspartate tripeptide (cRGD) (Alipour et al., 2020; Park et al., 2012). The targeting property of RGD integrin is influenced by its conformation, while cRGD has a higher affinity and specificity for $\alpha_{IIb}\beta_3$, an integrin that is overexpressed on platelet surfaces during thrombus formation (Huang et al., 2008). For example, Zhang et al. used cRGD as a targeting ligand for liposome encapsulated uPA NPs. When freshly prepared platelets were activated by the introduction of thrombin in vitro, the cRGD modified liposomes could interact well with activated platelets compared to control liposomes. The fucoidan polysaccharide has also been shown to be an effective targeting ligand since it can mimic the main ligand of P-selectin, a protein that is overexpressed by activated platelets localized in the thrombus (Bachelet et al., 2009). The enhancement of the thrombolytic efficiency of rt-PA delivered by NPs utilizing fucoidan-mediated targeting have been observed after only 30 minutes as shown by Juenet et al., which is faster than many targeted nanocarriers reported in literature (Juenet et al., 2018). Other targeting ligands such as the DAEWVDVS peptide have also been used to target P-selectin, as demonstrated by Pawlowski et al., who used both DAEWVDVS and RGD as heteromultivalent targeting ligands for streptokinase-delivering liposome NPs. $\alpha_{IIb}\beta_3$ integrins and P-selectin (platelet adhesion receptors), GPVI (collagen receptor), and other binding molecules are abundant on the surface of activated platelets, endowing them with an innate thrombus homing capability during their crucial role in thrombosis and hemostasis (Saboor et al., 2013). These properties have motivated the wide use of platelet membranes as a targeted surface coating for hybrid NPs (Li, Li, et al., 2020; Xu, Zhang, et al., 2020; Hu, Qian, et al., 2016). For instance, in the previously mentioned work, Xu et al. leveraged platelet membranes as a surface coating on polymer NPs to target platelet-rich thrombus sites and achieved rapid thrombolysis. The NPs could achieve a complete fibrin clearance after administering a dose of 5 mg/kg (Xu, Zhang, et al., 2020).

Recent efforts to improve the localization of NPs near the thrombus site have led to the implementation of dual targeting capabilities which are increasingly common amongst hybrid NPs. A typical dual targeted NP consists of a magnetic inorganic component for specific targeting of NPs guided by an external magnetic field in tandem with the contribution from a targeting ligand. For example, Zhang et al. modified dendrimer NPs with both Fe_3O_4 NPs and RGD for dual targeting (Zhang et al., 2020). These dendrimer NPs could surround and protect the thrombolytic drug nattokinase (NK). Clot lysis by 1 mg/mL of NK in vivo was the most effective when delivered by the dendrimer NPs under both magnetic guidance through the vascular system and RGD-mediated preferential accumulation on the thrombus. Finally, the stroke homing peptide (termed SHp or CLEVSRKNC peptide) has also been used (albeit relatively infrequently) as a targeting ligand due to its specificity to ischemic stroke tissues, which can enhance the local concentration of drugs while avoiding nonischemic tissues. In a study by Zhao et al., the SHp was used alongside with the T7 peptide for a liposome NP, delivering the ZL006 neuroprotectant to the ischemic brain (Zhao et al., 2016). The results showed that this dual targeted liposome could selectively deliver ZL006 to the ischemic tissue, enhancing the local concentration of the drugs while reducing their side effects. The NPs could reduce the brain infarct volume to less than 30% and decrease the rate of neuronal apoptosis to $11.3\% \pm 1.8\%$ (Zhao et al., 2016).

Finally, the efficient release of the drug from the PNPs at the clot site are critical. During the transport of the PNPs, chemisorption or encapsulation of drug molecules is preferred in order to prevent premature drug release. Therefore a stimulus mechanism is needed to activate the drug release once the PNPs arrive at the target site. There are two main mechanisms to trigger drug release: physical impetus and chemical stimulus, as shown in Fig. 15.5C. Physical impetus includes local temperature and pressure change as well as the motion of PNPs, while chemical stimulus is caused by the change of local chemical environment, such as ROS generation, pH environment variation, or chemical bond cleavage due to the presence of other reactive molecules or light.

Depending on the drug-loading mechanisms, the increase of local temperature could have different effects on targeted drug release: For drug-loading via surface adsorption, the raising of local temperature can increase the drug desorption rate; for drug encapsulated by polymer membranes, such as liposomes, the polymer membrane can be expanded at a higher temperature so that holes can be formed to release the drug; or if the drug molecules are entrapped by temperature-sensitive polymers, the change in temperature can alter the polymer chain configuration from the densely packed configuration to extended porous configuration, so that the drug can be released. The mechanisms to increase local temperature of the PNPs will be discussed in Section 15.4. For example, biomimetic magnetic microrobots, an assembly of magnetic NP chains embedded in microgel, were developed to load rt-PA into hydrogel (Xie et al., 2020). Under a 375 kHz alternating magnetic field (AMF), the rt-PA-loaded gel particles could achieve a local temperature increment of about 22°C and induce a significant hyperthermia effect to increase the release rate of rt-PA. A temperature-sensitive phase transmission material, 1-tetradecanol, was used to cover the gold@mesoporous silica core–shell nanospheres in order to encapsulate the uPA inside the pore of silica shell (Wang et al., 2017). The drug was released when the particles were irradiated by a laser and the local temperature exceeded the phase changing (from solid to liquid) temperature of 38°C of 1-tetradecanol (Wang et al., 2017).

For pressure-induced drug release, two different mechanisms have been developed: ultrasound breakable drug-capsules and shear-activated deagglomeration of NP clusters. Ultrasound is well-known to generate uniform dispersions of particles in liquid due to its cavitation effect, that is, the alternating high-pressure (compression) and low-pressure (rarefaction) cycles in liquid (Gallo et al., 2018). It has also been used for drug delivery and release (Pitt et al., 2004). As an example, Correa-Paz et al. synthesized rt-PA encapsulated polymer microparticles using the porous $CaCO_3$ microparticles (diameter ~ 620 nm) as the template (Correa-Paz et al., 2019). The porous $CaCO_3$ particles were first coated with rt-PA, then covered by polymer layers via layer-by-layer deposition, and finally the $CaCO_3$ cores were removed chemically. The rt-PA from such capsules, would be released from the porous shells upon application of ultrasound, and the activity of the released rt-PA after a 40-minutes treatment was approximately fourfold higher than the same amount of non-encapsulated rt-PA (1 mg/kg) (Correa-Paz et al., 2019). Furthermore, fluid shear stress can increase locally by one to two orders of magnitude in the stenotic and thrombosed blood vessels compared to normal vasculature, making it a viable physical impetus for triggering drug release. Microscale aggregates of nanoparticles can break up into nanoscale components to release thrombolytics when exposed to abnormally high fluid shear stress (Juenet et al., 2018; Korin et al., 2012). Korin et al. developed a biocompatible and biodegradable PLGA microaggregates around 3.8 mm to demonstrate such a mechanism (Korin et al., 2012). When coated with rt-PA and administered intravenously in mice, the shear-activated nanotherapeutics induced rapid clot dissolution in a mesenteric injury model, restored normal flow dynamics, and increased survival in an otherwise fatal mouse pulmonary embolism model (Korin et al., 2012).

In addition, if the drug molecules are physically adsorbed on the surfaces of NPs, even the motion of NPs can help to release the drug. For example, Hu et al. demonstrated that 1 mg/mL of rt-PA loaded Fe_3O_4 nanorods (with a rt-PA mass loading ratio of 6%) could release 4.4 ± 0.2 µg/mL rt-PA after 2 hours, while upon applying a rotation magnetic field (RMF, 20 Hz and 3 mT), the released rt-PA concentration became 6.8 ± 0.4 µg/mL, nearly a 50% increment (Hu, Huang, et al., 2016). Similar results were reported in low-intensity ultrasound-based rt-PA delivery. Wang et al. designed a multifunctional magnetic microbubble to incorporate rt-PA for accelerating thrombolysis (Wang, Guo, et al., 2020). When ultrasound was applied for five cycles at an acoustic pressure of 0.05 bar, approximately 5% of the rt-PA was released. If the ultrasound was continually applied for another 60 cycles, almost 90% of the rt-PA was released. Thus a stepwise release of rt-PA was achievable by increasing the number of cycles of applied ultrasound. Furthermore, by increasing the acoustic pressure to 0.1 and 0.15 bar, more rapid release kinetics were observed due to more intense microbubble oscillations (Wang, Guo, et al., 2020).

The microenvironment of the thrombus and pathological tissues are acidic compared to normal tissues, with pH values between 6.73 and 7.29 (Cui et al., 2016). Thus the pH triggering mechanism has been commonly implemented as a release mechanism with rt-PA encapsulating NPs. Polymer NPs have been utilized frequently for pH triggered release. For example, Li et al. used an oxidized dextran polymer NP with uPA conjugated by the pH sensitive imine linkage to form a protein-polymer conjugate with RGD as a targeting ligand (Li et al., 2018). This entangled polymer network could protect the uPA from enzymatic hydrolysis under normal physiological pH and degrade under acidic conditions. In vitro experimentation showed that the activity of these NPs was highest at a pH of 6.8, and decreased with smaller pH values with the smallest activity observed under a normal pH of 7.4 (Gallo et al., 2018). Similar drug release strategies have also been applied for other polymer NPs (Gallo et al., 2018; Cui et al., 2016; Mei et al., 2019). For example, liposome NPs have shown great potential for pH triggered release, since their amphiphilic nature allows them to undergo protonation, bilayer destabilization, and release of encapsulated content (Bruch et al., 2019). In the previously discussed dual targeted liposomes implemented by Zhao et al., the cumulative release curves of ZL006-loaded liposomes displayed similar release behaviors under two distinct pH conditions over 48 hours, reaching a cumulative release of 66.8% $\pm$ 1.9% at a pH of 5.5 and 70.3% $\pm$ 5.8% at a pH of 7.4 (Zhao et al., 2016).

The other environmental change due to ischemic stroke is the upregulated toxic ROS near the clot site (Ferreira et al., 2018), which has the potential to be used as a smart and sensitive trigger for controlled drug release (Yao et al., 2018). ROS includes hydrogen peroxide (H_2O_2), hydrogen radicals (*OH), hydroxyl ions (OH^-), superoxide anions ($*O_2^-$), nitric oxide (NO*), peroxynitrites ($ONOO^-$), hypochlorite (OCl^-), and others. For example, during the formation of a thrombus, H_2O_2 plays an essential role in the platelet activation, stimulates additional platelet recruitment (Kim et al., 2015; Freedman Jane, 2008), and is the most abundant in signaling in the vasculature (Del Principe et al., 1985; Iuliano et al., 1997). Depletion of H_2O_2 and prevention of platelet activation would be a promising antithrombotic therapeutic strategy for the treatment of various vascular thrombotic diseases (Dayal et al., 2013; Jung et al., 2018). Kang et al. developed a fibrin-targeted imaging and antithrombotic nanomedicine, termed FTIAN, as a theranostic system for obstructive thrombosis (Kang et al., 2017). FTIAN inhibited the generation of H_2O_2 and suppressed the expression of tumor necrosis factor-alpha (TNF-α) and soluble CD40 ligand in activated platelets, demonstrating its intrinsic antioxidant, anti-inflammatory, and antiplatelet activity. In a mouse model of ferric chloride-induced carotid thrombosis, FTIAN specifically targeted the obstructive thrombus and suppressed thrombus formation (Kang et al., 2017). Other ROS responsive materials have also been used for thrombolysis agent release, as in the case of the previously discussed dextran polymer core/RBC membrane shell nanocarrier developed by Lv et al. (2018).

Surface conjugated drugs often make use of linker molecules that are cleavable by the thrombin enzymes abundant in the thrombus area, while drug-encapsulating NPs utilize shells that are degradable by thrombin. Li et al. developed a thrombin degradable hydrogel nanocapsule for encapsulating rt-PA, which was prepared by in situ adsorption of the acrylamide and methacrylamide monomers onto the surface of rt-PA followed by cross-linking to a thrombin-cleavable peptide with acrylate end groups (Li et al., 2017). At the rt-PA to monomer ratio of 1/3000, 100% of rt-PA activity could be recovered after 4.5 hours, and in vitro experiments showed that rt-PA released from the nanocapsules had the same fibrinolytic activity as unmodified free rt-PA. The rate of recovery of the activity was found to be linearly correlated with thrombin concentration (Li et al., 2017). Alternatively, thrombin-cleavable linker peptides can be used for surface adsorbed rt-PA release. Xu et al. encapsulated ZL006 within a dextran polymer core coated by a platelet membrane, and rt-PA was linked to the platelet membrane surface by the thrombin-cleavable Tat peptide (Rao et al., 2019). A cumulative rt-PA release of 90% was observed at a thrombin concentration of 1 U/mL, followed by the subsequent release of ZL006 from the dextran polymer core by the previously discussed pH triggering mechanism. The NP retained the thrombolytic activity of rt-PA and achieved a significantly higher thrombolytic rate after circulation in blood for 20 minutes compared to free rt-PA (Rao et al., 2019).

Lastly, triggered and controlled drug release can be effectively achieved by using light from the "phototherapeutic window" (620−850 nm), which is harmless and has maximal tissue penetration (Cheng, Doane, et al., 2014). However, this has been done sparingly with PNPs, as many NPs reliant on external stimuli fall within the realm of active NPs (Section 15.4). In one example, Yang et al. used platelet membrane-coated gold nanorods to encapsulate and deliver drugs for thrombolysis (Yang et al., 2018). Using lysozyme as the encapsulated drug, release tests showed that under the NIR irradiation (808 nm, 2 W/cm^2), 80.9% of the lysozyme could be released after 3 hours. In vitro blood clot lysis capabilities using uPA as the encapsulated drug showed that the NIR triggered release could decrease blood clot weight from ~75 mg to ~15 mg in 3 hours (Yang et al., 2018).

15.4 Improved stroke treatment with enhanced mass transport

There are some limitations for PNP-based stroke treatment. First, because the transportation of drug carriers depends solely on the blood circulation, different targeting/releasing strategies need to be developed. For example, if the blood vessel is completely blocked, the shear-flow method is unimplementable. Second, the clot lysis rate is determined only by the local rt-PA concentration. If the local concentration fails to increase significantly, the lysis efficiency cannot be improved notably. Clearly, based on the discussion in Section 15.2, an alternative way to enhance stroke treatment is to enhance the mass transport during the thrombolysis in addition to increasing the local concentration. Below we will review some recent significant works based on this strategy.

15.4.1 Increasing the local temperature

According to Eqs. (15.2) and (15.3), the clot lysis rate increases monotonically with the local temperature T, and thus both magnetic hyperthermia and photothermal therapy have been employed to improve stroke treatment. The commonly used materials for magnetic hyperthermia are ferrite MNPs [magnetite (Fe_3O_4) or maghemite (γ-Fe_2O_3)] with a size of 10−100 nm. When exposed to a high-frequency AMF, the MNPs produce heat and increase temperature locally. Decuzzi et al. reported that the superparamagnetic Fe_3O_4 nanocube clusters coated with rt-PA and bovine serum albumin could achieve ~100-fold increase in lysis rate as compared to free rt-PA, and a further ~10-fold enhancement by using a 295-kHz AMF (Voros et al., 2015). A similar effect has been observed in the biomimetic magnetic microrobot, an assembly of magnetic NP chains embedded in microgel (Xie et al., 2020). Under a 375 kHz AMF, the rt-PA-loaded gel particles could achieve a local temperature increment of about 22°C and induce a significant hyperthermia effect, achieving an 12.8-fold thrombolytic effect compared to that without the AMF treatment. In addition to magnetic hyperthermia, the PTT has also been adopted to enhance stroke treatment. PTT has been adapted in work discussed previously by Wang et al., where gold@mesoporous silica core−shell nanospheres and a temperature-sensitive phase material, 1-tetradecanol, were used to encapsulate uPA, and the NIR laser was used to increase the local temperature and trigger the uPA release (Wang et al., 2017). A similar work has also been reported to deliver urokinase uPA for deep venous thrombosis therapy by the same group (Xu, Zhou, et al., 2020). Clearly, in most hyperthermia-based strategies, the change of the local temperature serves two purposes: one is to trigger drug release either through temperature-dependent desorption or via the coating of a temperature-responsive material; and the other is to enhance the local diffusion of the released drug molecules. Thus a significantly improved thrombolytic efficiency can be achieved.

15.4.2 Increasing mass transport by active particles

As suggested in Section 15.2, the other method to improve the stroke treatment is to enhance the mass transport of the reactants and the products via hydrodynamic principles by agitating the liquid locally around the blood clot. Such a hydrodynamic effect can be achieved by introducing NPs moving locally near the blood clot. The active NPs can perform either translational or rotatory motion, driven by remote and external interactions, thus introducing fast local flow around the particles so that the thrombolytic agents can transport with an enhanced diffusion coefficient. So far, three different interactions have been implemented for stroke treatment: magnetic field, ultrasound, and light.

Using Ni magnetic nanorods (MNRs) and an external magnetic field, the principle of improved stroke treatment by enhanced mass transport was demonstrated in in vitro experiments by Cheng et al., both quantitatively with a microfluidic system and qualitatively via an animal test (Cheng, Huang, et al., 2014). Based on an eight-channel plate, the authors showed that the clot lysis speed v_{R+T} was a function of MNR concentration C_R (Fig. 15.6A). In fact, when $C_R = 7$ mg/mL, the thrombolytic speed of low-concentration rt-PA (50 μg/mL) can be enhanced up to twofold, to the maximum lysis speed at high rt-PA concentration (Fig. 15.2B). The validity and efficiency of this enhanced treatment has also been demonstrated in a rat embolic model (Cheng, Huang, et al., 2014). Compared with PNP treatment strategies that focus on the loading rate of drug molecules to nanocarriers, active motion of NPs accelerate the thrombolysis by elevating drug transport through a hydrodynamic convection, which can be tuned by many parameters such as MNR concentration, magnetic field strength and frequency, which make the method more flexible and accessible (Cheng, Huang, et al., 2014). To further validate this idea, Hu et al. developed biocompatible rt-PA-functionalized Fe_3O_4 nanorods loaded with 6% rt-PA that was guided to the blood clot and to release rt-PA locally (Hu, Huang, et al., 2016). Two important observations were revealed: first they found that the rt-PA release rate was enhanced by the RMF, as shown in Fig. 15.6B, that is, an external RMF can stimulate the release of rt-PA from the functionalized Fe_3O_4 nanorods; second, the release time of rt-PA was found to be within ∼30 minutes, and thus the nanorods could act as capsule and create a safe time window for the targeted delivery of rt-PA (Hu, Huang, et al., 2016). In addition, using rt−PA-functionalized self-assembled magnetic colloidal microwheels (microwheels), Tasci et al. showed that the fibrinolysis rate depends strongly on the motion configuration of the microwheels (Tasci et al., 2017). As shown in Fig. 15.6C, by changing the magnetic field configuration, two different motions of microwheels

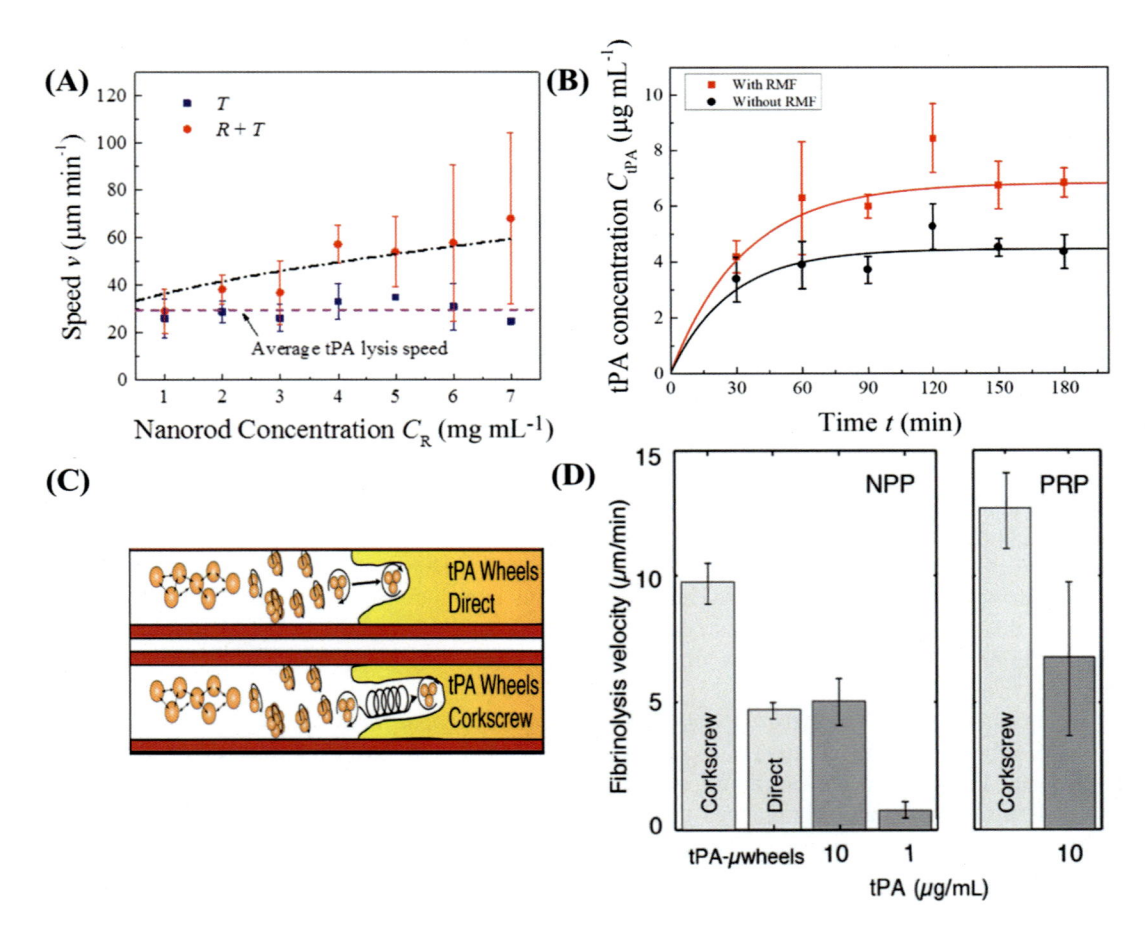

FIGURE 15.6

(A) Clot lysis speed v_T (rt-PA only) and v_{R+T} (rt-PA with rotating MNRs) versus nanorod concentration C_R (Cheng, Huang, et al., 2014). (B) The plot of the accumulated rt-PA concentration C_{tPA} released from functionalized Fe_3O_4 nanorods versus release time t with or without a RMF (Hu, Huang, et al., 2016). (C) Illustrations of direct motion and corkscrew (helical) motion of tPA-μwheels. (D) Fibrinolysis velocity of tPA versus tPA-μwheels for plasma-derived fibrin gels and gels formed with platelet rich plasma (PRP) (Tasci et al., 2017).

(A) Reprint with permission from Cheng, R., Huang, W., Huang, L., Yang, B., Mao, L., Jin, K., ZhuGe Q., & Zhao, Y. (2014). Acceleration of tissue plasminogen activator-mediated thrombolysis by magnetically powered nanomotors. ACS Nano, 8(8), 7746–7754. Copyright 2014 American Chemical Society. (B) Reprint with permission from Hu, J., Huang, W., Huang, S., ZhuGe, Q., Jin, K., & Zhao, Y. (2016). Magnetically active Fe3O4 nanorods loaded with tissue plasminogen activator for enhanced thrombolysis. Nano Research, 9(9), 2652–2661. Copyright 2016 Springer. (C) and (D) Reprint with permission from Tasci, T. O., Disharoon, D., Schoeman, R. M., Rana, K., Herson, P. S., Marr, D. W., & Neeves, K. B. (2017). Enhanced fibrinolysis with magnetically powered colloidal microwheels. Small, 13(36), 1700954. Copyright 2017 John Wiley & Sons, Inc.

were introduced: direct motion (the trajectory of the microwheels was a straight line) and corkscrew motion (the trajectory was a helical curve). The authors showed that the corkscrew motion induced a significant mechanical penetration of microwheels into the fibrin gels and platelet-rich thrombi and greatly improved the fibrinolysis rate. The fibrinolysis rate of the microwheels with corkscrew trajectories is twice the speed of the microwheels with direction motion and is fivefold faster than that with 1 μg/mL rt-PA (Fig. 15.6D). Thus the combination of increased rt-PA release, enhanced mass transport as well as mechanical destruction could all play a role to greatly improve the thrombolytic efficiency (Tasci et al., 2017).

The enhanced thrombolysis by magnetically activated and rt$-$PA-functionalized NPs is further demonstrated via in vivo animal experiments. Hu et al. functionalized rt-PA into porous Fe_3O_4-microrods with a mass loading ratio $\sim 12.9\%$ to achieve targeted thrombolytic therapy in ischemic stroke induced by distal middle cerebral artery occlusion (Hu et al., 2018). Once the desired amount of rt$-$PA-functionalized Fe_3O_4-microrod was injected via the vein of the mouse, a permanent magnetic bar was placed around the clot location to concentrate the microrods to the desired location. Both the magnetic concentration (targeted delivery) and microrod rotation by external magnetic could significantly reduce the total doses of rt-PA while achieving a similar clot lysis efficiency. The results showed that with only 1 mg/kg of rt-PA-microrods, it took 25 minutes to completely lyse the thrombus (the actual rt-PA concentration was only about 0.13 mg/kg). Compared to a rt-PA-only experiment (10 mg/kg, 85-minute lysis time), the amount of rt-PA was reduced by almost 100 times while the lysis speed increased more than three times. Such an effect could dramatically reduce the risk of rt-PA-mediated hemorrhagic complications. In vivo toxicity analysis indicated that the biomarkers for hepatic toxicity were not significantly altered compared from normal levels up to 7 days after intravenous administration of the rt-PA-microrods (10 mg/kg). The microrods were barely detectable in spleen, liver, and kidney, and they were discharged from the kidneys through urine (Hu et al., 2018).

Ultrasound radiation can also be used to induce active motion of nanoparticles (Hansen-Bruhn et al., 2018) and has recently been applied to enhance mass transport for stroke treatment. Ultrasound irradiation can activate stable oscillations of micro-/nanobubbles, that is, inertial expansion and shrinking as well as explosion, which facilitates local motion of the bubbles and induces bubble penetration into clots to improve the clot lysis rate. Ma et al. used ultrasound for the deep penetration of targeted nanobubbles with "cavitation effect" to enhance the transport of uPA into deep areas of the clot (Ma et al., 2019). However, this cavitation effect requires a high acoustic-driven force, which could introduce the risk of tissue injury. In the work by Wang, Guo, et al. (2020) low-intensity ultrasound was used to introduce stable oscillations of magnetic microbubbles incorporated with rt-PA. Such oscillations facilitated deeper penetration of rt-PA into the thrombi. When irrigated by low-intensity ultrasound (0.2 bar) for stable oscillations, it was found that the microbubbles could penetrate up to 1 cm into agarose-fibrin gel. In a mouse model of venous thrombosis, the residual thrombus decreased by 67.5% when compared to conventional injection of rt-PA. The improved penetration was evidenced by the histological images, where a relatively homogenous distribution of NPs within the clots was observed. The penetration of rt-PA by ultrasound was up to several hundred micrometers in thrombi, and the lysis happened not only at the interface of a clot but also at many interior locations, resulting in accelerated thrombolysis (Wang, Guo, et al., 2020).

Light has also been reported to activate micro-/nanomotors and is beneficial for active drug delivery (Xu et al., 2017; Wang et al., 2018; Ou et al., 2020). Wan et al. developed the porous silica/platinum nanomotors with platelet membrane modification for sequentially targeted delivery of thrombolytic and anticoagulant drugs for thrombus treatment (Wan et al., 2020). Due to the thermophoresis caused by the resonance absorption of NIR irradiation by Pt NPs asymmetrically located in the nanomotors, the motion speeds of the nanomotors increased with the NIR irradiation power, which could enhance the mass transport of thrombolytic drug and effectively promote drug penetration in thrombus site. In addition, after the nanomotors targeted the thrombus site, the platelet membranes were ruptured under NIR irradiation to achieve desirable sequential drug release, including rapid release of uPA (3 hours) and slow release of anticoagulant heparin (>20 days). Results from animal experiments confirmed that these functionalized nanomotors had good thrombolytic performance in vivo and could dissolve thrombus volume to less than 5% within 7 days (Wan et al., 2020).

15.5 Other strategies

While drug delivery-based thrombolysis is the primary clinical therapy for stroke treatment, due to their increased side effects, other drug-free-based treatment strategies have also been developed. In this section, nanotechnology-based drug-free thrombolytic therapies, including mechanical destruction, bubble burst, photothermal thrombus ablation, and ROS-mediated photodynamic therapy, will be discussed, and their corresponding principles are illustrated in Fig. 15.7.

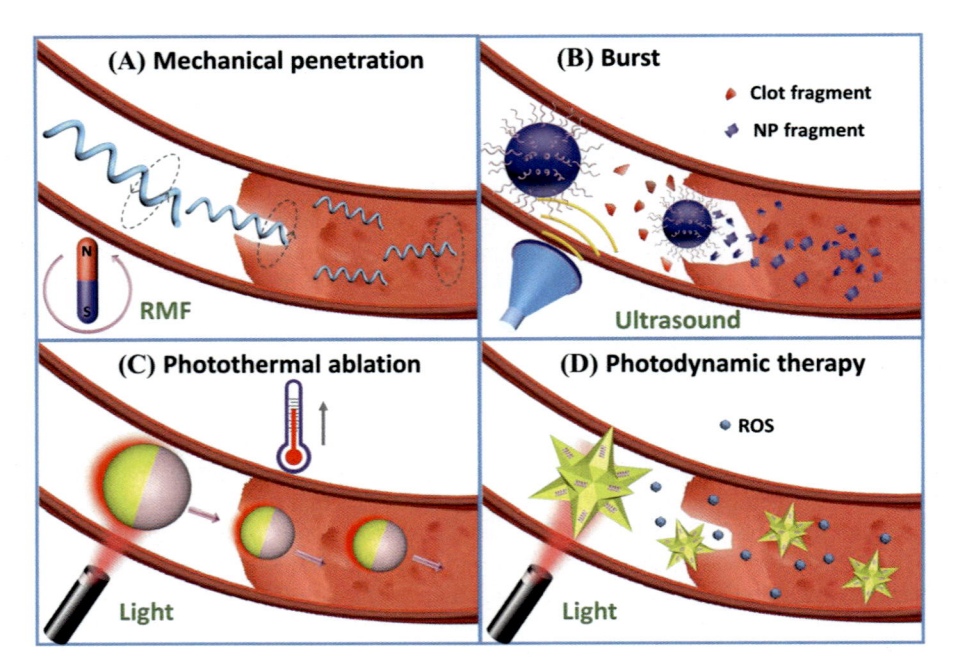

FIGURE 15.7

Schematic representation of drug-free stroke treatment: (A) mechanical penetration and destruction, (B) ultrasound-induced micro-/nanobubble burst, (C) photothermal ablation, and (D) photodynamic therapy.

15.5.1 **Mechanical destruction**

In the experiment by Hu et al., blood clot lysis experiments were compared in catheter for Fe_3O_4 MNRs with and without an RMF. They concluded that the rotating MNRs could mechanically destruct the clot network, which provided an additional mechanism for improving the lysis efficiency (Hu, Huang, et al., 2016). In addition, Tasci et al. demonstrated that the corkscrew motion of rt−PA-functionalized microwheels induced a significant mechanical penetration of microwheels into the fibrin gels and platelet-rich thrombi and greatly improved the fibrinolysis rate (Tasci et al., 2017). The microwheels following helical trajectories could penetrate through 100-μm-sized platelet-rich thrombi in a microfluidic model of hemostasis in ≈ 5 minutes. These results show that moved NPs can mechanically destruct the clot network, facilitating thrombolysis. Based on this principle, Khalil et al. designed a closed-loop motion control system to position and propel a magnetic helical robot to remove superficial blood clots, and the principle is illustrated in Fig. 15.7A (Khalil et al., 2019). They showed that the magnetic localization of a helical robot with a diameter of 1 mm could achieve a mean absolute position error of 2.35 ± 0.4 mm. In an in vitro catheter experiment with a 1-h-old blood clot, the rotation helical robot could achieve an average clot removal rate of 0.67 ± 0.47 mm^3/minutes, while for an ex vivo model inside a rabbit aorta, the helical robot could attain an average translational speed of 4.4 mm/second against a flow with a flow rate of 10 mL/hour, and a mean absolute position error of 2.6 mm (Khalil et al., 2019).

15.5.2 **NP/bubble burst**

As discussed in Section 15.3, ultrasound can be used for targeted drug release based on its cavitation effect. The periodic pressure changes around a NP or a microbubble can induce a local explosion, thus achieving drug release. Similar principles can be applied for clot removal if the NP or microbubble is located inside a blood clot, as shown in Fig. 15.7B. Wei et al. demonstrated that the NPs with a perfluorohexane (PFH) core and an Au NPs shell could absorb a 1064-nm laser pulse with a fluence under 70 mJ/cm^2, and create cavitation (Wei et al., 2014). The ultrasound images showed that at the NP suspension and blood clot interface, bubbles were generated under laser irradiation, and cavitation energy release ruptured the clot surface and broke fibrin structures, resulting in small pores at the interface which helped the suspension to further penetrate into the clot (Wei et al., 2014). The PFH is also a phase transition material for ultrasound. When PFH-containing NPs are triggered by ultrasound, the material changes from a liquid phase to a gas phase, inducing a significant volume expansion, or even a subsequent explosion of the NPs. Zhong et al. designed PFH encapsulated magnetic NPs functionalized with CREKA peptide (Cys-Arg-Glu-Lys-Ala) which can specifically target fibrin, for thrombolysis (Zhong et al., 2019). Under a low-intensity focused ultrasound, the volume of the NPs increased nearly 300-fold, and eventually vaporized. The thrombolytic rates were significantly improved for NPs with PFH compared to similar NPs without PFH encapsulation under the ultrasound power (Zhong et al., 2019). Such an improvement could be due to enhanced microstreaming around the thrombus, possibly loosening the structure, allowing additional NPs permeate into the thrombus, and continuing to disrupt the surrounding clot.

15.5.3 Photothermal ablation

Since blood clots are networked by fibrin fibers and the fibrin monomers are linked by noncovalent bonds to form the thrombus (Mosesson, 2005), not only can the direct mechanical destruction or NP/bubble burst remove or destroy fibrin network, but exposure to photothermal heating could also lyse the noncovalent interactions in the thrombus, resulting in loosening and downsizing of the clot. Singh et al. pioneered this idea and demonstrated the photothermal thrombus ablation with Au nanorods exposed to NIR irradiation, both in vitro from purified fibrinogen and in vivo in murine blood vessels (Singh et al., 2016). By NIR laser irradiation for 45 minutes in the presence of 18 μg/mL Au nanorods, the local temperature could increase up to $50°C-55°C$ and induce a significant photothermal fibrinolysis up to $16\% \pm 4\%$. Under arterial and venous shear, $18\% \pm 3\%$ and $7\% \pm 3\%$ fibrinolysis were achieved, respectively. With the addition of 50 U streptokinase, the lysis rate increased to $41\% \pm 5\%$ and $20\% \pm 3\%$, respectively (Singh et al., 2016). Dong et al. further confirmed the results by using Au nanorods irradiated with an 808 nm NIR laser (Dong et al., 2018). Shao et al. developed a rather complicated erythrocyte membrane-cloaked Janus polymeric motor asymmetrically coated with Au NPs which could propelled by NIR laser irradiation due to thermophoresis (Shao et al., 2018). The motion of the motors could be controlled by the NIR laser intensity, and could be navigated through the biological environments such as cell culture, serum, and blood. In addition, upon NIR irradiation, the motors were disrupted to release heparin, resulting in enhanced photothermal-mediated thrombolysis (Shao et al., 2018).

15.5.4 Photodynamic therapy

As discussed in Section 15.3, during the formation of a thrombus, local ROS levels will be upregulated. Thus by changing the local ROS environment around the thrombus, it is possible to destabilize the fibrin network or even break the fibrin skeleton of the blood clot. The generation of ROS species is well-known in the photocatalyst community and is therefore used to lyse blood clots. Such a process is called photodynamic therapy (PDT). Zhang et al. demonstrated this idea using mesoporous carbon nanospheres prepared from a metal—organic framework precursor, which possessed both PTT and PDT properties due to their porphyrin-like metal centers (Zhang, Liu, et al., 2019). When these NPs were intravenously injected into a mouse model of thrombus, the NPs actively targeted the glycoprotein IIb/IIIa receptors on the thrombus and accumulated in the thrombosed area. Irradiation by an 808-nm laser not only increased the temperature of the thrombotic sites noticeably, but also revealed ROS generation by a 2,7-dichlorodihydrofluorescein diacetate probe, indicating both the PTT and PDT effects. During these combined treatments, platelet factor 3 was damaged, the red cells underwent apoptosis, and the fibrin skeleton of the blood clot was broken. Secondary embolism was prevented due to the full breakage of the blood clots. The experiments show that for 10 minutes and 2 W/cm^2 NIR laser irradiation, a penetration length of 26.3 mm into the thrombus was achieved under the PTT/PDT synergetic therapy (Zhang, Liu, et al., 2019).

15.6 Conclusions and perspectives

Clearly, with the rapid development of biomedical nanotechnology and materials science, remarkable progress has been made for nanotechnology-based stroke treatment. From a systematic point of view, an ideal nanotechnology-based stroke treatment technology should include four major steps: (1) nanocarrier/NP preparation; (2) nanocarrier/NP transport or navigation and concentration in human body; (3) on-site thrombolysis; and (4) the discharge or biodegradation of the nanocarrier/NP or their fragments in the human body. This chapter only partially discusses Step 1 and Step 3. Step 2 is partially discussed in Chapters 4, 5, and 7, while most researchers are only concerned with the toxicity in Step 4 (Hoet et al., 2004; Chenthamara et al., 2019). Even for Steps 1 and 3, despite the rapid development in bionanotechnology, there are still many challenges that remain for real clinical application. First, the synthesis of drug-loaded NPs needs to meet the following criteria: (1) high drug-loading ratio; (2) ability to be concentrated or specifically target the thrombus; (3) stimulated release capability after targeting or concentration; and (4) the remains of the NPs after drug release should be biodegradable and biocompatible, or be easily discharged from the human body. In these regards, drug encapsulated particles functionalized with targeting ligands are the choice of selection. The second challenge is how to better control the motion of these NPs inside a human body. As of now, additional interactions via magnetic guidance, light propulsion, or ultrasound control are used and NPs made of materials possessing these functionalities can be added to the drug-loaded NPs via superparticle assembly. Third, an understanding of the stability and transport behaviors of superparticle NPs in complicated blood fluid and confined blood vessels becomes increasingly important. The blood fluid is a complicated environment consisting of RBCs, white blood cells, platelets, proteins, ions, etc., and bounded by the walls of blood vessels. Multiple physical and biochemical interactions, including mechanical, electrostatic/ionic, hydrodynamic, and chemical interactions, will influence the stability and trajectories of the superparticles, which eventually determines the amount of the particles that accumulates at the thrombus site. Fourth, exploration of the physical and chemical reaction mechanisms between the functionalized NPs and the porous clot network is critical to design external interaction or drug release mechanisms. Fifth, investigation of the biocompatibility and possible toxicity of nanocarrier materials, especially the possible long-term toxicity caused by accumulation in the body, is another important issue related to Step 4. Finally, the scale-up industrial production of these nanomedicines and associated equipment or accessories is also one of the biggest obstacles for successful clinical translation.

References

Abdalla, A., & Mäder, K. (2009). ESR studies on the influence of physiological dissolution and digestion media on the lipid phase characteristics of SEDDS and SEDDS pellets. *International Journal of Pharmaceutics*, *367*(1), 29–36.

Adams Harold, P., del Zoppo, G., Alberts Mark, J., Bhatt Deepak, L., Brass, L., Furlan, A., Grubb Robert, L., Higashida Randall, T., Jauch Edward, C., Kidwell, C., Lyden Patrick, D., Morgenstern Lewis, B., Qureshi Adnan, I., Rosenwasser Robert, H., Scott Phillip, A., & Wijdicks Eelco, F. M. (2007). Guidelines for the early management of adults with ischemic stroke. *Stroke*, *38*(5), 1655–1711.

Alexandrov, A. V., Molina, C. A., Grotta, J. C., Garami, Z., Ford, S. R., Alvarez-Sabin, J., Montaner, J., Saqqur, M., Demchuk, A. M., Moyé, L. A., Hill, M. D., & Wojner, A. W.CLOTBUST Investigators. (2004). Ultrasound-enhanced systemic thrombolysis for acute ischemic stroke. *New England Journal of Medicine*, *351*(21), 2170–2178.

Alipour, M., Baneshi, M., Hosseinkhani, S., Mahmoudi, R., Jabari Arabzadeh, A., Akrami, M., Mehrzad, J., & Bardania, H. (2020). Recent progress in biomedical applications of RGD-based ligand: From precise cancer theranostics to biomaterial engineering: A systematic review. *Journal of Biomedical Materials Research Part A, 108*(4), 839–850.

Bachelet, L., Bertholon, I., Lavigne, D., Vassy, R., Jandrot-Perrus, M., Chaubet, F., & Letourneur, D. (2009). Affinity of low molecular weight fucoidan for P-selectin triggers its binding to activated human platelets. *Biochimica et Biophysica Acta, 1790*(2), 141–146.

Balasubramanian, V., Domanskyi, A., Renko, J. M., Sarparanta, M., Wang, C. F., Correia, A., Makila, E., Alanen, O. S., Salonen, J., Airaksinen, A. J., Tuominen, R., Hirvonen, J., Airavaara, M., & Santos, H. A. (2020). Engineered antibody-functionalized porous silicon nanoparticles for therapeutic targeting of pro-survival pathway in endogenous neuroblasts after stroke. *Biomaterials, 227*119556.

Bannish, B. E., Chernysh, I. N., Keener, J. P., Fogelson, A. L., & Weisel, J. W. (2017). Molecular and physical mechanisms of fibrinolysis and thrombolysis from mathematical modeling and experiments. *Scientific Reports, 7*(1), 6914.

Bao, Q., Hu, P., Xu, Y., Cheng, T., Wei, C., Pan, L., & Shi, J. (2018). Simultaneous blood-brain barrier crossing and protection for stroke treatment based on edaravone-loaded ceria nanoparticles. *ACS Nano, 12*(7), 6794–6805.

Bhattacharjee, P., Bhattacharyya, D., & Kolev, K. J. F. (2014). Thrombolysis. *An insight into the abnormal fibrin clots—Its pathophysiological roles*, 1–29.

Biener, J., Wittstock, A., Baumann, T., Weissmüller, J., Bäumer, M., & Hamza, A. (2009). Surface chemistry in nanoscale materials. *Materials, 2*(4), 2404–2428.

Brott, T., & Bogousslavsky, J. (2000). Treatment of acute ischemic stroke. *New England Journal of Medicine, 343*(10), 710–722.

Bruch, G. E., Fernandes, L. F., Bassi, B. L., Alves, M. T. R., Pereira, I. O., Frézard, F., & Massensini, A. R. (2019). Liposomes for drug delivery in stroke. *Brain Research Bulletin, 152*, 246–256.

Cesarman-Maus, G., & Hajjar, K. A. (2005). Molecular mechanisms of fibrinolysis. *British Journal of Haematology, 129*(3), 307–321.

Chen, K., Wang, Y., Liang, H., Xia, S., Liang, W., Kong, J., Liang, Y., Chen, X., Mao, M., Chen, Z., Bai, X., Zhang, J., Li, J., Chang, Y. N., Li, J., & Xing, G. (2020). Intrinsic biotaxi solution based on blood cell membrane cloaking enables fullerenol thrombolysis in vivo. *ACS Applied Materials Interfaces, 12*(13), 14958–14970.

Cheng, R., Huang, W., Huang, L., Yang, B., Mao, L., Jin, K., ZhuGe, Q., & Zhao, Y. (2014). Acceleration of tissue plasminogen activator-mediated thrombolysis by magnetically powered nanomotors. *ACS Nano, 8*(8), 7746–7754.

Cheng, Y., Doane, T. L., Chuang, C. H., Ziady, A., & Burda, C. (2014). Near infrared light-triggered drug generation and release from gold nanoparticle carriers for photodynamic therapy. *Small, 10*(9), 1799–1804.

Chenthamara, D., Subramaniam, S., Ramakrishnan, S. G., Krishnaswamy, S., Essa, M. M., Lin, F.-H., & Qoronfleh, M. W. (2019). Therapeutic efficacy of nanoparticles and routes of administration. *Biomaterials Research, 23*(1), 20.

Chou, L. Y., Ming, K., & Chan, W. C. (2011). Strategies for the intracellular delivery of nanoparticles. *Chemical Society Reviews, 40*(1), 233–245.

Colasuonno, M., Palange, A. L., Aid, R., Ferreira, M., Mollica, H., Palomba, R., Emdin, M., Del Sette, M., Chauvierre, C., & Letourneur, D. (2018). Erythrocyte-inspired discoidal polymeric nanoconstructs carrying tissue plasminogen activator for the enhanced lysis of blood clots. *ACS Nano, 12*(12), 12224–12237.

Collet, J. P., Park, D., Lesty, C., Soria, J., Soria, C., Montalescot, G., & Weisel, J. W. (2000). Influence of fibrin network conformation and fibrin fiber diameter on fibrinolysis speed. *Arteriosclerosis, Thrombosis, and Vascular Biology, 20*(5), 1354–1361.

Correa-Paz, C., Poupard, M. F. N., Polo, E., Rodríguez-Pérez, M., Taboada, P., Iglesias-Rey, R., Hervella, P., Sobrino, T., Vivien, D., & Castillo, J. (2019). In vivo ultrasound-activated delivery of recombinant tissue plasminogen activator from the cavity of sub-micrometric capsules. *Journal of Controlled Release*, *308*, 162−171.

Cui, W., Liu, R., Jin, H., Lv, P., Sun, Y., Men, X., Yang, S., Qu, X., Yang, Z., & Huang, Y. (2016). pH gradient difference around ischemic brain tissue can serve as a trigger for delivering polyethylene glycol-conjugated urokinase nanogels. *Journal of Controlled Release*, *225*, 53−63.

Dayal, S., Wilson Katina, M., Motto David, G., Miller Francis, J., Chauhan Anil, K., & Lentz Steven, R. (2013). Hydrogen peroxide promotes aging-related platelet hyperactivation and thrombosis. *Circulation*, *127*(12), 1308−1316.

Del Principe, D., Menichelli, A., De Matteis, W., Di Corpo, M. L., Di Giulio, S., & Finazzi-Agro, A. (1985). Hydrogen peroxide has a role in the aggregation of human platelets. *FEBS Letters*, *185*(1), 142−146.

Dong, L., Liu, X., Wang, T., Fang, B., Chen, J., Li, C., Miao, X., Wei, C., Yu, F., & Xin, H. (2018). Localized light-Au-hyperthermia treatment for precise, rapid, and drug-free blood clot lysis. *ACS Applied Materials Interfaces*, *11*(2), 1951−1956.

Drozdov, A. S., Vinogradov, V. V., Dudanov, I. P., & Vinogradov, V. V. (2016). Leach-proof magnetic thrombolytic nanoparticles and coatings of enhanced activity. *Scientific Reports*, *6*, 28119.

Eustis, S., & El-Sayed, M. A. (2006). Why gold nanoparticles are more precious than pretty gold: Noble metal surface plasmon resonance and its enhancement of the radiative and nonradiative properties of nanocrystals of different shapes. *Chemical Society Reviews*, *35*(3), 209−217.

Ferreira, C. A., Ni, D., Rosenkrans, Z. T., & Cai, W. (2018). Scavenging of reactive oxygen and nitrogen species with nanomaterials. *Nano Research*, *11*(10), 4955−4984.

Freedman Jane, E. (2008). Oxidative stress and platelets. *Arteriosclerosis, Thrombosis, and Vascular Biology*, *28*(3), s11−s16.

Gallo, M., Ferrara, L., & Naviglio, D. (2018). Application of ultrasound in food science and technology: A perspective. *Foods*, *7*(10), 164.

Gao, W., Hu, C.-M. J., Fang, R. H., & Zhang, L. (2013). Liposome-like nanostructures for drug delivery. *Journal of Materials Chemistry B*, *1*(48), 6569−6585.

Group, N. I. O. N. D. S. R.-P. S. S. (1995). Tissue plasminogen activator for acute ischemic stroke. *New England Journal of Medicine*, *333*(24), 1581−1588.

Gu, B., Piebalgs, A., Huang, Y., Longstaff, C., Hughes, A. D., Chen, R., Thom, S. A., & Xu, X. Y. (2019). Mathematical modelling of intravenous thrombolysis in acute ischaemic stroke: Effects of dose regimens on levels of fibrinolytic proteins and clot lysis time. *Pharmaceutics*, *11*, 3.

Hansen-Bruhn, M., de Ávila, B. E.-F., Beltrán-Gastélum, M., Zhao, J., Ramírez-Herrera, D. E., Angsantikul, P., Vesterager Gothelf, K., Zhang, L., & Wang, J. (2018). Active intracellular delivery of a Cas9/sgRNA complex using ultrasound-propelled nanomotors. *Angewandte Chemie International Edition*, *57*(10), 2657−2661.

Harold, P., Adams, J., Zoppo, G., & Alberts, M. J. S. (2007). Guidelines for the early management of adults with ischemic stroke. *Stroke*, *38*(5), 1655−1711.

Hoet, P. H. M., Brüske-Hohlfeld, I., & Salata, O. V. (2004). Nanoparticles—Known and unknown health risks. *Journal of Nanobiotechnology*, *2*(1), 12.

Hsu, B. B., Conway, W., Tschabrunn, C. M., Mehta, M., Perez-Cuevas, M. B., Zhang, S., & Hammond, P. T. (2015). Clotting mimicry from robust hemostatic bandages based on self-assembling peptides. *ACS Nano*, *9*(9), 9394−9406.

Hu, J., Huang, S., Zhu, L., Huang, W., Zhao, Y., Jin, K., & ZhuGe, Q. (2018). Tissue plasminogen activator-porous magnetic microrods for targeted thrombolytic therapy after ischemic stroke. *ACS Applied Materials Interfaces*, *10*(39), 32988−32997.

Hu, J., Huang, W., Huang, S., ZhuGe, Q., Jin, K., & Zhao, Y. (2016). Magnetically active Fe_3O_4 nanorods loaded with tissue plasminogen activator for enhanced thrombolysis. *Nano Research*, *9*(9), 2652−2661.

Hu, Q., Qian, C., Sun, W., Wang, J., Chen, Z., Bomba, H. N., Xin, H., Shen, Q., & Gu, Z. (2016). Engineered nanoplatelets for enhanced treatment of multiple myeloma and thrombus. *Advanced Materials*, *28*(43), 9573−9580.

Huang, G., Zhou, Z., Srinivasan, R., Penn, M. S., Kottke-Marchant, K., Marchant, R. E., & Gupta, A. S. (2008). Affinity manipulation of surface-conjugated RGD peptide to modulate binding of liposomes to activated platelets. *Biomaterials*, *29*(11), 1676−1685.

Huang, J., Sun, C., Yao, D., Wang, C. Z., Zhang, L., Zhang, Y., Chen, L., & Yuan, C. S. (2018). Novel surface imprinted magnetic mesoporous silica as artificial antibodies for efficient discovery and capture of candidate nNOS-PSD-95 uncouplers for stroke treatment. *Journal of Material Chemistry B*, *6*(10), 1531−1542.

Huang, M., Zhang, S. F., Lu, S., Qi, T., Yan, J., Gao, C., Liu, M., Li, T., & Ji, Y. (2019). Synthesis of mesoporous silica/polyglutamic acid peptide dendrimer with dual targeting and its application in dissolving thrombus. *Journal of Biomedical Materials Research Part A*, *107*(8), 1824−1831.

Huang, X., Jain, P. K., El-Sayed, I. H., & El-Sayed, M. A. (2007). Plasmonic photothermal therapy (PPTT) using gold nanoparticles. *Lasers in Medical Science*, *23*(3), 217.

Huang, Y., Yu, L., Ren, J., Gu, B., Longstaff, C., Hughes, A. D., Thom, S. A., Xu, X. Y., & Chen, R. (2019). An activated-platelet-sensitive nanocarrier enables targeted delivery of tissue plasminogen activator for effective thrombolytic therapy. *Journal of Controlled Release*, *300*, 1−12.

Iuliano, L., Colavita, A. R., Leo, R., Praticò, D., & Violi, F. (1997). Oxygen free radicals and platelet activation. *Free Radical Biology and Medicine*, *22*(6), 999−1006.

Juenet, M., Aid-Launais, R., Li, B., Berger, A., Aerts, J., Ollivier, V., Nicoletti, A., Letourneur, D., & Chauvierre, C. (2018). Thrombolytic therapy based on fucoidan-functionalized polymer nanoparticles targeting P-selectin. *Biomaterials*, *156*, 204−216.

Jung, E., Kang, C., Lee, J., Yoo, D., Hwang, D. W., Kim, D., Park, S.-C., Lim, S. K., Song, C., & Lee, D. (2018). Molecularly engineered theranostic nanoparticles for thrombosed vessels: H_2O_2-activatable contrast-enhanced photoacoustic imaging and antithrombotic therapy. *ACS Nano*, *12*(1), 392−401.

Kang, C., Gwon, S., Song, C., Kang, P. M., Park, S.-C., Jeon, J., Hwang, D. W., & Lee, D. (2017). Fibrin-targeted and H_2O_2-responsive nanoparticles as a theranostics for thrombosed vessels. *Acs Nano*, *11*(6), 6194−6203.

Khalil, I. S., Adel, A., Mahdy, D., Micheal, M. M., Mansour, M., Hamdi, N., & Misra, S. (2019). Magnetic localization and control of helical robots for clearing superficial blood clots. *APL Bioengineering*, *3*(2), 026104.

Kim, W., Haller, C., Dai, E., Wang, X., Hagemeyer, C. E., Liu, D. R., Peter, K., & Chaikof, E. L. (2015). Targeted antithrombotic protein micelles. *Angewandte Chemie International (Edition)*, *127*(5), 1481−1485.

Korin, N., Kanapathipillai, M., Matthews, B. D., Crescente, M., Brill, A., Mammoto, T., Ghosh, K., Jurek, S., Bencherif, S. A., & Bhatta, D. (2012). Shear-activated nanotherapeutics for drug targeting to obstructed blood vessels. *Science*, *337*(6095), 738−742.

Kumar, C. S. S. R., & Mohammad, F. (2011). Magnetic nanomaterials for hyperthermia-based therapy and controlled drug delivery. *Advanced Drug Delivery Reviews*, *63*(9), 789−808.

Laverdure, A. M., Surbeck, J., North, M. O., & Tritto, J. (2001). Growth, development, reproduction, physiological and behavioural studies on living organisms, human adults and children exposed to radiation from video displays. *Indoor and Built Environment*, *10*(5), 306−309.

Li, B., Chen, R., Zhang, Y., Zhao, L., Liang, H., Yan, Y., Tan, H., Nan, D., Jin, H., & Huang, Y. (2018). RGD modified protein−polymer conjugates for pH-triggered targeted thrombolysis. *ACS Applied Bio Materials*, *2*(1), 437−446.

Li, C., Du, H., Yang, A., Jiang, S., Li, Z., Li, D., Brash, J. L., & Chen, H. (2017). Thrombosis-responsive thrombolytic coating based on thrombin-degradable tissue plasminogen activator (t-PA) nanocapsules. *Advanced Functional Materials*, *27*(45)1703934.

Li, M., Li, J., Chen, J., Liu, Y., Cheng, X., Yang, F., & Gu, N. (2020). Platelet membrane biomimetic magnetic nanocarriers for targeted delivery and in situ generation of nitric oxide in early ischemic stroke. *ACS Nano*, *14*(2), 2024−2035.

Li, Y., Zhang, X., Qi, Z., Guo, X., Liu, X., Shi, W., Liu, Y., & Du, L. (2020). The enhanced protective effects of salvianic acid A: A functionalized nanoparticles against ischemic stroke through increasing the permeability of the blood-brain barrier. *Nano Research*, *13*(10), 2791−2802.

Lu, Y., Li, C., Chen, Q., Liu, P., Guo, Q., Zhang, Y., Chen, X., Zhang, Y., Zhou, W., Liang, D., Zhang, Y., Sun, T., Lu, W., & Jiang, C. (2019). Microthrombus-targeting micelles for neurovascular remodeling and enhanced microcirculatory perfusion in acute ischemic stroke. *Advanced Materials*, *31*(21), e1808361.

Lv, W., Xu, J., Wang, X., Li, X., Xu, Q., & Xin, H. (2018). Bioengineered boronic ester modified dextran polymer nanoparticles as reactive oxygen species responsive nanocarrier for ischemic stroke treatment. *ACS Nano*, *12*(6), 5417−5426.

Ma, L., Wang, Y., Zhang, S., Qian, X., Xue, N., Jiang, Z., Akakuru, O. U., Li, J., Xu, Y., & Wu, A. (2019). Deep penetration of targeted nanobubbles enhanced cavitation effect on thrombolytic capacity. *Bioconjugate Chemistry*, *31*(22), 369−374.

Mauricio, M. D., Guerra-Ojeda, S., Marchio, P., Valles, S. L., Aldasoro, M., Escribano-Lopez, I., Herance, J. R., Rocha, M., Vila, J. M., & Victor, V. M. (2018). Nanoparticles in medicine: A focus on vascular oxidative stress. *Oxidative Medicine and Cellular Longevity*, *2018*, 6231482.

Mei, T., Kim, A., Vong, L. B., Marushima, A., Puentes, S., Matsumaru, Y., Matsumura, A., & Nagasaki, Y. (2019). Encapsulation of tissue plasminogen activator in pH-sensitive self-assembled antioxidant nanoparticles for ischemic stroke treatment—Synergistic effect of thrombolysis and antioxidant. *Biomaterials*, *215*, 119209.

Mosesson, M. W. (2005). Fibrinogen and fibrin structure and functions. *Journal of Thrombosis and Haemostasis*, *3*, 1894−1904.

Niu, Y., Tan, H., Li, X., Zhao, L., Xie, Z., Zhang, Y., Zhou, S., & Qu, X. (2020). Protein−carbon dot nanohybrid-based early blood−brain barrier damage theranostics. *ACS Applied Materials Interfaces*, *12*(3), 3445−3452.

Ou, J., Liu, K., Jiang, J., Wilson, D. A., Liu, L., Wang, F., Wang, S., Tu, Y., & Peng, F. (2020). Micro-/nanomotors toward biomedical applications: The recent progress in biocompatibility. *Small*, *16*(27), 1906184.

Park, J., Singha, K., Son, S., Kim, J., Namgung, R., Yun, C. O., & Kim, W. J. (2012). A review of RGD-functionalized nonviral gene delivery vectors for cancer therapy. *Cancer Gene Therapy*, *19*(11), 741−748.

Pawlowski, C. L., Li, W., Sun, M., Ravichandran, K., Hickman, D., Kos, C., Kaur, G., & Gupta, A. S. (2017). Platelet microparticle-inspired clot-responsive nanomedicine for targeted fibrinolysis. *Biomaterials*, *128*, 94−108.

Pitt, W. G., Husseini, G. A., & Staples, B. J. (2004). Ultrasonic drug delivery—A general review. *Expert Opinion on Drug Delivery*, *1*(1), 37−56.

Qin, Z., & Bischof, J. C. (2012). Thermophysical and biological responses of gold nanoparticle laser heating. *Chemical Society Reviews*, *41*(3), 1191−1217.

Rao, S., Chen, R., LaRocca, A. A., Christiansen, M. G., Senko, A. W., Shi, C. H., Chiang, P.-H., Varnavides, G., Xue, J., Zhou, Y., Park, S., Ding, R., Moon, J., Feng, G., & Anikeeva, P. (2019). Remotely controlled chemomagnetic modulation of targeted neural circuits. *Nature Nanotechnology*, *14*, 967−973.

Rizvi, S. A. A., & Saleh, A. M. (2018). Applications of nanoparticle systems in drug delivery technology. *Saudi Pharmaceutical Journal*, *26*(1), 64−70.

Saboor, M., Ayub, Q., Ilyas, S., & Moinuddin. (2013). Platelet receptors: An instrumental of platelet physiology. *Pakistan Journal of Medical. Science*, *29*(3), 891−896.

Shao, J., Abdelghani, M., Shen, G., Cao, S., Williams, D. S., & van Hest, J. C. (2018). Erythrocyte membrane modified janus polymeric motors for thrombus therapy. *ACS Nano*, *12*(5), 4877−4885.

Singh, N., Varma, A., Verma, A., Maurya, B. N., & Dash, D. (2016). Relief from vascular occlusion using photothermal ablation of thrombus with a multimodal perspective. *Nano Research*, *9*(8), 2327−2337.

Sun, T., Zhang, Y. S., Pang, B., Hyun, D. C., Yang, M., & Xia, Y. (2014). Engineered nanoparticles for drug delivery in cancer therapy. *Angewandte Chemie International (Edition)*, *53*(46), 12320−12364.

Tasci, T. O., Disharoon, D., Schoeman, R. M., Rana, K., Herson, P. S., Marr, D. W., & Neeves, K. B. (2017). Enhanced fibrinolysis with magnetically powered colloidal microwheels. *Small*, *13*(36), 1700954.

Voros, E., Cho, M., Ramirez, M., Palange, A. L., De Rosa, E., Key, J., Garami, Z., Lumsden, A. B., & Decuzzi, P. (2015). TPA immobilization on iron oxide nanocubes and localized magnetic hyperthermia accelerate blood clot lysis. *Advanced Functional Materials*, *25*(11), 1709−1718.

Wan, M., Wang, Q., Wang, R., Wu, R., Li, T., Fang, D., Huang, Y., Yu, Y., Fang, L., & Wang, X. (2020). Platelet-derived porous nanomotor for thrombus therapy. *Scientific Advances*, *6*(22), eaaz9014.

Wang, J., Xiong, Z., Zheng, J., Zhan, X., & Tang, J. (2018). Light-driven micro/nanomotor for promising biomedical tools: Principle, challenge, and prospect. *Accounts of Chemical Research*, *51*(9), 1957−1965.

Wang, M., Zhang, J., Yuan, Z., Yang, W., Wu, Q., & Gu, H. (2012). Targeted thrombolysis by using of magnetic mesoporous silica nanoparticles. *Journal of Biomedical Nanotechnology*, *8*(4), 624−632.

Wang, S., Guo, X., Xiu, W., Liu, Y., Ren, L., Xiao, H., Yang, F., Gao, Y., Xu, C., & Wang, L. (2020). Accelerating thrombolysis using a precision and clot-penetrating drug delivery strategy by nanoparticle-shelled microbubbles. *Scientific Advances*, *6*(31), eaaz8204.

Wang, S., Wang, R., Meng, N., Guo, H., Wu, S., Wang, X., Li, J., Wang, H., Jiang, K., Xie, C., Liu, Y., Wang, H., & Lu, W. (2020). Platelet membrane-functionalized nanoparticles with improved targeting ability and lower hemorrhagic risk for thrombolysis therapy. *Journal of Controlled Release*, *328*, 78−86.

Wang, X., Wei, C., Liu, M., Yang, T., Zhou, W., Liu, Y., Hong, K., Wang, S., Xin, H., & Ding, X. (2017). Near-infrared triggered release of uPA from nanospheres for localized hyperthermia-enhanced thrombolysis. *Advanced Functional Materials*, *27*(40)1701824.

Wei, C. W., Xia, J. J., Lombardo, M., Perez, C., Arnal, B., Larson-Smith, K., Pelivanov, I., Matula, T., Pozzo, L., & O'Donnell, M. (2014). Laser-induced cavitation in nanoemulsion with gold nanospheres for blood clot disruption: In vitro results. *Optical Letters*, *39*(9), 2599−2602.

Xie, M., Zhang, W., Fan, C., Wu, C., Feng, Q., Wu, J., Li, Y., Gao, R., Li, Z., Wang, Q., Cheng, Y., & He, B. (2020). Bioinspired soft microrobots with precise magneto-collective control for microvascular thrombolysis. *Advanced Materials*, *32*, 2000366.

Xu, J., Wang, X., Yin, H., Cao, X., Hu, Q., Lv, W., Xu, Q., Gu, Z., & Xin, H. (2019). Sequentially site-specific delivery of thrombolytics and neuroprotectant for enhanced treatment of ischemic stroke. *ACS Nano*, *13*(8), 8577−8588.

Xu, J., Zhang, Y., Xu, J., Liu, G., Di, C., Zhao, X., Li, X., Li, Y., Pang, N., & Yang, C. (2020). Engineered nanoplatelets for targeted delivery of plasminogen activators to reverse thrombus in multiple mouse thrombosis models. *Advanced Materials*, *32*(4), 1905145.

Xu, J., Zhou, Y., Nie, H., Xiong, Z., OuYang, H., Huang, L., Fang, H., Jiang, H., Huang, F., Yang, Y., Ding, X., Wang, X., & Zhou, W. (2020). Hyperthermia-triggered UK release nanovectors for deep venous thrombosis therapy. *Journal of Material Chemistry B*, *8*(4), 787−793.

Xu, L., Mou, F., Gong, H., Luo, M., & Guan, J. (2017). Light-driven micro/nanomotors: From fundamentals to applications. *Chemical Society of Review*, *46*(22), 6905−6926.

Yang, T., Ding, X., Dong, L., Hong, C., Ye, J., Xiao, Y., Wang, X., & Xin, H. (2018). Platelet-mimic uPA delivery nanovectors based on Au rods for thrombus targeting and treatment. *ACS Biomaterial Science and Engineering*, *4*(12), 4219−4224.

Yao, J., Cheng, Y., Zhou, M., Zhao, S., Lin, S., Wang, X., Wu, J., Li, S., & Wei, H. (2018). ROS scavenging Mn_3O_4 nanozymes for in vivo anti-inflammation. *Chemical Science*, *9*(11), 2927−2933.

Zhang, F., Liu, Y., Lei, J., Wang, S., Ji, X., Liu, H., & Yang, Q. (2019). Metal−organic-framework-derived carbon nanostructures for site-specific dual-modality photothermal/photodynamic thrombus therapy. *Advanced Science*, *6*(17), 1901378.

Zhang, H., Wu, J., Wu, J., Fan, Q., Zhou, J., Wu, J., Liu, S., Zang, J., Ye, J., Xiao, M., Tian, T., & Gao, J. (2019). Exosome-mediated targeted delivery of miR-210 for angiogenic therapy after cerebral ischemia in mice. *Journal of Nanobiotechnology*, *17*(1), 29.

Zhang, N., Li, C., Zhou, D., Ding, C., Jin, Y., Tian, Q., Meng, X., Pu, K., & Zhu, Y. (2018). Cyclic RGD functionalized liposomes encapsulating urokinase for thrombolysis. *Acta Biomaterialia*, *70*, 227−236.

Zhang, S. F., Lü, S., Gao, C., Yang, J., Yan, X., Li, T., Wen, N., Huang, M., & Liu, M. (2018). Multiarm-polyethylene glycol-polyglutamic acid peptide dendrimer: Design, synthesis, and dissolving thrombus. *Journal of Biomedical Material Research Part A*, *106*(6), 1687−1696.

Zhang, S.-F., Lü, S., Yang, J., Huang, M., Liu, Y., & Liu, M. (2020). Synthesis of multiarm peptide dendrimers for dual targeted thrombolysis. *ACS Macro Letters*, *9*(2), 238−244.

Zhang, T., Li, F., Xu, Q., Wang, Q., Jiang, X., Liang, Z., Liao, H., Kong, X., Liu, J., Wu, H., Zhang, D., An, C., Dong, L., Lu, Y., Cao, H., Kim, D., Sun, J., Hyeon, T., Gao, J., & Ling, D. (2019). Ferrimagnetic nanochains-based mesenchymal stem cell engineering for highly efficient post-stroke recovery. *Advanced Functional Materials*, *29*(24), 1900603.

Zhao, Y., Jiang, Y., Lv, W., Wang, Z., Lv, L., Wang, B., Liu, X., Liu, Y., Hu, Q., Sun, W., Xu, Q., Xin, H., & Gu, Z. (2016). Dual targeted nanocarrier for brain ischemic stroke treatment. *Journal of Controlled Release*, *233*, 64−71.

Zhong, Y., Zhang, Y., Xu, J., Zhou, J., Liu, J., Ye, M., Zhang, L., Qiao, B., Wang, Z.-G., & Ran, H.-T. (2019). Low-intensity focused ultrasound-responsive phase-transitional nanoparticles for thrombolysis without vascular damage: A synergistic nonpharmaceutical strategy. *ACS Nano*, *13*(3), 3387−3403.

Nanofiltration for blood products

16

Ibrahim M. Alarifi[1,2]

[1]*Department of Mechanical and Industrial Engineering, College of Engineering, Majmaah University, Al-Majmaah, Saudi Arabia* [2]*Engineering and Applied Science Research Center, Majmaah University, Al-Majmaah, Saudi Arabia*

16.1 Introduction

Blood safety needs to be monitored while conducting blood transfusions. Pathogen inactivation ensures blood safety by strengthening donor screening using different chemical, physical, and photochemical methods for removing the blood-borne pathogens. Nanofiltration is one of the pathogen inactivation methods, besides photochemical inactivation using riboflavin and methylene blue. Transmission of viral infections is among the most common complications associated with blood transfusion. Nanofiltration involves viral removal filtration systems that use nanotechnology to manufacture plasma-derived coagulation factors. Since the 1990s, nanofiltration has been used as a supplement of viral reduction treatment that improves safety of viruses (Jain & Jain, 2008). The viral reduction treatments including the solvent detergent treatment and heat treatment deactivate hepatitis B virus, hepatitis C virus, and HIV virus. The simple steps included in the process of nanofiltration include filtering of protein solution via membranes with pores sized between 15 and 40 nm. These conditions help in retaining viruses via a mechanism that is based on size exclusion. The typical membranes used for nanofiltration are illustrated in Fig. 16.1, which shows tight and loose nanofiltration membranes.

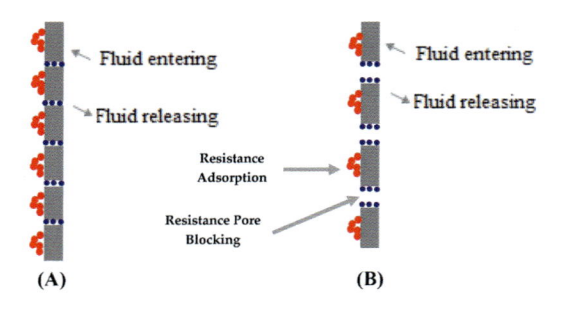

FIGURE 16.1

Typical nanofiltration membrane: (A) tight membrane structure; (B) loose membrane structure.

Nanotechnology for Hematology, Blood Transfusion, and Artificial Blood. DOI: https://doi.org/10.1016/B978-0-12-823971-1.00006-4

The need to incorporate virus contamination control strategies is underscored as virus contamination is dangerous for plasma-derived biopharmaceutical products. The enhancement of product safety improves the plasma donor selection and testing procedures for infectious agents are improved after the implementation of viral inactivation treatments and improved manufacturing practices (Burnouf & Radosevich, 2000). These procedures have eliminated the transmittance of hepatitis C virus, hepatitis B virus, and human immunodeficiency virus; however, hepatitis A virus and parvovirus B19 are not completely eliminated. The reason is that continuous experiments are conducted by manufacturers to develop plasma products that increase safety from viruses. Filtration of plasma protein solutions is among the new procedures that help in retaining viruses through a sieving mechanism. This technology is known as nanofiltration, in which the membranes have pores of a few nanometers that are used to remove viruses. The first application of this technique was observed for filtering concentrates of plasma-derived coagulation factors. However, now this technique is widely used and included as a complementary safety measure for various biopharmaceutical products.

Nanofilters used in nanofiltration process are sterilized and only used once. The mean pore size of these nanofilters is either 15, 35, or 72 nm. Recently, a new filter with mean pore size of 19 nm was introduced, which is capable of clearing 3−7 logs of nonenveloped small viruses from plasma. Straightforward scale-up is permitted through various filter surface areas with a range of 0.001−4 m^2. The structure of the membrane comprises three-dimensional networked pores that are capable of removing viruses through a sieving mechanism (Table 16.1). The process of nanofiltration is validated based on the size of plasma-borne and model viruses. The interconnection of capillaries and voids acting as multilayer (100−200 layers) filters clearly explain the three-dimensional networked pore structure of the sieve membrane (Burnouf & Radosevich, 2003).

An increase in throughput of greater than 95% is achieved for the majority of the plasma proteins due to the hydrophilic nature of cellulose that inhibits the adsorption of proteins. The recovery of protein is based on the surface area of filter, protein concentration, pH, temperature, and

Table 16.1 Sizes of viruses used to validate nanofiltration process (Burnouf & Radosevich, 2003).

Viruses	Size in (nm)
Human immunodeficiency virus	100−120
Hepatitis B virus	50
Hepatitis C virus	40−45
Hepatitis A virus	25−30
Parvovirus B19	18−24
Pseudorabies virus	120−200
Herpes simplex virus	150
Reovirus	60−80
Bovine viral diarrhea virus	40−60
Poliovirus	25−30
Animal parvoviruses	18−24
Encephalomyocarditis virus	25−30

presence of high MW aggregates. Two different types of integrity tests are performed by the users; a gold particle test and a leakage test. The gold particle test uses checks of the distribution of the pores on the filter using specifically calibrated gold particle solutions with specific surface tensions that are linked with virus removal capability. The leakage test checks the absence of membrane defects following the product filtration to remove small nonenveloped viruses. For instance, 20 N filter can remove over 4 logs of nonenveloped viruses; while, fibrinogen and immunoglobulin G are retained. The size of the pore determines the difference in protein recovery and virus removal as it provides flexibility. Fig. 16.2A and B shows the structure of a single device and an enhanced device that assembled several filters of same and different pore sizes. The majority of the devices

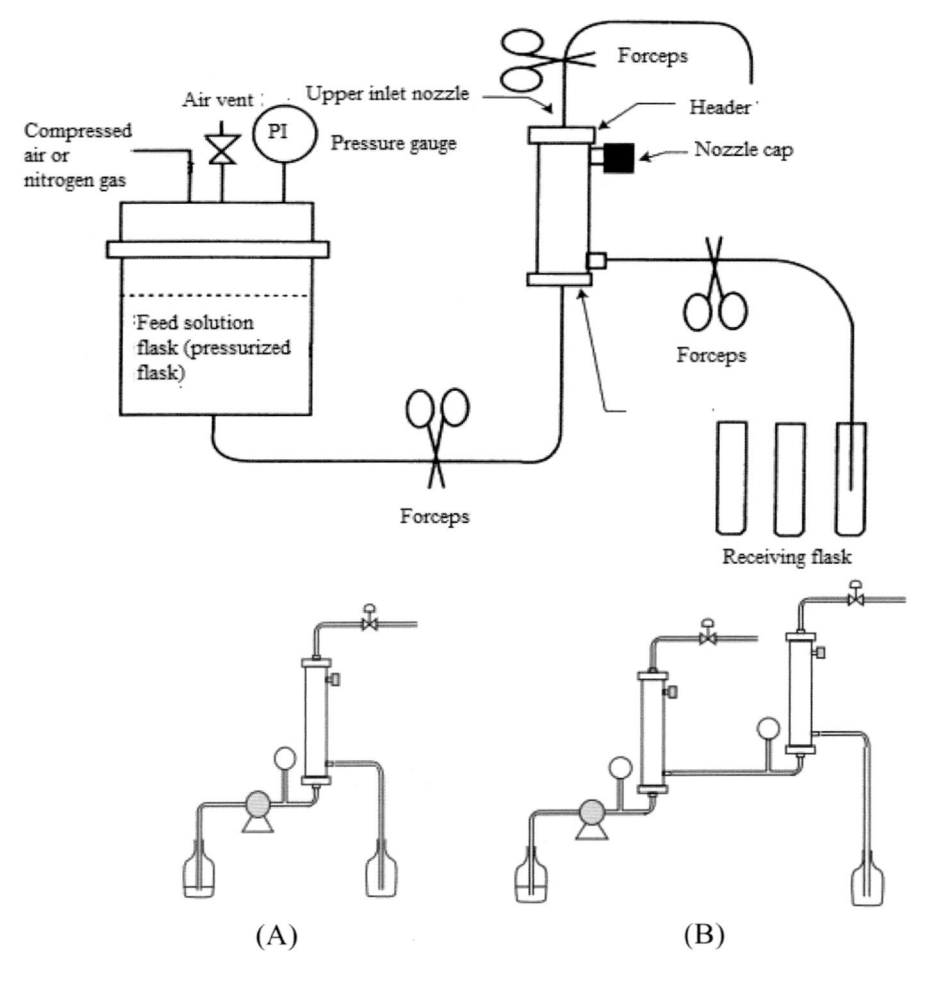

(A)   (B)

FIGURE 16.2

Nanofiltration modes: (A) single; (B) double (Burnouf & Radosevich, 2003).

are manufactured with a preference to the flow-through mode; while others even prefer the cross-flow mode.

16.2 Literature review

The introduction of the nanofiltration procedure daters back to the late 1980s where it was primarily used for removing viruses from biological products. Compromised throughputs and recovery, and extended retention of virus are the main technical issues faced in nanofiltration. It has been shown that the first application of nanofiltration was done by the plasma biopharmaceutical industry for purifying the concentrates of plasma-derived coagulation factors (Factor IX and XI). The rationale for choosing these factors was high filterability and difference of size between the virus and proteins. However, the application of nanofiltration has broadened after recent developments and advancements in qualified nanofiltration membranes. The routine application of this procedure is observed in different plasma products like factor VIII (high-molecular-weight protein fractions), immunoglobulin G, and von Willebrand factor (Burnouf & Radosevich, 2003).

Nanofiltration does not alter protein structure, which is one of the major advantages of this mild processing step. It complements the core viral reduction treatment to improve safety margin of plasma products as it is considered to be a viral removal method following a viral reduction step. It can remove 4−6 logs of enveloped and nonenveloped viruses. Nanofiltration is capable of either reducing or removing prions, which shows that pores of 15−35 nm can remove spiked scrapie agent from the albumin solution (Tateishi et al., 2001). It is not possible to nanofiltrate the whole plasma as it has a complex protein content and composition. The basic mechanism of the plasma nanofiltration mechanism is illustrated in Fig. 16.3.

The systematic analysis of samples obtained during filtration help in evaluating the impact of the filtration process on plasma proteins. There is a need to explore the content of global protein, concentration of relevant plasma proteins, electrophoretic profiles, properties of global coagulation, and activation of coagulation cascade. There is no change in the electrophoretic protein profile; however, the plasma proteins are either diluted or lost during the process of filtration. It is observed that total protein content is pronounced after filtration via 35 nm pore size membrane; whereas total protein content decreases after filtration through 75 nm pore size membrane during leukoreduction. After filtration, the decrease in protein content might be due to the diluted ACD solution used to wash out the entire system. The final protein content is likely to be greater than 50 g/L of the plasma fractionation (European Pharmacopoeia, 2002).

The D-dimer concentrations are set at 0.5 n/mL during filtration (Burnouf et al., 2003). Plasma inactivation was not induced by contact of whole plasma with filtration membranes, although this was confirmed by conducting some tests like fibrinopeptide A, thrombin− antithrombin complex, and prothrombin fragment 1 + 2. Previously, evidence for alteration or activation of protein during nanofiltration was not shown for purified plasma products. Plasma nanofiltration is applied to plasma and blood connection centers under standard conditions. Burnouf et al. (2003) showed that the entire procedure of nanofiltration for 200 mL of plasma took approximately 20 minutes, and so 40 minutes would be required for filtering 400 mL plasma. However, <40 minutes are required for further processing and optimized filtration of 650−800 mL of plasma.

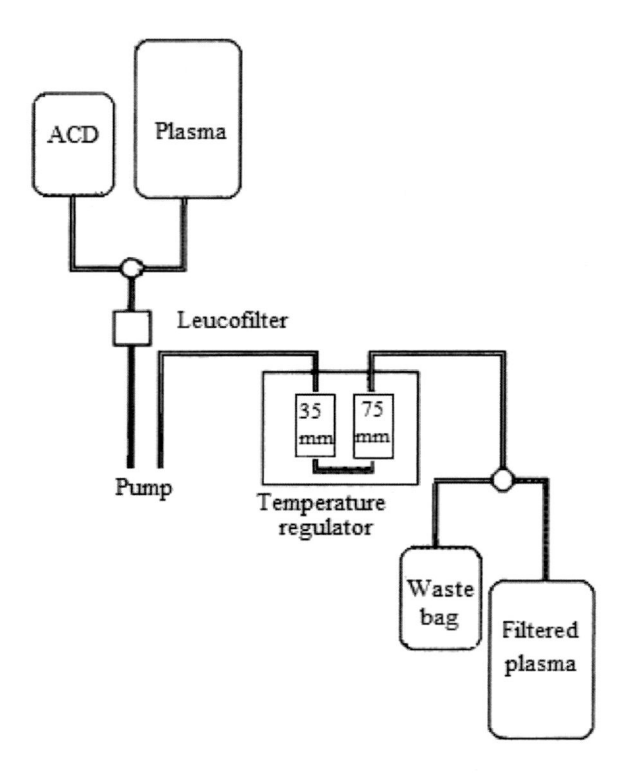

FIGURE 16.3

Plasma nanofiltration mechanism.

There is a significant impact of plasma nanofiltration on blood transfusion for plasma fractionation and transfusion. Nanofiltration taking place in the clinical plasma results in immunization of white blood cells antigens and human leucocyte antigen, immunosuppression, and severe pulmonary dysfunction (Brand, 2000; Hiruma & Okuyama, 2001). The risk of developing Creutzfeldt−Jakob disease can be minimized by reducing the removal of cell fragments and blood cell from plasma and its derivatives. The risk of developing this disease corresponds to the lymphocytes and platelets because prion proteins are linked with these blood components (Bessos et al., 2001). During leukoreduction process, the infection caused by Creutzfeldt−Jakob disease is aggravated via cell fragmentation; although there is no confirmed evidence regarding this. Experimentally, nanofiltration contributes to the direct removal of prions by using albumin solutions that are spiked with scrapie agent (Tateishi et al., 2001). The risk of transmission can be reduced by plasma nanofiltration for fractionation. The assessment of polymerase chain reaction and testing DNA from six polymerase chain reaction HCV-positive donations highlighted the benefits of nanofiltration for the viral safety of the plasma.

Viruses including human immunodeficiency virus and hepatitis B virus are larger than 35−40 nm that allow a nanofiltration sequence based on its particle size. However, viruses smaller than 35−40 nm (parvovirus B19), if present in aggregated form of antibodies within the plasma, can be removed the antigen from the body (Takeda et al., 2001). The robustness of nanofiltration is useful for plasma products as a virus-removal procedure (Burnouf & Radosevich, 2003). The plasma protein data is compared to obtained proteins using other procedures like illumination, SD treatment, and methylene blue. During the development process, the labile plasma proteins are affected to different extents (Goubran et al., 2000). For example, there is loss of 26%−32% factor VIII, 23%−26% fibrinogen, 16% factor XIII, 13%−39% factor XI, 10% factor V, and 18% activity of von Willebrand factor. It is shown that there is 43% loss of fibrinogen in cryoprecipitate by treating plasma (Hornsey et al., 2000). The SD-treated plasma using nanofiltration reveals a decrease in the activity of protein S and alpha2 antiplasmin with partial removal of the multimers of von Willebrand factors with high molecular weight.

A review study highlights the benefits of using nanofiltration during the manufacturing of biopharmaceutical products and plasma-derived coagulation factors from blood origin (Burnouf & Radosevich, 2003). Viral safety is improved by complementing viral reduction treatments like heat treatment and solvent detergent treatment to inactivate hepatitis B virus, hepatitis C virus, and human immunodeficiency virus, hepatitis C. The main reason behind the quick acceptance of nanofiltration corresponds to its simple steps of manufacturing that include filtration of plasma solution through a small pore size of 15−40 nm in the membranes. The nanofiltration is carried out under specific conditions that assist retaining viruses by efficient removal of a wide range of viruses greater than 4−6 logs without denaturing the plasma proteins. The viral markers including nucleic acids and antigens are removed through nanofiltration, unlike the treatment used for viral inactivation that kills the viruses without affecting these viral markers. Recently, nanofiltration has gained immense success after showing its ability to remove prions, which has opened up new perspectives and made this process part of a routine step in the manufacture of biopharmaceutical products. The success of nanofiltration also corresponds to minimized toxicity issues as the majority of the viral inactivation treatments need postelimination from the protein solution using mutagenic or toxic materials.

16.3 Learning objectives

The size exclusion of high-throughput via low-resolution requirements is possible by implementing membrane filtration in medical applications. The protein concentration and buffer exchange are offered through applications such as sterile filtration, microfiltration, and ultrafiltration. However, significant improvements were required in current research and development endeavors to maintain the inherent high-throughput characteristics (Doodeji & Zerafat, 2018). Membrane filtration is just confined to plasma pool clarification, bacterial filtration of buffers, and dialysis. Virus safety is significantly contributed by stringent viral reduction through the introduction of additional steps that enhance viral safety margins. The viruses are likely to be retained on sieving by filtering plasma protein solutions using membranes that are commercially available. The technology using membranes characterized by mean pore size is termed as nanofiltration, which makes it different from other filtration methods that are not primarily designed for removing viruses. Nanofiltration was initially applied in the field of biopharmacy for effective removal of viruses from plasma manufacturing. However, nanofiltration is used in various biopharmaceutical products because of its increased effectiveness.

The clinical immunoglobulin preparations demand viral safety as a significant prerequisite. The safety of human intravenous immunoglobulin is ensured by the common manufacturing practice that involves the utilization of virus inactivation processes. The high-resolution separation of protein and virus is made possible through the engineering of nanofiltration membrane structures. Therefore this process in extensively recognized and applied to medicinal products prepared from animal and human cells, and also the products derived from plasma. In the 1990s, the nanofiltration of plasma products was recognized at the industrial level as it complemented other treatments that reduce viruses. These treatments include the heat treatments and solvent detergent method that inactivates hepatitis B virus, hepatitis B virus, and human immunodeficiency virus.

Nanofiltration aims to increase the safety of plasma products against nonenveloped viruses and safeguard against the attack of infectious agents entering human plasma. The basic procedure followed during nanofiltration is explained in Fig. 16.4. The separation efficiency at different pH values for removing antibiotics is controlled through electrostatic interaction between membrane surface and antibiotics dissociation species. The membrane pores are approached as the antibiotic's dipole affects retention that further affects the molecular orientation (Zhao et al., 2018). Based on the materials, the nanofiltration membranes are categorized into two groups that are polymer and ceramic. The hybrid nanofiltration membranes have gained maximum attention as the result of working in harsh conditions; although nanofiltration has more benefits as compared to another removal techniques. The polymers used in hybrid membranes' top layer include:

- Polydimethylsiloxane
- Polysulfone
- Polyethersulfone
- Polyvinylidene fluoride

The polymers used in preparing nanofiltration membranes modify the size of pores and alter the chemical properties of membrane surfaces. A polymer like PDMS is nontoxic, flexible, chemically stable, thermal, and possesses antibacterial characteristics; therefore it is used in thin selective layers. The effect of molecular weight of PDMS on rejection can be investigated through fabrication of alumina−PDMS hybrid nanofiltration membranes. This is only possible by increasing the molecular weight of the polymer, which reduces the pore size that is further effective in fabricating the engineered membranes. These engineered membranes can be applied in a special application of virus removal with the desired pore size. The purification applications also implement the nanofiltration membranes.

In the medicine industries, high purity was obtained for xylitol that is an important stone corner. Initially, some of the traditional techniques such as adsorption and crystallization were used to

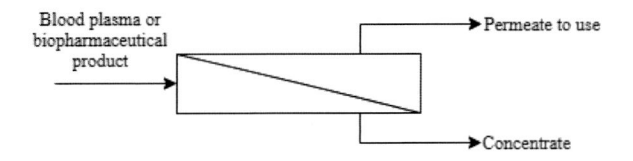

FIGURE 16.4

Basic mechanism of nanofiltration.

purify xylitol because of its high performance. Moreover, among the famous co-polymers used to blend polymers pluronic has received immense attention because of its strong hydrophilic nature and it is used as a surface modifier and pore former (Faneer et al., 2017).

Nanofiltration offers the removal of biological infectious agents that are larger than the nominal filter porosity, which helps in retaining these agents on the membrane and the smaller plasma proteins easily pass through the membrane. Nanofiltration, among the other viral reduction methods, permits efficient removal of nonenveloped and enveloped viruses. This technique is simple and performed using commercially available cartridges without any sophisticated equipment, that is, aseptic filtration that removes bacteria. The plasma products should go through membranes with porosity ranging between 15 and 40 nm. The viral clearance of small nonenveloped viruses is likely to increase, although a membrane with 35 nm pores has been proved to be successful in removing virus aggregates (Rezvan et al., 2006). However, recently the process of nanofiltration has showed the successful removal of prions; therefore, it is now used in the manufacturing of biopharmaceutical products.

The majority of the nanofiltration membranes are polymers for solvent filtration, whereas there is wide usage of ceramic nanofiltration membranes and combined ceramic and polymeric membranes (Dutczak et al., 2011). Nanofiltration membranes are divided into two categories, characterized as loose and tight nanofiltration membranes depending on the material and synthesis method. There is a direct relation of tight nanofiltration membrane with the loose ones. Both the membranes are synthesized in a similar way with a composite structure comprising a polysulfide sublayer and polyamide makes up its topmost layer. The topmost layers are made through interfacial polymerization by immersing the sublayer in an aqueous amine solution, followed by dipping in acyl chloride solution. A thin polymeric layer provides efficient separation capacity that is yielded as the result of a reaction between amine and acyl chloride. The optimization of permeability of the topmost layers is comparable to the concentration of monomers, choice of monomers, and the reaction time that helps in determining the eventual membrane performance (Schaefer et al., 2005).

The ceramic nanofiltration membranes possess an asymmetric structure that comprises three layers with different pore sizes (Fig. 16.5):

- Macroporous support layer
- Mesoporous intermediate layer
- Thin top layer with pores

The macroporous support layer is made by sintering of a powder that provides mechanical strength. α-Al_2O_3 possesses a pore size of 1 μm or smaller, and it is considered as the most

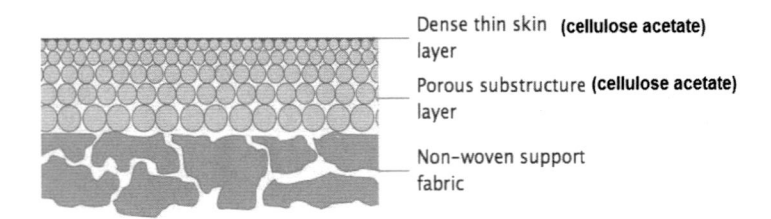

Dense thin skin **(cellulose acetate)** layer

Porous substructure **(cellulose acetate)** layer

Non-woven support fabric

FIGURE 16.5

Asymmetric structure of nanofiltration membrane (Nanofiltration Separations).

common material supporting all the layers of the membranes. It is important to check the quality of the support layer because any irregularity in the support layers eventually causes defects in the top layer. This can be avoided by reducing the surface roughness as a mesoporous intermediate layer is added in the sol−gel process (Van der Bruggen, 2013). The sol is a colloidal suspension that is manufactured by either a polymeric route or a colloidal route. The polymeric route allows mixing of a small amount of water in an organic solvent that yields branched polymeric modules. Whereas, the colloidal route allows mixing of the precursor in excess water to form colloids measuring a few nanometers. Usually, the colloidal route is followed for the intermediate layer. The oxides of titanium and zirconium possess more stability as compared to the oxides of silicon and aluminum. In the nanofiltration membrane, specifically, all these four oxides are used. There is often the appearance of intermediate γ-Al$_2$O$_3$ layers; however, they are only stable between pH 3 and 11. On the contrary, stability of TiO$_2$ and ZrO$_2$ is observed at lower as well as higher pH values. The oxides are hydrolyzed to form colloids owing to an excess of water, such as:

$$M(OR)_4 + H_2O \rightarrow HO-M(OR)_3 + ROHM = Ti, Zr$$

The hydroxides further undergo a condensation reaction:

$$(OR)_3M-OH + HO-M(OR)_3 \rightarrow (OR)_3M-O-M(OR)_3 + H_2OM = Ti, Zr$$

The addition of HNO$_3$ or HCl helps in achieving stabilization of colloids to avoid any aggregation. The process of dip coating is used to add the intermediate layer, here dipping into means dispersion. A gel is yielded as the result of drying below 100°C resulting in the formation of a network of structure of particles. The asymmetric drying that is starting parallel and moving toward the support layer causes the organic binders like polyvinyl alcohol to prevent cracks. The calcination of membrane is conducted at a specific temperature, which is the main parameter determining the pore size of the membrane (Van Gestel et al., 2002). The desired phase in anatase in the case of TiO$_2$ layers is achieved during thermal treatment between 250°C and 500°C. An amorphous structure is formed at a temperature below 250°C, with reduced corrosion resistance. Similarly, a phase transformation from anatase to rutile takes place above 500°C that yields large pores in the top layer of nanofiltration membrane. An amorphous phase is achieved like ZrO$_2$ at room temperature, whereas transition to a cubic or tetragonal monoclinic phase takes place at higher temperature. The temperature between 360°C and 550°C provides the desired tetragonal phase for calcinations. Transition in the monoclinic phase takes place at 550°C, with no open structure that shows a similarity to titanium. The method of formation of a thin top layer is like the one used for intermediate layers, with the only difference in the use of polymeric sols. The interpenetration of polymeric structures results in the formation of three-dimensional networks that eventually result in the pore formation. The calcination temperature is important for polymeric sols as the top layer is made based on special requirements. The best results are attained at a moderate calcination temperature, that is at 300°C.

The hydrolysis of oxides results in the formation of colloids, owing to excess water, shown as follows:

$$M(OR)_4 + H_2O \rightarrow HO-M(OR)_3 + ROHM = Ti, Zr$$

The condensation reaction takes place after hydrolysis, which is shown as follows:

$$(OR)_3M-OH + HO-M(OR)_3 \rightarrow (OR)_3M-O-M(OR)_3 + H_2OM = Ti, Zr$$

16.4 **Methods and techniques**

The size of pathogens and the size of membrane pores characterize the removal ability of nanofiltration. This process is successful in removing a wide range of nonenveloped as well as enveloped viruses because these viruses measure up to 20–200 nm, and the process guarantees protein activity recovery of 90%–95%. The consistency and robustness of nanofiltration in removing viruses is confirmed by a demonstration of 99.9% removal of viruses that include human immunodeficiency virus, hepatitis S virus, and von Willebrand virus (Caballero et al., 2014). Nanofiltration, being identified as a virus removal tool, can be differentiated from specific viral inactivation treatments and viral removal methods.

16.4.1 **Dedication method**

The main aim of the nanofiltration process is to remove the viruses. The efficient and reproducible removal of viruses is possible under controlled operating conditions such as temperature, pressure, rate of flow, ratio of filter area, and protein load (Fig. 16.6). The process of chromatography and centrifugation develop working conditions to purify proteins that might even remove the viruses, unlike nanofiltration which is completely different from these purification tools. The removal of viruses validates all the conditions characterizing this chromatographic step by avoiding any batch to batch contamination that consumes much time and tends to be costly.

16.4.2 **Predictability of virus removal**

Relying on the sieving mechanism, the viruses are removed in a predictable and reproducible way under specific process conditions. There is a chance of robustness for the sieving mechanism for

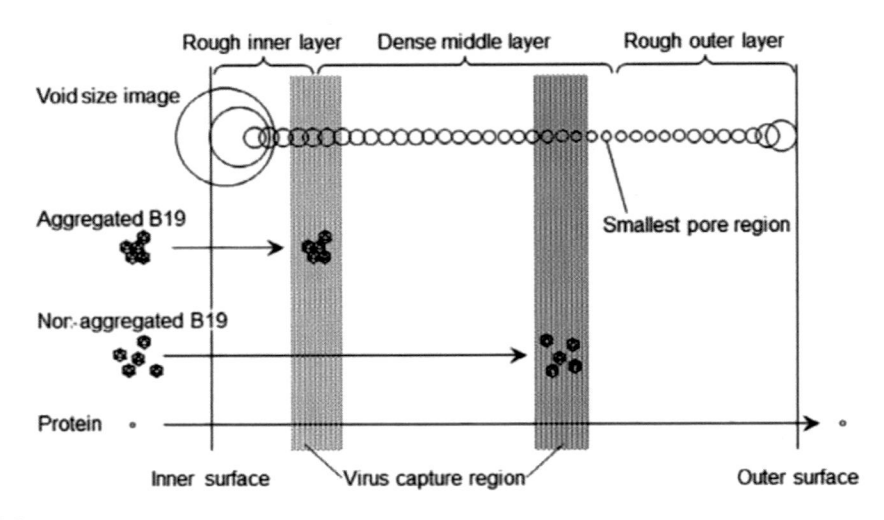

FIGURE 16.6

Dedication method (Adan-Kubo et al., 2019).

removing viruses larger than the membrane's pore size. Moreover, other factors affecting the extent of viral removal for viruses with small pore size are as follows;

- Virus presented in aggregates
- Protein content and adsorption of viruses in the solution as the result of the charge effect
- Forming complexes with antibodies

The reliability of nanofiltration is guaranteed through integrity tests that control the potential removal of viruses using a filter.

16.4.3 Robustness and efficacy

The increased removal of enveloped and nonenveloped viruses from plasma depends on the size of pores on the filtration membrane and size of virus. Apart from nanofiltration, no other method has been established yet that ensures the removal of nonenveloped and enveloped viruses from blood plasma. Through the establishment of worst-case conditions, the robustness of nanofiltration is demonstrated on a case by case basis.

16.4.4 Ease of use and flexibility

The degree of flexibility that nanofiltration affords is offered by a few methods. Various steps during the production process utilize nanofilters; however, this potential is significantly reduced in other viral reduction methods (Alarifi et al., 2018; Alharbi et al., 2016). The selection of an appropriate system to prepare protein is possible using commercially available devices with various pore sizes. The process stability is made as a relatively straightforward operation and it increases the availability of various surface areas. The existing devices are limited as they face problems in filtering protein solutions of 150 L or more. Therefore systems with larger filtration area should be made available to allow the predicted application of nanofiltration for albumin.

16.4.5 Viral markers

Nanofiltration can remove viral markers, which include nucleic acids and antigens, along with the protein mixture, unlike other viral inactivation treatments that destroy viruses but fail to remove these markers. This provides practical advantages for validation.

16.4.6 Toxicity

The majority of the inactivation treatments use mutagenic or toxic chemicals that need to be removed afterward. These types of toxicity issues are not linked with nanofiltration as no such chemicals are added. The process of validation is associated with the risk of leachables from the filter, which needs to be considered.

16.4.7 Risks of downstream contamination

The principles of good manufacturing practices are likely to be followed after implementing nanofiltration procedures (World Health Organization, 2004). The devices are just used once under the majority of the circumstances, whereas nanofiltration is conducted before bacterial aseptic filtration and filling at the end of the production process to minimize the risk of recontamination.

16.4.8 Protein integrity and product characteristics

There is 90%−95% chance of recovering protein antigens and their activity. Heat treatment is a viral inactivation method that results in the loss of 20% of protein. On the contrary, no protein content is lost and no neoantigens are generated through nanofiltration during in vitro analysis because the process is carried out at standard pH, temperature, and osmolarity. The experimental design of nanofiltration taking place at standard temperature is presented in Fig. 16.7. It is known that heat viral inactivation treatments and purification procedures use plasma coagulation factor VIII which has been reported to have increased incidence of inhibitors among the patients. However, there is no change in product characteristics after undergoing nanofiltration as it facilitates product registration processes and facilitates regulatory acceptance.

16.4.8.1 Adapting to a large product range

The application of nanofiltration is observed on different classes of plasma products, thus allowing the rationalization of regulatory efforts and product developments.

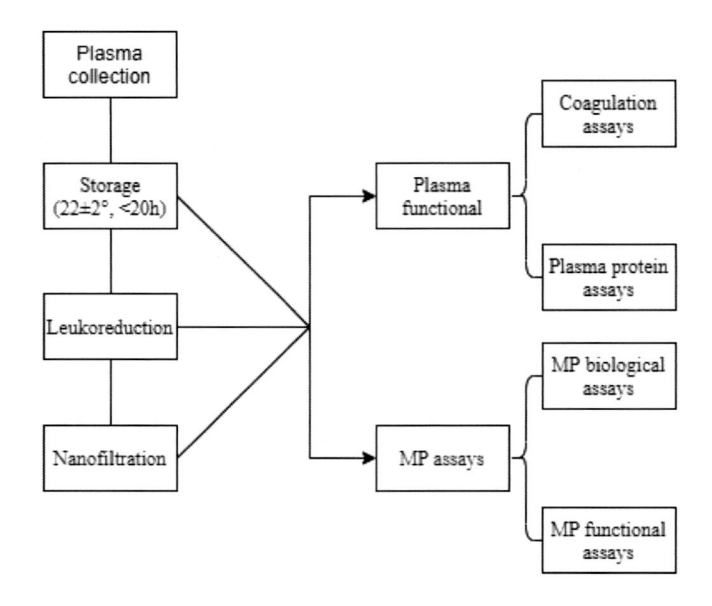

FIGURE 16.7

Experimental design of nanofiltration.

16.5 Applications

The nanofiltration of plasma products was implemented for improving the margin of viral safety, especially against the nonenveloped viruses entering the human plasma pool. It has evolved as a routine step in the manufacturing of biopharmaceutical products. The majority of the applications of nanofiltration refer to the products used clinically in the licensing processes. Different applications of nanofiltration process are described below.

16.5.1 Coagulation factors

The easy implementation of nanofiltration was proven technically at its first attempt at being applied with high-purity plasma protein coagulation factor of approximately 60 kDa to maintain the level of purity along with low protein content. In later steps, nanofiltration was used for larger products. The products that were successfully nanofiltered in the first attempt include factor IX (FIX) as it has low MW (57 kDa) with development preparations of higher purity, in comparison to prothrombin complex concentrates. The possibility of using 35 nm pore size membrane, along with newly developed 15 nm pore size membrane was evaluated in 1994. Three chromatographic steps were followed to prepare the FIX product; while, solvent−detergent treatment was performed for viral inactivation (Burnouf and Radosevich, 2003). The material used in the process of nanofiltration is the final FIX eluate. Single-step filtration was used to compare both the filters in the dead-ened mode. Virus validations were conducted that showed that nanomembranes can remove 7−7.8 log of simian virus 40, human immunodeficiency virus, bovine viral diarrhea virus, reovirus type, and porcine pseudorabies virus. It has also shown success in removing poliovirus Sabin type 1, bovine parvovirus, and small nonenveloped viruses. The nonfiltered product showed no alteration in the in vitro data, along with no change in the specific activity of FIX and product protein profile. The nanofiltration of coagulation factors is represented in Fig. 16.7.

Nanoinfiltration did not affect the vitro markers of activation as revealed by the rat hypotension model and rabbit Wessler model of thrombogenicity. FIX solution is filtered on 15 nm filter within 2−3 hours after being purified from 1000−2000 L of plasma, which shows the large-scale adaptability of nanofiltration. The hemophilia B patients demonstrated the absence of thrombogenicity with normal half-life and recovery (Burnouf and Radosevich, 2003). Usually, complete removal of the viruses is achieved when the mean pore size of virus is larger than the filtration membrane. While conducting 15 nm Planova filtration, approximately 5.2 canine parvovirus were removed with a mean size of 18−26 nm. Fifteen-nanometer Planova filtration also nanofiltered high-purity FIX concentrates that were either purified by chelate chromatography or viral inactivation (Table 16.2). Viruses including Herpes simplex virus 1, lipid-enveloped viruses, Semliki Forest virus, and Sindbis virus were removed in 7 logs. The nonenveloped viruses were removed in the following quantities (Burnouf and Radosevich, 2003);

- 6.0 logs of bovine parvovirus
- 5.6 logs of poliovirus
- 5.7 logs of encephalomyocarditis virus

Previously, it was found that the Planova 35 nm membrane was effective in removing 5.7 logs of Bovine papillomavirus (Burnouf & Radosevich, 2003). No detectable impact, including

Table 16.2 Virus removal through nanofiltration on Planova 15 nm membrane (Burnouf-Radosevich et al., 1994).

	Factor IX		Factor XI	
Virus	Virus challenge	Reduction factor	Virus challenge	Reduction factor
Polio Sabin type 1	8.5	>6.7	8.6	>6.9
Bovine parvovirus	8.0	>6.3	8.1	>6.4

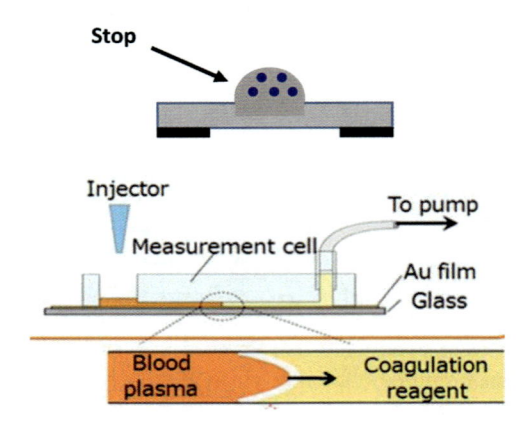

FIGURE 16.8

Blood coagulation monitor design and measurement cell.

thrombogenicity, was revealed after preclinical and clinical trials. Nanofiltration was effective in retaining high MW proteins present in Austrian FIC products, using Viresolve 70 tangential flow procedure. The high-purity FIX licensed in Australia with a combination of Planova 35- and 15 nm membrane filters was provided with sequential nanofiltration. The FIX is either purified by viral inactivation induced by incubating in sodium thiocyanate or by monoclonal antibody (mAb) chromatography. A Swedish nanofiltered FIX showed no significant side effect of this procedure; rather it showed its efficacy during continuous blood transfusions carried out on patients with hemophilia B (Burnouf & Radosevich, 2003). Therefore the use of small pore size nanofilters in the process of nanofiltration has gained popularity, as shown in Fig. 16.8 and it is widely accepted in the manufacturing of FIX concentrates because it is easy to use, provides great outcomes, and possesses no side effects

16.5.2 Prothrombin complex

Prothrombin complex concentrates possess high protein content, that is, 25−40 g L, proteins of high MW, and several coagulation factors; therefore their nanofiltration is not easy to achieve.

It has been shown that the first successful nanofiltration of prothrombin complex concentrates was conducted in Germany. Planova 35 nm was set like a filter that is capable of filtering 4 logs of hepatitis B virus, Bovine viral diarrhea virus (medium-sized enveloped viruses), and 7 logs of Herpes simplex virus-1 and human immunodeficiency virus-1 (large enveloped viruses) for conducting nanofiltration of prothrombin complex concentrates. Poliovirus was not retained by the filter and there was no impact on product profile with any increase in thrombogenicity. Solvent—detergent is used for the inactivation of prothrombin complex concentrates that are likely to be nanofiltered (Josić et al., 2000). During clinical analysis, no thrombogenicity was achieved in vitro or animal models. The virus removal capacity can be increased by combining two serially connected Planova 15 nm for the nanofiltration of prothrombin complex concentrates, which can remove 5.9 logs of canine parvovirus (Fig. 16.9).

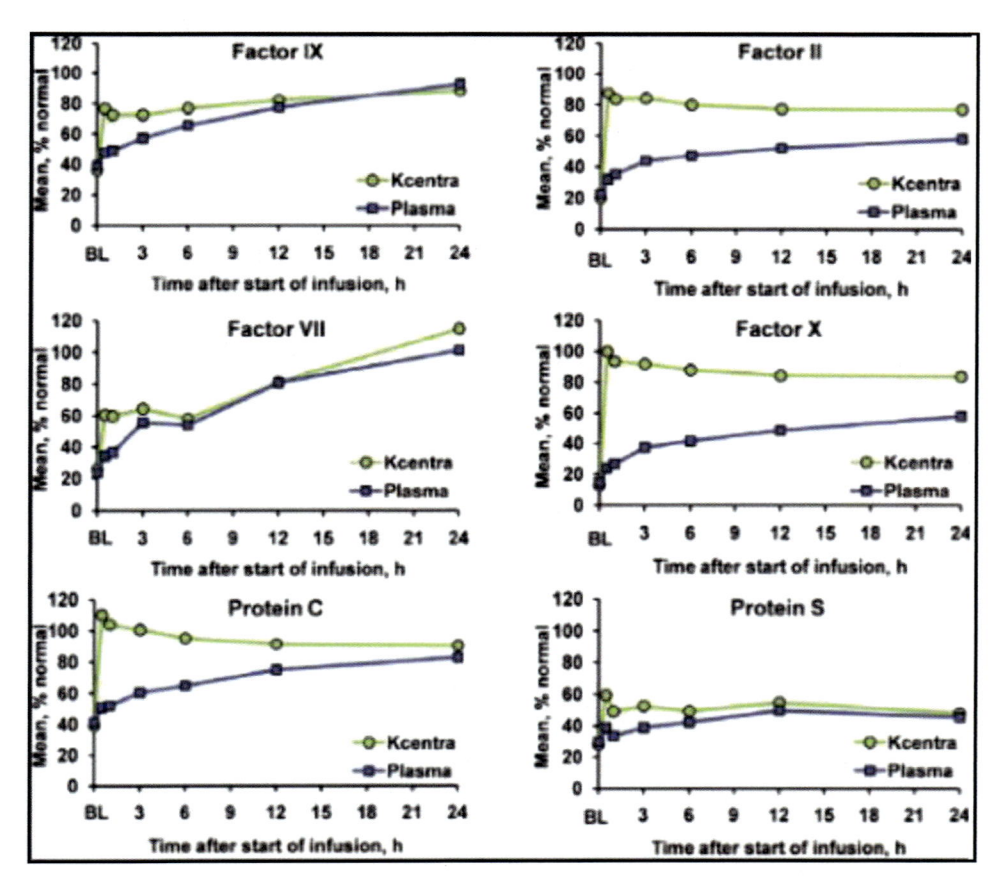

FIGURE 16.9

Nanofiltration of prothrombin complexes (Ismail, 2017).

16.5.3 **Factor VIII**

Chromatography methods, including affinity, anion-exchange, and size exclusion, have been used for purifying the concentrates of factor VIII since the 1990s (Burnouf & Radosevich, 2001). These steps are effective in removing allogenic plasma proteins inducing immunosuppression among the patients suffering from hemophilia A. It facilitates the treatment by eliminating the virus-killing agents like Triton X-100, Tween 80, and tri(n-butyl) phosphate. Plasma protein in bulk containing an earlier generation of factor VIII concentrates are likely to be removed by chromatography that increases its purity by 100- or 1000-fold. The nanofiltration of FVIII concentrates is made achievable for rendering high purity even for protein content equal to 1 g; despite its large molecular weight. Solvent—detergent is used to inactivate monoclonal purified factor VIII and later subjected to dead-end single filtration using Planova 35 nm (Borel-Derlon, 2000). There is much similarity in the in vitro characteristics presented by nanofiltered factor VIII and the nonnanofiltered product. One batch of product takes approximately 1.5 hours to go through the nanofiltration step. Nanofiltration showed success in removing medium and large size viruses; however, it failed to remove some of the small unenveloped viruses that include parvovirus B19, Encephalomyocarditis virus, and hepatitis A virus. The introduction of nanofiltration in the routine showed a significant contribution of the 35 nm filter membrane as there was a significant reduction in the DNA of parvovirus B19 in the final concentrates of factor VIII. All the parvovirus B19 strands were removed from factor VIII concentrates after introducing receptor-mediated hemagglutination screening of the patients' plasma units (Takeda et al., 2001). A sequence of Planova 35- and 15 nm was used to conduct nanofiltering under product licensed in France. Activation of this product was carried out through the solvent—detergent, whereas ion exchange chromatography assisted in its purification.

There is a need to dissociate von Willebrand factor using a buffer with a high concentration of $CaCl^2$ during anion-exchange chromatography to achieve successful nanofiltration on a 15 nm membrane. Later, factor VIII eluates were nanofiltered for several hours at 35°C. Removal of bovine viral diarrhea virus >4.06 log, Human immunodeficiency virus >3.77 log, Hepatitis A virus >3.72 log, and porcine pseudorabies virus >6.08 log was shown in viral validation during the dual step filtration. The final product in the form of von Willebrand content was half that of the product, which was not nanofiltered. The nanofiltration process does not degrade or activate factor VIII, as revealed through in vitro analysis. There was similarity between nonnanofiltered products and binding to phospholipids, activated factor X, and kinetics of proteolysis with thrombin.

The formation of neoantigens was not revealed through the immunogenicity tests conducted among the rabbits. There was a similarity in the nanofiltered product and pharmacokinetic profile in hemophilia A patients. Clinical efficacy of 100% for surgery and 94% for bleeding was achieved, which was rated either good or excellent. In particular, there was no viral seroconversion and appearance of antifactor VIII inhibitor against parvovirus B19, although it is not easy to investigate the reason for B19 seroconversion's absence. Its absence might be due to nucleic acid testing, nanofiltration, or both (Aubin et al., 2000). It is suspected that significant loss of factor VIII is induced by nanofiltration on Planova 15 nm, although there is no evidence to support this statement. The performance of virus removal is compromised in any case either by filtration membrane of similar pore size (like DV20) or Planova 20 nm. The safety of products corresponding to small plasma-borne viruses is ensured by the combined benefits of NAT plasma screening and nanofiltration, without compromising the supply of factor VIII. Here, it is also important to consider the increased number of viruses among the infected plasma donations.

16.5.4 Factor XI

Chromatographic procedure is followed for preparing licensed highly-purified Factor XI with molecular weight of 160 kDa. Factor XI is virally inactivated by solvent−detergent. The single-stage dead-end nanofiltration on Planova 15 nm ensures efficient removal of lipid-enveloped and nonenveloped viruses.

16.5.5 Von Willebrand factor

The use of new technologies, including affinity chromatography and anion-exchange, is introduced to produce concentrates of von Willebrand factor of high purity. It is possible to conduct nanofiltration on von Willebrand solution on 35 nm membrane to improve the concentration of low proteins and increase the purity (Fig. 16.10). At 80°C the nanofiltered product sustains additional viral

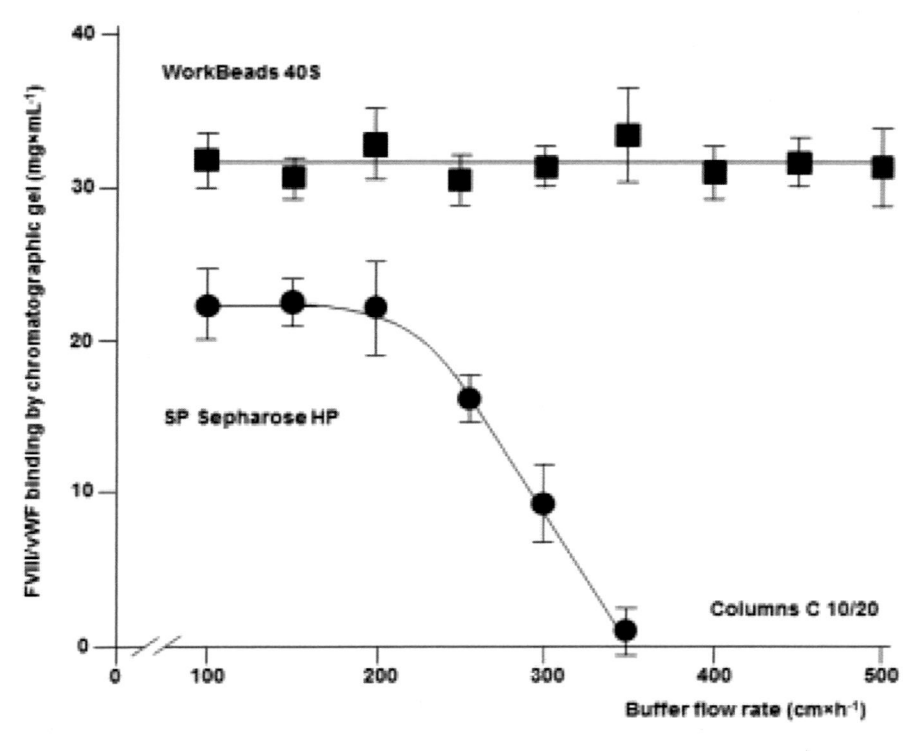

FIGURE 16.10

Nanofiltration of von Willebrand factor (Sergiy et al., 2017).

inactivation by dry-heat for approximately 72 hours and shows stability. There is a bioequivalence of this product with the nonnanofiltered product, rather than heat-treated product (Borel-Derlon, 2000).

16.5.6 **Plasma products**

The concentrated immunoglobulins (160 kDa) are known as viral safe products; however, there are chances of hepatitis C virus transmission as the result of specific viral inactivation treatment. Incubation at pH 4, solvent—detergent, or pasteurization procedures are likely to be used for deactivating the intravenous immunoglobulin G preparations. Recently, the use of nanofiltration has increased for enhancing viral safety of the products of immunoglobulin G. the use of membranes with 35 nm pores has been shown to achieve good reduction for some of the nonenveloped and enveloped viruses. There is consistent removal of nonenveloped viruses with no harm to the antibody binding activity and molecular characteristics of immunoglobulin G by conducting nanofiltration on 15 nm membranes. The 7% solution of intravenous immunoglobulin G is validated for the efficacy of Planova filtration to remove nonenveloped as well as enveloped viruses. At first, the prefiltration of intravenous immunoglobulin G was performed on 75 nm filter and later passed through 35 nm filters that are connected in series. Connecting two types of filters increase the possible aggregation of viruses as the 75 nm filter increases capacity, whereas 35 nm filters remove the viruses. Almost all the viruses typically over 5.4 logs and size larger than 35 nm are removed on Planova 35 at the first stage of filtration (Fig. 16.11). The logs of 4.3 encephalomyocarditis virus and >4.7 hepatitis Porcine pseudorabies virus of 2.6 logs were efficiently removed by filtration membrane with diameter of greater than 35 nm. Filtration of another immunoglobulin G at 35 nm showed incomplete removal of hepatitis C virus. Sufficient margin of viral safety is likely to be achieved by combining the process of nanofiltration with another viral reduction step (Table 16.3).

The fractionation of cold ethanol, at a concentration of 3% on Planova 75- and 35 nm, and plasma-derived intravenous immunoglobulin preparation, and inactivation through b-propiolactone results in the production of plasma-derived immunoglobulin G concentrate. Viral reduction greater than 7.1 for pseudorabies, 6.4 for bovine viral diarrhea virus, 6.6 for reovirus, and 4.8 for human immunodeficiency virus was achieved. However, it failed to remove the bovine parvovirus. The robustness of the nanofiltration step was presented with lots of intravenous immunoglobulins and other membranes (Alarifi et al., 2020). This step was followed with proper maintenance of various subclasses, no significant loss of immunoglobulin G, and preserved antibody activity. It is possible to nanofilter the intravenous immunoglobulin preparation on Planova 15 nm that was made by fractionation of ethanol, treated with solvent—detergent at pH 4, and chromatographed. This helped in establishing removal of >5.58 logs of encephalomyocarditis virus and >6.25 logs of canine parvovirus. Approximately, 200—300 g of protein per 1 m^2 of 15 nm Planova per hour is nanofiltered efficiently. Hyperimmune anti-D preparations were also reported for successful nanofiltration. Nanofiltration of a chromatographically purified preparation (protein concentration as low as <1 g at 22°C temperature) is performed on Planova 35 nm. There is no change in the immunoglobulin G subclass with no evidence about the neoantigens. The bovine parvovirus (>4.97 log) and lipid-enveloped viruses are completely removed. On the contrary, nanofiltration was not able to remove the unenveloped viruses including 4.3 log poliovirus, 3.2 log Theiler's murine encephalomyelitis virus, and 4.14 v log porcine parvovirus. The 5% preparation of anti-HBs was nanofiltered through

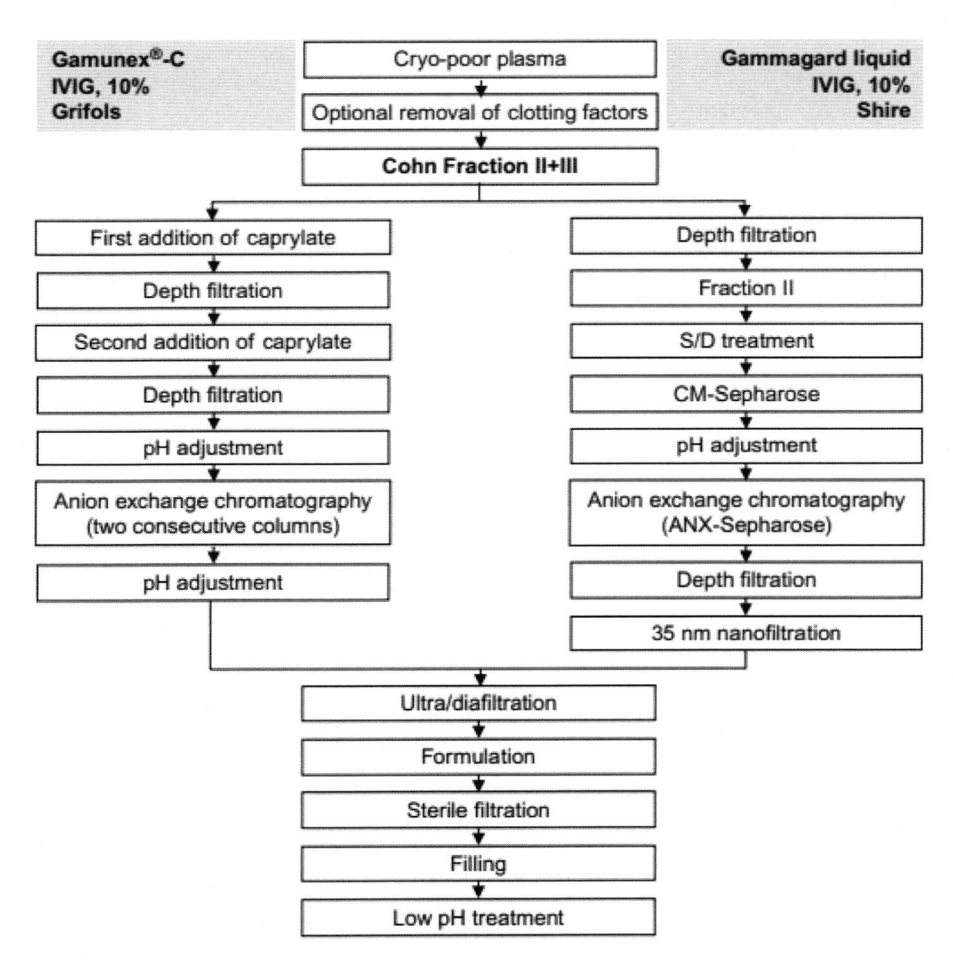

FIGURE 16.11

Nanofiltration of plasma immunoglobulins (Buchacher & Curling, 2018).

Table 16.3 Hepatitis C virus filtration through 35 nm membrane (Burnouf et al., 2003).

Plasma samples	Quantitative HCV assay		Qualitative HCV assay	
	Prefiltration	Postfiltration	Prefiltration	Postfiltration
1	20−50	<0−2	+	−
2	50−10	<0−2	+	−
3	>10	<0−2	+	−
4	20−50	<0−2	+	−
5	50−10	<0−2	+	−
6	>10	<0−2	+	−

dead-end nanofiltration on a single Planova 35 nm with excellent recovery (Aubin et al., 2000). Modest 2.5 log of B19 and 2.3 log of hepatitis A virus are likely to be achieved under such conditions that require a 1 m^2 module for each batch, taking 2.5 hours.

16.5.7 **Protease inhibitors**

The concentrate of C1-inhibitor (104 kDa) is subjected to nanofiltration that is prepared through a two-step chromatographic procedure from cryoprecipitate poor plasma. The dead-end nanofiltration on Planova 15 nm and Planova 35 nm helps to achieve specific viral inactivation by applying solvent detergent, along with specific viral removal. This procedure is also applied to an underdeveloped solvent−detergent treated 1-antitrypsin (Fig. 16.12). The total capacity of virus removal through 15 nm membrane was <8.6 logs for hepatitis A virus and <5.8 logs for porcine parvovirus.

16.5.8 **Transferrin**

The apo-transferrin from the Cohn fraction IV is purified using viral inactivation by the solvent−detergent procedure and two ion exchange chromatography steps. The purified form of apo-transferrin is nanofiltered on Planova 15 nm. The polymerase chain reaction assay removes less

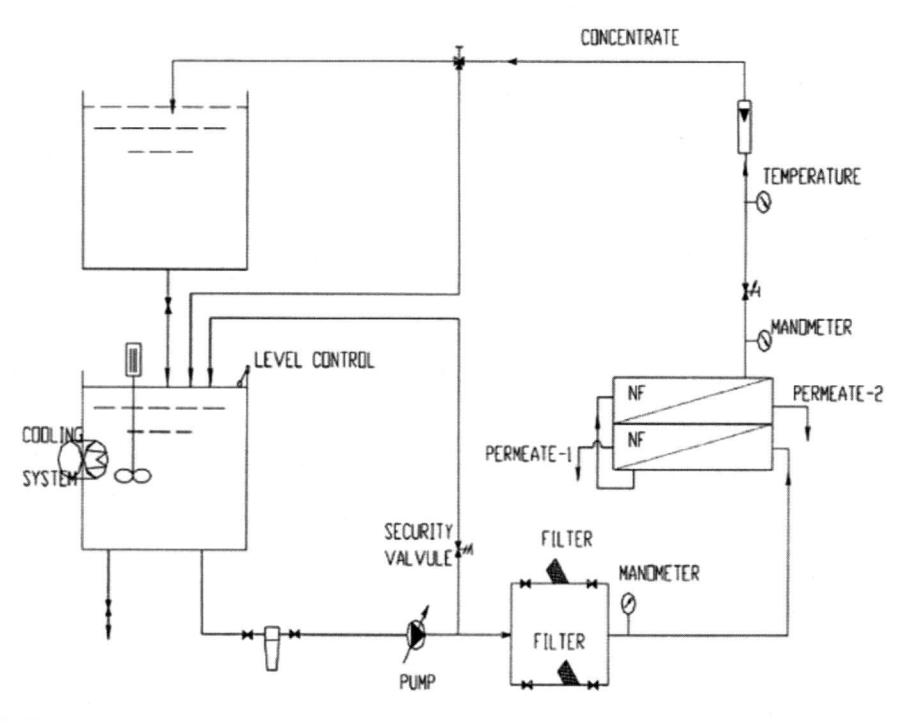

FIGURE 16.12

Schematic presentation of nanofiltration of protease inhibitors (Cuartas et al., 2004).

than 4.5 logs of human B19 positive, which was revealed through the spiking experiments. The robust virus removal capacity of >6.8 logs for Bovine viral diarrhea virus, >5.8 for Encephalomyocarditis virus, >7.7 logs for Human immunodeficiency virus, >7.8 for Pseudorabies virus, and 5.0 for porcine pseudorabies virus is likely to be achieved through nanofiltration. The clinical trials have shown safe and increased tolerance of the Finnish products in iron-binding therapies.

16.5.9 Other blood products

Experiments have been conducted to remove B19 from hemoglobin (64 kDa) solution through nanofiltration. In the first step, hemoglobin is spiked with B19 and later nanofiltration was conducted sequentially on Planova 35 nm filter and Planova 15 nm in the tangential mode. Greater than 6 logs of B19 are likely to be removed through 15 nm membrane; however, Planova 35 nm showed no reduction in the B19 titer. There was no change in methemoglobin ratio, whereas 70.4% ± 3.4% of hemoglobin was recovered. The filtration of red blood cell hemolysates is performed to retain virus through a reusable ultrafiltration system to produce diaspirin cross-linked hemoglobin. There is an increase in the retention of phiX174 phage at high transmembrane pressures and membrane crossflow rate in the presence of red blood cell hemolysate. Reductions of 4.2 logs for porcine pseudorabies virus, and 7.6 logs for hepatitis A virus, >7.9 logs for human immunodeficiency virus, pseudorabies, Bovine viral diarrhea virus were found under optimized conditions. The application of nanofiltration on Planova 15 nm on leucocyte-derived interferon showed efficient results, with no changes in leucocyte-derived interferon alfa.

16.5.10 Prion removal

The classical form of Creutzfeldt–Jakob disease is not transmitted through blood products, as shown through the epidemiological and experimental studies. In humans, related variants of Creutzfeldt–Jakob disease and epizootic bovine spongiform encephalopathy result in a new form of prion disease. There is no transmission of sporadic Creutzfeldt–Jakob disease in either hemophilic patients exposed to blood transfusions or recurrent injections of plasma-derived products (Farrugia, 2002). Possible concerns in transmission of Creutzfeldt–Jakob disease are reported only when there is increased prevalence of this disease in a population (Hilton et al., 2002). The disease can enter and infect healthy mice via the intracerebral route as blood components from mice infected with human and animal transmissible spongiform encephalopathy (Uddin et al., 2020). Considering sheep, transmissible spongiform encephalopathy can develop as the result of transfusing nonleucodepleted whole blood from a sheep fed with BSE-infected bovine brain material (Hunter & Houston, 2002). The presence and transmission of Creutzfeldt–Jakob disease variant is concerning because efficient deferrals of blood donors at risk are prohibited due to long asymptomatic incubation periods. At present, the potentially infectious donations cannot be detected through an appropriate test. Moreover, the viral inactivation processes do not affect prions as they are highly resistant. The ability of manufacturing process to remove the infectious agent can be ensured by observing the safety of the plasma products. The partitioning of prion proteins is significantly contributed by the manufacturing process of plasma products that are used at present. These processes include depth filtration, ethanol fractionation, and chromatographic processes (Lee et al., 2001).

Prions need to be removed using some specific removal tool, although viral removal was initially targeted by nanofiltration. The human transmissible spongiform encephalopathy pathogens present in diluted brain homogenate can be removed using a Millipore screen-type 0.025 lm membrane filter (Millipore) that allows a small amount of diluted brain homogenate to be removed. The Planova cartridges with mean pore sizes ranging between 75 and 10 nm are used to filter brain homogenate from infected mice. Planova 35 nm filtrate was detected with no infectivity. The approximate size or pathogenic agent was 40 nm; however, residual infectivity develops from Planova 10 nm filtrate after adding 1% Sarkosyl to the homogenate.

The removal of scrapie agent ME7 is possible by nanofilteration of albumin solution spiked with brain homogenate, which is a mouse adapted strain of scrapie for a variant of Creutzfeldt−Jakob disease and bovine spongiform encephalopathy. On the large scale, the recovery of albumin was observed to be 90%. The type of filter and adding an anionic detergent (Sarkosyl) in the protein solution affect the extent of viral removal. Planova 35 nm was used to remove infectivity of 4.93 logs and 1.61 logs without and with detergent, respectively. Similarly, Planova 15 nm helped in removing >5.87 logs without detergent and >4.21 logs with detergent. The use of 15 nm or even smaller pores in the membrane resulted in no residual infectivity in any of the filtrates. The prion spikes have the tendency to aggregate under specific conditions, therefore careful analysis is important. Some of the prions with minimum mean pore size of 35 nm or less can be removed from the biological solutions. However, effective removal of prions is possible through filters of 15 nm or less. Prions removal either corresponds to adsorption on the membrane or the sieving mechanism. The prion removal depends on its type, its aggregation, conformation, and the physicochemical nature of filtered solution. Therefore it would not be wrong to say that the process of nanofiltration is safe to be used on human plasma products and other biopharmaceutical products. These products include the recombinant proteins during the production, using human- or animal-derived materials or medicinal products derived from bovine sources.

16.6 Conclusion

The viral safety of plasma-derived products is improved, particularly, the concentrates of coagulation factor that are highly susceptible to being contaminated with infectious agents. The introduction of solvent detergent treatment was observed in the 1980s, which was most effective on enveloped viruses. The absence of denaturing effect of proteins provides a major significance of solvent and detergent treatment; whereas its major limitation is that it is not effective on nonenveloped viruses. The genomic amplification tests conducted on plasma helped in eliminating contamination that further reduced the viral load of plasma pools in the 1990s. During the process of production, this approach has some exacting benefits for the viruses, which neither tested for individual blood donations, inactivated them, nor removed them. At present, the combination of nanofiltration with robust virus inactivation method and other viral safety measures provide major progress to prepare safer plasma products.

Nanofiltration is a well-known virus removal tool, which is validated with defined pore size. In order to meet the above criteria, the term nanofiltration needs to be restricted to membrane filtration. The pores on membranes ranging between 10 and 40 nm help in removing 4−6 logs of viruses and save the essential plasma products. Maintaining balance between protein sizes, pore size, concentration of solution, and structure of used filter is important to attain positive results of applying nanofiltration. The small pore size of the nanofiltration membrane, that is, around 10−20 nm, ensures viral clearance of the smallest nonenveloped viruses <30 mm (hepatitis A virus and B19 virus). The filtering system used at present is resistant to the majority of the viral inactivation treatments because it is capable of removing nonenveloped viruses such as Reovirus 3. This indicates the effective removal of other viruses that consist of similar features to viruses like Reovirus 3 that contaminate the plasma pool. The benefits of nanofiltration can also be extended toward removing variants of Creutzfeldt−Jakob disease as it is easily transmitted through the plasma products. Nanofiltration is not only effective in robust virus removal, rather it also inhibits any sort of protein alterations to provide excellent recovery of protein. However, this is only possible under a specific molecular weight of protein and membrane cutoff limits.

The chapter has clearly shown that nanofiltration adds further to the concept of the sterilization of biopharmaceutical products from certain viruses and it is expected that this procedure would become routine practice with increased acceptance.

References

Adan-Kubo, J., Tsujikawa, M., Takahashi, K., Hongo-Hirasaki, T., & Sakai, K. (2019). Microscopic visualization of virus removal by dedicated filters used in biopharmaceutical processing: Impact of membrane structure and localization of captured virus particles. *Biotechnology Progress, 35*(6), e2875.

Alarifi, I., Prasad, B., & Uddin, M. K. (2020). *Conducting polymer membranes and their applications. Self-standing Substrates* (pp. 147−176). Cham: Springer.

Alarifi, I. M., Alharbi, R. A., Khan, M. N., Khan, W. S., Usta, A., & Asmatulu, R. (2018). Water Treatment using electrospun PVC/PVP nanofibers as filter medium. *International Journal of Material Science Research, 1*(2), 43−49.

Alharbi, A. R., Alarifi, I. M., Khan, W. S., & Asmatulu, R. (2016). Highly hydrophilic electrospun polyacrylonitrile/polyvinypyrrolidone nanofibers incorporated with gentamicin as filter medium for dam water and wastewater treatment. *Journal of Membrane and Separation Technology, 5*(2), 38−56.

Aubin, J. T., Defer, C., Vidaud, M., Montreuil, M. M., & Flan, B. (2000). Large−scale screening for human parvovirus B19 DNA by PCR: Application to the quality control of plasma for fractionation. *Vox Sanguinis, 78*(1), 7−12.

Bessos, H., Drummond, O., Prowse, C., Turner, M., & MacGregor, I. (2001). The release of prion protein from platelets during storage of apheresis platelets. *Transfusion, 41*(1), 61−66.

Borel-Derlon, A. (2000). vWF SD-35-DH: Pharmacokinetic data and clinical efficacy. *Plasma-Derived Medical Products and Haemostasis: Current Issues and Perspectives*, 2000.

Brand, A. (2000). Immunological aspects of blood transfusions. *Blood Reviews, 14*(3), 130−144.

Buchacher, A., & Curling, J. M. (2018). *Current manufacturing of human plasma immunoglobulin G. Biopharmaceutical Processing* (pp. 857−876). Elsevier.

Burnouf, T., & Radosevich, M. (2000). Reducing the risk of infection from plasma products: Specific preventative strategies. *Blood Reviews, 14*(2), 94−110.

Burnouf, T., & Radosevich, M. (2001). Affinity chromatography in the industrial purification of plasma proteins for therapeutic use. *Journal of Biochemical and Biophysical Methods, 49*(1−3), 575−586.

Burnouf, T., & Radosevich, M. (2003). Nanofiltration of plasma-derived biopharmaceutical products. *Haemophilia: The Official Journal of the World Federation of Hemophilia, 9*(1), 24−37.

Burnouf, T., Radosevich, M., El-Ekiaby, M., Satoh, S., Sato, T., Amin, S. N., & Goubran, H. A. (2003). Nanofiltration of single plasma donations: Feasibility study. *Vox Sanguinis, 84*(2), 111–119.

Burnouf-Radosevich, M., Appourchaux, P., Huart, J. J., & Burnouf, T. (1994). Nanofiltration, a new specific virus elimination method applied to high-purity factor IX and factor XI concentrates. *Vox Sanguinis, 67*(2), 132–138.

Caballero, S., Diez, J. M., Belda, F. J., Otegui, M., Herring, S., Roth, N. J., & Jorquera, J. I. (2014). Robustness of nanofiltration for increasing the viral safety margin of biological products. *Biologicals: Journal of the International Association of Biological Standardization, 42*(2), 79–85.

Cuartas, B., Alcaina, M. I., & Soriano, E. (2004). Separation of mineral salts and lactose solutions through nanofiltration membranes. *Food Science and Technology International, 10*(4), 255–262.

Doodeji, M. S., & Zerafat, M. M. (2018). A review on the applications of nanofiltration in virus removal and pharmaceutical industries.

Dutczak, S. M., Luiten-Olieman, M. W., Zwijnenberg, H. J., Bolhuis-Versteeg, L. A., Winnubst, L., Hempenius, M. A., & Stamatialis, D. (2011). Composite capillary membrane for solvent resistant nanofiltration. *Journal of Membrane Science, 372*(1–2), 182–190.

European Pharmacopoeia. (2002). Human plasma for fractionation. Monograph 01/2002: 0853. Strasbourg: Council of Europe, 1928–1929.

Faneer, K. A., Rohani, R., & Mohammad, A. W. (2017). Polyethersulfone/pluronic F127 blended nanofiltration membranes for xylitol purification. *Malaysian Journal of Analytical Sciences, 21*(1), 221–230.

Farrugia, A. (2002). Risk of variant Creuzfeldt–Jakob disease from factor concentrates: Current perspectives. *Haemophilia: The Official Journal of the World Federation of Hemophilia, 8*(3), 230–235.

Goubran, H. A., Burnouf, T., & Radosevich, M. (2000). Virucidal heat-treatment of single plasma units: A potential approach for developing countries. *Haemophilia: The Official Journal of the World Federation of Hemophilia, 6*(6), 597–604.

Hilton, D. A., Ghani, A. C., Conyers, L., Edwards, P., McCardle, L., Penney, M., & Ironside, J. W. (2002). Accumulation of prion protein in tonsil and appendix: Review of tissue samples. *BMJ (Clinical Research ed.), 325*(7365), 633–634.

Hiruma, K., & Okuyama, Y. (2001). Effect of leucocyte reduction on the potential alloimmunogenicity of leucocytes in fresh-frozen plasma products. *Vox Sanguinis, 80*(1), 51–56.

Hornsey, V. S., Krailadsiri, P., MacDonald, S., Seghatchian, J., Williamson, L. M., & Prowse, C. V. (2000). Coagulation factor content of cryoprecipitate prepared from methylene blue plus light virus-inactivated plasma. *British Journal of Haematology, 109*(3), 665–670.

Hunter, N., & Houston, F. (2002). Can prion diseases be transmitted between individuals via blood transfusion: Evidence from sheep experiments. *Developments in Biologicals, 108*, 93–98.

Ismail, A. (2017). Kcentra®(Prothrombin Complex Concentrate [Human]): Its indications, side effects, and contraindications. *The Southwest Respiratory and Critical Care Chronicles, 5*(21), 11–15.

Jain, K. K., & Jain, K. K. (2008). *The handbook of nanomedicine* (Vol. 404). Totowa: Humana Press.

Josić, D., Hoffer, L., Buchacher, A., Schwinn, H., Frenzel, W., Biesert, L., & Hellstern, P. (2000). Manufacturing of a prothrombin complex concentrate aiming at low thrombogenicity. *Thrombosis Research, 100*(5), 433–441.

Lee, D. C., Stenland, C. J., Miller, J. L., Cai, K., Ford, E. K., Gilligan, K. J., & Petteway, S. R., Jr (2001). A direct relationship between the partitioning of the pathogenic prion protein and transmissible spongiform encephalopathy infectivity during the purification of plasma proteins. *Transfusion, 41*(4), 449–455.

What-When-how. (2020). Nanofiltration Separations Part 1 (Nanotechnology). Retrieved from: <http://what-when-how.com/nanoscience-and-nanotechnology/nanofiltration-separations-part-1-nanotechnology/ >.

Rezvan, H., Motallebi, Z., Jalili, M. A., Mousavi, H. K., & Pourfath, E. A. (2006). Safety of blood and plasma derivatives: Pathogen reducing technologies. *Medical Journal of the Islamic Republic Iran, 20,* 2006, 2.

Schaefer, A., Fane, A. G., & Waite, T. D. (Eds.), (2005). *Nanofiltration: Principles and applications*. Elsevier.

Sergiy, P. H., Ievgenia, M. K., Olena, S. H., & Georgii, L. V. (2017). The simultaneous human FVIII/vWF purification and virus inactivation combined in chromatographic column. *Journal of Biomolecular Research & Therapeutics, 6*, 2.

Takeda, Y., Wakisaka, A., Noguchi, K., Murozuka, T., Katsubayashi, Y., Matsumoto, S., & Nishioka, K. (2001). Receptor-mediated haemagglutination screening and reduction in the viral load of parvovirus B19 DNA in immunopurified Factor VIII concentrate (Cross Eight M®). *Vox Sanguinis, 81*(4), 266−268.

Tateishi, J., Kitamoto, T., Mohri, S., Satoh, S., Sato, T., Shepherd, A., & Macnaughton, M. R. (2001). Scrapie removal using Planova® virus removal filters. *Biologicals: Journal of the International Association of Biological Standardization, 29*(1), 17−25.

Uddin, S., Mohamad, M., Rahimi-Gorji, M., Roslan, R., & Alarifi, I. M. (2020). Fractional electro-magneto transport of blood modeled with magnetic particles in cylindrical tube without singular kernel. *Microsystem Technologies, 26*(2), 405−414.

Van der Bruggen, B. (2013). Nanofiltration. *Encyclopedia of Membrane Science and Technology*, 1−23.

Van Gestel, T., Vandecasteele, C., Buekenhoudt, A., Dotremont, C., Luyten, J., Leysen, R., & Maes, G. (2002). Alumina and titania multilayer membranes for nanofiltration: Preparation, characterization and chemical stability. *Journal of Membrane Science, 207*(1), 73−89.

World Health Organization. (2004). Guidelines on viral inactivation and removal procedures intended to assure the viral safety of human blood plasma products. *WHO Technical Report*, Series, *924*, 150−224.

Zhao, S., Ba, C., Yao, Y., Zheng, W., Economy, J., & Wang, P. (2018). Removal of antibiotics using polyethylenimine cross-linked nanofiltration membranes: Relating membrane performance to surface charge characteristics. *Chemical Engineering Journal, 335*, 101−109.

Artificial red blood cells

17

Katja Ferenz, Ozan Karaman and Shah Bahrullah Shah
Institut für Physiologie, Universität Duisburg-Essen Universitätsklinikum Essen, Essen, Deutschland

17.1 Introduction

17.1.1 Historical introduction to transfusion history

Blood transfusion has a very long history dating back a couple of centuries. In Europe, bloodletting is more than two centuries old. Blood transfusion evolved through stages, initially from one animal to another and later on from animal to man. Andreas Libavius (1546–1616) of Halle in Saxony (Germany) stated that blood would flow from the artery of a young man to an aged man. A physician at Florence (Italy), Francesco Folli (1624–85), presented his transfusion from a donor animal via a cannula made of bone connected by an isolated artery's segment in the presence of Grand Duke Ferdinand II on August 13, 1654 with missing evidence confirming the success. The author of 'History of Royal Society' (1667), Thomas Sprat considered Christopher Wren as "the first author of the noble anatomical experiment of injecting liquors into the vein of animals … hence arising many new experiments" (Gibson, 1970), as the pioneer of transfusing blood. Forming the *Oxford Experimental Philosophy Club* during the 1650s with scientists such as Robert Hooke, Robert Boyle, Thomas Willis, and Christopher Wren, resulted in the first written confirmation of blood transfusion in 1666. The discovery of the blood groups A, B, O, and AB by Karl Landsteiner in 1901 took the so far failed practice into success (Giangrande, 2000).

17.1.2 Demand of blood transfusions

Currently hundreds of thousands of people receive blood and blood products on a daily basis all over the word. The vital role of the blood in our survival is to fulfill the continuous supply of oxygen to all tissue cells for their normal cellular activities. Blood and its cellular components are associated with important immune functions and are essential for the balance of blood clot formation and degradation. Furthermore, blood mediates the transport of hormones and sustains the physiological pH of 7.4. Blood volumes in human beings vary depending on age, weight, height, gender, and many other factors (Kron et al., 2016). Reduction in blood volume (e.g., from injury or surgery) below critical limits can lead to various complications, known as hypovolemic shock, which can result in organ failure or even death.

Nanotechnology for Hematology, Blood Transfusion, and Artificial Blood. DOI: https://doi.org/10.1016/B978-0-12-823971-1.00018-0

17.1.2.1 *Current state and limitations of conventional blood transfusions*

Red blood cell (RBCs) concentrates, mostly utilized in surgery or emergency medicine to ensure the maintenance of essential body functions of patients, represent an indispensable tool in different clinical scenarios. Statistical analysis prognosticates that with every passing year the number and quality (due to environmental impact and lifestyle) of blood donations and therefore the amount of generated RBCs decreases while the need is actually increasing (Giangrande, 2000; Kron et al., 2016). The portrayed situation is caused by several factors. Over the span of the last decades, a demographic shift has led to an increase of the high aged segment of the society, which represents the prime user population. Patients older than 65 years receive over 50% of blood transfusions and due to the demographic trend, it is expected that this group will double in number within the course of 30 years. Simultaneously, the counterpart, which comprises young people, hence the prime donor age population, will slowly diminish (Henkel-Hanke & Oleck, 2007). The available quantity of RBCs, especially in emergencies, such as wars and natural disasters, will be highly limited. In the current pandemic of SARS-CoV-2 the American Red Cross faced "severe blood shortage" and many blood drives got canceled (Pagano et al., 2020). Furthermore, blood donation from a number of asymptomatic donors entered the blood supply chain (Chang et al., 2020). The highly regulated process of blood collection, processing, and supply in such a scenario is hard to preserve. In consequence, after the initial surge of the current pandemic, there was higher blood demand and collection difficulties due to the restrictions, when some US health institutions finally restored the elective surgeries. Expected financial and logistic stress will build hugely on top of civilian blood collection and processing facilities in the case of bigger disasters such as nuclear detonation etc. Moreover, shortage of blood is associated with economic problems as costs for a unit of blood will drastically increase. Humanity will be confronted with a shortfall of 4 million units of packed RBCs by the year 2030 (Henkel-Hanke & Oleck, 2007).

In addition to the points mentioned above, conventional blood donations or RBCs exhibit properties that impair their utilization and cause undesired side effects. Common and grave complications, which follow a treatment with RBCs span from immune modulations, acute transfusion reactions, transfusion-related lung injury to volume overload and hemolytic reactions (Ferenz & Steinbicker, 2019; Jahr et al., 2019). Further, transmission of bacterial and viral infections as well as general contaminations remain a concern, especially in developing countries (Ferenz & Steinbicker, 2019; Jahr et al., 2019). Due to new emerging diseases, which can be transmitted via transfusion, the cost of a blood unit increases, as additional screenings need to be developed and performed prior to use (Cabrales & Intaglietta, 2013). In addition to the hygienic concerns, the storage of RBCs over a longer period of time results in the so-called storage lesion effect (Brunskill et al., 2015; Henkel-Hanke & Oleck, 2007). The flexibility of the RBC membrane diminishes over time. The erythrocytes begin to leak potassium and are increasingly prone to hemolysis. After storage beyond 42 days 2,3-bisphosphoglycerate becomes undetectable and the ability to offload oxygen is extremely reduced (Henkel-Hanke & Oleck, 2007). Ultimately, mistransfusions still represent a possible scenario, which can be responsible for severe complications (Ferenz & Steinbicker, 2019). All the factors and limitations of conventional blood donations/RBCs stated above provide the impetus for the development of artificial red blood cells as an alternative to blood donations in order to reduce the dependency on and the demand for RBCs.

17.1.3 Solving the limitations of conventional blood transfusions with artificial red blood cells

A perfect alternative or substitute covering all functions of allogenic blood is far from reality, as artificial counterparts of the blood share only specific aspects such as (1) delivery and evacuation of gases, (2) coagulation, and (3) immunological functions, whereas blood exhibits a higher complexity in functions and composition (Ferenz & Steinbicker, 2019). Within this chapter, we will address only gas transport functions of red blood cells. Importantly the term "blood substitutes" needs definition, as conventional "blood substitutes" are merely designed to fulfill the function of oxygen transport and are therefore no complete substitute for blood. More precise are terms such as "artificial red blood cells," "oxygen therapeutic agents," or "artificial oxygen carriers" (AOCs). AOCs aim to generate an adequate and practical alternative to RBCs in order to lower the dependency on blood donations and offer physicians an additional tool in certain problematic clinical scenarios (rare blood groups, immunogenic reactions, religious reasons, oxygen delivery during transplantation, and organ storage) (Ferenz & Steinbicker, 2019; Jahr et al., 2019).

17.2 General requirements for an ideal artificial oxygen carriers

The following criteria need to be fulfilled by a product to consider it as an ideal AOC:

Capability to efficiently operate at physiological gas tensions and to supply the tissue with sufficient oxygen. Ideally, similar behavior as natural hemoglobin concerning oxygen- and carbon dioxide-transport and -delivery (Ferenz & Steinbicker, 2019; Jahr et al., 2019).

Reduction or complete lack of harmful and toxic effects. Properties that can lead to an overload of the reticuloendothelial system (RES), cause nephrotoxicity, impair the cardiovascular system, induce immunologic reactions or other defense mechanisms of the body should be avoided (Henkel-Hanke & Oleck, 2007; Jahr et al., 2007; Nouwairi, 2004).

Similar or lower viscosity as blood to avoid the probability of turbulent flow to occur (Henkel-Hanke & Oleck, 2007; Nouwairi, 2004).

A short organ retention time. Metabolization and elimination of the components should preferentially be mediated by the recipient's natural metabolic system. Sufficient intravascular half-life that allows conveying an impactful physiological effect and universal compatibility (independent of the patient's blood type) (Henkel-Hanke & Oleck, 2007; Jahr et al., 2007).

Long shelf life without quality change to allow for preproduction. Further, ideal AOCs should exhibit convenient usage, which involves stability at room temperature and immediate availability in large quantities (Henkel-Hanke & Oleck, 2007; Nouwairi, 2004).

From an economical perspective, costs to produce AOC should not exceed those of RBCs (Henkel-Hanke & Oleck, 2007).

17.2.1 **Types of artificial oxygen carriers**

The following section will cover the different classes of AOCs. One can distinguish between three different types of AOCs:

1. Hemoglobin-based artificial oxygen carriers (HBOCs)
2. Perfluorocarbon-based artificial oxygen carriers (PFCOCs)
3. Stem cell-derived artificial oxygen carriers (cultured red blood cells, cRBCs).

17.2.2 **Hemoglobin-based artificial oxygen carriers**

As the name already hints, this category of AOCs comprises natural hemoglobin derived from different sources. Hemoglobin can be obtained from blood (outdated RBC or animal blood) or synthesized in bacteria and plants (Ferenz & Steinbicker, 2019). In the 1930s Anderson demonstrated the oxygen transport capability of hemoglobin solutions derived from lysed RBCs and laid the foundation for HBOCs (Ferenz & Steinbicker, 2019; Jahr et al., 2019). However, focused research and development was first conducted in the mid-1980s following the concerns of HIV transmissions via blood transfusions (Henkel-Hanke & Oleck, 2007).

Development and efficient usage of HBOCs are associated with some challenges and drawbacks: HBOCs often exhibit a significantly shorter in vivo half-life of 18−23 hours in comparison to erythrocytes (120 days). Therefore, to adequately replace the function of erythrocytes and maintain oxygen delivery for a longer period, repetitive administration of HBOCs is required. Additionally, necessary hemoglobin still is mostly harvested from RBCs or animal blood. Hence, the generation of HBOCs still depends to a large part on the willingness to donate blood and the risks associated with conventional blood donations remain (Ferenz & Steinbicker, 2019; Henkel-Hanke & Oleck, 2007; Jahr et al., 2019). Furthermore, hemoglobin, as the fundamental component of HBOCs, turns the HBOCs into a double-edged sword. Hemoglobin needs to be highly purified and processed to avoid unwanted side effects; stromal debris can cause renal failure or activate the immune system. However, even the utilization of purified acellular hemoglobin is associated with several complications. Free tetrameric hemoglobin exhibits low stability in circulation and rapidly dissociates into pairs of $\alpha\beta$ dimers. These dimers are extremely prone to oxidation and show increased renal excretion. The oxidation of free hemoglobin to methemoglobin results in nonfunctional hemoglobin and in destabilization of the globin protein chains with subsequent release of cytotoxic heme into the blood circulation. High concentrations of free heme damage the tubule system of the kidney and lead to renal failure. In addition, methemoglobin induces oxidative stress on cellular level (damage of cell membranes, oxidization of nucleic acids and proteins) by producing harmful reactive oxygen species (ROS). Acellular hemoglobin is also able to mediate vasoconstrictive effects and to induce systemic hypertension by scavenging nitric oxide (NO) generated by endothelial cells resulting in reduced blood flow due to local hyperoxia (Cabrales & Intaglietta, 2013; Ferenz & Steinbicker, 2019; Jahr et al., 2019).

To reduce the severe side effects associated with the use of acellular hemoglobin different approaches have been developed. Most common are modifications directly on the hemoglobin molecule itself such as cross-linking. Cross-linking can be performed on the intramolecular (connecting individual or between two adjacent globin chains within a hemoglobin tetramer) or intermolecular level (linking several whole hemoglobin tetramers). This modification increases the stability of the quaternary structure and prevents the dissociation of hemoglobin into αβ dimers, which lowers the risk of renal damage (Tsai et al., 2015). Additionally, hemoglobin can also be linked to adenosine triphosphate (ATP)-adenosine and glutathione (Cabrales & Intaglietta, 2013). While the ATP conveys a stabilizing effect on the tetramer and prevents dimerization, adenosine allows for the creation of hemoglobin oligomers. Further, adenosine and ATP counteract vasoconstrictive and proinflammatory properties of acellular hemoglobin. Lastly, glutathione protects heme from oxidative stress and introduces an electronegative charge. The charge results in an increase of circulation time by reducing renal excretion, extravasation, and phagocytosis (Cabrales & Intaglietta, 2013).

Besides engineering the hemoglobin itself, the strategy holding the most potential to advance HBOC development is the encapsulation of the hemoglobin in any kind of membrane or nanocapsule. Encapsulation reduces toxicity and eradicates many undesired, harmful effects of HBOCs. Hemoglobin engulfed in the protective layer of a nanoparticle exhibits higher stability as well as lower vulnerability to oxidative stress and interaction with NO. Further, nanoparticles enable the integration of modifications to counteract the negative aspects of HBOC application (Ferenz & Steinbicker, 2019; Jahr et al., 2019).

Additional examples of some modifications and their benefits are presented below and in Table 17.1:

The addition of polyethylene glycol to the surface leads to a hydrophilic neutral charge, which decreases opsonization and elimination by the mononuclear phagocytic system. Thereby immunogenicity is reduced and the half-life is prolonged. The half-life in circulation increases proportionally with the percentage of polyethylene glycol coating (Sheng et al., 2009; Tsai et al., 2015; Zolog et al., 2011).

By coencapsulation of enzymes and antioxidative compounds the formation of methemoglobin and the associated toxic effects are avoided (Jahr et al., 2019). Examples are hemoglobin cross-linked with superoxide dismutase and catalase (D'Agnillo & Chang, 1998) or the integration of ferulic acid (Qi et al., 2017).

Choosing the amount of incorporated hemoglobin allows control of the oxygen-carrying capacity, effects on viscosity, and the oncotic pressure (Chang et al., 2020).

A very recent approach is the use of silica nanoparticles. By adsorption to silica, hemoglobin preserves its quaternary structure and enhances its oxygen affinity (Devineau et al., 2018).

17.3 Perfluorocarbon-based oxygen carriers

Perfluorocarbons (PFCs) are compounds in which hydrogen atoms are replaced by halogen atoms, mainly fluorine. The biggest advantage of PFCs lays in the strength of the carbon−fluorine bond. Hence, PFCs are metabolically and chemically inert and their use in vivo will not result in the formation of toxic metabolites (Riess & Riess, 2001). PFCs exhibit a high gas solubility of respiratory gases that linearly depends on their partial pressure (Henry's Law). As they physically dissolve the gases and do not chemically bind them, no saturation of the system occurs (Ferenz & Steinbicker, 2019; Jägers et al., 2020). In contrast to hemoglobin the association and dissociation of oxygen to PFCs happens two times faster. van der Waals interactions between oxygen and PFC molecules are significantly weaker than bonds between hemoglobin and oxygen, resulting in higher extraction rates. Additionally, PFCs unload more than 90% of the dissolved oxygen to the tissue which is three times higher than the physiological oxygen extraction rate (Faithfull, 1992). Further, the oxygen release of PFCs is virtually unaffected by external circumstances. Oxygen release remains effective at physiologically relevant partial pressures, rendering a cooperative effect unnecessary

Table 17.1 Tabular presentation of relevant HBOCs.

Compound/ company	P_{50} (mmHg)	Characteristics	Exemplifying applications/ studies
Hemopure/ Hemoglobin Oxygen Therapeutics	38 (Winslow, 2006)	• Purified bovine hemoglobin with an increased stability due to polymerization with glutaraldehyde.	• Clinically approved in South Africa (Mer et al., 2016) and Russia (Ortiz et al., 2014). • Successfully utilized during hyperhemolysis syndrome of sickle cell anemia (Epstein & Hadley, 2019). • A structural equivalent, OxyGlobin (Biopure; P_{50} 36 mmHG (Izbicki & Ameis, 2006a, 2006b)), is the only FDA approved HBOC for veterinarian use (Liu & Silverstein, 2014).
Sanguinate, PP-007/Prolong	7−16 (Abuchowski, 2017)	• Bovine hemoglobin covalently attached with PEG. -additionally acts as carbon monoxide transporter; functions as vasodilative and antiinflammatory agent and supports oxygen offload in ischemic tissue (Abuchowski, 2017). • In addition, carbon monoxide prevents methemoglobin formation (Abuchowski, 2017).	• Successfully used in severe clinical situations like liver transplantation (Holzner et al., 2018) and anemia following hemorrhage (Thenuwara et al., 2017). • Investigated as a therapeutic option for vasoocclusive crisis of sickle cell patients (NCT02411708).
HemO$_2$Life, HemoxyCarrier, HemoxCell/ Hemarina	7 (Mallet et al., 2014)	• Hemoglobin originated from the marine invertebrate *Arenicola marina* (Mallet et al., 2014). • Structure composes of a hexagonal bilayer formed by 156 linked globin molecules. One molecule is able to bind 156 molecules of O_2 (Mallet et al., 2014). • Exhibits an internal superoxide dismutase activity (Mallet et al., 2014).	• Successfully utilized as an additive for organ preservation solutions (Le Meur et al., 2020) (NCT02652520). • A study, focusing on efficacy of HemO$_2$life in preservation solutions for kidney transplantations is currently ongoing (NCT04181710).
OxyVita/ OXYVITA	6 (Harrington & Wollocko, 2011)	• Globular shaped compound achieved by intra- and intermolecular cross-linking of bovine hemoglobin molecules (Ferenz & Steinbicker, 2019). • Cross-linking via amide bonds (Wollocko et al., 2017) without the use of common toxic agents like glutaraldehyde.	• Studies with respect to use in hemorrhagic shock (Jahr et al., 2012). • Surpassed RBCs in stroke models in respect of reoxygenation and rescue of ischemic brain tissue (Mito et al., 2009).

Table 17.1 Tabular presentation of relevant HBOCs. *Continued*

Compound/ company	P_{50} (mmHg)	Characteristics	Exemplifying applications/ studies
HbVesicles/Terumo Corp.	9−30 (Sakai, 2017)	• Human hemoglobin encased in biocompatible liposomes (Azuma et al., 2017). The outer layer is additionally equipped with PEG$_{5000}$ and p$_{50}$ of HbVesicles is adjustable by using pyridoxalphosphate (Sakai, 2017). • One can distinguish between HbVesicles transporting oxygen (HbO$_2$) or carbon monoxide (Hb-CO) (Ferenz & Steinbicker, 2019).	• Successfully tested in isolated organ perfusion, extra corporeal membrane oxygenation priming and cell culture (Kohno et al., 2017; Sakai, 2017). • Hb-CO exhibits antiinflammatory properties by modulating macrophage and neutrophil activity and was able to reduce inflammation of an acute pancreatitis (Taguchi et al., 2018).
ErythroMer/ KaloCyte Inc.	21 (Pan et al., 2016)	• Human hemoglobin encapsulated in a synthetic polymer shell with integrated 2,3-bisphosphoglycerate shuttle and the antioxidant leukomethylene blue enable the sensing of fluctuation in pH and hence pH dependent changes in oxygen affinity (Pan et al., 2016).	• Reestablished oxygen transport and physiological hemodynamics in anemia/major bleeding models (Pan et al., 2016).
HemoAct/Japan Blood Products Organization	9 (Haruki et al., 2015)	• Human hemoglobin as a core encased in human serum albumin as a protective shell. • One hemoglobin molecule is surrounded by three albumin molecules (Funaki et al., 2019; Haruki et al., 2015).	• Animal models proofed preclinical safety (Haruki et al., 2015). • Exhibited neuroprotective and antioxidative properties in cerebral ischemia (Gekka et al., 2018). • By using recombinant canine and feline serum albumin, Hemoact was customized for cats (HemoAct-F) (Yokomaku et al., 2018) and dogs (HemoAct-C) (Yamada et al., 2016).
VitalHeme (PNPH)/SynZyme Technologies LLC	11 (Cao et al., 2017)	• Pegylated bovine hemoglobin. • The polyethylene glycol is additionally covalently labeled with catalytic antioxidative nitroxides and contributes to colloid pressure (Hsia & Ma, 2012).	• Showed neurovascular protective effects in animal models for traumatic brain injury (Hsia & Ma, 2012). • Showed promising protective effects during arterial occlusion by maintaining vascular function and increasing collateral blood flow (Cao et al., 2017).

(Continued)

Table 17.1 Tabular presentation of relevant HBOCs. *Continued*

Compound/company	P$_{50}$ (mmHg)	Characteristics	Exemplifying applications/studies
YQ23/New B Innovation Limited	40 (Li et al., 2014)	• Bovine, nonpolymerized hemoglobin with cross-linked tetramers and free from stromal remnants (Li et al., 2014).	• Effectively prevented liver tumor metastasis formation after hepatectomy (Li et al., 2014). • First-in-man study to test safety, tolerability, pharmacokinetics started in 2019 (NCT03802292). – studies for the use in solid tumors is still recruiting in 2020 (NCT04513067).
RRBC—rebuilt red blood cells	–	• A multistep protocol was utilized to generate RRBCs (bioreplication of native RBCs by silification, layer-by-layer annealing of biocompatible biopolymers, followed by silica etching and membrane vesicle fusion of RBC ghosts) (Guo et al., 2020)-native RBCs' shape is preserved as well as unique flexibility, deformability and oxygen transport properties (Guo et al., 2020). • RRBCs can be designed with different shells and integrated molecules thanks to the layer-by-layer assembly process and the resulting modular structure (Guo et al., 2020).	– RRBCs can be equipped with several nonnative functionalities for example, loading with MRI imaging agents, therapeutic drugs and luciferase-luciferin ATP biosensor for toxin detection (Guo et al., 2020).
Metal–organic framework (MOF)-based hemoglobin nanoparticles	–	• MOF comprising aluminum (Al) show a crystalline structure with differently sized cavities (Liu et al., 2020). • Al-based MOFs display the highest void volumes and therefore load high amounts of hemoglobin (Liu et al., 2020). • Functionality and structure of hemoglobin preserved (Liu et al., 2020).	• Larger caves occupied by hemoglobin molecules, smaller pores allow diffusion of smaller compounds (oxygen, reducing agents) (Liu et al., 2020).

(Cabrales & Intaglietta, 2013). Additionally, oxygen release by PFCOCs is not regulated by pH and not impaired by temperature (Ignarro et al., 1987). PFCOCs experience no negative alterations of oxygen transport by storage lesions or during circulation as they, in contrast to hemoglobin, are not subjected to oxidization (Cabrales & Intaglietta, 2013). PFCs are not only suitable for oxygen delivery but are also able to dissolve carbon monoxide (CO) and nitrogen (N_2), which might be relevant for the treatment of gas poising or decompression sickness (Ferenz & Steinbicker, 2019; Mayer & Ferenz, 2019). Due to their extraordinary properties, PFCs exhibit strong hydro- and lipophobicity. Therefore PFCs require emulsification or encapsulation for intravascular administration. The development of intravascularly applicable PFCOCs is associated with two challenges:

The selection of the appropriate PFC: the compound is required to exhibit fast elimination from the body, high purity, and good emulsifiability (Giraudeau et al., 2012; Zarif et al., 1994).

The preparation of stable emulsions: PFCOC need to be small-sized, sterile and contain only biocompatible surfactants. The preparation should counteract molecular diffusion and Oswald ripening, two mechanisms responsible for particle growth over time (Giraudeau et al., 2012; Zarif et al., 1994).

The number of PFCs complying with these prerequisites is heavily restricted. The excretion rate of a PFC and the stability of its respective emulsions are dictated by its molecular weight. A rapid excretion is associated with a low molecular weight and therefore with a high lipid solubility. In contrast to that, emulsion stability benefits from high molecular weight, which corresponds to high water solubility. Molecular weight, furthermore, influences the vapor pressure. Low molecular weight PFCs can promote air retention in alveoli, which can lead to increased pulmonary residual volume (Riess & Riess, 2001). Due to their small particle size, PFCOCs appear within the cell-free layer of every blood vessel in contrast to erythrocytes, which are restricted to the center by the Fåhræus—Lindqvist effect. Spatial proximity to the endothelium guarantees a more balanced distribution of oxygen carriers during vessel branching and enhanced oxygen delivery to the microcirculation (Jägers et al., 2020). This gets even more important in an ischemic microcirculation, where PFCOCs perform both cellular oxygenation and restoration of the flexibility of acidotic, stiffened erythrocytes by reinstituting aerobic metabolism (Cabrales & Intaglietta, 2013). From an economical point of view PFC are available in high amounts at modest cost (Cabrales & Intaglietta, 2013). Therefore PFCs hold the potential to enable a cost-efficient AOC production. Yet, although PFCOCs appear to be attractive compounds, they are associated with some drawbacks. Especially the first generation of PFCOCs struggled with significant limitations such as low oxygen delivery and labor-intensive preparation, as well as long organ retention times, low intravascular half-life, and short shelf life (Ferenz & Steinbicker, 2019; Jahr et al., 2019). Due to their ability to sequestrate NO, PFCOCs can cause in general, vasoconstriction, reduction of RBC velocity in postcapillary venules and increase venous leukocyte sticking, when applied in a large dose (Rafikova et al., 2002, 2004).

Table 17.2 briefly introduces relevant PFCOCs.

17.4 Stem cell-derived oxygen carriers

Groundbreaking advances in the area of stem cells research in the last 15 years (recently reviewed by Lee et al., (2018) as well as Christaki et al. (2019)) lay the foundation for efficient in vitro generation of stem cell-derived oxygen carriers also named cultured red blood cells (cRBCs). The production process mimics the natural human erythropoiesis, a highly complex and precisely regulated process, via successive addition of defined transcription and growth factors (Giarratana et al., 2011; Hirose et al., 2013; Lu et al., 2008; Olivier et al., 2012). Hence, physicochemical properties of

Table 17.2 Relevant PFCOCs.

Compound/ company	Characteristics/composition	Oxygen-carrying capacity (mL O_2)/ (dL × mm Hg)	Exemplifying applications/ studies
Fluosol-DA/Fluosol-DA, Green cross crop., and Alpha Therapeutic	PFD, egg-yolk lecithin 0.4% (Hill, 2019)	0.08×10^{-2} (Riess & Riess, 2001)	• Complete myocardial protection (Magovern et al., 1982) • Cerebral-hypoxia (Sutherland et al., 1984) • Angioplasty (Kerins, 1994) • Occlusion (Kent et al., 1990) • Anemia (Tremper et al., 1982)
Oxygent/AF0144, Alliance Pharmaceutical Corporation Double Crane Pharm Co.	58% perfluorooctylbromide, 2% perfluorodecylbromide, egg-yolk phospholipids (Ferenz & Steinbicker, 2019), a-Tocopherol, ethylenediaminetetraacetic acid, sodium chloride, phosphate buffer	3.1×10^{-2} (Faithfull, 1994)	• hemorrhagic Shock (Paxian et al., 2003) • orthopedic surgery with preoperative hemodilution (Spahn et al., 1999) • Decreased allogenic blood transfusion in high blood loss noncardiac surgery (Spahn et al., 2002)
Oxyfluor/HemaGen/ PFC	78% perfluorodichlorooctane, safflower oil, egg-yolk phospholipds (Riess & Krafft, 2006)	0.18×10^{-2} (Kaufman RJ et al., 1994)	• Ardiopulmonary bypass (Briceño et al., 1999) • Hemorrhagic shock (Kaufman RJ et al., 1994)
Oxycyte, ABL-101/ Tenax Therapeutics Inc. Oxygen Biotherapeutics Inc. Formerly: Synthetic Blood International Inc.	60% of perfluorotertbutylcyclohexane, egg-yolk phospholipids (Cabrales et al., 2004)	1.7×10^{-2} (Cabrales et al., 2004)	• Hemodilution (Yang et al., 2008) • Lungs Injury (Haque et al., 2016) • Ischemia (Cabrales et al., 2004) • An ongoing study of utilization of Oxycyte in acute ischemic stroke. (NCT03463551).
OxyPherol/formerly Fluosol-43, Green cross crop. and Alpha Therapeutic	20% perfluorotributylamine, Pluronic F-68 (Riess & Riess, 2001)	0.6×10^{-2} (Faithfull, 1994)	• Angioplasty (Young et al., 1990) • Coagulation and inflammation (Von Der Hardt et al., 2004) • Vasodilation (Southworth et al., 2005) • Isolated heart (Bito et al., 2000)

Table 17.2 Relevant PFCOCs. *Continued*

Compound/ company	Characteristics/composition	Oxygen-carrying capacity (mL O_2)/ (dL $\times$ mm Hg)	Exemplifying applications/ studies
Perftoran, Vidaphor (OJCS SPF Perftoran, Russian Academy of Sciences/Perftoran Company)	12% perfluorodecalin, 3% perfluoromethylcyclohexylpiperidine, 6.5% proxanol 268 (Durnovo et al., 2008)	0.08×10^{-2} (Durnovo et al., 2008)	• Air embolism (Kozhura et al., 2005) • Artherosclerosis (Leskova et al., 2003) • Ischemia (Kozhura et al., 2005) • Inflammation in peritonitis (Lykova et al., 2004)
PFC@PLGA-RBCM	perfluorooctylbromide, polylactide-co-glycolide, RBC membrane	-	• Tumor hypoxia/cancer radiotherapy (Gao et al., 2017)
Dodecafluoropentane/ NuvOx	2% dodecafluoropentane, 5% human serum albumin	2.94 (gas, 37°C)	• Short-term oxygenation of tissue after stroke (Culp et al., 2019) • Tissue preconditioning prior to surgery to decrease ischemia/ reperfusion tissue damage (Strom et al., 2014)
A-AOC	17% perfluorodecalin, 5% albumin	1.43×10^{-2}	• Prevented hypoxic tissue damage in massive hemodilution (Wrobeln et al., 2020) • Effective against decompression illness (Mayer & Ferenz, 2019) • Preserving functionality of ex situ perfused organs (Wrobeln et al., 2017)

cRBCs exhibit the highest similarity to natural erythrocytes when compared to their artificial counterparts (HBOCs, PFCOCs) (Giarratana et al., 2011; Lu et al., 2008). So far five different concepts have been realized: (1) human hematopoietic stem and progenitor cells (hHSPCs), (2) human embryonic stem cells (hESCs) (Giarratana et al., 2011; Lu et al., 2008); (3) induced pluripotent stem cells (iPSCs) (Arora & Daley, 2012; Doulatov et al., 2013; Lengerke et al., 2009), (4) mesenchymal stem cells (Lau et al., 2017), and (5) immortalized cell lines (Lee et al., 2018). Early research was focused on hHSPCs which can be harvested from peripheral blood (Giarratana et al., 2005; Neildez-Nguyen et al., 2002), umbilical cord blood, bone marrow (Baek et al., 2008), mobilized apheresis products, or remnants from leukoreduction of RBCs (Zhou et al., 2020); an expensive and time-consuming process when used in an autologous manner. Today differentiation requirements for hHSPCs are well understood but researchers face problems with availability and upscaling (Zhou et al., 2020). By using peripheral blood mononuclear cells the latest protocols exploit sources that are easier to access (Heshusius et al., 2019) or try to increase yield by improving culture conditions; for example, by developing serum-free media (Giarratana et al., 2011; Hirose et al., 2013; Lu et al., 2008; Olivier et al., 2012) or by mimicking hHSCPs' natural environment such as the bone marrow niche (Severn et al., 2019).

HESCs harvested from the inner layer of blastocysts generated after in vitro fertilization show potential to divide up to 300 times in culture without changing important properties such as pluripotency, telomer length, or appropriate karyotype (Christaki et al., 2019). Completed nucleation and expression of adult β-globin chains indicate mature phenotype (Lu et al., 2008) but remain still an issue. Importantly because of ethical and immunological concerns clinical relevance of hESCs is still limited today (Giarratana et al., 2011; Zhou et al., 2020). Adult somatic cells such as skin fibroblasts reprogrammed to embryonic-like cells are named iPSCs and avoid immunological issues when autologous cells are used (Zhou et al., 2020). Widespread use of iPSC-derived RBCs in clinical practice is limited by inefficient maturation of cells (lack of enucleation, expression of fetal/embryonic globin chains) and problems with upscaling to reach the high numbers of erythrocytes (10^{12}) needed for one blood unit (Focosi & Pistello, 2016). Low expression rates of, for example, Krüppel-like factor 1 and B-cell lymphoma/leukemia 11 A, two factors required for the switch from fetal to adult globin expression, might be responsible for the immature phenotype of iPSC-derived blood cells (Siatecka & Bieker, 2011; Xu et al., 2010). In addition, high fluctuations in transduction efficiencies of required viral vectors impair the homogeneity and reproducibility of the production (Tao & Ghoroghchian, 2014). The cultivation of a robust iPSCs culture required the usage of feeder cells and serum, which pose a potential risk for xenogeneic contamination, until some groups developed feeder-free and xenobiotic-free protocols (Bernecker et al., 2019; Hamada et al., 2020; Olivier et al., 2019). To generate cRBCs using protocols via embryonic body formation is cheaper and more physiological than standard culture conditions (Zhou et al., 2020). However, there is still a need for more studies addressing the physicochemical behavior of iPSC-derived RBCs in vivo (Focosi & Pistello, 2016). Most recent activities in the cRBCs field are the exploitation of mesenchymal stem cells or the use of immortalized erythroid precursor cell lines (Christaki et al., 2019; Lee et al., 2018; Zhou et al., 2020).

17.5 Fate/biodistribution of artificial oxygen carriers in the body

Pharmacokinetics including half-life in the blood as well as organ retention time are important parameters for evaluating AOCs efficiency. Both immediate and late effects of AOCs on the body need to be considered. The ideal AOC should stay in the circulation for at least 24 hours and avoid retention until it is finally metabolized. The biodistribution and metabolism of the AOCs are greatly dependent on their nature (e.g., particle size) and on the material used for their chemical or genetic engineering, which are used to increase the stability (e.g., encapsulation) of the product and to improve the oxygen-carrying efficiency (McGoron et al., 1994). In the case of PFCOCs, organ retention time additionally depends exponentially on the molecular weight of the chosen PFC. Using a combination of a low molecular weight PFC for oxygen transport (e.g., perfluorooctylbromide) with only a small percentage of a slightly lipophilic secondary PFC (e.g., perfluorodecylbromide) can increase PFCOC stability and reduce the excessive organ retention (Riess, 2006). Properties of AOCs should ideally be similar or close to the blood colloid which is exchanged, as circulatory time in the bloodstream is important and should be long enough to preserve an adequate circulating volume (Jägers et al., 2020). For maintaining the blood volume, it is necessary to preserve isoosmotic pressure for attracting body fluids sufficiently. The product should induce neither sludging nor thrombosis by sedimentation of red blood cells due to the difference in viscosity (Hartman, 1952).

Different classes of AOCs show different pharmacokinetic profiles in the body. Radioactive-labeled HBOC "liposomes encapsulating hemoglobin" (Leh) were found up to 50% in blood, 15% in liver, 14% in spleen, 3% in muscles, and some traces appeared in brain, kidney, and heart (Phillips et al., 2009). Carbon clearance measurement and histopathological studies of the HBOC hemoglobin vesicles (HbV) revealed removal by the RES $\sim 40\%$ already after 24 hours and were no longer detectable at day 14 (Sakai et al., 2001). Interestingly, phagocytosis activity decreased at day 3, which was partly attributed to the suppressed defensive function of the body, while a following increase in phagocytic activity leading to HbV elimination at day 14 might been caused by the increased metabolism of HbV (Phillips et al., 2009). This was confirmed in another study of the same group demonstrating significant but minor increases in plasma activities of enzymes such as aminotransferases, alkaline phosphatase, or leucine aminopeptidase but no signs of deteriorative damage to the liver; the main organ for the HbV entrapment and metabolism (Sakai et al., 2001). One year follow-up studies showed elevated plasma lipid levels after HbV administration as a symptom of hyperlipidemia attributed to obesity and aging (Sakai et al., 2007).

Using chemically and metabolically inert PFCs for PFCOCs synthesis is advantageous as no toxic metabolites are formed within the body. Monitoring pharmacokinetics of PFCOCs in real time is possible under ^{19}F-MRI as fluorine atoms are physiologically not present within the body. Distribution of PFC in various organs is dependent on the tissue-blood partition coefficient, the tissue blood flow, and the tissue compartment volume. Diagnostically interesting perfluoro-15-crown-5 ether displays an ideal candidate for a [19]F —MRI contrast agent as it exhibits 20 magnetically equivalent fluorine atoms (Colotti et al., 2017). However, its deposition in liver and spleen remains unchanged for months as compared to other PFCs such as perfluorooctylbromide (PFOB) or perfluorodecalin whose half-life is less than 2 weeks (Jacoby et al., 2014). Various studies (Holaday et al., 1972; Liu & Long, 1976; Modell et al., 1973) demonstrated that vessel-rich organs saturated quickly while fat-rich organs accumulated more PFC. PFOB-based polylactide-co-glycolide (PLGA)-PEG/PFOB spread toward many organs, such as liver heart kidney, lungs, and bowel, in the first hour after the emulsion had been administered through the trachea (Wang et al., 2018). Strategies such as biomimicking RBCs by encapsulating PFC using RBC membrane as in RBC-PFC or PFC@PLGA-RBCM improved the circulatory half-life, prevented cellular uptake, aided immune compatibility, as well as enhanced the outcome in cases of hemorrhagic shock and cancer radiotherapy (Gao et al., 2017; Zhuang et al., 2018). The key route for PFC elimination is expiration through the lungs. Studies suggest PFC uptake with RES, primarily in the liver and spleen (Shaffer, 1996). After uptake of PFCOCs by the RES, pure PFC molecules reach the lung via the bloodstream while bound to lipoproteins (Ferenz & Steinbicker, 2019).

AOCs based on natural giant extracellular hemoglobin from polychaete annelids (HEMOXYCarrier by Hemarina SA, France) showed quick diffusion of fluorescently-labeled product within the whole animal including lungs, liver, brain, and ovaries, but showed no striking side effects (Zhang et al., 2019).

17.6 Biomedical applications of artificial oxygen carriers beyond emergency medicine

Besides the use of AOCs in acute clinical settings like severe hemorrhage, shock, or trauma to minimize the usage of allogenic RBC, AOCs exhibit a broad spectrum of further applications (Table 17.3), and therefore offer additional benefits in comparison to conventional RBCs.

17.7 Role of artificial oxygen carriers in viral infections, for example, in lung diseases

Apart from serving the basic physiological function of blood, AOCs have also enough room to serve a dual function to fight against various pathogen or assist a therapy. In the case of viral infection, nanoparticles in general can serve as a vehicle for a drug or antibiotic. Various potential avenues for nanomaterials, ranging from diagnostics, treatment opportunities, vaccines, and virus-deactivating opportunities in the context of Covid-19, are discussed by Chakhalian et al. (2020). Woods et al. (2015) demonstrated well-tolerated potential use of an albumin vehicle as a drug carrier for pulmonary use and for protease activity in human lungs. These strategies can be adopted by AOCs for drug delivery to virus-infected organs including lungs, for example, by using previously successful albumin-derived perfluorodecalin-based oxygen supplying nanocapsules by Wrobeln et al. (2020).

Blast lung injury accounts for a number of morbidities and mortalities (Zhang et al., 2019). PFC, the main component of PFCOC, has the potential to attenuate acute respiratory distress syndromes both in animals and human models. In a study with dogs Zhang et al. (Zhang et al., 2020) demonstrated how vaporized PFC protected dogs against the severe effects of blast lung injury by suppressing the expression of mitogen-activated protein kinase (MAPK) and nuclear factor 'kappa-light-chain-enhancer' of activated B-cells (NF-κB). The dogs treated with PFC showed low expressions of interleukin (IL)-6 and tumor necrosis factor alpha (TNF-α). PFC regulates the signaling pathways of IL-6 and TNF-α via NF-κB and MAPK pathways, which in turn inhibit the inflammatory factors leading to injury causing reactive oxygen species production.

One of the main concerns in viral infection is thrombosis or thrombotic bodies formation as demonstrated in a case study of critically ill intensive care unit-admitted patients conducted by Klok et al. (2020), where 49% of the study group developed thrombosis. PFCOC have the potential to inhibit thrombosis: PFC nanoparticles functionalized with phenylalanine-proline-arginine-chloromethylketone (PPACK) (PFC-PPACK), as previously shown by Vemuri et al. (2018), improved the allograft function, decreased renal damage, protected vasculature, and improved longevity of renal grafts. The same approach can be used to treat viral infections, especially in the

Table 17.3 Further fields of applications of AOCs.

Field of application	Examples
Transplantation medicine	Preservation of organs destined for transplantation (Funakoshi et al., 1997; Skibba et al., 1985). Organs perfused with AOCs over prolonged periods were successfully transplanted in humans (Fuchinoue et al., 2020).
Oxygenation of cell culture	AOCs can shorten oxygen diffusion distances and improve culture conditions in cell culture, especially on large scale. The only AOC commercially available so far is HEMOXCell by Hemarina (France) (Mouré et al., 2019; Rodriguez-Brotons et al., 2016).
Ultrasound contrast agents	PFCOCs can enhance ultrasound contrast and thus improve diagnostic and allow for targeted therapy (Jägers et al., 2020; Scheer et al., 2017; Schutt et al., 2003).
Sickle cell disease	Patients suffering from sickle cell disease were safely treated with hemoglobin solutions. In comparison to saline-treated patients, infusion of hemoglobin solution resulted in a significantly lower increase in heart rate by exercise. Further, vasoocclusive episodes of sickle cell disease might be an indication for AOC administration (Gonzalez et al., 1997).
Carcinoma therapy/ targeted tumor therapy	Due to their elevated metabolism tumors generate a hypoxic microenvironment. Hypoxia causes resistance against therapy as radiotherapy and the majority of anticancer drugs require an oxygenated tumor environment to cause maximum cytotoxicity. In order to enhance therapy effectivity, AOCs can increase local oxygenation of hypoxic and therapeutically resistant tumor regions (Teicher, 1992; Yu et al., 2019). As PFCOCs can be visualized with MRI or ultrasound, PFCOCs also increase selectivity of tumor therapies as cytotoxic agents such as 5-aminolevulinic acid can be selectively photo-activated in tumor cells only (Scheer et al. 2017).
Microcirculatory support/ perfusion of ischemic tissues	Microcirculatory disturbances can be caused by global ischemia (induced by hemorrhagic shock) or focal ischemia (induced by arterial vessel obstructions). Alterations of the microcirculation comprise leukocyte adhesion and activation, endothelial cell swelling, capillary plugging by RBCs and other blood components. PFCOCs are capable to reduce adhesion and activation of leukocytes in ischemic tissue. Additionally, being smaller than erythrocytes, allows acellular AOCs to pass barriers in microcirculation and supply ischemic tissue (Jägers et al., 2020; Riess & Krafft, 1998).
Visualization of tissue oxygenation	The flour isotope ^{19}F displays magnetic properties and therefore can be detected by magnetic resonance imaging. Tagging PFCOCs with ^{19}F allows for visualization and an displays an additional option for observation of tissue oxygenation (Bellemann et al., 2002).
Reduction of gas embolism	As PFCs present a high solubility for any kind of gas, the usage of PFCOCs to prevent or dissolve gaseous emboli was successfully demonstrated (Mayer et al., 2020). Possible indications are decompression illness (Mayer & Ferenz, 2019) and reduction of air embolism during cardiopulmonary bypass (Reah et al., 1997).
Septic shock	HBOCs administration is associated with systemic hypertension resulting from NO scavenging and induction of norepinephrine by hemoglobin. Clinical trials revealed that HBOC application in patients suffering from septic shock reduced the norepinephrine consumption by 50% (Reah et al., 1997; Waschke & Lenz,1999).
Deep space exploration	Eventual possibility of making secure transfusion of the crew in space by lyophilizing blood products before space crew departure into space, thus minimizing the higher risk of traumatic injuries (Nowak et al., 2019).

case of lung-related viral infections. Another similar approach was to pack antifibrin monoclonal antibodies and urokinase onto PFC nanoparticles' surfaces, as was studied in dogs (Marsh et al., 2011).

17.8 Outlook

Despite the effort and resources invested in the development of AOCs, the transfusion of conventional RBCs still remains the standard procedure to treat high amounts of blood loss. Apparently, major challenges remain to be solved to enable a safe and efficient usage of AOCs. Yet, research will continue since AOCs are urgently awaited to meet the demands of an aging population. Over the last decades many studies, approaches and products failed, but researchers capitalized from these drawbacks by drawing the right conclusions and learning from mistakes. The gained knowledge paved the way to promising and innovative agents, which have been already successfully utilized in in vitro and in vivo scenarios as well as in clinical trials or already approved (e.g., Hemopure, Sanguinate, $HemO_2Life$, Perftoran, Oxygent). AOC are superior to RBCs in many aspects. They offer universal compatibility, longer shelf life convenient storage and the possibility of modifications to transform the compounds to more than oxygen transporting vehicles. The field of AOCs has still not revealed its full potential and will profit from advances in bionanotechnology that open a whole new dimension of novel approaches. An example for innovative ideas is Hb-Scuba, a chimeric hemoglobin created by transferring unique allosteric effects from crocodile to human hemoglobin. The allosteric effect consists of reduction of oxygen affinity through accumulation of bicarbonate ions, the final product of respiration. These ions cause the release of large fractions of hemoglobin-bound oxygen, what allows crocodiles to stay longer under water. However, in humans high amounts of bicarbonate ions to effectively deliver oxygen with Hb-Scuba are not encountered physiologically (Komiyama et al., 1995). Yet, Hb-Scuba with its special properties can be utilized to improve oxygenation in regions of highest need and strong metabolic carbon dioxide production (e.g., tumors).

Like Hb-Scuba, each AOC product, with its unique characteristics, can be considered as a specialized tool that can be utilized in specific clinical scenario and in diagnostic/therapeutic concepts. AOC do not aim to replace RBC transfusions completely or to be considered as blood substitutes. AOC contribute to Patient Blood Management and lower the consumption of allogenic RBCs. Considering the advances and the innovative spirit in the field of artificial oxygen carrier it is only a matter of time until broadly accepted products will be clinically available.

Further reading referring to Tables 26.1, 26.2 and 26.3

1. (Gibson, 1970)
a. (Giangrande, 2000)
b. (Kron et al., 2016)
c. (Pagano et al., 2020)

d. (Chang et al., 2020)

e. (Ferenz & Steinbicker, 2019)

f. (Jahr et al., 2019)

g. (Cabrales & Intaglietta, 2013)

h. (Henkel-Hanke & Oleck, 2007)

i. (Brunskill et al., 2015)

j. (Tissot et al., 2017)

k. (Nouwairi, 2004)

l. (Jahr et al., 2007)

m. (Harris & Palmer, 2008)

n. (Tsai et al., 2015)

o. (Zolog et al., 2011)

p. (Sheng et al., 2009)

q. (D'Agnillo & Chang, 1998)

r. (Qi et al., 2017)

s. (Bäumler et al., 2014)

t. (Chang, 2010)

u. (Devineau et al., 2018)

v. (Winslow, 2006)

w. (Mer et al., 2016)

x. (Ortiz et al., 2014)

y. (Epstein & Hadley, 2019)

z. (Izbicki & Ameis, 2006a, 2006b)

aa. (Holaday et al., 1972; Liu & Long, 1976; Modell et al., 1973)

ab. (Abuchowski, 2017)

ac. (Holzner et al., 2018)

ad. (Thenuwara et al., 2017)

ae. (Mallet et al., 2014)

af. (Le Meur et al., 2020)

ag. (Harrington & Wollocko, 2011)

ah. (Wollocko et al., 2017)

ai. (Jahr et al., 2012)

aj. (Mito et al., 2009)

ak. (Hiromi, 2017)

al. (Hiroshi et al., 2017)

am. (Kohno et al., 2017)

an. (Taguchi et al., 2018)

ao. (Dipanjan et al., 2016)

ap. (Haruki et al., 2015)

aq. (Funaki et al., 2019)

ar. (Gekka et al., 2018)

as. (Yokomaku et al., 2018)

at. (Yamada et al., 2016)

au. (Cao et al., 2017)
av. (Hsia & Ma, 2012)
aw. (Li et al., 2014)
ax. (Guo et al., 2020)
ay. (Liu et al., 2020)
az. (Riess & Riess, 2001)
ba. (Jägers et al., 2020)
bb. (Faithfull, 1992)
bc. (Keipert et al., 1996)
bd. (Ignarro et al., 1987)
be. (Mayer & Ferenz, 2019)
bf. (Zarif et al., 1994)
bg. (Giraudeau et al., 2012)
bh. (Riess, 2005)
bi. (Rafikova et al., 2004)
bj. (Rafikova et al., 2002)
bk. (Hill, 2019)
bl. (Magovern et al., 1982)
bm. (Sutherland et al., 1984)
bn. (Kerins, 1994)
bo. (Kent et al., 1990)
bp. (Tremper et al., 1982)
bq. (Faithfull, 1994)
br. (Paxian et al., 2003)
bs. (Spahn et al., 1999)
bt. (Spahn et al., 1999)
bu. (Riess & Krafft, 2006)
bv. (Goodin et al., 1994)
bw. (Briceño et al., 1999)
bx. (Cabrales et al., 2004)
by. (Cabrales et al., 2004)
bz. (Yang et al., 2008)
ca. (Haque et al., 2016)
cb. (Young et al., 1990)
cc. (Von Der Hardt et al., 2004)
cd. (Southworth et al., 2005)
ce. (Bito et al., 2000)
cf. (Durnovo et al., 2008)
cg. (Kozhura et al., 2005)
ch. (Leskova et al., 2003)
ci. (Lykova et al., 2004)
cj. (Gao et al., 2017)
ck. (Culp et al., 2019)

cl. (Strom et al., 2014)
cm. (Wrobeln et al., 2020)
cn. (Wrobeln et al., 2017)
co. (Lee et al., 2018)
cp. (Christaki et al., 2019)
cq. (Lu et al., 2008)
cr. (Giarratana et al., 2011)
cs. (Olivier et al., 2012)
ct. (Hirose et al., 2013)
cu. (Lengerke et al., 2009)
cv. (Arora & Daley, 2012)
cw. (Doulatov et al., 2013)
cx. (Lau et al., 2017)
cy. (Lee et al., 2016)
cz. (Trakarnsanga et al., 2017)
da. (Neildez-Nguyen et al., 2002)
db. (Giarratana et al., 2005)
dc. (Baek et al., 2008)
dd. (Zhou et al., 2020)
de. (Heshusius et al., 2019)
df. (Kim et al., 2019)
dg. (Olivier et al., 2019)
dh. (Severn et al., 2019)
di. (Focosi & Pistello, 2016)
dj. (Xu et al., 2010)
dk. (Siatecka & Bieker, 2011)
dl. (Tao & Ghoroghchian, 2014)
dm. (Bernecker et al., 2019)
dn. (Hamada et al., 2020)
do. (McGoron et al., 1994)
dp. (Riess, 2006)
dq. (Hartman, 1952)
Dr. (Phillips et al., 2009)
ds. (Sakai et al., 2001)
dt. (Sakai et al., 2007)
du. (Colotti et al., 2017)
dv. (Jacoby et al., 2014)
dw. (Liu & Long, 1976)
dx. (Modell et al., 1973)
dy. (Holaday et al., 1972)
dz. (Wang et al., 2018)
ea. (Zhuang et al., 2018)
eb. (Shaffer, 1996)

ec. (Zhuang et al., 2018)
ed. (Woods et al., 2015)
ee. (Zhang et al., 2020)
ef. (Klok et al., 2020)
e.g., (Vemuri et al., 2018)
eh. (Marsh et al., 2011)
ei. (Skibba et al., 1985)
ej. (Funakoshi et al., 1997)
ek. (Fuchinoue et al., 1986)
el. (Mouré et al., 2019)
em. (Rodriguez-Brotons et al., 2016)
en. (Scheer et al., 2017)
eo. (Schutt et al., 2003)
ep. (Gonzalez et al., 1997)
eq. (Teicher, 1992)
er. (Peng et al., 2019)
es. (Riess & Krafft, 1998)
et. (Bellemann et al., 2002)
eu. (Mayer et al., 2020)
ev. (Reah et al., 1997)
ew. (Waschke & Lenz, 1999)
ex. (Nowak et al., 2019)
ey. (Komiyama et al., 1995)

References

Abuchowski, A. (2017). Sanguinate (PEGylated Carboxyhemoglobin Bovine): Mechanism of action and clinical update. *Artificial Organs*, *41*(4), 346−350. Available from https://doi.org/10.1111/aor.12934.

Arora, N., & Daley, G. Q. (2012). Pluripotent stem cells in research and treatment of hemoglobinopathies. *Cold Spring Harbor Perspectives in Medicine*, *2*(4), a011841. Available from https://doi.org/10.1101/cshperspect.a011841.

Azuma, H., Fujihara, M., & Sakai, H. (2017). Biocompatibility of HbV: Liposome-encapsulated hemoglobin molecules-liposome effects on immune function. *Journal of Functional Biomaterials*, *8*(3), 24. Available from https://doi.org/10.3390/jfb8030024.

Baek, E. J., Kim, H. S., Kim, S., Jin, H., Choi, T. Y., & Kim, H. O. (2008). In vitro clinical-grade generation of red blood cells from human umbilical cord blood CD_3^{4+} cells. *Transfusion*, *48*(10), 2235−2245. Available from https://doi.org/10.1111/j.1537-2995.2008.01828.x.

Bäumler, H., Xiong, Y., Liu, Z. Z., Patzak, A., & Georgieva, R. (2014). Novel hemoglobin particles-promising new-generation hemoglobin-based oxygen carriers. *Artificial Organs*, *38*(8), 708−714. Available from https://doi.org/10.1111/aor.12331.

Bellemann, M. E., Brückner, J., Peschke, P., Brix, G., & Mason, R. P. (2002). Quantification and visualization of oxygen partial pressure in vivo by19f NMR imaging of perfluorocarbons. *Biomedizinische Technik*, *47*, 451−454. Available from https://doi.org/10.1515/bmte.2002.47.s1a.451.

Bernecker, C., Ackermann, M., Lachmann, N., Rohrhofer, L., Zaehres, H., Araúzo-Bravo, M. J., Van Den Akker, E., Schlenke, P., & Dorn, I. (2019). Enhanced ex vivo generation of erythroid cells from human induced pluripotent stem cells in a simplified cell culture system with low cytokine support. *Stem Cells and Development*, *28*(23), 1540−1551. Available from https://doi.org/10.1089/scd.2019.0132.

Bito, A., Inoue, K., Asano, M., Ando, S., & Takaba, T. (2000). Experimental myocardial preservation study of adding perfluorochemicals (FC$_{43}$) in lidocaine cardioplegia. *The Japanese Journal of Thoracic and Cardiovascular Surgery : Official Publication of the Japanese Association for Thoracic Surgery = Nihon Kyōbu Geka Gakkai Zasshi*, *48*(5), 280−290. Available from https://doi.org/10.1007/BF03218140.

Briceño, J. C., Rincón, I. E., Vélez, J. F., Castro, I., Arcos, M. I., & Velásquez, C. E. (1999). Oxygen transport and consumption during experimental cardiopulmonary bypass using oxyfluor. *ASAIO Journal*, *45*(4), 322−327. Available from https://doi.org/10.1097/00002480-199907000-00013.

Brunskill, S. J., Wilkinson, K. L., Doree, C., Trivella, M., & Stanworth, S. (2015). Transfusion of fresher vs older red blood cells for all conditions. *Cochrane Database of Systematic Reviews*, *2015*(5). Available from https://doi.org/10.1002/14651858.CD010801.pub2.

Cabrales, P., & Intaglietta, M. (2013). Blood substitutes: Evolution from noncarrying to oxygen- and gas-carrying fluids. *ASAIO Journal*, *59*(4), 337−354. Available from https://doi.org/10.1097/MAT.0b013e318291fbaa.

Cabrales, P., Tsai, A. G., Frangos, J. A., Briceño, J. C., & Intaglietta, M. (2004). Oxygen delivery and consumption in the microcirculation after extreme hemodilution with perfluorocarbons. *American Journal of Physiology: Heart and Circulatory Physiology*, *287*. Available from https://doi.org/10.1152/ajpheart.01166.2003. (1 56−1).

Cao, S., Zhang, J., Ma, L., Hsia, C. J. C., & Koehler, R. C. (2017). Transfusion of polynitroxylated pegylated hemoglobin stabilizes pial arterial dilation and decreases infarct volume after transient middle cerebral artery occlusion. *Journal of the American Heart Association*, *6*(9). Available from https://doi.org/10.1161/JAHA.117.006505.

Chakhalian, D., Shultz, R. B., Miles, C. E., & Kohn, J. (2020). Opportunities for biomaterials to address the challenges of COVID-19. *Journal of Biomedical Materials Research - Part A*, *108*(10), 1974−1990. Available from https://doi.org/10.1002/jbm.a.37059.

Chang, L., Zhao, L., Gong, H., Gong, H., Gong, H., Wang, L., Wang, L., Wang, L., Yang, Z., & Xu, B. (2020). Severe acute respiratory syndrome coronavirus 2 RNA detected in blood donations. *Emerging Infectious Diseases*, *26*(7), 1631−1633. Available from https://doi.org/10.3201/eid2607.200839.

Chang, T. M. S. (2010). Blood replacement with nanobiotechnologically engineered hemoglobin and hemoglobin nanocapsules. *Wiley Interdisciplinary Reviews: Nanomedicine and Nanobiotechnology*, *2*(4), 418−430. Available from https://doi.org/10.1002/wnan.95.

Christaki, E. E., Politou, M., Antonelou, M., Athanasopoulos, A., Simantirakis, E., Seghatchian, J., & Vassilopoulos, G. (2019). Ex vivo generation of transfusable red blood cells from various stem cell sources: A concise revisit of where we are now. *Transfusion and Apheresis Science*, *58*(1), 108−112. Available from https://doi.org/10.1016/j.transci.2018.12.015.

Colotti, R., Bastiaansen, J. A. M., Wilson, A., Flögel, U., Gonzales, C., Schwitter, J., Stuber, M., & van Heeswijk, R. B. (2017). Characterization of perfluorocarbon relaxation times and their influence on the optimization of fluorine-19 MRI at 3 tesla. *Magnetic Resonance in Medicine*, *77*(6), 2263−2271. Available from https://doi.org/10.1002/mrm.26317.

Culp, W. C., Onteddu, S. S., Brown, A., Nalleballe, K., Sharma, R., Skinner, R. D., Witt, T., Roberson, P. K., & Marsh, J. D. (2019). Dodecafluoropentane emulsion in acute ischemic stroke: A phase Ib/II randomized and controlled dose-escalation trial. *Journal of Vascular and Interventional Radiology*, *30*(8), 1244−1250. Available from https://doi.org/10.1016/j.jvir.2019.04.020, e1.

Devineau, S., Kiger, L., Galacteros, F., Baudin-Creuza, V., Marden, M., Renault, J. P., & Pin, S. (2018). Manipulating hemoglobin oxygenation using silica nanoparticles: A novel prospect for artificial oxygen carriers. *Blood Advances*, *2*(2), 90−94. Available from https://doi.org/10.1182/bloodadvances.2017012153.

Dipanjan, P., Stephen, R., Santosh, M., Gururaja, V., Lisa, G., Albert, T., Nikhil, M., Ahmed, S., Philip, S., Greg, H., Greg, L., & Allan, D. (2016). Erythromer (EM), a nanoscale bio-synthetic artificial red cell: Proof of concept and in vivo efficacy results. *Blood*, 1027. Available from https://doi.org/10.1182/blood. v128.22.1027.1027.

Doulatov, S., Vo, L. T., Chou, S. S., Kim, P. G., Arora, N., Li, H., Hadland, B. K., Bernstein, I. D., Collins, J. J., Zon, L. I., & Daley, G. Q. (2013). Induction of multipotential hematopoietic progenitors from human pluripotent stem cells via respecification of lineage-restricted precursors. *Cell Stem Cell*, *13*(4), 459−470. Available from https://doi.org/10.1016/j.stem.2013.09.002.

Durnovo, E. A., Furman, I. V., Pushkin, S. Y., Maslennikov, I. A., Bondar, O. G., & Ivanitsky, G. R. (2008). Clinical results of the application of perftoran for the treatment of odontogenous abcesses and phlegmons in the maxillofacial region. *Journal of Cranio-Maxillofacial Surgery*, *36*(3), 161−172. Available from https://doi.org/10.1016/j.jcms.2007.07.012.

D'Agnillo, F., & Chang, T. M. S. (1998). Polyhemoglobin-superoxide dismutase catalase as a blood substitute with antioxidant properties. *Nature Biotechnology*, *16*(7), 667−671. Available from https://doi.org/10.1038/nbt0798-667.

Epstein, S. S., & Hadley, T. J. (2019). Successful management of the potentially fatal hyperhaemolysis syndrome of sickle cell anaemia with a regimen including bortezomib and Hemopure. *Journal of Clinical Pharmacy and Therapeutics*, *44*(5), 815−818. Available from https://doi.org/10.1111/jcpt.12998.

Faithfull, N. S. (1992). Oxygen delivery from fluorocarbon emulsions—Aspects of convective and diffusive transport. *Biomaterials, Artificial Cells and Immobilization Biotechnology*, *20*(4), 797−804. Available from https://doi.org/10.3109/10731199209119721.

Faithfull, N. S. (1994). Mechanisms and efficacy of fluorochemical oxygen transport and delivery. *Artificial Cells, Blood Substitutes, and Biotechnology*, *22*(2), 181−197. Available from https://doi.org/10.3109/10731199409117413.

Ferenz, K. B., & Steinbicker, A. U. (2019). Artificial oxygen carriers—Past, present, and future—A review of the most innovative and clinically relevant concepts. *Journal of Pharmacology and Experimental Therapeutics*, *369*(2), 300−310. Available from https://doi.org/10.1124/jpet.118.254664.

Focosi, D., & Pistello, M. (2016). Effect of induced pluripotent stem cell technology in blood banking. *Stem Cells Translational Medicine*, *5*(3), 269−274. Available from https://doi.org/10.5966/sctm.2015-0257.

Fuchinoue, S., Takahashi, K., & Teraoka, S. (1986). Clinical experience in kidney preservation with a new fluorocarbon emulsion perfusate. *Transplantation Proceedings*, *18*(3), 566−570, Accessed November 11, 2020. Available from https://www.researchgate.net/publication/283940089_Clinical_experience_in_kidney_preservation_with_a_new_fluorocarbon_emulsion_perfusate.

Funaki, R., Kashima, T., Okamoto, W., Sakata, S., Morita, Y., Sakata, M., & Komatsu, T. (2019). Hemoglobin-albumin clusters prepared using N-succinimidyl 3-maleimidopropionate as an appropriate cross-linker. *ACS Omega*, *4*(2), 3228−3233. Available from https://doi.org/10.1021/acsomega.8b03474.

Funakoshi, Y., Fujita, S., Fuchinõue, S., Agishi, T., & Ota, K. (1997). Neo red cell as an organ preservation solution. *Artificial Cells, Blood Substitutes, and Immobilization Biotechnology*, *25*(4), 407−416. Available from https://doi.org/10.3109/10731199709118930.

Gao, M., Liang, C., Song, X., Chen, Q., Jin, Q., Wang, C., & Liu, Z. (2017). Erythrocyte-membrane-enveloped perfluorocarbon as nanoscale artificial red blood cells to relieve tumor hypoxia and enhance cancer radiotherapy. *Advanced Materials*, *29*(35). Available from https://doi.org/10.1002/adma.201701429.

Gekka, M., Abumiya, T., Komatsu, T., Funaki, R., Kurisu, K., Shimbo, D., Kawabori, M., Osanai, T., Nakayama, N., Kazumata, K., & Houkin, K. (2018). Novel hemoglobin-based oxygen carrier bound with albumin shows neuroprotection with possible antioxidant effects. *Stroke: A Journal of Cerebral Circulation*, *49*(8), 1960−1968. Available from https://doi.org/10.1161/STROKEAHA.118.021467.

Giangrande, P. L. F. (2000). The history of blood transfusion. *British Journal of Haematology*, *110*(4), 758−767. Available from https://doi.org/10.1046/j.1365-2141.2000.02139.x.

Giarratana, M. C., Kobari, L., Lapillonne, H., Chalmers, D., Kiger, L., Cynober, T., Marden, M. C., Wajcman, H., & Douay, L. (2005). Ex vivo generation of fully mature human red blood cells from hematopoietic stem cells. *Nature Biotechnology*, *23*(1), 69−74. Available from https://doi.org/10.1038/nbt1047.

Giarratana, M. C., Rouard, H., Dumont, A., Kiger, L., Safeukui, I., Le Pennec, P. Y., François, S., Trugnan, G., Peyrard, T., Marie, T., Jolly, S., Hebert, N., Mazurier, C., Mario, N., Harmand, L., Lapillonne, H., Devaux, J. Y., & Douay, L. (2011). Proof of principle for transfusion of in vitro-generated red blood cells. *Blood*, *118*(19), 5071−5079. Available from https://doi.org/10.1182/blood-2011-06-362038.

Gibson, W. C. (1970). The bio-medical pursuits of Christopher Wren. *Medical History*, *14*(4), 331−341. Available from https://doi.org/10.1017/S0025727300015787.

Giraudeau, C., Djemaï, B., Ghaly, M. A., Boumezbeur, F., Mériaux, S., Robert, P., Port, M., Robic, C., Bihan, D. L., Lethimonnier, F., & Valette, J. (2012). High sensitivity 19F MRI of a perfluorooctyl bromide emulsion: Application to a dynamic biodistribution study and oxygen tension mapping in the mouse liver and spleen. *NMR in Biomedicine*, *25*(4), 654−660. Available from https://doi.org/10.1002/nbm.1781.

Gonzalez, P., Hackney, A. C., Jones, S., Strayhorn, D., Hoffman, E. B., Hughes, G., Jacobs, E. E., & Orringer, E. P. (1997). A phase I/II study of polymerized bovine hemoglobin in adult patients with sickle cell disease not in crisis at the time of study. *Journal of Investigative Medicine*, *45*(5), 258−264. Available from http://jim.bmj.com/content/about-us.

Goodin, T. T., Grossbard, E. B., Kaufman, R. J., Richard, T. J., Kolata, R. J., Allen, J. S., & Layton, T. E. (1994). A perfluorochemical emulsion for prehospital resuscitation of experimental hemorrhagic shock. *Critical Care Medicine*, 680−689. Available from https://doi.org/10.1097/00003246-199404000-00026.

Guo, J., Agola, J. O., Serda, R., Franco, S., Lei, Q., Wang, L., Minster, J., Croissant, J. G., Butler, K. S., Zhu, W., & Brinker, C. J. (2020). Biomimetic rebuilding of multifunctional red blood cells: Modular design using functional components. *ACS Nano*, *14*(7), 7847−7859. Available from https://doi.org/10.1021/acsnano.9b08714.

Hamada, A., Akagi, E., Obayashi, F., Yamasaki, S., Koizumi, K., Ohtaka, M., Nishimura, K., Nakanishi, M., Toratani, S., & Okamoto, T. (2020). Induction of Noonan syndrome-specific human-induced pluripotent stem cells under serum-, feeder-, and integration-free conditions. *In Vitro Cellular and Developmental Biology - Animal*. Available from https://doi.org/10.1007/s11626-020-00515-9.

Haque, A., Scultetus, A. H., Arnaud, F., Dickson, L. J., Chun, S., McNamee, G., Auker, C. R., McCarron, R. M., & Mahon, R. T. (2016). The emulsified PFC Oxycyte® improved oxygen content and lung injury score in a swine model of oleic acid lung injury (OALI). *Lung*, *194*(6), 945−957. Available from https://doi.org/10.1007/s00408-016-9941-9.

Harrington, J. P., & Wollocko, H. (2011). Molecular design properties of oxyvita hemoglobin, a new generation therapeutic oxygen carrier: A review. *Journal of Functional Biomaterials.*, *2*(4), 414−424. Available from https://doi.org/10.3390/jfb2040414.

Harris, D. R., & Palmer, A. F. (2008). Modern cross-linking strategies for synthesizing acellular hemoglobin-based oxygen carriers. *Biotechnology Progress*, *24*(6), 1215−1225. Available from https://doi.org/10.1002/btpr.85.

Hartman, F. W. (1952). Fate and disposal of plasma substitutes. *Annals of the New York Academy of Sciences*, *55*(3), 504−512. Available from https://doi.org/10.1111/j.1749-6632.1952.tb26569.x.

Haruki, R., Kimura, T., Iwasaki, H., Yamada, K., Kamiyama, I., Kohno, M., Taguchi, K., Nagao, S., Maruyama, T., Otagiri, M., & Komatsu, T. (2015). Safety evaluation of hemoglobin-albumin cluster

\hemoact\as a red blood cell substitute. *Scientific Reports*, *5*. Available from https://doi.org/10.1038/srep12778.

Henkel-Hanke, T., & Oleck, M. (2007). Artificial oxygen carriers: A current review. *AANA Journal*, *75*(3), 205−211, Accessed November 11, 2020. Available from https://pubmed.ncbi.nlm.nih.gov/17591302/.

Heshusius, S., Heideveld, E., Burger, P., Thiel-Valkhof, M., Sellink, E., Varga, E., Ovchynnikova, E., Visser, A., Martens, J. H. A., von Lindern, M., & van den Akker, E. (2019). Large-scale in vitro production of red blood cells from human peripheral blood mononuclear cells. *Blood Advances*, *3*(21), 3337−3350. Available from https://doi.org/10.1182/bloodadvances.2019000689.

Hill, S. E. (2019). Perfluorocarbons. *Shock (Augusta, Ga.)*, *52*(1S), 60−64. Available from https://doi.org/10.1097/SHK.0000000000001045.

Hiromi, S. (2017). Overview of potential clinical applications of hemoglobin vesicles (HbV) as artificial red cells, evidenced by preclinical studies of the academic research consortium. *Journal of Functional Biomaterials*, *10*. Available from https://doi.org/10.3390/jfb8010010.

Hirose, S. I., Takayama, N., Nakamura, S., Nagasawa, K., Ochi, K., Hirata, S., Yamazaki, S., Yamaguchi, T., Otsu, M., Sano, S., Takahashi, N., Sawaguchi, A., Ito, M., Kato, T., Nakauchi, H., & Eto, K. (2013). Immortalization of erythroblasts by c-MYC and BCL-XL enables large-scale erythrocyte production from human pluripotent stem cells. *Stem Cell Reports*, *1*(6), 499−508. Available from https://doi.org/10.1016/j.stemcr.2013.10.010.

Hiroshi, A., Mitsuhiro, F., & Hiromi, S. (2017). Biocompatibility of HbV: Liposome-encapsulated hemoglobin molecules-liposome effects on immune function. *Journal of Functional Biomaterials*, *24*. Available from https://doi.org/10.3390/jfb8030024.

Holaday, D. A., Fiserova-Bergerova, V., & Modell, J. H. (1972). Uptake, distribution, and excretion of fluorocarbon FX-80 (perfluorobutyl perfluorotetrahydrofuran) during liquid breathing in the dog. *Anesthesiology*, *37*(4), 387−394. Available from https://doi.org/10.1097/00000542-197210000-00005.

Holzner, M. L., DeMaria, S., Haydel, B., Smith, N., Flaherty, D., & Florman, S. (2018). Pegylated bovine carboxyhemoglobin (SANGUINATE) in a Jehovah's witness undergoing liver transplant: A case report. *Transplantation Proceedings*, *50*(10), 4012−4014. Available from https://doi.org/10.1016/j.transproceed.2018.09.006.

Hsia, C. J. C., & Ma, L. (2012). A hemoglobin-based multifunctional therapeutic: Polynitroxylated pegylated hemoglobin. *Artificial Organs*, *36*(2), 215−220. Available from https://doi.org/10.1111/j.1525-1594.2011.01307.x.

Ignarro, L. J., Byrns, R. E., Buga, G. M., & Wood, K. S. (1987). Endothelium-derived relaxing factor from pulmonary artery and vein possesses pharmacologic and chemical properties identical to those of nitric oxide radical. *Circulation Research*, *61*(6), 866−879. Available from https://doi.org/10.1161/01.RES.61.6.866.

Izbicki, M. J. R., & Ameis, H. M. (2006a). Effekte der therapeutischen Gabe der zellfreien Hämoglobinlösung HBOC-301 (Oxyglobin TM) auf die Gewebsoxygenierung des Pankreas bei schwerer akuter Pankreatitis im Großtiermodell. https://ediss.sub.uni-hamburg.de/handle/ediss/1654.

Izbicki, M. J. R., & Ameis, H. M. (2006b). Universitätsklinikum Hamburg-Eppendorf Klinik Und Poliklinik Für Allgemein-, Viszeral-Und Thoraxchirurgie.

Jacoby, C., Temme, S., Mayenfels, F., Benoit, N., Krafft, M. P., Schubert, R., Schrader, J., & Flögel, U. (2014). Probing different perfluorocarbons for in vivo inflammation imaging by 19F MRI: Image reconstruction, biological half-lives and sensitivity. *NMR in Biomedicine*, *27*(3), 261−271. Available from https://doi.org/10.1002/nbm.3059.

Jägers, J., Wrobeln, A., & Ferenz, K. B. (2020). Perfluorocarbon-based oxygen carriers: From physics to physiology. *Pflugers Archiv European Journal of Physiology*. Available from https://doi.org/10.1007/s00424-020-02482-2.

Jahr, J. S., Walker, V., & Manoochehri, K. (2007). Blood substitutes as pharmacotherapies in clinical practice. *Current Opinion in Anaesthesiology*, *20*(4), 325−330. Available from https://doi.org/10.1097/ACO.0b013e328172225a.

Jahr, J. S., Akha, A. S., & Holtby, R. J. (2012). Crosslinked, polymerized, and peg-conjugated hemoglobin-based oxygen carriers: Clinical safety and efficacy of recent and current products. *Current Drug Discovery Technologies*, *9*(3), 158−165. Available from https://doi.org/10.2174/157016312802650742.

Jahr, J. S., Guinn, N. R., Lowery, D. R., Shore-Lesserson, L., & Shander, A. (2019). Blood substitutes and oxygen therapeutics. *Anesthesia and Analgesia*, 1−9. Available from https://doi.org/10.1213/ane.0000000000003957, Publish Ah(Xxx).

Keipert, P. E., Faithfull, N. S., Roth, D. J., Bradley, J. A. D., Batra, S., Jochelson, P., & Flaim, K. E. (1996). *Supporting tissue oxygenation during acute surgical bleeding using a perfluorochemical-based oxygen carrier. Advances in experimental medicine and biology* (Vol. 388, pp. 603−609). New York LLC: Springer. Available from https://doi.org/10.1007/978-1-4613-0333-6_77.

Kent, K. M., Cleman, M. W., Cowley, M. J., Forman, M. B., Jaffe, C. C., Kaplan, M., King, S. B., Krucoff, M. W., Lassar, T., McAuley, B., Smith, R., Wisdom, C., & Wohlgelernter, D. (1990). Reduction of myocardial ischemia during percutaneous transluminal coronary angioplasty with oxygenated Fluosol®1 1 Fluosol® (20% Intravascular Perfluorochemical Emulsion) is a registered trademark of The Green Cross Corporation, Osaka, Japan. *The American Journal of Cardiology*, *66*(3), 279−284. Available from https://doi.org/10.1016/0002-9149(90)90836-P.

Kerins, D. M. (1994). Role of the perfluorocarbon fluosal-DA in coronary angioplasty. *American Journal of the Medical Sciences*, *307*(3), 218−221. Available from https://doi.org/10.1097/00000441-199403000-00009.

Kim, S. H., Lee, E. M., Han, S. Y., Choi, H. S., Ryu, K. Y., & Baek, E. J. (2019). Improvement of red blood cell maturation in vitro by serum-free medium optimization. *Tissue Engineering - Part C: Methods*, *25*(4), 232−242. Available from https://doi.org/10.1089/ten.tec.2019.0023.

Klok, F. A., Kruip, M. J. H. A., van der Meer, N. J. M., Arbous, M. S., Gommers, D., Kant, K. M., Kaptein, F. H. J., van Paassen, J., Stals, M. A. M., Huisman, M. V., & Endeman, H. (2020). Confirmation of the high cumulative incidence of thrombotic complications in critically ill ICU patients with COVID-19: An updated analysis. *Thrombosis Research*, *191*, 148−150. Available from https://doi.org/10.1016/j.thromres.2020.04.041.

Kohno, M., Ikeda, T., Hashimoto, R., Izumi, Y., Watanabe, M., Horinouchi, H., Sakai, H., Kobayashi, K., & Iwazaki, M. (2017). Acute 40% exchange-transfusion with hemoglobin-vesicles in a mouse pneumonectomy model. *PLoS One*, *12*(6). Available from https://doi.org/10.1371/journal.pone.0178724.

Komiyama, N. H., Miyazaki, G., & Tamef, J. (1995). Transplanting a unique allosteric effect from crocodile into human haemoglobin. *Nature*, *373*(6511), 244−246. Available from https://doi.org/10.1038/373244a0.

Kozhura, V. L., Basarab, D. A., Timkina, M. I., Golubev, A. M., Reshetnyak, V. I., & Moroz, V. V. (2005). Reperfusion injury after critical intestinal ischemia and its correction with perfluorochemical emulsion "perftoran.". *World Journal of Gastroenterology: WJG*, *11*(45), 7084−7090. Available from https://doi.org/10.3748/wjg.v11.i45.7084.

Kron, S., Schneditz, D., Leimbach, T., Czerny, J., Aign, S., & Kron, J. (2016). Determination of the critical absolute blood volume for intradialytic morbid events. *Hemodialysis International*, *20*(2), 321−326. Available from https://doi.org/10.1111/hdi.12375.

Lau, S. X., Leong, Y. Y., Ng, W. H., Ng, A. W. P., Ismail, I. S., Yusoff, N. M., Ramasamy, R., & Tan, J. J. (2017). Human mesenchymal stem cells promote CD_3^{4+} hematopoietic stem cell proliferation with preserved red blood cell differentiation capacity. *Cell Biology International*, *41*(6), 697−704. Available from https://doi.org/10.1002/cbin.10774.

Le Meur, Y., Badet, L., Essig, M., Thierry, A., Büchler, M., Drouin, S., Deruelle, C., Morelon, E., Pesteil, F., Delpech, P. O., Boutin, J. M., Renard, F., & Barrou, B. (2020). First-in-human use of a marine oxygen carrier (M101) for organ preservation: A safety and proof-of-principle study. *American Journal of Transplantation, 20*(6), 1729−1738. Available from https://doi.org/10.1111/ajt.15798.

Lee, E., Sivalingam, J., Lim, Z. R., Chia, G., Shi, L. G., Roberts, M., Loh, Y. H., Reuveny, S., & Oh, S. K. W. (2018). Review: In vitro generation of red blood cells for transfusion medicine: Progress, prospects and challenges. *Biotechnology Advances, 36*(8), 2118−2128. Available from https://doi.org/10.1016/j.biotechadv.2018.09.006.

Lee, S. A., Kim, J. Y., Choi, Y., Kim, Y., & Kim, H. O. (2016). Application of mutant JAK2V617F for in vitro generation of red blood cells. *Transfusion, 56*(4), 837−843. Available from https://doi.org/10.1111/trf.13431.

Lengerke, C., Grauer, M., Niebuhr, N. I., Riedt, T., Kanz, L., Park, I. H., & Daley, G. Q. (2009). *Hematopoietic development from human induced pluripotent stem cells. Annals of the New York Academy of Sciences* (Vol. 1176, pp. 219−227). Blackwell Publishing Inc. Available from https://doi.org/10.1111/j.1749-6632.2009.04606.x.

Leskova, G. F., Michunskaya, A. B., Koboz eva, L. P., & Klimenko, E. D. (2003). Influence of perftoran on structural and metabolic disturbances in the liver during experimental atherosclerosis. *Bulletin of Experimental Biology and Medicine, 136*(4), 340−343. Available from https://doi.org/10.1023/B:BEBM.0000010946.82543.42.

Li, C. X., Wong, B. L., Ling, C. C., Ma, Y. Y., Shao, Y., Geng, W., Qi, X., Lau, S. H., Kwok, S. Y., Wei, N., Tzang, F. C., Ng, K. T. P., Liu, X. B., Lo, C. M., & Man, K. (2014). A novel oxygen carrier \YQ$_{23}$\suppresses the liver tumor metastasis by decreasing circulating endothelial progenitor cells and regulatory T cells. *BMC Cancer, 14*(1). Available from https://doi.org/10.1186/1471-2407-14-293.

Liu, D. T., & Silverstein, D. C. (2014). *Crystalloids, colloids, and hemoglobin-based oxygen-carrying solutions. Small animal critical care medicine* (2nd (Ed.), pp. 311−316). Elsevier Health Sciences. Available from https://doi.org/10.1016/B978-1-4557-0306-7.00058-1.

Liu, M. S., & Long, D. M. (1976). Biological disposition of perfluoroctylbromide. *Investigative Radiology*, 479−485. Available from https://doi.org/10.1097/00004424-197609000-00154.

Liu, X., Jansman, M. M. T., & Hosta-Rigau, L. (2020). Haemoglobin-loaded metal organic framework-based nanoparticles camouflaged with a red blood cell membrane as potential oxygen delivery systems. *Biomaterials Science, 8*(21), 5859−5873. Available from https://doi.org/10.1039/d0bm01118e.

Lu, S. J., Feng, Q., Park, J. S., Vida, L., Lee, B. S., Strausbauch, M., Wettstein, P. J., Honig, G. R., & Lanza, R. (2008). Biologic properties and enucleation of red blood cells from human embryonic stem cells. *Blood, 112*(12), 4475−4484. Available from https://doi.org/10.1182/blood-2008-05-157198.

Lykova, O. F., Romanova, T. V., Moiseeva, L. M., Konysheva, T. V., Grigor'ev, E. V., & Zorin, N. A. (2004). Effect of perftoran on macroglobulin content in the plasma and peritoneal exudate of rats with acute exudative inflammation (peritonitis). *Bulletin of Experimental Biology and Medicine, 138*(4), 351−353. Available from https://doi.org/10.1007/s10517-005-0097-5.

Magovern, G. J., Flaherty, J. T., Gott, V. L., Bulkley, B. H., & Gardner, T. J. (1982). Optimal myocardial protection with fluosol cardioplegia. *Annals of Thoracic Surgery, 34*(3), 249−257. Available from https://doi.org/10.1016/S0003-4975(10)62493-9.

Mallet, V., Dutheil, D., Polard, V., Rousselot, M., Leize, E., Hauet, T., Goujon, J. M., & Zal, F. (2014). Dose-ranging study of the performance of the natural oxygen transporter HEMO2life in organ preservation. *Artificial Organs, 38*(8), 691−701. Available from https://doi.org/10.1111/aor.12307.

Marsh, J. N., Hu, G., Scott, M. J., Zhang, H., Goette, M. J., Gaffney, P. J., Caruthers, S. D., Wickline, S. A., Abendschein, D., & Lanza, G. M. (2011). A fibrin-specific thrombolytic nanomedicine approach to acute ischemic stroke. *Nanomedicine: Nanotechnology, Biology, and Medicine, 6*(4), 605−615. Available from https://doi.org/10.2217/nnm.11.21.

Mayer, D., & Ferenz, K. B. (2019). Perfluorocarbons for the treatment of decompression illness: How to bridge the gap between theory and practice. *European Journal of Applied Physiology*, *119*(11−12), 2421−2433. Available from https://doi.org/10.1007/s00421-019-04252-0.

Mayer, D., Guerrero, F., Goanvec, C., Hetzel, L., Linders, J., Ljubkovic, M., Kreczy, A., Mayer, C., Kirsch, M., & Ferenz, K. B. (2020). Prevention of decompression sickness by novel artificial oxygen carriers. *Medicine and Science in Sports and Exercise*, *52*(10), 2127−2135. Available from https://doi.org/10.1249/MSS.0000000000002354.

McGoron, A. J., Pratt, R., Zhang, J., Shiferaw, Y., T, S., & Millard, R. (1994). Perfluorocarbon distribution to liver, lung and spleen of emulsions of perfluorotributylamine (FTBA) in pigs and rats and perfluorooctyl bromide (PFOB) in rats and dogs by 19F NMR spectroscopy - PubMed. *Artificial Cells, Blood Substitutes, and Biotechnology*, *22*(4), 1243−1250. Available from https://doi.org/10.3109/10731199409138822.

Mer, M., Hodgson, E., Wallis, L., Jacobson, B., Levien, L., Snyman, J., Sussman, M. J., James, M., van Gelder, A., Allgaier, R., & Jahr, J. S. (2016). Hemoglobin glutamer-250 (bovine) in South Africa: Consensus usage guidelines from clinician experts who have treated patients. *Transfusion*, *56*(10), 2631−2636. Available from https://doi.org/10.1111/trf.13726.

Mito, T., Nemoto, M., Kwansa, H., Sampei, K., Habeeb, M., Murphy, S. J., Bucci, E., & Koehler, R. C. (2009). Decreased damage from transient focal cerebral ischemia by transfusion of zero-link hemoglobin polymers in mouse. *Stroke: A Journal of Cerebral Circulation*, *40*(1), 278−284. Available from https://doi.org/10.1161/STROKEAHA.108.526731.

Modell, J. G., Tham, M. K., Modell, J. H., Calderwood, H. W., & Ruiz, B. C. (1973). Distribution and retention of fluorocarbon in mice and dogs after injection or liquid ventilation. *Toxicology and Applied Pharmacology*, *26*(1), 86−92. Available from https://doi.org/10.1016/0041-008X(73)90088-4.

Mouré, A., Bacou, E., Bosch, S., Jegou, D., Salama, A., Riochet, D., Gauthier, O., Blancho, G., Soulillou, J. P., Poncelet, D., Olmos, E., Bach, J. M., & Mosser, M. (2019). Extracellular hemoglobin combined with an O_2—Generating material overcomes O_2 limitation in the bioartificial pancreas. *Biotechnology and Bioengineering*, *116*(5), 1176−1189. Available from https://doi.org/10.1002/bit.26913.

Neildez-Nguyen, T. M. A., Wajcman, H., Marden, M. C., Bensidhoum, M., Moncollin, V., Giarratana, M. C., Kobari, L., Thierry, D., & Douay, L. (2002). Human erythroid cells produced ex vivo at large scale differentiate into red blood cells in vivo. *Nature Biotechnology*, *20*(5), 467−472. Available from https://doi.org/10.1038/nbt0502-467.

Nouwairi, N. S. (2004). The risks of blood transfusions and the shortage of supply leads to the quest for blood substitutes. *AANA Journal*, *72*(5), 359−364.

Nowak, E. S., Reyes, D. P., Bryant, B. J., Cap, A. P., Kerstman, E. L., & Antonsen, E. L. (2019). Blood transfusion for deep space exploration. *Transfusion*, *59*(10), 3077−3083. Available from https://doi.org/10.1111/trf.15493.

Olivier, E., Qiu, C., & Bouhassira, E. E. (2012). Novel, high-yield red blood cell production methods from CD34-positive cells derived from human embryonic stem, yolk sac, fetal liver, cord blood, and peripheral blood. *Stem Cells Translational Medicine*, *1*(8), 604−614. Available from https://doi.org/10.5966/sctm.2012-0059.

Olivier, E. N., Zhang, S., Yan, Z., Suzuka, S., Roberts, K., Wang, K., & Bouhassira, E. E. (2019). PSC-RED and MNC-RED: Albumin-free and low-transferrin robust erythroid differentiation protocols to produce human enucleated red blood cells. *Experimental Hematology*, *75*, 31. Available from https://doi.org/10.1016/j.exphem.2019.05.006, 52.e15.

Ortiz, D., Barros, M., Yan, S., & Cabrales, P. (2014). Resuscitation from hemorrhagic shock using polymerized hemoglobin compared to blood. *American Journal of Emergency Medicine*, *32*(3), 248−255. Available from https://doi.org/10.1016/j.ajem.2013.11.045.

Pagano, M. B., Hess, J. R., Tsang, H. C., Staley, E., Gernsheimer, T., Sen, N., Clark, C., Nester, T., Bailey, C., & Alcorn, K. (2020). Prepare to adapt: Blood supply and transfusion support during the first 2 weeks of the 2019 novel coronavirus (COVID-19) pandemic affecting Washington State. *Transfusion*, *60*(5), 908−911. Available from https://doi.org/10.1111/trf.15789.

Pan, D., Rogers, S., Misra, S., et al. (2016). Erythromer (EM), a nanoscale bio-synthetic artificial red cell: Proof of concept and in vivo efficacy results. *Blood*, *128*(22), 1027. Available from https://doi.org/10.1182/blood.v128.22.1027.1027.

Paxian, M., Keller, S. A., Huynh, T. T., & Clemens, M. G. (2003). Perflubron emulsion improves hepatic microvascular integrity and mitochondrial redox state after hemorrhagic shock. *Shock (Augusta, Ga.)*, *20* (5), 449−457. Available from https://doi.org/10.1097/01.shk.0000090601.26659.87.

Peng, Y., Xiaoxue, H., Lining, Y., Kangyu, H., Yunfei, G., Ahu, Y., Yiqiao, H., & Jinhui, W. (2019). Artificial red blood cells constructed by replacing heme with perfluorodecalin for hypoxia-induced radiore-sistance. *Advanced Therapeutics*, 1900031. Available from https://doi.org/10.1002/adtp.201900031.

Phillips, W. T., Rudolph, A. S., Coins, B., & Klipper, R. (2009). Biodistribution studies of liposome encapsu-lated hemoglobin (Leh) studied with a newly developed 99m-technetium liposome label. *Biomaterials, Artificial Cells and Immobilization Biotechnology*, 757−760. Available from https://doi.org/10.3109/10731199209119715.

Qi, D., Li, Q., Wang, P., & Wang, X. (2017). Haemoglobin site-specifically modified with ferulic acid to sup-press the autoxidation. *Artificial Cells, Nanomedicine and Biotechnology*, *45*(6), 1077−1081. Available from https://doi.org/10.1080/21691401.2017.1309659.

Rafikova, O., Rafikov, R., & Nudler, E. (2002). Catalysis of S-nitrosothiols formation by serum albumin: The mechanism and implication in vascular control. *Proceedings of the National Academy of Sciences of the United States of America*, *99*(9), 5913−5918. Available from https://doi.org/10.1073/pnas.092048999.

Rafikova, O., Sokolova, E., Rafikov, R., & Nudler, E. (2004). Control of plasma nitric oxide bioactivity by perfluorocarbons: Physiological mechanisms and clinical implications. *Circulation*, *110*(23), 3573−3580. Available from https://doi.org/10.1161/01.CIR.0000148782.37563.F8.

Reah, G., Bodenham, A. R., Mallick, A., Daily, E. K., & Przybelski, R. J. (1997). Initial evaluation of diaspirin cross-linked hemoglobin (DCLHb(TM)) as a vasopressor in critically ill patients. *Critical Care Medicine*, *25*(9), 1480−1488. Available from https://doi.org/10.1097/00003246-199709000-00014.

Riess, J. G. (2005). Understanding the fundamentals of perfluorocarbons and perfluorocarbon emulsions rele-vant to in vivo oxygen delivery. *Artificial Cells, Blood Substitutes, and Immobilization Biotechnology*, *33* (1), 47−63. Available from https://doi.org/10.1081/BIO-200046659.

Riess, J. G. (2006). Perfluorocarbon-based oxygen delivery. *Artif Cells, Blood Substitutes, Biotechnol*, *34*(6), 567−580. Available from https://doi.org/10.1080/10731190600973824.

Riess, J. G., & Krafft, M. P. (1998). *Fluorinated materials for in vivo oxygen transport (blood substitutes), diagnosis and drug delivery*. *Biomaterials* (Vol. 19, pp. 1529−1539). Elsevier Sci Ltd. Issue 16. Available from https://doi.org/10.1016/S0142-9612(98)00071-4.

Riess, J. G., & Krafft, M. P. (2006). *Fluorocarbon emulsions as in vivo oxygen delivery systems. Background and chemistry*. *Blood substitutes* (pp. 259−275). Elsevier Ltd. Available from https://doi.org/10.1016/B978-012759760-7/50033-0.

Riess, J. G., & Riess, J. G. (2001). Oxygen carriers ("blood substitutes")—Raison d'etre, chemistry, and some physiology. *Chemical Reviews*, *101*(9), 2797−2919. Available from https://doi.org/10.1021/cr970143c.

Rodriguez-Brotons, A., Bietiger, W., Peronet, C., Langlois, A., Magisson, J., Mura, C., Sookhareea, C., Polard, V., Jeandidier, N., Zal, F., Pinget, M., Sigrist, S., & Maillard, E. (2016). Comparison of perfluorodecalin and HEMOXCell as oxygen carriers for islet oxygenation in an in vitro model of encapsulation. *Tissue Engineering - Part A*, *22*(23−24), 1327−1336. Available from https://doi.org/10.1089/ten.tea.2016.0064.

Sakai, H. (2017). Overview of potential clinical applications of hemoglobin vesicles (HbV) as artificial red cells, evidenced by preclinical studies of the academic research consortium. *Journal of Functional Biomaterials*, *8*(1), 10. Available from https://doi.org/10.3390/jfb8010010.

Sakai, H., Horinouchi, H., Tomiyama, K., Ikeda, E., Takeoka, S., Kobayashi, K., & Tsuchida, E. (2001). Hemoglobin-vesicles as oxygen carriers: Influence on phagocytic activity and histopathological changes in reticuloendothelial system. *American Journal of Pathology, 159*(3), 1079–1088. Available from https://doi.org/10.1016/S0002-9440(10)61783-X.

Sakai, H., Tsuchida, E., Horinouchi, H., & Kobayashi, K. (2007). One-year observation of Wistar rats after intravenous infusion of hemoglobin-vesicles (artificial oxygen carriers). *Artificial Cells, Blood Substitutes, and Biotechnology, 35*(1), 81–91. Available from https://doi.org/10.1080/10731190600974582.

Scheer, A., Kirsch, M., & Ferenz, K. B. (2017). Perfluorocarbons in photodynamic and photothermal therapy. *Journal of Nanoscience and Nanomedicine, 10*(16), 21–27. Available from https://doi.org/10.7150/thno.46288.

Schutt, E. G., Klein, D. H., Mattrey, R. M., & Riess, J. G. (2003). Injectable microbubbles as contrast agents for diagnostic ultrasound imaging: The key role of perfluorochemicals. *Angewandte Chemie - International Edition, 42*(28), 3218–3235. Available from https://doi.org/10.1002/anie.200200550.

Severn, C. E., Eissa, A. M., Langford, C. R., Parker, A., Walker, M., Dobbe, J. G. G., Streekstra, G. J., Cameron, N. R., & Toye, A. M. (2019). Ex vivo culture of adult CD_3^{4+} stem cells using functional highly porous polymer scaffolds to establish biomimicry of the bone marrow niche. *Biomaterials, 225.* Available from https://doi.org/10.1016/j.biomaterials.2019.119533.

Shaffer, T. H. (1996). Liquid ventilation in premature lambs: Uptake, biodistribution and elimination of perfluorodecalin liquid. *Reproduction, Fertility, and Development, 8*(3), 409–416. Available from https://doi.org/10.1071/RD9960409.

Sheng, Y., Yuan, Y., Liu, C., Tao, X., Shan, X., & Xu, F. (2009). In vitro macrophage uptake and in vivo biodistribution of PLA-PEG nanoparticles loaded with hemoglobin as blood substitutes: Effect of PEG content. *Journal of Materials Science: Materials in Medicine, 20*(9), 1881–1891. Available from https://doi.org/10.1007/s10856-009-3746-9.

Siatecka, M., & Bieker, J. J. (2011). The multifunctional role of EKLF/KLF1 during erythropoiesis. *Blood, 118*(8), 2044–2054. Available from https://doi.org/10.1182/blood-2011-03-331371.

Skibba, J. L., Somalia, J., Petroff, R. J., Jr., & Denor, P. (1985). Canine liver isolation-perfusion at normo- and hyperthermic temperatures with perfluorochemical emulsion (Fluosol-43). *European Surgical Research*, 301–309. Available from https://doi.org/10.1159/000128482.

Southworth, R., Blackburn, S. C., Davey, K. A. B., Sharland, G. K., & Garlick, P. B. (2005). The low oxygen-carrying capacity of Krebs buffer causes a doubling in ventricular wall thickness in the isolated heart. *Canadian Journal of Physiology and Pharmacology, 83*(2), 174–182. Available from https://doi.org/10.1139/y04-138.

Spahn, D. R., Van Brempt, R., Theilmeier, G., Reibold, J. P., Welte, M., Heinzerling, H., Birck, K. M., Keipert, P. E., & Messmer, K. (1999). Perflubron emulsion delays blood transfusions in orthopedic surgery. *Anesthesiology, 91*(5), 1195–1208. Available from https://doi.org/10.1097/00000542-199911000-00009.

Spahn, D. R., Waschke, K. F., Standl, T., Motsch, J., Van Huynegem, L., Welte, M., Gombotz, H., Coriat, P., Verkh, L., Faithfull, S., & Keipert, P. (2002). Use of perflubron emulsion to decrease allogeneic blood transfusion in high-blood-loss non-cardiac surgery. *Anesthesiology*, 1338–1349. Available from https://doi.org/10.1097/00000542-200212000-00004.

Strom, J., Swyers, T., Wilson, D., Unger, E., Chen, Q. M., & Larson, D. F. (2014). Dodecafluoropentane emulsion elicits cardiac protection against myocardial infarction through an ATP-sensitive K^+ channel dependent mechanism. *Cardiovascular Drugs and Therapy, 28*(6), 541–547. Available from https://doi.org/10.1007/s10557-014-6557-2.

Sutherland, G. R., Farrar, J. K., & Peerless, S. J. (1984). The effect of fluosol-da on oxygen availability in focal cerebral ischemia. *Stroke; a Journal of Cerebral Circulation, 15*(5), 829–835. Available from https://doi.org/10.1161/01.STR.15.5.829.

Taguchi, K., Nagao, S., Maeda, H., Yanagisawa, H., Sakai, H., Yamasaki, K., Wakayama, T., Watanabe, H., Otagiri, M., & Maruyama, T. (2018). Biomimetic carbon monoxide delivery based on hemoglobin vesicles ameliorates acute pancreatitis in mice via the regulation of macrophage and neutrophil activity. *Drug Delivery*, *25*(1), 1266−1274. Available from https://doi.org/10.1080/10717544.2018.1477860.

Tao, Z., & Ghoroghchian, P. P. (2014). Microparticle, nanoparticle, and stem cell-based oxygen carriers as advanced blood substitutes. *Trends in Biotechnology*, *32*(9), 466−473. Available from https://doi.org/10.1016/j.tibtech.2014.05.001.

Teicher, B. A. (1992). Use of perfluorochemical emulsions in cancer therapy. *Biomaterials, Artificial Cells and Immobilization Biotechnology*, *20*(4), 875−882. Available from https://doi.org/10.3109/10731199209119734.

Thenuwara, K., Thomas, J., Ibsen, M., Ituk, U., Choi, K., Nickel, E., & Goodheart, M. J. (2017). Use of hyperbaric oxygen therapy and PEGylated carboxyhemoglobin bovine in a Jehovah's Witness with life-threatening anemia following postpartum hemorrhage. *International Journal of Obstetric Anesthesia*, *29*, 73−80. Available from https://doi.org/10.1016/j.ijoa.2016.10.006.

Tissot, J. D., Bardyn, M., Sonego, G., Abonnenc, M., & Prudent, M. (2017). The storage lesions: From past to future. *Transfusion Clinique et Biologique: Journal de la Societe Francaise de Transfusion Sanguine*, *24*(3), 277−284. Available from https://doi.org/10.1016/j.tracli.2017.05.012.

Trakarnsanga, K., Griffiths, R. E., Wilson, M. C., Blair, A., Satchwell, T. J., Meinders, M., Cogan, N., Kupzig, S., Kurita, R., Nakamura, Y., Toye, A. M., Anstee, D. J., & Frayne, J. (2017). An immortalized adult human erythroid line facilitates sustainable and scalable generation of functional red cells. *Nature Communications*, *8*. Available from https://doi.org/10.1038/ncomms14750.

Tremper, K. K., Friedman, A. E., Levine, E. M., Lapin, R., Camarillo, D., Tremper, K. K., Friedman, A. E., Levine, E. M., Lapin, R., Camarillo, D., Tremper, K. K., Friedman, A. E., Levine, E. M., Lapin, R., & Camarillo, D. (1982). The preoperative treatment of severely anemic patients with a perfluorochemical oxygen-transport fluid, Fluosol-DA. *New England Journal of Medicine*, *307*(5), 277−283. Available from https://doi.org/10.1056/NEJM198207293070503.

Tsai, A. G., Cabrales, P., Young, M. A., Winslow, R. M., & Intaglietta, M. (2015). Effect of oxygenated polyethylene glycol decorated hemoglobin on microvascular diameter and functional capillary density in the transgenic mouse model of sickle cell anemia. *Artificial Cells, Nanomedicine and Biotechnology*, *43*(1), 10−17. Available from https://doi.org/10.3109/21691401.2014.936063.

Vemuri, C., Upadhya, G. A., Arif, B., Jia, J., Lin, Y., Gaut, J. P., Fazal, J., Pan, H., Wickline, S. A., & Chapman, W. C. (2018). Antithrombin perfluorocarbon nanoparticles improve renal allograft function in a murine deceased criteria donor model. *Transplantation Direct*, *4*(9). Available from https://doi.org/10.1097/TXD.0000000000000817.

Von Der Hardt, K., Kandler, M. A., Brenn, G., Scheuerer, K., Schoof, E., Dötsch, J., & Rascher, W. (2004). Comparison of aerosol therapy with different perfluorocarbons in surfactant-depleted animals. *Critical Care Medicine*, *32*(5), 1200−1206. Available from https://doi.org/10.1097/01.CCM.0000124876.31138.F6.

Wang, J., Wang, R., Li, N., Shen, X., Huang, G., Zhu, J., & He, D. (2018). High-performance reoxygenation from PLGA-PEG/PFOB emulsions: A feedback relationship between ROS and HIF-1α. *International Journal of Nanomedicine*, *13*, 3027−3038. Available from https://doi.org/10.2147/IJN.S155509.

Waschke, K. F., & Lenz, C. (1999). Biomedical applications of artificial oxygen carriers. *Transfusion Medicine and Hemotherapy*, *26*(2), 31−36. Available from https://doi.org/10.1159/000053536.

Winslow, R. M. (2006). Current status of oxygen carriers ('blood substitutes'): 2006. *Vox Sanguinis*, *91*(2), 102−110. Available from https://doi.org/10.1111/j.1423-0410.2006.00789.x.

Wollocko, H., Wollocko, B. M., Wollocko, J., Grzegorzewski, W., & Smyk, L. (2017). OxyVita®C, a next-generation haemoglobin-based oxygen carrier, with coagulation capacity (OVCCC). Modified lyophilization/spray-drying process: Proteins protection. *Artificial Cells, Nanomedicine and Biotechnology*, *45*(7), 1350−1355. Available from https://doi.org/10.1080/21691401.2017.1339052.

Woods, A., Patel, A., Spina, D., Riffo-Vasquez, Y., Babin-Morgan, A., De Rosales, R. T. M., Sunassee, K., Clark, S., Collins, H., Bruce, K., Dailey, L. A., & Forbes, B. (2015). In vivo biocompatibility, clearance, and biodistribution of albumin vehicles for pulmonary drug delivery. *Journal of Controlled Release, 210*, 1−9. Available from https://doi.org/10.1016/j.jconrel.2015.05.269.

Wrobeln, A., Schlüter, K. D., Linders, J., Zähres, M., Mayer, C., Kirsch, M., & Ferenz, K. B. (2017). Functionality of albumin-derived perfluorocarbon-based artificial oxygen carriers in the Langendorff-heart†. *Artificial Cells, Nanomedicine and Biotechnology, 45*(4), 723−730. Available from https://doi.org/10.1080/21691401.2017.1284858.

Wrobeln, A., Jägers, J., Quinting, T., Schreiber, T., Kirsch, M., Fandrey, J., & Ferenz, K. B. (2020). Albumin-derived perfluorocarbon-based artificial oxygen carriers can avoid hypoxic tissue damage in massive hemodilution. *Scientific Reports, 10*(1). Available from https://doi.org/10.1038/s41598-020-68701-z.

Xu, J., Sankaran, V. G., Ni, M., Menne, T. F., Puram, R. V., Kim, W., & Orkin, S. H. (2010). Transcriptional silencing of γ-globin by BCL11A involves long-range interactions and cooperation with SOX6. *Genes and Development, 24*(8), 783−789. Available from https://doi.org/10.1101/gad.1897310.

Yamada, K., Yokomaku, K., Kureishi, M., Akiyama, M., Kihira, K., & Komatsu, T. (2016). Artificial blood for dogs. *Scientific Reports, 6*. Available from https://doi.org/10.1038/srep36782.

Yang, Z. J., Price, C. D., Bosco, G., Tucci, M., El-Badri, N. S., Mangar, D., & Camporesi, E. M. (2008). The effect of isovolemic hemodilution with oxycyte®, a perfluorocarbon emulsion, on cerebral blood flow in rats. *PLoS One, 3*(4). Available from https://doi.org/10.1371/journal.pone.0002010.

Yokomaku, K., Akiyama, M., Morita, Y., Kihira, K., & Komatsu, T. (2018). Core-shell protein clusters comprising hemoglobin and recombinant feline serum albumin as an artificial O_2 carrier for cats. *Journal of Materials Chemistry B, 6*(16), 2417−2425. Available from https://doi.org/10.1039/c8tb00211h.

Young, L. H., Jaffe, C. C., Revkin, J. H., McNulty, P. H., & Cleman, M. (1990). Metabolic and functional effects of perfluorocarbon distal perfusion during coronary angioplasty. *The American Journal of Cardiology, 65*(15), 986−990. Available from https://doi.org/10.1016/0002-9149(90)91001-M.

Yu, P., Han, X., Yin, L., et al. (2019). Artificial red blood cells constructed by replacing heme with perfluorodecalin for hypoxia-induced radioresistance. *Advances in Therapy, 2*(6), 1900031. Available from https://doi.org/10.1002/adtp.201900031.

Zarif, L., Postel, M., Septe, B., Trevino, L., Riess, J. G., Mahé, A. M., & Follana, R. (1994). Biodistribution of mixed fluorocarbon−hydrocarbon dowel molecules used as stabilizers of fluorocarbon emulsions: A quantitative study by fluorine nuclear magnetic resonance (NMR). *Pharmaceutical Research: An Official Journal of the American Association of Pharmaceutical Scientists, 11*(1), 122−127. Available from https://doi.org/10.1023/A:1018914215345.

Zhang, F., Zhuang, J., Esteban Fernández De Ávila, B., Tang, S., Zhang, Q., Fang, R. H., Zhang, L., & Wang, J. (2019). A nanomotor-based active delivery system for intracellular oxygen transport. *ACS Nano, 13*(10), 11996−12005. Available from https://doi.org/10.1021/acsnano.9b06127.

Zhang, Z., Li, H., Liang, Z., Li, C., Yang, Z., Li, Y., Cao, L., She, Y., Wang, W., Liu, C., & Chen, L. (2020). Vaporized perfluorocarbon inhalation attenuates primary blast lung injury in canines by inhibiting mitogen-activated protein kinase/nuclear factor-κB activation and inducing nuclear factor, erythroid 2 like 2 pathway. *Toxicology Letters, 319*, 49−57. Available from https://doi.org/10.1016/j.toxlet.2019.10.019.

Zhou, P., Ouchari, M., Xue, Y., & Yin, Q. (2020). In vitro generation of red blood cells from stem cell and targeted therapy. *Cell Transplantation, 29*, 1−5. Available from https://doi.org/10.1177/0963689720946658.

Zhuang, J., Ying, M., Spiekermann, K., Holay, M., Zhang, Y., Chen, F., Gong, H., Lee, J. H., Gao, W., Fang, R. H., & Zhang, L. (2018). Biomimetic nanoemulsions for oxygen delivery in vivo. *Advanced Materials, 30*(49). Available from https://doi.org/10.1002/adma.201804693.

Zolog, O., Mot, A., Deac, F., Roman, A., Fischer-Fodor, E., & Silaghi-Dumitrescu, R. (2011). A new polyethyleneglycol-derivatized hemoglobin derivative with decreased oxygen affinity and limited toxicity. *Protein Journal, 30*(1), 27−31. Available from https://doi.org/10.1007/s10930-010-9298-5.

Platelet substitutes

Mohammad Feroz Alam[1] and Khaliqur Rahman[2]

[1]*Department of Pathology, JN Medical College, Aligarh, India* [2]*Department of Hematology, Sanjay Gandhi Post Graduate Institute of Medical Sciences, Lucknow, India*

18.1 Platelet physiology

Blood is an amazing fluid made up of formed cellular elements (45%) and plasma (55%). The cellular elements are erythrocytes (red blood cells/RBCs), leukocytes (white blood cells/WBCs), and, thrombocytes (platelets) (Fig. 18.1). Platelets are anucleate cells/cellular fragments about $2-3\ \mu m$ in size and they have a major role in blood clotting (hemostasis). Platelets are formed by budding off the cytoplasm of the megakaryocytes in the bone marrow (Fig. 18.2). Megakaryocytes are the precursors of platelets. These are large cells with multilobated nuclei found in bone marrow, which extends long, branching processes, known as proplatelets. These proplatelets finally yield platelets that are structurally and functionally similar to the circulating blood platelets (Italiano et al., 1999). Human adults contain nearly one trillion blood platelets in circulation, each with an average life span of only $8-10$ days. The normal platelet counts of $150-400 \times 10^9$ platelets per liter of whole blood are maintained by roughly adding 100 billion new platelets daily from bone marrow megakaryocytes.

18.1.1 Platelet structure

Platelets have a unique structure, although anucleate, they contain secretory granules, mitochondria, and a large variety of other cellular organelles. The secretory granules are vesicles that release their contents either to the platelet surface or in the extracellular fluid. Platelets contain at least three major types of granules—α-granules, dense granules, and lysosomes, along with some peroxisomes and recently described T granules. α-Granules are specific to platelets and are the most abundant of different granules; there are about $50-80$ α-granules/platelet (White, 1998). These granules measure $200-500$ nm in diameter and account for about 10% of platelet volume. The α-granules contain mainly proteins, both membrane associated receptors (e.g., $\alpha IIb\beta3$ and P-selectin) and soluble cargo (e.g., fibrinogen). More than 300 soluble α-granule proteins are involved in a wide variety of hemostatic, inflammatory, and wound regenerative functions (Maynard et al., 2010).

Dense granules (also known as δ-granules) are the second most abundant type of platelet granules, there are about $3-8$ δ-granules/platelet. Each δ-granule measures about 150 nm in diameter and mainly contains bioactive amines (for example, serotonin and histamine), adenine nucleotides,

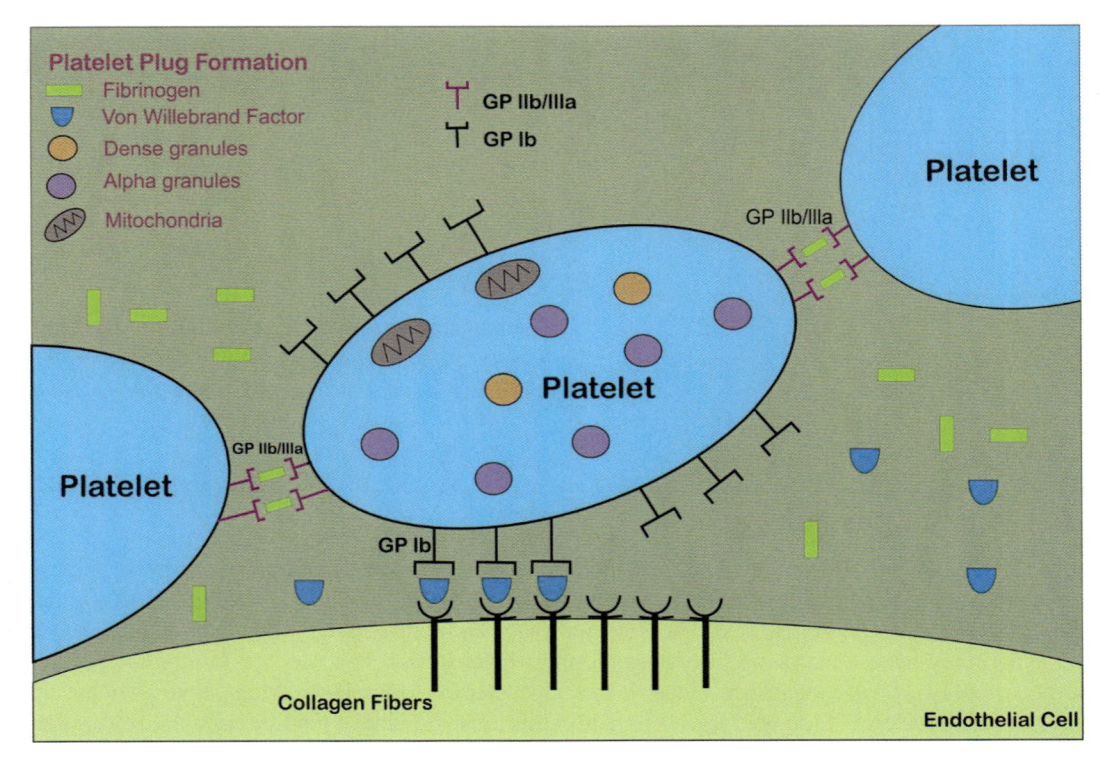

FIGURE 18.1

Peripheral blood smear showing erythrocytes, leukocytes and the small thrombocytes (arrows) in the background (Leishman's stain × 1000).

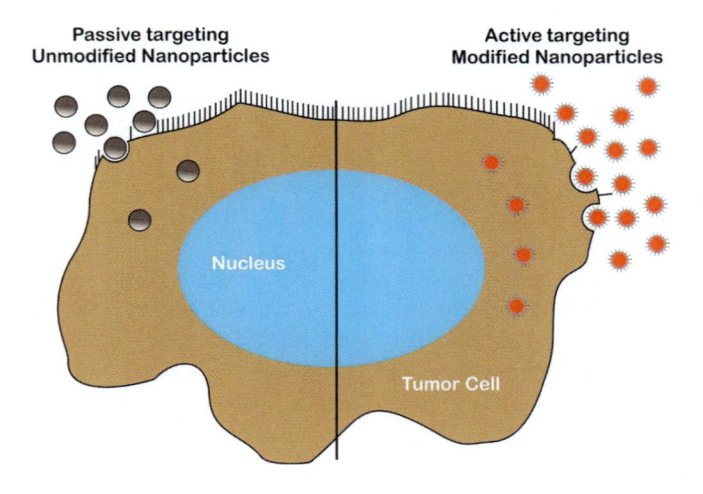

FIGURE 18.2

Bone marrow biopsy showing budding of platelets (arrows) from the cytoplasm of megakaryocyte (Leishman's stain × 1000).

polyphosphates, and pyrophosphates, as well as cations, mainly calcium in high concentrations. These granules derive their name from their electron-dense appearance on electron microscopy, due to their high cation concentrations (Gerrard et al., 1977). Other platelet granules have also been described; the platelet lysosomes and peroxisomes have ill-defined functions. The platelet granules are unloaded via exocytosis, upon stimulation by an agonist, the cargo stored in platelet granules is released, and the rate and extent of degranulation are dependent on the stimulation strength. Dense granule exocytosis is fastest and most sensitive to agonists, whereas lysosome exocytosis is slow. α-Granule exocytosis is considered to be intermediate of the two. The platelet exocytosis kinetics is governed by the concentration and potency of the agonist used (Jonnalagadda et al., 2012).

Due to the absence of a nucleus-based DNA transcriptional regulation, the platelets are pre-loaded with megakaryocytic RNA and organelles to confront any stress (e.g., vascular injury). To identify and mount an adequate response to stress, platelets have high-energy-producing mitochondria. Platelets usually have 5−8 mitochondria, which serve a variety of purposes, as seen in nucleated cells, like metabolism, activation, ATP production, and hemostasis (Hayashi et al., 2011). Maintenance of calcium homeostasis for optimal functioning of platelets requires a constant energy supply which is maintained by the high production of ATP inside mitochondria. Platelet activation is mediated by several agonists: collagen, thrombin, and ADP, and these are also implicated in the regulation of hemostasis (Davì & Patrono, 2007). The activity of these agonists is mediated by a common increase in intracellular calcium, which is maintained by the mitochondrial ATP. Platelets also have metabolic flexibility that allows the platelet to adapt to different situations, such as hypoxia or the presence of mitochondrial inhibitory agents (Melchinger et al., 2019). After vascular injury platelets get activated and adhere to the vascular wall initiating adhesion which leads to a coagulation cascade. In addition to the well-known function of energy production, mitochondria have recently been demonstrated to contribute to several platelet functions during activation such as the mitochondrial permeability transition, increased ROS generation, and collapse of mitochondrial membrane potential (Melchinger et al., 2019).

18.1.2 Platelet surface receptors

A large array of mobile transmembrane receptors are present on the platelet membrane, including many integrins (αIIbβ3, α2β1, α5β1, α6β1, αVβ3), leucine-rich repeated receptors (glycoprotein GP Ib/IX/V, Toll-like receptors), G-protein coupled seven transmembrane receptors (PAR-1 and PAR-4 thrombin receptors, P2Y1 and P2Y12 ADP receptors, TPα and TPβ TxA2 receptors), proteins belonging to the immunoglobulin superfamily (GP VI, FcγRIIA), C-type lectin receptors (P-selectin), tyrosine kinase receptors (thrombopoietin receptor, Gas-6, ephrins and Eph kinases), and a miscellany of other types (CD63, CD36, P-selectin ligand 1, TNF receptor type, etc.). Many of these receptors are present on other cells, but some are exclusive to platelets. The platelet surface receptors have an important role in hemostasis, allowing specific interactions, as well as functional responses of vascular adhesive proteins and platelet agonists. In addition, platelet surface receptors are involved in other less well-understood platelet functions such as inflammation, tumor growth and metastasis, and immunological host defense (Rivera et al., 2009). Vascular injury initiates the process of thrombus formation when blood platelets interact with the extracellular matrix components of the vessel (particularly VWF, collagen, fibronectin, thrombospondin, and laminin) which are exposed due to injury. Collagens I and III present in the vessel wall are considered the most important in causing platelet adhesion to the damaged vasculature. The GP Ib/IX/V complex is the major platelet receptor which mediates interaction with vWF. This GP Ib/IX/V complex consists of leucine-rich repeat glycoproteins: GP Ibα and GP Ibβ that are linked by disulfide and are noncovalently associated with GP IX and GP V (Rivera et al., 2000). The GP Ib/IX/V complex also binds to other adhesive proteins (collagen, thrombospondin-1), α-thrombin, and coagulation factors (kininogen, FXI, FXII). The α2β1 integrin, commonly referred to as GP Ia/IIa, VLA-2, or CD49b/CD29, also plays an important role in the adhesion of platelets to collagen and further subsequent activation.

18.1.3 Hemostatic platelet functions

Hemostasis is the mechanism by which the body prevents blood loss from an injured vessel. It is a complex process that is dependent upon a complex interaction between platelets, plasma coagulation mechanisms, fibrinolytic proteins, blood vasculature, and cytokines. Hemostasis encompasses three different subsets, primary hemostasis, secondary hemostasis, and fibrinolysis. Primary hemostasis refers to the physiological process which halts bleeding from an injured blood vessel, by the formation of a so-called "platelet plug," but simultaneously also maintaining normal blood flow elsewhere in the circulation. Secondary hemostasis is the activation of the coagulation cascade leading to the deposition of insoluble fibrin at the site of injury. Tertiary hemostasis is the dissolution of fibrin clot due to activation of the plasminogen. An uninjured normal endothelium is nonadhesive to the flowing platelets, however, in the setting of vascular injury, the subendothelium gets exposed and platelets start adhering to the exposed extracellular matrix components, and finally form a localized platelet plug, sealing the blood vessel. The formation of platelet plug is achieved through three distinct but interrelated platelet mechanisms—platelet adhesion, platelet activation and secretion, and platelet aggregation (Clemetson, 2012).

Platelet adhesion to the damaged vessel is a result of binding of various platelet receptors to the damaged and exposed subendothelium. Subendothelial extracellular matrix (ECM) elements that bind the platelets are collagen, vWF, fibronectin, laminin, fibrinogen, etc. The collagens type I and III are highly thrombogenic and they are the most powerful anchors of platelets because of their strong activity and affinity for vWF (Farndale et al., 2003). The initial platelet anchorage is mediated by the interaction between the vWF present in the subendothelial matrix, and the GPIbα fraction on the platelet GPIb-IX-V receptor. After the initial platelet anchorage, platelet collagen receptors interact with exposed collagen increasing cytosolic Ca^{2+} and leading to a platelet shape change, degranulation, and integrin activation. In the final steps of platelet adhesion there is binding of platelets to other ECM components such as fibronectin, laminin, and immobilized vWF (Peyvandi et al., 2011). *Platelet activation* follows platelet adhesion and is necessary for hemostasis to continue. This activation reaction is amplified by soluble agonists recruiting more circulating platelets at the site of injury. These agonists primarily are ADP, Thromboxane A2 (TxA2), epinephrine, and thrombin. ADP is secreted from platelet dense granules and binds to its relevant receptors. Binding of ADP to the platelets initiates increase of intracellular platelet Ca^{2+}, TxA2 synthesis, protein phosphorylation, shape change, granule release, and most importantly, activation of αIIbβ3 or the GPIIb/IIIa receptor (Andrews & Berndt, 2004). *Platelet aggregation* is the final step of primary hemostasis and occurs because of cross-linking of αIIbβ3 on adjacent platelets by the fibrinogen molecule. Although platelet aggregation is a complex process involving different receptors and their ligands, the main process is mediated via the integrin αIIbβ3. The resting plate-lets have low affinity for αIIbβ3 binding to its ligand but this affinity markedly increases upon platelet activation. Finally, the fibrinogen bridges the nearby platelets and platelet aggregates are formed (Fig. 18.3) (Ruggeri & Mendolicchio, 2007).

18.1.4 **Nonhemostatic platelet functions**

The central role of platelets in hemostasis and thrombosis is well-known; however, there are a wide range of other nonrelated conditions which are continually establishing the importance of platelets in their pathogenesis (Leslie, 2010). Varied conditions ranging from inflammation to atherosclero-sis, microbial infections to tumor growth and metastasis are being influenced by platelets. Platelets influence the inflammation by expressing pattern recognition receptors, toll-like receptors, which initiate the innate immune response. Also, there exists a platelet/leukocyte and platelet/monocyte axis where specific platelet receptors and counter receptors on the mentioned leucocytes facilitate their interaction in the bloodstream. Platelets can, upon activation, release many inflammatory mediators, such as interleukin-1 (IL-1) that can exacerbate the immune response. The IL-1β, has been

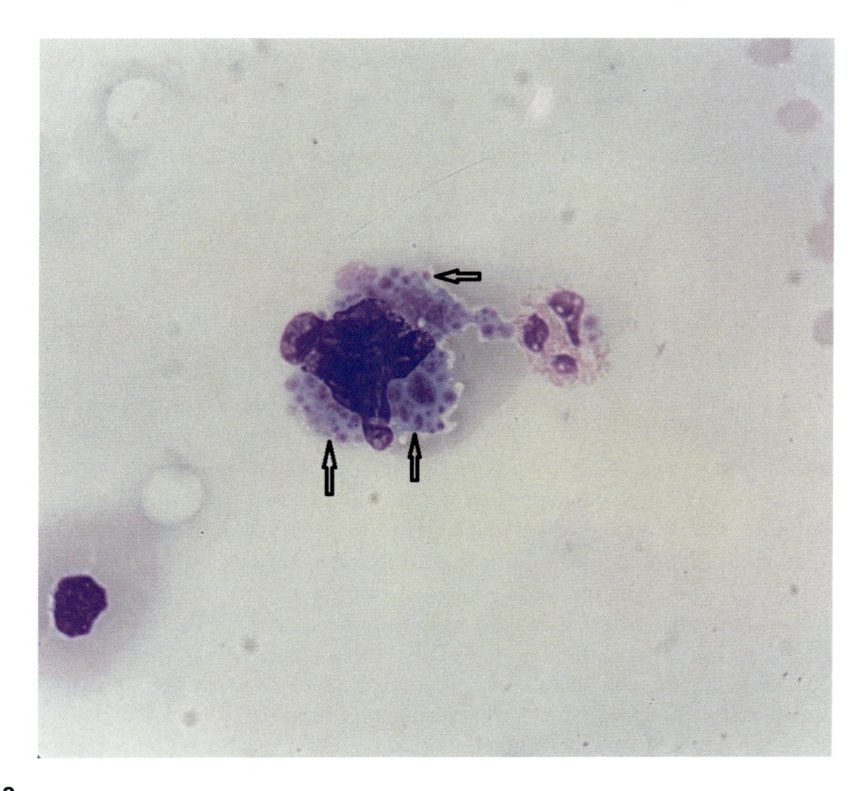

FIGURE 18.3

Diagrammatic representation of platelet plug formation.

specifically linked to the pathogenesis of arthritis and systemic lupus erythematosus. Nonclassical activation of platelets can release platelet microparticles and these microparticles have been linked to the inflammatory pathways associated with rheumatoid arthritis. Hence, platelets can influence inflammation by multiple dynamic processes (Ware et al., 2013). Neurodegenerative diseases, atherosclerosis, and transfusion-related lung injury, are just few of the diseases showing underlying inflammatory pathways that have strong association with altered platelet activity (Ware et al., 2013).

Platelets and the coagulation pathways have been linked to various stages of tumor growth and metastasis (Yapijakis et al., 2012). Platelet-influenced tumor growth can occur at many levels but most important is the tumor cell hijacked normal wound healing of platelets leading to uncontrolled cellular growth, angiogenesis, and metastasis (Labelle & Hynes, 2012). Local tumor expansion is enhanced by inflammation and angiogenesis, but in cancer the normal physiological process of dialing-down angiogenesis is prevented by a continuous, oncogene-mediated induction of tumor angiogenesis. There are several proteins that are preferentially expressed in platelets of tumor-bearing animals and cancer patients, many of them were angiogenesis regulators such as VEGF, bFGF, platelet-derived growth factor (PDGF), PF4, TSP1, MMP9, endostatin, angiopoietin-1 and -2. This angiogenic "proteome," as well as the concentrations of the individual proteins, is fairly stable under normal physiological conditions, although it is altered very early in tumor growth (Klement et al., 2009). The integrity of the endothelium in the tumor microenvironment having tumor cell-platelet heteroaggregates may be directly, or indirectly, affected by growth factors released from platelet α-granules, including PDGF, TGF-β, EGF, insulin-like growth factor -1 (IGF-1), and VEGF-A (Smyth et al., 2009). The circulating tumor cells (CTCs) along with the platelet-leukocyte heteroaggregates, at the vascular endothelium, might enhance their extravasation and invasion of distant target tissues. However, these hypotheses have yet to be examined.

18.2 Artificial platelets

Blood transfusion annually saves millions of lives globally, however, the quantity and quality of blood units available for transfusion are still a major concern, and more so in the developing countries. A recent study (Roberts et al., 2019) showed that in 2017 out of 195 countries, 119 countries (61%) did not have sufficient blood supply to meet their need. Across these 119 countries with insufficient supply, the total shortfall amounted to more than 100 million units, which was equal to around 1849 units per 100,000 people. In most high-income countries, supply was able to meet the demand; the scarcity of blood was marked in low and middle income countries. India had the largest absolute shortage, being short of nearly 41 million units in 2017 (52.5 million needed vs. 11.3 million supplied), creating a huge unmet gap. The WHO recommends that all blood donations should be screened for infections prior to use. Screening for HIV, hepatitis B, hepatitis C, and syphilis should be mandatory. 99.8% of the donations in high-income countries and 99.9% in upper-middle-income countries are screened following basic quality procedures, as compared to 82% in lower-middle-income countries and 80.3% in low-income countries. The prevalence of transfusion-transmissible infections in blood donations in high-income countries is considerably lower than in low- and middle-income countries (WHO, 2018). Acute shortage of blood and a definite risk of

TTIs along with a danger of transfusion-related complications bound us to think for some other alternative, although this idea appears to be unpractical and impossible, or at least very difficult.

Nanotechnology is revolutionizing medicine, nanomedicine deals with the concept of manipulation and assembly of the matter at the nanoscale for applications at the clinical level of medical sciences. In a broad sense, nanomedicine is the application of nanoscale technologies to the practice of medicine. This enables the miniaturization of many current devices, resulting in faster operation or integration of several operations. Furthermore, at this scale, man-made structures match typical sizes of natural functional units in living organisms. This allows them to interact with the biology of living organisms at the smallest structural level. Nanobiotechnology-based artificial cells have attracted much attention in recent times as substitutes for natural cells. The bio-inspired artificial cells may be defined and designed in many ways, they can be integral biological cell imitators with cell-like structures and exhibit some of the key characteristics of living cells. Alternatively, they can be engineered materials that only mimic some of the properties of cells, such as surface characteristics, shapes, morphology, or a few specific functions (Xu et al., 2016). These artificial cells can have applications in many fields from medicine, and hemostasis being one such field. Various micro- and nanosystems with bio-inspired surface fabrication and intracellular loading have been developed to mimic the function of platelets, especially in the event of vascular injury.

18.2.1 Surface fabrication of fibrinogen and its segments

A core of albumin polymers having an average diameter of 1020 ± 250 nm was prepared by disulfide polymerization of recombinant human serum albumin by controlling pH and temperature. Fibrinogen was coated on the surface of this albumin polymer. Under experimental flow conditions, the fibrinogen-conjugated albumin polymers were irreversibly attached to the platelet-immobilized surface in the reconstituted blood, even at a low platelet concentration. This attachment of fibrinogen-conjugated albumin polymers to platelets was suppressed when anti-GPIIb/IIIa monoclonal antibodies were added. It was therefore confirmed that these synthetic fibrinogen-albumin polymers specifically interacted with GPIIb/IIIa expressed on the surface of the activated platelets. Although platelets with a low platelet concentration were hardly attached to the platelet-immobilized surface under the flow conditions, the addition of fibrinogen-albumin polymers enhanced the attachment of the remaining platelets to the surface, indicating that the fibrinogen-albumin polymers would help the hemostatic ability of platelets at the site of vascular injury of patients with thrombocytopenia also. However, the use of complete fibrinogen molecules to coat platelet substitutes has some disadvantages in terms of storage stability, immunogenicity, and off-target activity. It also raises the potential for transmitting infectious diseases (Takeoka et al., 2001).

Albumin-Fibrinogen/RGD/H-12 complex: The human fibrinogen molecule contains three putative binding sites for platelet GPIIb/IIIa, namely a tetrapeptide containing an arginine-glycine-aspartic acid (RGD) sequence, for example RGDF and RGDS at α 95−98 and α 572−575, respectively, and a dodecapeptide (HHLGGAKQAGDV, H12) at the γ-chain carboxyterminal sequence (γ 400−411). The RGD sequence is the cell attachment site of a large number of other adhesive extracellular matrix, blood, and cell surface proteins, and nearly half of the over 20 known integrins recognize this sequence in their adhesion protein ligands. Covalent attachment of a RGD-containing peptide instead of fibrinogen, to erythrocytes was done, but nonselective attachment of RGD to other nonplatelet receptors and a potential risk of indiscriminate thrombus formation due to attachment to nonactivated platelets was noted (Coller, 1980). The H12 sequence of the fibrinogen molecule can prevent the nonspecific binding of RGD sequence. Therefore the H12 sequence was preferred instead of full-length fibrinogen. The H12 conjugates also showed minimal interaction with nonstimulated platelets compared with RGD conjugates, based on results obtained from flow cytometric analyses of platelet agglutination (Takeoka et al., 2003). Poly(ethylene glycol) (PEG) modification of the H12- polyAlb complex (H12−PEG−polyAlb) was shown to increase the hemostatic effects from a few hours to about 6 hours. Platelets were also shown to interact with inert beads coated with fibrinogen and that the interactions were more rapid if the platelets were activated with ADP. H12−PEG−polyAlb particles promote thrombus formation by accelerating and enhancing the aggregation of the flowing platelets. H12−PEG−polyAlb particles even contribute for the thrombus by their own mass effect, and hence can also work in patients with severe thrombocytopenia (Okamura et al., 2007).

Thromboerythrocytes: Agam and Livne (1992) used formaldehyde to cross-link fibrinogen molecules to the surface membrane of erythrocytes. The formaldehyde-dependent cross-linking was successful in attaching 58 fibrinogen molecules per erythrocyte. These coated erythrocytes were indistinguishable from untreated erythrocytes in properties like osmotic fragility, bound hemoglobin concentration, sedimentation rate, acetylcholinesterase activity, and phagocytosis by macrophages. Although, these fibrinogen-coated erythrocytes had the capability to enhance agonist-induced platelet aggregation in vitro, the platelet-dependent aggregation induced by ADP or thrombin was proportional to the fibrinogen density on the red cell membrane. The major limitation was that formaldehyde is a cytotoxic agent that may have carcinogenic potential and so it should not be used as a cross-linking reagent for in vivo use. Thromboerythrocytes were also made by covalent attachment of peptide to erythrocytes via a heterobifunctional cross-linking reagent where approximately $0.5-1.5 \times 10^6$ peptide molecules could be bound per erythrocyte after 2 hours of incubation. Thromboerythrocytes were able to selectively interact with platelets activated with ADP, epinephrine, or thrombin to produce large aggregates containing mixtures of platelets and erythrocytes. Studies with monoclonal antibodies to GPIIb/IIIa and fluid-phase RGD peptides indicate that the RGD peptides on the erythrocytes bind to the activated GPIIb/IIIa receptors on the platelets (Coller et al., 1992). The surface fabrication of erythrocytes with fibrinogen or some of its sequence demonstrated hemostatic properties in certain in vivo preclinical models, but their actual clinical translation was limited due to issues of antigen-matching of erythrocytes.

Fibrinogen-coated albumin microcapsules/microspheres (FAMs): The fibrinogen molecule was coated to albumin microspheres instead of the antigenic red cells. The potential ability of large microspheres to pass through the pulmonary microcirculation necessitated that the size of the microsphere must be kept small. Synthocytes were albumin microcapsules (FAMs) produced by spray-drying a 10% solution of human albumin to form albumin microcapsules. Next, the human fibrinogen molecule was immobilized to the surface of albumin microcapsules under specific ionic conditions and pH. The resultant coated microcapsules had a fibrinogen content of $<2\%$ of the total protein and the median diameter of coated microcapsules was $3.5-4.5\ \mu m$ with only $<2\%$ coated microcapsules having a diameter $>6\ \mu m$. A single bolus infusion of 1.5 (or 0.75) $\times$ 10^9 FAMs/kg shortened the prolonged microvascular ear-bleeding time in a thrombocytopenic rabbit (busulfan-induced) from a mean of 21.7 to 5.2 minutes. Similarly, FAMs were also seen to decrease the blood loss from a standard abdominal wall surgical wound in the thrombocytopenic rabbit. The hemostatic effect of FAMs was measurable at 3h postinfusion, but not 8h postinfusion.

No thrombogenicity was observed and no cardiopulmonary toxicity was seen in single-dose toxicity studies. Also, experimental studies using platelet deficient and normal whole blood in a parallel perfusion chamber were performed which suggested that FAMs enhance the adhesion of platelet-containing aggregates to activated endothelial cells. On evaluation these aggregates were found to contain FAMs, platelets, and fibrin, indicating an interaction of FAMs with platelets. Such an interaction was also supported by finding FAMs surrounded by degranulated platelets in biopsies from the bleeding X wounds (Levi et al., 1999). The study showed an enhanced hemostatic activity in vivo, but its translation into actual clinical bleeding scenario was unpredictable. A preparation similar to that of FAMs was the thrombospheres, which also showed in vivo hemostatic capabilities. Thrombospheres decreased the microvascular ear-bleeding time in thrombocytopenic rabbits, but their hemostatic effects were seen at least after 72 hours of a single bolus injection. The mechanism of action was also not elucidated because the hemostatic effect of thrombospheres persisted even after they were not detectable in the circulating blood (Yen et al., 1995). A major disadvantage of both the agents described above was that they both relied essentially on the presence of natural platelets.

Coated Liposomes: The attachment of platelet membrane glycoproteins onto lipid vesicles was described by several groups in the 1980s. A liposome-based platelet substitute called the platelet-some was produced by incorporating an extract of platelet membranes into unilamellar lipid vesicles. Plateletsomes had no effect on platelet aggregation in vitro, however, infusion of plateletsomes decreased the tail blood loss in thrombocytopenic rats. No evidence of consumptive coagulopathy was seen after intravenous infusion of plateletsomes into rabbits, nor was there any evidence of pathological thrombosis in the treated rats (Lee & Blajchman, 2001). In a study done by Nishiya et al. (2002) liposomes were covalently bound with recombinant GPIa/IIa (rGPIa/IIa) and/or recombinant fragments of GPIbα (rGPIbα) and their interaction was evaluated with the collagen or vWF surface under flow conditions in the absence of other platelet components. It was seen that liposomes carrying rGPIbα (rGPIbα-liposomes) reversibly interact with the vWF surface under flow conditions. The rGPIa/IIa supports immediate arrest of flowing liposomes onto the collagen surface, hence, rGPIa/IIa is essential for the stability of liposome adhesion to the collagen surface. rGPIa/IIa, expressed by its ability to act in concert with the rGPIbα−vWF interaction to promote stable adhesion of rGPIa/IIa-Ibα-liposomes to the collagen surface. The two receptors, rGPIa/IIa and rGPIbα, therefore, have complementary roles, and the corresponding adhesive substrates, collagen and vWF, are also complementary in the adhesion of rGPIa/Iia-Ibα-liposomes. Hemostatic effects were seen when these liposomes were conjugated with H12 sequence on their surface. The H12-conjugated liposomes interacted with activated platelets by binding to the GPIIb/IIIa platelet receptor. This interaction triggered a release reaction in which the liposome released encapsulated material that was kept within them. The rate of content release from the liposome was dependent upon the surface density of H12 molecules present on the liposome. In addition, other modes of interaction between the liposome and platelets may occur, such as internalization of the liposomes by the platelets upon interaction. When the liposomes are conjugated with octaarginine instead of H12, the liposomes were internalized, hence releasing their contents inside the activated platelets (Nishiya & Toma, 2004).

Okamura, Fukui et al. (2009) and Okamura, Takeoka et al. (2009) developed synthetic platelet substitutes based on a strategy of using polymerized liposomes (mean diameter 0.22−0.26 μm) as a carrier vehicle and synthetic H12 peptides as a surface-coating ligand to target activated platelets.

H12-coated liposomes with polyethyleneglycol-surface modifications showed specific interaction with activated platelets and augmented effects on platelet thrombus formation onto collagen immobilized surfaces under flow conditions in vitro, and prolonged hemostatic ability in vivo to correct bleeding time in a dose-dependent manner in a thrombocytopenic rat model. To strengthen the hemostatic ability of H12-coated particles as a platelet substitute, they exploited installation of drug delivery function by encapsulating potent platelet agonist ADP into liposomes. This lipid nanovesicles (H12-ADP-liposomes) designed for targeting activated platelets at sites of vascular injury, helped reinforce primary hemostasis by residual platelets in thrombocytopenia. They also visualized the specific accumulation of the liposomes at the sites of vascular injury by encapsulating the contrast dye iopamidol into the H12-conjugated liposomes (H12—iopamidol—liposome complex), which provided the first visual evidence that H12-conjugated particles specifically interact with activated platelets (Okamura, Eto et al., 2010). To further modulate the release reaction and enhance their activity, the H12-(ADP)-vesicles with different lamellarities and membrane flexibilities were made. It was shown that H12-(ADP)-vesicles were capable of augmenting platelet aggregation by releasing ADP in an aggregation-dependent manner. The amount of ADP released from the vesicles was dependent on their membrane properties. The mechanism of the ADP release and its regulation depends on the fact that vesicles incorporated in the platelet aggregates are strongly bound to neighboring platelets and are subjected to physical forces, which pull on the vesicle and continuously change its shape. This deformability depends on the surface flexibilities and lamellarities, which are dependent on the composition of the vesical membrane. The deformability (including a possible disruption) is correlated with the amount of ADP released from the vesicle. Specifically, the amount of ADP released increased with decreasing lamellarity and tended to increase with increasing the membrane flexibility. The in vivo results clearly demonstrated that H12-(ADP)-vesicles with the ability to release ADP exert considerable hemostatic action in terms of correcting prolonged bleeding time in a busulfan-induced thrombocytopenic rat model. Hence, the ADP release reaction can efficiently be tuned by modulating the membrane properties (Okamura, Katsuno et al., 2010).

Nanosheets: Using nanotechnology principles biodegradable poly(D,L-lactide-*co*-glycolide) (PLGA) disk-shaped nanosheets were made instead of the usual spherical carriers. The main advantage of nanosheets over spherical carriers was that they had a large contact area for the targeting site, compared to the small contact area of spherical carriers. The nanosheets were coated with the H12 dodecapeptide and similar surface area spherical H12-coated PLGA microparticles were also evaluated for hemostatic activity comparison. Both nanosheets and spherical carriers conjugated with a fluorescent marker were flown over thrombocytopenic blood attached to the collagen surface. It was seen that the nanosheets adhered to the collagen surface at twice the rate of the spherical carriers. Hence, the ellipsoidal nanosheets adhered more effectively than classical spherical carriers because of the larger contact area of the nanosheets. More binding sites were available, supporting the adhesive strength. The adhesive rate of the carriers on a collagen surface can be controlled by the change of their shape and the best adherence is obtained with ellipsoidial nanosheets with a medium aspect ratio. These nanosheets induced only a two-dimensional spread of platelet thrombi, however, the spherical microparticles allowed for a three-dimensional spread of the growing thrombus. The microparticles thrombi piled up dramatically, in contrast to the evenly spreading thrombi of the nanosheets (Okamura, Fukui et al., 2009; Okamura, Takeoka et al., 2009). Piled up thrombi may lead to blood vessel occlusion and can have severe consequences in vivo.

Platelet-derived microparticles: Platelet microparticles are microvesicles of platelet membranes which are formed spontaneously during storage of platelets and are found in platelet concentrates, fresh-frozen plasma, and cryoprecipitate. Platelet microparticles possess procoagulant activity like intact platelets and adhere to the injury-exposed vascular subendothelium initiating primary hemostasis. Cypress Bioscience (San Diego, CA, USA) synthesized microparticulate human platelet preparation from outdated platelet concentrates, the infusible platelet membranes (IPMs). The IPMs consist of spherical vesicles with a diameter of approximately 0.6 mm, and contain various procoagulant phospholipids. In the in vivo studies on thrombocytopenic rabbits, the IPMs shortened the prolonged ear-bleeding time for up to 6 hours posttransfusion; however, by 24 hours their hemostatic activity was no longer detectable. Toxicity studies in experimental animals have not demonstrated any pathological thrombogenicity, or potentiation of disseminated intravascular coagulation in endotoxin-treated rabbits (Blajchman, 2003). IPMs have been successfully administered in normal human volunteers and thrombocytopenic patients in phase I and II clinical trials and have provided some improvement leading to cessation of bleeding in some patients with a single dose of IPMs, although, the results of phase III clinical trials have not been reported (Nasiri, 2019).

18.2.2 Platelet-mimetic nanoparticles

Artificial platelet-mimetic technologies address the shortage, storage, and immunogenicity related issues of natural platelets, and also artificial materials do not need to be matched prior to therapeutic use. Still, biocompatibility, complement activating, and other potential adverse effects like thrombosis have to be evaluated. Clearance of nanoparticles can be finely tuned by controlling the size or material used for synthesis. Synthetic materials have longer shelf life, scale-up potential, and ease of manufacturing and greater reproducibility. The physicochemical properties of artificial platelet-mimetic devices can also be controlled to simulate the in vivo properties of natural platelets in both their resting as well as active conformations, helping to obtain desirable results for clinical applications. Polymer engineering led to the development of one such nanospherical synthetic platelets consisting of poly(lactic-co-glycolic acid)-poly-L-lysine (PLGA-PLL) block copolymer cores conjugated with PEG arms terminating with RGD functionalities. Nanosphere cores were approximately 170 nm in diameter, but the overall diameter increased with increasing PEG molecular weight. The incorporation of the RGD moiety could influence cellular interactions with biomaterials; a longer PEG spacer between the PLGA surface and the RGD moiety led to a greater probability that the RGD functionality is available to bind with the activated platelets. The RGD-functionalized polymer interacted specifically with activated platelets and the binding of activated platelets to the functionalized PLGA-PLL-PEG was specific. The cores of the synthetic platelets were degradable polyester nanoparticles, and degradable polyester nanoparticles have been shown to induce clot formation. However, PEGylation of the particles reduced this effect dramatically as PEGylated nanoparticles do not alter clotting behavior. This lack of alteration in the coagulation cascade with PEGylated nanoparticles is attributed mainly to the hydrophilic PEG corona which reduces the surface adsorption of plasma proteins and lipids. The PEG corona also acts as a shield for nanoparticles, increasing their circulation time which allows the particles to reach the site of injury before being cleared. These functionalized nanoparticles halve the bleeding time after intravenous administration in a rat model of major trauma, and may be useful for early intervention in trauma (Bertram et al., 2009).

Platelet like nanoparticles (PLNs): PLNs were synthesized via the layer-by-layer (LbL) method for precise control over their size, shape, and material composition. Similar to natural platelets, PLNs were discoidal in shape and were functionalized with platelet-specific peptides, so as to mimic the biochemical interactions of natural platelets with both injured endothelium and with other platelets. Their discoidal shape and flexible exterior showed individual and aggregated PLN binding to targeted surfaces under flow. PLNs were then independently assessed for their specific adhesion to activated platelets under physiologically relevant flow conditions. PLNs functionalized with fibrinogen interacting peptides (GRGDS) showed high specific adhesion to activated platelet-coated surfaces. While PLNs having no peptide functionalization showed minimal adhesion to activated platelet-coated surfaces. Furthermore, it was proposed that, in vitro, the PLNs marginate to the wall under shear due to their physical properties (e.g., shape and flexibility) and are likely to have higher local concentration at the wall under flow. This increase in local PLN concentration can potentially cause PLNs to interact with each other and clump together if shear conditions permit. In the case of normal hemostatic plug formation, circulating platelets become activated and bind to the damaged endothelium due to exposure of collagen and release of VWF from the wound site. In the case of hemostatic plug formation following in vivo injection of PLNs in tail amputated mice, activated circulating platelets and PLNs both bind to injured endothelium, as well as to each other, effectively forming the hemostatic plug much faster than in the absence of PLNs. However, an increase in microthrombi was seen in histological lung sections in some cases following treatment with PLNs. This was likely due to increased overall activity of circulating platelets following tail transaction, compounded with the observed lung accumulation. To minimize these complications, a low PLN dose was used and no such complications were observed with the lower dose (Anselmo et al., 2014).

SynthoPlate: natural platelet mechanisms led to the development of a synthetic platelet technology (SynthoPlate) that integrates platelet-mimetic site-selective adhesion and aggregation functionalities via heteromultivalent surface-decoration of lipid vesicles with von Willebrand Factor-binding, collagen-binding, and active platelet integrin GPIIb-IIIa-binding peptides. SynthoPlate vesicles do not have systemic prothrombotic or procoagulant risks, but can amplify activated platelet-mediated primary and secondary hemostatic mechanisms selectively at a bleeding site, even at low platelet concentrations. In vivo studies further confirmed that prophylactic administration of SynthoPlate can dose-dependently improve hemostasis in a tail bleeding model in severely thrombocytopenic mice, with higher doses capable of correcting bleeding times to levels comparable to that of normal mice. Immunofluorescence microscopy and immunoblot analyses confirmed striking colocalization of murine platelets and SynthoPlate, as well as enhanced formation of fibrin in hemostatic foci of SynthoPlate injected mice. SynthoPlate vesicles showed reasonably long circulation periods and were primarily cleared by the liver and spleen (Shukla et al., 2017). Building on these hemostatic results in mice, the effects of SynthoPlate were investigated in emergency management of traumatic exsanguinating hemorrhage in large animal (pig) model. The in vitro and in vivo studies demonstrated that the SynthoPlate dose did not have any cytotoxicity and systemic risks in the experimental conditions tested. Following arterial injury and bolus administration of SynthoPlate (or control) particles, the results showed that compared to saline (0% survival), SynthoPlate treatment renders 100% survival during the "golden hour." Interestingly, unmodifed (control) particles showed ~50% survival during this first hour, however, survival for this control treatment group fell to 25% during the second hour of the experimental window. Since, unmodified (control) particles were incapable of biofunctionally enhancing hemostasis, these animals continued to bleed at a higher rate and succumbed. In contrast, a combination of hemodynamic stabilization and hemostatic enhancement significantly improved the survival of the SynthoPlate-treated pigs within and beyond the "golden hour." Superior hemostatic capability of SynthoPlate can also be potentially combined with externally applied hemostatic dressings, to further improve the hemostatic outcomes in prehospital scenarios. However, the treatments in the study were administered within 1-minute after injury and hence, such an instant initiation of treatment in traumatic patients is not possible in actual clinical settings (Hickman et al., 2018).

18.3 **Platelet membrane cloaking for drug delivery**

Nanotechnology is increasingly employed for coating natural cell membranes on nanoconstructs to achieve desired functionalization for theranostics. Targeted drug delivery to unreachable cellular or subcellular locations can be achieved through membrane cloaking. Drug targeting can be passive, where nanoparticles in blood extravasate through endothelial gaps of the leaky diseased vasculature and accumulate in the surrounding tissue via the enhanced permeability and retention (EPR) effect. Passive targeting is limited in utility for its low target specificity, thereby, lower than desired concentration at the target site. Specificity and concentration requires "active targeting" with moieties such as small ligands, antibodies, and natural membranes having surface biomarkers capable of specific binding to target site, facilitating efficient uptake, internalization, and receptor-mediated endocytosis resulting in elevated concentrations in tumor cells (Fig. 18.4) (Alam et al., 2014). Platelet membranes can be used for passive targeting, as the membrane is recognized as self (CD 47) by the immune system, increasing the circulation half-life. Platelet membrane surface receptor-mediated specific interactions with damaged endothelium, tumor cells, and pathogenic bacteria can be utilized for active targeting.

18.3.1 **Vascular targeting**

Platelet adhesion, activation, and aggregation at sites of vascular injury and their role in primary hemostasis are well-known. The platelet surface receptors have an important role in hemostasis, allowing specific interactions and functional responses. In addition, these receptors are also involved in other less well-understood platelet functions such as inflammation, tumor growth, and metastasis (Rivera et al., 2009). Platelets also interact with injured and inflamed endothelium in multiple ways, such as in an atheroma. They are also capable of attaching to immune cells and regulating their recruitment and activity at the site of inflammation. Also, their binding at sites of plaque rupture can lead to the deadly thrombus formation. Hence, to target atherosclerosis, platelet membrane-coated nanoparticles (PNPs) were fabricated by coating the membrane-derived from repeated freeze—thawing of fresh platelets onto preformed poly(lactic-co-glycolic acid) (PLGA) cores loaded with a far-red fluorescent dye, which was used to demonstrate payload delivery of the nanoparticles to target atherosclerotic plaque. Finally, cores and the membrane were mixed together and sonicated, resulting in membrane-coated nanoparticles. In vitro evaluations showed that the

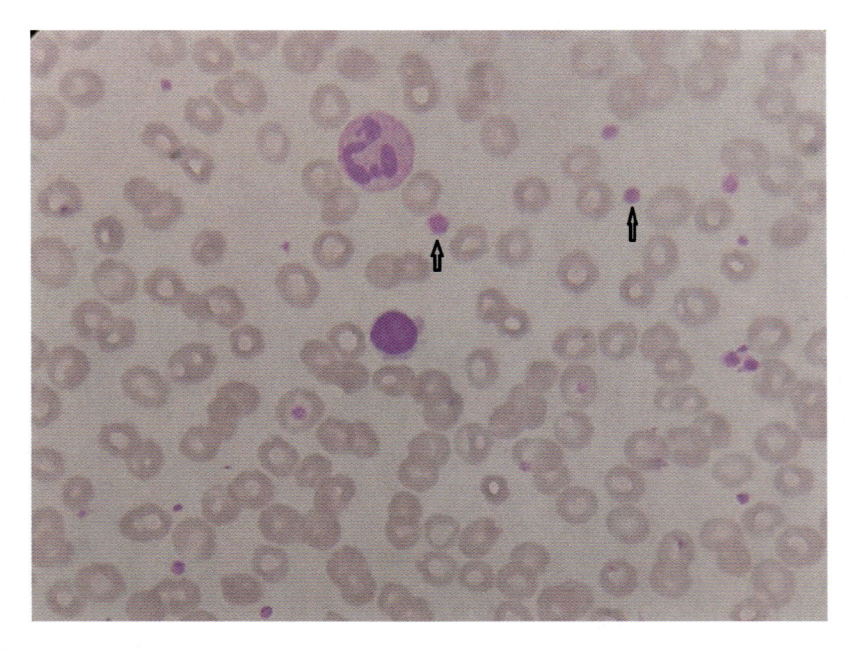

FIGURE 18.4

Nanoparticle active and passive targeting.

PNPs can bind to foam cells, collagen, and activated endothelium (all present in plaque) and it was also demonstrated that PNPs possess strong affinity to bind with these components. Similar, targeting of the plaque by PNPs was seen in in vivo studies using mice with atherosclerotic plaque model. Histological sections were also taken to obtain a more detailed microscopic view of the in vivo PNP targeting of atherosclerotic and preatherosclerotic regions. In the first instance, a tissue sample with significant plaque development and high degree of luminal occlusion was studied for the distribution of the far-red dye in the PNP formulation; it was seen that the nanoparticle signal could be found in significant amounts throughout the plaque region. Sections were then stained with separate antibodies to probe for various elements of atherosclerosis development; this included ICAM-1 for activated endothelial cells, CD68 for macrophages, as well as type IV collagen. Each of the three was found to be present, and they displayed high degrees of proximity or colocalization with the nanoparticle signal on the sections. In the second instance, a section was taken from a vessel with significantly less disease burden, and the region of endothelium proximal to the plaque stained for ICAM-1. It was confirmed that there was activated endothelial cells in this region. This signal colocalized well with PNP signal, confirming that the particles are capable of binding to regions that are plaque-free but susceptible to future plaque development. In all, these in vivo targeting experiments further indicated the potential of the platelet membrane-coated formulation to bind with atherosclerotic sites at different stages of development, including preatherosclerotic regions with no visible sign of disease. Finally, the utility of the PNP formulation for live imaging was confirmed, and it was established that PNPs could be used to successfully target and deliver a sufficient quantity of contrast agent to provide observable contrast at the site of the plaque (Wei et al., 2018).

In another study PNPs were used as a targeted drug delivery platform to treat atherosclerosis through mimicking platelets targeting plaques. PNPs displayed 4.98-fold greater radiant efficiency than control nanoparticles in atherosclerotic arterial trees, indicating their effective homing to atherosclerotic plaques in vivo. In an atherosclerosis model established in apolipoprotein E-deficient mice, PNPs encapsulating rapamycin significantly attenuated the progression of atherosclerosis and stabilized atherosclerotic plaques. These results demonstrated the perfect efficacy and proresolving potential of PNP as a targeted drug delivery platform for atherosclerosis treatment (Song et al., 2019). Again, inspired by natural platelets (PLTs) and their role in targeting adhesion to the damaged blood vessel during thrombus formation, a biomimetic nanocarrier comprising a platelet membrane envelope loaded with L-arginine and γ-Fe_2O_3 magnetic nanoparticles (PAMNs) for thrombus-targeted delivery of L-arginine and in situ generation of nitric oxide (NO), were fabricated. It was demonstrated that the engineered 200-nm PAMNs inherit the natural properties of the platelet membrane and achieve rapid targeting to ischemic stroke lesions under the guidance of an external magnetic field. Subsequent to the release of L-arginine at the thrombus site, endothelial cells produce NO, which promotes vasodilation to disrupt the local platelet aggregation. Rapid targeting of PAMNs to stroke lesions as well as in situ generation of NO prompts vasodilation, recovery of blood flow, and reperfusion of the stroke microvasculature. Thus, these platelet membranes-derived nanocarriers were diagnostically beneficial for localizing stroke lesions and a promising modality for executing therapies (Li et al., 2020).

Platelets have an important role in the formation of thrombus. Hence, platelet membrane-based nanoparticles have been tried recently for thrombolytic therapy. Xu et al. (2019) modified recombinant tissue plasminogen activator (rtPA) on the surface of platelet membrane for sequentially site-specific delivery of rtPA for ischemic stroke treatment. Mediated by the cloak of the platelet membrane, the nanoformulation targets the thrombus site and rtPA is triggered to release by the upregulated thrombin. From the in vitro and in vivo evaluation, of the nanoformulation it was demonstrated that it significantly enhanced the antiischemic stroke efficacy in the rat model with middle cerebral artery occlusion, showing a decrease in ischemic area and reactive oxygen species level compared to that with free drug combination. Similarly, urokinase plasminogen activator (uPA) and gold nanorods (AuNRs) were loaded in a platelet membrane nanoparticle. When the nanoparticles targeted the thrombosis, the inner uPA could be released from the nanoparticles in a sustained manner, resulting in an obvious thrombus decrease (Yang et al., 2018). Hence, PNPs are showing promise in targeting atherosclerotic plaque and thrombus for therapeutic recanalization of the affected blood vessel; still there is a long way to go before actual clinical application.

18.3.2 Tumor targeting

CD 47 on the surface of the platelets (marker of self), allows for the escape of immune recognition of the platelet membrane-coated nanoparticles (PNPs). The nanodimension of PNPs prevents macrophage engulfment and increases circulation half-life, and they preferentially accumulate at the tumor due to leakage through abnormal tumor vasculature. However, the main reason for the increasing use of PNPs in tumor-targeted drug delivery is the specific binding interactions between the platelets and the cancer cells. Platelet-influenced tumor growth can occur at many levels but most important is the tumor cell hijacked normal wound healing of platelets leading to uncontrolled cellular growth, angiogenesis, and metastasis (Labelle & Hynes, 2012). During the metastasis, tumor cells shield themselves by attaching to platelets for protection against shear forces and immune attack, therefore increasing the chance of metastasis formation. Tumor cells achieve such binding interactions by overly expressing ligands such as podoplanin, P-selectin PSGL-1, A disintegrin and metalloproteinase domain-containing protein 9 (ADAM-9), and fibrinogen/$\alpha v\beta 3$. These ligands bind with platelet receptors, including C-type lectin-like receptor 2 (CLEC-2), P-selectin, integrin $\alpha 6\beta 1$, and integrin $\alpha IIb\beta 3$, respectively. Although, tumor cells utilize such binding interactions to gain advantages for their metastasis, PNPs can exploit these interactions for active drug targeting (Wang et al., 2020). Poly(lactic-co-glycolic acid) (PLGA) nanoparticles were developed to deliver bufalin, an active ingredient which exhibits strong antineoplastic activities, sparing normal cellular replication. Biodegradable PLGA was firstly conjugated to chitosan oligosaccharide (CS), followed by self-assembly with the pore producing, vitamin E polyethylene glycol succinate (TPGS) and bufalin by the nanoprecipitation method, resulting in porous bufalin-loaded nanoparticles (CS-pPLGA/Bu NPs) with positive surface charges. These were functionalized by absorbing the negatively charged platelet membrane by layer-by-layer (LbL) self-assembly, and the therapeutic effects of the NPs were explored in a murine H-22 hepatoma tumor model. A high level of uptake of platelet membrane-coated NPs by cancer cells were observed by confocal microscopy and flow cytometry analysis. In vivo imaging showed accumulation of the NPs in the tumor site, demonstrating their active targeting ability as a result of targeted binding of platelet membrane P-selectin to the CD44 receptors overexpressed on the surface of cancer cells. The platelet membrane-coated NPs were also found to be effective in reducing tumor growth in vivo, and did not cause off-target effects on the major organs (Wang et al., 2019).

RNA interference (RNAi) is a naturally occurring mechanism for gene downregulation. Since its discovery in the late 1990s, it has been widely leveraged as a tool for biological studies. Through cytosolic RNA-induced silencing complex, target genes can be posttranscriptionally silenced via degradation of the corresponding mRNA. Small interfering RNAs (siRNAs) are short and well-defined double-stranded RNA molecules that can be synthetically manufactured to take advantage of the RNAi pathway. They have also been widely explored as therapeutics for human diseases, especially cancer. However, there are several hurdles that have prevented the more widespread approval of siRNA therapies, and these include their inherent vulnerability in biological environments, suboptimal uptake and knockdown efficiency, rapid blood clearance, and possible immunogenicity. The formulation of siRNA into nanoscale delivery vehicles represents an effective strategy for enhancing bioavailability and biocompatibility. Zhuang et al. (2020) loaded synthetic siRNAs inside porous metal−organic framework (MOF) nanoparticles and further coated them by a layer of naturally derived platelet membrane. The impact of membrane coating on nanoparticle interaction with macrophages was evaluated, and both bare MOF-siRNA and platelet membrane-coated MOF-siRNA nanoparticles were incubated with murine J774 macrophages. Flow cytometric analysis revealed that the platelet membrane coating was able to significantly reduce interaction of the MOF-siRNA nanoparticles with the macrophages. Next, membrane P-selectin-based targeting of the CD-24 was studied in SK-BR-3 breast cancer cells line using fluorescence imaging, showing increased interaction between the two receptors. After confirming successful targeting of the tumor cells, internalization of the siRNA payload was assessed using fluorescently labeled siRNA, in either, free form or platelet membrane (P-MOF-siRNA) or red cell membrane (R-MOF-siRNA). After 24 hours of incubation with SK-BR-3 breast cancer cells, the amount of uptake was quantified by flow cytometry. Free siRNA exhibited a negligible amount of uptake due to its inability to cross the cell membrane. On the other hand, cells incubated with the two membrane-coated nanoformulations, displayed a substantial amount of endocytosis induced uptake. However, because of the specific interactions of cancer cells with platelets, the P-MOF-siRNA nanoparticles were better at delivering the siRNA compared with R-MOF-siRNA nanoparticles. Next was to demonstrate the internalized payloads release from the endosomal compartment, which is essential for the silencing effect of siRNA since RNAi machinery is located in the cytosol. To accomplish this, fluorescently labeled siRNA was delivered via P-MOF-siRNA, and the intracellular localization was visualized by fluorescence microscopy over the course of 24 hours. At 1 hour after the start of incubation, most of the payload could still be found on the periphery of the cells, but a notable portion had already been endocytosed after 4 hours. At 8 hours, the siRNA signal could be seen throughout the cell, and this effect was even more pronounced after one full day of incubation. Overall, these data indicated that the siRNA could successfully escape or release into the cytosol after dissociation of the MOF scaffold in the more acidic pH of the endosomes. Next, the gene silencing efficiency of the nanoformulation was evaluated using green fluorescent protein (GFP) as a model target. Both the P-MOF-siRNA[GFP] and R-MOF-siRNA[GFP] groups showed gene silencing effects, the effect of the former was considerably stronger than the latter. The P-MOF-siRNA[GFP] formulation reduced the level of expression to near the baseline established by wild-type SK-BR-3 cells. The same study was repeated using fluorescence imaging, and the results corroborated that P-MOF-siRNA[GFP] was better at knocking down gene expression compared with R-MOF-siRNA[GFP]. The results from these studies validated the utility of MOF-based nanoparticulate delivery for improving siRNA activity, and the increased cancer cell—specific interactions afforded by the platelet membrane helped to further boost silencing efficiency.

One of the most important functions of platelets is that they adhere to CTCs and create a cellular shield which protects CTCs from hemodynamic force and cytotoxic effects of immune cells. This platelet—tumor cell interaction further enhances tumor cell arrest in the vasculature and eventually results in metastasis. As a key regulator of the metastatic process, platelets represent a potential target as well as an efficient vehicle to deliver novel cancer therapeutic agents. Recent studies suggest that functionalizing platelet membranes with cytotoxic drugs is more efficient in killing CTCs and reducing cancer metastasis. Inspired by the adhesion of activated platelets to CTC-associated microthrombi, biocompatible silica (Si) particles were functionalized with membrane-derived vesicles from activated platelets. This biomimetic coating allows for targeting of synthetic particles to CTCs. Additionally, the major tumor-killing cytokine, Tumor necrosis factor—related apoptosis-inducing ligand (TRAIL) was conjugated to the platelet membrane-coated Si particles. Mice injected with MDA-MB-23 cells, a human breast cancer cell line, followed by the injection of platelet membrane-coated particles developed less lung metastasis than those injected with saline buffer, particles without conjugated TRAIL, or soluble TRAIL (Li et al., 2016). Overall, many recent studies have demonstrated PNPs as a versatile platform for broad cancer targeting. Platelet membrane cloaking reduces immunogenicity, attenuates macrophage uptake, and enhances pharmacokinetic profiles, all contributing to higher anticancer efficacy in vivo.

18.4 **Limitations of artificial hemostats**

Natural platelet based micro- and nanoparticles offer many advantages over other artificial hemostatic agents, still many limitations have to be overcome before their actual clinical utilization starts. Adsorption of proteins onto any surface or particle that comes in contact with blood is believed to cause rapid uptake of the particle by the RES system. Evasion of the RES system is considered necessary for any successful nanoformulation. PEGylation of nanoparticles is the most common method used to overcome protein adsorption. Even with the use of PEGylation, there will still be some protein adsorption, and these nanoparticles will have a relatively short circulation time. Also, complement activation-related pseudoallergy (CARPA) can be elicited during administration of certain liposomes and several polymeric-based nanoparticles. The hallmark symptoms of CARPA appear within 1−3 minutes following nanoparticle administration and include cardiopulmonary distress and a characteristic erythema upon reperfusion of the tissues. Usually, these issues spontaneously resolve within minutes, still, the consequences of a severe CARPA episode during administration of intravenous hemostatic agents could be catastrophic. Therefore the development of robust models to ensure a minimized risk for CARPA during nanoparticle administration will be of paramount importance in clinical translation of intravenous hemostatic technologies. Further, nanoparticles are stored in suspension, frozen, or as a lyophilized product, depending on their stability in each phase. For biodegradable polymer nanoparticles, they are generally lyophilized for long-term storage. However, due to crystallization during processing, aggregates are formed, which then require sonication in order to recover prelyophilization size distributions (Lashof-Sullivan et al., 2013).

To conclude, artificial platelet substitute still remain highly sought-after, which can potentially circumvent several limitations associated with natural blood products. Newer scientific discoveries elaborating the role of platelets in physiological and pathological conditions, along with nanotechnological advancements in producing and interacting subcellular molecules have raised the hope of developing synthetic platelets for clinical utilization soon.

References

Agam, G., & Livne, A. (1992). Erythrocytes with covalently bound fibrinogen as a cellular replacement for the treatment of thrombocytopenia. *European Journal of Clinical Investigation, 22,* 105−112.

Alam, F., Naim, M., Aziz, M., & Yadav, N. (2014). Unique roles of nanotechnology in medicine and cancer. *Indian Journal of Cancer, 51,* 506−510.

Andrews, R. K., & Berndt, M. C. (2004). Platelet physiology and thrombosis. *Thrombosis Research, 114* (5−6), 447−453, 447.

Anselmo, A. C., Modery-Pawlowski, C., Menegatti, S., Kumar, S., Vogus, D. R., Tian, L. L., et al. (2014). Platelet-like nanoparticles: Mimicking shape, flexibility, and surface biology of platelets to target vascular injuries. *ACS Nano, 8,* 11243−11253.

Bertram, J. P., Williams, C. A., Robinson, R., Segal, S. S., Flynn, N. T., & Lavik, E. B. (2009). Synthetic platelets: Nanotechnology to halt bleeding. *Science Translational Medicine, 1,* 11−22.

Blajchman, M. A. (2003). Substitutes and alternatives to platelet transfusions in thrombocytopenic patients. *Journal of Thrombosis and Haemostasis: JTH, 1,* 1637−1641.

Clemetson, K. J. (2012). Platelets and primary haemostasis. *Thrombosis Research*, *129*(3), 220−224.

Coller, B. S. (1980). Interaction of normal, thrombasthenic, and Bernard−Soulier platelets with immobilized fibrinogen: Defective platelet−fibrinogen interaction in thrombasthenia. *Blood*, *55*, 169−178.

Coller, B. S., Springer, K. T., Beer, J. H., Mohandas, N., Scudder, L. E., Norton, K. J., et al. (1992). Thromboerythrocytes. Invitro studies of a potential autologous, semi-artificial alternative to platelet transfusions. *The Journal of Clinical Investigation*, *892*, 546−555.

Davì, G., & Patrono, C. (2007). Platelet activation and atherothrombosis. *The New England Journal of Medicine*, *357*, 2482−2494.

Farndale, R. W., Siljander, P. R., Onley, D. J., Sundaresan, P., Knight, C. G., & Barnes, M. J. (2003). Collagen platelet interactions: Recognition and signalling. *Biochemical Society Symposium*, *70*, 81−94. Available from https://doi.org/10.1042/bss0700081.

Gerrard, J. M., Rao, G. H., & White, J. G. (1977). The influence of reserpine and ethylenediaminetetraacetic acid (EDTA) on serotonin storage organelles of blood platelets. *The American Journal of Pathology*, *87*(3), 633−646.

Hayashi, T., Tanaka, S., Hori, Y., Hirayama, F., Sato, E. F., & Inoue, M. (2011). Role of mitochondria in the maintenance of platelet function during in vitro storage. *Transfusion Medicine*, *21*, 166−174.

Hickman, D. A., Pawlowski, C. L., Shevitz, A., Luc, N. F., Kim, A., Girish, A., et al. (2018). Intravenous synthetic platelet (SynthoPlate) nanoconstructs reduce bleeding and improve 'golden hour' survival in a porcine model of traumatic arterial hemorrhage. *Scientific Reports*, *8*(1), 3118.

Italiano, J. E., Jr., Lecine, P., Shivdasani, R. A., & Hartwig, J. H. (1999). Blood platelets are assembled principally at the ends of proplatelet processes produced by differentiated megakaryocyte. *The Journal of Cell Biology*, *147*, 1299−1312.

Jonnalagadda, D., Izu, L. T., & Whiteheart, S. W. (2012). Platelet secretion is kinetically heterogeneous in an agonist-responsive manner. *Blood*, *120*(26), 5209−5216.

Klement, G. L., Yip, T. T., Cassiola, F., Kikuchi, L., Cervi, D., Podust, V., et al. (2009). Platelets actively sequester angiogenesis regulators. *Blood*, *113*(12), 2835−2842.

Labelle, M., & Hynes, R. O. (2012). The initial hours of metastasis: The importance of cooperative host-tumor cell interactions during hematogenous dissemination. *Cancer Discovery*, *2*(12), 1091−1099.

Lashof-Sullivan, M., Shoffstall, A., & Lavik, E. (2013). Intravenous hemostats: Challenges in translation to patients. *Nanoscale*, *5*(22), 10719−10728.

Lee, D. H., & Blajchman, M. A. (2001). Novel treatment modalities: New platelet preparations and subsititutes. *British Journal of Haematology*, *114*, 496−505.

Leslie, M. (2010). Cell biology. Beyond clotting: The powers of platelets. *Science (New York, N.Y.)*, *328*(5978), 562−564.

Levi, M., Friederich, P. W., Middleton, S., De Groot, P. G., Wu, Y. P., Harris, R., et al. (1999). Fibrinogen-coated albumin microcapsules reduce bleeding in severely thrombocytopenic rabbits. *Nature Medicine*, *5*, 107−111.

Li, J., Ai, Y., Wang, L., Bu, P., Sharkey, C. C., Wu, Q., et al. (2016). Targeted drug delivery to circulating tumor cells via platelet membrane-functionalized particles. *Biomaterials*, *76*, 52−65.

Li, M., Li, J., Chen, J., Liu, Y., Cheng, X., Yang, F., et al. (2020). Platelet membrane biomimetic magnetic nanocarriers for targeted delivery and in situ generation of nitric oxide in early ischemic stroke. *ACS Nano*, *14*(2), 2024−2035, 25.

Maynard, D. M., Heijnen, H. F., Gahl, W. A., & Gunay-Aygun, M. (2010). The α-granule proteome: Novel proteins in normal and ghost granules in gray platelet syndrome. *Journal of Thrombosis and Haemostasis: JTH*, *8*(8), 1786−1796.

Melchinger, H., Jain, K., Tyagi, T., & Hwa, J. (2019). Role of platelet mitochondria: Life in a nucleus-free zone. *Frontier in Cardiovascular Medicine*, *6*, 153. Available from https://doi.org/10.3389/fcvm.2019.00153.

Nasiri, S. (2019). Infusible platelet membrane may be effective and safe: A mini review. *Thrombosis & Haemostasis Research*, *3*(1), 1021.

Nishiya, T., Kainoh, M., Murata, M., Handa, M., & Ikeda, Y. (2002). Reconstitution of adhesive properties of human platelets in liposomes carrying both recombinant glycoproteins Ia/IIa and Iba under flow conditions: Specific synergy of receptor—ligand interactions. *Blood, 100,* 136—142.

Nishiya, T., & Toma, C. (2004). Interaction of platelets with liposomes containing dodecapeptide sequence from fibrinogen. *Thrombosis and Haemostasis, 91,* 1158—1167.

Okamura, Y., Eto, K., Maruyama, H., Handa, M., Ikeda, Y., & Takeoka, S. (2010). Visualization of liposomes carrying fibrinogen g-chain dodecapeptide accumulated to sites of vascular injury using computed tomography. *Nanomedicine: Nanotechnology, Biology, and Medicine, 6,* 391—396.

Okamura, Y., Fujie, T., Maruyama, H., Handa, M., Ikeda, Y., & Takeoka, S. (2007). Prolonged hemostatic ability of poly(ethylene glycol)-modified polymerized albumin particles carrying fibrinogen g chain dodecapeptide. *Transfusion, 47,* 1254—1262.

Okamura, Y., Fukui, Y., Kabata, K., Suzuki, H., Handa, M., Ikeda, Y., et al. (2009). Novel platelet substitutes: Disk-shaped biodegradable nanosheets and their enhanced effects on platelet aggregation. *Bioconjugate Chemistry, 20,* 1958—1965.

Okamura, Y., Katsuno, S., Suzuki, H., Maruyama, H., Handa, M., Ikeda, Y., et al. (2010). Release abilities of adenosine diphosphate from phospholipid vesicles with different membrane properties and their hemostatic effects as a platelet substitute. *Journal of Controlled Release: Official Journal of the Controlled Release Society, 148,* 372—379.

Okamura, Y., Takeoka, S., Eto, K., Maekawa, I., Fujie, T., Maruyama, H., et al. (2009). Development of fibrinogen gamma-chain peptide-coated, adenosine diphosphate-encapsulated liposomes as a synthetic platelet substitute. *Journal of Thrombosis and Haemostasis: JTH, 7,* 470—477.

Peyvandi, F., Garagiola, I., & Baronciani, L. (2011). Role of von Willebrand factor in the haemostasis. *Blood Transfusion, 9*(S2), 3—8. Available from https://doi.org/10.2450/2011.002S.

Rivera, J., Lozano, M. L., Corral, J., González-Conejero, R., Martínez, C., & Vicente, V. (2000). Platelet GP Ib/IX/V complex: Physiological role. *Journal of Physiology and Biochemistry, 56,* 355—365.

Rivera, J., Lozano, M. L., Navarro-Núñez, L., & Vicente, V. (2009). Platelet receptors and signaling in the dynamics of thrombus formation. *Haematologica, 94,* 700—711.

Roberts, N., James, S., Delaney, M., & Fitzmaurice, C. (2019). The global need and availability of blood products: A modelling study. *The Lancet Haematology, 6*(12), e606—e615.

Ruggeri, Z. M., & Mendolicchio, G. L. (2007). Adhesion mechanisms in platelet function. *Circulation Research, 100*(12), 1673—1685.

Shukla, M., Sekhon, U. D., Betapudi, V., Li, W., Hickman, D. A., Pawlowski, C. L., et al. (2017). In vitro characterization of SynthoPlate™ (synthetic platelet) technology and its in vivo evaluation in severely thrombocytopenic mice. *Journal of Thrombosis and Haemostasis: JTH, 15*(2), 375—387.

Smyth, S. S., McEver, R. P., Weyrich, A. S., Morrell, C. N., Hoffman, M. R., Arepally, G. M., et al. (2009). Platelet functions beyond hemostasis. *Journal of Thrombosis and Haemostasis, 7,* 1759—1766.

Song, Y., Huang, Z., Liu, X., Pang, Z., Chen, J., Yang, H., et al. (2019). Platelet membrane-coated nanoparticle-mediated targeting delivery of Rapamycin blocks atherosclerotic plaque development and stabilizes plaque in apolipoprotein E-deficient (ApoE$^{-/-}$) mice. *Nanomedicine: Nanotechnology, Biology, and Medicine, 15*(1), 13—24.

Takeoka, S., Okamura, Y., Teramura, Y., Watanabe, N., Suziki, H., Tsuchida, E., et al. (2003). Function of fibrinogen γ -chain dodecapeptide-conjugated latex beads under flow. *Biochemical and Biophysical Research Communications, 312,* 773—779.

Takeoka, S., Teramura, Y., Okamura, Y., Handa, M., Ikeda, Y., & Tsuchida, E. (2001). Fibrinogen-conjugated albumin polymers and their interaction with platelets under flow conditions. *Biomacromolecules, 2,* 1192—1197.

Wang, H., Wu, J., Williams, G. R., Fan, Q., Niu, S., Wu, J., et al. (2019). Platelet-membrane-biomimetic nanoparticles for targeted antitumor drug delivery. *Journal of Nanobiotechnology, 17*(1), 60.

Wang, S., Duan, Y., Zang, Q., Komarla, A., Gong, H., Gao, W., et al. (2020). Drug targeting via platelet membrane—coated nanoparticles. *Small Structure*. Available from https://doi.org/10.1002/sstr.202000018.

Ware, J., Corken, A., & Khetpal, R. (2013). Platelet function beyond hemostasis and thrombosis. *Current Opinion in Hematology, 20*(5), 451—456. Available from https://doi.org/10.1097/MOH.0b013e32836344d3.

Wei, X., Ying, M., Dehaini, D., Su, Y., Kroll, A. V., Zhou, J., et al. (2018). Nanoparticle functionalization with platelet membrane enables multifactored biological targeting and detection of atherosclerosis. *ACS Nano, 12*(1), 109—116.

White, J. G. (1998). Use of the electron microscope for diagnosis of platelet disorders. *Seminars in Thrombosis and Hemostasis, 24*(2), 163—168.

WHO. (2018). Blood safety and availability. Available at https://www.who.int/news-room/fact-sheets/detail/blood-safety-and-availability

Xu, C., Hu, S., & Chen, X. (2016). Artificial cells: From basic science to applications. *Materials Today, 19*(9), 516—532.

Xu, J., Wang, X., Yin, H., Cao, X., Hu, Q., Lv, W., et al. (2019). Sequentially site-specific delivery of thrombolytics and neuroprotectant for enhanced treatment of ischemic stroke. *ACS Nano, 13*(8), 8577—8588, 27.

Yang, T., Ding, X., Dong, L., Hong, C., Ye, J., Xiao, Y., et al. (2018). Platelet-mimic uPA delivery nanovectors based on Au rods for thrombus targeting and treatment. *ACS Biomaterials Science and Engineering, 4*(12), 4219—4224.

Yapijakis, C., Bramos, A., Nixon, A. M., Ragos, V., & Vairaktaris, E. (2012). The interplay between hemostasis and malignancy: The oral cancer paradigm. *Anticancer Research, 32*(5), 1791—1800.

Yen, R. C., Ho, T. W., & Blajchman, M. A. (1995). A novel approach to correcting the bleeding associated with thrombocytopenia. *Transfusion, 35*, 41S.

Zhuang, J., Gong, H., Zhou, J., Zhang, Q., Gao, W., Fang, R. H., et al. (2020). Targeted gene silencing in vivo by platelet membrane-coated metal-organic framework nanoparticles. *Science Advance, 6*(13), eaaz6108. Available from https://doi.org/10.1126/sciadv.aaz6108.

Artificial white blood cells—WBC substitute

19

Khaliqur Rahman[1] and Mohammad Feroz Alam[2]

[1]*Department of Hematology, Sanjay Gandhi Post Graduate Institute of Medical Sciences, Lucknow, India*
[2]*Department of Pathology, JN Medical College, Aligarh, India*

19.1 Leukocyte physiology

Leukocytes or white blood cells (WBCs) are cells involved in the immune surveillance required for defending the human body from infectious agents or foreign bodies. These are found throughout the body but mainly in the peripheral blood, bone marrow, and lymphatic systems. The different types of leukocytes originate from a common precursor, the pleuripotent hematopoietic stem cells (HSCs). They further differentiate into myeloid and lymphoid progenitors and give rise to all the major forms of leukocytes. Five major kinds of leukocytes can be identified microscopically on a stained peripheral blood smear. These are the neutrophils, eosinophils, basophils, monocytes, and the lymphocytes. The first four are the myeloid series cells, while the last one is a lymphoid cell. In addition to these, there are some other leukocytes which are mainly fixed in tissues and are very rare in peripheral circulation. These include the histiocytes, dendritic cells, plasma cells, and mast cells. Mature myeloid series cells are larger in size with size ranging from 12 to 18 μm, while the lymphoid series cells are smaller with size ranging from 8 to 12 μm, in diameter. Their number in peripheral blood ranges from 4 to 11 $\times$ 10^9 cells/L.

Based on the presence or absence of cytoplasmic granules, the leukocytes can be divided into granulocytes (neutrophils, eosinophils, basophils, mast cells) or agranulocytes (lymphocytes, monocytes, plasma cells). These granules are membrane-bound enzymes which are released and used in the digestion of endocytosed particles. Though this is a naming convention, some lymphocytes and monocytes can display the presence of cytoplasmic granules in certain conditions.

The primary role of leukocytes is in the body's defense mechanism against pathogens, and each type has a specific function. Some of them are involved in innate immunity, while others are involved in the specific adaptive immunity. There is also a continuous cross talk between these cells with the help of cytokines which leads to successful defense.

19.1.1 Normal functions of the leukocytes

Neutrophils: neutrophils migrate toward the site of inflammation through a process known as chemotaxis, where the cell surface receptors detect the chemical gradients of inflammatory molecules like IL6, IFN-γ, C3a, C5a, etc. They also have receptors to detect and adhere to endothelial surface

Nanotechnology for Hematology, Blood Transfusion, and Artificial Blood. DOI: https://doi.org/10.1016/B978-0-12-823971-1.00015-5

and Fc receptors for opsonin (Serhan et al., 2010). They move to the site of inflammation via ameboid movement and in certain conditions in the form of a neutrophil swarm (Lämmermann et al., 2013). They quickly reach the site of inflammation and release cytokines, which further amplify inflammatory reactions by several other cell types. Neutrophils, being the front line of defense, attack the microorganism directly by three methods:

- *Phagocytosis*: where they internalize the opsonized microbe, resulting in the formation of a phagosome followed by secretion of reactive oxygen species (ROS) and hydrolytic enzyme digestion.
- *Degranulation* (release of cytoplasmic granules): neutrophils have three types of granules with different contents and have antimicrobial properties. These include the primary or azurophilic granules containing myeloperoxidase, defensins, serine proteases, elastases, cathepsin G; secondary or specific granules which contain alkaline phosphatases, lysozyme, NADPH oxidase, collagenases; and tertiary granules which contain collagenase, gelatinase, and cathepsin.
- *Neutrophil extracellular traps*: these are web-like structures of DNA and serine proteases that trap and kill the extracellular organisms. These act as a concentrated mechanism to kill the microbes and in addition also act as a physical barrier to prevent the spread of infection (Brinkmann et al., 2004; Urban et al., 2009).

Eosinophil: these leukocytes are the principle cells involved in combating parasitic infection. These are also involved in controlling the mechanism associated with asthma and allergy. Although phagocytic, eosinophils are less efficient than neutrophils in killing intracellular bacteria. Upon activation, they produce and release granules containing basic protein and eosinophil cationic proteins, which are toxic to several parasites and mammalian cells (Hogan et al., 2008). These proteins bind heparin and neutralize its anticoagulant activity. They also release ROS, such as hypobromite, superoxide, and peroxide; eicosanoids from the leukotriene (e.g., LTC_4, LTD_4, LTE_4); prostaglandin (e.g., PGE_2) families[18]; enzymes, such as elastase; growth factors, such as TGF beta, VEGF, and PDGF[19][20]; and cytokines, such as IL-1, IL-2, IL-4, IL-5, IL-6, IL-8, IL-13, and TNF alpha. (Bandeira-Melo et al., 2002; Hogan et al., 2008; Kato et al., 2005). Eosinophils also have some antiviral properties as is evident by the presence of RNAses in their granules.

Basophils: these have an important role in mechanisms involved in allergic reactions. Their granules contain heparin, which has an anticoagulant property; histamine, a potent vasodilator; and proteolytic enzymes like elastases. The release of their granular content is mediated by IgE binding to the receptor on the cell surface of basophils. Like eosinophils, they also have a role in combating parasitic and helminthic infestations (Voehringer, 2009). Recent studies in mice have suggested that basophils may also regulate the T cells' behavior and mediate the magnitude of the secondary immune response (Nakanishi, 2010).

Monocytes/macrophages: monocytes are noted in peripheral circulation. Once they reach the tissue, they differentiate into macrophages and just like neutrophils these cells phagocytize pathogens. They also generate toxic oxygen metabolites that have a role in killing microorganisms. Monocytes are essential in antigen presentation to T lymphocytes. They also are necessary for tissue repair and remodeling.

Dendritic cells (DCs): DCs are the cells which are important for both innate and adaptive immunity. They activate the T cells. The organized contact between a T-cell and a dendritic cell during

T-cell activation is known as an immunological synapse. DCs capture the fragments of pathogens from the environment and display the peptides of foreign material to T cells in the cleft of the major histocompatibility complex (MHC) molecules. This interaction helps the T cells to mount an appropriate immune response via the type of costimulatory ligands presented by the DCs to the T-cell along with cytokines released at the synapse.

19.1.1.1 Lymphocytes: (B cells, T cells, and natural killer cells)

B cells are activated with the help of antigenic stimulus in the secondary lymphoid organs in either T cell-dependent or T cell-independent mechanisms. The protein antigens activate B cells in T cell-dependent mechanisms, while the polysaccharide does the same in a T cell-independent mechanism (Nutt et al., 2015). Once activated, B cells differentiate into either plasmablasts, for an immediate protection or long-lived plasma and memory B cells for prolonged protection. The antibodies produced by plasmablasts are usually weak antibodies of class IgM (MacLennan et al., 2003). Plasma cells are the class switched cells that secrete more potent antibodies and provide long-lived protection. Memory B cells circulate through the body and initiate a strong and rapid antibody response against the antigen for which they have been exposed earlier (known as the anamnestic secondary antibody response) (Kurosaki et al., 2015). Owing to the production of antibodies in the serum, the immunity provided by B cells is also known as humoral immunity.

T cells provide adaptive cell-mediated immunity. There are several important subsets of T cells which have different functions. Some of them help in activation and differentiation of other cells, especially B cells (helper T cells, CD4 +), while the others are cytotoxic and destroy virus-infected cells and tumor cells (cytotoxic T cells, CD8 +). Some T cells retain the memory of previous infections. These are long-lived and quickly expand to a large number of effector T cells upon reexposure to an antigen. Yet another crucial subset of T cells is the regulatory T cells (T regs), which are important for immune tolerance. Their major function is to put an end to the T cell-mediated immune reaction once their job is done. This is helpful in preventing the autoimmune phenomenon.

Natural killer (NK) cells are lymphoid cells that are involved in innate immunity. They kill the target cells primarily through three different mechanisms. The first mechanism is through the release of perforin and granzyme from cytoplasmic granules to kill the cells by perforating the target cell's membrane (Lieberman, 2003). The second mechanism is through the release of cytokines that help other immune cells. And lastly, by the expression of ligands for the engagement of NK cells to their targets required for the cell death (Zamai et al., 1998).

19.2 Artificial leukocytes

The last two decades have rapidly witnessed the progress in the development of artificial blood cells. By artificial cells, we mean engineered particles having the size of biological cells with engineered functionality that mimics normal physiological properties. The primary functions of leukocytes are protection against harmful foreign agents by rendering the cytotoxic destruction and phagocytic clearance of these agents. They reach the specific site of infection/inflammation with the help of receptors on their surface. By combining properties of conventional synthetic nanoparticles with biomimetic materials, new solutions in the field of drug delivery are being offered (Doshi et al., 2012; Hammer et al., 2008).

Though the complex structure of leukocytes has not been successfully simulated by the nanotechnology, several interesting synthetic approaches for the functional mimicry of leukocytes have been explored. The majority of these approaches are used for the drug delivery in order to avoid the biological barriers which limit their optimal biodistribution. Some of these important bio-inspired synthetic components are discussed below.

19.2.1 Leukosomes

These are the proteolipid vesicles made by incorporating proteins derived from the leukocyte plasma membrane into lipid nanoparticles (Molinaro et al., 2016) (Fig. 19.1). These leukosomes retain the versatility and physicochemical properties of liposomal formulations in terms of loading and release of therapeutic preparations. They also retain the physiological tropism of leukocytes toward inflamed vasculature and preferential accumulation in the inflamed tissue.

Using a mouse model, Molinaro et al. successfully demonstrated the advantages of the leukosomes manufactured by them. The cell membrane was extracted from both primary and immortalized immune cells. A mixture of cholesterol and synthetic choline-based phospholipids was assembled which mimicked the physiologic composition of the plasmalemma and it was enriched with purified proteins using the thin layer evaporation. They loaded these leukosomes with antiinflammatory glucocorticoids. This therapeutic loading did not affect the surface properties of the leukosome. In the flow chamber assay using the restructured monolayer of activated human umbilical vein endothelial cells (HUVEC) and physiologically relevant shear stress, adherent properties of conventional liposomal preparations were compared with leukosomes. Compared with conventional liposomes, leukosomes preferentially recognized the inflamed endothelium. The leukosomes were capable of selective and effective deliver of dexamethasone to the inflamed tissue. The success of treatment was demonstrated by reduced expression of proinflammatory markers and endothelial

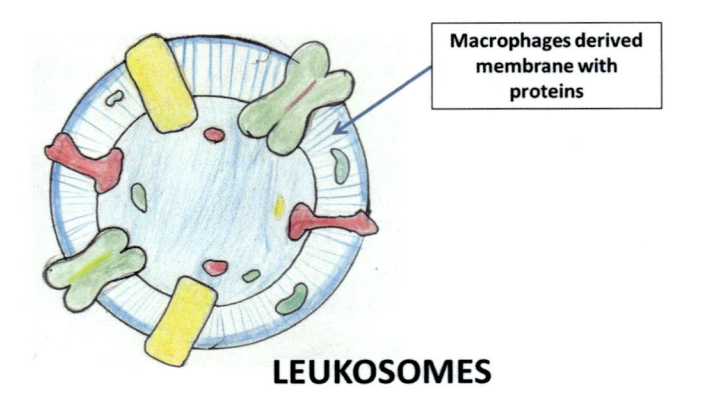

LEUKOSOMES

FIGURE 19.1

Leukosome. Membrane extracted from macrophages covering the lipid nanoparticle.

adhesion molecules as well as increased levels of the antiinflammatory gene using a polymerase chain reaction. It was also noted that leukosomes did not initiate any significant immune response or antibody production in the injected mice.

In an attempt to target the inflamed vasculature, Boada et al. (2020) used leukosomes with proteins derived from liposaccharide-stimulated macrophages. These leukosomes were loaded with rapamycin and were called Leuko-Rapa. The Leuko-Rapa were delivered for a relatively short therapeutic course (once a day for 7 days) into hypercholesterolemic, ApoE$^{-/-}$, mice. It led to reduced production of inflammatory cytokines and decreased matrix metalloproteinase activity along with a decreased proliferation of macrophage in atherosclerotic aorta. These loaded leukosomes have very aptly been termed as "smart bombs" (Csányi et al., 2020).

19.2.2 Leukopolymerosomes

Leukopolymerosomes are the polymerosomes with adhesive properties of leukocytes. The technology utilizes modular avidin—biotin chemistry to attach adhesion ligands that mimic selectins and integrins of leukocytes (Hammer et al., 2008). Polymersomes have a spectrum of advantages over phospholipid vesicles, which include a tougher and more stable structure with critical strains before rupture, being as much as seven times higher than those for lipid vesicles, even if the polymer is in the liquid state (Bermudez et al., 2002). These have excellent adhesive properties. These particles support the adhesion effectively at the shear rate at which inflammation typically occurs. Another unique advantage of polymerosomes is their ability to effectively encapsulate both aqueous and nonaqueous components (Fig. 19.2).

Leukocyte adhesion to the vessel wall requires two adhesion pathways: the selectin and the integrins pathways. A successful and functional adhesion requires both these to work in coherence. The selectin pathway causes the initial deceleration and leukocyte-endothelial rolling. Even in the

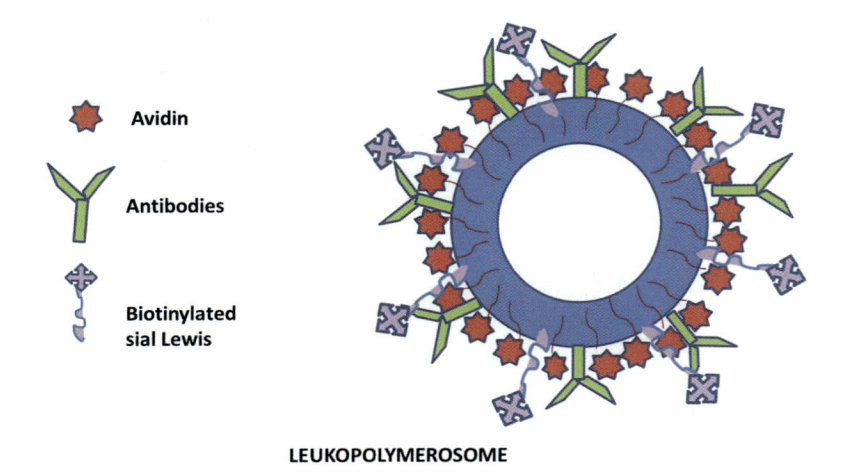

Avidin

Antibodies

Biotinylated
sial Lewis

LEUKOPOLYMEROSOME

FIGURE 19.2

Leukopolymerosome. Polymer based nanovesicles coated with adhesion ligands.

highest concentration of selectins, leukocytes are not able firmly adhere. Engagement of selectins leads to the activation of integrins which are the principle force for a firm adhesion of leukocytes. The selectin interactions are kinetically fast but mechanically weak, whereas integrin interactions are kinetically slow but mechanically strong. Hence, one should incorporate both selectin mimics (to support rolling interactions) and integrin mimics (to support firm adhesion) to make effective constructs of leukopolymerosomes. When loaded with imaging or therapeutic agents, these can be successfully used for diagnosis and management of inflammatory disease and allow us to understand the physical chemistry of cell adhesion.

19.2.3 Leukolike vectors

In an attempt to evade the immune system, cross the biological barrier and selectively target the tissue, nanoparticles composed of nanoporous silicon (NPS) coated with cellular membranes derived from leukocytes were manufactured. These hybrid particles were known as Leukolike vectors (LLV) (Parodi et al., 2013) (Fig. 19.3). Both in vitro as well as in vivo studies have suggested that LLVs may be well suited for chemotherapeutics delivery to the cancer cells. The LLVs have a potential to recognize and bind to tumor endothelium in a nondestructive manner. In a trans-well model comprising a HUVEC monolayer in an upper chamber and confluent MDA-MB-231 breast cancer cell monolayer in a lower chamber, the LLV with doxorubicin (DOX) payload was found to be superior to free DOX and DOX-NPS in reducing the tumor cells' viability while maintaining the endothelial layer viability (Parodi et al., 2013). The proteomic profiling of LLV has demonstrated the ability to transfer more than 150 proteins associated with leukocyte membranes with at least a portion of them maintaining the correct orientation. About 50% of the proteins were associated to the plasma membrane with most of them involved in cellular transportation, adhesion, and signaling. The remaining 50% belonged to various subcellular compartments such as the endoplasmic reticulum, mitochondria, ribosomes, and nucleus (Corbo et al., 2015). The cellular source of membrane has been shown to impact the immunogenic response of these LLV, with syngeneic cells having more favorable properties as compared to xenogeneic source (Evangelopoulos et al., 2016).

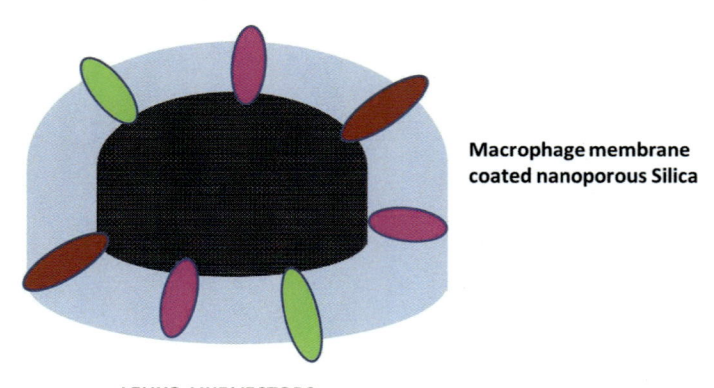

Macrophage membrane coated nanoporous Silica

LEUKO-LIKE VECTORS

FIGURE 19.3

Leuko like vectors. Nanoporus silicon core coated with membranes derived from leucocytes.

19.2.4 Nanoghosts

These are the biomimetic nanovesicles which are made by homogenizing the cytoplasm free, intact cell membranes (Ghosts cells) into nanosized vesicles while entrapping a therapeutic agent of choice (Toledano Furman et al., 2013) (Fig. 19.4). Briefly, the cells undergo hypotonic treatment followed by a mild homogenization process in order to remove cytosolic content. After centrifugation of these cytoplasm free cell membranes (cell ghosts), these are stabilized in sucrose solution. These membranes are then fractionated by a short sonication to yield 100−200 nm nanoghosts (NG). NG Further purification and ultracentrifugation leads to formation of nanovesicles composed of cell membrane. These are then PEGylated and finally loaded with therapeutic payloads (Oieni et al., 2020). Most of the studies have shown the ghost cells derived from the mesenchymal sem cells (MSC) owing to their hypoimmunogenecity and their inherent cancer targeting properties (Studeny et al., 2004; Toledano Furman et al., 2013; Uccelli et al., 2008). These MSC-derived nanoghosts have opened a new avenue for drug delivery for the treatment of refractory cancers and other diseases. These have been found superior to synthetic liposome in terms of specific tumor targeting. One of the studies has utilized the monocyte-derived nanoghost and shown that doxorubicin-loaded nanoghosts had greater cellular uptake and cytotoxicity compared to noncoated nanoparticle controls in metastatic MCF-7 breast cancer cell lines (Krishnamurthy et al., 2016).

19.2.5 Artificial antigen presenting cells

Artificial antigen presenting cells (aAPCs) are cell-sized polystyrene beads conjugated with immobilized peptide-MHC and costimulatory receptor ligands (Fig. 19.5). These are being used in immunology to tune the immunity in experimental settings (Hellstrom et al., 2001; Tham et al., 2001; Trickett & Kwan, 2003). Multifunctional aAPCs are being engineered with more of the physiological features of genuine APCs. For example, aAPCs composed of biodegradable polylactide-co-glycolide (PLGA) and surface-modified with avidin−palmitate conjugates to anchor peptide-MHC

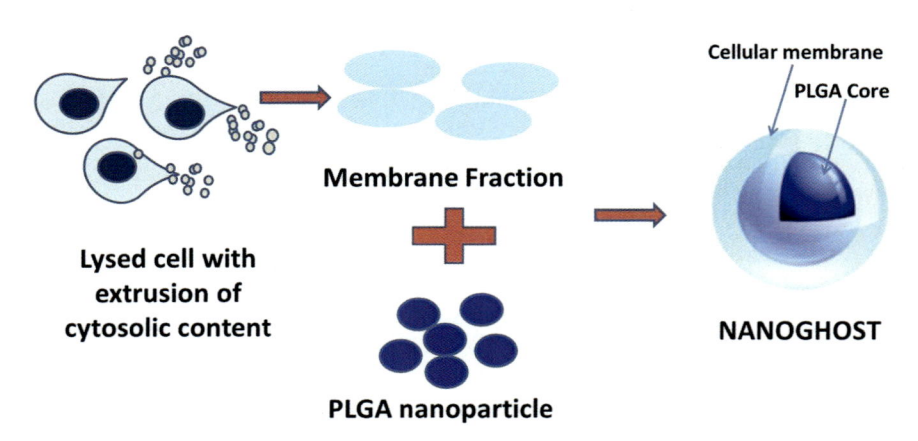

FIGURE 19.4

Nanoghost. Cytosol free membrane derived from leukocytes covering the core of PLGA nanoparticle.

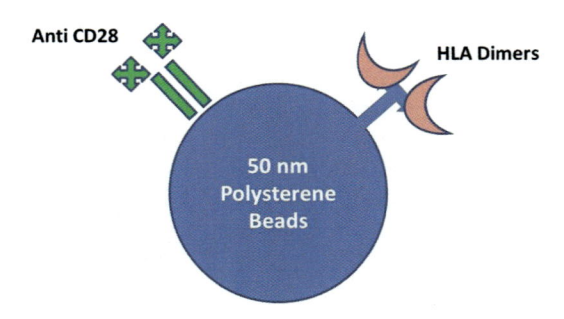

Artificial Antigen presenting cells (aAPC)

FIGURE 19.5

Artificial Antigen Presenting cells. Nano sized polystyrene beads conjugated with immobilized peptide-MHC and co-stimulatory receptor ligand.

and costimulatory ligands to the particle surfaces have been generated (Steenblock & Fahmy, 2008). These aAPCs when loaded with cytokines have been found to have a much higher stimulating effects on T cells as compared to the PLGA particles with costimulatory ligands alone (Steenblock & Fahmy, 2008). Additionally, the remodeling of T cell—aAPC interaction suggested that the concentration of IL-2 developing in the synapse near an IL-2-releasing aAPCs exceeds that obtained with a 1000-fold-greater concentration of cytokine added in the coculture solution. Not only in vitro experiments have shown the advantages of aAPC but certain in vivo experiments have also demonstrated their utility. In an experiment by Ugel et al. (2009), aAPCs made by covalent coupling of MHC-Ig dimers and B7.1-Ig molecules to magnetic beads were injected into mice bearing lung melanoma metastases to assess the efficacy of melanoma-specific cytotoxic T lymphocytes (CTLs). Significant tumor reduction was noted in these experimental mice. aAPCs also augmented the activity of transferred CTLs in a subcutaneous melanoma model. It was summarized that an in vivo aAPC administration had a potential as a novel approach to improve cancer immunotherapy (Ugel et al., 2009). Similarly, in another experiment, i.v. administration of aAPC has been shown to activate the Ag-specific CD4 + T cells and their proliferation in secondary lymphoid organs. This conferred partial protection against subcutaneous tumors, and prevented the establishment of lung metastasis (Caserta et al., 2008).

19.2.6 Microbivores

Microbivores are a speculative large class of medical nanorobots intended to be used in human patients for destroying microbial pathogens found in blood using a digest and discharge protocol (Freitas, 2005). It was proposed to be an artificial WBC or the phagocyte needed to combat the microbial infections. They were planned to be administered intravenously, and they would leave the bloodstream through the intestine. However, in spite of its proposed detailed design and mechanics, we still lack the facility to synthesize it.

19.3 Limitations of artificial leukocytes

Despite the overall positive findings, clinical translation of these leukomimetic particles has several challenges, which include the availability of technology, production yields, homogeneity of product, and purification difficulties to make a clinical grade compound.

Although it is assumed that these biomimetics have selective targeted actions, other off-target tissues, like lungs, have also been shown to be affected (Csányi et al., 2020). The dynamics of drug delivery also need to be better controlled. An unusually rapid release may lead to systemic side effects, while very slow release may be noneffective.

19.4 Conclusion

Technological advances had led to a paradigm change in the practice of medicine. Artificial blood has been a possibility. However, compared to the platelets and RBCs, the production and usage of artificial WBCs has not been very well established. Most of the efforts in the field of manufacturing of functional leukocyte mimics (artificial leukocytes) have explored the efficient and targeted drug delivery owing to physiological tropism of leukocytes toward inflamed and infected tissue, as well as capabilities of immune evasion. There has been success in demonstration of their utility using both in vitro and in vivo animal models. Some of the studies have also successfully demonstrated the use of biomimetic nanoparticles to tune the immunity.

Extensive in vitro and in vivo evaluation of different approaches is required to check the scalability, ensure manufacturing reproducibility, consistent composition, and safety and therapeutic advantages. A concerted effort from biomedical and materials engineers, hematologist, immunologists, and appropriate regulatory agencies is needed to reach the desired goal.

References

Bandeira-Melo, C., Bozza, P. T., & Weller, P. F. (2002). The cellular biology of eosinophil eicosanoid formation and function. *The Journal of Allergy and Clinical Immunology, 109*, 393–400.

Bermudez, H., Brannon, A. K., Hammer, D. A., Bates, F. S., & Discher, D. E. (2002). Molecular weight dependence of polymersome membrane structure, elasticity, and stability. *Macromolecules, 35*, 8203–8208.

Boada, C., Zinger, A., Tsao, C., Zhao, P., Martinez, J. O., Hartman, K., Naoi, T., Sukhoveshin, R., Sushnitha, M., Molinaro, R., Trachtenberg, B., Cooke, J. P., & Tasciotti, E. (2020). Rapamycin-loaded biomimetic nanoparticles reverse vascular inflammation. *Circulation Research, 126*(1), 25–37.

Brinkmann, V., Reichard, U., Goosmann, C., Fauler, B., Uhlemann, Y., Weiss, D. S., Weinrauch, Y., & Zychlinsky, A. (2004). Neutrophil extracellular traps kill bacteria. *Science (New York, N.Y.), 303*, 1532–1535.

Caserta, S., Alessi, P., Guarnerio, J., Basso, V., & Mondino, A. (2008). Synthetic CD^{4+} T cell-targeted antigen-presenting cells elicit protective antitumor responses. *Cancer Research, 68*(8), 3010–3018.

Corbo, C., Parodi, A., Evangelopoulos, M., Engler, D. A., Matsunami, R. K., Engler, A. C., Molinaro, R., Scaria, S., Salvatore, F., & Tasciotti, E. (2015). Proteomic profiling of a biomimetic drug delivery platform. *Current Drug Targets, 16*(13), 1540–1547.

Csányi, G., Lucas, R., & Annex, B. H. (2020). Loaded leukosome. A smart bomb to halt vascular inflammation. *Circulation Research, 126*, 38−40.

Doshi, N., Orje, J. N., Molins, B., Smith, J. W., Mitragotri, S., & Ruggeri, Z. M. (2012). Platelet mimetic particles for targeting thrombi in flowing blood. *Advanced Material, 24*(28), 3864−3869.

Evangelopoulos, M., Parodi, A., Martinez, J. O., Yazdi, I. K., Cevenini, A., van de Ven, A. L., & Tasciotti, E. (2016). Cell source determines the immunological impact of biomimetic nanoparticles. *Biomaterials, 82*, 168−177.

Freitas, R. A., Jr (2005). Microbivores: Artificial mechanical phagocytes using digest and discharge protocol. *Journal of Evolution and Technology, 14*(1), 55−92.

Hammer, D. A., Robbins, G. P., Haun, J. B., Lin, J. J., Qi, W., Smith, L. A., Ghoroghchian, P., Therien, M. J., & Bates, F. S. (2008). Leuko-polymersomes. *Faraday Discussions, 139*, 129−141.

Hellstrom, I., Ledbetter, J. A., Scholler, N., Yang, Y., Ye, Z., Goodman, G., Pullman, J., Hayden-Ledbetter, M., & Hellstrom, K. E. (2001). CD3-mediated activation of tumor-reactive lymphocytes from patients with advanced cancer. *Proceedings of the National Academy of Sciences, 98*(12), 6783−6788.

Hogan, S. P., Rosenberg, H. F., Moqbel, R., Phipps, S., Foster, P. S., Lacy, P., Kay, A. B., & Rothenberg, M. E. (2008). Eosinophils: Biological properties and role in health and disease. *Clinical and Experimental Allergy, 38*(5), 709−750.

Kato, Y., Fujisawa, T., Nishimori, H., Katsumata, H., Atsuta, J., Iguchi, K., & Kamiya, H. (2005). Leukotriene D4 induces production of transforming growth factor-beta1 by eosinophils. *International Archives of Allergy and Immunology, 137*, 17−20.

Krishnamurthy, S., Gnanasammandhan, M. K., Xie, C., Huang, K., Cui, M. Y., & Chan, J. M. (2016). Monocyte cell membrane-derived nanoghosts for targeted cancer therapy. *Nanoscale, 8*(13), 6981−6985.

Kurosaki, T., Kometani, K., & Ise, W. (2015). Memory B cells. *Nature Reviews Immunology, 15*(3), 149−159.

Lämmermann, T., Afonso, P. V., Angermann, B. R., Wang, J. M., Kastenmüller, W., Parent, C. A., & Germain, R. N. (2013). Neutrophil swarms require LTB4 and integrins at sites of cell death in vivo. *Nature, 498*, 371−375.

Lieberman, J. (2003). The ABCs of granule-mediated cytotoxicity: New weapons in the arsenal. *Nature Reviews Immunology, 3*(5), 361−370.

MacLennan, I. C. M., Toellner, K.-M., Cunningham, A. F., Serre, K., Sze, D. M.-Y., Zúñiga, E., Cook, M. C., & Vinuesa, C. G. (2003). Extrafollicular antibody responses. *Immunological Reviews, 194*, 8−18.

Molinaro, R., Corbo, C., Martinez, J. O., Taraballi, F., Evangelopoulos, M., Minardi, S., Yazdi, I. K., Zhao, P., De Rosa, E., Sherman, M. B., De Vita, A., Toledano Furman, N. E., Wang, X., Parodi, A., & Tasciotti, E. (2016). Biomimetic proteolipid vesicles for targeting inflamed tissues. *Nature Materials, 15*(9), 1037−1046.

Nakanishi, K. (2010). Basophils as APC in Th$_2$ response in allergic inflammation and parasite infection. *Current Opinion in Immunology, 22*(6), 814−820.

Nutt, S. L., Hodgkin, P. D., Tarlinton, D. M., & Corcoran, L. M. (2015). The generation of antibody-secreting plasma cells. *Nature Reviews. Immunology, 15*(3), 160−171.

Oieni, J., Levy, L., Letko Khait, N., Yosef, L., Schoen, B., Fliman, M., Shalom-Luxenburg, H., Malkah Dayan, N., D'Atri, D., Cohen Anavy, N., & Machluf, M. (2020). Nano-Ghosts: Biomimetic membranal vesicles, technology and characterization. *Methods (San Diego, Calif.), 177*, 126−134.

Parodi, A., Quattrocchi, N., van de Ven, A. L., Chiappini, C., Evangelopoulos, M., Martinez, J. O., Brown, B. S., Khaled, S. Z., Yazdi, I. K., Enzo, M. V., Isenhart, L., Ferrari, M., & Tasciotti, E. (2013). Synthetic nanoparticles functionalized with biomimetic leukocyte membranes possess cell-like functions. *Nature Nanotechnology, 8*(1), 61−68.

Serhan, C. N., Ward, P. A., & Gilroy, D. W. (2010). *Fundamentals of inflammation* (pp. 53−54). Cambridge University Press, ISBN 978−0−521−88729-8.

Steenblock, E. R., & Fahmy, T. M. (2008). A comprehensive platform for ex vivo T-cell expansion based on biodegradable polymeric artificial antigen-presenting cells. *Molecular Therapy: The Journal of the American Society of Gene Therapy*, *16*(4), 765–772.

Studeny, M., Marini, F. C., Dembinski, J. L., Zompetta, C., Cabreira-Hansen, M., Bekele, B. N., Champlin, R. E., & Andreeff, M. (2004). Mesenchymal stem cells: Potential precursors for tumor stroma and targeted-delivery vehicles for anticancer agents. *Journal of the National Cancer Institute*, *96*(21), 1593–1603.

Tham, E. L., Jensen, P. L., & Mescher, M. F. (2001). Activation of antigen-specific T cells by artificial cell constructs having immobilized multimeric peptide-class I complexes and recombinant B7-Fc proteins. *Journal of Immunological Methods*, *249*(1–2), 111–119.

Toledano Furman, N. E., Lupu-Haber, Y., Bronshtein, T., Kaneti, L., Letko, N., Weinstein, E., Baruch, L., & Machluf, M. (2013). Reconstructed stem cell nanoghosts: A natural tumor targeting platform. *Nano Letters*, *13*(7), 3248–3255.

Trickett, A., & Kwan, Y. L. (2003). T cell stimulation and expansion using anti-CD3/CD28 beads. *Journal of Immunological Methods*, *275*(1–2), 251–255.

Uccelli, A., Moretta, L., & Pistoia, V. (2008). Mesenchymal stem cells in health and disease. *Nature Reviews Immunology*, *8*, 726–736.

Ugel, S., Zoso, A., De Santo, C., Li, Y., Marigo, I., Zanovello, P., Scarselli, E., Cipriani, B., Oelke, M., Schneck, J. P., & Bronte, V. (2009). In vivo administration of artificial antigen-presenting cells activates low-avidity T cells for treatment of cancer. *Cancer Research*, *69*(24), 9376–9384.

Urban, C. F., Ermert, D., Schmid, M., Abu-Abed, U., Goosmann, C., Nacken, W., Brinkmann, V., Jungblut, P. R., & Zychlinsky, A. (2009). Neutrophil extracellular traps contain calprotectin, a cytosolic protein complex involved in host defense against *Candida albicans*. *PLoS Pathogens*, *5*, e1000639.

Voehringer, D. (2009). The role of basophils in helminth infection. *Trends in Parasitology*, *25*(12), 551–556.

Zamai, L., Ahmad, M., Bennett, I. M., Azzoni, L., Alnemri, E. S., & Perussia, B. (1998). Natural killer (NK) cell-mediated cytotoxicity: Differential use of TRAIL and Fas ligand by immature and mature primary human NK cells. *The Journal of Experimental Medicine*, *188*(12), 2375–2380.

Index

Note: Page numbers followed by "*f*" and "*t*" refer to figures and tables, respectively.

Printed in the United States
by Baker & Taylor Publisher Services